TAURINE 2

Basic and Clinical Aspects

ADVANCES IN EXPERIMENTAL MEDICINE AND BIOLOGY

Recent Volumes in this Series

Volume 400A
EICOSANOIDS AND OTHER BIOACTIVE LIPIDS IN CANCER, INFLAMMATION, AND RADIATION INJURY, Part A
Edited by Kenneth V. Honn, Santosh Nigam, and Lawrence J. Marnett

Volume 400B
EICOSANOIDS AND OTHER BIOACTIVE LIPIDS IN CANCER, INFLAMMATION, AND RADIATION INJURY, Part B
Edited by Kenneth V. Honn, Santosh Nigam, and Lawrence J. Marnett

Volume 401
DIETARY PHYTOCHEMICALS IN CANCER PREVENTION AND TREATMENT
Edited under the auspices of the American Institute for Cancer Research

Volume 402
AIDS, DRUGS OF ABUSE, AND THE NEUROIMMUNE AXIS
Edited by Herman Friedman, Toby K. Eisenstein, John Madden, and Burt M. Sharp

Volume 403
TAURINE 2: Basic and Clinical Aspects
Edited by Ryan J. Huxtable, Junichi Azuma, Kinya Kuriyama, Masao Nakagawa, and Akemichi Baba

Volume 404
SAPONINS USED IN TRADITIONAL AND MODERN MEDICINE
Edited by George R. Waller and Kazuo Yamasaki

Volume 405
SAPONINS USED IN FOOD AND AGRICULTURE
Edited by George R. Waller and Kazuo Yamasaki

Volume 406
MECHANISMS OF LYMPHOCYTE ACTIVATION AND IMMUNE REGULATION VI: Cell Cycle and Programmed Cell Death in the Immune System
Edited by Sudhir Gupta and J. John Cohen

Volume 407
EICOSANOIDS AND OTHER BIOACTIVE LIPIDS IN CANCER, INFLAMMATION, AND RADIATION INJURY, Volume 2
Edited by Kenneth V. Honn, Lawrence J. Marnett, Santosh Nigam, Robert L. Jones, and Patrick Y-K. Wong

Volume 408
TOWARD ANTI-ADHESION THERAPY FOR MICROBIAL DISEASES
Edited by Itzhak Kahane and Itzhak Ofek

TAURINE 2

Basic and Clinical Aspects

Edited by

Ryan J. Huxtable
University of Arizona
Tucson, Arizona

Junichi Azuma
Osaka University
Osaka, Japan

Kinya Kuriyama
Kyoto Prefectural University of Medicine
Kyoto, Japan

Masao Nakagawa
Kyoto Prefectural University of Medicine
Kyoto, Japan

and

Akemichi Baba
Osaka University
Osaka, Japan

PLENUM PRESS • NEW YORK AND LONDON

Library of Congress Cataloging-in-Publication Data

Taurine 2 : basic and clinical aspects / edited by Ryan J. Huxtable
... [et al.].
p. cm. -- (Advances in experimental medicine and biology ; v.
403)
"Proceedings of the International Taurine Symposium '95, an
officialsatellite symposium of the 15th Biennial Meeting of the
International Society for Neurochemistry, held June 27-July 1, 1995,
in Osaka, Japan"--T.p. verso.
Includes bibliographical references and index.
ISBN 0-306-45385-1
1. Taurine--Physiological effect--Congresses. 2. Taurine-
-Therapeutic use--Congresses. I. Huxtable, Ryan J.
II. International Society for Neurochemistry. Meeting (15th : 1995
: Osaka, Japan) III. InternationalTaurine Symposium '95 (1995 :
Osaka, Japan) IV. Series.
[DNLM: 1. Taurine--congresses. W1 AD559 v.403 1996 / QU 60 T2287
1996]
QP801.T3T37 1996
612'.0157--dc20
DNLM/DLC
for Library of Congress 96-32064
CIP

Proceedings of the International Taurine Symposium '95, an official Satellite Symposium of the 15th Biennial Meeting of the International Society for Neurochemistry,
held June 27 – July 1, 1995, in Osaka, Japan

ISBN 0-306-45385-1

A Division of Plenum Publishing Corporation
233 Spring Street, New York, N. Y. 10013

10 9 8 7 6 5 4 3 2 1

Printed in the United States of America

PREFACE

This volume comprises the edited proceedings of the International Taurine Symposium held in Osaka, Japan, in June 1995, as a Satellite Symposium of the 15th Biennial Meeting of the International Society for Neurochemistry. This Taurine Symposium was the latest in a series held since 1975 at approximately two-year intervals by an informal group of international researchers. It attracted contributions from 20 countries, ranging from Armenia via Finland and Spain to the United States. Some 121 participants attended.

The Symposium was organized and chaired by Junichi Azuma, University of Osaka. Other members of the Organizing Committee in Japan consisted of Kinya Kuriyama and Masao Nakagawa, both from the Kyoto Prefectural University of Medicine, and Akemichi Baba, from Osaka University. The Committee had to contend with the disaster of the Kobe earthquake, which struck on January 21. The epicenter was only around 25 miles from the meeting site, and the quake demolished the home of one Committee member. Despite this unnaturally natural handicap, the participants experienced a superbly organized meeting, one which more than maintained the high social and scientific standards established for this series. In his Welcome Message, Dr. Azuma listed a threefold objective for the Symposium: To provide a forum for the interdisciplinary exchange of information on taurine; to give an opportunity for renewing old friendships and making new friends; and to promote cooperation among participants from around the world. The meeting easily met all three objectives, and provided yet again an object lesson in the uniqueness of science in its ability to transcend borders, politics, and culture.

The next meeting is a return to the roots, to the town where the first two symposia were held in 1975 and 1977. It is planned for Tucson, Arizona, July 15–19, 1997. Considering the summer climate of southern Arizona, this promises to be the hottest taurine symposium ever. We look forward to seeing both familiar faces and new faces, and to writing a further chapter in the exciting but unfinished story of taurine research that has been assembled over the last 20 years.

We are indebted once more to Taisho Pharmaceutical Company for their continued generous support of the taurine symposia. In particular, we appreciate the help of Dr. Yasuhide Tachi and Ms. Etsuko Tamura of the Taisho company. We are also grateful to the International Society for Neurochemistry for their assistance. Ms. Emi Yoneya provided secretarial and organizational support for many aspects of the meeting. Finally, I thank Carla Hayes for helping in the retyping into a database of literally hundreds of references.

Ryan J. Huxtable
Editor-in-Chief

CONTENTS

Part 1. Chemistry, Biochemistry, Metabolism of Taurine

Part 2. Taurine and the Liver

Part 3. Taurine and the Kidney

Part 5. Taurine and the Audiovisual Systems

Part 6. Taurine and the Central Nervous System

Part 7. Taurine and Diabetes

Part 8. Taurine – Clinical Studies

Part 1

CHEMISTRY, BIOCHEMISTRY, METABOLISM OF TAURINE

1

IN VITRO REACTIONS OF HYPOTAURINE

Silvestro Duprè,[1] Mario Fontana,[1] Giuseppina Pitari,[2] and Doriano Cavallini[1]

[1] Dipartimento di Scienze Biochimiche "A. Rossi Fanelli"
Università di Roma "La Sapienza", Roma, Italy
and Centro di Biologia Molecolare
CNR, Roma, Italy
[2] Dipartimento di Biologia di Base ed Applicata
Università di L'Aquila, L'Aquila, Italy

INTRODUCTION

Hypotaurine and cysteine sulfinic acid are natural sulfinates found in relatively low amounts in some mammalian tissues[5]. They are metabolic intermediates in the biosynthetic pathways leading to the corresponding sulfonates. The function of hypotaurine as a biological antioxidant in protecting some cells from oxidative damage, by scavenging free radical forms of oxygen, has been proposed in the past[3,9] and recently demonstrated *in vitro.*[1,4] Chemically, sulfinates are not very stable compounds; oxidation is the most important but not the only reaction they are involved in. The low content of natural sulfinates in physiological tissues or fluids makes it often necessary, to perform quantitative determinations, to submit the biological samples to enrichment procedures which could give rise to artefactual results. The presence of an acidic milieu was supposed to slow down the oxidation of hypotaurine to taurine, which is enhanced at alkaline pH. However, it seems that oxidation is also facilitated in the presence of deproteinizing acids such as sulfosalycilic acid[3] or trichloroacetic acid.

Dilute hypotaurine solutions submitted to temperatures higher than room temperature, as during a concentration step in the rotary evaporator or when spotting the solution on chromatographic plates, may also show artefacts which should be kept under control. We present data that demonstrate that, under relatively mild conditions, sulfinates may undergo not only oxidation but also other reactions leading to partial disappearance of the investigated compound and to the appearance of some new derivatives in detectable amounts.

Taurine 2, edited by Huxtable et al.
Plenum Press, New York, 1996

MATERIALS AND METHODS

Products

Hypotaurine, 5,5'-dithio-bis(2-nitrobenzoic acid), cystamine dihydrochloride, and dithiothreitol were from Sigma Chemicals. Monobromobimane was from Calbiochem.

Procedures

Hypotaurine (10 mg) is dissolved in 150 ml HCl (from 0.1 N to 6 N), and taken to dryness in a rotary evaporator under reduced pressure at water bath temperature ranging from 40° to 80°. The residue is dissolved in 150 ml water and taken to dryness again. The residue is dissolved in 1.0 ml water and analyzed.

TLC on 0.2 mm thick silica gel plates on aluminium sheets (Merck) is developed in *n*-butanol:methanol:acetic acid:water (25:40:10:25) and compounds are detected with ninhydrin. Cysteamine plus cystamine after reduction by dithiothreitol was determined by HPLC after reaction with monobromobimane[8]. Amino acid analysis is performed on a Carlo Erba 3A30 analyzer.

The determination of the content of thiol and disulfide groups of the solution after the concentration step was performed with Ellman's reagent, in 1.0 ml 0.2 M phosphate buffer, pH 8.0[2]. The determination of disulfide groups was performed after reduction with $NaBH_4$ (20 mg for 5 min) followed by careful destruction of borohydride at acid pH. Standard curves are obtained under the same conditions with cysteine and cystine.

Hypotaurine hydrochloride was prepared by adding a suitable amount of hypotaurine in water to a 2:1 molar excess of HCl and precipitating with ethanol. The oily residue was

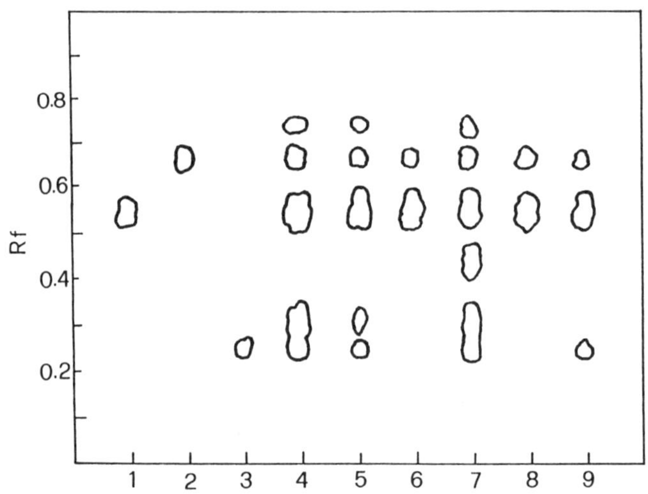

Figure 1. Thin layer chromatogram (silica gel on aluminium sheet) developed in n-butanol:methanol:acetic acid:water (25:40:10:25). Ninhydrin-positive spots are shown. Lane 1-3: standards of hypotaurine, taurine and cystamine respectively; lane 4: hypotaurine in 1 N HCl, concentrated under vacuo at 70°; lane 5: hypotaurine in 6 N HCl, concentrated under vacuo at 70°; lane 6: hypotaurine in water, concentrated under vacuo at 70°; lane 7: hypotaurine in 1 N HCl, concentrated under vacuo at 70° in the presence of excess dithiothreitol; lane 8: hypotaurine in 6 N HCl, heated at 70° for 2 h in stoppered vial; lane 9: hypotaurine hydrochloride, prepared as described in Methods, heated for 1.5 h at 70° in dry state.

collected and desiccated under vacuo over P_2O_5 and NaOH. A vitrous, hygroscopic mass was obtained which was not further characterized.

RESULTS AND DISCUSSION

The hypotaurine solution, taken to dryness as described in the methods section, has been analyzed by TLC and by amino acid analysis; the amounts of cyst(e)amine and of thiol or disulfides groups present are evaluated respectively by HPLC separation of monobromobimane derivatives and with Ellman's reagent before and after reduction with $NaBH_4$.

Fig. 1 shows schematically the TLC chromatogram of the solution after the concentration procedure. Qualitatively the chromatogram pattern does not change with initial HCl concentration, water bath temperature or time taken to concentrate the solution. Some ninhydrin-positive compounds are detected, and the comparison with standards suggests the presence of hypotaurine, taurine and cystamine as the main components (R_f values respectively 0.53, 0.66 and 0.26) together with minor amounts of other compounds. In vacuo concentration of hypotaurine solutions in water or in 0.2 M NaOH shows that only low or partial oxidation to taurine takes place (Fig. 1, lane 6). Other compounds are not detected. In the presence of ten molar excess of dithiothreitol or mercaptoethanol during the concentration step, new spots on TLC are detected, probably due to the presence of unidentified mixed disulfides (Fig. 1, lane 7). Similar chromatographic analyses performed on the hypotaurine solution before the sample is fully evaporated or on hypotaurine solutions in 0.1-6 N HCl heated in stoppered vials for various time intervals (up to 2 h) show the presence of hypotaurine and taurine but not of cystamine.

Concentration in vacuo of diluted HCl solutions leads to the formation of an azeotrope at 6.5 N. At the end of the concentration step hypotaurine will be exposed to 6.5 N HCl for a while and the dry compound obtained should contain hypotaurine in the hydrochloride form. This compound, prepared as described and heated in the dry state at 70° for 2 h, shows on TLC the presence of hypotaurine and a certain amount of cystamine and of taurine (Fig. 1, lane 9).

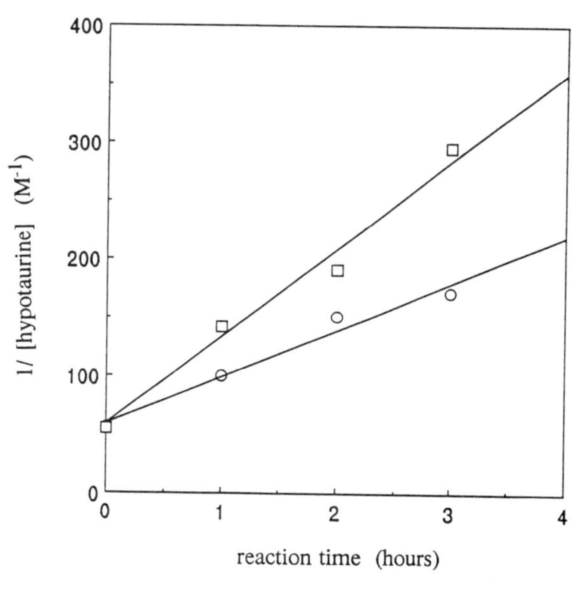

Figure 2. Plot of 1/[hypotaurine] vs. time for disappearance of hypotaurine in acetic acid at 70°. ○: concentrated acetic acid; □: concentrated acetic acid + 0.18 M H_2SO_4. Initial hypotaurine concentration 18.5 mM.

Table 1 collects quantitative data obtained by amino acid analysis, HPLC determination and dosage with Ellman's reagent.

Taurine is the product of an oxidation reaction. The presence of cystamine as one of the products obtained at the end of the concentration procedure may be explained by a disproportionation reaction of hypotaurine in acid conditions, followed by other reactions involving the unstable sulfenate derivative and/or other intermediate compounds, according to the reaction scheme shown below.

The disproportionation reaction of sulfinates has been studied in the past with aromatic sulfinates[7] and takes place in strong anhydrous acid conditions, as in glacial acetic acid and concentrated sulfuric acid at 70°. The time course of the same reaction performed on hypotaurine, following the decrease of hypotaurine concentration with amino acid analysis, shows that the reaction is second order in hypotaurine. The second order rate constant of disappearance (k_2) in acetic acid is $1.1 \times 10^{-2}\ M^{-1}s^{-1}$ and in acetic acid at 0.18 M sulfuric acid is $2.1 \times 10^{-2}\ M^{-1}s^{-1}$ (Fig. 2) which is one order higher than the corresponding constant ($1.1 \times 10^{-3}\ M^{-1}s^{-1}$ in acetic acid) obtained with *p*-toluene-sulfinic acid[7]. As already known[10], aliphatic sulfinic acids such as hypotaurine are less stable than aromatic sulfinates and we were able under these conditions to detect on TLC (not shown) and by disulfide group titration the presence of cystamine (Table 1). Amino acid analysis of a sample of hypotaurine in concentrated acetic acid after 2 h at 70° (Fig. 3) shows the presence of two other more alkaline ninhydrin-positive compounds, one of which has been identified as ethanolamine.

Cysteine sulfinic acid, when submitted to the same reactions as described above, does not present the same reactivity. The degree of oxidation to cysteic acid and of disproportionation is less than 2%.

A reaction sequence which may explain these results is shown in the Scheme 1. Cystamine may be formed (Scheme 1A) as one of the final products after the first disproportionation reaction of hypotaurine and the disproportionation of the intermediate thiol sulfonate[6]. The sum of this sequence of reactions would give a stoichiometric ratio hypo-

Table 1. Percentages of various compounds detected after different reaction conditions

Conditions	Amino Acid Analyzer Hypotaurine	Amino Acid Analyzer Taurine	Cyst(e)amine Bimane-HPLC	Equivalent (SH + SS) DTNB + $NaBH_4$	Equivalent (SH) DTNB
In 6 N HCl 70° dried under vacuo	46	12	5	8	<0.1
In 2 N HCl 70° dried under vacuo	65	25	2	5	0.2
In water 70° dried under vacuo	>95	<2	n.t.	<0.1	<0.1
In 6 N HCl 70° + dithiothreitol dried under vacuo	45	<1	n.t.	n.t.	n.t.
In 6 N HCl 70° + mercaptoethanol dried under vacuo	80	19	n.t.	n.t.	n.t.
Hypotaurine hydrochloride heated dry at 70°	80	5	7	6	0.1
In conc. acetic acid + H_2SO_4 heated at 70°	85	8	n.t.	0.7	n.t.
In conc. acetic acid - H_2SO_4 heated at 70°	80	13	n.t.	12	n.t.

Data of representative experiments are given as the percentage of the tested compound compared to the starting molar amount of hypotaurine. n.t. = not tested.

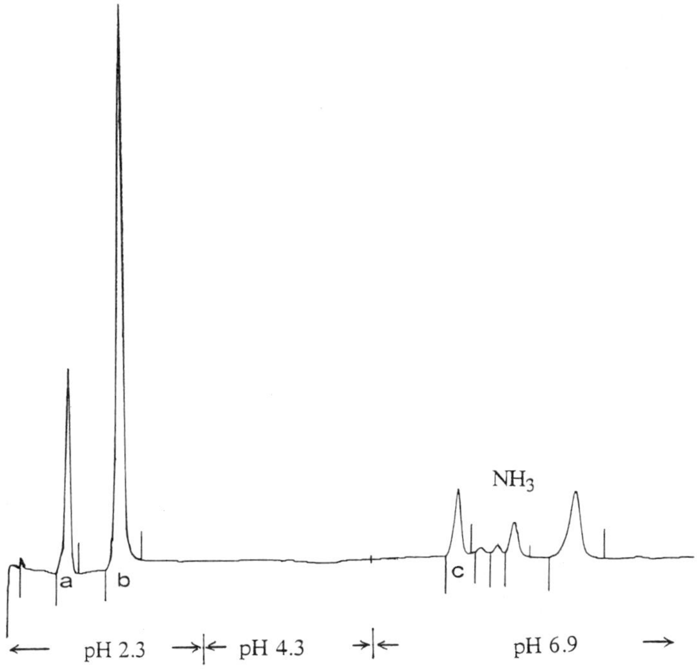

Figure 3. Amino acid elution profile of hypotaurine solution in concentrated acetic acid heated at 70° for 2 h in stoppered vial. Identified peaks are: a, taurine; b, hypotaurine; c, ethanolamine.

taurine:taurine:cystamine of 5:3:1. Furthermore the possible presence at the end of the process of some other intermediates (such as the more stable disulfone and thiolsulfinic derivatives or the less stable thiolsulfonic derivative) or the presence of other reaction pathways running side by side or alternatively to the outlined reaction sequence (as for instance a disproportionation reaction of the very unstable sulfenate) (Scheme 1B) may

A.

$$2\ R\text{-}SO_2H \longrightarrow R\text{-}SO_3H + R\text{-}SOH$$
$$R\text{-}SO_2H + R\text{-}SOH \longrightarrow R\text{-}SO_2\text{-}S\text{-}R + H_2O$$
$$3\ R\text{-}SO_2\text{-}S\text{-}R \longrightarrow 4\ R\text{-}SO_2H + R\text{-}S\text{-}S\text{-}R$$

Overall Reaction:

$$5\ R\text{-}SO_2H \longrightarrow 3\ R\text{-}SO_3H + R\text{-}S\text{-}S\text{-}R + H_2O$$

B.

$$2\ R\text{-}SO_2\text{-}S\text{-}R + 2\ H_2O \longrightarrow 3\ R\text{-}SO_2H + R\text{-}SH$$
$$R\text{-}SO_2\text{-}S\text{-}R + R\text{-}SH \longrightarrow R\text{-}SO_2H + R\text{-}S\text{-}S\text{-}R$$
$$3\ R\text{-}SOH \longrightarrow R\text{-}SO_2H + R\text{-}S\text{-}S\text{-}R$$
$$2\ R\text{-}SOH \longrightarrow R\text{-}SO_2H + R\text{-}SH$$

Scheme 1. Proposed mechanism for the disproportionation reaction of hypotaurine. A: partial reactions and final sum of disproportionation reaction showing a stoichiometric ratio hypotaurine:taurine:cystamine 5:3:1. B: some possible side reactions.

explain the experimentally found variable final ratios of the products (see Table 1). In the presence of excess thiols, cystamine may also be formed through the reduction of thiol sulfonate. Again a different stoichiometry would be obtained. The amount of taurine which has been found at the end of the concentration procedure of HCl solutions is however not related to the amount of cystamine and exceeds the amount expected from the disproportionation reaction. The oxidation runs at the same time as the disproportionation reaction.

The conclusion we can draw from these data is that in experiments concerned with sulfinates, and especially when hypotaurine is involved, great care must be exerted to avoid artefacts due to unexpected reaction pathways, such as the disproportionation reaction of sulfinates. The presence of some compounds found *in vivo* may, at least in part, originate from artefactual reactions during the isolation steps or could even due to a nonenzymatic general acid-catalyzed disproportionation reaction.

ACKNOWLEDGMENTS

We express our gratitude to Dr. C. Foppoli for the amino acid analyses and to Dr. A. Antonucci for the HPLC determination of cystamine. This work was supported by funds of M.U.R.S.T (40%).

REFERENCES

1. Aruoma, O.I., Halliwell, B., Hoey, B.M., and Butler, J. 1988, The antioxidant action of taurine, hypotaurine and their metabolic precursors, *Biochem.J.* 256:251-255.
2. Ellman, G.L. 1959, Tissue sulfhydryl groups, *Arch.Biochem.Biophys.* 82:70-77.
3. Fellman, J.H. and Roth, E.S. 1985, The biological oxidation of hypotaurine to taurine: hypotaurine as an antioxidant, in: "Taurine: Biological Actions and Clinical Perspectives", Oja, S.S., Ahtee, L., Kontro, P. and Paasonen, M.K. eds., Alan R. Liss, New York, pp. 71-82.
4. Green, T.R. and Fellman, J.H. 1994, Effect of photolytically generated riboflavin radicals and oxygen on hypotaurine antioxidant free radical scavenging activity, in: "Taurine in Health and Disease", Huxtable, R.J. and Michalk, D.V. eds., Plenum Press, New York, pp. 19-29.
5. Huxtable, R.J. 1986, " The Biochemistry of Sulfur", Plenum Press, New York, p. 134.
6. Jocelyn, P.C. 1972, " Biochemistry of the SH Group", Academic Press, London, pp. 107-110.
7. Kice, J.L. and Bowers, K.W. 1962, The mechanism of the disproportionation of sulfinic acids, *J.Amer.Chem.Soc.* 84:605-610.
8. Newton, G.L., Dorian, R., and Fahey, R.C. 1981, Analysis of biological thiols: Derivatization with monobromobimane and separation by reverse-phase high-performance liquid chromatography, *Anal.Biochem.* 114:383-387.
9. Ricci, G., Duprè, S., Federici, G., Spoto, G., Matarese, R.M., and Cavallini, D. 1978, Oxidation of hypotaurine to taurine by ultraviolet irradiation, *Physiol.Chem.&Physics*, 10:435-441.
10. Wellisch, E., Gipstein, E., and Sweeting, O.J. 1960, Thermal decomposition of sulfinic acids, *J.Org.Chem.* 27:1810-1812.

2

THE REACTIVITY OF TAURINE WITH HYPOCHLOROUS ACID AND ITS APPLICATION FOR EYE DROPS

Ikuo Koyama, Katsu Nakamori, Tomomi Nagahama, Mizuho Ogasawara, and Masami Nemoto

Research Center
Taisho Pharmaceutical Co., Ltd.
No.403, Yoshino-Cho, 1-Chome Ohmiya-Shi, Saitama, 330, Japan

INTRODUCTION

Taurine plays an important role in protecting neutrophils from oxidative attack by excessive HOCl.[8,10] This protection is attributed to taurine functioning as a trap for HOCl or a competing amine that dose not yield toxic N-Cl derivatives. Wright *et al.* reported the role of taurine as a scavenger of HOCl in biological systems.[11]

HOCl is a potent oxidizing agent whose biocidal properties are well known, particularly in the disinfection of municipal water systems. It can directly oxidize a variety of biologically significant substances such as carbohydrates, nucleic acids, peptides and amino acids. Kitano *et al.*[7] reported that HOCl used to disinfect water in swimming pools caused corneal damage. In the present study, the ability to protect the ocular surface against HOCl was investigated using canine erythrocytes and rabbit eyes for the purpose of developing eye drops containing taurine.

MATERIALS AND METHODS

Materials

The taurine used in this study was a standardized product (51 A.M,No.796) recognized by the Ministry of Welfare which had been synthesized in our laboratory. Sodium Hypochlorite solution (NaClO) and N,N-diethyl-*p*-phenylene-diamine sulfate (DPD) were purchased from WAKO Pure Chemical Industries. Sepharose gel with free amino groups (EAH-Sepharose gel 4B) was purchased from Pharmacia LKB. The chemicals used in this study were reagent grade commercial products.

Taurine 2, edited by Huxtable et al.
Plenum Press, New York, 1996

Cells

Fresh blood from a healthy beagle dog was drawn into heparinized tubes. Red blood cells were then separated by centrifugation at 550 x g for 5 min and washed three times with 40 mM phosphate-buffered isotonic saline solution (PBS; pH 7.4). The packed cells were then diluted with PBS to prepare a red blood cell suspension of 4 x 10^8 cells/ml. This erythrocyte suspension was kept at 4°C and used as the stock solution for experiments.

Quantitative Determination of Hemolytic Behavior

The amount of oxidized chlorine (Cl^+) was determined using the DPD Ferrous Titrimetric Method[1]. Free available chlorine reacts instantly with DPD indicator to produce a red color. Subsequent addition of a small amount of iodide ion acts catalytically to cause monochloramine (-NHCl) to produce color. Further addition of iodide ion to excess evokes a rapid response from dichloramine (-NCl_2). In the titrimetric procedure, decolorization by standard ferrous ammonium sulfate (FAS) titrant is instantaneous, thereby enabling each step to be performed more rapidly. One ml of HOCl solution at various concentrations was added to 1 ml of erythrocyte suspension (4 x 10^8 cells/ml) in a test tube at 4°C.

After treatment for 5 s, the erythrocytes were then separated by centrifugation at 550 x g for 1 min and washed once with PBS in order to remove free HOCl and oxidized soluble substances. The washed erythrocytes were then diluted with PBS to 4 x 10^7 cells/ml. HOCl-treated erythrocytes were stable for at least 1 h at 4°C. HOCl-treated erythrocytes (10^7 cells/ml) were incubated with various concentrations of taurine or other amino acids over a 4 h period at 37°C. HOCl-induced hemolysis was assayed by measuring the absorbance of released hemoglobin at 416 nm. The percentage (I) of inhibition of lysis by taurine and other amino acids after 4 h incubation was calculated using the formula:

$$I(\%) = 100 \times (T-E)/(T-C)$$

where T and E are the percentage of lysis of HOCl-treated erythrocytes with and without amino acids, respectively, and C is the percentage of lysis of untreated erythrocytes.

Determination of Cl^+ Removed from HOCl-Treated Erythrocytes and from HOCl-Treated Sepharose Gel by Taurine

To prepare the HOCl-treated Sepharose gel, 1 ml of the gel with between 7 μmol and 11 μmol of free amino groups was used. The Sepharose gel was washed 3x and diluted 10x with PBS. HOCl (2.5 mM; 1 ml) was added to 1 ml of the Sepharose gel suspension in a test tube at 4°C. Since the amount of amino groups in the Sepharose gel was greatly in excess of HOCl, a portion of each amino group was oxidized by HOCl, thereby producing monochloroamino groups (-NHCl). HOCl-treated Sepharose gel was then separated by centrifugation at 550 x g for 5 min and washed 3x with PBS. HOCl-treated Sepharose gel was diluted with PBS to produce the same Sepharose gel concentration as that before HOCl-treatment. The HOCl-treated Sepharose gel suspension was kept at 4°C. HOCl-treated erythrocyte suspension (4 x 10^8 cells/ml) or HOCl-treated Sepharose gel suspension of 2 ml was incubated with 2 ml of PBS and 4 ml of various concentrations of taurine for a 4 h period at 37°C. The amount of TauNHCl formed in the supernatant by taurine scavenging Cl^+ was determined using the DPD Ferrous Titrimetric Method[1] or by measuring the absorbance of TauNHCl at 250 nm[11]. TauNHCl was stable enough that by measuring it we were able to determine the amount of Cl^+ moiety removed.

Meniscus Tear Sampling Method and Measurement of LDH Activity in Meniscus Tears

Meniscus tear samples from normal albino rabbits (males, 2.5-4.0 kg) were collected by the saline pool method. In the saline pool method, 150 μl of saline was poured into the cul-de-sac of a rabbit eye (the pouch between the cornea and conjunctiva), and tears were collected from the meniscus tear. The LDH activity in 80 μl of the tear sample was measured with a Hitachi Automatic Analyzer (7150 type).

Measurement of LDH Release Into Meniscus Tears Induced by HOCl Treatment

Various HOCl solutions were prepared by diluting NaClO solution with 40 mM phosphate-buffered isotonic saline solution (pH 7.4, PBS). Concentrations were confirmed using the DPH ferrous titrimetric method.[1] The meniscus tear was collected by the saline pool method as described above. The LDH activity in meniscus tears of each albino rabbit before HOCl-treatment was measured, and this value was used as the initial value of LDH activity. After serial instillation of 150 μl of HOCl solution at various concentrations, the ocular surface was washed once with 150 μl of saline to remove the free HOCl and certain oxidized soluble substances, and then meniscus tear samples were collected at 120 min. A 30 μl portion of the meniscus tear sample was used for measuring LDH activity.

Effect of Post-Application of Taurine

After serial instillations of 150 μl of 0.7 mM HOCl solution for 60 s, the ocular surface was washed once with 150 μl of saline. Then 150 μl of 240 mM taurine, or saline as a negative control, was instilled in the cul-de-sac of rabbit eyes. Thirty minutes later, a meniscus tear sample was collected and LDH activities were measured.

Effect of Pre-Application of Taurine

Thirty minutes after serial instillation of 150 μl of isotonic taurine solution (pH 7.0) at various concentrations or saline as a negative control, 150 μl of HOCl solution (0.7 mM) was instilled into the cul-de-sac of rabbit eyes. At a designated time after HOCl treatment, meniscus tear samples were collected using the saline pool method and LDH activities were measured.

RESULTS AND DISCUSSION

The lysis of HOCl-treated erythrocytes depended on the concentration of HOCl and the incubation time (Fig. 1). Although the highest concentration of HOCl (423 μM) caused rapid lysis of erythrocytes, the hemoglobin inside the erythrocytes was not oxidized. At concentrations less than 213 μM, HOCl attacked the erythrocyte membranes, causing gradual lysis. The lysis of HOCl-treated erythrocytes followed swelling due to increase in membrane permeability to water and ions. HOCl is an exceedingly powerful oxidant, known to react rapidly with amines ($R\text{-}NH_2$) and thiols (RSH).[2] These reactions yield chloramines (R-NHCl) and disulfides (R-SS-R). Thus, the reaction of HOCl with amines such as taurine, glycine or alanine resulted in formation of the respective chloramines, as detected by uv spectrophotometry (λ_{max} 250 nm). In the same fashion, HOCl may attack and decompose

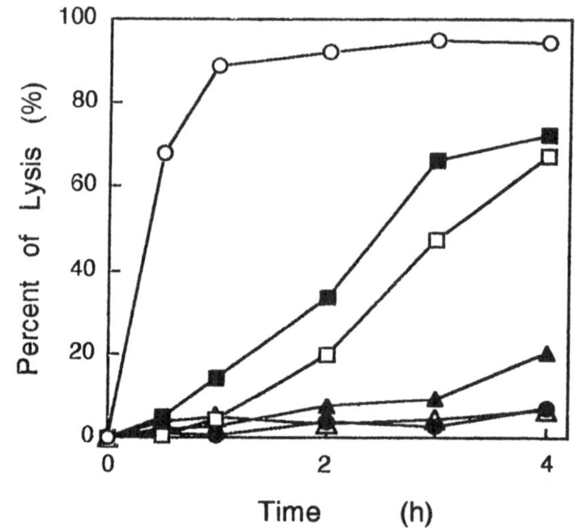

Figure 1. HOCl-induced lysis of erythrocytes. Δ, 0 μM HOCl (final concentration); ●, 70 μM HOCl; ▲, 105 μM HOCl; □, 140 μM HOCl; ■, 213 μM HOCl; ○, 423 μM HOCl. The erythrocytes (10^7 cells/ml) were treated with varying concentrations of HOCl at 37 ºC.

-NH_2 and -SH groups at the membrane of the target cell, leading to lysis due to membrane denaturation. The lysis may be due mainly to the gradual decomposition of amine- and sulfhydryl-containing molecules of membrane components, and to the gradual degradation of peptide bonds. Lipid peroxidation of the membrane may also contribute to the lysis.

The gradual lysis of erythrocytes that had been previously treated with HOCl was inhibited to a remarkable extend by the addition of taurine (Fig. 2). The inhibition depended

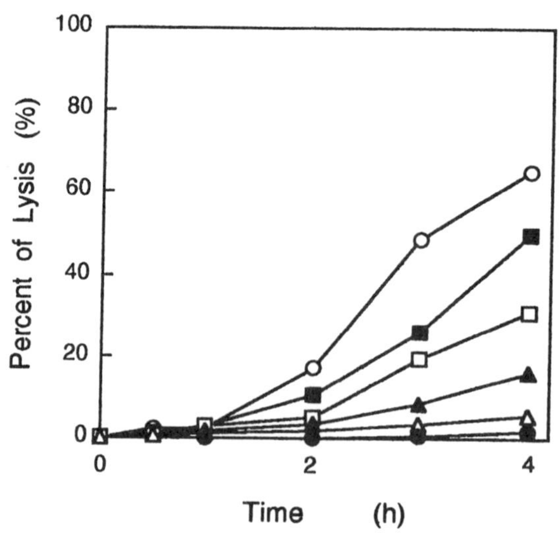

Figure 2. Effect of taurine on inhibition of HOCl-induced lysis. ○, 0 mM (final concentration); ■, 5mM taurine; □, 20 mM taurine; ▲, 40 mM taurine; Δ, 120 mM taurine; ▼, 160 mM taurine; ●, control. HOCl-treated erythrocyte (10^7 cells/ml) were incubated at 37 ºC in PBS with varying concentrations of taurine. Intact erythrocytes were incubated at 37 ºC in PBS as the control. The final concentration of HOCl used in HOCl-treatment was 140 μM.

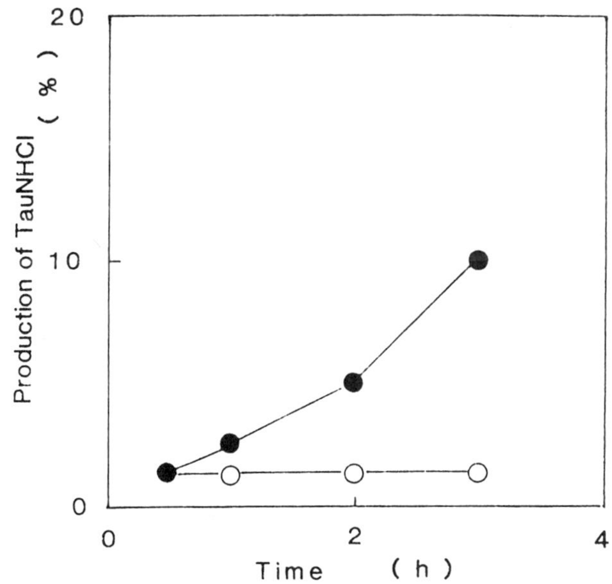

Figure 3. Effect of taurine on removal of the oxidized chlorine (Cl^+) from HOCl-treated erythrocytes. ○, 0 mM taurine; ●, 160 mM taurine. HOCl-treated erythrocytes (10^8 cells/ml) were incubated in PBS with taurine. The final concentration of HOCl used in HOCl-treatment was 0.175 mM. Percentage of removed Cl^+ moiety was calculated by determination of the amount of TauNHCl using the DPD ferrous titrimetric method.

on the concentration of taurine, complete inhibition of lysis being attained at 160 mM. There are some reports that taurine efficiently inhibited the lysis of human erythrocytes (HRBC) caused by HOCl generated in neutrophils, as well as the lysis caused by HOCl in a cell-free system.[4,9] This effect of taurine was employed mainly to protect HRBC from the attack of HOCl by scavenging HOCl in the medium. In our study, however, HOCl and other oxidized substances in the medium were removed by washing HOCl-treated erythrocytes. Accordingly, our findings suggest that taurine effectively inhibits lysis by suppressing denaturing of membranes attacked by HOCl.

Fig. 3 shows that taurine removed oxidized chlorine (Cl^+) from HOCl-treated erythrocytes, resulting in the formation of TauNHCl. In the supernatant of the suspension of HOCl-treated erythrocytes with taurine during incubation at 37°C, TauNHCl was formed gradually at a rate depending on the incubation time, but the concentration of the removed Cl^+ moiety was low. It has been reported that ascorbate, methionine, hypotaurine, $FeSO_3$, KCN and Na_2SO_3 all reduced TauNHCl with equivalent stoichiometry, and that cysteine, glutathione and KI reduced TauNHCl at a ratio of 2 mol of reductant to 1 mol of TauNHCl[5]. Thomas *et al.* reported that TauNHCl in the medium was transported into erythrocytes by the anion-transport system.[9] Steady-state concentrations of TauNHCl are difficult to determine with accuracy in complex biological systems. However, our findings indicate that removal of the Cl^+ moiety from HOCl-treated erythrocytes by taurine resulted in the production of TauNHCl and inhibition of lysis of HOCl-treated erythrocytes.

Fig. 4 shows the effect of removal of the Cl^+ moiety from HOCl-treated Sepharose gel by taurine. Transfer of the electrophilic Cl^+ moiety from monochloramine groups (-NHCl) in HOCl-treated Sepharose gel to the nucleophilic acceptor, taurine, occurred, thereby producing TauNHCl. The yield of TauNHCl depended on the concentration of taurine and the incubation time. This finding supports the hypothesis that taurine removes

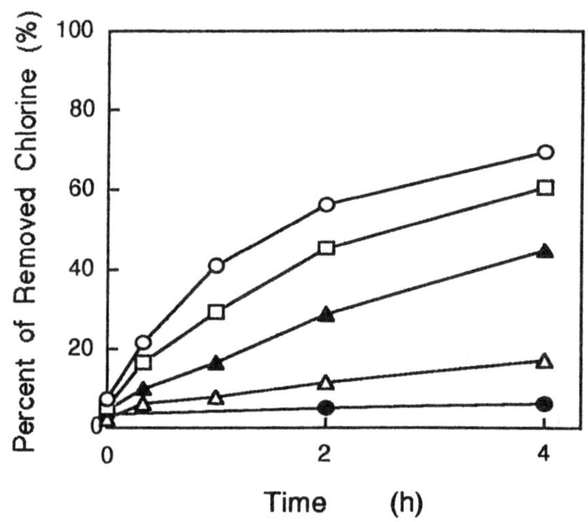

Figure 4. Effect of taurine on removal of the oxidized chlorine (Cl^+) from HOCl-treated sepharose gel. ●, 0 mM (final concentration); △, 10 mM taurine; ▲, 20 mM taurine; □, 40 mM taurine; ○, 80 mM taurine. HOCl-treated sepharose gel was incubated with varying concentrations of taurine at 37 ºC. The final concentration of HOCl used in HOCl-treatment was 0.7 mM. Percentage of removed Cl^+ moiety was calculated from absorbance of TauNHCl at 252 nm.

the Cl^+ moiety from monochloroamine groups in the membrane proteins and lipids of erythrocytes attacked by HOCl.

Amino acids other than taurine also inhibited the lysis of HOCl-treated erythrocytes, and the extent of inhibition of lysis depended on the concentration of amino acids (Fig. 5). The concentrations of taurine, α-alanine, β-alanine and glycine required for 50 % inhibition of the lysis of erythrocytes were 18, 30, 34, and 40 mM, respectively. If the inhibition of lysis of HOCl-treated erythrocytes by amino acids is mainly due to removal of the Cl^+ moiety, then taurine was the most potent reagent of all amino acids tested here.

The accurate sampling of meniscus tears is required for quantitative estimation of ocular surface damage by measurement of the amounts of tear components released from the ocular surface tissues during eye irritation. Imayasu *et al.* reported that LDH activity in meniscus tears collected with a 2 µl glass micropipette is a good index for estimating corneal damage.[6] In this method, 2 µl of a tear sample must be collected carefully within 30 s in order to avoid inclusion of LDH produced by stimulation of, or damage to, the ocular surface at tear sampling. Because meniscus tear volume is small,[3] 7.5 ± 2.5 µl, it is difficult to collect meniscus tears within 3 s. We developed the saline pool method, in which all meniscus tears are collected within 10 s after pooling the tears in the cul-de-sac of a rabbit eye with 150 µl of saline. This method has the following advantages over the conventional micropipette method: first, it is easy to collect meniscus tears, and secondly, the LDH value is thought to be the least sensitive to dilution by tear secretion induced by eye irritation and the presence of residual sample in the cul-de-sac. The mean LDH value obtained by the saline pool method was 58.5 ± 32.3 U/L (n=42), which corresponds to an actual LDH concentration of 1229 U/L, since in the saline pool method, the LDH value in meniscus tears was diluted approximately 21 times (157.5/7.5 = 21) based on the normal resident volume (7.5 µl) in the cul-de-sac of rabbit eyes. This value is close to that obtained by the conventional micropipette method, showing that the mean LDH value obtained by the saline pool method reflects the real LDH concentration in meniscus tears.

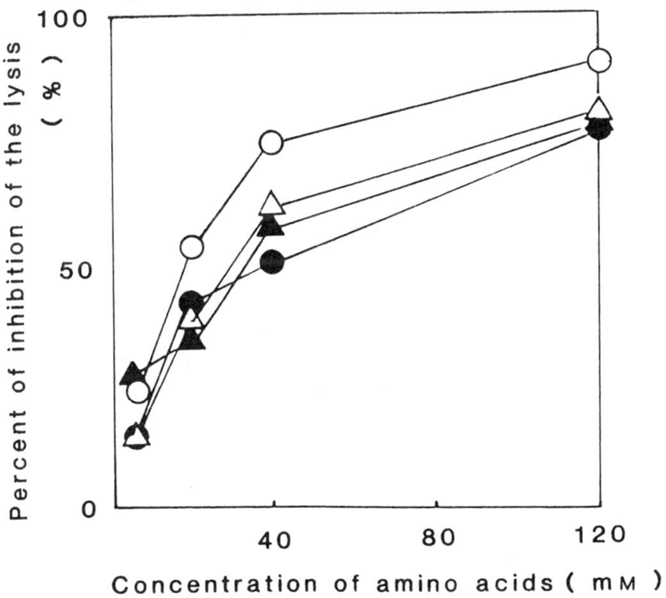

Figure 5. Effect of Various amino acids on inhibition of the HOCl-induced lysis. ○, taurine; ●, glycine; Δ, α-alanine; ▲, β-alanine. HOCl-treated erythrocytes (10^7 cells/ml) were incubated in PBS with various amino acids at 37 ºC. Concentration of HOCl used in HOCl treatment was 0.7 mM.

The LDH activity 30 min after serial instillations of HOCl for 30 s increased dose-dependently with increasing HOCl concentration, indicating that LDH activity is a good index of ocular surface damage due to HOCl (Fig. 6).

Fig. 7 shows the effect of application of taurine on ocular surface damage induced by HOCl. Immediately after serial instillations of 0.7 mM HOCl for 30 s, ocular surface tissues were washed once with saline in order to remove the free HOCl and the oxidized meniscus tear components, and then 240 mM taurine and saline were instilled in the same fashion. The LDH activity 30 min after post-application of 240 mM taurine was significantly ($p < 0.01$, n=4-5) lower than that in the case of saline application. This finding indicates that taurine is effective in suppressing further damage of ocular surface tissues already attacked by HOCl. The finding of our *in vitro* study indicated that taurine effectively inhibited the lysis of canine erythrocytes which had been pre-treated with HOCl, due to scavenging of the oxidized chlorine moiety from the HOCl-treated erythrocytes and a direct protective effect on erythrocyte membranes. In addition to protecting canine erythrocyte membranes against HOCl, our findings indicated that taurine is effective in protecting ocular surface tissues against HOCl by suppressing increase in damage.

At 30 min after serial instillations of 240 mM taurine for 1 min, 0.7 mM HOCl was instilled in the same fashion (Fig. 8). LDH activity at 30 min after HOCl treatment was significantly ($p < 0.01$, n=4-5) lower than that in the case of saline application. Ocular surface tissues can be protected against HOCl as long as an amount of taurine sufficient for scavenging HOCl remains in ocular surface tissues. In a histochemical and morphological examination of enucleated rabbit eyes, Yoshimura *et al.* found that taurine is effective in healing corneal damage caused by high concentration of HOCl (5.6 mM) in rabbits.[12] Using our quantitative method of evaluation, in which LDH activity in meniscus tears used as an index of ocular surface damage, we showed that the application of taurine is effective in protecting against ocular surface damage caused by a low concentration of HOCl (0.7 mM)

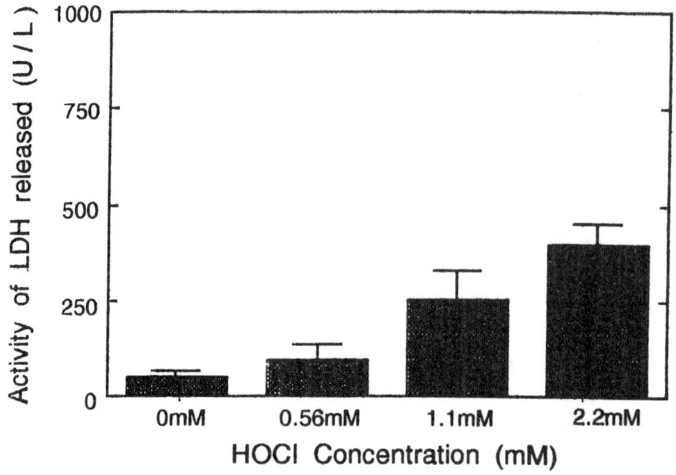

Figure 6. Effect of HOCl concentration on HOCl-induced release of LDH from ocular surface tissues. Each value was LDH activity after 30 min from HOCl-treatment. Each value represents the means ± S.E. (n = 6 ~ 7).

actually used in swimming pools. These findings suggest that the use of taurine in swimming pools may significantly decrease ocular discomfort.

SUMMARY

The ability of taurine to protect biomembranes attacked by HOCl was examined using canine erythrocytes which had been pre-treated with HOCl. Of the amino acids tested, taurine

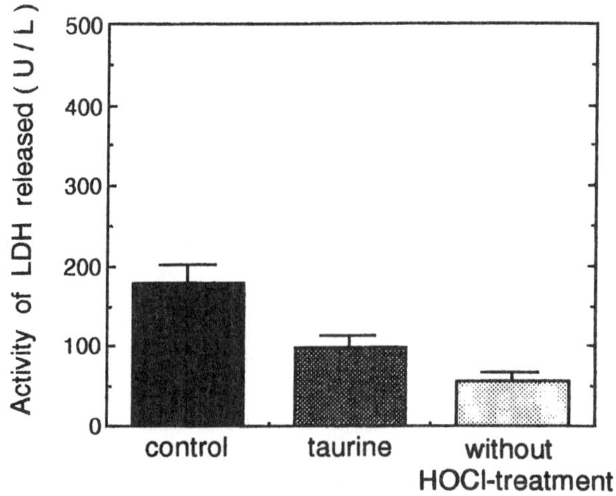

Figure 7. Effect of post-application of taurine on the ocular surface damage caused by HOCl. 0.9% NaCl was used as a control and the concentration of taurine was 240 mM. HOCl used in HOCl-treatment was 0.7 mM. Each sample was dropped to the ocular surface after 30s from HOCl-treatment. Each value was LDH activity after 30 min from HOCl-treatment. Each value represents the means ± S.E. (n = 10).

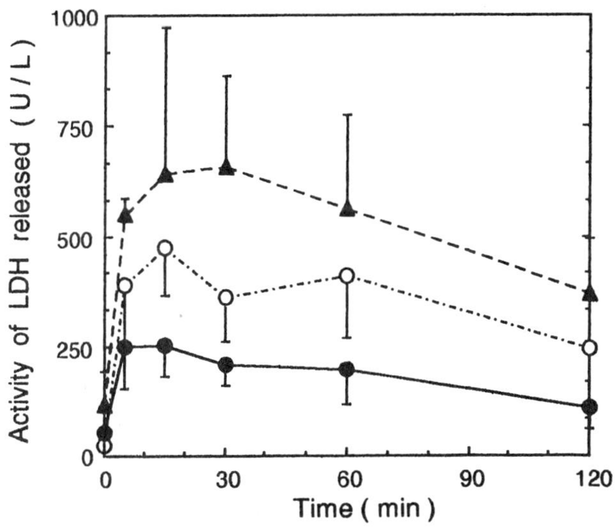

Figure 8. Effect of pre-application of taurine on the HOCl-induced release of LDH from ocular surface tissues. ▲, 0.9% NaCl; ○, 80 mM taurine; ●, 240 mM taurine. HOCl used in HOCl-treatment was 4.2 mM. Each sample was dropped to the ocular surface before 10 min from HOCl-treatment. Each value represents the mean ± S.E. (n = 4 ~ 5).

was the most effective in inhibiting attack by HOCl, followed by glycine, α-alanine and β-alanine. During the incubation of HOCl-treated erythrocytes with taurine, an appreciable amount of monochlorotaurine (TauNHCl) was detected in the supernatant. This suggests that taurine may remove the oxidized chlorine from HOCl-treated erythrocytes, resulting in the production of TauNHCl. The effect of taurine on the removal of Cl^+ moiety was examined using Sepharose gel with amino groups. Taurine removed Cl^+ from HOCl-treated Sepharose gel. The yield of TauNHCl depended on the concentration of taurine.

The effectiveness of taurine in preventing ocular surface damage caused by HOCl was investigated in albino rabbits. The activity of lactate dehydrogenase (LDH) released from ocular tissues into tears during eye irritation was used as an index of ocular surface damage. Taurine effectively protected ocular surface tissues from damage induced by HOCl, and arrested the progression of tissue damage that had already been initiated by HOCl. These finding suggest that taurine may be clinically useful in the treatment of ocular surface damage caused by oxidants such as HOCl.

REFERENCES

1. A.P.H.A., A.W.W.A., and W.P.C.F. 1975, Standard Methods for the Examination of Water and Wastewater", 14th edition, Washington, D.C. pp. 329-332.
2. Albrich, J.M., McCarthy, A., and Hurst, J.K. 1981, Biological reactivity of hypochlorous acid. *Proc.Natl.Acad.Sci.USA*, 78:210-214.
3. Charai, S.S., Patton, T.F., Metha, A., and Robinson, J.R. 1973, Lacrimal and instilled flud dynamics in rabbit eyes. *Pharm.Sci.* 62:1112-1121.
4. Dallegri, F., Ballestrero, A., Frumento, G., and Patrone, F. 1985, Erythrocyte lysis by PMA-triggered neutrophil polymorphonuclears. *Immunology*, 55:639-645.
5. Dilley, J.V. 1972, The origin of urinary taurine excretion during chronic radiation injury, *Radiation Res.* 50:191-196.

6. Imayasu, M., Hirata, T., Mitsunaga, S., Kotani, S., and Hamano, H. 1990, Quantitative evaluation of corneal lesion by measuring the lactate dehydrogenase activity in rabbit. *Atarashii Ganka*, 7:297-302.
7. Kitano, S. and Yoshimura, H. 1985, Corneal and conjunctival damage due to chlorine. *Nihon no Ganka*, 56:539-546.
8. Thomas, E.L., Grisham, M.B., and Jefferson, M.M. 1983, Myeloperoxidase-dependent effect of amines on functions of isolated neutrophils, *J.Clin.Invest.* 72:441-454.
9. Thomas, E.L., Grisham, M.B., Melton, D.F., and Jefferson, M.M. 1985, Evidence for a role of taurine in the *in vitro* oxidative toxicity of neutrophils toward erythrocytes, *J.Biol.Chem.* 260:3321-3329.
10. Weiss, S.J., Klein, R., Slivka, A., and Wei, M. 1982, Chlorination of taurine by human neutrophils - evidence for hypochlorous acid generation, *J.Clin.Invest.* 70:598-607.
11. Wright, C.E., Lin, T.T., Lin, Y.Y., Sturman, J.A., and Gaull, G.E. 1985, Taurine scavenges oxidized chlorine in biological systems, in: "Taurine: Biological Actions and Clinical Perspectives", Oja, S.S., Ahtee, L., Kontro, P. and Paasonen, M.K. eds., Alan R. Liss, New York, pp. 137-147.
12. Yoshimura, H., Shoji, J., and Kitano, S. 1989, Therapeutic effect of taurine on corneal damage due to chlorine. *Folia Ophthalmol.Jpn.* 40:1797-1804.

3

DEPLETION OF FELINE TAURINE LEVELS BY β-ALANINE AND DIETARY TAURINE RESTRICTION

J. A. Sturman, P. Lu, J. M. Messing, and H. Imaki

New York State Institute for Basic Research in Developmental Disabilities
Department of Developmental Biochemistry
1050 Forest Hill Road, Staten Island, New York 10314

INTRODUCTION

It has long been known that cats are dependent on a dietary source of taurine to maintain their body pools because they have a limited capacity for biosynthesis. Taurine deficiency results in retinal degeneration, myocardial failure, decreased immune responsiveness, and a profound adverse effect on feline pregnancy and outcome of the progeny[30,32,33]. Even with zero dietary taurine intake, tissue levels of taurine are only reduced to the point that equilibrium with the biosynthesized taurine is reached, no matter how long cats are maintained on such a diet[25,34]. A number of abnormalities have been documented in the brains of surviving kittens from taurine-deprived mothers[33], but, to date, no abnormalities have been reported in the brains of taurine-deprived adult cats, other than the decreased concentration of taurine. In this study we have attempted to reduce taurine levels even further by using β-alanine, which competes with taurine for the same transport systems, in the drinking water. β-Alanine was used rather than the more frequently used competitive inhibitor of taurine transport, guanidinoethanesulfonic acid (GES), because cats are able to extensively metabolize GES to form taurine using a transamidinase or amidinohydrolase[14,15,31]. We report the concentrations of taurine and β-alanine in tissues and fluids of cats following 20 weeks of β-alanine ingestion, and document the resulting morphological changes in the cerebellum and profound degenerative changes in the retina.

EXPERIMENTAL PROCEDURES

Female domestic cats raised in the colony of the Institute for Basic Research in Developmental Disabilities and vaccinated against rhinotracheitis, panleukopenia, calici virus [FVR-C-P (MLV) Pitman-Moore, Washington Crossing, NJ, USA] were fed a completely defined, taurine-free purified diet (Bioserve, Frenchtown, NJ, USA) alone or containing 0.05% taurine for at least two years before being assigned to this study. Five cats in

Taurine 2, edited by Huxtable et al.
Plenum Press, New York, 1996

each group were caged individually and fed the same diets and provided with 5% β-alanine in the drinking water for 20 weeks. At this time the cats were killed by exsanguination following an overdose of sodium pentobarbital (Nembutal, Abbott Laboratories, North Chicago, IL, USA). Other taurine-supplemented and taurine-deprived cats were killed in a similar way after drinking water alone, for comparison. The concentrations of taurine in tissues of these cats have been reported previously[34] and some of the values are included in this report for comparison; some immersion-fixed brains and syringe-perfused eyes from these same animals were used for morphological and immunohistochemical studies.

From each cat the brain was removed and the right half immersed in cold fixative (1% paraformaldehyde and 2.5% glutaraldehyde in 0.1 M Sorensen's phosphate buffer, pH 7.4) for several days prior to processing for routine histology and immunohistochemistry, and the right eyes enucleated and perfused with the same fixative using syringes and needles, before immersing in the same fixative for storage.

The left halves of brain, the left eyes, and other body tissues were rapidly dissected and frozen on dry ice prior to extraction for determination of taurine and β-alanine concentrations. Tissues were homogenized in 10 vol 20% trifluoroacetic acid (TFA) and centrifuged at 20,000 x g for 30 min.

The clear supernatants were passed through a 0.45 μm μStar filter (Costar, Cambridge, MA, USA) and stored at -70°C until used for taurine and β-alanine measurements. Analysis was accomplished by derivatizing with phenylisothiocynate and the taurine and β-alanine derivatives separated by reverse-phase HPLC. The apparatus consisted of a Spectra Physics 8800 Ternary HPLC pump (Piscataway, NJ, USA) and a 4.6 mm X 25 cm Baker Bond C-18 column (Baker, Phillipsburg, NJ, USA) maintained at 34°C. The taurine and β-alanine derivatives were detected at 254 nm with an LDC SpectroMonitor D (Milton Roy, Riviera Beach, FL, USA) and quantified using Nelson Analytical 2600 chromatography software (Cupertino, CA, USA) with an IBM-compatible 486 PC (Gateway, Sioux City, SD, USA). A refrigerated Waters 712 WISP (Milford, MA, USA) maintained at 5°C allowed the automatic analysis of up to 96 samples.

Standard deviations and significance of difference between groups were calculated using Student's t test using a standard computer program (STATA, Computing Resource Center, Santa Monica, CA, USA).

Immunohistochemistry

Immunogens were synthesized by the method of Campistron *et al.*,[1] starting with 100 mg of taurine or β-alanine, and coupling with glutaraldehyde to bovine serum albumin (BSA) or poly-L-lysine (PL; MW 30 to 70 kD, Sigma Chemical Co., St. Louis, MO, USA).

Rabbits were immunized by first injecting 500 μg of immunogen emulsified in Freund's incomplete adjuvant. Injections alternated amino acid-glutaraldehyde-BSA and amino acid-glutaraldehyde-PL. Two rabbits were injected s.c. and two i.m. Blood was collected 7 and 21 days after each injection and serum frozen in 1 ml batches.

The serum was characterized for titer and cross-reactivity using the ELISA method[19]. High, stable titers of reactivity against amino acid-glutaraldehyde-BSA were reached after three injections (> 1:64,000 dilution). The anti-taurine serum had a small cross-reactivity to GABA-glutaraldehyde-BSA and to β-alanine-glutaraldehyde-BSA. This reactivity was removed by preadsorption with both conjugates and resulted in little loss of taurine-glutaraldehyde-BSA activity. Similarly, the anti-β-alanine serum was preadsorbed with taurine-glutaraldehyde-BSA and GABA-glutaraldehyde-BSA to ensure the fidelity of the immune reactions.

Tissue was embedded in paraffin and serial 6 μm sagittal sections cut and mounted on glass slides. Sections were deparaffined and hydrated by warming at 60°C for 65 min in a histoclear bath, followed by graded ethanol washes. Staining with the antisera (1 to 2,000 dilution) was carried out at 4°C overnight and visualized using the peroxidase-conjugated avidin/DAB method (Dako Corporation, Carpinteria, CA, USA). Control sections were included in every batch, those in which the antisera were preadsorbed with the immunogen prior to processing, and others replacing the primary antiserum with preimmune serum at similar dilutions. Control slides showed no visible staining after processing, whereas slides prepared with the antisera showed brown reaction products. Counterstaining, when employed, was with hematoxylin. Other histological stains employed were cresyl violet, toluidine blue, hematoxylin and eosin, and Luxol fast blue. Photomicrographs were taken on a Zeiss Axiophot using Kodak 40 color reversal film and Kodak Panatomic X film.

Electron Microscopy

The fixed right eyes were hemisected and the posterior eye cups were dissected radially around the area centralis after securing the optic nerve and separated into the superior, temporal, nasal and inferior sectors, rinsed overnight with the buffer, postfixed in 2% OsO_4 for 1-2 h and dehydrated in an ascending series of acetone solutions before embedding in EM-bed 812. Semithin sections (0.75 μm) were stained with toluidine blue and micrographs were taken by Zeiss Photomicroscope III using Kodak TMX-100 film. Thin sections were stained with uranyl acetate and lead citrate in Ultrostainer (LKB, Bromma, Sweden), and examined in a Philips EM 300 operated at 80 kv.

RESULTS

The cats tolerated the β-alanine in the drinking water well, each cat consuming approximately 500 g of β-alanine during the 20 week period. There was an overall 10% loss of weight during this time. There were no significant differences in routine blood chemistry parameters, previously reported for taurine-supplemented and taurine-deprived cats[34]. The concentration of taurine in tissues of both taurine-supplemented and taurine-deprived cats was reduced globally by feeding β-alanine (Table 1). Large amounts of β-alanine accumulated in non-neural tissues of all cats and in neural tissues of taurine-deprived cats (Table 2). β-Alanine conjugated to cholic acid predominantly and to a lesser extent to chenodeoxycholic acid was found in bile from all taurine-deprived cats. Only traces were noted in bile from some taurine-supplemented cats. No glycine-conjugated bile acids were detected in bile from any cats.

Microscopic examination of the cerebellum of cats treated with β-alanine showed a reduced number of granule cells compared to the cerebellum of cats treated only with water and many of those remaining were pyknotic (Fig. 1). There was also a clear reduction in the number of Purkinje cells and many of those remaining were dead or dying. The cerebellar white matter contained numerous long and swollen fibers resembling Rosenthal fibers[22] which are not present in cats treated only with water (Fig. 2). These fibers also showed staining with an antibody to ubiquitin localized to the periphery (not shown). There was also prominent gliosis in the cerebellum of cats treated with β-alanine, shown by GFAP staining, compared to cats treated only with water (Fig. 3).

Immunohistochemical staining of cerebellum of cats treated with β-alanine with an antibody to β-alanine showed prominent localization in Purkinje cells and their dendrites (Fig. 4). Golgi II cells also contained β-alanine. Some granule cells contained β-alanine, especially in taurine-deprived cats treated with β-alanine while others were devoid. Cerebellum of cats treated only with water were negative to this antibody. Using an antibody to taurine, this compound appears to have been virtually eliminated from cerebellar Purkinje

Table 1. Concentration of taurine in tissues of taurine-supplemented and taurine-deprived cats drinking water or 5% β-alanine. (μmol/g wet weight)

	Taurine-Deprived		Taurine-Supplemented	
Tissue	Water[a]	5% β-Alanine[b]	Water[a]	5% β-Alanine[b]
Retina	15.9 ± 5.5	6.48 ± 3.42	42.4 ± 4.4	21.8 ± 8.9
Tapetum	5.78 ± 2.94	1.20 ± 0.23	11.1 ± 3.6	3.43 ± 1.57
Lens	1.68 ± 0.88	0.69 ± 0.36	6.64 ± 2.30	1.32 ± 0.56
Liver	0.60 ± 0.48	0.15 ± 0.06	8.50 ± 3.33	0.77 ± 0.56
Kidney	0.92 ± 0.45	0.41 ± 0.13	5.15 ± 1.91	0.97 ± 0.65
Lung	2.11 ± 1.63	0.34 ± 0.18	8.28 ± 2.60	2.08 ± 1.18
Spleen	1.46 ± 0.70	0.40 ± 0.19	7.34 ± 2.44	1.17 ± 0.76
Adrenal	3.63 ± 1.87	1.03 ± 0.53	12.5 ± 3.6	3.01 ± 1.58
Heart	1.67 ± 1.04	0.40 ± 0.26	12.0 ± 2.7	1.85 ± 0.83
Gastrocnemius	0.82 ± 0.38	0.42 ± 0.08	5.84 ± 1.02	1.10 ± 0.68
Biceps	1.26 ± 0.94	0.32 ± 0.08	6.35 ± 1.62	0.89 ± 0.34
Triceps	1.47 ± 1.41	0.46 ± 0.25	7.83 ± 2.96	1.03 ± 0.61
Diaphragm	0.67 ± 0.57	0.14 ± 0.01	5.49 ± 2.33	0.35 ± 0.13
Plasma, μmol/L	7.6 ± 6.1	84 ± 67	127 ± 53	49 ± 51
Whole blood, μmol/L	3.85 ± 32.5	62 ± 24	252 ± 105	70 ± 69
Occipital lobe	0.56 ± 0.48	0.17 ± 0.07	2.28 ± 0.88	0.27 ± 0.14
Frontal lobe	0.50 ± 0.37	0.19 ± 0.08	2.19 ± 0.49	0.46 ± 0.24
Temporal lobe	0.43 ± 0.35	0.15 ± 0.06	1.92 ± 0.88	0.38 ± 0.21
Parietal lobe	0.57 ± 0.48	0.15 ± 0.09	2.32 ± 0.76	0.31 ± 0.15
Cerebellum	0.45 ± 0.24	0.16 ± 0.05	3.14 ± 0.80	0.45 ± 0.28
Superior colliculus	0.27 ± 0.13	0.12 ± 0.04	1.65 ± 0.44	0.24 ± 0.08
Inferior colliculus	0.36 ± 0.28	0.11 ± 0.03	1.56 ± 0.47	0.26 ± 0.09
Hippocampus	0.42 ± 0.22	0.19 ± 0.07	2.19 ± 0.64	0.54 ± 0.33
Corpus callosum	0.40 ± 0.29	0.12 ± 0.07	2.67 ± 0.57	0.36 ± 0.18
Thalamus	0.38 ± 0.44	0.10 ± 0.03	1.55 ± 0.29	0.26 ± 0.11
Pons	0.25 ± 0.25	0.10 ± 0.04	1.48 ± 0.38	0.19 ± 0.14
Medulla	0.36 ± 0.35	0.08 ± 0.03	1.47 ± 0.47	0.19 ± 0.14
Olfactory bulb	3.19 ± 1.57	1.06 ± 0.41	7.43 ± 1.89	3.55 ± 1.17
Lateral geniculate nucleus	0.39 ± 0.30	0.12 ± 0.04	1.95 ± 0.71	0.25 ± 0.09
Optic tract	1.07 ± 0.48	0.16 ± 0.07	2.25 ± 0.84	0.37 ± 0.19
Optic nerve	1.37 ± 0.58	0.63 ± 0.42	3.29 ± 1.08	1.00 ± 0.57
Spinal cord	0.39 ± 0.44	0.13 ± 0.08	1.03 ± 0.41	0.19 ± 0.10
Sciatic nerve	0.27 ± 0.29	0.16 ± 0.10	1.14 ± 0.40	0.18 ± 0.14

[a]Values published in Sturman and Messing[34]; Messing and Sturman[25]; [b]Mean of 5 cats ± SD.

and granule cells of cats treated with β-alanine and concentrated in Golgi II and glial cells (Fig. 5). β-Alanine was also localized in neurons in other brain regions: in pyramidal cells and granule cells of the hippocampus, and in the visual cortex (not shown).

Light microscopic examination of the retina showed a range of degenerative changes, from moderate to severe in the taurine-supplemented cats, and from quite severe to extremely severe in the taurine-deprived cats. Sections from superior and inferior retina are illustrated in Fig. 6 and sections from temporal and nasal retina are illustrated in Fig. 7. In all cases sections from different cats in each group are presented in order of decreasing retinal taurine concentration which invariably corresponds with increasing severity of changes. In general, the severity of changes in a given cat was nasal > temporal > superior > inferior sector of retina. The morphology of retinas from taurine-supplemented cats treated with β-alanine ranged from small abnormalities (Figs. 6g, 6h) to greater changes including a reduced thickness of the outer nuclear layer and reductions in length of both the inner and outer segments (Figs. 6b, 6c, 7g) and sectors with reduced numbers of photoreceptors with large

Table 2. Concentration of β-alanine in tissues of taurine-supplemented and taurine-deprived cats drinking 5% β-alanine. (μmol/g wet weight)

Tissue	Taurine-deprived[a]	Taurine-supplemented[a]
Retina	9.44 ± 3.03	5.52 ± 1.42
Tapetum	8.71 ± 2.67	3.42 ± 1.50
Lens	3.03 ± 0.53	3.66 ± 1.95
Liver	3.13 ± 1.25	1.88 ± 0.77
Kidney	6.59 ± 4.19	3.24 ± 1.71
Lung	2.25 ± 1.27	1.28 ± 0.53
Spleen	1.37 ± 1.02	1.03 ± 0.58
Adrenal	3.83 ± 2.40	2.81 ± 1.29
Heart	1.11 ± 0.68	1.41 ± 0.66
Gastrocnemius	6.17 ± 3.73	7.19 ± 3.34
Biceps	5.80 ± 3.62	6.21 ± 3.32
Triceps	6.87 ± 4.31	7.54 ± 3.18
Diaphragm	2.58 ± 1.88	3.18 ± 2.42
Plasma, μmol/L	2930 ± 2480	286 ± 262
Whole blood, μmol/L	1660 ± 1330	555 ± 290
Occipital lobe	1.98 ± 0.68	0.14 ± 0.08
Frontal lobe	2.40 ± 0.73	0.14 ± 0.07
Temporal lobe	2.45 ± 0.77	0.16 ± 0.07
Parietal lobe	2.52 ± 0.87	0.18 ± 0.08
Cerebellum	1.82 ± 0.78	0.12 ± 0.08
Superior colliculus	1.83 ± 0.52	0.12 ± 0.07
Inferior colliculus	1.31 ± 0.75	0.18 ± 0.07
Hippocampus	1.72 ± 1.33	0.36 ± 0.15
Corpus callosum	1.76 ± 0.73	0.21 ± 0.07
Thalamus	2.64 ± 1.22	0.19 ± 0.05
Pons	1.73 ± 1.17	0.16 ± 0.12
Medulla	2.17 ± 1.05	0.11 ± 0.04
Olfactory bulb	1.46 ± 0.76	0.52 ± 0.22
Lateral geniculate nucleus	1.63 ± 0.56	0.10 ± 0.04
Optic tract	2.65 ± 1.86	0.09 ± 0.05
Optic nerve	2.32 ± 1.23	0.21 ± 0.10
Spinal cord	5.90 ± 3.90	0.35 ± 0.25
Sciatic nerve	2.68 ± 1.91	0.51 ± 0.28

[a]Mean of 5 cats ± SD.

intercellular spaces between surviving photoreceptor nuclei associated with vestigial remnants of inner and outer segments (Figs. 7b, 7h). These changes were comparable to those observed in retinas of cats fed taurine-depleting diets for prolonged periods. In contrast, the morphology of the retinas from taurine-deprived cats treated with β-alanine generally showed more severe abnormalities than produced by a taurine-free diet alone. The retinas from these cats were all thinner than normal, in some case less than half the normal thickness due primarily to the depletion of photoreceptors. The left retinas which were dissected for measurement of taurine and β-alanine concentrations were extremely fragile. Sections showing the least severe changes are illustrated in Figs. 6d, 6i, 7c and 7i. These sections showed many abnormalities, nevertheless, including severe disruption of the photoreceptor layer with large intercellular spaces between remaining nuclei and barely discernible inner and outer segments (Figs. 6e, 7d). The most severely affected cats had retinas virtually devoid of photoreceptors (Figs. 6f, 6j, 7e, 7j).

Other changes noted by light microscopy include a marked decrease in the number and size of vascular elements in the inner retina, especially in the outer plexiform layer, and in the choriocapillaris, and apparent thinning of retinal epithelium in the retinas of some

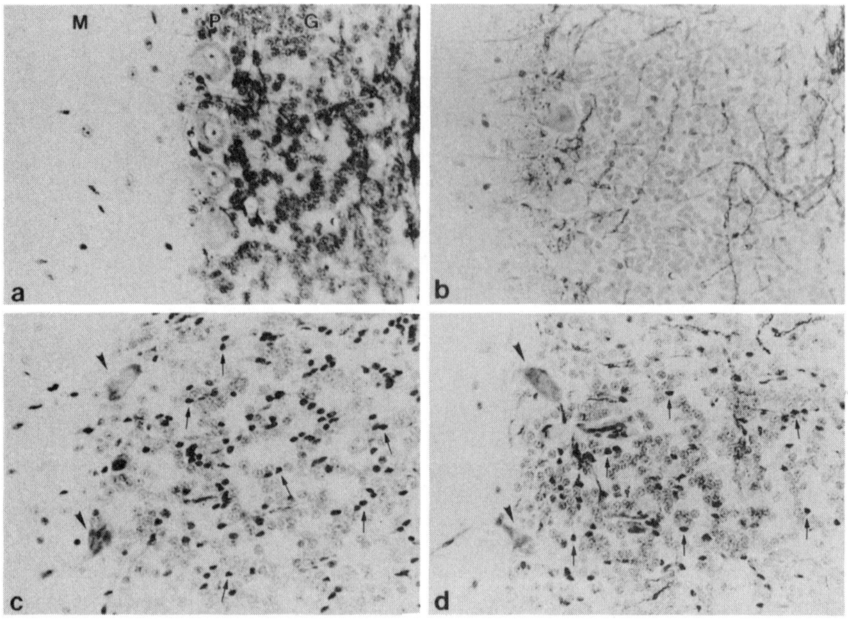

Figure 1. Sagittal sections of cerebellum stained with Luxol fast blue. Cats were raised as follows: a: 0.05% taurine; b: 0% taurine; c. 0.05% taurine + 5% β-alanine; d: 0% taurine + 5% β-alanine. M: molecular layer; P: Purkinje cell layer; G: granule cell layer. Arrowheads indicate pyknotic and dying Purkinje cells; Arrows, pyknotic and dying granule cells. X 230.

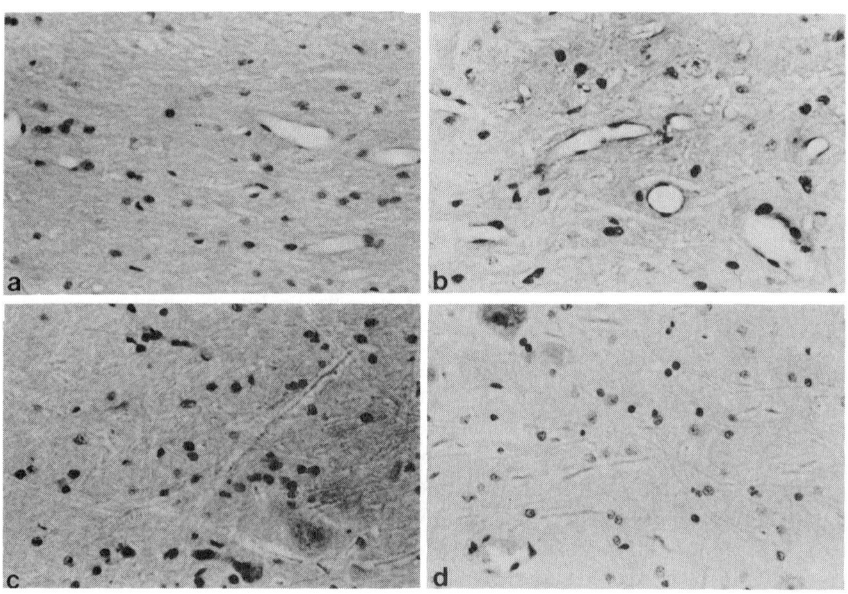

Figure 2. Sagittal sections of cerebellar white matter stained with hematoxylin and eosin. a: 0.05% taurine; b: 0% taurine; c: 0.05% taurine + 5% β-alanine; d: 0% taurine + 5% β-alanine. The swollen fibers are clearly seen in c and d but not in a and b. X 240.

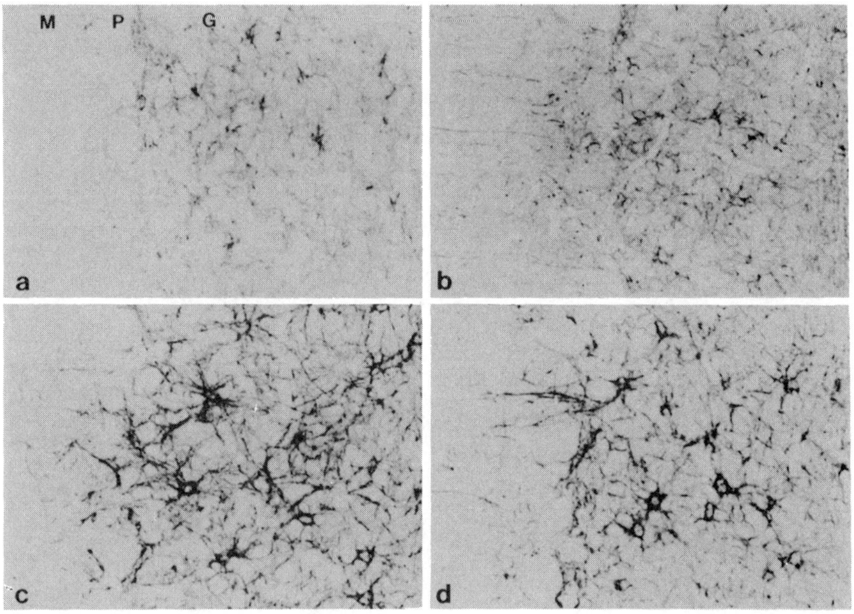

Figure 3. Sagittal sections of cerebellum stained with GFAP. a: 0.05% taurine; b: 0% taurine; c: 0.05% taurine + 5% β-alanine; d: 0% taurine + 5% β-alanine. Note the greatly increased staining in c and d compared with a and b. M: molecular layer; P: Purkinje cell layer; G: granule cell layer. X 270.

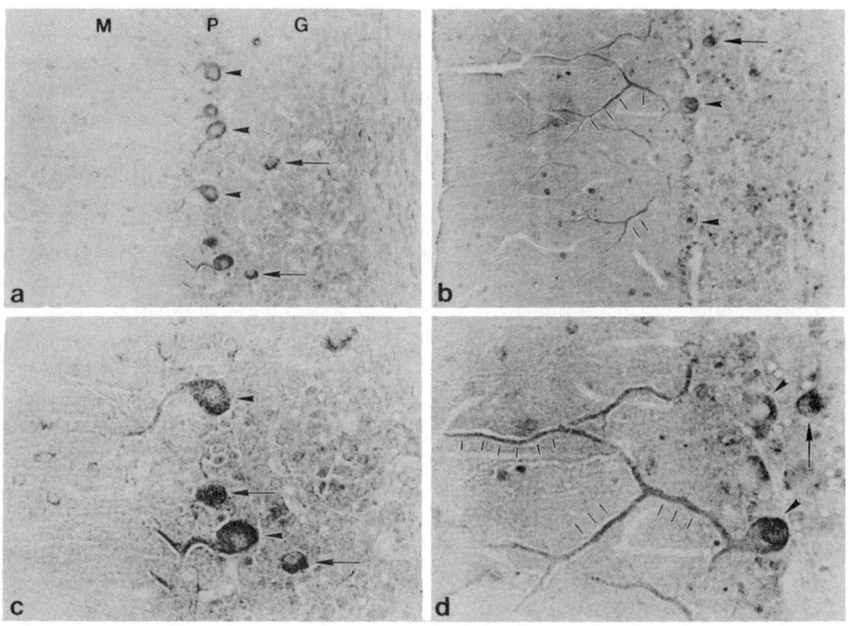

Figure 4. Sagittal sections of cerebellum stained with antibody to β-alanine. a, c: 0.05% taurine + 5% β-alanine; b, d: 0% taurine + 5% β-alanine. M: molecular layer; P: Purkinje cell layer; G: granule cell layer. Arrowheads indicate Purkinje cells; Arrows, Golgi II cells; bars, Purkinje cell dendrites. Sections from cats raised on 0.05% taurine and 0% taurine do not stain with this antibody. a, b X 130; c, d X 260.

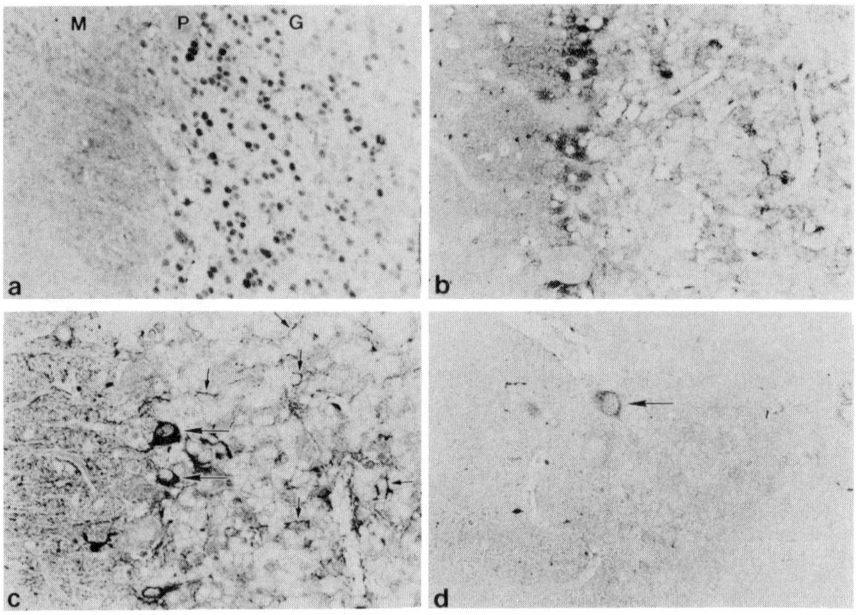

Figure 5. Sagittal sections of cerebellum stained with antibody to taurine. a: 0.05% taurine; b: 0% taurine; c: 0.05% taurine + 5% β-alanine; d: 0% taurine + 5% β-alanine. M: molecular layer; P: Purkinje cell layer; G: granule cell layer. Large arrows indicate Golgi II cells; small arrows, glial cell fibers. X 260.

β-alanine treated cats. The tapetum of these cats were relatively intact, although there appeared to be more swollen or compressed tapetal cells in some cases. Examination of transverse sections of optic nerves revealed that while their overall diameter and number of axons seemed unchanged, the thickness and organization of myelin sheaths were significantly affected in some β-alanine treated cats (see below).

The observations made in toluidine blue-stained sections described above were confirmed and finer morphological changes were revealed by electron microscopy. The soma of rods and cones in the retina of a taurine-supplemented cat treated with β-alanine which showed relatively minor changes at the light microscopic level (Fig. 7g) were, for example, swollen to varying degrees in comparison to the corresponding structures in the control cats (Figs. 8a, 8b). The inner and outer segments of the same retina were also swollen or disorganized, although their density and overall dimension were not significantly different from those in the control retina (Figs. 9a and 9b). The inner and outer segments in the temporal retina of a taurine-deprived cat treated with β-alanine (same area as Fig. 7d) were distinctively shorter, deformed and diminished in number in comparison to elongated and tightly-packed counterparts in the control cat retina (Figs. 9a, 10a and 10b). In a more severely affected retinal area of a taurine-deprived cat treated with β-alanine (same as Fig. 6e), both the inner and outer segments were further reduced in length and number, and, occasionally, macrophages containing irregular inclusion bodies in their rather electron-lucent cytoplasm were found between them (Fig. 10c). Similar macrophages also were encountered between crenated or hyperchromatic nuclei of photoreceptor cells in the outer nuclear layer (Fig. 10d). In the temporal retina from a taurine-supplemented cat was a more severely affected area (same as Fig. 7b) in which the outer nuclear layer consisted of only one or two rows of photoreceptor nuclei and was frequently encroached by macrophages

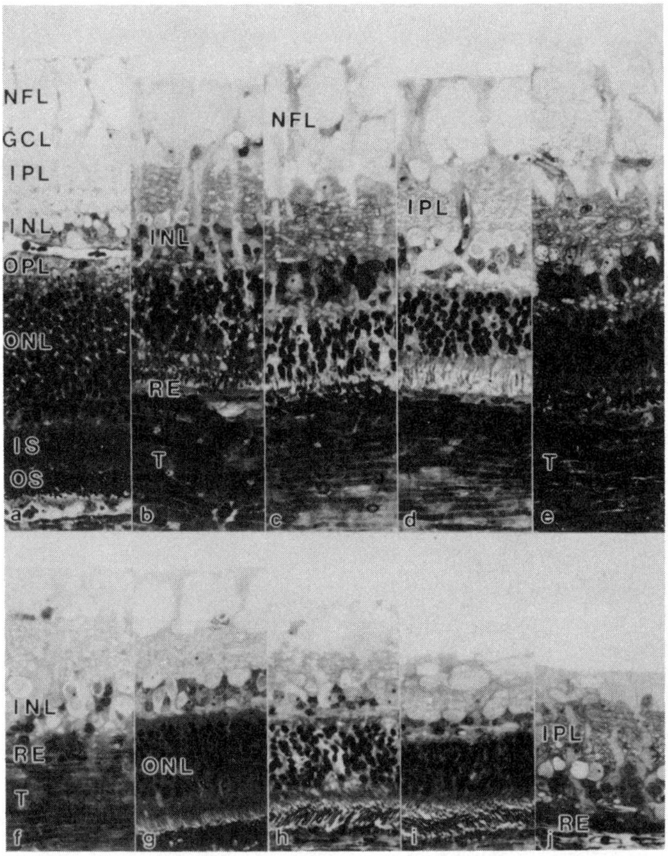

Figure 6. a-f. Light micrographs (LM) of superior retinas from a: a control cat, b: and c: taurine-supplemented cats treated with β-alanine, d:, e: and f: taurine-deprived cats treated with β-alanine. All stained with toluidine blue, X 350. g-j: LM of inferior retinas from g: and h: taurine-supplemented cats treated with β-alanine, i: and j: taurine-deprived cats treated with β-alanine.

and expanded processes of Muller cells, and the inner and outer segments were vestigial or missing (Fig. 11a). Photoreceptors were completely absent from any areas of retina examined in one of the taurine-deprived cats which had the lowest retinal taurine (Figs. 6f, 6j, 7j). In such areas, cells of the inner nuclear layer were in direct apposition with the retinal epithelial cells which displayed profuse apical microvilli entwined with equally delicate microvilli of Muller cells in some foci but otherwise were flattened and condensed (Fig. 11b). The retinal epithelial cells abutting the tips of outer segments which showed exceptionally abundant fuzzy membrane discs in the retina of another taurine-deprived cat (same area as Fig. 7c), by contrast, were swollen with numerous vacuoles of varying sizes occupying the basal cytoplasm (Fig. 12a). More commonly, however, the retinal epithelial cells of most cats given β-alanine showed relatively large, obtuse mitochondria and phagosomes of similar size with varying electron densities, as those of the control cats, although their cytoplasm often appeared somewhat condensed as shown here for a taurine-deprived cat (Fig. 12b, similar area as in Fig. 6e). Tapetal cells of these cats usually displayed characteristic tapetal rods arranged in orderly parallel arrays, comparable to those seen in the control cats, but there seemed to be greater number of tapetal cells with either more or less tightly-packed rods in

Figure 7. a-e. LM of temporal retinas from a: a control cat, b: taurine-supplemented cat treated with β-alanine. c:, d: and e: taurine-deprived cats treated with β-alanine., f-j. LM of nasal retinas from f: a control cat, g: and h: taurine-supplemented cats treated with β-alanine, i: and j: taurine-deprived cats treated with β-alanine.

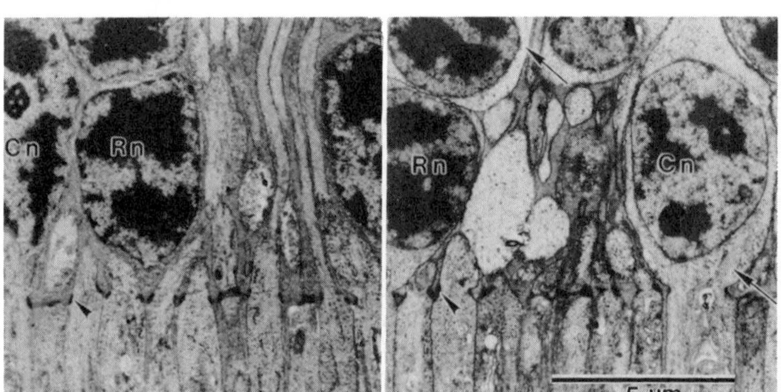

Figure 8. Electron micrographs (EM) showing photoreceptors located in the outermost level of the outer nuclear layer in the retinas of a) a control cat, and b) a taurine-supplemented cat treated with β-alanine. Cn: cone nuclei; Rn: rod nuclei; Arrowheads point to the terminal bars forming external limiting membrane. Arrows in b) point the swollen cytoplasm of rod and cone somata.

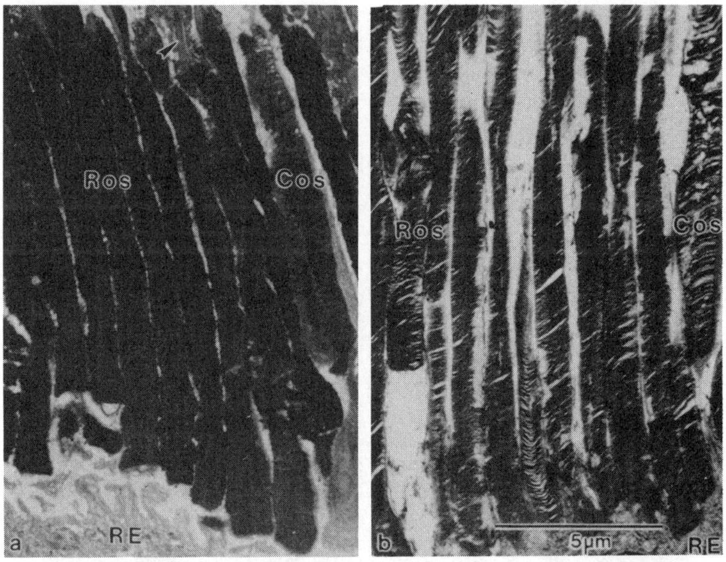

Figure 9. EM showing outer segments of rods and cones (Ros and Cos) in the retinas of a: a control cat and b: a taurine-supplemented cat treated with β-alanine (same as 8b). Proximally, each outer segment is connected to inner segment by a connecting cilium (arrowhead), and distally enclosed by apical microvilli of retinal epithelium (RE). The scale bar in 9b also applies to 8a, b and 9a.

β-alanine-treated cats (Fig. 12b). Large electron-dense globular bodies which appeared to replace some tapetal rods in outer tapetal cells described previously in adult cats fed a taurine-free diet alone and in kittens from female cats consuming a taurine-free diet[16,35,36] were not detected in these cats.

The cells in the inner nuclear layer showed little changes. In the moderately affected nasal retina of a taurine-supplemented cat (superior and temporal areas shown in Figs. 6c, 7b) both bipolar and horizontal cells showed somewhat angular or crenated nuclei and swollen mitochondria, while Muller cells were hypertrophied and less electron-dense in both nuclear and cytoplasmic portions relative to corresponding cells in the control cat retina (Figs. 8a and 8b). Amacrine cells showed even smaller changes, although they appeared swollen in a number of retinas. Similarly, ganglion cells were largely intact, only occasional retinas exhibiting swollen axons contributing to increased thickness of the nerve fiber layer. Finally, electron microscopic examination of optic nerves demonstrated that although the number and density of axons did not seem to differ much between different regions or nerves, their configuration and dimension and associated structures were substantially altered by β-alanine in the drinking water. Thus in contrast to the rather spherical profiles of axons with varying diameters in normal feline optic nerves, axons in the optic nerves of β-alanine-treated cats, illustrated here in a taurine-deprived cat with a moderately affected retina (temporal and nasal areas shown in Figs. 7c, 7i), showed irregular, angular profiles with smaller diameters especially in the core region (Figs. 14a and 14b). Furthermore, these axons were surrounded by a markedly thinner, fraying myelin sheath, often loosely-fitted, or focally detached from axolemma, in contrast to the axons in control cats which were invested by rather solid myelin which seemed to show the five thickness frequency modes described in normal cat optic nerves. In addition to spaces between axons and myelin, there were numerous intercellular spaces between shrunken axons, either myelinated or unmyelinated,

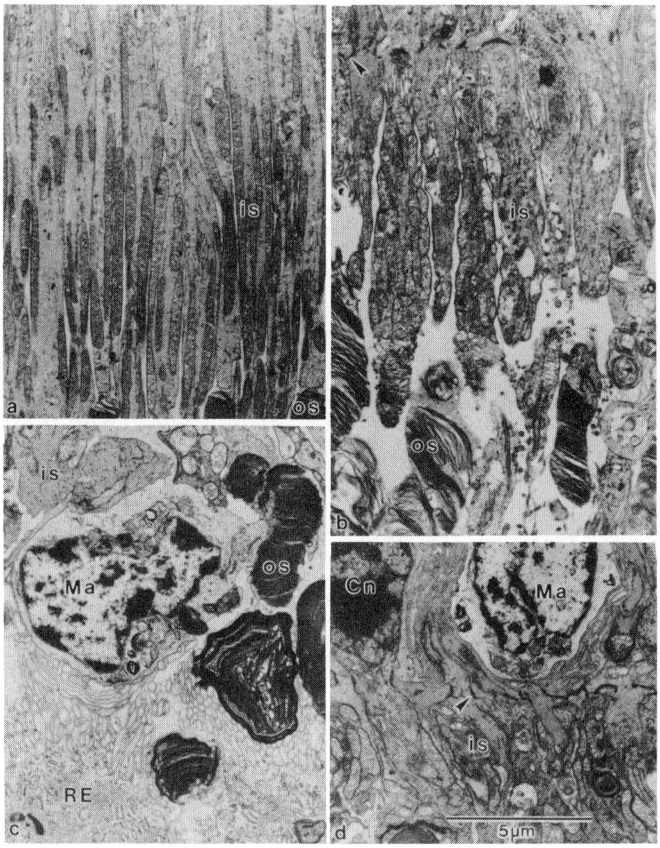

Figure 10. EM showing parts of inner and outer segments (is and os) of retinas from a: control cat, b:, c: and d: taurine-deprived cats treated with β-alanine. Note that both the inner and outer segments are greatly reduced in length and deformed in these cats. In c), a macrophage (Ma) is found in the subretinal space adjacent to retinal epithelial cell (RE) and irregularly oriented, disfigured inner and outer segments (is and os), and in d), between photoreceptor nuclei (Cn) in the outer nuclear layer. Arrowheads mark the external limiting membrane.

and other cellular elements such as astrocytes, oligodendrocytes and their processes, rendering axon bundles less demarcated by indistinct trabeculae which seemed to contain smaller or collapsed blood vessels and increased numbers of oligodendrocytes and astrocytes.

DISCUSSION

β-Alanine has been used to modify tissue taurine concentrations in several species, including mice, rats, guinea pigs and ferrets[6,18,29]. This study is the first to report its use in a species with a dietary dependency on taurine. As found in other species, β-alanine reduced tissue taurine concentrations in cats, even in cats whose taurine stores had been severely depleted by prolonged feeding of a taurine-free diet. The use of this species produced intriguing results: visceral tissues of both taurine-supplemented and taurine-deprived cats had taurine concentrations greatly reduced by β-alanine treatment accompanied by accumulation of β-alanine. The concentration of taurine in brain of taurine-supplemented cats was

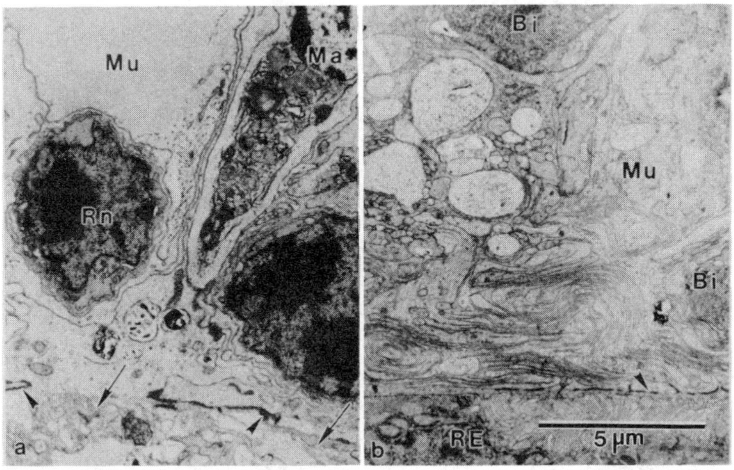

Figure 11. EM showing outer retinal parts from a: a taurine-supplemented cat treated with β-alanine, and b: a taurine-deprived cat treated with β-alanine. In a) the outer nuclear layer is made of one or two rod nuclei (Rn) encroached by a macrophage (Ma) and a thick Muller cell process (Mu). Both the inner and outer segments are missing but membranous debris (arrows) immediately outside the external limiting membrane (arrowheads) may represent vestigial elements. In b) photoreceptors are completely missing in the area, and sheathlike Muller cell processes (Mu) form intricate whirls inside the external limiting membrane (arrowhead) separating bipolar cells (Bi) from the retinal epithelial cells (RE).

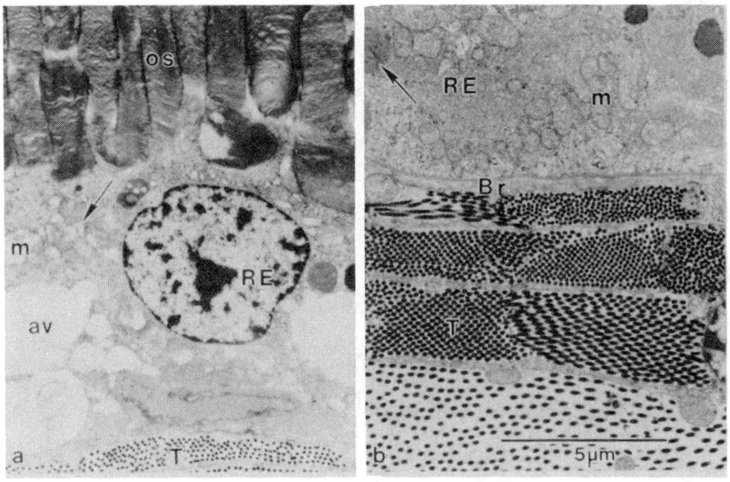

Figure 12. EM showing parts of retinal epithelial cells (RE) and tapetum (T) from taurine-deprived cats treated with β-alanine. The RE in a) contains large autophagic vacuoles (av) and distended cisternae of ER (arrow), swollen mitochondria (m), and its apical surface tightly wraps tips of outer segments (os) which show unusually frazzled or fuzzy disc membranes. In b) the RE shows typically large mitochondria (m), some phagosomes (arrow) and fine tubular ER. Melanosomes are found only in RE of inferior retina where tapetum is absent (see Fig. 6j). Three tapetal cells (T) subjacent RE are slightly compressed and the 4th unusually distended. Br: Bruch's membrane.

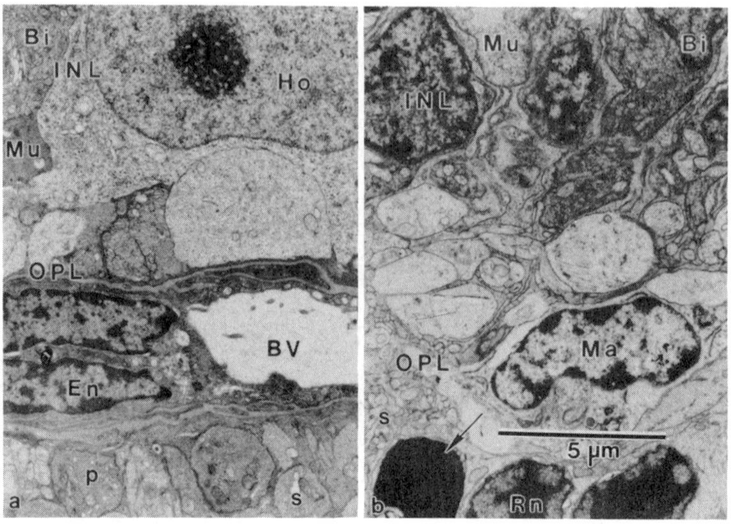

Figure 13. EM showing parts of inner nuclear layer (INL) and outer plexiform layer (OPL) from a: a normal (control) cat, and b: a taurine-supplemented cat treated with β-alanine. In a), despite intervening blood vessels (BV) and endothelial cells (En), pedicles and pherules (p and s) display typical profiles as do the horizontal, bipolar and Muller cells (Ho, Bi and Mu). In b), a macrophage (Ma) is seen in the outer plexiform layer, adjacent to somewhat misshapen spherules (s) and a pyknotic rod nucleus (arrow). Neighboring rod nuclei (Rn) show irregularly crenated contour as do the nuclei of bipolar and muller cells (Bi and Mu) in the inner nuclear layer.

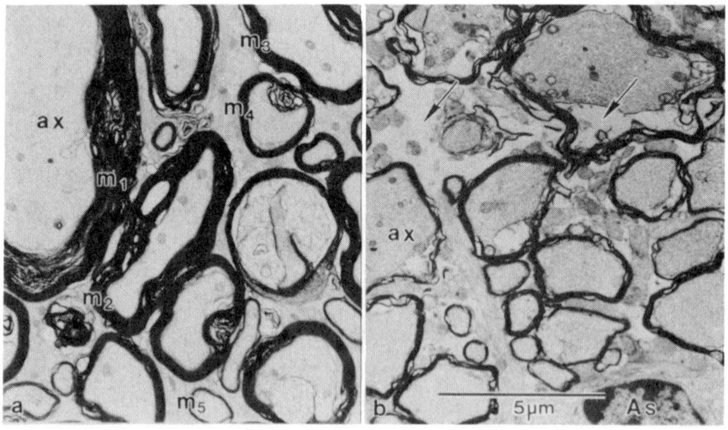

Figure 14. EM showing central portion of optic nerve from a: a control cat, and b: a taurine-deprived cat treated with β-alanine. In a), axons (ax) of varying diameters are invested by rather solid myelin sheath, displaying 5 modes of thickness distribution (m_1 - m_5). Axons in b) are surrounded by thinner, frazzled or loosened myelin. Clear spaces are found between such myelin sheaths and axons inside, as well as outside (arrows) which in the control optic nerve is filled with processes of fibrous astrocytes, oligodendrocytes and unmyelinated axons. As: astrocyte.

reduced from approximately 2 μmol/g wet wt to 0.2 to 0.5 μmol/g wet wt by β-alanine treatment accompanied by the accumulation of only 0.1 to 0.2 μmol/g wet wt of β-alanine. Taurine-deprived cats, on the other hand, had brain taurine concentrations reduced from approximately 0.4 μmol/g wet wt to 0.1 to 0.2 μmol/g wet wt, but accompanied by the accumulation of approximately 2 μmol/g wet wt of β-alanine, tenfold the amount in the taurine-supplemented cat brains. These results suggest that the intracellular concentration of taurine in cat brain has much less effect on the accumulation of β-alanine than the extracellular concentration of taurine, and may provide a partial explanation, at least, of the puzzling results obtained with taurine and taurine analogues, including β-alanine, on the cat cerebellar granule cell cultures[37,38]. Another possibility is that prolonged treatment with β-alanine and/or prolonged dietary depletion of taurine changes the kinetic characteristics of the taurine transporter in cat brain, enabling more active accumulation of β-alanine, which uses the same transporter, in taurine-deprived cats. Such an effect of dietary taurine has been elegantly demonstrated to occur in the rat kidney[2-5,8].

The observation of β-alanine-conjugated bile acids in this study is the first, to our knowledge. Cats normally conjugate bile acids only with taurine and not with glycine, even in conditions of taurine depletion. The presence of large amounts of β-alanine as in this study suggests that β-alanine can subserve the bile acid conjugating functions of taurine in the face of a taurine deficiency.

Dietary β-alanine resulted in reduced numbers of granule and Purkinje cells in the cerebellum. Many of those remaining appeared pyknotic and in the process of dying. These degenerative changes were accompanied by extensive gliosis, much more severe than has been noted in taurine-deprived kittens[23,39]. Also of interest was the presence of numerous long and swollen fibers in the cerebellar white matter. Such fibers are not seen in taurine-supplemented or taurine-deprived cats not treated with β-alanine. These structures resemble the Rosenthal fibers described in several human disease states of the cerebellum, generally accompanied by reactive or neoplastic astrocytosis. As with Rosenthal fibers, these fibers in the cat cerebellar white matter also stained on the periphery with an antibody to ubiquitin.

Using antibodies specific for β-alanine and taurine, we have shown the accumulation of substantial amounts of β-alanine in Purkinje cells and their dendrites. This is true for all Purkinje cells examined, including those with pyknotic appearance and apparently dying. Some granule cells stained positively for β-alanine particularly in taurine-deprived cats treated with β-alanine while others appeared devoid of β-alanine. None of the granule cells with pyknotic appearance were stained by the antibody to β-alanine (or taurine). β-Alanine and taurine have been shown repeatedly to be transported by a single carrier in neural cells, so it is not surprising that the prolonged treatment with β-alanine in this study resulted in displacement of taurine from many neural components[11,12,21,24,27]. These *in vivo* results for the localization of β-alanine in cat cerebellum differ from those reported in cultured rat cerebellar cells in that we found predominantly neuronal localization whereas it was present only in glial cells in the cultures[13]. Dietary β-alanine appears to have completely eliminated any taurine from cerebellar Purkinje and granule cells. The remaining small amounts of taurine are localized to Golgi II cells and glia. A brief examination of β-alanine staining in hippocampus and visual cortex, in which β-alanine is accumulated by various classes of neurons, suggest that β-alanine accumulation within neurons is global.

The morphological changes in the retinas of the taurine-supplemented cats treated with β-alanine closely resemble those documented for cats deprived of taurine for much longer periods of time[9]. These results suggest that retinal photoreceptors, their outer segments especially, are exquisitely dependent on taurine to maintain their fine structure, and can rapidly degenerate if sufficient taurine is not present. They also indicate that this structural function of taurine cannot be subserved by β-alanine. The morphological changes in the retinas of the taurine-deprived cats treated with β-alanine were extreme and greater than has been noted in cats deprived of taurine for many years. In some of the cats, the retinas were devoid of photoreceptors in large areas, a phenomenon that has never been reported in cats deprived of taurine. These results support and emphasize the conclusion reached from the results from the taurine-supplemented cats treated with β-alanine; namely that retinal photoreceptors are exquisitely dependent on large amounts of taurine to maintain their structure. We have located only one other report of retinal changes, in rats injected with β-alanine for four days, but no data for tissue concentrations were presented[26].

The degenerative changes noted in the optic nerves of cats treated with β-alanine may be the result of taurine depletion rather than a neurotoxic effect of β-alanine. This suggestion is supported by the fact that the degeneration in optic nerves appeared to reflect the degree of taurine depletion. As in the retina, the greatest effects were present in taurine-deprived cats treated with β-alanine. A previous report has shown that rats treated with GES have a reduced number of optic nerve fibers in addition to reduced retinal taurine concentrations, a reduction of axon diameters and thinning of myelin sheaths, resembling the present observations in the cat[17,20]. In addition, offspring from such treated mothers have a higher proportion of small diameter optic nerve fibers.

We conclude that although dietary treatment with β-alanine results in great reductions in the concentration of taurine in cerebellum (and other brain regions) of taurine-supplemented and taurine-deprived cats, the cerebellar pathology most likely results from toxic actions of β-alanine itself because the taurine concentrations in taurine-deprived cat cerebellum are no different from those in taurine-supplemented cats treated with β-alanine. Yet taurine-deprived cats do not show the unusual pathology described here. In this connection, there have been two published reports of hyper-β-alaninemia in young children, both associated with severe neurological symptoms, which supports the neurotoxic nature of β-alanine[10,28]. The retinal degeneration, on the other hand, is most likely the result of severe taurine depletion.

ACKNOWLEDGMENTS

We are grateful to Anna Parese for secretarial assistance and to Michael Natelli and other members of the IBR Animal Colony Facility for practical help throughout this study. This work was supported by the New York State Office of Mental Retardation and Developmental Disabilities. Figures 1-5 are reproduced with the permission of Wiley-Liss, Inc., and Figures 6-14 are reproduced with the permission of The Faculty Press.

REFERENCES

1. Campistron, G., Geffard M. and Buijs, R. M. 1986, Immunological approach to the detection of taurine and immunocytochemical results, *J. Neurochem.* 46:862-868.
2. Chesney, R.W., Gusowski, N. and Dabbagh, S. 1985, Renal cortex taurine content regulates the adaptive response to altered dietary intake of sulfur amino acids, *J. Clin. Invest.* 76:2213-2221.
3. Chesney, R.W., Gusowski, N. and Friedman, A.L. 1983, Renal adaptation to altered dietary sulfur amino acid intake occurs at the luminal brush border membrane, *Kidney Int.* 24:588-584.
4. Chesney, R.W., Gusowski, N. and Theissen, M. 1984, Developmental aspects of renal β-amino acid transport. IV. Brush border membrane response to altered intake of sulfur amino acids, *Pediatr. Res.* 18:611-618.
5. Chesney, R.W., Lippincott, S., Gusowski, N., Padilla, M. and Zelikovic, I. 1986, Studies on renal adaptation to altered dietary amino acid intake: Tissue taurine responses in nursing and adult rats, *J. Nutr.* 116:1965-1976.

6. Davies, W.E., Kay, I.S. and Birnso, O.V. 1990, Taurine function in the auditory system, in: "Taurine: Functional Neurochemistry, Physiology, and Cardiology", Pasantes-Morales, H., Shain, W., Martin, D.L. and Martin del Rio, R. eds., Wiley-Liss, New York, pp. 397-405.
7. Freeman, B. 1978. Myelin sheath thickness and conduction latency groups in the cat optic nerve. *J.Comp.Neurol.* 181:183-196.
8. Han, X. and Chesney, R.W. 1994, Expression of taurine transport and its regulation by diet in xenopus laevis oocytes following injection of rat kidney cortex mRNA, in: "Taurine in Health and Disease", Huxtable, R.J. and Michalk, D. eds., Plenum Press, New York, pp. 121-130.
9. Hayes, K.C., Carey, R.E. and Schmidt, S.Y. 1975, Retinal degeneration associated with taurine deficiency in the cat, *Science* 188:949-951.
10. Higgins, J.J., Kaneski, C.R., Bernardini, I., Brady, R.O. and Barton, N.W. 1994, Pyridoxine-responsive hyper-β-alaninemia associated with Cohen's syndrome, *Neurol.* 44:1728-1732.
11. Holopainen, I. 1988, Taurine and β-alanine uptake in primary astrocytes differentiating in culture: Effects of ions, *Neurochem. Res.* 13:853-858.
12. Holopainen, I., Malminen, O. and Kontro, P. 1987, Sodium-dependent high-affinity uptake of taurine in cultured cerebellar granule cells and astrocytes, *J. Neurosci. Res.* 18:479-483.
13. Hosli, E. and Hosli, L. 1980, Cellular localization of the uptake of [^{3}H]taurine and [^{3}H]β-alanine in cultures of the rat central nervous system, *Neurosci.* 5:145-152.
14. Huxtable, R.J. and Lippincott, S.E. 1981, Comparative metabolism and taurine-depleting effects of guanidinoethanesulfonate in cats, mice, and guinea pigs, *Arch. Biochem. Biophys.* 210:698-709.
15. Huxtable, R.J., Laird, H.E. and Lippincott, S.E. 1979, The transport of taurine in the heart and the rapid depletion of tissue taurine content by guanidinoethyl sulfonate, *J. Pharmacol. Exper. Therapeutics* 211:465-471.
16. Imaki, H., Moretz, R.C., Wisniewski, H.M. and Sturman, J.A. 1986, Feline maternal taurine deficiency: Effects on retina and tapetum of the offspring, *Devel. Neurosci.* 8:160-181.
17. Lake, N. 1992, Localization of taurine and glial fibrillary acidic protein in human optic nerve using immunocytochemical techniques, in "Taurine: Nutritional Value and Mechanisms of Action", Lombardini, J.B., Schaffer, S.W. and Azuma, J. eds., Plenum Press, New York, pp. 303-307.
18. Lake, N. and De Marte, L. 1988, Effects of β-alanine treatment on the taurine and DNA content of the rat heart and retina, *Neurochem. Res.* 13:1003-1006.
19. Lake, N. and Verdone-Smith, C. 1989, Immunocytochemical localization of taurine in the mammalian retina, *Curr. Eye Res.* 8:163-173.
20. Lake, N., Malik, N. and De Marte, L. 1988, Taurine depletion leads to loss of rat optic nerve axons, *Vis. Res.* 28:1071-1076.
21. Larsson, O.M., Griffiths, R., Allen, I.C. and Schousboe, A. 1986, Mutual inhibition kinetic analysis of γ-aminobutyric acid, taurine, and β-alanine high-affinity transport into neurons and astrocytes: Evidence for similarity between the taurine and β-alanine carriers in both cell types, *J. Neurochem.* 47:426-432.
22. Lowe, J., Morrell, K., Lennox, G., Landon, M. and Mayer, R.J. 1989, Rosenthal fibers are based on the ubiquitination of glial filaments, *Neuropathol. Appl. Neurobiol.* 15:45-53.
23. Lu, P., Schuller-Levis, G. and Sturman, J.A. 1991, Distribution of taurine-like immunoreactivity in cerebellum of kittens from taurine-supplemented and taurine-deficient mothers, *Int. J. Dev. Neurosci.* 9:621-629.
24. Martin, D.L. and Shain, W. 1979, High affinity transport of taurine and β-alanine and low affinity transport of Ó-aminobutyric acid by a single transport system in cultured glioma cells, *J. Biol. Chem.* 254:7076-7084.
25. Messing, J.M. and Sturman, J.A. 1993, Evaluation of taurine status in cats consuming diets containing different amounts of taurine by determination of plasma and whole blood taurine concentrations, *J. Nutr. Biochem.* 4:168-171.
26. Pasantes-Morales, H., Quesada, O., Carabez, A. and Huxtable, R.J. 1983, Effects of the taurine transport antagonist, guanidinoethane sulfonate, and β-alanine on the morphology of rat retina, *J. Neurosci. Res.* 9:135-143.
27. Saransaari, P. and Oja, S.S. 1993, Uptake and release of β-alanine in cerebellar granule cells in primary culture: Regulation of release by glutamatergic and GABAergic receptors, *Neurosci.* 53:475-481.
28. Scriver, C.R., Pueschel, S. and Davies, E. 1966, Hyper β-alaninemia associated with β-aminoaciduria and gamma-aminobutyric-aciduria, somnolence and seizures, *N. Engl. J. Med.* 274:635-643.
29. Shaffer, J.E. and Kocsis, J.J. 1981, Taurine mobilizing effects of beta alanine and other inhibitors of taurine transport, *Life Sci.* 28:2727-2736.
30. Sturman, J.A. 1986, Nutritional taurine and central nervous system development, in "Mental Retardation: Research, Education and Technology Transfer", Annals New York Academy of Sciences, pp. 196-213.

31. Sturman, J.A., 1990, Taurine deficiency, in: "Taurine: Functional Neurochemistry, Physiology, and Cardiology", Pasantes-Morales, H., Shain, W., Martin, D.L. and Martin del Rio, R. eds., Wiley-Liss, New York, pp. 385-395.
32. Sturman, J.A., 1992, Review: taurine deficiency and the cat, in "Taurine: Nutritional Value and Mechanisms of Action", Lombardini, J.B., Schaffer, S.W. and Azuma, J. eds., Plenum Press, New York, pp. 1-5.
33. Sturman, J.A. 1993, Taurine in development, *Physiol. Rev.* 73:119-147.
34. Sturman, J.A. and Messing, J.M. 1991, Dietary taurine content and feline reproduction and outcome, *J. Nutr.* 121:1195-1203.
35. Sturman, J.A., Wen, G.Y., Wisniewski, H.M. and Hayes, K.C. 1981, Histochemical localization of zinc in the feline tapetum: Effect of taurine depletion. *Histochemistry* 72:341-350.
36. Sturman, J.A., Wen, G.Y., Wisniewski, H.M., Niemann, W.H., and Hayes, K.C., 1982, Taurine and tapetum structure, in "Taurine in Nutrition and Neurology, Vol. 139", Huxtable, R.J. Pasantes-Morales, H. eds., Plenum Publishing Corporation, New York, pp. 65-78.
37. Trenkner, E. and Sturman, J.A. 1991, The role of taurine in the survival and function of cerebellar cells in cultures of early postnatal cat, *Int. J. Devl. Neurosci.* 9:77-88.
38. Trenkner, E., Gargano, A., Scala, P. and Sturman, J. 1992, Taurine synthesis in cat and mouse in vivo and in vitro, in "Taurine: Nutritional Value and Mechanisms of Action", Lombardini, J.B., Schaffer, S.W. and Azuma, J. eds., Plenum Press, New York, pp. 7-14.
39. Xu, Y., Lu, P., Imaki, H. and Sturman, J.A. 1993, Feline maternal taurine deficiency: a quantitative morphometric and immunohistochemical study of 8-week-old kitten visual cortex, *Biomed. Letts.* 48:329-344.

4

THE EFFECTS OF TAURINE IN A RODENT MODEL OF AGING

Ralph Dawson, Jr., Baerbel Eppler, Tucker A. Patterson, Debbie Shih, and Steve Liu

Department of Pharmacodynamics
College of Pharmacy, University of Florida
JHMHC Box 100487, Gainesville, Florida 32610

INTRODUCTION

A number of excellent review articles have examined the various physiological roles of taurine in adult and developing organisms[19,20,39]. A conclusion that can be drawn from these various reviews is that there is a dearth of information on the role of taurine in aging and senescence. A few studies have examined the tissue content of taurine in aging animal models. In general taurine content seems to decline modestly with advanced age. What is unclear is whether the aging process may increase the demand for the protective and regulatory actions of taurine in cellular homeostasis. This article will briefly review what is known about changes in taurine content and function during senescence and describe our recent studies of taurine supplementation using a common rodent model of aging. We will also discuss the potential consequences of a diminished cytoprotective role of taurine in advanced aging due to a age-related decrement in taurine homeostasis.

The male Fischer 344 (F344) rat has served as a major model of aging for a number of years. A number of legitimate concerns have been raised about this model due to specific pathological changes that occur in this model. The major concerns with the male F344 model are the high incidence of nephropathy and testicular interstitial cell tumors[5,38]. These age-associated pathologies can confound some studies on the basic biology of aging. Human renal function also declines in a roughly linear function with age and this decline begins at about 30 years of age. Hepatic function also diminishes with age. The liver and kidney are both key organ systems in whole body taurine homeostasis. Thus, the ability to regulate the biosynthesis and the renal conservation of taurine may be compromised with advancing age. Intracellular Ca^{2+} regulation[15], protein phosphorylation[24], and antioxidant defense systems[17,22] are all compromised in aging and are also biochemical processes which involve taurine. Therefore, there is a need to have a better understanding of how whole body taurine homeostasis is maintained in advanced aging and if there is an increased functional demand for taurine in the aging organism.

Taurine 2, edited by Huxtable et al.
Plenum Press, New York, 1996

BRIEF REVIEW OF TAURINE AND AGING

Taurine content has been measured in brain and peripheral tissues of aged rodents. Timeras *et al.*[40] measured brain regional content of taurine in Long-Evans rats that were 2, 14, 22 and 30 months of age. Taurine content was reported to decline 25-43% in the brain regions examined (spinal cord, brainstem, cerebellum and cerebral cortex) when compared to the 2 month time point. The decline in taurine content was less marked when comparing the 14 to the 30 month old group (11-37% decline)[40]. Tyce and Wong[44] found no difference in taurine content in the cerebrum and hindbrain when they compared 3 month and 22 month old female F344 rats. Banay-Schwartz *et al.*[2] examined taurine content in 53 microdissected brain regions in 3 and 29 month old male F344 rats. In general, taurine content declined about 12% when all brain areas were averaged[34]. Taurine depletion was greatest in the ventromedial nucleus of the hypothalamus where taurine content was reduced 45%. Interestingly, Donzanti and Ung[12] examined microdissected subregions of the striatum in 6 and 20 month old male F344 rats and found an average increase of 23% in taurine content in the aged rats in anterior regions of the striatum. This is in contrast to Banay-Schwartz *et al.*[2,34] who found significant reductions in taurine in the caudate, putamen and globus pallidus. Benedetti *et al.*[3] found significant reductions in taurine content in the striatum, accumbens, cerebellum and cortex of 21-22 month-old male Wistar rats when compared to 3-month-old controls. We previously examined brain taurine content in 6 month and 24 month old male F344 rats and found a consistent 6-8% depletion of taurine in cortex, striatum, and cerebellum[10,46]. Taurine was unchanged in the hippocampus and midbrain, but increased in the brainstem (18%)[46]. Subcellular fractionation of cerebral cortex from 8 month and 30 month old male F344 rats did not uncover any age-related changes in the distribution or content of taurine[9]. We have also examined taurine content in the brain regions from female Long-Evans rats that were 6 and 30 months of age[8]. Taurine was reduced in the entorhinal cortex (16%), amygdala-piriform cortex (14%), striatum (3%) and mediobasal hypothalamus (20%) and increased in the dorsal (16%) and ventral (12%) hippocampus of aged rats[8]. Massie *et al.*[25] examined whole brain taurine content in male C57BL/6J mice ranging in age from 53-932 days of age. Taurine content tended to decrease with increasing age, yielding a -0.43 correlation coefficient that did not quite reach statistical significance. Kirzinger and Fonda[21] found no difference in whole brain taurine content between 12 month and 32 month old male C57BL/6J mice.

Oja *et al.*[32] examined K^+-stimulated taurine release in brain slices from 3, 6, 12, and 18 month old mice. The stimulated release of taurine appeared to increase in the hippocampus and decline in the striatum with age. Taurine release was not substantially altered in the brainstem, cerebellum or cerebral cortex[32]. We have examined taurine efflux stimulated by the neurotoxin trimethyltin in slices of cerebral cortex from 6 month and 24 month old male F344 rats[7]. There was no age-related difference in taurine efflux stimulated by trimethyltin. Taurine content of the cortical slices was, however, significantly lower in the aged F344 slices treated with trimethyltin than the adult controls[7]. We also examined kainic acid-stimulated amino acid release in cortical slices from adult (6 month) and aged (30 month) female Long-Evans rats[8]. There was no effect of age on kainic acid-stimulated taurine release despite an age-related increase in aspartate release[8]. Ooboshi *et al.*[33] reported that ischemia-induced release of taurine was attenuated in the hippocampus of aged spontaneously hypertensive rats (SHR) (19-23 month old) when compared to adult controls (5-7 month old).

We examined the content of taurine in a number of peripheral and cardiovascular tissues[9]. Taurine content was found to be decreased in the atria, caudal artery and kidney of male 30 month old male F344 rats when compared to 8 month old controls[9]. Renal content of taurine was constant at 6, 8, and 22 months of age, but declined in 30 month old male

Table 1. Taurine concentrations in the blood of aged rodents

Animal Model (age in months)	Sample Type	Concentration (nmol/ml)	Ref.
Female Wistar Rats	Plasma (C)		
6		199 ± 31	6
18		220 ± 30	
30		220 ± 41	
Male Sprague-Dawley	Plasma (?)		
2		102 ± 10	13
6		269 ± 22	
24		275 ± 20	
Male F344 Rats	Plasma (C)		
3		36 ± 3	28
24		37 ± 6	
Male F344 Rats	Serum (DC)		
6		333 ± 22	9,46
8		322 ± 15	
24		239 ± 13	
30		255 ± 20	
Female Long-Evans	Serum (DC)		
6		315 ± 28	8
24		184 ± 23	
Female F344 Rats	Serum (CP)		
3		137 ± 7	Unpublished
24		138 ± 16	findings

C: catheter; DC: decapitation; CP: cardiac puncture.

F344 rats[9]. Corman *et al.*[6] reported that aged, female Wistar rats had a significant reduction in urinary taurine excretion. The reduction in urinary taurine excretion was present at 18 months of age and was sustained until 30 months of age[6]. Massie *et al.*[25] found that taurine content in the mouse heart increased with age, whereas liver and kidney was unchanged and leg muscle declined.

Serum and plasma levels of taurine from a number of studies of aged rodents are presented in Table 1. Our laboratory has consistently found that trunk blood collected from decapitated aged rats yields serum taurine concentrations significantly lower than adult rats. Massie *et al.*[25] found no significant age-related change in the taurine content of mouse blood. Other studies including our own that used plasma or serum collected via catheterization or cardiac puncture did not show an age-related decline in taurine concentration (Table 1). Several potential explanations are possible for these findings which include: sympathoadrenal activation may alter circulating taurine concentrations in aged rats, taurine may be released from platelets, lymphocytes, or red cells differentially in aged rats or rapid postmortem changes occur in serum from decapitated aged rats. Decapitation has previously been shown to elevate plasma taurine concentrations[27]. The aged F344 rat has a number of hematological changes that would be consistent with decreased serum taurine derived from lysed cells[5]. Nishio *et al.*[31] reported that taurine augmented mitogen-stimulated proliferative responses to a greater degree in T cells from aged mice (21-28 months old) than from adult controls (2-3 months old). Taurine was also found by Nishio *et al.*[31] to increase intracellular Ca^{2+} concentrations in T cells of aged mice in response to mitogen stimulation to levels comparable to adult mice.

It is difficult to summarize simply the effects of aging on whole body taurine homeostasis or tissue content. In general, tissue content of taurine tends to decline somewhat with advanced age. Differences in the species, strain, sex and age of the animal models used in aging research often lead to problems in the interpretation of the literature.

EFFECTS OF TAURINE IN AGING MALE F344 RATS

Animal Model and Experimental Design

Male F344 rats, 18 months of age, were obtained from Harlan Sprague-Dawley's NIA-supported colony. A group (n=12) of adult rats that were 3 months old at the start of the study were also included. The 18 month old F344 rats (n=90) were divided into 3 dietary conditions (n=30 per group). The groups included: a normal rat chow group (Purina 5012), taurine-supplemented group (5012 chow + 1.5% taurine in the drinking water) and a taurine-deficient diet group (Purina 5729C-M diet). The taurine-deficient diet was identical to the 5012 chow except plant-derived protein was substituted for animal protein to eliminate taurine. All rats were maintained in their own colony room on a 12 h light-dark cycle and food and fluids were available *ad libitum*. Food and fluid intake was monitored at weekly intervals along with body weights. The rats were maintained on these diets for 320 days or until they died. An outline of the experimental design and schedule of the parameters assessed are given in Fig. 1. Data was analyzed using analysis of variance (parametric or nonparametric) followed by appropriate post-hoc group comparisons.

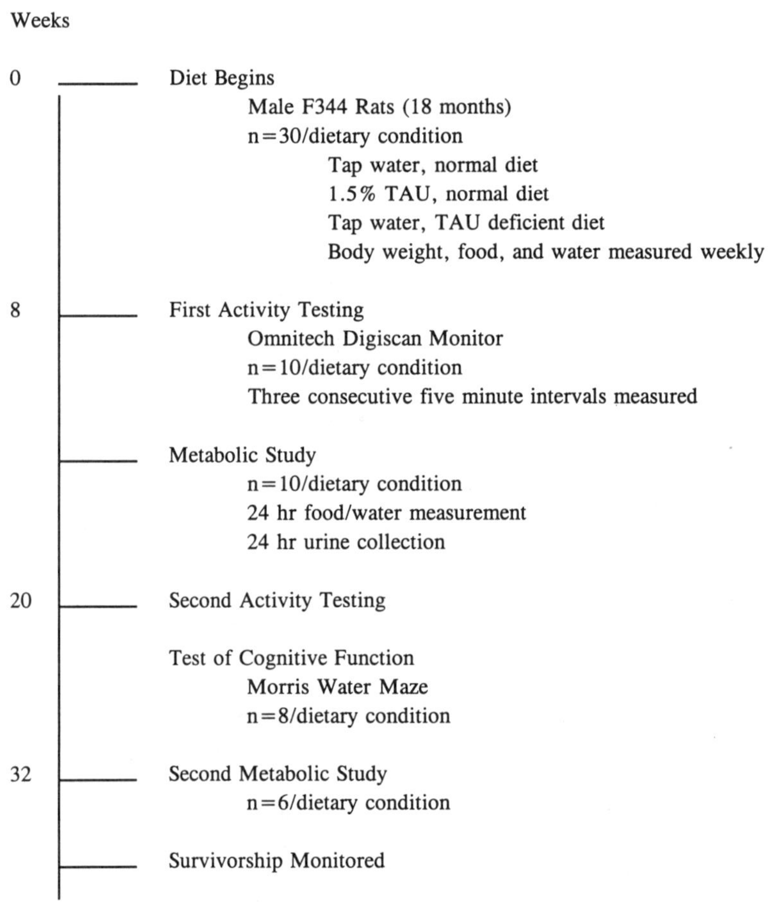

Figure 1. Experimental design.

Metabolic Studies

Metabolic studies were conducted after 12 (n=10 per group) and 32 (n=6 per group) weeks on the diets. The rats were allowed 48 h to acclimate themselves to the metabolic cages prior to collection of data. Food and fluid intake were measured for two consecutive days along with 24 h urine output. These data are presented in Table 2. The urinary excretion of Na^+, K^+, Ca^{2+}, protein, creatinine, urea nitrogen, and glucose was also measured. Urinary excretion of taurine, alanine and glycine was measured by HPLC with electrochemical detection of the o-pthalaldehyde derivatives[9].

Experimental Findings

The effects of diets supplemented with 1.5% taurine in the drinking water or devoid of taurine had no significant effect on survival when compared to a standard rodent diet (Purina 5012) (Fig. 2). The median days of survival on the diets were 241, 245 and 274 for the control, taurine-supplemented and taurine-deficient groups respectively. Taurine supplementation resulted in a greater than 5x elevation in serum taurine in the aged F344 rats

Table 2. Metabolic measures in adult and aged F344 rats

Measure	Adult Control	Aged Control	No TAU	1.5% TAU
Body Weight (g)				
I	240 ± 5	441 ± 11[a,b]	441 ± 7[a,b]	401 ± 10[a]
II	336 ± 22	430 ± 8[a]	413 ± 3[a]	430 ± 2[a]
Food Intake (g/24 h)				
I	19 ± 1	17 ± 1	16 ± 1	17 ± 1
II	25 ± 3	38 ± 2[a]	34 ± 4	39 ± 4[a]
Fluid Intake (g/24 h)				
I	23 ± 1	32 ± 3	32 ± 3	33 ± 4
II	24 ± 4	77 ± 16[a]	67 ± 13[a]	78 ± 5[a]
Urine Output (ml/24 h)				
I	10 ± 1	16 ± 2[a]	12 ± 1	17 ± 2[a]
II	11 ± 1	34 ± 4[a]	30 ± 5[a]	36 ± 5[a]
Glucose Excretion (mg/24 h)				
I	2.92 ± 0.39	5.33 ± 0.57[a]	4.74 ± 0.42	6.35 ± 0.73[a]
II	4.15 ± 1.14	8.62 ± 1.84	7.64 ± 0.95	11.85 ± 3.32
Sodium Excretion (mEq/24 h)				
I	1.63 ± 0.09	1.86 ± 0.12[c]	1.08 ± 0.09[a,b]	1.72 ± 0.16
II	1.27 ± 0.29	1.68 ± 0.20	1.57 ± 0.19	1.67 ± 0.22
Potassium Excretion (mEq/24 h)				
I	0.81 ± 0.10	1.52 ± 0.11[a,c]	0.99 ± 0.07[b]	1.54 ± 0.17[a]
II	2.97 ± 0.47	4.38 ± 0.26[a]	4.23 ± 0.24	4.15 ± 0.30
Calcium Excretion (mg/24 h)				
I	1.25 ± 0.31	1.03 ± 0.14	0.71 ± 0.10[b]	1.79 ± 0.29
II	0.38 ± 0.10	2.42 ± 0.88[a]	2.09 ± 0.41	3.69 ± 0.70[a]
Creatinine (mg/24 h)				
I	15.5 ± 0.9	18.0 ± 1.0	15.7 ± 0.8	17.2 ± 0.8
II	17.6 ± 1.6	19.0 ± 1.2	18.3 ± 1.1	17.6 ± 0.6
UUN (mg/24 h)				
I	252 ± 16	281 ± 17	246 ± 22	243 ± 25
II	228 ± 44	244 ± 57	258 ± 24	193 ± 55

I = First metabolic study: Adult = 4 month old, Aged = 20 month old; II = Second metabolic study: Adult = 10 month old, Aged = 26 month old. a = Significantly different from adult control ($p < 0.05$); b = Significantly different from 1.5% TAU ($p < 0.05$); c = Significantly different from no TAU ($p < 0.05$).

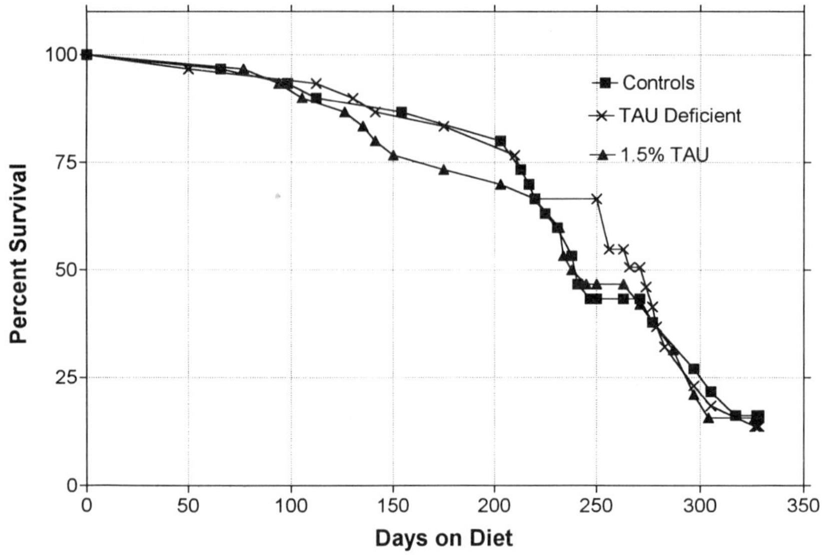

Figure 2. Effects of taurine on life span in F344 rats.

(Fig. 3). As previously reported[9,46], serum taurine was significantly ($p < 0.05$) lower in aged controls (26-29 month old) when compared to adult controls (10-11 month old) (Fig. 3). Serum taurine was also significantly ($p < 0.05$) lower in the aged rats on the taurine-deficient diet compared to the adult controls, but was not different from age matched controls (Fig. 3). Urinary taurine excretion in adult and aged rats is presented in Fig. 4. Taurine excretion was significantly ($p < 0.05$) reduced in 26 month old F344 rats on both normal and taurine-deficient diets when compared to adult (10 month old) or 20 month old F344 rats. As expected, taurine supplementation resulted in highly significant ($p<0.01$) increases in urinary taurine

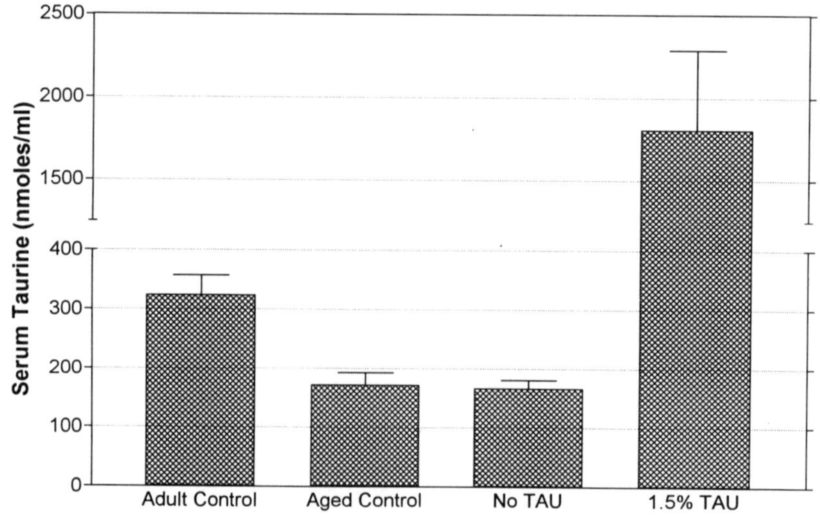

Figure 3. Effects of aging and dietary modifications on serum taurine.

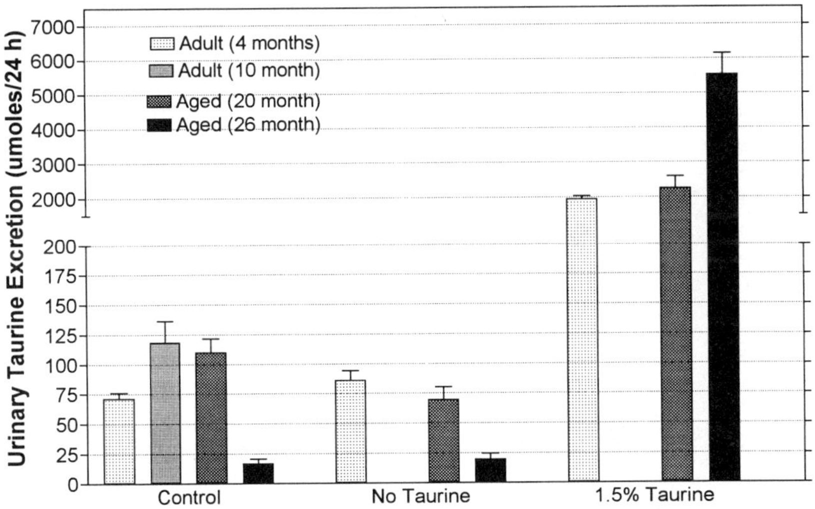

Figure 4. The effects of aging on urinary taurine excretion.

excretion in both adult and aged F344 rats. Urinary excretion of glycine and alanine was also measured in rats on the control and taurine-deficient diets. The high levels of taurine in the urine of the taurine-supplemented group necessitated diluting the urine for analysis to the point that alanine and glycine were not detectable in this group. The pattern of excretion of glycine and alanine with increasing age was distinct from that of taurine. Alanine excretion declined in a roughly linear fashion with age (data not shown). Glycine excretion exhibited a more complex change over time with excretion being stable from 4-10 months (32 μmol/24 h), falling significantly ($p < 0.05$) at 20 months (16 μmol/24 h) and increasing significantly ($p < 0.05$) at 26 months (65 μmol/24 h). Glycine excretion was not affected by the taurine-deficient diet.

There was a pronounced age-related increase in urine output and fluid intake (Table 2). Electrolyte excretion was altered by age and to a modest degree by the dietary manipulations (Table 2). We had previously found that elevated dietary intake of taurine enhanced urinary excretion of K^+ and Ca^{2+} in spontaneously hypertensive rats (SHR)[26]. Calcium excretion increased in the 26 month old rats on the control diet and the taurine-supplemented diet compared to the 10 month old rats on the control diet (Table 2). Calcium excretion was also lower in 20 month old rats on the taurine-deficient diet compared to age matched taurine-supplemented rats (Table 2). Potassium excretion was elevated in the aged rats on the control diet for both metabolic studies when compared to the adults on the control diet (Table 2). Potassium excretion in the 20 month old rats on the taurine-deficient diet was lower than the age-matched rats on the taurine-supplemented diet. However, K^+ excretion did not differ between these groups at 26 months of age (Table 2). Excretion of creatinine and urea nitrogen were not altered by age or diet (Table 2).

Taurine supplementation did not prevent or retard the age-related increase in urinary protein excretion (Fig. 5). The increase in urinary protein excretion is a good marker of the progressive nephropathy that occurs with advancing age in the F344 rat. Taurine supplementation also did not improve the heart weight to body weight ratio for the aged rats. The ratios (heart weight (mg)/100 g body weight) for the groups were: adult control (273 ± 10), aged control (366 ± 18), aged no taurine (400 ± 26) and aged taurine supplemented (369 ± 22).

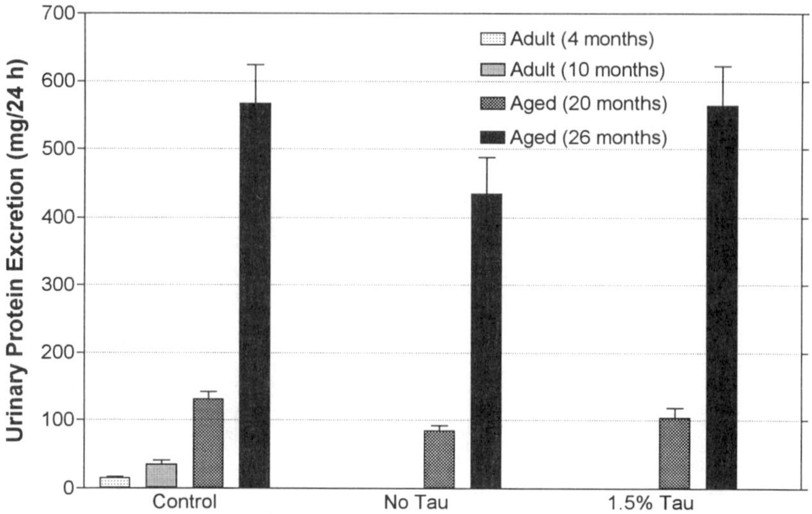

Figure 5. Urinary protein excretion in adult and aged F344 rats.

Serum creatinine and urea nitrogen were significantly ($p < 0.05$) elevated in the aged control group and the no taurine group when compared to the adult controls (Figs. 6 and 7). The aged rats on the taurine-supplemented diet had serum creatinine and urea nitrogen values that did not differ significantly from the adult controls (Figs. 6 and 7). Serum growth hormone, Ca^{2+}, glucose and Na^+ were not affected by either age or dietary treatment (Table 3). Total serum protein levels were decreased in aged rats (Table 3).

Locomotor activity was assessed after 8 and 20 weeks on the diets. There was no significant effect of the diets on locomotor activity (data not shown). Cognitive function was assessed using the Morris Water Maze. The aged F344 rats (n=8 per dietary condition) selected for water maze training were in good general health and had no visible signs of

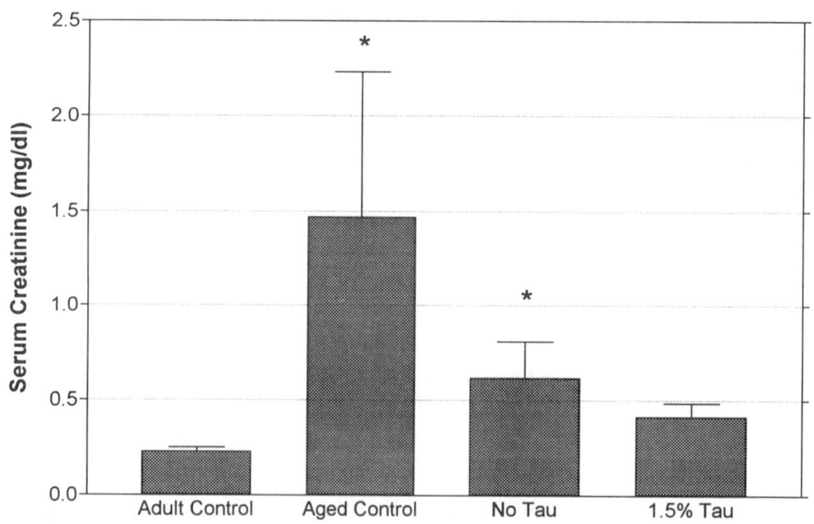

Figure 6. Serum creatinine in adult and aged F344 rats.

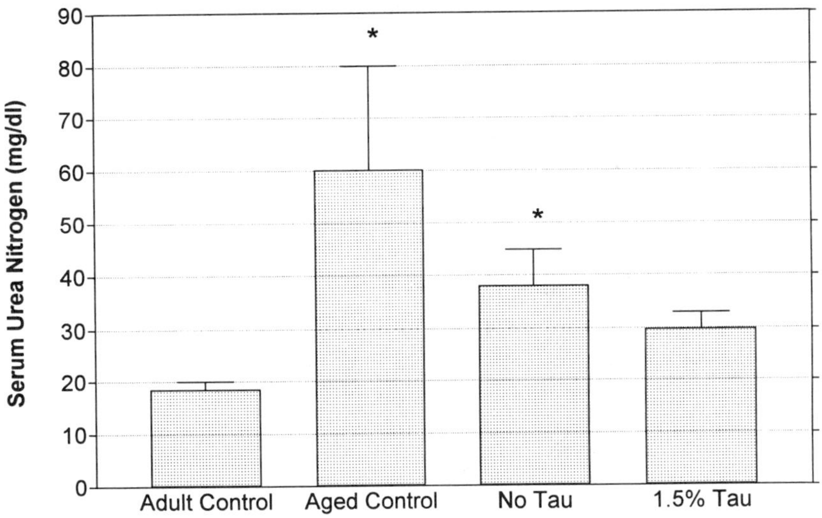

Figure 7. Serum urea nitrogen in adult and aged F344 rats.

cataracts. The aged F344 rats generally exhibited very poor maze learning performance as has previously been reported[23]. Taurine supplementation or restriction had no significant effect on acquisition or retention of the maze learning task (data not shown).

Tissue taurine content has been determined in 3 peripheral tissues (Table 4). Adrenal taurine content was not significantly altered by taurine restriction or supplementation when compared to aged rats on normal diets or adult rats on normal diets. Taurine supplementation resulted in a significant increase in liver taurine content relative to the aged rats eating the normal diet or the taurine free diet. Liver taurine content was variable within treatment groups and this resulted in the aged control group not statistically differing from the adult control group despite a 67% decrease in taurine content. Taurine content of the spleen was significantly decreased in all the aged groups when compared to the adult controls.

Table 3. Serum values for adult and aged F344 rats

Measure	Adult Control	Aged Control	No TAU	1.5% TAU
Growth Hormone (ng/ml)	13.5 ± 8.8	14.1 ± 3.4	29.0 ± 11.1	15.1 ± 4.7
Calcium (n = 3) (mg/dl)	ND	9.3 ± 0.5	10.9 ± 0.2	10.4 ± 0.5
Sodium (mEq/l)	123 ± 2	125 ± 2	129 ± 3	126 ± 3
Glucose (mg/dl)	151 ± 8	120 ± 11	120 ± 7	125 ± 7
Protein (mg/ml)	98 ± 5	70 ± 5^{a,b}	61 ± 3^{a}	52 ± 3^{a}

ND = Not determined; a = Significantly different from adult control ($p< 0.05$);
b = Significently different from 1.5% taurine ($p< 0.05$)

Table 4. Tissue content of taurine in aged F344 rats

Group	Adrenal	Liver	Spleen
Adult Control (n = 6)	6.85 ± 0.52	9.62 ± 2.79	14.14 ± 0.38[2]
Aged Control (n = 11)	7.00 ± 0.62	3.13 ± 0.55	8.55 ± 0.70
Aged No Taurine (n = 14)	7.83 ± 0.52	6.53 ± 1.47	10.09 ± 0.79
Aged 1.5% Taurine (n = 10)	9.33 ± 1.04	20.04 ± 2.20[1]	11.66 ± 0.76

All data expressed as μmol taurine/g tissue weight ± SEM. [1]$p<0.01$ versus aged control or no taurine group; [2] $p<0.05$ versus aged control, aged no taurine or aged 1.5% taurine groups.

GENERAL DISCUSSION

This is the first comprehensive study of the long term effects of taurine supplementation in a rodent model of aging. Our previous findings of reduced serum taurine and reductions in tissue content of taurine provided the rationale for assessing the effects of taurine supplementation and restriction in the F344 model of aging. We replicated our previous findings of reduced serum taurine in the aged male F344 rat and also found that urinary excretion of taurine declines significantly with advancing age. The finding of reduced renal excretion of taurine in aged F344 rats is in agreement with Corman *et al.*[6] who reported an age-related decline in taurine excretion in aged female Wistar rats. These findings suggest that the aged F344 rat can conserve taurine via renal adaptive mechanisms[14,37] despite significant renal nephropathy. These findings also suggest that the aged F344 rat shows signs of taurine deficiency as indexed by reduced serum taurine and renal conservation of taurine. The reduction in urinary taurine could, however, reflect a decrease in liver taurine content (Table 4) as suggested by Waterfield *et al.*[47]. Restriction of the dietary intake of taurine did not greatly exacerbate the reduction in serum taurine or result in further reductions in urinary taurine excretion in the aged F344 rat. Biosynthesis of taurine from precursors (methionine and cysteine) appears normal in aged rats since elimination of a direct dietary source of taurine did not result in significant changes in serum, tissue, or urinary taurine concentrations. Studies are currently underway to complete the measurement of tissue content of taurine in these experimental groups. Thus, despite age-related impairment of renal and hepatic function, taurine biosynthesis and renal adaptive mechanisms appear to compensate for low dietary intake of taurine in the F344 rat. Detailed analysis of tissue content of taurine in specific brain regions, the kidney and cardiovascular tissues is needed to fully assess the impact of taurine restriction on whole body taurine homeostasis in this rat model. Our data on tissue content of taurine would suggest long-term dietary taurine restriction seems to blunt some of the age-related decline seen in tissue taurine content perhaps by stimulation of *de novo* synthesis of taurine. The severe renal damage in the aged rats resulted in high urinary output and excessive fluid intake suggesting a possible mild degree of dehydration (Table 2). We have previously shown in the Brattleboro rat which suffers from diabetes insipidus that brain content of taurine is increased 30-40%[11]. This increase in brain taurine content is thought to be an adaptive response to chronic dehydration[11]. If aged F344 rats do suffer mild dehydration, then the fact that taurine content in brain is not increased with advanced age may reflect an unappreciated age-related depletion of taurine. It is unclear what impact aging would have on species which do not have the high capacity that rats exhibit for taurine biosynthesis.

Table 5. Renoprotective effect of taurine

Strain or Disease State	TAU supplement	Urea	Creatinine	Urinary Protein	Ref.
SHR	1.5%	↓ 32%[1]	—	—	49
WKY	1.5%	↓ 31%[1]	—	—	
SHR	1%	—	—	↓ 58%	43
SP-SHR	1.5%	↓ 21%[3]	↓ 22%[3]	↓ 59%	unpublished findings
PAMN Nephropathy	1%	-	↓ 53%[2]	↑ 10%	41
Sprague-Dawley control	1%	—	↓ 5%[2]	NC	42
STZ Diabetes	1%	—	↓ 4%[2]	↓ 35%	
PAMN Nephropathy	1%	—	↓ 53%[2]	↓ 41%	

STZ = streptozocin; PAMN = puromycin aminonucleoside; NC = no change; 1 = plasma, 2 = serum, 3 = urine.

Taurine supplementation was expected to have a renoprotective effect in the aged F344 rat based on previous studies in a number of animal models of impaired renal function (Table 5). The elevated urinary protein excretion and other age-related changes in renal function (Table 2) were not greatly affected by taurine supplementation. Serum markers of renal nitrogenous waste removal (serum urea nitrogen and creatinine) were elevated to a greater extent in the aged control and aged rats on the taurine restricted diet when compared to the taurine supplemented group. Thus, taurine supplementation blunted the age-related increase in serum markers of nitrogenous waste. Taurine supplementation has previously been shown to reduce serum creatinine and urea nitrogen in other models of impaired renal function (Table 5). This also appears to be the case in aging since taurine supplementation did improve these measures. The age-related increase in urinary protein excretion is related to glomerular structural injury and increased glomerular basement membrane permeability[16]. The renal nephropathy and proteinuria seen in the aged F344 rat may occur by mechanisms that are unresponsive to the renoprotective effects of taurine. There is little experimental data on the influence of taurine on nitrogen balance and excretion. The biosynthesis of taurine can result in the urinary excretion of nitrogen in the form of taurine. Ammonia has been shown to stimulate the release of taurine from cultured astrocytes[1]. Aged rats exhibit deficits in ammonia metabolism[21,45]. Although renal excretion of urea nitrogen and creatinine was not increased by taurine supplementation (Table 2), serum levels were lower. This would suggest that the rate of nitrogenous waste production was diminished by taurine supplementation. The normalization of serum creatinine in taurine supplemental rats could be related to a recent report of improved skeletal muscle function in taurine treated aged rats[36]. Further studies are needed to explore the mechanism for the reduced level of serum nitrogenous waste in taurine supplemented rats.

We have recently completed a series of preliminary studies that suggest taurine may be capable of directly inhibiting iron-stimulated autoxidation of dopamine and l-dopa. Iron can act as a powerful catalysis for the autoxidation of catecholamines resulting in the generation of free radicals and cytotoxic quinones[29,30]. Quinone formation can be monitored spectrophotometrically and the rate of metal-stimulated catecholamine autoxidation can be measured[18]. The results of such an experiment in the presence and absence of taurine are presented in Fig. 8. An extensive series of control experiments has been performed suggesting that taurine and hypotaurine (HTau) can inhibit ($p < 0.05$) iron-stimulated l-dopa oxidation and taurine precursors and ß-alanine (B-Ala) either have no effect or enhance oxidation rates. While taurine can reduce iron-stimulated catecholamine autoxidation, iron-stimulated lipid peroxidation of brain homogenates was unaffected by 20 mM taurine (unpublished observations). We have also found that basal levels of lipid peroxidation in the

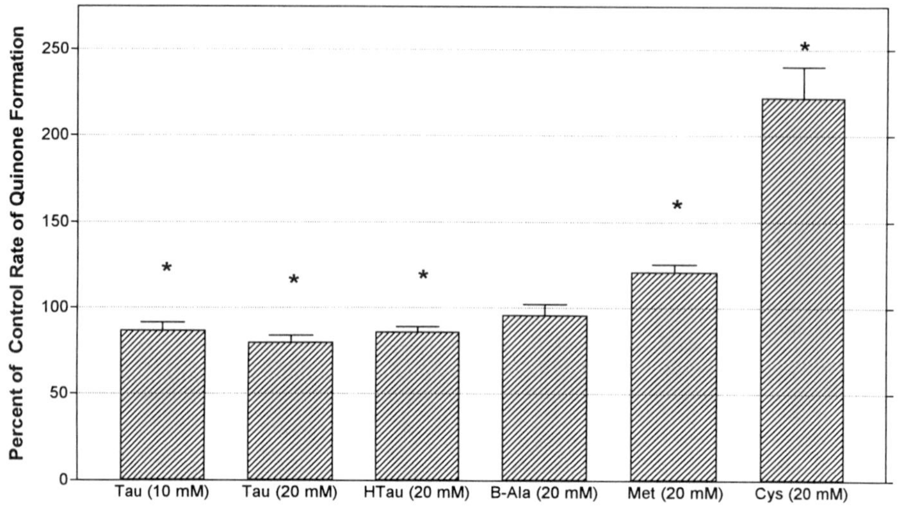

Figure 8. Effects of taurine and related amino acids on ferric chloride-stimulated l-DOPA oxidation.

adrenal glands from the rats used in this study were not altered by aging or dietary treatments. The free radical hypothesis of aging suggests that age-related cellular deterioration is related to the cumulative effects of oxidative damage and the failure to adequately repair such damage[17,22]. There is substantial evidence that oxidative stress increases with advanced age and antioxidant defense systems may not function optimally[17,22]. Taurine has been suggested to have antioxidant functions [20,48]. Specifically, taurine has been shown to protect lymphoblastoid cells from iron-induced cellular damage, but did not reduce iron-induced lipid peroxidation[35]. If taurine content declines with advanced age or whole body taurine homeostasis is altered as suggested by altered serum levels and decreased urinary excretion, then the reduced availability of taurine in advanced age may contribute to an impairment in antioxidant defense mechanisms. This may be particularly important in catecholamine rich brain regions in light of age-related increases in brain iron content[4]. Our laboratory is currently exploring the role of taurine in the aged brain and its potential antioxidant and neuroprotective functions.

Taurine has well established functions in osmoregulation, Ca^{2+} homeostasis, cardiovascular control, and the regulation of neural excitability[19,20,48]. With advancing age the ability for fine-tuned regulation of complex physiological functions diminishes and the ability to respond to metabolic and oxidative challenges also declines. An age-related decline in taurine content or a diminished ability to mobilize or regulate taurine availability could potentially contribute to compromised cellular function or lead to cell death. Clearly, the role taurine in "normal" cellular aging and age-related diseases needs further elucidation and clarification.

REFERENCES

1. Albrecht, J., Bender, A.S., and Norenberg, M.D. 1994, Ammonia stimulates the release of taurine from cultured astrocytes, *Brain Res.* 660:288-292.
2. Banay-Schwartz, M., Lajtha, A., and Palkovits, M. 1989, Changes with aging in the levels of amino acids in rat CNS structural elements. II. Taurine and small neutral amino acids, *Neurochem.Res.* 14:563-570.

3. Benedetti, M.S., Russo, A., Marrari, P., and Dostert, P. 1991, Effects of aging on the content in sulfur-containing amino acids in rat brain, *J.Neural Transm.* 86:191-203.
4. Benkovic, S.A. and Connor, J.R. 1993, Ferritin, transferrin, and iron in selected regions of the adult and aged rat brain, *J.Comp.Neurol.* 338:97-113.
5. Coleman, G.L., Barthold, S.W., Osbaldiston, G.W., Foster, S.J., and Jonas, A.M. 1977, Pathological changes during aging in barrier-reared Fischer 344 male rats, *J.Gerontol.* 32:258-278.
6. Corman, B., Pratz, J., and Poujeol, P. 1985, Changes in anatomy, glomerular filtration, and solute excretion in aging rat kidney, *Am.J.Physiol.* 248:R282-R287.
7. Dawson, R., Patterson, T.A., and Eppler, B. 1995, Endogenous excitatory amino acid release from brain slices and astrocyte cultures evoked by trimethyltin and other neurotoxic agents, *Neurochem.Res.* 20:847-858.
8. Dawson, R. and Wallace, D.R. 1992, Kainic acid-induced seizures in aged rats: neurochemical correlates, *Brain Res.Bull.* 29:459-468.
9. Dawson, R. and Wallace, D.R. 1992, Taurine content in tissues from aged Fischer 344 rats, *Age*, 15:73-81.
10. Dawson, R., Wallace, D.R., and Meldrum, M.J. 1989, Endogenous glutamate release from frontal cortex of adult and aged rats, *Neurobiol.Aging*, 10:665-668.
11. Dawson, R., Jr., Wallace, D.R., and King, M.J. 1990, Monoamine and amino acid content in brain regions of Brattleboro rats, *Neurochem.Res.* 15:755-761.
12. Donzanti, B.A. and Ung, A.K. 1990, Alterations in neurotransmitter amino acid content in the aging rat striatum are subregion dependent, *Neurobiol.Aging*, 11:159-162.
13. Fierabracci, V., Del Roso, A., Novelli, M., Masiello, P., Fosella, P.V., and Bergamini, E. 1990, Age-related changes in amino acid metabolism in the rat, in: "Protein Metabolism in Aging", Anonymous Wiley-Liss, New York.
14. Friedman, A.L., Albright, P.W., Gusowski, N., Padilla, M., and Chesney, R.W. 1983, Renal adaptation to alteration in dietary amino acid intake, *Am.J.Physiol.* 245:F159-F166.
15. Gibson, G.E. and Peterson, C. 1987, Calcium and the aging nervous system, *Neurobiol.Aging*, 8:329-343.
16. Goldstein, R.S., Tarloff, J.B., and Hook, J.B. 1988, Age-related nephropathy in laboratory rats, *FASEB J.* 2:2241-2251.
17. Harman, D. 1988, Free radicals and aging, *Mol.Cell.Biochem.* 84:155-161.
18. Heikkila, R.E. and Cabbat, F.S. 1978, The stimulation of 6-hydroxydopamine autoxidation by bivalent copper: potential importance in neurotoxic process, *Life Sci.* 23:33-38.
19. Huxtable, R.J. 1989, Taurine in the central nervous system and the mammalian actions of taurine, *Prog.Neurobiol.* 32:471-533.
20. Huxtable, R.J. 1992, Physiological actions of taurine, *Physiol.Rev.* 72:101-163.
21. Kirzinger, S.S. and Fonda, M.L. 1978, Glutamine and ammonia metabolism in the brains of senescent mice, *Exp.Gerontol.* 13:255-261.
22. LeBel, C.L. and Bondy, S.C. 1992, Oxidative damage and cerebral aging, *Prog.Neurobiol.* 38:601-609.
23. Luine, V., Bowling, D., and Hearns, M. 1990, Spatial memory deficits in aged rats: contributions of monoaminergic systems, *Brain Res.* 537:271-278.
24. Magnoni, M.S., Govoni, S., Battaini, F., and Trabucchi, M. 1991, The aging brain: protein phosphorylation as a target of changes in neuronal function, *Life Sci.* 48:373-385.
25. Massie, H.R., Williams, T.R., and DeWolfe, L.K. 1989, Changes in taurine in aging fruit flies and mice, *Exp.Gerontol.* 24:57-65.
26. Meldrum, M.J., Tu, R., Patterson, T.A., Dawson, R., and Petty, T. 1994, The effect of taurine on blood pressure, and urinary sodium, potassium and calcium excretion, in: "Taurine in Health and Disease", Huxtable, R. and Michalk, D.V. eds., Plenum Press, New York.
27. Milakofsky, L., Hare, T.A., Miller, J.M., and Vogel, W.H. 1984, Comparison of amino acid levels in rat blood obtained by catheterization and decapitation, *Life Sci.* 34:1333-1340.
28. Milakofsky, L., Harris, N., and Vogel, W.H. 1993, Effect of repeated stress on plasma catecholamines and taurine in young and old rats, *Neurobiol.Aging*, 14:359-366.
29. Miller, D.M., Buettner, G.R., and Aust, S.D. 1990, Transition metals as catalysts of "autoxidation" reactions, *Free Radical Biol.Med.* 8:95-108.
30. Monks, T.J., Hanzlik, R.P., Cohen, G.M., Ross, D., and Graham, D.G. 1992, Quinone chemistry and toxicity, *Toxicol.Appl.Pharmacol.* 112:2-16.
31. Nishio, S., Negoro, S., Hosokawa, T., Hara, H., Tanaka, T., Deguchi, Y., Ling, J., Awata, N., Azuma, J., Aoike, A., Kawai, K., and Kishimoto, S. 1990, The effect of taurine on age-related immune decline in mice: the effect of taurine on T cell and B cell proliferative response under costimulation with ionomycin and phorbol myristate acetate, *Mech.Aging Dev.* 52:125-139.

32. Oja, S.S., Holopainen, I., and Kontro, P. 1990, Stimulated taurine release from different brain preparations: Changes during development and aging, in: "Taurine: Functional Neurochemistry, Physiology, and Cardiology", Pasantes-Morales, H., Martin, D.L., Shain, W. and del Río, R.M. eds., Wiley-Liss, New York, pp. 277-287.
33. Ooboshi, H., Sadoshima, S., Yao, H., Ibayashi, S., Matsumoto, T., Uchimura, H., and Fujishima, M. 1995, Ischemia-induced release of amino acids in the hippocampus of aged hypertensive rats, *J.Cereb.Blood Flow Metab.* 15:227-234.
34. Palkovits, M., Banay-Schwartz, M., and Lajtha, A. 1990, Taurine levels in brain nuclei of young adult and aging rats, in: "Taurine: Functional Neurochemistry, Physiology, and Cardiology", Pasantes-Morales, H., Martin, D.L., Shain, W. and del Río, R.M. eds., Wiley-Liss, New York, pp. 45-51.
35. Pasantes-Morales, H., Wright, C.E., and Gaull, G.E. 1985, Taurine protection of lymphoblastoid cells from iron-ascorbate induced damage, *Biochem.Pharmacol.* 34:2205-2206.
36. Pierno, S., DeLuca, A., Roselli, A., Degiorgi, A., Huxtable, R.J., and Conte Camerino, D. 1996, Effects of chronic taurine treatment on the electrical and contractile properties of skeletal muscle fibers of aged rats, in: "Taurine: Basic and Clinical Aspects", Huxtable, R.J., Azuma, J., Kuriyama, K., Nakagawa, M. and Baba, A. eds., Plenum Publishing Co. New York.
37. Rozen, R. and Scriver, C.R. 1982, Renal transport of taurine adapts to perturbed taurine homeostasis, *Proc.Natl.Acad.Sci.USA*, 79:2101-2105.
38. Shimokawa, I., Higami, Y., Hubbard, G.B., McMahan, C.A., Masoro, E.J., and Yu, B.P. 1993, Diet and the suitability of the male Fischer 344 rat as a model for aging research, *J.Gerontol.* 48:B27-B32.
39. Sturman, J.A. 1993, Taurine in development, *Physiol.Rev.* 73:119-147.
40. Timiras, P.S., Hudson, D.B., and Oklund, S. 1973, Changes in central nervous system free amino acids with development and aging, *Prog.Brain Res.* 40:267-275.
41. Trachtman, H., Del Pizzo, R., Futterweit, S., Levine, D., Rao, P.S., Valderrama, E., and Sturman, J.A. 1992, Taurine attenuates renal disease in chronic puromycin aminonucleoside nephropathy, *Am.J.Physiol.Renal,Fluid Electrolyte Physiol.* 262:F117-F123.
42. Trachtman, H., Lu, P., and Sturman, J.A. 1993, Immunohistochemical localization of taurine in rat renal tissue: Studies in experimental disease states, *J.Histochem.Cytochem.* 41:1209-1216.
43. Trachtman, H., Pizzo, R.D., Rao, P., Rujikarn, N., and Sturman, J.A. 1989, Taurine lowers blood pressure in the spontaneously hypertensive rat by a catecholamine independent mechanism, *Am.J.Hyperten.* 2:909-912.
44. Tyce, G.M. and Wong, K. 1980, Conversion of glucose to neurotransmitter amino acids in the brains of young and aging rats, *Exp.Gerontol.* 15:527-532.
45. Wallace, D.R. and Dawson, R. 1992, Ammonia regulation of phosphate-activated glutaminase displays regional variation and impairment in the brain of aged rats, *Neurochem.Res.* 17:1113-1122.
46. Wallace, D.R. and Dawson, R., Jr. 1990, Decreased plasma taurine in aged rats, *Gerontology*, 36:19-27.
47. Waterfield, C.J., Turton, J.A., Scales, M.D.C., and Timbrell, J.A. 1993, The correlation between urinary and liver taurine levels and between pre-dose urinary taurine and liver damage, *Toxicology*, 77:1-5.
48. Wright, C.E., Tallan, H.H., Lin, Y.Y., and Gaull, G.E. 1986, Taurine: biological update, *Ann.Rev.Biochem.* 55:427-453.
49. Yamamoto, J., Akabane, S., Yoshimi, H., Nakai, M., and Ikeda, M. 1985, Effects of taurine on stress-evoked hemodynamic and plasma catecholamine changes in spontaneous hypertensive rats, *Hypertension*, 7:913-922.

5

TAURINE ADMINISTRATION RAISES PLASMA TAURINE LEVELS AND AFFECTS CERTAIN PLASMA AMINO ACIDS AND RELATED COMPOUNDS IN RATS

Karsten Korang,[1] Louis Milakofsky,[1,2] Theodor A. Hare,[1] James M. Hofford,[3] and Wolfgang H. Vogel[1]

[1] Departments of Pharmacology and Psychiatry and Human Behavior
Jefferson Medical College of Thomas Jefferson University
Philadelphia, Pennsylvania, 19107
[2] Department of Chemistry
Penn State University, Berks Campus
Reading, Pennsylvania 19107
[3] Department of Medicine
Jefferson Medical College of Thomas Jefferson University
Philadelphia, Pennsylvania, 19107

INTRODUCTION

In spite of the interest in the pharmacology of taurine, little information is available on its absorption into the blood stream after administration to the rat. For this reason, we conducted a pharmacokinetic study on the fate of injected taurine in rat plasma for 4 h and measured its effects on a large number of plasma amino acids and related compounds.

MATERIALS AND METHODS

Male Sprague-Dawley rats weighing about 250 g were housed for three days prior to the experiment in Polycarb containers with 3 animals per cage in the animal quarters at 24°C and 55% humidity. They were maintained on a 12 h light/dark cycle and allowed free access to food (Purina Chow) and water.

For the plasma-time study, animals were catheterized in the left jugular vein under anesthesia[3]. After 24 h, they were injected ip with saline or taurine (0.8 g/kg) and blood was obtained at different times in syringes and tubes containing 200 IU of heparin. After centrifugation, the plasma was deep frozen.

Taurine 2, edited by Huxtable et al.
Plenum Press, New York, 1996

At the time of the analysis, 0.32 N perchloric acid (PCA) was added to the plasma[3]. The PCA solution contained 2.632 nmol/ml of m-fluorophenylalanine (MFP) as an internal standard.

Amino acid levels of deproteinized plasma were determined by ion-exchange HPLC with fluorometric post-column detection with o-phthaldialdehyde. Lithium citrate elution buffers were used to separate the amino acids. The amino acid levels were quantitatively determined by relating the chromatographic peak heights to those obtained from a known physiological standard and the internal standard MFP. The technique has been shown to be reliable, sensitive, and valid[3].

A two-way analysis of variance (ANOVA) with repeated measures was used to evaluate plasma values. Spearman linear correlation analyses were performed to compare compounds within the plasma. Due to the large number of statistical evaluations, significance was set at $p < 0.01$.

RESULTS

The plasma levels of taurine and affected other amino acids and related amino compounds before and after injection of 0.8 g/kg of taurine are shown in Table 1. Plasma taurine increases rapidly to a maximum at 15 min after injection (about 70 times baseline levels) and declines steadily thereafter approaching baseline values at about 240 min. Concurrently, the levels of a large number of other amino compounds decrease either immediately or subsequently, generally paralleling changes in taurine levels. Major decreases in the plasma are seen with alanine, arginine, asparagine, citrulline, glutamine, glycine, ornithine, threonine, tryptophan, tyrosine and valine while ß-alanine and phosphoserine are increased.

No changes were seen with α-aminobutyric acid, GABA, aspartic acid, ethanolamine, glutamic acid, histidine, leucine, methionine, phenylalanine, and serine.

Table 1. Effect of taurine injection (0.8 g/kg; i.p.) on plasma levels (nmol/ml) of taurine and other amino acids

Amino Acid	T0	T15	T30	T60	T120	T240
Alanine	131 ± 10	99.2 ± 7.8	87.3 ± 7.5	78.7 ± 5.9**	81.1 ± 5.0**	120 ± 10
Ammonia	40.1 ± 2.6	46.7 ± 4.3	39.2 ± 3.9	36.6 ± 2.4	32.8 ± 2.0	27.6 ± 2.0*
Arginine	97.8 ± 3.4	72.2 ± 4.9**	62.9 ± 8.5**	56.2 ± 5.9**	52.3 ± 3.9**	64.2 ± 3.1*
Asparagine	25.6 ± 1.0	14.7 ± 0.7**	13.7 ± 1.3	13.7 ± 0.8**	16.3 ± 1.4**	21.7 ± 1.0
β-Alanine	1.84 ± 0.25	4.41 ± 0.46**	3.21 ± 0.26	2.01 ± 0.22	2.11 ± 0.23	1.89 ± 0.21
Citrulline	39.7 ± 2.2	26.6 ± 2.2*	24.8 ± 3.2*	28.0 ± 2.6	28.6 ± 2.1	28.8 ± 2.5
Glutamine	258 ± 20	146 ± 3	154 ± 14**	167 ± 9**	210 ± 12	244 ± 13
Glycine	128 ± 14	89.0 ± 2.8*	77.8 ± 9.3	101 ± 11	107 ± 8	109 ± 10
Isoleucine	43.5 ± 4.6	28.2 ± 1.5*	31.1 ± 1.8	35.4 ± 3.2	40.2 ± 1.5	37.2 ± 2.8
Lysine	141 ± 9	102 ± 8*	99.1 ± 11.6	106 ± 7	118 ± 12	128 ± 7
Ornithine	21.2 ± 1.7	23.9 ± 1.7	17.9 ± 0.9	15.2 ± 1.1	13.6 ± 0.3**	14.7 ± 0.9*
Phosphoserine	4.83 ± 0.48	9.23 ± 0.58**	9.19 ± 0.58**	6.24 ± 0.43	3.60 ± 0.32	3.82 ± 0.37
Taurine	66.4 ± 5.0	4893 ± 242**	2814 ± 439**	1789 ± 139**	223 ± 16**	107 ± 9*
Threonine	86.6 ± 5.2	57.4 ± 3.4**	59.0 ± 2.9**	54.5 ± 2.2**	55.4 ± 3.7**	60.7 ± 4.0*
Tryptophan	31.6 ± 2.1	25.2 ± 1.4	23.5 ± 1.5	23.3 ± 1.4	21.9 ± 1.6**	31.1 ± 2.8
Tyrosine	47.9 ± 3.7	37.5 ± 2.2	35.3 ± 3.4	27.3 ± 1.5**	32.7 ± 2.0**	32.7 ± 2.0*
Valine	92.4 ± 6.0	59.2 ± 3.4**	68.1 ± 4.2	82.3 ± 7.1	80.1 ± 3.1	68.1 ± 7.1

Values represent averages ± SEM. Number of animals was 5-6.T0 represents values before and T15 to T240 values after injection in min. *$p < 0.001$, **$p < 0.005$.

We also looked for correlations between the levels of taurine with those of some of the other major compounds in plasma. Taurine correlated inversely with serine, glutamine, valine, leucine and histidine (all $r=-0.8$, $p < 0.05$). These correlations generally disappeared after taurine administration.

DISCUSSION

The levels of taurine and other amino compounds in the plasma of this study are similar to those reported earlier[3]. Our rat plasma levels of about 60 nmol/ml are similar to those of 44 nmol/ml found for human plasma[1,4]. All the other compounds show the expected differential distribution in the plasma.

The ip administration of taurine increases plasma levels rapidly with a peak level 70-fold above baseline values after 15 min. Thereafter, levels decline steadily. After 4 h, baseline levels are almost reached. This results in a half-life of about 40 min.

The increased plasma levels of taurine are associated with significant changes in the concentrations of several amino acids or related amino compounds. Most amino compounds decrease in the plasma. At present, we do not know how taurine affects these other amino compounds but interference with the uptake, storage and/or release of some of the amino compounds is possible. Interactions of taurine with certain amino acids have indeed been reported[2]. Of the compounds linked metabolically to arginine, methionine shows no change, citrulline decreases briefly at the beginning and ornithine decreases at the end of the period studied. However, these changes are minor.

A significant number of correlations between taurine and certain amino compounds is found in control rats which might suggest special relationships such as shared transport mechanisms. Interestingly, they are mostly lost after taurine injection indicating that the large amounts of taurine might overwhelm certain physiological mechanisms.

In conclusion, taurine is rapidly absorbed into the bloodstream after ip administration. Here, it has a half-life of about 40 min. The increased taurine levels affect a number of amino compounds and disturb some of the relationships found in the resting state. Pharmacological results obtained with taurine administration can be due to raised taurine levels per se but also be caused by taurine-induced changes in some other amino compounds.

ACKNOWLEDGMENT

Financial support by the Anesthesia Research Foundation, Wilmington, is gratefully acknowledged.

REFERENCES

1. Fukuda, K. and Usui, T. 1983, Characteristic patterns of free amino acid content in plasma, erythrocytes, lymphocytes, and granulocytes in man, *Hiroshima J. Med. Sci.,* 32:163-166.
2. Huxtable, R.J. 1992, Physiological Actions of Taurine, *Physiol. Rev.* 72:101-163.
3. Milakofsky, L., Hare, T.A., Miller, J.M. and Vogel, W.H. 1983, Comparison of amino acid levels in rat blood obtained by decapitation or catherization, *Life Sci.,* 34:1333-1340.
4. Paauw, J.D. and Davis, T.A. 1990, Taurine supplementation at three different dosages and its effect on trauma patients, *Am. J. Nutr.,* 60:203-206.

6

TAURINE AND KYNURENINASE

Y. Shibata,[1,4] T. Ohta,[2] M. Nakatsuka,[1] H. Ishizu,[3] Y. Matsuda,[4]
T. Shindo,[4] F. Takeuchi,[1] M. Yoshino,[1] S. Hirano,[5] and T. Noguchi[5]

[1] Department of Biochemistry
Aichi Medical University
Aichi, Japan
[2] Japan Food Research and Development Institute Co. Ltd.
Kyoto, Japan
[3] Ueki Hospital
Sakai, Osaka, Japan
[4] Japan Clinic Co. Ltd.
Kyoto, Japan
[5] Department of Physiology
Toho University, Tokyo, Japan

INTRODUCTION

Kynureninase is the most important enzyme in tryptophan metabolism. In the main pathway of tryptophan metabolism, kynureninase forms 3-OH-anthranilic acid from 3-OH-kynurenine. In this reaction vitamin B_6 is a co-enzyme. An important relationship between -SH groups and liver kynureninase apoenzyme has been reported by Takeuchi[8,9] and Izuta[1]. We have experimented on kynureninase and S-involved compounds as follows: (i) Kynureninase activity was inhibited in vitamin B_6-deficient rats and in streptozotocin (STZ)-induced diabetic rats; (ii) Following tryptophan administration, in rats given taurine or oysters (rich in taurine), the excretion of xanthurenic acid (an index of vitamin B_6 deficiency) decreased compared with an excess methionine-administered group; and (iii) The active center of kynureninase apoenzyme (464 amino acid residues) is located near cysteine.

METHODS AND RESULTS

The activity of liver kynureninase was decreased in STZ-induced diabetic rats (Table 1) and in vitamin B_6-deficient rats (Table 2).

The low molecular fraction of oyster extract improved the glucose tolerance test of alloxan diabetic rats (Table 3).

Taurine 2, edited by Huxtable et al.
Plenum Press, New York, 1996

Table 1. Liver kynureninase activity of STZ-induced diabetic rats

Pyridoxal Phosphate	Control	STZ
+	18.3 ± 2.1	9.4 ± 0.3
—	16.8 ± 2.5	7.7 ± 0.9

STZ: Streptozotocin-induced diabetic rats.

Table 2. Kynureninase activity of 100,000 g supernatant of 25% rat liver homogenate (nmol/ml/min)

Pyridoxal Phosphate	Control	Vitamin B_6-Deficient
+	39.6 ± 7.5	32.3 ± 5.5
—	20.6 ± 4.7	12.7 ± 8.5

Control: Complete diet fed group

In excess methionine-administered rats, the excretion of xanthurenic acid increased in urine. Administration of taurine or oyster extract (rich in taurine) did not increase xanthurenic acid excretion (Table 4).

DISCUSSION

Recently Takeuchi *et al.*[9] reported the primary structure of kynureninase protein. This protein (apoenzyme) consists of 464 amino acids. The active center of kynureninase apoenzyme is located near cysteine (C) and the coenzyme, pyridoxal phosphate, binds with lysine (K) at position 276. The N-terminal is CH_3CO-Met and the C-terminal is Asn. This enzyme also acts as the apoenzyme of cysteine ß-lyase. It is also related to cystathionine synthetase and cystathionine lyase, both of which are also vitamin B_6-dependent enzymes (Fig. 1).

Table 3. Effect of low molecular fraction of oyster extract on glucose tolerance test in alloxan diabetic rats (Glucose (mg/dl))

Time	Control	Alloxan-Diabetic	Alloxan-Diabetic rats fed Oyster Extract
0	138	224	252
10 min	319	350	364
30 min	257 ± 30	457 ± 38	371 ± 38
1 h	198 ± 23	448 ± 26	352 ± 33
2 h	219 ± 20	439 ± 23	312 ± 27
3 h	149 ± 22	452 ± 26	307 ± 24
4 h		410 ± 33	300 ± 24
24 h	133 ± 19	373 ± 20	318 ± 21
48 h	138 ± 20	401 ± 21	306 ± 29

Rats were given 50 g glucose at time 0.

Table 4. Excretion of xanthurenic acid in urine of rats after oral tryptophan administration

Dietary Enrichment	Urinary Xanthurenic Acid (μg/day)
Taurine	2689.1
Serine + Methionine	3667.1
Oyster Extract	1851.1
Control	2681.4

	271										282
Kynureninase	C	W	C	S	Y	-	K*	Y	L	N	S	G
Rat CGL	C	M	C	S	A	T	K*	Y	M	N	-	G
E. coli CGS	Y	L	H	S	C	T	K*	Y	L	N	-	G

Figure 1. Comparison of amino acid sequence at pyridoxal phosphate-binding site. CGL: cystathionine γ-lyase; CGS: cystathionine γ-synthase; *pyridoxal phosphate binding site.

SUMMARY OF LITERATURE

i. In vitamin B_6 deficient rats, xanthurenic acid shows a diabetogenic action[3]. In diabetes induced by the Zn^{2+} chelating agent, 8-hydroxyquinoline oxine[5], proinsulin synthesis is inhibited. The cytosolic enzyme, kynureninase is inhibited, but not the mitochondrial enzyme, kynurenine aminotransferase[4].
ii. Xanthurenic acid excretion increases in vitamin B_6 deficiency, and xanthurenic acid also inhibits kynureninase[8,9].
iii. In our experiments, taurine had a beneficial action in diabetes mellitus patients[2].
iv. Zinc can improve the disturbance of taste in diabetes mellitus patients. Sometimes, zinc content in such patients is decreased. Zinc, *in vitro*,inhibits kynureninase activity[8,9]. In vitamin B_6 deficient rats, Zn^{2+} content in the brain stem is increased[6]. *In vivo*, administration of Zn^{2+} inhibits DOPA decarboxylase activity in liver and brain stem.
v. Hypertension and hypercholesterolemia develops in rats given excess methionine, but not in rats given excess taurine.
vi. In STZ diabetic rats, vitamin B_6 deficiency was not observed, but the formation of pyridoxal from pyridoxine decreased[7].

REFERENCES

1. Izuta, J.A. 1991, *Aichi Med.Univ.Assoc.* 19:637-647.
2. Kotake, Y. and Hattori, M. 1996, *Proc.Trace Nutrients Research, Japan Trace Nutrients Research Society, Kyoto, Japan*, (in press).
3. Kotake, Y., Inada, T., and Matsumura, Y. 1954, Experiments of protracted accumulation effects of xanthurenic acid causing chronic diabetic symptoms in white rats, *J.Biochem.* 41:255.
4. Kotake, Y., Ogasawara, N., and Hagino, Y. 1962, Kynurenine-transaminase, kynureninase and the elevated amount of excretion of xanthurenic acid, *Proc.Japan Acad.* 138:223-226.
5. Okamoto, K. 1960, Diabetes Mellitus, *Shindan & Chiryou-Sha Co.Ltd.* 59.

6. Shibata, O., Kimura, M., Itokawa, Y., Kikuchi, H., and Ishikawa, M. 1996, *Proc. Trace Nutrients Research, Japan Trace Nutrents Research Society, Kyoto, Japan*, (in press).
7. Shibuya, M., Kagosaki, S., Kadoki, M., Tanaka, M., and Okada, M. 1995, *Abstract of 49th meeting of Japanese Society of Nutrition and Food Science, May 19-21, Gifu, Japan*, 162.
8. Takeguchi, F., Tsubouchi, R., Nakatsuka, M., and Shibata, Y. 1992, *Abstract of 7th Meeting on International Study Group for Tryptophan Research, June 6-10, Nagoya*, 105.
9. Takeuchi, F., Tsubouchi, R., Yoshino, M., and Shibata, Y. 1995, Amino-acid sequence of rat liver kynureninase, *Biochim.Biophys.Acta*, 1252:185-188.

7

DEXAMETHASONE RECOVERS PHORBOL-ESTER INDUCED REDUCTION OF TAURINE TRANSPORTATION IN MOUSE MACROPHAGE CELL LINE, RAW 264.7

Ha Won Kim,[1] Mi Ja Shim,[1] Won Bae Kim,[2] and Byong Kak Kim[3]

[1] Department of Life Science
College of Liberal Arts and Natural Science
Seoul City University, Jeonnong-Dong, Seoul 130-743, Korea
[2] Research Laboratories
Dong-A Pharmaceutical Co., Ltd.
Kiheung-Up, Yongin-Kun, Kyungki-Do 449-900, Korea
[3] College of Pharmacy
Seoul National University
Seoul 151-742, Korea

ABSTRACT

The human taurine transporter gene has 12 transmembrane domains and its molecular weight is 69.6 kDa. The long cytoplasmic carboxy and amino termini might function as regulatory attachment sites for other proteins. Six potential protein kinase C phosphorylation sites have been reported in human taurine transporter. In this report, we studied the effects of phorbol 12-myristate 13-acetate (PMA) and glucocorticoid hormone on the taurine transportation in RAW 264.7, a mouse macrophage cell line. When the cells were incubated with [^{3}H]taurine in the presence or absence of Na^+ ion for 40 min at 37°C, [^{3}H]taurine uptake rate was 780-times higher in the Na^+-containing buffer than in the Na^+-deficient buffer, indicating that this cell line expresses taurine transporter protein on the cell surface. THP1, human promonocyte cell line, also showed a similar property. The [^{3}H]taurine uptake rate was not influenced by the inflammatory-inducing cytokines such as interleukin-1, gamma-interferon or interleukin-1 + gamma-interferon, but was decreased by PMA in the RAW 264.7 cell line. This suggests that activation of protein kinase C inhibits taurine transporter activity directly or indirectly. The inhibition of [^{3}H]taurine uptake by PMA was time-dependent. Maximal inhibition occurred with 1 h stimulation with PMA. Increasing the treatment time beyond 1 h reduced the inhibition of [^{3}H]taurine uptake, due to the depletion or inactivation of protein kinase C. The cell line also showed concentration-dependent [^{3}H]taurine uptake by PMA stimulation. Phorbol ester (1 μM) caused 23% inhibition. The

Taurine 2, edited by Huxtable et al.
Plenum Press, New York, 1996

inhibition was significant even at a concentration as low as 10 nM PMA. The reduced [^{3}H]taurine uptake could be reversed by treatment with glucocorticosteroid hormone. Dexamethasone recovered the reduced taurine uptake induced by phorbol ester, the maximum effect being seen at 1 h. This may suggest that macrophage cells require higher taurine concentration in a stressed state, for the secretion of glucocorticoid hormone is increased by hypothalamo-pituitary-adrenocortical axis activation in the blood stream.

INTRODUCTION

Taurine is one of the major intracellular amino acids in mammals. It is involved in a number of important physiological processes such as membrane stabilization, detoxification, antioxidation, osmoregulation, modulation of Ca^{2+} flux and neuroregulation.

Recently there has been increasing interest in the mechanisms of taurine transportation in various tissues. Ehrlich cells increase taurine influx as a function of extracellular pH[12]. Rat brain synaptic membrane vesicles required Na^+ and Cl^- ions for taurine uptake[7]. Taurine could be transported above the concentration of plasma osmolarity in toads[1]. Therefore these properties support the presence of a taurine transporter in various tissues.

Cloning of cDNA encoding the transporter of taurine has been done in mouse brain cell[13], rat brain cells[19], canine kidney cells[21] and human thyroid cells[8]. Taurine transporter gene was found to be located in chromosome 3 p24 - p26 in human placenta[18]. This gene has a molecular weight of 69.6 kDa. The long cytoplasmic carboxy and amino termini are variable and probably function as regulatory domain attachment sites for other proteins. Six potential sites for protein kinase C (PKC) phosphorylation have been reported in human taurine transporter. PKC activation was reported to play some important roles in the regulation of taurine uptake in cultured cells[11]. TPA is a tumor-promoter and can activate PKC[2]. It was known as an inhibitor for monocyte differentiation[6].

In this report, we demonstrate that stimulation of PKC reduces taurine uptake in macrophage cells and that dexamethasone reverses the reduced taurine uptake induced by PKC stimulation in this cell line.

METHODS

Cell Culture and Taurine Uptake

Taurine transport studies were performed on THP1 (human monocyte, ATCC TIB 202) and RAW 264.7 (mouse macrophage, ATCC TIB 71) cell lines that were obtained from the American Type Culture Collection, Maryland, USA). Cells were cultured in DMEM medium containing 10% heat-inactivated fetal calf serum (FCS) and penicillin/streptomycin in 90 mm culture dishes for maintaining and in 6-well plates for taurine uptake experiments. When the cells were confluent on the dish, the medium was removed from the monolayer culture and replaced with 2 ml of Na^+-containing buffer which contained 25 mM HEPES/Tris (pH 7.5), 140 mM NaCl, 5.4 mM KCl, 1.8 mM $CaCl_2$, 0.8 mM $MgSO_4$, 5 mM D-glucose or Na^+-deficient buffer which contained 140 mM sucrose instead of 140 mM NaCl. After stabilizing the cells for 2 h at 37°C, 2 μCi of [2-^{3}H(N)]taurine (21.9 Ci/mmol, Dupont, NEN) were added to each culture and incubation continued for 40 min. The cells were washed thrice with ice-cold Na^+-deficient buffer and lyzed with one ml of 0.01 mM NaOH and 1 mM EDTA. The lysate was mixed with 8 ml cocktail solution and was used in measuring radioactivity for three min.

Cell Stimulation

RAW 264.7 cells were stimulated with IL-1β (100 ng/ml), IFN-γ (100 IU/ml), PMA (10 ng/ml) and dexamethasone (100 nM) for 6 h in 10% FCS-DMEM in 6-well plates. The medium was then replaced with Na^+ buffer and the cells were stabilized for 1 h. Then 20 μl (2 μCi) of [^{3}H]taurine in 2 ml buffer were added and cultured for 40 min. Uptaken [^{3}H]taurine was measured by liquid scintillation counter after the cells were lysed. To see the effects of treatment time with PMA on taurine uptake, mouse macrophage cells, RAW 264.7 line, were stimulated for various times with PMA (10 ng/ml) or PMA (10 ng/ml) + dexamethasone (100 nM) in 40 mm culture dishes in 10% DMEM medium. The medium was replaced with Na^+ buffer and stabilized for 10 min and then 2.5 μCi [^{3}H]taurine were added and incubated for 40 min. The cells were thoroughly washed with buffer and lyzed with 1 ml of 20 μM NaOH containing 1 mM EDTA for 1 h at room temperature. The lysates were used for measuring taurine uptake in this cell line.

For the pretreatments of Na^+-deficient buffer and cold taurine in the human macrophage cells, THP1 cells were grown in RPMI 1640 medium containing 10% heat-inactivated FCS, penicillin and streptomycin, and L-glutamine. The cells were washed with Na^+-containing or Na^+-deficient buffer and each 3 x 10^6 cell aliquot in one ml medium stabilized in 24-well plates for 30 min at 37°C in 7% CO_2 incubator. Cold taurine (100 μl of 1 M) or PBS were added to cells and further incubated for 30 min. Then were added 2 μCi (20 μl) of [^{3}H]taurine to each well and incubated for 30 min. The cells were washed with ice-cold PBS 3x and the cells were lyzed with 0.2 M NaOH. The lysates were used for measurement of uptaken [^{3}H]taurine into the cells by adding 2 ml of the cocktail solution.

RESULTS

Macrophage Cells Expressed Taurine Transporter

We measured the uptake of [^{3}H]taurine in human and mouse macrophage cell lines to examine whether macrophage cells express taurine transporter on their cell membrane. When we compared the effects of Na^+ on taurine uptake in RAW 264.7 cell line, the cells in Na^+-deficient buffer did not transport taurine into the cells (Table 1). The Na^+-containing buffer induced 780-times higher taurine transportation than Na^+-deficient buffer did in the mouse macrophage cells.

Similar experiments were conducted in the human macrophage cell line, THP1. In this cell line, the Na^+-containing buffer transported 33x higher [^{3}H]taurine than the Na^+-deficient buffer (Table 2). In addition to this, when the THP1 cells were pretreated with 100 μM cold-taurine for 30 min, [^{3}H]taurine was uptaken in amounts similar to that seen in the Na^+-deficient buffer. These results showed that human and mouse macrophage cells expressed taurine transporter on their cell membranes.

Table 1. Effect of Na^+ ion for taurine transportation in macrophage cell line, RAW264.7

Cell	Na^+-Containing Buffer (cpm x 10^{-6})	Na^+-Deficient Buffer (cpm x 10^{-6})
RAW	6.570 ± 0.473	0.008 ± 0.001

Cells were pulsed with [^{3}H]taurine for 40 min in Na^+-containing buffer (25 mM-HEPES/Tris (pH 7.5), 140 mM NaCl, 5.4 mM KCl, 1.8 mM $CaCl_2$, 0.8 mM $MgSO_4$, 5 mM D-glucose) or Na^+-deficient buffer which contained 140 mM sucrose instead of 140 mM NaCl. The lysate was mixed with 8 ml cocktail solution and was used in measuring β-count for 3 min.

Table 2. Pretreatment with cold taurine prevented [^{3}H]taurine uptake in the human promonocytic cell line, THP1

Buffer	Pretreatment (30 min)	[^{3}H]Taurine Uptake (30 min; cpm)
Na^+-Deficient	PBS	313 ± 24
Na^+-Containing	PBS	10465 ± 1354
Na^+-Containing	Taurine (100 μM)	239 ± 58

THP1 cells were grown in RPMI 1640 medium containing 10% heat inactivated FCS, pen/str, L-glutamine. The cells were washed with Na^+ or Na^+-deficient buffer and each 3 x 10^6 cells/ml medium stabilized in 24-well plate for 30 min at 37°C in a 7% CO_2 incubator. Cold taurine (100 μl, 1 M) or PBS was added to cells which were incubated for a further 30 min. [^{3}H]Taurine (20 μl; 2 μCi) was then added to each well and incubated for 30 min. The cells were washed with ice-cold PBS three times and lysed with 0.2 M NaOH. Data are means ± SEM.

PKC Stimulation Blocked Taurine Uptake in Macrophage

When the RAW 264.7 cells were grown confluently on 6-well culture plates, they were treated with various concentrations of PMA, which is known to be a PKC stimulator. After 1 h, the medium was replaced with Na^+-containing buffer, stabilized for 1 h, and incubated with 2 μCi [^{3}H]taurine for 30 min. Low concentrations of PMA, such as 0.1 or 1.0 ng/ml PMA, did not affect [^{3}H]taurine uptake. However the concentration of 10 ng/ml PMA reduced [^{3}H]taurine uptake in the cell. The reduction of taurine uptake was gradual as the PMA concentration increased. At concentrations of 10 and 100 ng/ml PMA, [^{3}H]taurine uptake was 85% and 82% that of the control, respectively (Table 3).

Dexamethasone Abolished PMA Effect on Taurine Transportation

RAW 264.7 cells were stimulated with 100 ng/ml IL-1β, 100 IU/ml IFN-γ, 10 ng/ml PMA or 100 nM dexamethasone for 6 h and cultured in the presence of [^{3}H]taurine for 40 min. The single treatment or co-stimulation of the cells with IL-1β and IFN-γ did not affect [^{3}H]taurine uptake by the cell line in the presence of Na^+-containing buffer. However, addition of PMA to the culture system of IL-1β + IFN-γ reduced [^{3}H]taurine uptake to 82% of that of the control. Interestingly, the addition of dexamethasone abolished the effect of PMA when the cell line was stimulated with IL-1β + IFN-γ + PMA + dexamethasone. These results showed that inflammatory cytokines such as IL-1β or IFN-γ did not affect taurine transporter activity, but stimulation of protein kinase C with PMA reduced the activity of the transporter. When the cells were stimulated with PMA and dexamethasone simultaneously,

Table 3. [^{3}H]Taurine uptake was dose-dependently blocked by PMA stimulation in RAW 264.7

PMA (ng/ml; 1 h)	[^{3}H]Taurine Uptake (cpm)	% Control
PBS	513700 ± 31600	100
0.1	507900 ± 12000	99
1.0	506400 ± 71700	98
10	435000 ± 84600	85
100	422900 ± 20600	82
1000	395100 ± 16900	77

Cells were cultured and stimulated in 6-well plate. [^{3}H]Taurine (2 μCi) was incubated for 30 min. Values are mean ± SE of duplicates.

Table 4. Dexamethasone recovers reduced taurine uptake by PMA in the RAW cell line

Stimulation	[^{3}H]Taurine Uptake (cpm)
PBS	454,100 ± 1,700
IL-1β	431,600 ± 9,700
IFN-γ	423,200 ± 2,200
IL-1β + IFN-γ	461,100 ± 8,300
IL-1β + IFN-γ + PMA	373,400 ± 4,600
IL-1β + IFN-γ + PMA + Dex	485,500 ± 1,800

RAW 264.7 cells were stimulated for 6 h in 10% FCS-DMEM in a 6-well plate. The medium was then replaced with Na^+ buffer and cells were stabilized for 1 h. [^{3}H]Taurine (20 μl; 2 μCi) was then added in 2 ml buffer and cells cultured for 40 min. Uptake of [^{3}H]taurine was measured by β-scintillation counter after the cells were lysed. The final concentrations were as follows: IL-1β 100 ng/ml, IFN-γ 100 IU/ml, PMA 10 ng/ml, and dexamethasone 100 nM. Data are shown as mean ± SEM.

dexamethasone blocked signal transduction of PKC to the taurine transporter in the cytoplasm.

In the next experiment, the cell line was stimulated with 10 ng/ml PMA for 5 min, 30 min, 1 h and 3 h in absence or presence of 100 nM dexamethasone. The results showed that 5 and 30 min stimulations with PMA could not reverse the effects of dexamethasone on [^{3}H]taurine transport. At least 1 h stimulation was required for the blocking of the signal transduction of PKC activity in this system (Table 5).

DISCUSSION

In the present study, we demonstrated that macrophage cell lines express taurine transporter protein on their cell surface and that phorbol ester inhibits the activity of the transporter. Moreover, this study is the first evidence for the regulation by dexamethasone of signal transduction of protein kinase C in taurine transporter activity in macrophage cells. The fact that human and mouse macrophage cell lines showed Na^+-dependent taurine uptake

Table 5. Dexamethasone recovers reduced taurine uptake induced by PKC stimulation after 1 h

Stimulation	Without Dexamethasone (dpm)	With Dexamethasone (dpm)
PBS	764,700 ± 20,500	685,300 ± 28,100
PMA (5 min)	762,000 ± 14,500	650,700 ± 14,100
PMA (30 min)	656,500 ± 17,600	610,200 ± 9,900
PMA (1 h)	577,700 ± 20,900	744,200 ± 29,100
PMA (3 h)	674,200 ± 16,100	ND

Cells of the mouse macrophage line, RAW 264.7, were stimulated for various times with PMA (10 ng/ml) or PMA (10 ng/ml) + dexamethasone (100 nM) in 40 mm culture dishes in 10% DMEM medium. The medium was replaced with Na^+ buffer and stabilized for 10 min and then 2.5 μCi [^{3}H]taurine was added and incubated for 40 min. The cells were lyzed with 1 ml of 0.2 M NaOH containing 1 mM EDTA for 1 h at room temperature. The data are expressed as mean ± SE. ND (not determined).

strongly indicates that these cells expressed taurine transporter on the cell surface. In addition, macrophage cells did not take up [^{3}H]taurine when they were pretreated with cold taurine. Most of the taurine transporters of human, mouse, rat and dog showed Na^+ and Cl^- dependence. Therefore dependence of taurine uptake on these ions can be a parameter indicating the presence of the taurine transporter. Stoichiometrically, the ratios of Na^+:taurine and Cl^-:taurine during taurine uptake were 2:1 and 1:1, respectively[17,20]. Activation of RAW 264.7 cells by interferon-γ inhibited synthesis of nitric oxide (NO) and tumor necrosis factor by treatment with taurine chloramine[14].

PMA is known to be a PKC activator. Upon activation, PKC moves from the cytosol to the intracellular membrane. Taurine transporter activity was dose-dependently inhibited by PMA (Table 3). The inhibition was observed only above 10 ng/ml PMA, which is a typical concentration for the expression of its activity. PKC was found to inhibit the activity of taurine transporter in various tissues such as placental choriocarcinoma cell line[11], renal epithelial cell[9] and colon carcinoma cell line[3]. Most of the taurine transporter molecules from various species have 12 transmembrane domains and their sizes are about 70 kDa[8]. The long cytoplasmic carboxy and amino termini of the transporter are variable and may function as regulatory domain attachment sites for other proteins. Six potential sites for PKC phosphorylation have been reported in human taurine transporter[8]. PKC activation was reported to play some important roles in the regulation of taurine uptake in cultured cells[11]. It is still unknown whether PKC acts on the taurine transporter directly or indirectly in the macrophage. Furthermore it needs to be clarified which isotype of the PKC is involved in the inhibition of the activity of the taurine transporter. Transport mechanisms for taurine uptake and release in macrophages should be studied in detail. The signal pathways for the regulation of the activity of the transporter by various stimulators also require elucidation.

Inflammatory cytokines such as IL-1β and IFN-γ did not affect inhibitory effects on the transporter of RAW264.7. Combination treatment of the cells with IL-1β, IFN-γ and PMA inhibited the transporter activity due to PMA. Addition of dexamethasone to this system recovered the activity of the transporter (Table 4). Dexamethasone is a glucocorticoid hormone with a cytoplasmic receptor. The complex of dexamethasone and its receptor moves to the nucleus and binds to DNA resulting in modulation of some protein expression. The activity of the transporter that was inhibited by PKC might be recovered by dexamethasone in three ways as shown in Fig. 1. The first possibility is that protein(s) induced by dexamethasone may block the signal pathway of activated PKC. This can be verified by the time-course recovery experiment showing that dexamethasone could recover [^{3}H]taurine uptake in 1 h in PMA-treated RAW cells (Table 5). The cells that were treated with dexamethasone for short times such as 5 or 30 min failed to recover the activity of taurine transporter. The effects of dexamethasone can appear either in a short time or in a delayed fashion. The rapid response by the glucocorticoid hormone can be mediated by the constituent protein(s) in the cell. Delayed response, however, can be mediated by synthesized second messenger protein(s)[15]. The second possibility is that the recovery can be mediated by NO whose production can be blocked by dexamethasone. NO increase induced by PKC activation may inhibit the activity of taurine transporter in RAW cells. NO production could be inhibited by the concomitant incubation of the murine microglial cell with PMA and dexamethasone when compared with single incubation with PMA[10]. The third possibility is that the PKC signal may be blocked by the heat shock protein 90 (hsp 90) that was detached from the glucocorticosteroid receptor upon binding of the dexamethasone with its receptor[4]. Hsp 90 is a dimeric protein, and hormone-free glucocorticoid receptor is bound with both hsp 56 and hsp 90 in cytosol and intact cells[16]. The precise functions of the released hsp 90 are not well characterized. However, it is thought that hsp 90 performs a general chaperone function in the cell[16].

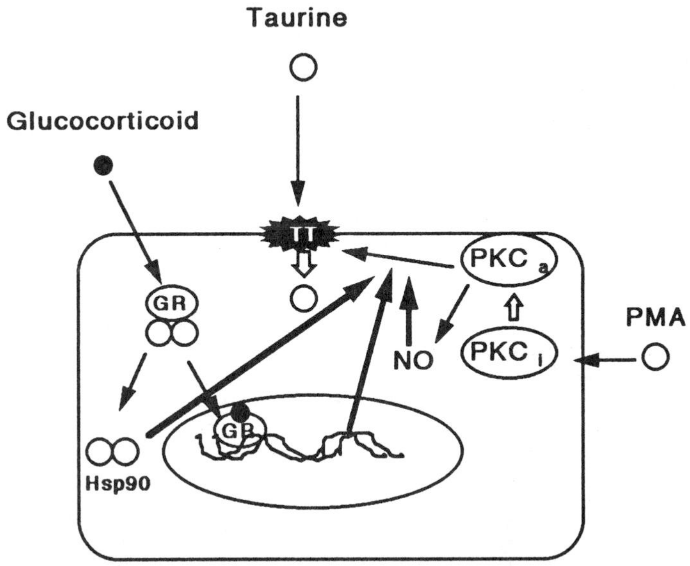

Figure 1. Hypothetical signal pathway affecting taurine transporter activity in macrophage cells. Activated PKC inhibits the activity of the taurine transporter. Glucocorticoid hormone may block the signal pathway of the PKC to the transporter through reduced nitric oxide, released hsp 90, or expressed second messenger protein. GR: glucocorticoid hormone; hsp 90: heat shock protein 90; NO: nitric oxide; PKC_i: inactive protein kinase C; PKC_a: active PKC; PMA: phorbol myristate acetate; TT: taurine transporter.

The fact that dexamethasone recovered the reduced taurine uptake in the PMA-stimulated cells may suggest that macrophage cells require higher taurine concentrations in a stressed state, for the glucocorticoid hormone could be secreted into the blood stream by the activation of hypothalamo-pituitary-adrenocortical (HPA) axis due to stress or injury[5].

ACKNOWLEDGMENT

Supported in part from a grant of the Korea Taurine Society.

REFERENCES

1. Baxter, C.F., Baldwin, R.A., and Oh, C.C. 1990, Acclimation to hyper- and hypo-osmotic changes: The anomalous behavior of taurine in the nervous system and in the blood of vertebrates, in: "Taurine: Functional Neurochemistry, Physiology, and Cardiology", Pasantes-Morales, H., Martin, D.L., Shain, W. and del Río, R.M. eds., Wiley-Liss, New York, pp. 329-336.
2. Blumberg, P.M. 1988, Protein kinase C as the receptor for the phorbol ester tumor promoters: Sixth Rhoads memorial award lecture, *Cancer Res.* 48:1-8.
3. Brandsch, M., Miyamoto, Y., Ganapathy, V., and Leibach, F.H. 1993, Regulation of taurine transport in human colon carcinoma cell lines (HT-29 and Caco-2) by protein kinase C, *Am.J.Physiol.Gastrointest.Liver Physiol.* 264:G939-G946.
4. Evans, R.M. 1988, The steroid and thyroid hormone receptor superfamily, *Science*, 240:889-895.
5. Goulding, N.J. and Guyre, P.M. 1993, Glucocorticoids, lipocortins and the immune response, *Curr.Opinion Immunol.* 5:108-1113.

6. Hass, R., Brach, M., Kharbanda, S., Giese, G., Traub, P., and Kufe, D. 1991, Inhibition of phorbol ester-induced monocytic differentiation by dexamethasone is associated with down-regulation of c-fos and c-jun (AP-1), *J.Cellular Physiol.* 149:125-131.
7. Iwata, H., Koyama, Y., Uemura, A., Tanaka, H., Noda, M., and Baba, A. 1990, Chloride-dependent amino acid transport in synaptic membrance vesicle, in: "Taurine: Functional Neurochemistry, Physiology and Cardiology", Pasantes-Morales, H., Martin, D.L., Shain, W. and Martín del Río, M. eds., Wiley-Liss, New York, pp. 259-266.
8. Jhiang, S.M., Fithian, L., Smanik, P., McGill, J., Tong, Q., and Mazzaferri, E.L. 1993, Cloning of the human taurine transporter and characterization of taurine uptake in thyroid cells, *FEBS Lett.* 318:139-144.
9. Jones, D.P., Miller, L.A., Dowling, C., and Chesney, R.W. 1991, Regulation of taurine transporter activity in LLC-PK1 cells: role of protein synthesis and protein kinase C activation, *J.Am.Soc.Nephrol.* 2:1021-1029.
10. Jun, C.D., Hoon-Ryu, Um, J.Y., Kim, T.Y., Kim, J.M., Kang, S.S., Kim, H.M., and Chung, H.T. 1994, Involvement of protien kinase C in the inhibition of nitric oxide production from murine microglial cells by glucocorticoid, *Biochem.Biophys.Res.Comm.* 199:633-638.
11. Kulanthaivel, P., Cool, D.R., Ramamoorthy, S., Mahesh, V.B., Leibach, F.H., and Ganapathy, V. 1991, Tyrosine residues are essential for the activity of the human placental taurine transporter, *Biochem.J.* 277:53-58.
12. Lambert, I.H. and Hoffmann, E.K. 1990, Taurine transport and cell volume regulation in a mammalian cell, in: "Taurine: Functional Neurochemistry, Physiology, and Cardiology", Pasantes-Morales, H., Martin, D.L., Shain, W. and del Río, R.M. eds., Wiley-Liss, New York, pp. 267-276.
13. Liu, Q.-R., López-Corcuera, B., Nelson, H., Mandiyan, S., and Nelson, N. 1992, Cloning and expression of a cDNA encoding the transporter of taurine and β-alanine in mouse brain, *Proc.Natl.Acad.Sci.USA*, 89:12145-12149.
14. Park, E., Quinn, M.R., Wright, C.E., and Schuller-Levis, G. 1993, Taurine chloramine inhibits the synthesis of nitric oxide and the release of tumor necrosis factor in activated RAW 264.7 cells, *J.Leukocyte Biol.* 54:119-124.
15. Peers, S.H., Smillie, F., Elderfield, A.J., and Flower, R.J. 1993, Glucocorticoid and non-glucocorticoid induction of lipocortins (annexins) 1 and 2 in rat peritoneal leucocytes *in vivo*, *Br.J.Pharmacol.* 108:66-72.
16. Pratt, W.B. 1993, The role of heat shock proteins in regulating the function, folding and trafficking of the glucocorticoid receptor, *J.Biol.Chem.* 268:21455-21458.
17. Ramamoorthy, S., Kulanthaivel, P., Leibach, F.H., Mahesh, V.B., and Ganapathy, V. 1993, Solubilization and functional reconstitution of the human placental taurine transporter, *Biochim.Biophys.Acta Bio-Membr.* 1145:250-256.
18. Ramamoorthy, S., Leibach, F.H., Mahesh, V.B., Han, H., Yang-Feng, T., Blakely, R.D. and Ganapathy, V. 1994, Functional characterization and chromosomal localizatioin of a cloned taurine transporter from human placenta, *Biochem. J.* 300:893-900.
19. Smith, K.E., Borden, L.A., Wang, C.-H.D., Hartig, P.R., Branchek, T.A., and Weinshank, R.L. 1992, Cloning and expression of a high affinity taurine transporter from rat brain, *Mol.Pharmacol.* 42:563-569.
20. Tiruppathi, C., Brandsch, M., Miyamoto, Y., Ganapathy, V., and Leibach, F.H. 1992, Constitutive expression of the taurine transporter in a human colon carcinoma cell line, *Am.J.Physiol.Gastrointest.Liver Physiol.* 263:G625-G631.
21. Uchida, S., Kwon, H.M., Yamauchi, A., Preston, A.S., Marumo, F., and Handler, J.S. 1992, Molecular cloning of the cDNA for an MDCK cell Na^+- and Cl^--dependent taurine transporter that is regulated by hypertonicity, *Proc.Natl.Acad.Sci.USA*, 89:8230-8234.

Part 2

TAURINE AND THE LIVER

8

TAURINE PRODUCTION IN RAT PRIMARY HEPATOCYTES

Jun Ohta[1,2] and Martha H. Stipanuk[1]

[1] Division of Nutritional Sciences
Cornell University
Ithaca, New York 14853
[2] Department of Biochemistry
Okayama University Medical School
Okayama 700, Japan

INTRODUCTION

Cysteine metabolism in rat hepatocytes is highly regulated in response to the dietary levels of protein and sulfur amino acids[1,2]. Primary cultures of rat hepatocytes might be useful for studies of the regulation of cysteine metabolism, including taurine production. However, conventional monolayer cultures are known to lose many of the differentiated functions of liver cells. Recently, it has been reported that hepatocytes cultured as spheroids (multicellular aggregates with well-defined cellular morphology) maintain certain highly differentiated functions of liver parenchymal cells such as secretion of albumin[7,5].

In this paper, we describe taurine formation from L-cysteine by hepatocytes cultured as a monolayer or as spheroids under two L-methionine levels. Our results suggest that hepatocytes in primary culture may be used as a model for studies for cysteine metabolism including studies of taurine production and its regulation.

MATERIALS AND METHODS

Male Sprague-Dawley rats (Harlan Sprague-Dawley, Indianapolis, IN) were housed in the Division of Nutritional Sciences animal facility and given ad libitum access to a nonpurified diet (Prolab RMH 1000, Agway, Syracuse, NY). Rats weighed approximately 250 g when they were used for hepatocyte preparation. Primaria dishes and plates (Becton Dickenson) were coated with calf skin collagen (10 μg/cm^2 of surface) or poly(2-hydroxyethyl methacrylate) (poly-HEMA) (5 mg/cm^2 of surface) for monolayer and spheroid cultures, respectively, and sterilized under uv light for 16 h before use.

Hepatocytes were isolated aseptically by collagenase perfusion as described by Berry *et al.*[3] and cultured on collagen- or poly-HEMA-coated vessels. The plating medium used

Taurine 2, edited by Huxtable et al.
Plenum Press, New York, 1996

was a modification of serum-free hormonally defined medium[4]; that is, Williams' E medium (WE medium) supplemented with 10 μg/ml insulin, 50 ng/ml epidermal growth factor (EGF), 3 nM Na_2SeO_3, 100 units/ml penicillin G sodium, 100 μg/ml streptomycin sulfate, 0.25 μg/ml amphotericin B, and L-methionine (0 or 2 mM). The basal WE medium itself contained 0.1 mM methionine and 0.5 mM cysteine (cysteine + half cystine). Hepatocyte suspension (5 x 10^5/ml of the plating medium) was plated at the cell density of 1 x $10^5/cm^2$ and cultured at 37°C in 5% CO_2, 95% air. Half of the medium was changed 4 h after plating and then every 24 h.

Taurine production from [^{35}S]cysteine was measured 3 days after plating. Monolayer (~4 mg DNA) was washed twice with sulfur amino acid-free Williams' E medium (SAFWE medium) and was incubated with 0.2 mM [^{35}S]cysteine in 0.4 ml of SAFWE medium (37°C, pH 7.4, equilibrated with 5% CO_2/95% air) supplemented with 10 μg/ml insulin, 50 ng/ml EGF, 3 nM Na_2SeO_3, 100 unit/ml penicillin G sodium, 100 μg/ml streptomycin sulfate, 0.25 μg/ml amphotericin B, and 0.5 mM bathocuproine disulfonate. Spheroids formed in poly-HEMA-coated dishes were collected by centrifugation at 100 x g for 30 s and washed twice with SAFWE medium and, then, a portion (~1.6 mg DNA) of the spheroids obtained was incubated with [^{35}S]cysteine in a poly-HEMA-coated well as described above for monolayers. Incubations were stopped at 0 or 6 h by the addition of 0.8 ml of 6% (w/v) sulfosalicylic acid, and [^{35}S]taurine in the sulfosalicylate extracts was measured by anion-exchange HPLC after oxidation of hypotaurine to taurine[8]; the term taurine is used in this paper to indicate the sum of taurine and its precursor hypotaurine. Zero-time incubations served as blanks. DNA content was measured by the Hoechst 33258 method[6] using calf thymus DNA as the standard. Taurine production was expressed as pmol product formed/min/μg DNA.

RESULTS AND DISCUSSION

As shown in Figure 1, for each preparation of hepatocytes, taurine production from 0.2 mM cysteine was higher in hepatocytes cultured with 2 mM methionine supplementation than in those cultured without the supplemental methionine. The ratio of taurine production by hepatocytes cultured in basal WE medium to that by hepatocytes cultured in 2 mM methionine-supplemented medium was 0.35 ± 0.13 and 0.29 ± 0.12 for monolayer and spheroid cultures, respectively. Both ratios were significantly lower than 1 ($p < 0.02$), clearly

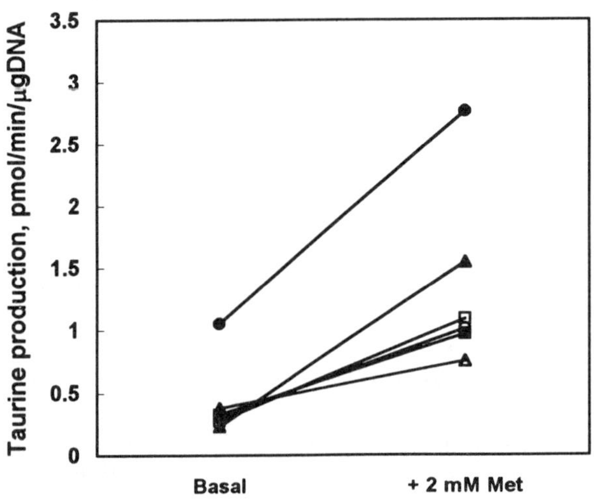

Figure 1. Taurine production by primary hepatocytes cultured as monolayer (open symbol) or spheroids (closed symbol). The shape of symbols indicates the rat used for hepatocyte preparation. Basal, without L-methionine supplementation; + 2 mM Met, with 2 mM supplemental L-methionine. For details, see Materials and Methods.

showing higher capacity for taurine production in hepatocytes cultured under higher methionine concentration. No statistically significant ($p > 0.05$, Student's t-test) difference in the basal/supplemented ratios for taurine production by monolayer and spheroid cultures was detected. These results seem to indicate that L-methionine and/or its metabolites may act directly on cultured hepatocytes to affect the capacity for taurine formation. Our previous study has shown that taurine production by freshly isolated hepatocytes is increased in response to dietary methionine supplementation prior to hepatocyte isolation[2]. In the present study, taurine formation by hepatocytes cultured *in vitro* was increased in response to methionine supplementation of culture medium. This similarity of response of hepatocytes *in vitro* to those *in vivo* suggests that primary cultures can be a useful *in vitro* system for studies of cysteine metabolism.

ACKNOWLEDGMENTS

We gratefully acknowledge the assistance of Lawrence L. Hirschberger with these experiments and the funding provided through NRICCP/USDA Competitive Research Grant #92-37200-7583 and USDA/CSREES Grant #94-34324-0987.

REFERENCES

1. Bagley, P.J. and Stipanuk M.H. 1994, The activities of rat hepatic cysteine dioxygenase and cysteinesulfinate decarboxylase are regulated in a reciprocal manner in response to dietary casein level. *J. Nutr.* 124:2410-2421.
2. Bagley, P.J. and Stipanuk M.H. 1995, Rats fed a low protein diet supplemented with sulfur amino acids have increased cysteine dioxygenase activity and increased taurine production in hepatocytes. *J. Nutr.* 125:933-940.
3. Berry, M.N., Edwards, A.M. and Barritt, G.J. 1991, Isolated Hepatocytes: Preparation, Properties and Applications, Elsevier, New York, NY. pp. 16-32.
4. Enat, R., Jefferson, D.M., Ruiz-Opazo, N., Gatmaitan, Z., Leinwand, L.A. and Reid, L.M. 1984, Hepatocyte proliferation *in vitro*: its dependence on the use of serum-free hormonally defined medium and substrata of extracellular matrix. *Proc. Natl. Acad. Sci. USA* 81:1411-1415.
5. Koide, N., Sakaguchi, K., Koide, Y., Asano, K., Kawaguchi, M., Matsushima, H., Takenami, T., Shinji, T., Mori, M. and Tsuji, T. 1990, Formation of multicellular spheroids composed of adult rat hepatocytes in dishes with positively charged surfaces and under other nonadherent environments. *Exp. Cell Res.* 186:227-235.
6. Labarca, C. and Paigen, K. 1980, A simple, rapid, and sensitive DNA assay procedure. *Anal. Biochem.* 102:344-352.
7. Landry, J., Bernier, D., Quellet, C., Goyette, R. and Marceau, N. 1985, Spheroidal aggregate culture of rat liver cells: histotypic reorganization, biomatrix deposition, and maintenance of functional activities. *J. Cell Biol.* 101:914-923.
8. Stipanuk, M.H., Hirschberger, L.L. and Bagley, P.J. 1992, Anion-exchange HPLC of taurine, cysteinesulfinate and cysteic acid, in: "Taurine - Nutritional Value and Mechanisms of Action", Lombardini, J.B., Schaffer, S.W. and Azuma, J. eds. Plenum Press, New York, pp. 429-435.

9

HIGH LEVELS OF DIETARY PROTEIN OR METHIONINE HAVE DIFFERENT EFFECTS ON CYSTEINE METABOLISM IN RAT HEPATOCYTES

Deborah L. Bella and Martha H. Stipanuk

Division of Nutritional Sciences
Cornell University, Ithaca, New York 14853

INTRODUCTION

Cysteine catabolism in mammalian liver can occur by both cysteine sulfinate-dependent and cysteine sulfinate-independent pathways. Both pathways lead to the production of pyruvate and sulfate, but only the cysteine sulfinate-dependent pathway leads to taurine production. Cysteine dioxygenase (CDO; EC 1.13.11.20) catalyzes the first reaction in the cysteine sulfinate-dependent catabolic pathway, by which cysteine is oxidized to cysteine sulfinate. Cysteine sulfinate has two possible catabolic fates. It can be transaminated by aspartate aminotransferase (AAT; EC 2.6.1.1) which ultimately yields pyruvate and sulfate, or it can be decarboxylated by cysteine sulfinate decarboxylase (CSAD; EC 4.1.1.29) to hypotaurine. Hypotaurine is then presumably non-enzymatically oxidized to taurine. Both taurine and sulfate are important metabolites required by the body for essential functions as well as being end-products of cysteine catabolism that are excreted in the urine.

The activities of both CDO and CSAD are highly affected by dietary protein and methionine level,[2,3] and both are potential regulatory enzymes in cysteine catabolism and taurine synthesis.[10,14,15,18,19,21,32] Although both cysteine sulfinate-dependent and -independent pathways of cysteine catabolism occur in mammalian liver cells, cysteine sulfinate-dependent catabolic enzymes undergo the largest changes in response to dietary protein or sulfur amino acid supplementation.[2,3,10,29]

We have reported that rat hepatic CDO and CSAD are regulated in a reciprocal manner in response to dietary casein level over the range of 100 to 600 g casein per kg diet.[2] High levels of dietary casein were associated with higher levels of CDO activity and with lower levels of CSAD activity. In addition, we have reported that rat hepatic CDO activity increased with dietary sulfur amino acid supplementation (3 or 10 g L-methionine per kg diet or the sulfur equivalent as L-cystine)[3] but that CSAD activity was unaffected by these levels of supplementation with sulfur amino acids. Thus, sulfur amino acids may be less effective than protein in decreasing CSAD activity and may result in a pattern of cysteine

Taurine 2, edited by Huxtable et al.
Plenum Press, New York, 1996

catabolism that favors taurine production. However, in the previous studies, the effects of methionine and protein were determined in separate experiments and the total sulfur amino acid content of the diets used for the two studies was different.

The purpose of this study was to compare the effects of dietary protein versus methionine supplementation on the hepatic and renal activities of key enzymes in the cysteine sulfinate-dependent pathway of cysteine metabolism. We also evaluated the effects of these dietary treatments on the distribution of cysteine among pathways of cysteine catabolism in isolated hepatocytes. In particular, we compared results for rats fed a basal low protein diet (LP) to those for rats fed high protein diets (HP or HPM) or excess methionine diets (LP+L-Met or LP+DL-Met). In addition, we compared results for rats fed a high protein diet not supplemented with methionine to improve the protein quality (HP) with those for rats fed a diet that was supplemented with methionine (HPM). Excess methionine diets containing excess L-methionine (LP+L-Met) were compared with those for rats fed diets containing excess DL-methionine (LP+DL-Met); although rats are able to utilize DL-methionine efficiently, we wanted to ensure that apparent differences in results of previous studies were not due to whether L- or DL-methionine was used.

The distribution of cysteine catabolism between pathways leading to sulfate versus taurine production is significant for several reasons. Both are essential metabolites or nutrients for cells, and the distribution between taurine and sulfate affects the availability of cysteine carbon for energy and the metabolic acid load that is generated from sulfur amino acid metabolism.

The results presented in this paper are a portion of a large experiment which addressed several experimental questions; other data and comparisons have been published in a separate paper.[4]

MATERIALS AND METHODS

Materials

L-cysteine, the disodium salt of bathocuproine disulfonate and α-ketoglutarate were purchased from Sigma Chemical (St. Louis, MO). L-Cysteine sulfinate was purchased from Aldrich Chemical (Milwaukee, WI). L-[^{35}S]Cysteine was purchased from Amersham (Product #SJ141, Arlington Heights, IL) as the hydrochloride.

L-[^{35}S]Cysteine was purified prior to use by applying it to a 0.6 x 4 cm Dowex 1-X8 column (200-400 mesh, acetate form) and eluting it with approximately 20 ml of deionized water. The eluate was aliquoted into portions (0.37 MBq/tube), lyophilized and stored at -70°C until used.

Animals and Treatments

Male Sprague-Dawley rats (Harlan Sprague-Dawley, Indianapolis, IN) were housed in stainless steel mesh cages in a room maintained at 20°C and 60-70% humidity with light from 2000 to 0800 h. Rats had free access to diet and water for the duration of the experiment. Animals were blocked into groups by initial body weight. Rats within each block were randomly assigned to receive one of five semi-purified experimental diets based on the AIN-76A formulation.[1] Modifications were made to prepare diets that contained 100 g casein + 1.5 g L-methionine per kg diet (LP), 600 g casein + 1.5 g L-methionine per kg diet (HP), 600 g casein + 9 g L-methionine per kg diet (HPM), 100 g casein + 1.5 g L-methionine + 21.2 g DL-methionine per kg diet (LP+DL-Met), or 100 g casein + 22.7 g L-methionine per kg diet (LP+L-Met). The LP diet provided 4.25 g sulfur amino acids/kg diet, with 2.75 g

Table 1. Composition of semi-purified diets containing various levels of casein and methionine[1]

	Diet				
Ingredients	LP	HP	HPM	LP+DL-Met	LP+L-Met
Vitamin Free Casein (g/kg)	100	600	600	100	100
L-Methionine (g/kg)	1.5	1.5	9.0	1.5	22.7
DL-Methionine (g/kg)	0	0	0	21.2	0
Cornstarch (g/kg)	367	117	113.25	356.4	356.4
Sucrose (g/kg)	367	117	113.25	356.4	356.4
Cellulose (g/kg)	50	50	50	50	50
Corn Oil (g/kg)	50	50	50	50	50
Vitamin Mix (AIN 76A(g/kg)	10	10	10	10	10
Choline Bitartrate (g/kg)	2	2	2	2	2
Mineral Mix (AIN 76[2](g/kg)	52.5	52.5	52.5	52.5	52.5

[1]Diets were prepared by Dyets, Inc. (Bethlehem, PA) in pelleted form. [2]The AIN-76 mineral mix was modified by replacement of $CaHPO_4$ and $K_3C_6H_5O_7.H_2O$ with $CaCO_3$ and KH_2PO_4 to prepare a basal mineral mix to facilitate manipulation of the chloride content of a diet discussed in a separate paper{{18000}}. In all cases, the weight of the mineral mix was maintained constant at 1.5-times the original AIN-76 formulation by addition or omission of sucrose such that 52.5 g of mineral mix was required per kg diet instead of the usual 35 g/kg of the standard AIN-76 mineral mix.

from casein and 1.5 g from free L-methionine. The HP diet provided 18 g sulfur amino acids per kg diet, and the HPM, LP+DL-Met and LP+L-Met diets all provided approximately 25.5 g sulfur amino acids per kg diet. The diets were designed to provide equivalent amounts of protein (HP and HPM; 600 g casein per kg) or equivalent amounts of total sulfur amino acids (HPM, LP+DL-Met, LP+L-Met; 25.5 g sulfur amino acids per diet) or a constant ratio of casein to L-methionine (LP and HPM; 1.5 g methionine per 100 g casein). Methionine was the major sulfur amino acid in all of the diets because approximately 85% of the total sulfur amino acids in casein are present as methionyl residues and methionine was used for sulfur amino acid supplementation of these diets as described. The composition of the diets is shown in Table 1.

Rats were fed the experimental diets for 2-3 weeks. Hepatocytes were isolated between 1000 and 1400 h by the method of Berry *et al.*[6] Hepatocytes were isolated from three rats per day, with the rats being randomly selected from a particular block and with the heavier weight blocks being used first. Kidneys were removed and placed on ice immediately following the isolation of the liver for perfusion.

Isolated hepatocytes from one rat were suspended in 40 ml of oxygenated Krebs-Henseleit bicarbonate buffer at 37°C. The average cellular ATP level for hepatocytes for all groups was 2.6 mmol/kg wet cell weight. One hepatocyte preparation in the HP group was discarded due to an ATP value < 2.0 mmol per kg wet cell weight.

To determine the wet weight of cells, two 0.5 ml aliquots of cell suspension were centrifuged for 30 s in pre-weighed tubes in a Beckman Microfuge™ 12 Centrifuge (Fullerton, CA) at 10,000 g. The supernatant was decanted, the inside of the tube was blotted dry, and the tube and pellet were weighed. The wet weight was estimated by subtracting the weight of the tube from the weight of the tube and pellet.

Incubations

Hepatocytes (approximately 100 mg wet weight of cells) were incubated with 0.2 mmol/L L-[^{35}S]cysteine plus 0.05 mM bathocuproine disulfonate for 15 min in a final volume of 2.5 ml. Bathocuproine disulfonate, a copper chelator, decreases oxidation of cysteine to cystine.[8] Incubations were performed in duplicate and incubations stopped at zero-time served as blanks. To measure GSH, 0.5 ml of the incubation suspension was added to a tube that contained 0.2 ml of a solution which contained 6 M perchloric acid, 3.75 mM each of bathocuproine disulfonate and bathophenanthroline disulfonic acid (iron chelator) and 2.5 mM γ-glutamylglutamate (internal standard). Then, 0.4 ml of 1 M 5-sulfosalicylic acid (SSA) was immediately added to the remaining incubation suspension. The acidified samples were centrifuged at 2000 g for 10 min, and the acid supernatants were stored at -20°C for subsequent analysis of total [^{35}S]GSH or [^{35}S]sulfate and [^{35}S]taurine plus [^{35}S]hypotaurine production from [^{35}S]cysteine. [^{35}S]Taurine was measured after oxidation of hypotaurine to taurine; the term taurine is used throughout this paper to indicate the sum of taurine and its precursor hypotaurine. Protein precipitated with SSA was collected for measurement of [^{35}S]cysteine incorporation into protein. For all metabolites, values for zero-time blanks were subtracted from the 15 min values.

Separate duplicate incubations for determination of ATP were stopped with the addition of 0.5 ml of 1 M perchloric acid to each flask. The acidified mixture was centrifuged for 10 min at 2000 g. The supernatant was stored on ice for up to 3 h until ATP was measured.

An aliquot of the hepatocyte suspension used for the incubations was analyzed for protein concentration.

Enzyme Activities

CDO and CSAD activities were measured in isolated hepatocytes and in kidney tissue. CDO activity was assayed in hepatocyte and kidney homogenates by measuring production of [^{35}S]cysteine sulfinate from [^{35}S]cysteine.[2] CSAD activity was measured in supernatants of hepatocyte and kidney homogenates using the procedure described by Bagley and Stipanuk[2] except that unlabeled cysteine sulfinate was used as substrate and production of unlabeled taurine (plus hypotaurine) was measured.

AAT activity was measured in liver homogenates by the procedure described by Bergmeyer and Bernt[5] except 20 mM CSA was used as substrate instead of aspartate.[10] This assay was included near the end of the study and was done on homogenates of liver obtained from additional rats fed the various experimental diets and not on hepatocyte preparations.

Analytical Methods

For the determination of sulfate production, 0.5 ml of SSA supernatant was applied to a 3 x 0.5 cm column of Dowex-50 (H^+-form). The column was then washed with 1 ml deionized water. The eluate was collected and stored at -20°C until [^{35}S]sulfate was separated by anion-exchange HPLC on a Hamilton PRP X-100 column and measured by liquid scintillation spectrometry as described by Coloso *et al.*[9]

For analysis of [^{35}S]hypotaurine plus [^{35}S]taurine production, hypotaurine was oxidized to taurine by treatment of the supernatant with hydrogen peroxide as previously described.[30] A 0.5 ml aliquot of the oxidized supernatant was applied to a 3 x 0.5 cm column of Dowex-50 (H^+ form), and the column was washed with 2.0 ml of deionized water to elute the taurine. The eluate was collected and stored at -20°C until [^{35}S]taurine was measured by an anion-exchange HPLC method previously described.[31] This HPLC method (without the oxidation step) was also used to measure [^{35}S]cysteine sulfinate produced in the CDO assay and, using post-column o-phthalaldehyde derivatization, for analysis of unlabeled taurine produced in the CSAD assay.[30]

[^{35}S]GSH and GSH were measured using a modification of the method of Fariss and Reed[1] as previously described by Coloso *et al.*[30] Samples were treated with dithiothreitol to reduce disulfides prior to derivatization with 1-fluoro-2,4-dinitrobenzene and HPLC analysis of the dinitrophenyl-derivatives. Molar GSH content was determined spectrophotometrically whereas GSH production was calculated from ^{35}S-radioactivity in the dinitrophenyl-GSH peak.

To measure [^{35}S]cysteine incorporation into protein, the precipitated protein pellet from the hepatocyte incubation stopped with SSA was re-washed with 1 ml 1 M SSA and then 1 ml water and resuspended in 2 ml of water. One ml of 1 M hydrogen peroxide was added and incubated with the protein suspension for 15 min at room temperature to oxidize cysteine to cysteic acid, to break all disulfide bonds, and to remove cyst(e)ine not attached by peptide bonds. The pellet was washed two additional times with 2 ml of water and then resuspended with 1 ml of water. The resuspended pellet was mixed with 1.5 ml of Solvable™ (NEN Research Products, Boston, MA) in a scintillation vial and placed in a shaking water bath at 50°C for 60 min. Then Ecoscint liquid scintillation cocktail (National Diagnostics, Atlanta, GA) was added, and radioactivity was measured by liquid scintillation spectrophotometry.

All measurements of [^{35}S]-radioactivity were corrected for counting efficiency and decay. Production of [^{35}S]sulfate, [^{35}S]taurine, and [^{35}S]GSH and [^{35}S]cysteine incorporation into protein were calculated by dividing ^{35}S-radioactivity in the product by the specific radioactivity of the [^{35}S]cysteine added as substrate as discussed previously.[2,3]

ATP was assayed enzymatically by the method of Lamprecht and Trautschold.[22] Protein concentration of the hepatocyte suspensions and the tissue preparations used for enzyme assays was determined by a modification of the Lowry method.[24]

Statistics

Data were analyzed with the paired t-test or analysis of variance (Minitab 81.1, Pennsylvania State University) and Tukey's or Tukey-Kramer's ω-procedure.[28] When variances were unequal, values were transformed ($\log_{10}$) prior to statistical analysis as indicated in the table footnotes. Significance is reported in the tables for $p \leq 0.05$; higher levels of significance reported in the text were obtained by repeating the Tukey's or Tukey-Kramer's ω-procedure using p values of 0.025, 0.01 and 0.005.

RESULTS

As shown in Table 2, the average daily diet consumption of rats was similar for most groups and mean daily weight gain was not significantly different for most experimental groups; the LP+DL-Met group had a significantly lower food intake than did the LP group. Rats fed the LP diet (100 g casein + 1.5g L-Met/kg) grew at an average rate of 4.5 ± 0.8 g per day. No increase in this rate of growth was observed for rats fed the diets that contained additional protein and/or methionine, but this could be due to the very high levels of methionine and protein used. Food intake and weight gain tended to be lower in the groups fed the high sulfur-amino acid diets, and final body weight was significantly less ($p \leq 0.05$) in the rats fed the diets with excess methionine than in those fed the other diets. This difference was largely due to lower initial rates of weight gain for rats in the HPM, LP+DL-Met and LP+L-Met dietary groups (data not shown).

Also shown in Table 2 are the protein concentration and the GSH concentration of freshly isolated hepatocytes. The protein concentration of hepatocytes was similar for all five groups. The mean initial GSH level ranged from 2.2 to 4.3 μmol/g and was higher ($p \leq 0.05$) for hepatocytes from rats fed the LP+L-Met diet than for those fed the LP, HP, or HPM diets. The GSH content of isolated hepatocytes from rats fed the LP+DL-Met diet were not different from that of hepatocytes isolated from rats fed the LP+L-Met diet.

Table 2. Effects of diet on food intake and weight gain of rats (for second week of feeding period) and on protein concentration and initial GSH concentration of hepatocytes isolated from the rats

	Diet				
	LP	HP	HPM	LP+DL-Met	LP+L-Met
Daily diet consumption (g)	24 ± 2.2^{b}	19 ± 1.8^{ab}	17 ± 1.0^{ab}	14 ± 0.3^{a}	16 ± 0.5^{ab}
Daily weight gain (g)	4.5 ± 0.8	4.9 ± 0.5	3.5 ± 0.5	2.9 ± 0.5	3.4 ± 0.3
Body weight at end of feeding period (g)	290 ± 9.8^{c}	279 ± 7.8^{bc}	264 ± 9.3^{abc}	237 ± 5.2^{a}	245 ± 5.8^{ab}
Protein concentration (mg protein/g wet weight of hepatocytes)	149 ± 7.0	170 ± 10.0	161 ± 5.6	148 ± 7.8	156 ± 6.9
Initial GSH Concentration (μmol/g wet weight of hepatocytes)	2.6 ± 0.4^{ab}	2.2 ± 0.1^{a}	2.4 ± 0.1^{ab}	3.8 ± 0.5^{bc}	4.3 ± 0.2^{c}

Values are means ± SEM for n = 4-6 rats. Values within a row and not followed by the same superscript letter are significantly different ($p \leq 0.05$) by Tukey-Kramer's ω-procedure.

The specific activity of CDO (V_{max}) was the highest in homogenates of isolated hepatocytes from rats fed the methionine-supplemented diets, and intermediate levels of activity were observed in hepatocytes isolated from rats fed the protein-supplemented diets (Table 3). Rats fed diets supplemented with excess DL-methionine and L-methionine had hepatic CDO specific activities that were 12.7- and 17.3-times, respectively, ($p \leq 0.005$) the activity of hepatocytes from rats fed the LP diet. The specific activity of CDO for hepatocytes from rats fed the HP and HPM diets was 5- and 7-times, respectively, ($p \leq 0.005$) that of the LP group. CSAD specific activity in hepatocytes was significantly affected by protein- and methionine-supplementation. Specific activity for rats fed the HP and HPM diets was only 12.7% and 12.0%, respectively, ($p \leq 0.005$) of the CSAD specific activity in hepatocytes isolated from rats fed the LP diet. CSAD specific activity for hepatocytes from rats fed the

Table 3. Activities of cysteine dioxygenase and cysteine sulfinate decarboxylase in isolated hepatocytes and of aspartate aminotransferase in liver from rats fed diets with different casein and methionine levels

	Enzyme (nmol·min^{-1}·mg protein^{-1})		
Diet	Cysteine Dioxygenase	Cysteine sulfinate Decarboxylase	Aspartate (Cysteine sulfinate:α-Ketoglutarate) Aminotransferase
LP	0.3 ± 0.08^{a}	29.9 ± 2.3^{c}	753 ± 41
HP	1.6 ± 0.3^{b}	3.8 ± 0.5^{a}	1071 ± 63
HPM	2.1 ± 0.5^{bc}	3.6 ± 0.7^{a}	998 ± 71
LP+DL-Met	3.8 ± 0.7^{bc}	9.5 ± 1.1^{b}	845
LP+L-Met	5.2 ± 0.9^{c}	16.1 ± 1.7^{b}	802 ± 23

Values are means ± SEM for n = 4-6 rats for cysteine dioxygenase and cysteine sulfinate decarboxylase; n = 3 for aspartate (cysteine sulfinate: α-ketoglutarate) aminotransferase except n=1 for LP+DL-Met. Values within the same column not followed by the same superscript letter are significantly different ($p \leq 0.05$) by Tukey-Kramer's ω-procedure. Values are expressed as specific activity in the hepatocyte homogenate (cysteine dioxygenase), supernatant fraction of hepatocyte homogenate (cysteine sulfinate decarboxylase) or liver homogenate [aspartate (cysteine sulfinate: α-ketoglutarate) aminotransferase]. To express results for cysteine sulfinate decarboxylase as mg protein of hepatocyte homogenate rather than supernatant, multiply values by 0.68. The data for CDO and CSAD (nmol·min^{-1}·mg protein^{-1}) were transformed to $\log_{10}$ prior to statistical analysis.

DL-methionine- and L-methionine-supplemented diets was reduced to 32% and 54%, respectively, ($p \leq 0.005$) of the specific activity for the LP group. AAT activity did not change markedly in response to diet in this study (Table 3). However, using a t-test for analysis, the specific activity for hepatic AAT was significantly greater ($p \leq 0.02$) for the protein-supplemented groups (HP and HPM) than for the LP group. No difference in AAT specific activity was observed between the methionine-supplemented groups (LP+DL-Met and LP+L-Met) and the LP group.

The kidney has a high capacity for cysteine catabolism, so activities of renal cysteine catabolic enzymes were also measured. Renal CDO and CSAD specific activities were 0.42 ± 0.03 and 1.68 ± 0.21 nmol·min^{-1}·mg protein^{-1}, respectively, for rats fed all five experimental diets. No effect of dietary treatment ($p > 0.05$) was observed on the V_{max} or specific activity of either renal enzyme. Thus, the effects of dietary protein and methionine supplementation on specific activities of CDO and CSAD are probably limited to the liver.

The effects of dietary treatment on the metabolism of [^{35}S]cysteine by hepatocytes isolated from rats fed the protein- and methionine-supplemented diets are shown in Table 4. Both [^{35}S]taurine production and [^{35}S]sulfate production were significantly greater ($p \leq 0.005$) in hepatocytes from rats fed diets supplemented with either protein or methionine than in hepatocytes from rats fed the LP diet. Hepatocytes from methionine-supplemented rats had a significantly greater ($p \leq 0.005$) rate of taurine production but a significantly lower ($p \leq 0.005$) rate of sulfate production than did hepatocytes from rats fed the high protein diets.

The rate of [^{35}S]GSH production was highest in hepatocytes from rats fed the LP diet and was significantly ($p \leq 0.005$) lower in hepatocytes from rats fed either the protein- or methionine-supplemented diets. However, the decrease in the rate of [^{35}S]GSH production was greatest ($p \leq 0.005$) for the protein-supplemented groups. [^{35}S]Cysteine incorporation into protein was similar for hepatocytes isolated from rats in all five groups.

Hepatocytes isolated from rats fed the high protein diets (HP and HPM) had a significantly ($p \leq 0.025$) higher rate of total [^{35}S]cysteine utilization than hepatocytes from rats fed the LP diet. This higher rate of [^{35}S]cysteine utilization was largely due to the high rate of sulfate production in the protein-supplemented groups. Total [^{35}S]cysteine utilization by hepatocytes from rats fed either of the diets with excess methionine (LP+DL-Met or LP+L-Met) was intermediate and not significantly different ($p > 0.05$) from that for hepatocytes from rats fed either the LP diet or the high protein diets.

Total [^{35}S]cysteine catabolism ([^{35}S]sulfate + [^{35}S]taurine production) values for hepatocytes isolated from rats fed the five diets were as follows: LP, 58 pmol·min^{-1}·mg

Table 4. Comparison of [^{35}S]cysteine metabolism in hepatocytes from rats fed diets with different casein and methionine levels (pmol· min^{-1} · mg protein $^{-1}$)

Diet	[^{35}S]Taurine	[^{35}S]Sulfate	[^{35}S]GSH	[^{35}S]Cysteine incorporation into protein	Total [^{35}S]cysteine utilization
LP	2.6 ± 0.6^{a}	55 ± 4^{a}	230 ± 15^{c}	44 ± 3	331 ± 20^{a}
HP	24 ± 4^{b}	379 ± 35^{c}	34 ± 6^{a}	28 ± 4	460 ± 32^{b}
HPM	23 ± 3^{b}	396 ± 24^{c}	45 ± 6^{a}	31 ± 6	496 ± 17^{b}
LP+DL-Met	89 ± 11^{c}	187 ± 35^{b}	93 ± 5^{b}	42 ± 5	380 ± 44^{ab}
LP+L-Met	101 ± 6^{c}	173 ± 5^{b}	104 ± 10^{b}	36 ± 2	413 ± 15^{ab}

Values are means ± SEM for n = 4-6 rats. Values for [^{35}S]taurine are the sum of [^{35}S]hypotaurine and [^{35}S]taurine production. Values within the same column and not followed by the same superscript letter are significantly different ($p \leq 0.05$) by Tukey-Kramer's ω-procedure. The data for taurine and sulfate production were transformed to $\log_{10}$ prior to statistical analysis.

protein^{-1}; HP, 398 pmol·min^{-1}·mg protein^{-1}; HPM, 419 pmol·min^{-1}·mg protein^{-1}; LP+DL-Met, 276 pmol·min^{-1}·mg protein^{-1}; and LP+L-Met, 250 pmol·min^{-1}·mg protein^{-1}. In comparison to the LP group, total [^{35}S]cysteine catabolism by isolated hepatocytes was higher in hepatocytes from rats fed either the methionine- or protein- supplemented diets, but activity was greater for rats fed the high protein diets ($p \leq 0.05$) than for rats fed excess methionine alone.

[^{35}S]Taurine production by hepatocytes as a percentage of total [^{35}S]cysteine catabolism was as follows: LP, 5%; HP, 6%; HPM, 5%; LP+DL-Met, 33%; and LP+L-Met, 37%. Values for taurine production as a percentage of total cysteine catabolism were similar for rats fed the HP, HPM and the LP diets. Although CDO activity was high in hepatocytes from rats fed the high protein diets,[2] most of the cysteine sulfinate generated by the CDO-catalyzed reaction was apparently further metabolized to sulfate rather than to taurine. This interpretation of the data is supported by the low CSAD activity and high AAT activity in hepatocytes from rats fed the high protein diets. Taurine production as a percentage of total cysteine catabolism for the DL-methionine- or L-methionine-supplemented groups was 32% and 37%, respectively, and was significantly higher ($p \leq 0.005$) than that for either the LP or protein-supplemented groups. The apparent greater conversion of cysteine sulfinate to taurine in the methionine-supplemented groups is consistent with the high CDO activity and the higher level of CSAD activity compared to the protein-supplemented groups.

DISCUSSION

Relatively High Levels of Both CDO and CSAD Activities are Needed for High Rates of Taurine Production in Hepatocytes

We and other researchers have attempted to correlate either CDO or CSAD specific activity, but not both enzyme activities, with taurine production or urinary taurine excretion.[12-14,16-18,20,21,29] However, the results of this study demonstrate that both enzymes play an important role in the regulation of cysteine catabolism and of the rate of taurine synthesis. In Figure 1, the rate of taurine production by hepatocytes isolated from rats fed the various experimental diets is compared to their CDO and CSAD activities. Both protein and methionine supplementation of the diet resulted in an increase in CDO and a decrease in CSAD activity in hepatocytes. However, the magnitude of increase for CDO was greater and the magnitude of decrease for CSAD was smaller for the methionine-supplemented groups compared to the protein-supplemented groups, such that the methionine-supplemented rats

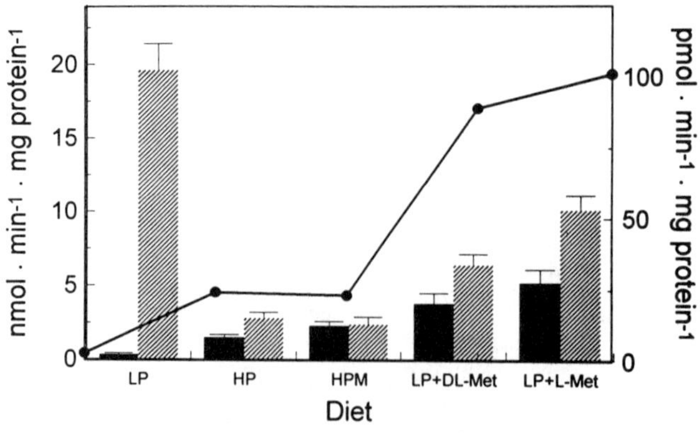

Figure 1. Comparison of cysteine dioxygenase (CDO) and cysteine sulfinate decarboxylase (CSAD) activities with the rate of taurine production (●) in hepatocytes isolated from rats fed various levels of protein or methionine. Both CDO (solid) and CSAD (shaded) activities are expressed as nmol·min^{-1} per mg protein in the total hepatocyte homogenate.

had higher levels of both CDO and CSAD than did the protein-supplemented rats. These findings support our previous studies in which methionine or cystine supplementation substantially elevated hepatic CDO activity but had little effect on hepatic CSAD activity relative to the effect of protein supplementation.[2,3] The differences between the effects of protein versus methionine observed in our studies are greater than would have been predicted from the work of Kohashi *et al.*[21] with CDO and of Jerkins and coworkers[18,19] with CSAD in which both high protein and high sulfur amino acid-containing diets seemed to have very similar effects. Our findings in this and previous studies[2,3] indicate that dietary methionine alone increases hepatic CDO activity but has little effect on hepatic CSAD activity unless it is provided at extremely high levels (> 10 g per kg diet). In contrast, a relatively modest increase in dietary protein (from 100 g to 300 g per kg diet) has a marked effect on both hepatic CDO and CSAD activities.

In this study, the net balance of the two enzyme activities resulted in significantly different rates of taurine synthesis. For example, the LP group had the lowest hepatic CDO specific activity but the highest CSAD specific activity of all the treatment groups. Despite the high CSAD activity, net production of taurine was low. With minimal production of cysteine sulfinate via oxidation of cysteine by CDO, an elevated CSAD activity did not lead to appreciable taurine synthesis. In comparison to the LP group, hepatocytes from rats fed the methionine-supplemented diets had higher CDO activity and reduced but substantial CSAD activity. Relatively high activities of both enzymes in animals fed these methionine-supplemented diets led to the greatest production of taurine. Hepatocytes from the rats fed high protein diets had more CDO activity than those from rats fed the LP diet, but these hepatocytes also had the lowest level of CSAD activity and a low rate of taurine production relative to total cysteine catabolism. Apparently the low CSAD activity limited further metabolism of cysteine sulfinate to taurine and, thus, favored conversion of cysteine sulfinate to sulfate. From these results, it is clear that neither of the two enzyme activities by themselves can be used to predict the rate of taurine production. It is necessary to evaluate both CDO and CSAD activity to assess the capacity of a particular tissue to produce taurine from cysteine.

Supplementation of Diets with Excess Methionine Favored High Rates of Taurine Production Whereas Supplementation of Diets with Excess Protein Favored High Rates of Sulfate Production

The HP and HPM diets had similar effects despite the higher methionine content in the HPM diet, and supplementation with either L-methionine or DL-methionine resulted in similar effects. However, although the HPM diet and the low protein diets supplemented with methionine provided equimolar amounts of sulfur amino acids, protein- versus methionine-supplementation clearly had different effects on hepatic cysteine metabolism. Hence, not all of the effects of the high protein diet on cysteine metabolism can be explained by the elevated supply of sulfur amino acids. Because the protein-supplemented diets were much lower in carbohydrate content (25% of total calories from carbohydrate) compared to the methionine-supplemented diets (76% of total calories from carbohydrate), the metabolic and hormonal state of the animals fed diets supplemented with protein versus methionine were undoubtedly different. The animals fed the protein-supplemented diets would be expected to have lower insulin to glucagon ratios, higher activities of hepatic enzymes involved in gluconeogenesis,[7] and higher levels of urea cycle enzymes[26,27] than rats in the methionine-supplemented groups. It is likely that some factor related to these differences in hormonal and metabolic state may have affected the regulation of CDO and CSAD activities, resulting in down-regulation of both activities and particularly of CSAD, so that hepatocytes of rats fed high protein diets had less CDO activity and much less CSAD activity than did those fed the methionine-supplemented diets. These observed changes in enzyme activities were associated with increased conversion of cysteine to sulfate plus pyruvate, a gluconeogenic precursor, with restriction of taurine production, which would have resulted in loss of the cysteine carbon chain as an energy source. Hence, taurine production as a percentage of total catabolism remained at the basal 5% level in hepatocytes from rats fed the high protein diets whereas it increased to 35% in hepatocytes from rats fed the high methionine diets.

Not only was the distribution of cysteine catabolism between taurine and sulfate different, but the rate of total catabolite production from cysteine was also different between hepatocytes from animals fed the protein- and methionine-supplemented diets. Total catabolite production from cysteine was significantly higher in hepatocytes from rats fed high protein diets than in those from rats fed the methionine-supplemented ($p \leq 0.05$) or low protein ($p \leq 0.005$) diets. However, CDO activity tended to be lower for the high protein groups than for the methionine-supplemented groups. The higher catabolic rate seen in animals fed the protein-supplemented diets appears to be due to the greater stimulation of hepatic sulfate production. This increase may be partially due to an increase in cysteine sulfinate-independent catabolism of cysteine to sulfate or to an overall stimulation of amino acid catabolism, including cysteine sulfinate transamination, due to consumption of a high protein diet.[2]

High Rates of Cysteine Catabolism to Taurine and Sulfate are Associated with Low Rates of GSH Synthesis by Hepatocytes

Increases in cysteine catabolism occurred in association with decreases in cysteine utilization for GSH synthesis. Results of a previous study[3] in which rats were fed diets with different levels of L-methionine or L-cystine also suggested that cysteine catabolism and GSH synthesis may be coordinately regulated. The basis for this apparent reciprocal relationship between the rate of cysteine catabolism and the rate of GSH production is not known. Comparison of the initial GSH concentration of isolated hepatocytes from animals fed the five experimental diets (Table 2) with rates of GSH synthesis (Table 4) shows that decreases in GSH synthesis did not consistently correspond to higher intracellular GSH concentrations, suggesting that feedback inhibition of GSH synthesis[25] was not responsible for the lower rates of GSH synthesis. Cysteine incorporation into protein was not significantly different ($p \leq 0.05$) across the five dietary treatment groups, so differences in cysteine utilization for protein synthesis were not responsible for differences in rates of GSH synthesis.

Preliminary studies of the effects of dietary methionine on the activity of γ-glutamylcysteine synthetase (E.C. 6.3.2.2), the rate limiting enzyme for GSH synthesis,[25] indicated that hepatic γ-glutamylcysteine synthetase activity decreased with methionine supplementation of a low protein diet (Stipanuk, Delans and Hirschberger, unpublished results). The γ-glutamylcysteine synthetase activity was assayed under conditions in which tissue GSH levels did not affect activity. These preliminary results suggest that dietary regulation of hepatic γ-glutamylcysteine synthetase may occur in the opposite direction to the dietary regulation of hepatic CDO.

In addition to this possible effect of sulfur amino acid intake, it is likely that the capacity of the hepatocyte for GSH synthesis would be affected by the hormonal state of the animal. Lu and Ge[23] showed that cAMP or glucagon administration inhibited γ-glutamylcysteine synthetase activity. Consistent with this hypothesis, the rate of GSH synthesis was lower in the animals fed the high protein (low carbohydrate) diets than in those fed the methionine-supplemented diets.

SUMMARY

This study clearly indicates that relatively high levels of both CDO and CSAD activity are needed for substantial taurine synthesis and that protein and methionine supplementation, at equimolar sulfur amino acid levels, are not equivalent in terms of their effects on cysteine catabolic enzyme activities and cysteine metabolism in hepatocytes. Evidence for a reciprocal regulation of cysteine catabolism (or CDO activity) and GSH synthesis (or γ-glutamylcysteine synthetase activity) in rat liver was also obtained. Although very high levels of protein and methionine were fed in this study, previous studies with lower levels of protein[2] or methionine[3] showed similar changes in cysteine metabolism.

Several questions regarding regulation of cysteine metabolism remain unanswered. Beyond sulfur amino acid availability, animals fed high protein diets appear to have other signals for regulation of CDO and CSAD activities. These signals may be related to the different hormonal and metabolic state of these animals. Furthermore, little is known about the molecular mechanisms involved in the observed changes in CDO and CSAD activities. The association between CDO activity and CDO protein has not been evaluated. Jerkins and Steele,[19] using immunochemical detection and quantification of CSAD protein in rat liver, showed that changes in CSAD protein concentration were correlated to changes in CSAD activity. The exact mechanisms or direct effectors which bring about changes in CDO and CSAD activities have yet to be determined. Further exploration of these potential regulatory mechanisms needs to be conducted to better understand the response of cysteine sulfinate-dependent cysteine catabolism to high levels of dietary protein or sulfur amino acids.

ACKNOWLEDGMENTS

We gratefully acknowledge the assistance of Lawrence L. Hirschberger and of Dr. Pamela J. Bagley with these studies. This work was supported by NRICCP/USDA Competitive Research Grant #92-37200-7583 and by USDA/CSRS Grant #89-34115-4498. This work was presented in preliminary form at Experimental Biology '94, April 24-28, 1994 at Anaheim, CA [Bella, D.L., Stipanuk, M.H., Hirschberger, L.H. and Bagley, P.J. (1994) The effect of dietary protein and sulfur amino acids on taurine vs sulfate production in cysteine metabolism. FASEB J. 8: A546 (abs.)].

REFERENCES

1. AIN Ad Hoc Committee on Standards for Nutrition Studies. Report of the AIN ad hoc committee on standards for nutritional studies. J. Nutr. 107:1340-1348, 1977.
2. Bagley, P. J., and M. H. Stipanuk. The activities of hepatic cysteine dioxygenase and cysteinesulfinate decarboxylase are regulated in a reciprocal manner in response to dietary casein level in rats. *J. Nutr.* 124:2410-2421, 1994.
3. Bagley, P. J., and M. H. Stipanuk. Rats fed a low protein diet supplemented with sulfur amino acids have increased cysteine dioxygenase activity and increased taurine production in hepatocytes. *J. Nutr.* 125: 933-940, 1995.
4. Bella, D. L., and M. H. Stipanuk. Comparison of effects of excess dietary protein, methionine or chloride on acid-base balance and on cysteine catabolism to taurine versus sulfate. Am. J. Physiol., 1995, in press.
5. Bergmeyer, H. U., and Bernt, E. Glutamate-oxaloacetate transaminawe. In: *Methods in Enzymatic Analysis*, edited by H. U. Bergmeyer, New York: Academic Press, Inc., vol. 2, 1974, p. 727-744.
6. Berry, M. N., A. M. Edwards, and G. J. Borritt. Isolated Hepatocytes: Preparation, Properties and Applications. Elsevier, New York, 1991, p.16-32, 56-57.
7. Claus, T. H., C. R. Park, and S. J. Pilkis. Glucagon and gluconeogenesis. In: *Glucagon I*, edited by P. Lefebvre, Berlin, Germany: Springer-Verlag, 1983, p. 361-382.

8. Coloso, R. M., M. R. Drake, and M. H. Stipanuk. Effect of bathocuproine disulfonate, a copper chelator, on cyst(e)ine metabolism in freshly isolated rat hepatocytes. *Am. J. Physiol.* 259:E443-E450, 1990.
9. Coloso, R. M., and M. H. Stipanuk. Metabolism of cyst(e)ine in rat enterocytes. *J. Nutr.* 119:1914-1924, 1989.
10. Daniels, K. M., and M. H. Stipanuk. The effect of dietary cysteine level on cysteine metabolism in rats. *J. Nutr.* 112(11):2130-2141, 1982.
11. Fariss, M. W., and D. J. Reed. High-performance liquid chromatography of thiols and disulfides: dinitrophenyl derivatives. *Methods Enzymol.* 143:101-109, 1987.
12. Gaull, G. E., D. K. Rassin, N.C.R. Raiha, and K. Heinonen. Milk protein quantity and quality in low-birth-weight infants. III. Effects on sulfur amino acids in plasma and urine. *J. Pediat.* 90:348-355, 1977.
13. Hardison, W.G.M., C. A. Wood, and J. H. Proffitt. Quantification of taurine synthesis in the intact rat and cat liver. *Proc. Soc. Exp. Biol. Med.* 155:55-58, 1977.
14. Hosokawa, Y., S. Nizeki, H. Tojo, I. Sato, and K. Yamaguchi. Hepatic cysteine dioxygenase activity and sulfur amino acid metabolism in rats: possible indicators in the evaluation of protein quality. *J. Nutr.* 118:456-461, 1988.
15. Hosokawa, Y., K. Yamaguchi, N. Kohashi, Y. Kori, and I. Ueda. Decrease of rat liver cysteine dioxygenase (cysteine oxidase) activity mediated by glucagon. *J. Biochem.* 84:419-424, 1978.
16. Jacobsen, J. G., and L. H. Smith, Jr. Comparison of decarboxylation of cysteine sulfinic acid-1-^{14}C and cysteic acid-1-^{14}C by human, dog and rat liver and brain. *Nature* 200:575-577, 1963.
17. Jacobsen, J. G., L. L. Thomas, and L. Smith. Properties and distribution of mammalian L-cysteine sulfinate carboxylyases. *Biochim. Biophys. Acta* 85:103-116, 1964.
18. Jerkins, A. A., L. E. Bobroff, and R. D. Steele. Hepatic cysteine sulfinic acid decarboxylase activity in rats fed various levels of dietary casein. *J. Nutr.* 119:1593-1597, 1989.
19. Jerkins, A. A., and R. D. Steele. Dietary sulfur amino acid modulation of cysteine sulfinic acid decarboxylase. *Am. J. Physiol.* 261:E551-E555, 1991.
20. Kohashi, N., K. Yamaguchi, Y. Hosokawa, Y. Kori, O. Fuji, and I. Ueda. Dietary control of cysteine dioxygenase in rat liver. *J. Biochem.* 84:159-168, 1978.
21. Knopf, K., J. A. Sturman, M. Armstrong, and K. C. Hayes. Taurine: an essential nutrient for the cat. *J. Nutr.* 108:773-778, 1978.
22. Lamprecht, W., and I. Trautschold. ATP determination with hexokinase and glucose-6-phosphate dehydrogenase. In: *Methods of Enzymatic Analysis*, edited by H. Bergmeyer. New York: Academic Press, vol. 4, 1974, p. 2101-2110.
23. Lu, S. C., and J-L. Ge. Loss of suppression of GSH synthesis at low cell density in primary cultures of rat hepatocytes. *Am. J. Physiol.* 263:C1181-C1189, 1992.
24. Markwell, M.A.K., S. M. Haas, L. L. Bieber, and N. E. Tolbert. A modification of the Lowry procedure to simplify protein determination in membrane and lipoprotein samples. *Anal. Biochem.* 87:206-210, 1978.
25. Richman, P. G., and A. Meister. Regulation of τ-glutamylcysteine synthetase by nonallosteric feedback inhibition of glutathione. *J. Biol. Chem.* 216:763-773, 1975.
26. Schimke, R. T. Adaptive characteristics of urea cycle enzymes in the rat. *J. Biol. Chem.* 237:459-468, 1962.
27. Snodgrass, P. J., R. C. Lin, W. A. Muller, and T. T. Aoki. Induction of urea cycle enzymes of rat liver by glucagon. *J. Biol. Chem.* 253:2748-2753, 1978.
28. Steel, R.G.D., and J. H. Torrie. *Principles and Procedures of Statistics*. New York: McGraw-Hill, 1960, p. 99-160.
29. Stipanuk, M. H. Effect of excess dietary methionine on the catabolism of cysteine in rats. *J. Nutr.* 109:2126-2139, 1979.
30. Stipanuk, M. H., L. L. Hirschberger, and P. J. Bagley. Anion-exchange HPLC of taurine, cysteinesulfinate and cysteic acid. In: *Taurine - Nutritional Value and Mechanisms of Action*, edited by J. B. Lombardini, S. W. Schaffer and J. Azume. New York: Plenum Press, 1992, p. 429-435.
31. Yamaguchi, K., S. Sakakibara, K. Kyoichiro, and U. Iwao. Induction and activation of cysteine oxidase of rat liver I. The effects of cysteine, hydrocortisone and nicotinamide injection on hepatic cysteine oxidase and tyrosine aminotransferase activities of intact and adrenalectomized rats. *Biochim. Biophys. Acta* 237:502-512, 1971.

10

TAURINE IN THE LIVER

The Function of Taurine Conjugated with Bile Acids

Toshiaki Nakashima, Toshihide Shima, Hironori Mitsuyoshi, Koji Inaba, Ninko Matsumoto, Yoshikuni Sakamoto, and Kei Kashima

Third Department of Internal Medicine
Kyoto Prefectural University of Medicine
Kawaramachi-Hirokoji, Kamigyo-ku, Kyoto 602, Japan

INTRODUCTION

In the enterohepatic circulation, bile acids have important roles in digesting dietary fat in the intestine. The detergent action of bile acids can be expected to influence biological membranes. In fact, feathery degeneration of hepatocytes and irregular-shaped erythrocytes are seen in chronic cholestatic liver diseases where the concentration of bile acids increases in hepatic tissue and serum.

There are five main bile acids in human: cholic acid (CA), deoxycholic acid (DCA), chenodeoxycholic acid (CDCA), ursodeoxycholic acid (UDCA) and lithocholic acid (LCA). Each bile acid can be conjugated with glycine or taurine at position C-24. As the conjugation of bile acids depends on the taurine concentration in the liver and is not influenced by the hepatic concentration of glycine, it has been observed that the levels of taurine-conjugated bile acids increase markedly after administration of taurine[2]. Therefore, the pharmacological effects of taurine-conjugated bile acids cannot be neglected after the administration of taurine[3,6,7].

In the present study, the effects of both taurine and taurine conjugated with bile acids on rat isolated hepatocytes and the biological membranes were determined *in vitro*.

MATERIALS AND METHODS

Isolated Rat Hepatocytes

Isolated hepatocytes were prepared from male Wistar rats (200 g) by the collagenase perfusion method. Hepatocytes were incubated for 3 h in Krebs-Henseleit buffer oxygenated with a stream of 95% O_2. Hypoxic conditions were established by evacuating the O_2 from the flasks with a stream of 95% N_2[4].

Taurine 2, edited by Huxtable et al.
Plenum Press, New York, 1996

Cell viability was determined by the exclusion test with trypan blue. The lipid peroxidation of hepatocytes was estimated according to the thiobarbituric acid method.

^{31}Phosphorous Nuclear Magnetic Resonance (^{31}P-NMR) Measurement

Human erythrocytes ghosts prepared according to the method by Dodge *et al.*[1] were resealed in isotonic buffer (154 mM NaCl with 5 mM Tris-HCl, pH 7.4) and then sonicated at 0°C, 100 W for 60 min by ultrasound sonicator. The precipitates (vesicles) which were obtained after ultracentrifugation at 100,000 g for 60 min were suspended and incubated in the isotonic buffer at 37°C for 10 min after the addition of taurine (10 mM) or bile acids (1 mM; 6 mM). The NMR spectra of the sample were obtained using JNM-PFT-100 NMR spectroscopy (JEOL, Japan).

Electron Spin Resonance (ESR) Measurement

Hepatic microsomes of rat were incubated with taurine or bile acids for 15 min and then the suspension was spin-labeled by incubation with 5-doxyl stearic acid (DSA) or 12-DSA for 15 min. The precipitates (microsomes) which were obtained after centrifuge were transferred to a capillary tube for use in an ESR spectrometer (FE2XG, JEOL, Japan). The flow properties of the shallow and deep layers of the membrane were determined with the spectra of 5-DSA and 12-DSA, respectively[1]. Order parameter(s) was calculated according to the Gaffney's equation.

RESULTS AND DISCUSSION

Effects of Taurine and Taurine-Conjugated Bile Acids on the Viability of Hepatocytes

When isolated rat hepatocytes were incubated for 180 min under oxygenation, the cell viability decreased and the lipid peroxidation (LPO) increased more in the Ca^{2+}-free medium than in the Ca^{2+}-containing medium. Taurine (10 mM) suppressed markedly both the reduction of viability and the increase of LPO in the Ca^{2+}-containing medium[4]. Under the hypoxic condition, the viability of hepatocytes was markedly reduced in both Ca^{2+}-containing and Ca^{2+}-free medium. Taurine prevented the hypoxia-induced hepatocyte death in the Ca^{2+}-containing medium, but not in the Ca^{2+}-free medium (Fig. 1)[5].

Bile acids (1 mM), on the other hand, significantly reduced the cell viability and increased LPO of hepatocytes under oxygenation. These effects were marked in the treatment of monohydroxy bile acid such as LCA and dihydroxy bile acids such as CDCA and DCA, compared with a trihydroxy bile acid such as CA. When the effects of free bile acids were compared with those of taurine-conjugated bile acids, no significant difference was found in either cell viability or LPO formation of hepatocytes (Fig. 2).

Although there was some difference in effects on hepatocytes between taurine and bile acids, there is little difference between effects of free bile acids and taurine-conjugated bile acids. Therefore, the direct effects of taurine and free or taurine-conjugated bile acids on the biological membranes were also examined.

Effects of Taurine and Taurine-Conjugated Bile Acids on the Mobility of Human Erythrocytes Ghosts

Biological membranes are generally composed of two layers of phospholipids that are moving continuously. The mobility or ordering of phosphorus at the head of phospholipids can

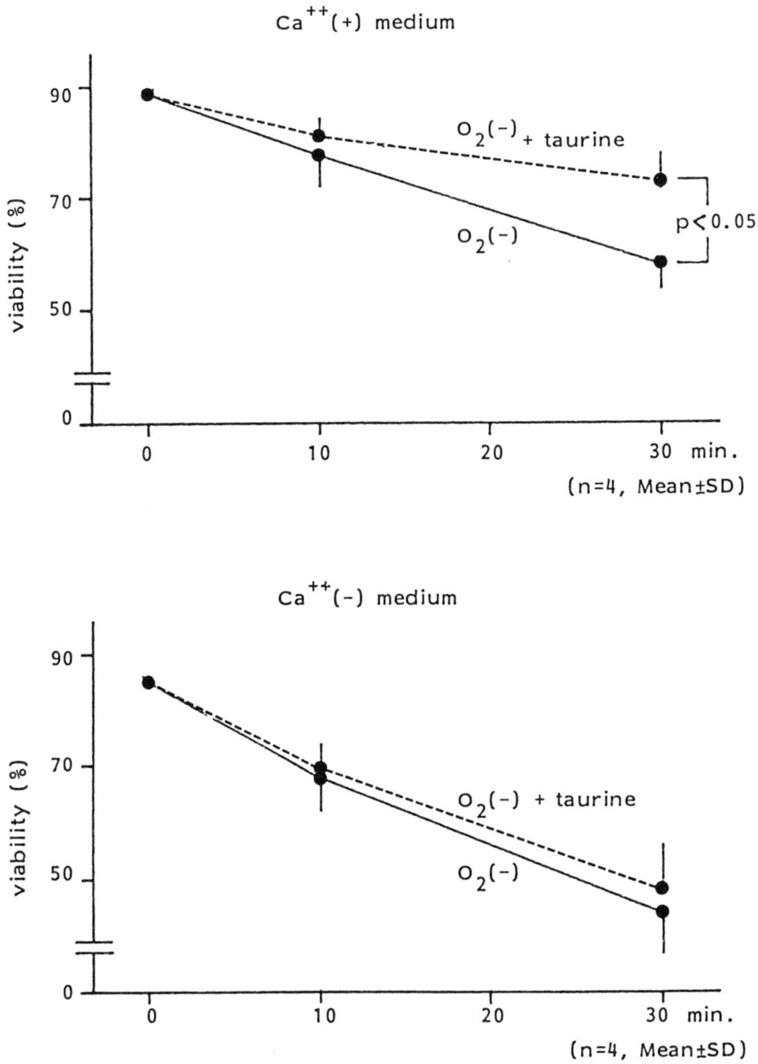

Figure 1. Effects of taurine on hypoxia-induced hepatocyte death in Ca^{2+}-containing (2.5 mM) or Ca^{2+}-free medium.

be analyzed using ^{31}P-NMR spectroscopy. As shown in Figure 3, the vesicles of erythrocytes ghosts revealed an NMR signal that had two components: the peak at higher magnetic field (10.4 ppm) that comes from the phospholipids arranged in order and the peak at lower magnetic field (0.6 ppm) that comes from the randomly moving phospholipids. The broadening of line width shows decrease of mobility and the narrowing of line width shows increase of mobility.

Taurine broadened the line width of the spectra, which indicates that taurine decreases membrane mobility (Fig. 3). The line width of the spectra from erythrocyte ghosts was not significantly changed by the treatment with free CA or UDCA, but DCA or CDCA broadened the line width, which indicates that some free bile acids decrease the mobility. After treatment with taurine-conjugated bile acids, narrowing of the component at the lower magnetic field was observed (Fig. 4). These data suggest that a part of the membrane was dissolved and the

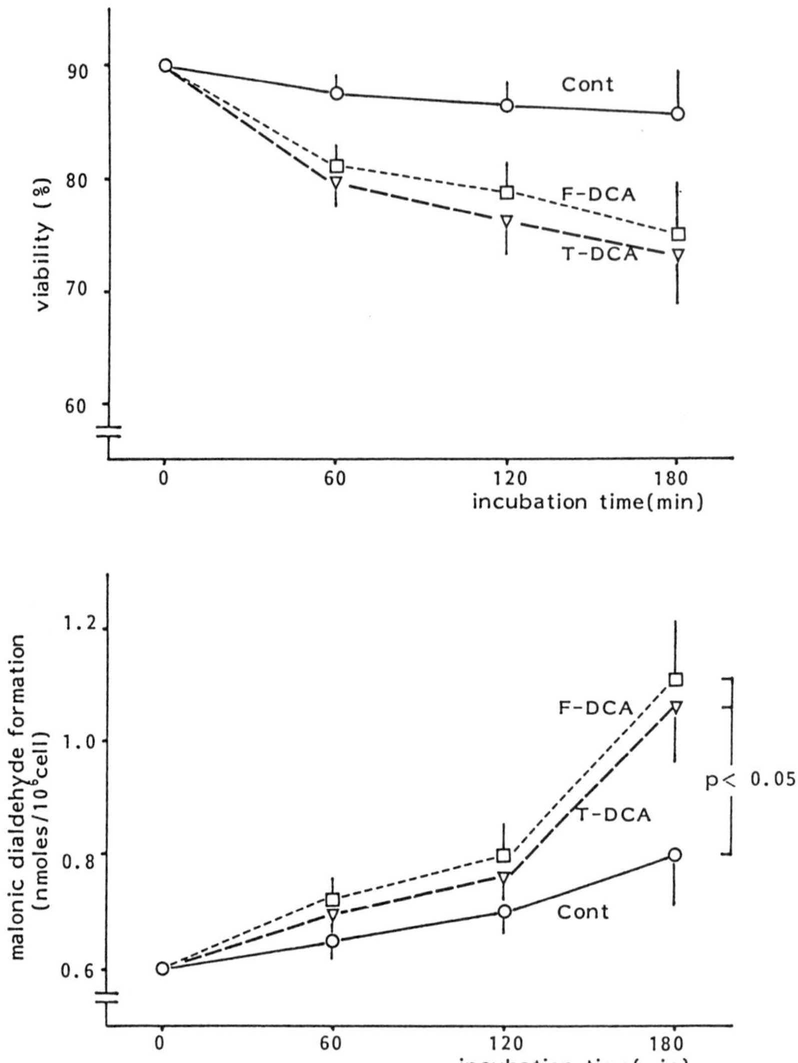

Figure 2. Effects of bile acids (1 mM) on cell viability and lipid peroxidation of hepatocytes in Ca^{2+}-containing and oxygenated medium: comparison between free bile acid and taurine-conjugated bile acid. F-DCA: free (unconjugated) deoxycholic acid-Na^{+} salt, T-DCA: taurine-conjugated deoxycholic acid-Na^{+} salt.

mobility was increased by treatment with taurine-conjugated bile acids. Narrowing of the spectra at the lower magnetic field was also observed in the erythrocyte ghosts treated with glycine-conjugated bile acids, but the degree of spectral change was less than that shown in the ghosts treated with taurine-conjugated bile acids.

Effects of Taurine and Taurine-Conjugated Bile Acids on the Fluidity of Hepatic Microsomal Membranes

ESR spectra arising from spin-labeling fatty acid probes incorporated into membranes yield information about flow properties of membranes. In the ESR spin-labeling

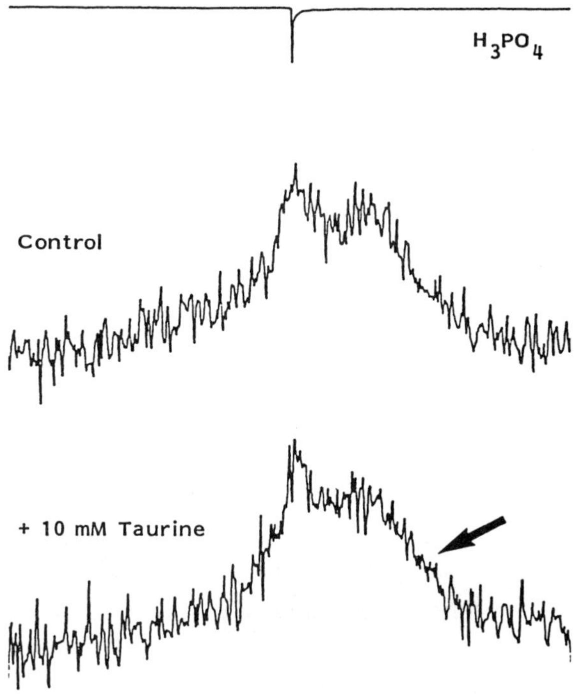

Figure 3. Effect of taurine on ^{31}P-NMR spectra of human erythrocyte ghosts. Proton noise decoupling was used and 85% H_3PO_4 was employed as an external standard for chemical shift.

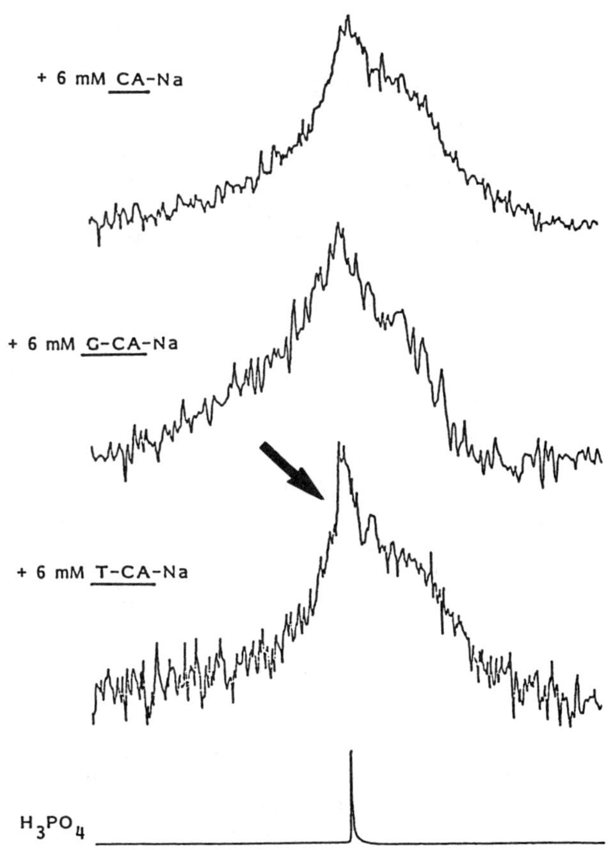

Figure 4. Effects of bile acids on ^{31}P-NMR spectra of human erythrocyte ghosts: comparison between free bile acid and taurine-conjugated bile acid. CA-Na: free (unconjugated) cholic acid-Na^+ salt, G-CA-Na: glycine-conjugated cholic acid-Na^+ salt, T-CA-Na: taurine-conjugated cholic acid-Na^+ salt.

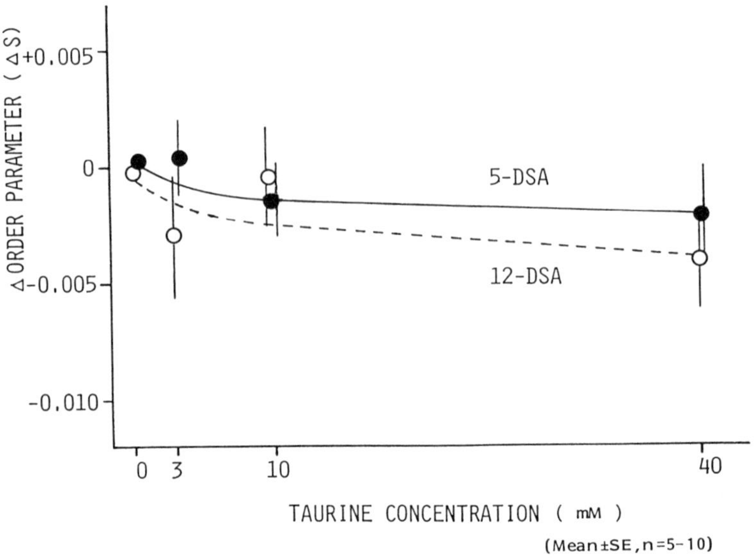

Figure 5. Effect of taurine on the order parameter (S) of 5-DSA or 12-DSA in hepatic microsomal membranes of rats. The deviation (ΔS) from the control value of S is shown in each experiment.

method, the increase of order parameter (S) indicates the decrease of fluidity and the decrease of S indicates the increase of fluidity. Taurine had no clear effect on the membrane fluidity estimated by 5-DSA or 12-DSA (Fig. 5). Bile acids mainly affected the fluidity of the shallow membrane layer in a dose-dependent manner, as estimated by 5-DSA. Free bile acids significantly decreased the fluidity. Taurine-conjugated bile acids, on the contrary, increased the fluidity (Fig. 6)[1]. There was no significant difference between the effect of taurine-conjugates and that of glycine-conjugates. Furthermore, taurine pretreatment did not influence the change in S induced by bile acids. Although taurine itself has no effect on the fluidity, it is suggested that taurine may affect the fluidity through the formation of taurine-conjugated bile acids.

SUMMARY

1. Taurine had cytoprotective effects on hepatocytes under hyperoxic or hypoxic condition in the presence of Ca^{2+}.
2. Bile acids had toxic effects on hepatocytes; there was no significant difference in the toxicity between free bile acids and taurine-conjugated bile acids.
3. The mobility of the membranes, examined by ^{31}P-NMR, decreased after treatment with taurine or free bile acids, but increased after treatment with taurine-conjugated bile acids.
4. The fluidity of the membranes, measured by ESR, did not change after treatment with taurine, but decreased after treatment with free bile acids and increased after treatment with taurine-conjugated bile acids.

In conclusion, because most bile acids are conjugated with taurine in the liver after administration of taurine, the influence of taurine-conjugated bile acids has to be taken into consideration in estimating the functions of taurine in the liver.

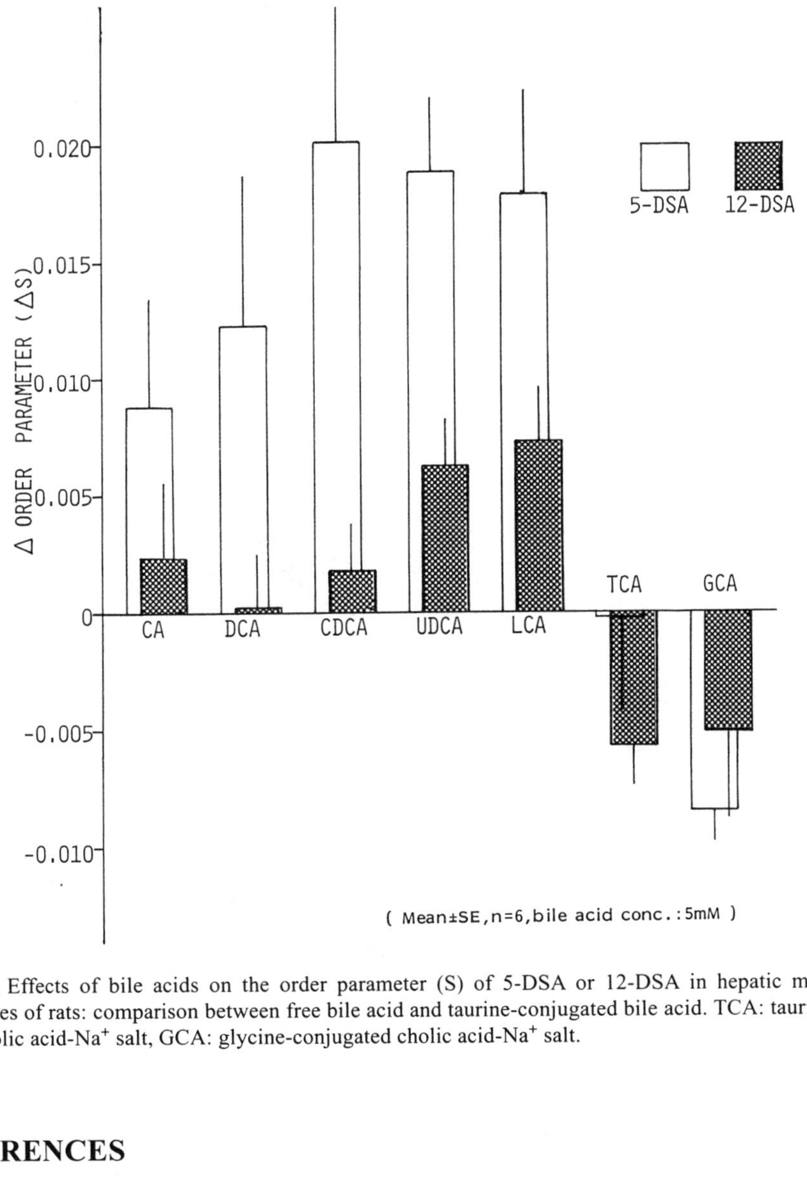

Figure 6. Effects of bile acids on the order parameter (S) of 5-DSA or 12-DSA in hepatic microsomal membranes of rats: comparison between free bile acid and taurine-conjugated bile acid. TCA: taurine-conjugated cholic acid-Na^+ salt, GCA: glycine-conjugated cholic acid-Na^+ salt.

REFERENCES

1. Dodge, J.T., Mitchell, C., and Hanahan, D.J. 1963, The preparation and chemical characteristics of hemoglobin-free ghosts of human erythrocytes, *Arch.Biochem.Biophys.* 110:119-130.
2. Nakashima, T., Nakagawa, Y., Sano, A., Sakamoto, Y., Shima, T., Nakajima, T., Seto, Y., Okuno, T., and Kashima, K. 1990, Regurgitation of bile acids in rat liver under bile drainage: quantitative analysis by taurine or ursodeoxycholate loading test, *J.Gastroenterol.Hepatol.* 5:121-125.
3. Nakashima, T., Sano, A., Nakagawa, Y., Okuno, T., Takino, T., and Kuriyama, K. 1983, Effect of taurine on the course of drug-induced chronic liver disease, in: "Sulfur Amino Acids: Biochemical and Clinical Aspects", Kuriyama, K., Huxtable, R.J. and Iwata, H. eds., Alan R. Liss, Inc., New York, pp. 472
4. Nakashima, T., Seto, Y., Nakajima, T., Shima, T., Sakamoto, Y., Cho, N., Sano, A., Iwai, M., Kagawa, K., Okanoue, T., and Kashima, K. 1990, Calcium-associated cytoprotective effects of taurine on the calcium and oxygen paradoxes in isolated rat hepatocytes, *Liver*, 10:167-172.
5. Nakashima, T., Shima, T., Sakamoto, Y., Nakajima, T., Seto, Y., Sano, A., Iwai, M., Okanoue, T., and Kashima, K. 1993, Effects of bile acids and taurine on the lipid fluidity of hepatic microsomes in normal and bile duct-ligated rats—A spin label study, *J.Hepatol.* 18:74-79.

6. Nakashima, T., Takino, T., and Kuriyama, K. 1982, Therapeutic effect of taurine administration on carbon tetrachloride-induced hepatic injury, *Jpn.J.Pharmacol.* 32:583-590.
7. Nakashima, T., Takino, T., and Kuriyama, K. 1983, Therapeutic and prophylactic effects of taurine administration on experimental liver injury, in: "Sulfur Amino Acids: Biochemical and Clinical Aspects", Kuriyama, K., Huxtable, R.J. and Iwata, H. eds., Alan R. Liss, Inc., New York, pp. 449-460.

11

COMBINED EFFECT OF TAURINE AND OX BILE ON BILIARY FLOW

Yukitaka Fukaya, Nobuko Senda, Akiko Fujita, Shoji Imai, and Ikuo Sawada

Matsuura Yakugyo Co., Ltd.
36, Torashinden, Odaka-cho, Midori-ku, Nagoya 459, Japan

INTRODUCTION

Taurine, similar to glycine, is an important amino acid for bile acid conjugation. It has been reported that taurine has an effect on metabolic function of the liver[1]. Animal bile has been used for a long time as a Chinese traditional medicine for stomach and intestinal disorders. We have previously examined various animal biles, such as ox, bear and pig bile, for their cholagogic effect in anesthetized rats. The results showed that the cholagogic effect of ox bile was significantly higher than that of other animal biles[2].

The purpose of the present study is to examine the combined effect of taurine and ox bile on biliary flow and bile acid composition in anesthetized rats and rabbits.

MATERIALS AND METHODS

Materials

Taurine was purchased from Nacalai Tesque, Inc., Kyoto, Japan. Bile Acid Standard Kit for HPLC was purchased from Japan Spectroscopic Co., Ltd., Kyoto, Japan. All other chemicals were of analytical reagent grade.

Biliary Flow Studies

Male Wistar rats (7-week old), after an overnight fast, were anesthetized with urethane (1.5 g/kg) by i.p. administration. Male albino rabbits (2.5 kg body weight), after an overnight fast, were anesthetized with sodium pentobarbital (25 mg/kg) by i.v. administration into the ear vein, and maintained under anesthesia for the duration of the experiment. The common bile duct of each rat or rabbit was cannulated with a polyethylene tube. After the bile was collected from each rat or rabbit for 30 min, taurine (100 mg/kg) and ox bile

Taurine 2, edited by Huxtable et al.
Plenum Press, New York, 1996

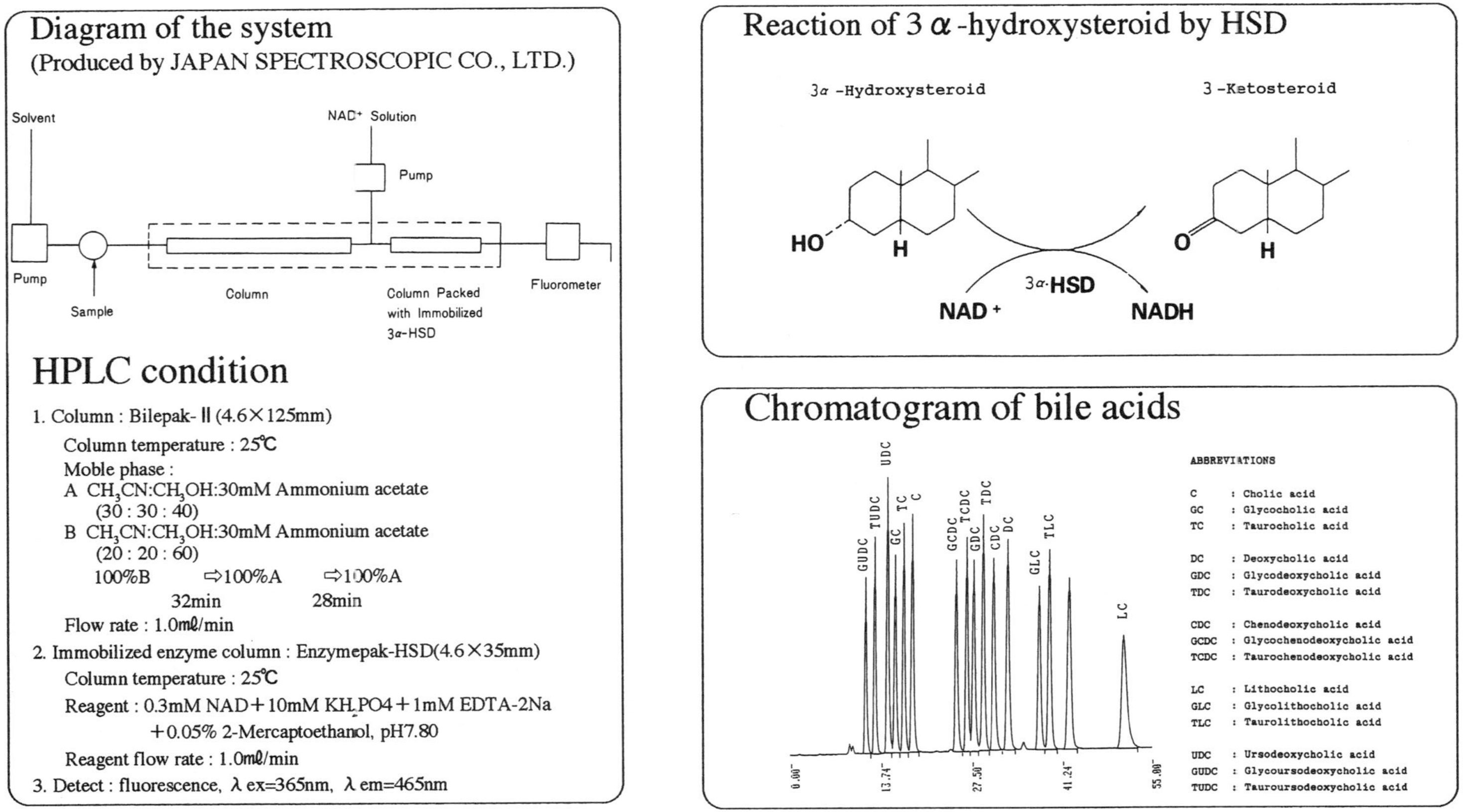

Figure 1. Analytical method for individual bile acids using HPLC and immobilized 3α-hydroxysteroid dehydrogenase (HSD) in column form.[3]

(1 g/kg) were co-administered to the duodenum. The bile of rat was collected for 4 h at 30 min intervals, and The bile of rabbit was collected for 7.5 h at 30 min intervals.

Measurement of Biliary Bile Acids

To a 50 μl portion of the collected bile in rats, in a 1.5 ml polypropylene tube, 950 μl of methanol were added. The mixture was agitated by vortex mixer and centrifuged for 15 min at 600 g. The supernatant was then passed through a membrane filter (0.45 μm). Aliquots (10 μl) of the filtrate were injected into a HPLC system.

Taurine and glycine conjugates and free bile acids in the bile of rats, rabbits and ox were measured by the method of individual bile acids using HPLC and immobilized 3α-hydroxysteroid dehydrogenase in column form (Fig. 1)[3]. The HPLC system was purchased from Japan Spectroscopic Co., Ltd., Kyoto, Japan.

Statistical Analysis

Data were summarized as means ± SEM and were analyzed using a Bonferroni's multiple comparison test.

RESULTS AND DISCUSSION

Ox bile (1 g/kg) produced an acceleration of biliary flow (Fig. 2). The cholagogic effect of ox bile was promoted by combined treatment with taurine (100 mg/kg) in rats, notwithstanding the finding that taurine alone had no effect on biliary flow (data not shown).

In addition, taurine, when combined with ox bile, enhanced both taurine-conjugated bile acid content and glycine-conjugated content of rat bile (Fig. 3). However, when taurine

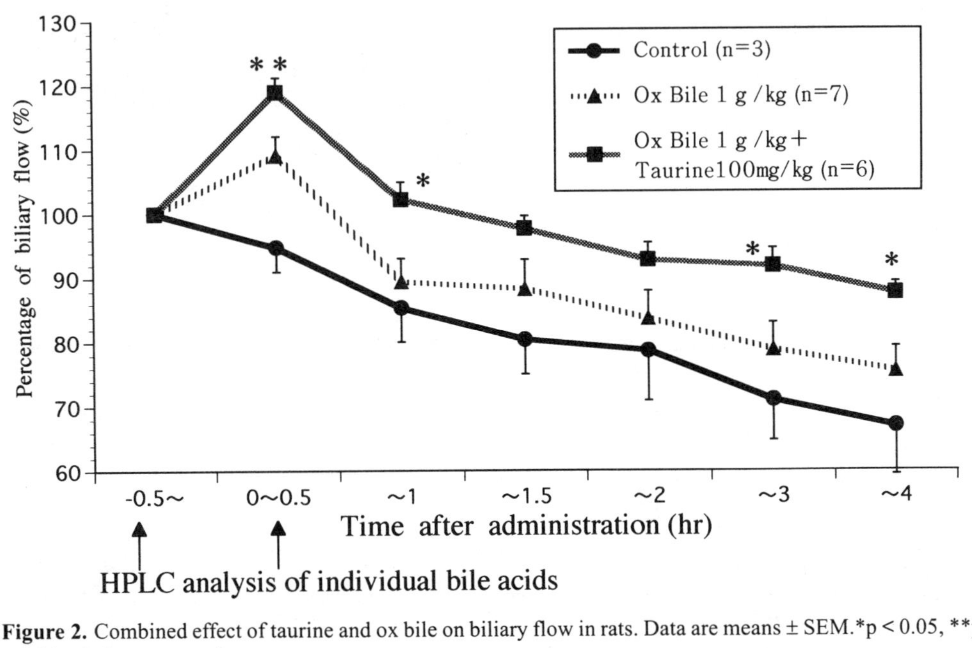

Figure 2. Combined effect of taurine and ox bile on biliary flow in rats. Data are means ± SEM.*$p < 0.05$, **$p < 0.01$ relative to control.

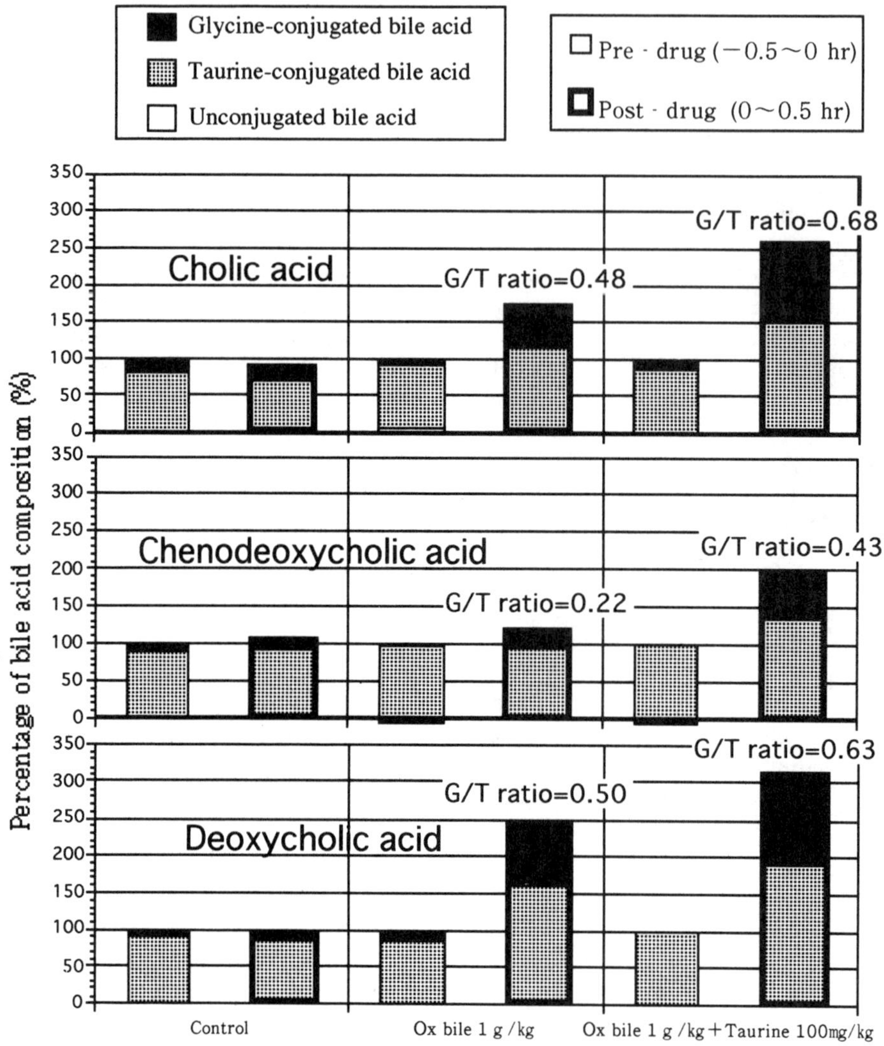

Figure 3. Combined effect of taurine and ox bile on bile acid composition in rats.

Table 1. Bile Acid composition and ratio between glycine- and taurine-conjugated bile acid (G/T ratio) of ox, rat and rabbit bile

	GC	GDC	GCDC	GLC	TC	TDC	TCDC	C	G/T ratio
Ox bile (lyophilized; mg/g)	265	50	18		193	46	18	13	1.30
Rat bile (mg/ml)	0.357	0.058	0.0347		2.885	0.517	0.505	0.121	0.12
Rabbit bile (mg/ml)	0.154	3.300		0.039					§
Human bile[5]									2~3.2

§: almost all glycine-conjugated
GC: Glycocholic acid; GDC: Glycodeoxycholic acid; GCDC: Glycochenodeoxycholic acid; GLC: Glycolithocholic acid; TC: Taurocholic acid; TDC: Taurodeoxycholic acid; TCDC: Taurochenodeoxycholic acid; C: Cholic acid.

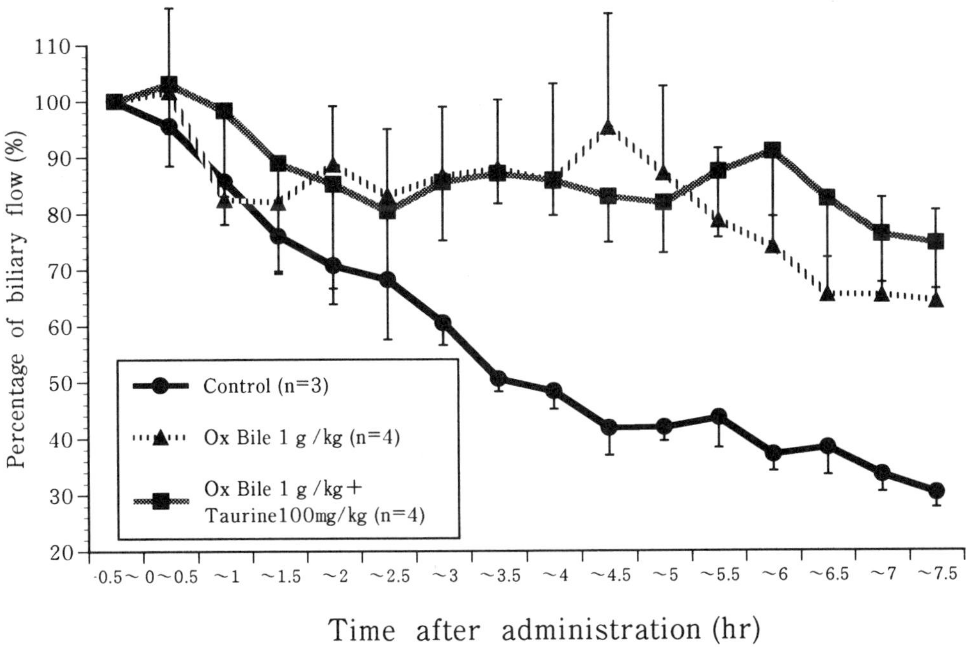

Figure 4. Combined effect of taurine and ox bile on biliary flow in rabbits. Data are means ± SEM.

was incubated with ox bile in an *in vitro* system, the contents of conjugated bile acids did not change (data not shown).

In rabbit bile, the ratio between glycine- and taurine-conjugated bile acids (G/T ratio) differs significantly from rat bile (Table 1.). Taurine did not promote the cholagogic effect of ox bile (Fig. 4).

It has been reported that taurine increased bile acid synthesis, as reflected by stimulation of cholesterol 7α-hydroxylase activity[4]. These results suggest that taurine may promote the increase in bile acid synthetic ability induced by ox bile in rats, but further investigations are required to understand the mechanism.

REFERENCES

1. Takino, T., Nakashima, T., Sano, A., Nakagawa, T. and Seto, Y. 1985, Liver diseases and sulfur amino acids, *Sulfur Amino Acids,* 8:1-11 (in Japanese).
2. Matsuura, K., Yamakawa, K., Fukaya, Y. and Hirata, Y. 1991, The effect of promoting the excretion of bile by using animal bile and bile acids, *Abstracts of the 5th International Symposium on Chinese Drugs (Nanjing,* p.36.
3. Okuyama, S., Kokubun, N., Higashidate, S., Uemura, D. and Hirata Y. 1979, A new analytical method of individual bile acids using high performance liquid chromatography and immobilized 3α-hydroxysteroid dehydrogenase in column form, *Chemistry Letters,* 1443-1446.
4. Nakamura, Y., Tsuji, K. and Ichikawa, T. 1987, Effect of dietary taurine on cholesterol 7α-hydroxylase activity in the liver of mice fed a lithogenic diet, *J. Nutr. Sci. Vitaminol.,* 33:239-243.
5. Ide, T. 1992, Dietary fiber - the biological metabolic response and physiological significance, *Shokubin Kougyo,* 46-54 (in Japanese).

12

SIGNIFICANCE OF TAURINE CONJUGATION OF BILE ACID IN THE BILIARY EXCRETION OF BILIRUBIN

Junichi Yamao, Yoshinobu Matsumura, Yoichi Hokaze, Masahide Yoshikawa, Norie Umemoto, Yuji Matsui, Hiroshi Fukui, and Tadasu Tsujii

Third Department of Internal Medicine
Nara Prefectural University of Medicine, Kashihara, Japan

INTRODUCTION

Almost all bile acids are conjugated with either taurine or glycine. The conjugates are more potent cholelogues than the unconjugated bile acids. Previously, we have confirmed that administration of taurine increased bile flow and biliary excretion of bile acids in rats with cholestasis induced by α-naphthylisothiocyanate[6]. Moreover, we found that administration of taurine shortened duration of jaundice in patients with acute hepatitis[4]. These results suggest a possibility that taurine may accelerate biliary excretion of bilirubin. In the present study, we investigated the effect of taurine on the biliary excretion of bilirubin in normal and mutant hyperbilirubinuric rats[4].

MATERIALS AND METHODS

Chemicals

Free sodium ursodeoxycholate (FUDC), sodium tauroursodeoxycholate (TUDC) and sodium glycoursodeoxycholate (GUDC) were kindly provided by Tokyo Tanabe Co. (Tokyo, Japan). [^{14}C]Cholic acid was obtained from Amersham Co. (Tokyo, Japan). Other chemicals were purchased from Nakarai Tesc Co. (Kyoto, Japan). All were analytical grade.

Animals

Male Sprague-Dawley (SD) rats weighing 200-250 g were purchased from Japan SLC Inc. (Hamamatsu, Japan). Mutant jaundiced rats (Eizai hyperbilirubinuria rat, EHBR) were kindly provided by Eizai Co. (Tokyo, Japan). All animals were fed commercial rat pellets. EHBR was given free access to distilled water. Half of the SD rats were also given

Taurine 2, edited by Huxtable et al.
Plenum Press, New York, 1996

distilled water, but the other half were given 3% β-alanine solution instead of water (β-alanine-fed rats). All animals were fasted for 24 h before experiments.

Bile Acid Conjugation

Bile acid conjugation in liver homogenates was analyzed by the method of Ekdahl *et al.*[1] to evaluate conjugating ability in SD rats and EHBR. [^{14}C]Cholate (3 μCi/mg) and unlabelled cholic acid (10-70 nmol) were added to liver homogenate from SD rats and EHBR and the homogenate was incubated at 37°C for 2 h. Thin layer chromatography was used for separation of cholic acid, taurocholic acid and glycocholic acid.

Hepatic Taurine and Glycine Concentration

Taurine and glycine concentrations in liver homogenate were measured in SD rats and β-alanine-fed rats. After the rat had been killed, the liver was immediately excised, weighed and homogenized in a buffer solution. The resultant homogenate was deproteinized with trichloroacetic acid. After centrifugation at 1425 rpm for 60 s, taurine and glycine were determined in the supernatant by HPLC.

Biliary Excretion

Biliary excretion of bilirubin was investigated under the continuous iv infusion of FUDC, TUDC or GUDC in SD, EHBR and β-alanine-fed rats. Under ether anesthesia, the common bile duct was cannulated with a PE-10 polyethylene catheter and bile was collected every 30 min during experiment. Infusion of bile salt was started after 30 min baseline sampling of bile. FUDC, TUDC or GUDC was continuously infused via an catheter inserted into the femoral vein at a constant rate of 0.6 or 1.2 μmol/min/100 g body weight. The bile flow was calculated from the weight of collected bile sample with an assumption that specific gravity of rat bile is 1.0. Biliary concentration of bilirubin was determined by Diazo method.

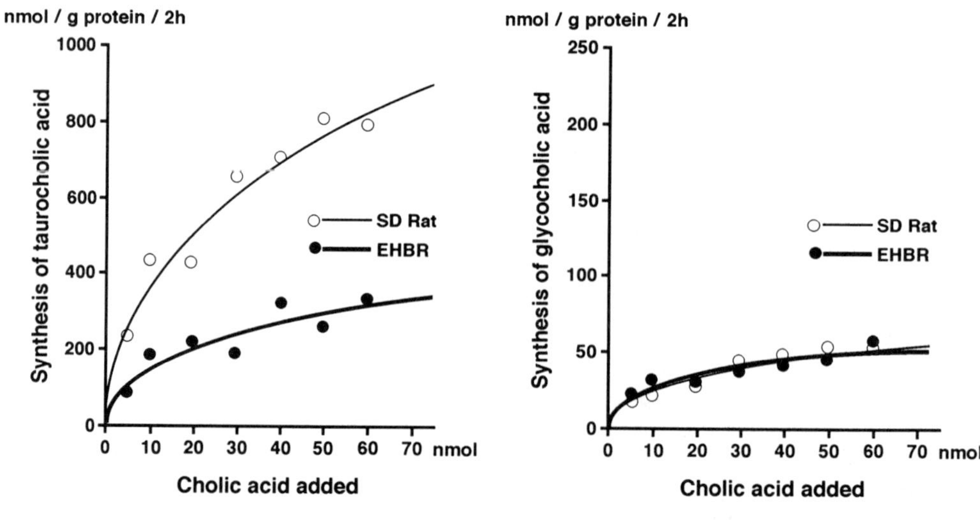

Figure 1. Conjugation of cholic acid in EHBR and SD rats.

RESULTS

Bile Acid Conjugation in EHBR and SD Rats

The synthesis of taurocholic acid was markedly reduced in EHBR compared with SD rats. The synthesis of glycocholic acid did not differ between EHBR and SD rats (Fig. 1).

Hepatic Taurine and Glycine Concentration in β-Alanine-Fed Rats and SD Rats

Hepatic taurine concentration in β-alanine-fed rats was reduced to approximately 25% of that in SD rats. β-Alanine feeding had no effect on hepatic glycine concentration (Fig. 2).

Effect of FUDC on Biliary Excretion of Bilirubin

When FUDC was continuously infused at a rate of 0.6 μmol/min/100 g body weight, the biliary excretion rate of bilirubin stayed higher than the control levels obtained by saline infusion throughout the 2 h experiment in SD rats (Fig. 3). In EHBR and β-alanine fed rats, biliary excretion rates of bilirubin were higher than the control levels until 30 min after the start of FUDC infusion, but thereafter they fell below the control levels (Figs. 4 and 5).

When the infusion rate of FUDC was doubled to 1.2 μmol/min/100 g body weight, the biliary excretion rate of bilirubin showed a rapid decrease to below control levels, similar to that seen in EHBR and β-alanine fed rats late in the experimental period (Fig. 6).

Effect of Taurine and Glycine on Biliary Excretion of Bilirubin

The decrement in biliary excretion rate of bilirubin induced by high dose (1.2 μmol/min/100 g body weight) infusion of FUDC in SD rats was attenuated by co-infusion of glycine. This kept biliary excretion of bilirubin higher than the control levels obtained by saline infusion throughout the 2 h experiment. When taurine was co-infused with high

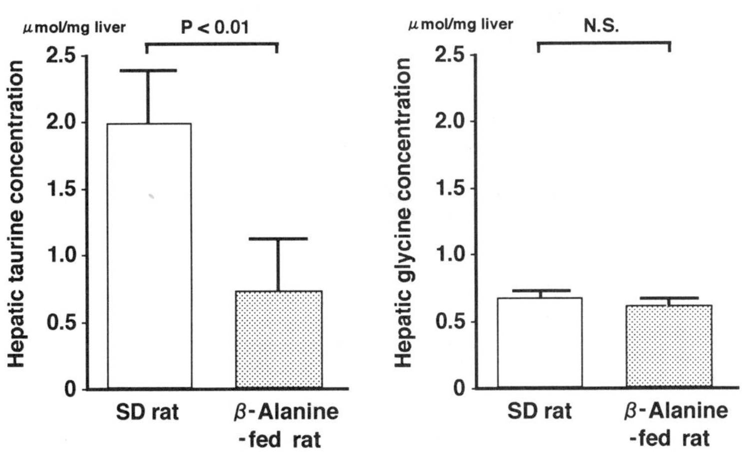

Figure 2. Hepatic taurine and glycine concentration in SD rats and β-alanine-fed rats.

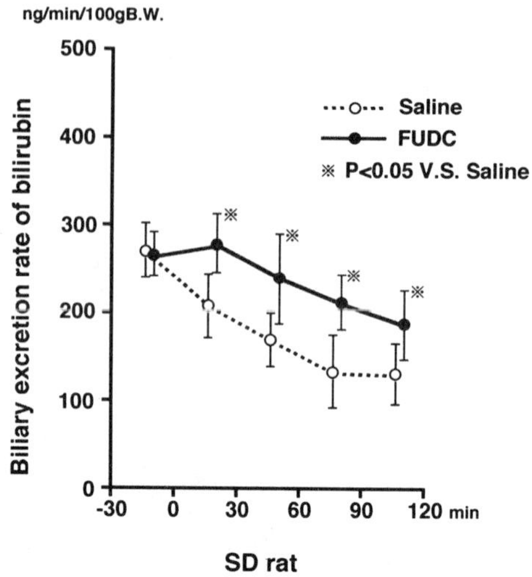

Figure 3. Biliary excretion rate of bilirubin under the continuous intravenous infusion of FUDC (0.6 μmol/min/100gB.W.) in SD rats.

dose FUDC, biliary excretion of bilirubin was augmented and continuously increased throughout the experiment (Fig. 7).

Effect of TUDC and GUDC on Biliary Excretion of Bilirubin

Continuous infusion of TUDC or GUDC at a rate of 0.6 μmol/min/100 g body weight led to augmentation of biliary excretion of bilirubin compared with control levels in SD rats, EHBR and β-alanine fed rats (Figs. 8-10).

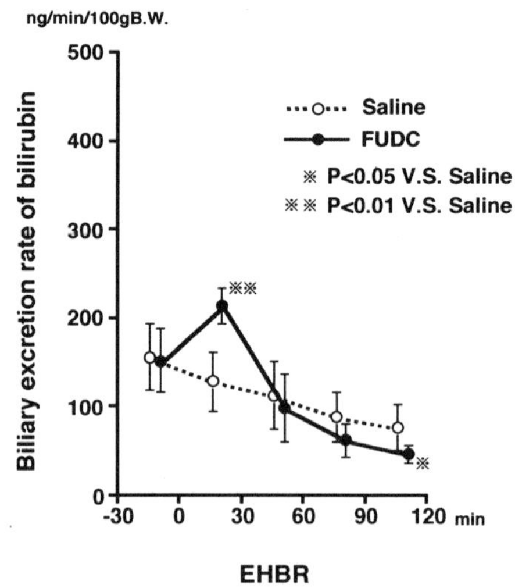

Figure 4. Biliary excretion rate of bilirubin under the continuous intravenous infusion of FUDC (0.6 μmol/min/100gB.W.) in EHBR.

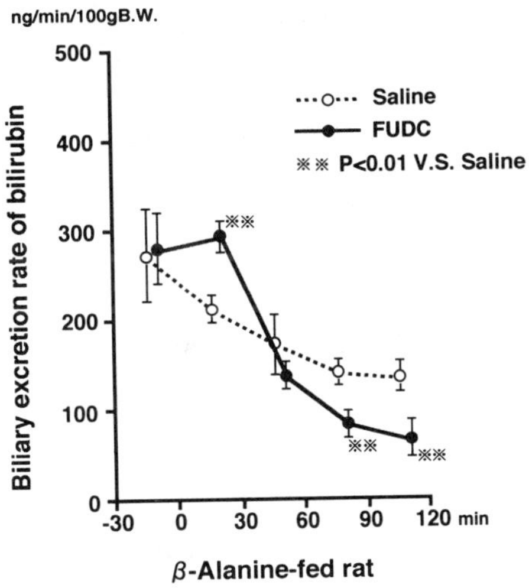

Figure 5. Biliary excretion rate of bilirubin under the continuous intravenous infusion of FUDC (0.6 μmol/min/100gB.W.) in β-alanine-fed rats.

DISCUSSION

The effect of continuous iv infusion of FUDC on the biliary excretion of bilirubin in EHBR was much less than in SD rats (Figs. 3 and 4). A similar phenomenon was seen in β-alanine-fed rats (Fig. 5) and even in SD rats when infusion rate of FUDCD was increased to 1.2 μmol/min/100 g body weight (Fig. 6). This difference in the effect of FUDC can be explained by impaired bile acid conjugation with taurine, as we found that EHBR were deficient in conjugating bile acid with taurine (Fig. 1) and confirmed β-alanine feeding lowered hepatic taurine in rats, as previously reported[5] (Fig. 2). Kitani and Kanai[3] reported

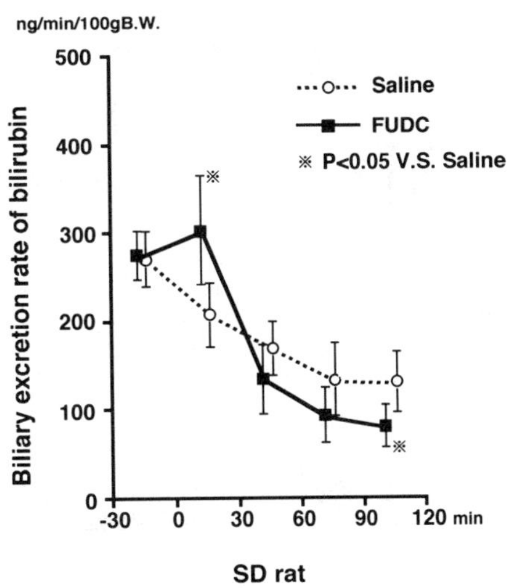

Figure 6. Biliary excretion rate of bilirubin under the continuous intravenous infusion of FUDC (1.2 μmol/min/100gB.W.) in SD rats.

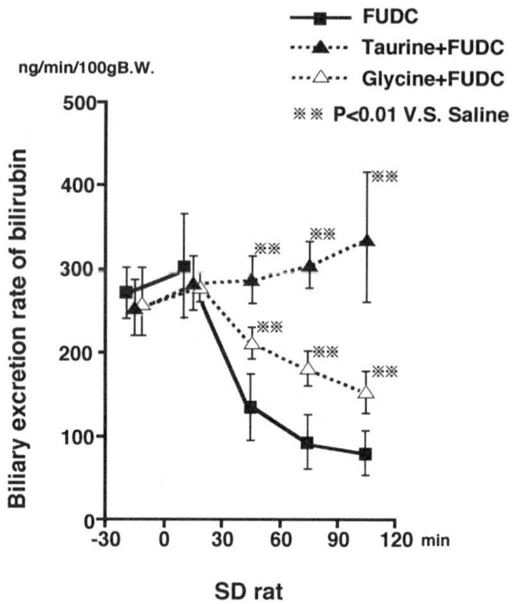

Figure 7. Effect of taurine and glycine on biliary excretion rate of bilirubin under the continuous intravenous infusion of FUDC (1.2 μmol/min/100gB.W.) in SD rats.

that continuous iv infusion of FUDC at the dose we used (1.2 μmol/min/100 g body weight) depleted hepatic taurine. Furthermore, co-infusion of taurine and glycine prevent the decrement in biliary excretion of bilirubin under the high dose infusion of FUDC in SD rats (Fig. 7). From these results, it is suggested that bile acid conjugation is necessary for the promoting effect of FUDC on biliary excretion of bilirubin. Hence, we investigated the effect

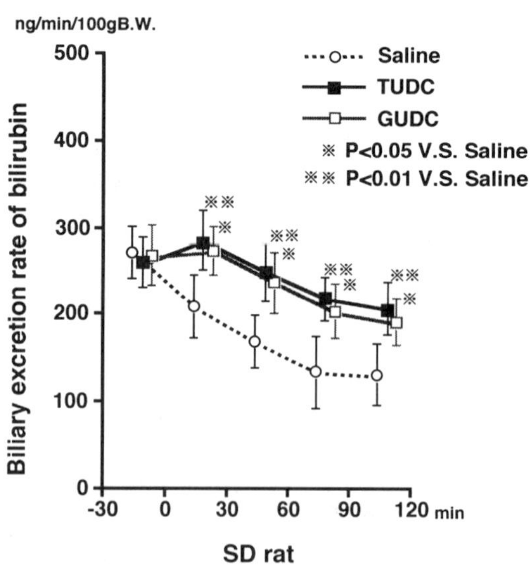

Figure 8. Biliary excretion rate of bilirubin under the continuous intravenous infusion of TUDC and GUDC (0.6 μmol/min/100gB.W.) in SD rats.

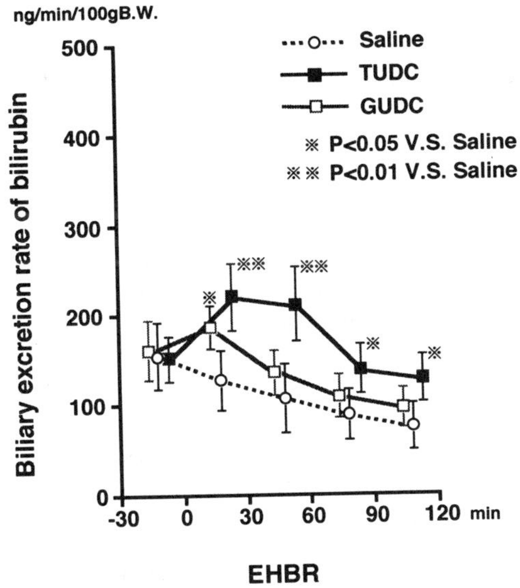

Figure 9. Biliary excretion rate of bilirubin under the continuous intravenous infusion of TUDC and GUDC (0.6 μmol/min/100gB.W.) in EHBR.

of infusion of TUDC and GUDC in SD rats, EHBR and β-alanine-fed rats and showed that TUDC and GUDC augmented the biliary excretion of bilirubin in all strains (Figs. 8-10).

The biliary excretion rate of bilirubin under simultaneous infusion of taurine and FUDC is much greater than occurs with glycine and FUDC (Fig. 7). This superiority of taurine may be associated with the fact[2] that taurine undergoes bile acid conjugation in preference to glycine.

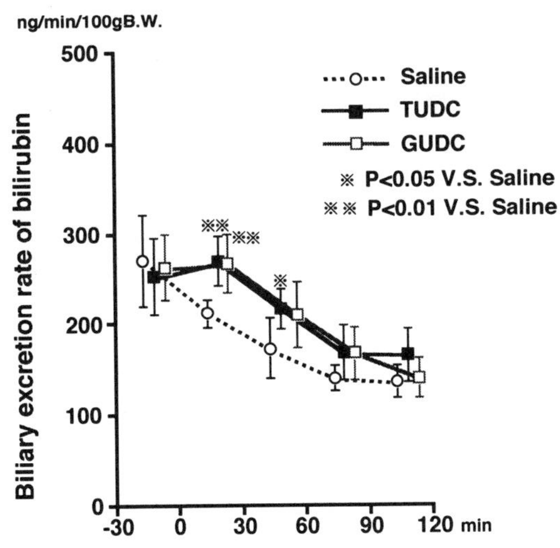

Figure 10. Biliary excretion rate of bilirubin under the continuous intravenous infusion of TUDC and GUDC (0.6 μmol/min/100gB.W.) in β-alanine-fed rats.

CONCLUSION

Bile acid conjugation, especially conjugation with taurine, plays an important role in the biliary excretion of bilirubin.

REFERENCES

1. Ekdahl, P.H., Gottfreis, A., and Scherstein, T. 1965, A sensitive method for the study of bile acid conjugation in homogenates of human liver biopsies, *Scandinav.J.Clin.Lab.Investigation*, 17:103-109.
2. Gaubutt, J.T., Heaton, K.W., Lack, L. and Tyor, M.P. 1969, Increased ratio of glycine-to-taurine-conjugated bile salts in patients with ileal disorders, *Gastroenterol.* 56:711-720.
3. Kitani, K. and Kanai, S. 1982, Effect of ursodeoxycholate on the bile flow in the rat, *Life Sci.* 31:1973-1985.
4. Matsuyama, Y., Morita, T., Higuchi, M., Tsujii, T. and Iwata, H. 1983, The effect of taurine administration on patients with acute hepatitis, in: "Sulfur Amino Acids: Biochemical and Clinical Aspects", Kuriyama, K., Huxtable, R.J. and Iwata, H. eds., Alan R. Liss, Inc., New York, pp. 461-468.
5. Shaffer, J.E. and Kocsis, J.J. 1981, Taurine mobilizing effects of beta alanine and other inhibitors of taurine transport, *Life Sci.* 28:2727-2736.
6. Tsujii, T., Matsuyama, Y., and Takagi, M. 1978, Effect of taurine and steroid on the biliary excretion of experimental intraheptic cholestasis in rats, *Sulfur Amino Acids*, 1:289-301.

13

EFFECT OF DIETARY TAURINE ON LIPID METABOLISM IN NORMCHOLESTEROLEMIC AND HYPERCHOLESTEROLEMIC STROKE-PRONE SPONTANEOUSLY HYPERTENSIVE RATS

Hiroshi Ogawa

Department of Hygiene
Kinki University School of Medicine
377-2 Ohno-Higashi, Osaka-Sayama City, Osaka 589, Japan

INTRODUCTION

It is well known that taurine plays an important role in conjugation of bile acids which are formed from cholesterol in the liver, suggesting that there is a close relationship between taurine and cholesterol metabolism. Kibe *et al.*[4] reported that taurine stimulates the activity of cholesterol 7α-hydroxylase, the rate-limiting enzyme of bile acid synthesis. Tsuji *et al.*[11] and Yamori *et al.*[12] reported that taurine shows a hypocholesterolemic effect in experimentally induced hypercholesterolemic animal models. These findings indicated a possibility that taurine is effective in diet-induced hypercholesterolemia.

Although rats are known to be resistant to atherosclerosis, stroke-prone spontaneously hypertensive rats (SHRSP) easily develop hypercholesterolemia and arterial fat deposition by cholesterol feeding, indicating that SHRSP may be an useful model for studies on lipid metabolism. Using SHRSP as a model, we[8] reported that taurine improves serum lipoprotein metabolism in SHRSP with hypercholesterolemia induced by cholesterol feeding. However, details of the hypocholesterolemic effect remain unclear.

In the present study, to elucidate the hypocholesterolemic effect of taurine, we examined the effect of dietary taurine on lipid metabolism in the serum and liver of normocholesterolemic and diet-induced hypercholesterolemic SHRSP.

MATERIALS AND METHODS

Tri-[1-^{14}C]oleate (4.1 GBq/mmol) was purchased from Dupont New England Nuclear (Boston, MA). All other chemicals and solvents were purchased from Sigma Chemical Co. (St. Louis, MO) and Wako Chemical Ind. (Osaka).

Taurine 2, edited by Huxtable et al.
Plenum Press, New York, 1996

In experiment I, male SHRSP were used at the age of 18 weeks and divided into two groups (5 rats in each group). The control group was given a regular diet (CE-2; Clea Japan Inc., Osaka) and tap water, and the experimental group was given the CE-2 diet and 3% taurine solution for 6 weeks ad libitum. In experiment II, male SHRSP were used at the age of 15 weeks and divided into two groups (6 rats in each group). The control group was given a high-fat and high-cholesterol diet (HFC diet) prepared by adding 5% lard, 2% cholesterol and 0.5% cholic acid by weight to the CE-2 diet and tap water, and the experimental group was given the HFC diet and 3% taurine solution for 6 weeks ad libitum.

After 5 weeks on the diet and drinking water, post-heparin plasma samples were obtained from the jugular vein under light ether anesthesia 3 min after heparin (300 units/kg) injection without fasting. Post-heparin plasma lipolytic activity (PHLA), hepatic triglyceride lipase (h-TGL) activity and lipoprotein lipase (LPL) activity were determined using tri-[1-^{14}C]oleate as a substrate by the combination of immunoprecipitation using anti-rat h-TGL sera and radioisotope method as reported previously[7]. One week later, whole blood was collected from the abdominal aorta under ether anesthesia after fasting overnight and serum was separated by centrifugation at 3,000 rpm, 4°C. In the case of experiment I, phase-shift in light-dark cycle was performed for a week before animals were killed.

Blood pressure was measured by the tail-pulse pickup method (USM-105-R, Ueda Electronic Works, Ltd., Tokyo). Various serum lipoprotein fractions (VLDL: very low density lipoprotein; d<1.006; IDL: intermediate density lipoprotein; d: 1.006~1.020; LDL: low density lipoprotein; d: 1.020~1.063; HDL: high density lipoprotein; d:1.063~1.210 and Free: lipoprotein-free; d>1.210) were separated by stepwise density-gradient ultracentrifugation (TL-100 with TL-100-2 rotor, Beckman, USA) according to the method of Hatch *et al.*[3]. Total liver lipids were extracted according to the method of Folch *et al.*[2] Serum cholesterol, phospholipid and triglyceride, and the various lipoprotein fractions and liver lipids were determined by commercially available kits (Cholesterol E-test Wako, Phospholipid C-test Wako and Triglyceride E-test Wako, Wako Chemical Ind., Osaka). The contents of various apolipoproteins (apoA-I, A-IV, B and E) were estimated by Rocket immunoelectrophoresis according to the method of Laurell[5], and modified by us[9]. Hepatic microsomal cholesterol 7α-hydroxylase activity was measured by HPLC according to the method of Ogishima *et al.*[10]

Statistical analysis was carried out by the MUSCOT statistical analysis program (YUKMUS Co., Tokyo) and $p < 0.05$ was defined as a significant difference. Values are expressed as the mean ± SEM, with n=5 in each group of experiment I and n=6 in each group of experiment II.

RESULTS

There was no significant difference in the growth, or intakes of food and drinking water between the two groups in both experiment I and experiment II.

Effects of taurine on systolic blood pressure, serum concentrations of total cholesterol, phospholipid, triglyceride and apolipoproteins (apoA-I, A-IV, B and E), and apoB/apoA-I ratio in the normocholesterolemic SHRSP (Experiment I) are shown in Table 1. In the normocholesterolemic SHRSP, taurine significantly increased serum concentrations of total cholesterol and phospholipid, while no significant change in the level of triglyceride was observed. In addition, taurine significantly decreased serum concentration of apoA-IV and tended to increase serum concentration of apoE.

To elucidate the alterations of serum lipid and apolipoproteins, distributions of serum lipid and apolipoproteins among various lipoprotein fractions were examined (Fig. 1 and 2). The increase of serum total cholesterol and phospholipid was mainly due to the increase in

Table 1. Effects of taurine on body weight, blood pressure, serum concentrations of total cholesterol, phospholipid, triglyceride, and various apolipoproteins (apo-A-I, A-IV, B and E), and apoB/apoA-I ratio in SHRSP

	Control (CE-2 diet)	Taurine
Body Weight (g)	305 ± 6	305 ± 7
Blood Pressure (mmHg)	230 ± 6	225 ± 3
Total Cholesterol (mg/dl)	68.1 ± 2.5	88.5 ± 5.1**
Phospholipid (mg/dl)	116 ± 2.4	130 ± 5
Triglyceride (mg/dl)	62.1 ± 4.7	59.7 ± 1.3
ApoA-I (mg/dl)	53.7 ± 3.5	55.2 ± 2.7
ApoA-IV (mg/dl)	10.0 ± 0.1	9.3 ± 0.2*
ApoB (mg/dl)	6.3 ± 0.2	6.2 ± 0.4
ApoE (mg/dl)	37.7 ± 1.0	43.1 ± 2.6
ApoB/ApoA-I(%)	11.5 ± 1.0	12.0 ± 0.3

Values are means ± SEM (N=5 in each group). Significant difference from the control group: $^*p < 0.05$; $^{**}p < 0.01$

the LDL fraction (d:1.006~1.063) where apoE content was also significantly elevated. On the other hand, the decrease of serum apoA-IV was due to the decrease in the HDL fraction (d:1.063~1.210).

Effects of taurine on PHLA, h-TGL activity and LPL activity in the normo-cholesterolemic SHRSP are shown in Table 2. Taurine significantly increased PHLA, which corresponded to the increase of LPL activity. Consequently, the ratio of h-TGL/PHLA was significantly decreased and the ratio of LPL/PHLA was significantly increased.

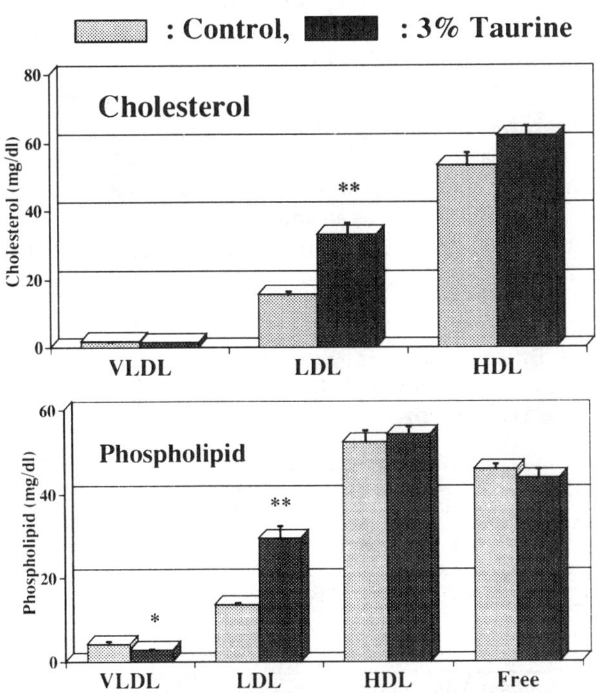

Figure 1. Distributions of serum cholesterol and phospholipid among various lipoprotein fractions in SHRSP. The vertical bars indicate the SEM. Significant difference from the control group: $^*p < 0.05$; $^{**}p < 0.01$.

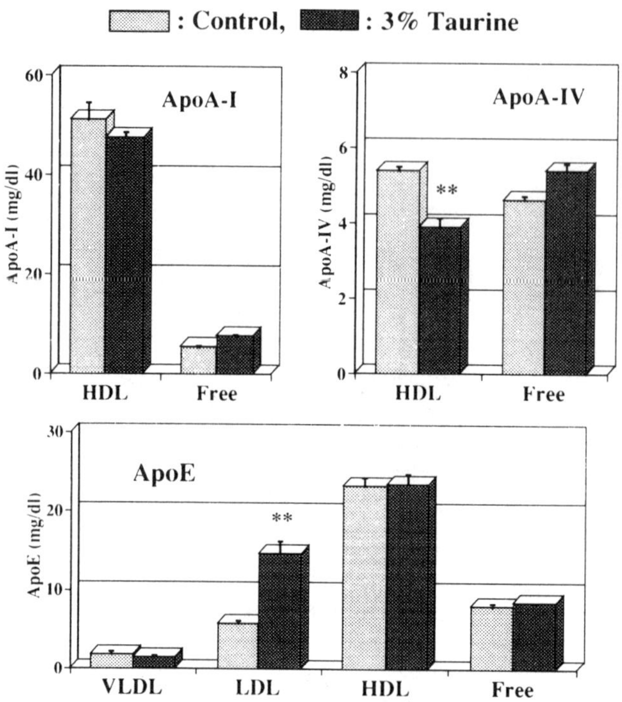

Figure 2. Distributions of serum apoA-I, A-IV and E among various lipoprotein fractions in SHRSP. The vertical bars indicate the SEM. Significant difference from the control group: $*p < 0.05$; $**p < 0.01$.

Effects of taurine on relative liver weight, hepatic cholesterol and phospholipid, and activity of hepatic microsomal 7α-hydroxylase in normocholesterolemic SHRSP are shown in Table 3. There was no significant difference in the relative weight and hepatic cholesterol content between the two groups. Phospholipid content of the experimental group was significantly lower than that of the control group. The activity of 7α-hydroxylase was much higher in comparison with the control group.

Next, to gain further understanding of the hypocholesterolemic effect of taurine, alterations of lipid metabolism in the serum and liver were examined using experimentally-induced hypercholesterolemic SHRSP as an animal model (Experiment II). Effects of taurine on body weight, systolic blood pressure, serum total cholesterol, phospholipid, triglyceride

Table 2. Effects of taurine on post-heparin plasma lipolytic activity (PHLA), hepatic triglyceride lipase (h-TGL) activity, lipoprotein lipase (LPL) activity, and the ratios of h-TGL/PHLA and LPL/PHLA in SHRSP

	Control (CE-2 diet)	Taurine
PHLA (FFA μmol/ml/h)	58.1 ± 2.2	68.6 ± 1.1**
h-TGL (FFA μmol/ml/h)	25.6 ± 1.0	24.5 ± 0.9
LPL (FFA μmol/ml/h)	32.6 ± 1.7	44.2 ± 1.6**
h-TGL/PHLA (%)	44.4 ± 1.4	36.4 ± 2.0**
LPL/PHLA (%)	55.6 ± 1.4	63.6 ± 2.3**

FFA μmol/ml/h: μmol of free fatty acid liberated per one ml of plasma per one hour. Values are means ± SEM (N=5 in each group). Significant difference from the control group: $*p < 0.05$; $**p < 0.01$

Table 3. Effects of taurine on liver weight, contents of cholesterol and phospholipid in the liver, and activity of hepatic microsomal cholesterol 7α-hydroxylase (7α-hydroxylase) in SHRSP

	Control (CE-2 diet)	Taurine
Weight (g/100g B.W.)	3.72 ± 0.12	4.07 ± 0.09
Cholesterol (mg/wet g)	1.50 ± 0.08	1.64 ± 0.05
Phospholipid (mg/wet g)	15.7 ± 0.24	14.1 ± 0.5*
7α-Hydroxylase (pmol/min/mg protein)	42.6 ± 1.4	130.9 ± 16.6**

Values are means ± SEM (N=5 in each group). Significant difference from the control group: *p < 0.05; **p < 0.01

and various apolipoproteins (apoA-I, A-IV, B and E), and the apoB/apoA-I ratio in the hypercholesterolemic SHRSP are shown in Table 4.

In the hypercholesterolemic SHRSP, although systolic blood pressure was significantly lower than that of the normocholesterolemic SHRSP fed a regular diet (CE-2), there was no significant difference between the HFC group and the taurine group. Taurine markedly suppressed the increase in the concentrations of serum total cholesterol and triglyceride. In addition, both the decrease in the concentration of serum apoA-I and the increase in the concentrations of serum apoB were also significantly suppressed by taurine administration, resulting in the reduction of apoB/apoA-I ratio, one of the atherogenic indexes.

Distribution of serum lipid and apolipoproteins among various lipoprotein fractions are shown in Fig. 3 and 4. In the VLDL and IDL fractions, taurine dramatically suppressed the marked increase in cholesterol and triglyceride contents, which was associated with suppression of the increase in apoE content. On the one hand, in the HDL fraction, taurine significantly suppressed the decrease in cholesterol content, which was associated with suppression of the decrease in apoA-I content. Moreover, taurine significantly suppressed the increase in apoA-IV content in the Free fraction.

Effects of taurine on PHLA, h-TGL activity, LPL activity, and the ratios of h-TGL/PHLA and LPL/PHLA in the hypercholesterolemic SHRSP are shown in Table 5. By the HFC feeding, h-TGL activity and h-TGL/PHLA ratio were significantly decreased, and

Table 4. Effects of taurine on body weight, blood pressure, serum concentrations of total cholesterol, phospholipid and various apolipoproteins (apoA-I, A-IV, B and E), and apoB/apoA-I ratio in the hypercholesterolemic SHRSP

	Regular diet (CE-2 diet)	HFC diet	HFC diet + 3% Taurine
Body weight (g)	310 ± 6	332 ± 10	339 ± 9
Blood Pressure (mmHg)	230 ± 1	216 ± 6#	213 ± 5
Total Cholesterol (mg/dl)	88.3 ± 2.6	243 ± 16##	163 ± 20**
Phospholipid (mg/dl)	105 ± 1.7	149 ± 6#	107 ± 6*
ApoA-I (mg/dl)	48.3 ± 1.1	36.8 ± 2.0##	42.6 ± 1.4*
ApoA-IV (mg/dl)	9.5 ± 0.4	13.5 ± 0.3##	12.2 ± 1.0
ApoB (mg/dl)	5.6 ± 0.1	24.0 ± 0.7##	15.8 ± 0.5**
ApoE (mg/dl)	31.6 ± 0.5	25.3 ± 1.4##	16.6 ± 1.3**
ApoB/ApoA-I (%)	12.0 ± 1.1	66.0 ± 2.1##	37.0 ± 1.0**

Values are means ± SEM (N=6 in each group). Significant difference from the regular diet (CE-2) group: #p < 0.05; ##p < 0.01; Significant difference from the HFC diet group: *p < 0.05; **p < 0.01

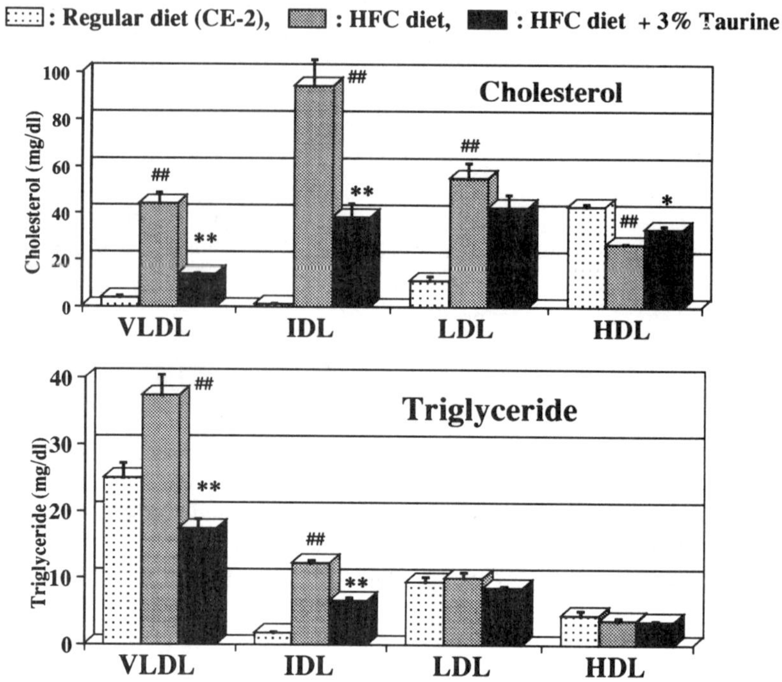

Figure 3. Distributions of serum cholesterol and triglyceride among various lipoprotein fractions in the hypercholesterolemic SHRSP. The vertical bars indicate the SEM. Significant difference from the regular diet (CE-2) group: #p < 0.05; ##p < 0.01); Significant difference from the HFC diet group: *p < 0.05; **p < 0.01).

LPL activity and LPL/PHLA ratio were significantly increased. However, no significant difference in any parameter was observed between the two groups in the hypercholesterolemic state.

Effects of taurine on relative liver weight, hepatic cholesterol and phospholipid, and activity of hepatic microsomal 7α-hydroxylase in the hypercholesterolemic SHRSP are shown in Table 6. Taurine significantly suppressed the marked increase in cholesterol content

Table 5. Effects of taurine on post-heparin plasma lipolytic activity (PHLA), hepatic triglyceride lipase (h-TGL) activity, lipoprotein lipase (LPL) activity, and the ratios of h-TGL/PHLA and LPL/PHLA in the hypercholesterolemic SHRSP

	Regular diet (CE-2 diet)	HFC diet	HFC diet + 3% Taurine
PHLA (FFA μmol/ml/h)	60.5 ± 2.5	70.8 ± 6.8	77.9 ± 4.8
h-TGL (FFA μmol/ml/h)	29.3 ± 1.8	16.2 ± 3.1#	22.0 ± 2.6
LPL (FFA μmol/ml/h)	31.2 ± 2.1	54.7 ± 6.3#	55.9 ± 5.6
h-TGL/PHLA (%)	48.5 ± 2.4	23.3 ± 4.4#	29.3 ± 4.6
LPL/PHLA (%)	51.6 ± 2.4	76.7 ± 4.4#	76.7 ± 4.4

FFA μmol/ml/h: μmol of free fatty acid liberated per ml of plasma per hour (cf Table 2). Values are means ± SEM (N=6 in each group). Significant difference from the regular diet (CE-2) group: #p < 0.05; ##p < 0.01; Significant difference from the HFC diet group: *p < 0.05; **p < 0.01.

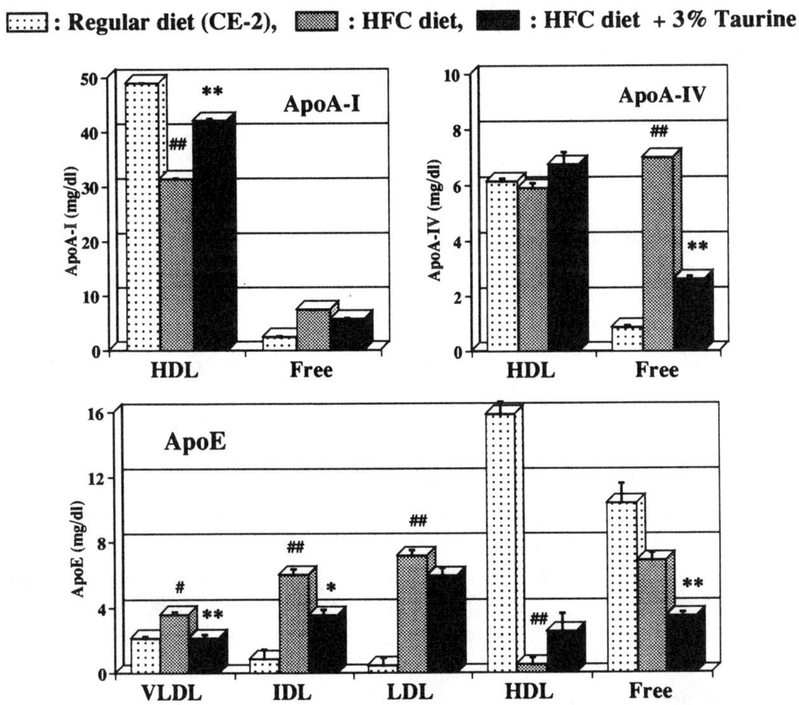

Figure 4. Distributions of serum apoA-I, A-IV and E among various lipoprotein fractions in the hypercholesterolemic SHRSP. The vertical bars indicate the SEM. Significant difference from the regular diet (CE-2) group: #p < 0.05; ##p < 0.01); Significant difference from the HFC diet group: *p < 0.05; **p < 0.01).

and the reduction of phospholipid content. The activity of hepatic microsomal 7α-hydroxylase, which was significantly increased by the HFC feeding, remained unchanged in the experimental group.

DISCUSSION

Among the widespread beneficial effects of taurine, in the present paper, we focussed on its hypolipidemic effect. In the normocholesterolemic SHRSP fed a regular diet, taurine

Table 6. Effects of taurine on liver weight, contents of cholesterol and phospholipid in the liver, and activity of hepatic microsomal cholesterol 7α-hydroxylase in the hypercholesterolemic SHRSP

	Regular diet (CE-2 diet)	HFC diet	HFC diet + 3% Taurine
Weight (g/100g B.W.)	4.31 ± 0.14	5.19 ± 0.09	5.28 ± 0.10
Cholesterol (mg/wet g)	1.49 ± 0.07	9.45 ± 0.73##	6.80 ± 0.38**
Phospholipid (mg/wet g)	15.0 ± 0.4	11.8 ± 0.70##	13.4 ± 0.4*
7α-Hydroxylase (pmol/min/mg protein)	3.70 ± 0.46	8.73 ± 0.92##	7.85 ± 0.68

Values are means ± SEM (N=6 in each group). Significant difference from the regular diet (CE-2) group: #p < 0.05; ##p < 0.01; Significant difference from the HFC diet group: *p < 0.05; **p < 0.01

significantly increased the content of cholesterol, phospholipid and apoE in the LDL fraction. In the LDL fraction of rat, both native LDL which contains only apoB as apolipoprotein and HDL_1 (referred to as apoE-HDL), one of HDL subfractions, which contains apoE as the major apolipoprotein, are present. Mahley *et al.*[6] also reported that HDL_1 encompasses both the LDL and HDL fractions in the rat. Accordingly, the increase of cholesterol, phospholipid and apoE in the LDL fraction of the taurine group is attributed to the increase in apoE-HDL. The apoE HDL transfers cholesterol from the peripheral tissues to the liver. In other words, it plays an important role in reverse cholesterol transport.

Moreover, taurine significantly increased LPL activity, while no change in h-TGL activity was observed. LPL which binds to endothelial cells catalyzes the hydrolysis of the triacylglycerol portion of chylomicrons and VLDL to produce free fatty acid and monoacylglycerol. This suggests that taurine may promote the hydrolysis of triacylglycerol-rich lipoproteins such as chylomicrons and VLDL, and it is anticipated to contribute to reduction of serum lipids in the hyperlipidemic state. In addition, the decrease in the content of apoA-IV in the HDL fraction of the taurine group may be due to the increase in the LPL activity since apoA-IV is one of major components of chylomicrons.

In the liver, taurine significantly increased the microsomal 7α-hydroxylase activity. This is the rate-limiting enzyme of hepatic cholesterol catabolism to bile acids. This result is in agreement with the result in guinea pig reported by Kibe *et al.*[4] and with that in the hamster reported by Belletani *et al.*[1] This increase in the microsomal 7α-hydroxylase activity indicates that taurine may promote hepatic taurine-conjugated bile acid synthesis, leading to reduction of cholesterol content when an excess amount of cholesterol is taken up in the liver, as the case of cholesterol feeding. Next, to investigate the details of the hypocholesterolemic effect of taurine, alterations of serum lipoproteins and hepatic lipids were studied in the diet-induced hypercholesterolemic SHRSP. Taurine dramatically suppressed the marked elevation of cholesterol and triglyceride contents in the VLDL and IDL fractions, which may partially be due to the increase in the LPL activity, as mentioned for the normocholesterolemic SHRSP. In addition. there is a possibility that taurine may also reduce hepatic synthesis and/or secretion of atherogenic and cholesterol-rich ß-VLDL, which is formed by cholesterol feeding.

In the liver of the hypercholesterolemic SHRSP, taurine markedly reduced the cholesterol content. This may be attributed to the increase in the microsomal 7α-hydroxylase activity, as mentioned for the normocholesterolemic SHRSP. However, there is a discrepancy that no significant difference in 7α-hydroxylase activity was observed between the control and experimental groups in the hypercholesterolemic condition, although the 7α-hydroxylase activity was significantly increased by the HFC feeding. This suggests that the 7α-hydroxylase activity in the light period may be maximally induced by the HFC feeding.

In conclusion, taurine increased apoE-HDL (HDL_1) which transfers the cholesterol from the peripheral tissues to the liver, resulting in the increase in serum cholesterol, phospholipid and apoE. Taurine also increased LPL activity, suggesting a promotion of the hydrolysis of triacylglycerol-rich lipoproteins such as chylomicrons and VLDL. Moreover, taurine increased the hepatic 7α-hydroxylase activity, indicating that taurine may stimulate hepatic bile acid synthesis. These effects result in a hypocholesterolemic response, leading to suppression of the marked elevation of atherogenic lipoproteins, and prevention of hepatic cholesterol accumulation in the hypercholesterolemic SHRSP.

REFERENCES

1. Bellentani. S., Pecorari, M., Cordoma, P., Marchegiano, P., Manenti, F., Bosisio, E., De Fabiani, E. and Galli, G. 1987, Taurine increases bile acid pool size and reduces bile saturation index in the hamster, *J. Lipid Res.* 28:1021-1027.

2. Folch, J., Lees, M.R. and Stanley, G.H.S. 1957, A simple method for the isolation and purification of total lipids from animal tissues, *J. Biol. Chem.* 226:497-509.
3. Hatch, F.T. and Lee, R.S. 1968, Practical methods for plasma lipoprotein analysis, *Adv. Lipid Res.* 6:1-68.
4. Kibe, A., Wake, C., Kuramoto, T. and Hoshita, T. 1980, Effect of dietary taurine on bile acid metabolism in guinea pigs, *Lipids* 15:224-229.
5. Laurell, C.B. 1966, Quantitative estimation of proteins by electrophoresis in agarose-gel containing antibodies, *Anal. Biochem.* 15:45-52.
6. Mahley, R.W. 1981, Cellular and molecular biology of lipoprotein metabolism in atherosclerosis, *Diabetes*, 30 (Suppl. 2):60-65.
7. Ogawa, H., Nishikawa, T. and Sasagawa, S. 1991, Effect of cholesterol feeding on the compositions of plasma lipoproteins and plasma lipolytic activities in SHRSP, *Clin. Exper. Hypertens.* A(13):999-1008.
8. Ogawa, H., Nishikawa, T., Fukushima, S. and Sasagawa, S. 1988, Studies on stroke-prone spontaneously hypertensive rats (SHRSP) fed a high-fat and high-cholesterol diet: Effects of taurine on serum lipoprotein metabolism in hypercholesterolemic rats, *Jpn. J. Hyg.* 14:648-658.
9. Ogawa, H., Shiota, C., Nishikawa, T., Fukushima, S. and Sasagawa, S. 1986, Age-related changes in serum concentrations of apolipoproteins A-I, E and A-IV in stroke-prone spontaneously hypertensive rats, *J. Hypertens.* 4 (Suppl. 3):429-431.
10. Ogishima, T. and Okuda, K. 1986, An improved method for assay of cholesterol 7α-hydroxylase activity, *Anal. Biochem.* 158:228-232.
11. Tsuji, K., Iwao, H., Nakagawa, Y. and Seki, T. 1980, Influence of dietary taurine on distribution and fecal excretion of cholesterol in hypercholesterolemic rats, *Sulfur-Containing Amino Acids* 3:165-171.
12. Yamori, Y., Nara, Y., Kihara, M., Horie, R., Oshima, A., Iritani, N. and Fukuda, E. 1980, Possible involvement of taurine in pathological cholesterol metabolism of arteriolipidosis-prone rats (ALR), *Sulfur-Containing Amino Acids* 3:107-113.

14

EFFECTS OF TAURINE AND STRUCTURALLY RELATED ANALOGUES ON Ca^{2+} UPTAKE AND RESPIRATION RATE IN RAT LIVER MITOCHONDRIA

M. Palmi,[1] F. Fusi,[1] G. Youmbi,[1] M. Frosini,[1] L. Bianchi,[2] L. Della Corte,[2] G. P. Sgaragli,[1] and K. F. Tipton[3]

[1] Istituto di Scienze Farmacologiche
Università di Siena, Siena, Italy
[2] Dipartimento di Farmacologia Preclinica e Clinica "M. Aiazzi-Mancini"
Università di Firenze, Firenze, Italy
[3] Department of Biochemistry
Trinity College, Dublin 2, Ireland

INTRODUCTION

Mitochondria from vertebrate sources possess an elaborate system for transporting Ca^{2+} across their inner membrane. This process includes accumulation of the cation into their matrix, via an uniporter mechanism. This is balanced by electroneutral Ca^{2+} release mediated by the functioning of a 2 Na^+:1 Ca^{2+} exchanger or "antiporter" in mitochondria from most tissues, including heart, brain and brown adipose tissue, or by a $2H^+/Ca^{2+}$ antiporter in liver[7]. In addition to these functions, mitochondria produce about 95% of the common cellular energy as ATP by means of oxidative phosphorylation. Under physiological conditions, Ca^{2+} transport and energy production seem to be strictly correlated and there is convincing evidence that the intramitochondrial Ca^{2+} concentration functions as a metabolic control in signalling the mitochondria to modify its metabolic rate in response to increased energy demand. According to the theory of flux control, the steps of Ca^{2+}-mediated metabolic control are distributed and include activation of the Ca^{2+}-sensitive dehydrogenases[2,9] and other Ca^{2+}-sensitive metabolic processes[8,16]. In this context, drugs that alter the flux of Ca^{2+} across mitochondrial membranes could, in theory, play a role in modulating the cellular energetic metabolism.

The available evidence indicates that taurine modulates numerous Ca^{2+}-dependent processes in several tissues, including heart, brain and retina[11,12,20] and that it has marked effects on the kinetics of Ca^{2+} movement across membranes, both in whole tissue as well as in subcellular preparations. Thus, high affinity (5-20 μM) Ca^{2+} uptake from the cytosol has been shown to be stimulated by taurine in several membrane systems, including cardiac

Taurine 2, edited by Huxtable et al.
Plenum Press, New York, 1996

sarcolemma, retinal rod outer segments and disk membranes, brain synaptosomes and mitochondria[13,15,21]. In mitochondria, the taurine-stimulated transport of Ca^{2+} has been shown to be inhibited by ruthenium red, oligomycin and atractyloside[14].

In order to investigate the possible role of taurine in modulating Ca^{2+} homeostasis and the respiration rate in mitochondria, we have studied its effects on these two functions in isolated, functionally-intact, mitochondria from rat liver. The specificity of taurine in these processes was investigated by determining the effects of a number of structurally-related analogues of taurine on mitochondrial Ca^{2+} transport and oxidative phosphorylation.

MATERIALS AND METHODS

Materials

2-Aminoethylarsonic acid and *N,N*-dimethyltaurine were synthesized and kindly provided by Dr. H.B.F. Dixon (Department of Biochemistry, University of Cambridge, Cambridge CB2 1QW, UK). Other reagents were obtained from Sigma chemical Co., (St. Louis, MO, USA). Livers from male Wistar rats (200-250 g) were homogenised in STE buffer (250 mM sucrose, 5 mM Tris, 1 mM EGTA, pH 7.4) and mitochondria were separated by differential centrifugation[3]. After centrifugation, the mitochondrial pellet was resuspended in either 1-2 ml KSH buffer (150 mM KCl, 25 mM sucrose, 5mM Mg^{2+}, 2.7 mM Hepes), for determinations of Ca^{2+} uptake, or in 3-4 ml of STE buffer for determination of respiration rates. The mitochondrial protein content was determined by the biuret method[5] using bovine serum albumin as a standard.

Determination of Ca^{2+} Uptake

Ca^{2+} uptake was determined spectrophotometrically by the procedure of Palade[18], using antipyrylazo III as indicator[19]. 1.9 ml of uptake buffer (92.5 mM KCl, 18.5 mM KMOPS, 7.5 mM sodium pyrophosphate, pH 7.0) containing 250 μM antipyrylazo III was added to a cuvette and 1 mM MgATP, creatine phosphokinase (20 μg/ml), 5 mM phoshocreatine and 1 mg of mitochondrial protein were then separately added to bring the final volume to 2.0 ml. This was equilibrated at 30°C, with stirring, before being placed in a Shimazdu model 160 spectrophotometer. Changes in Ca^{2+} concentration were monitored with time, after the addition to the cuvette of 50 μM $CaCl_2$, by measuring the absorbance at 710 nm and subtracting the absorbance at 790 nm. Calcium addition caused an initial rapid increase in the absorbance due to the formation of the calcium-antipyrylazo complex. This was followed by a time-resolvable decrease in absorbance as a result of the transport of Ca^{2+} into the mitochondria and its consequent disappearance from the extra-mitochondrial solution. In order to calibrate the absorbance signal under identical experimental conditions, 100 nmol of $CaCl_2$ was added to the cuvette after the absorbance decrease had stabilized, and the absorbance change was monitored. Compounds to be tested were dissolved in the buffered antipyrylazo III solution prior to the assay.

Determination of Respiration Rate in Mitochondria

Oxygen uptake was measured by using an O_2 electrode (YSI Model 5300, Yellow Springs, Ohio, USA) contained in the base of a Perspex chamber. An O_2 concentration of 475 nmol O_2/ml in the air-saturated reaction mixture was assumed[22]. The respiration rate was assessed under three sets of conditions: coupled respiration in the absence of ADP (state 4), respiration in phosphorylating mitochondria (state 3) and uncoupled respiration (state 3 uncoupled). All experiments were completed within 3 h of isolation of the mitochondria.

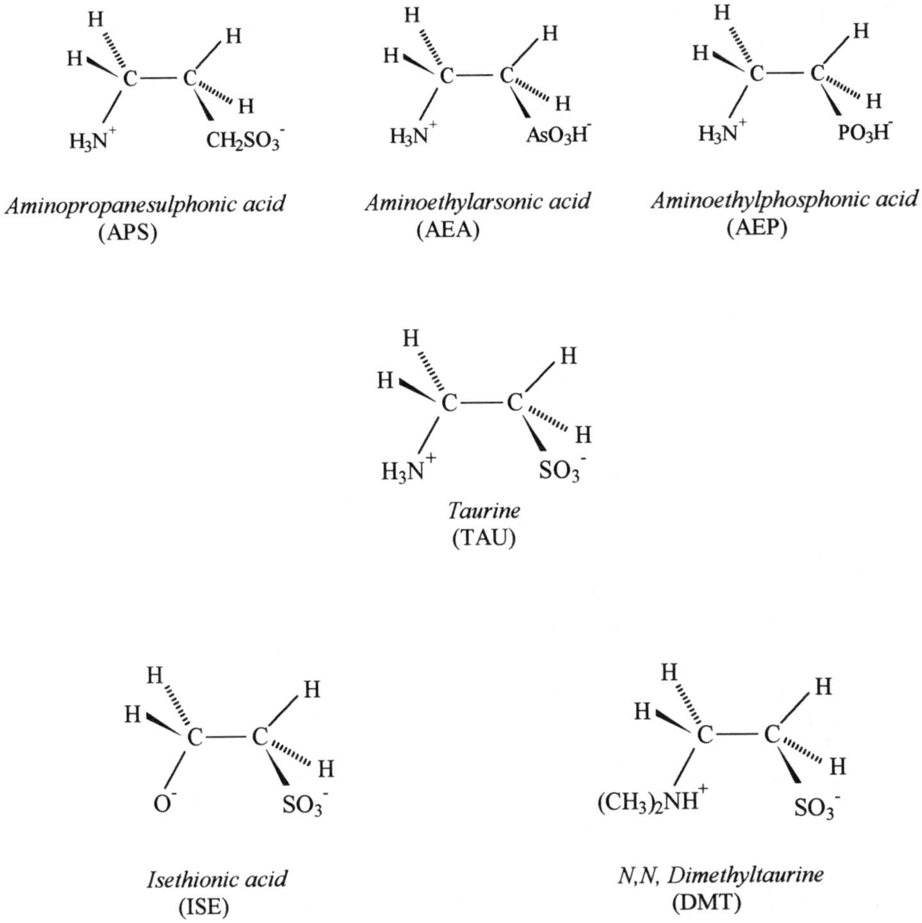

Figure 1. Structure of taurine and its analogues.

Isolated mitochondrial preparation (6 mg) was suspended in 3 ml of experimental medium (120 mM KCl, 5 mM KH_2PO_4, 2 mM $MgCl_2$, 10 mM HEPES, 1 mM EGTA, 13.3 μM rotenone, pH 7.2) in an air-tight and rapidly-stirred chamber thermostatted at 25°C. After 1 min, 10 μl of a stock solution of the compound of interest was added, followed 2 min later by 10 mM succinate (K^+ salt), and the rate of coupled respiration in the absence of ADP was then determined. Two minutes after succinate addition, 200 μM ADP (K^+ salt) was added and the respiration rate of the phosphorylating mitochondria was determined. After a further 2 min an excess (0.33 μM) of FCCP (carbonyl cyanide p-trifluoromethoxyphenylhydrazone) was added and the uncoupled rate of respiration was determined[4]. The mitochondrial preparations typically gave coupled respiration rates of 15-17 nmol $O_2.min^{-1}.mg^{-1}$ mitochondrial protein when respiring on succinate in the absence of added ADP under the above conditions. This rate was stimulated some 6-7 fold by the addition of excess FCCP.

Statistics

The mean values ± SEM of duplicate determinations from 3-5 separate mitochondrial preparations were determined and the percentage differences between experiment and control were then calculated. The statistical significance of the differences was estimated using Student's t test for unpaired data.

RESULTS

The structures and abbreviations of the taurine and the analogues used in this study are shown in Figure 1. Figure 2 shows the effects of different concentrations of taurine on the Ca^{2+} uptake by rat liver mitochondria.

The addition of taurine to the reaction medium, at concentrations varying between 1 and 10 mM, resulted in a significant ($p < 0.01$) increase of Ca^{2+} uptake. The concentration dependence tended to a plateau value at 10 mM taurine, at which point uptake was over twice the value determined in its absence.

The effects of the taurine analogues, each at 10 mM, on Ca^{2+} uptake are shown in Figure 3. Among the compounds studied only *N,N*-dimethyltaurine (DMT) showed a stimulating effect that was comparable to that given by 10 mM taurine. Isethionic acid (ISE), aminopropanesulphonic acid (APS) and aminoethylphosphonic acid (AEP) had no statistically significant effects on uptake whereas aminoethylarsonic acid (AEA) inhibited this process significantly, lowering the Ca^{2+} uptake rate by over 10% as compared to the control values.

The effects of taurine and its analogues on mitochondrial oxidative phosphorylation were also tested by determining the rate of respiration under coupled, uncoupled and phosphorylating conditions. The effects on coupled respiration in the absence of added ADP are shown in Figure 4b. A value greater than that of control indicates that the compound has stimulated the coupled respiration, suggesting it to be an uncoupler. By this criterion, taurine can be seen to have a substantial uncoupling effect whereas any stimulatory effects of ISE, APS and DMT were small and not statistically significant. In contrast, AEA inhibited coupled respiration, reducing it by over 29% as compared to the control values. The effects of these compounds on uncoupled respiration are shown in Figure 4c. AEA and to a lesser extent DMT were both inhibitors of uncoupled respiration whereas taurine, ISE and APS had no

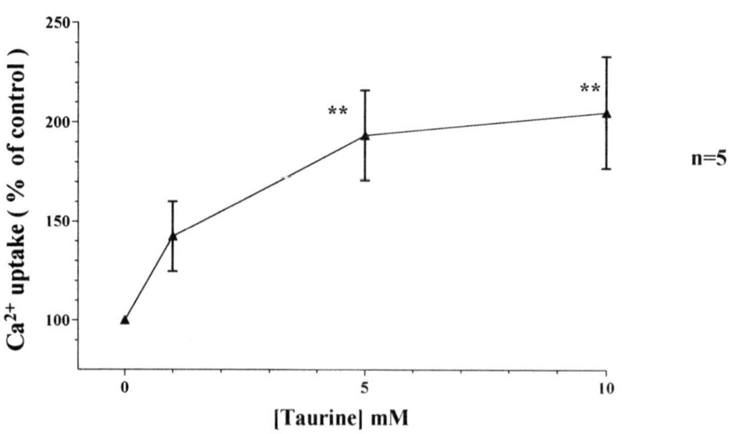

Figure 2. Effects of different doses of taurine on Ca^{2+} transport (nmol/min/mg protein) in rat liver mithocondria, recorded by differential (710-790 nm) absorbance changes of antipyrylazo III. Values represent the mean ± SEM of duplicate determinations from 3-5 separate experiments. Asterisks show statistical significance by Student's t test of the differences between experiments and controls.

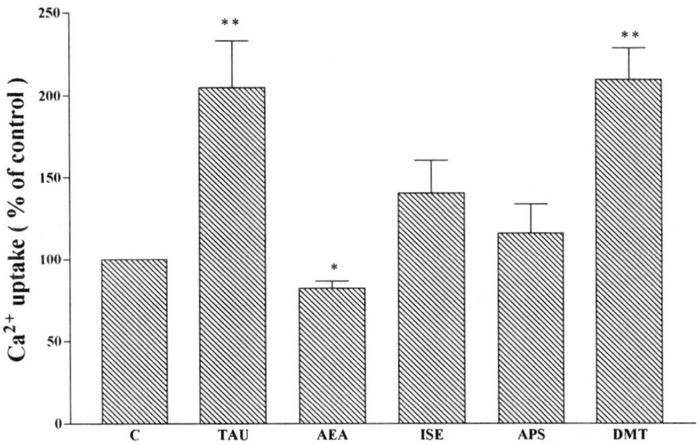

Figure 3. Effects of a single dose (10 mM) of taurine (TAU), aminoethylarsonic acid (AEA), isethionic acid (ISE), aminopropanesulphonic acid (APS) and *N,N*-dimethyltaurine (DMT) on Ca^{2+} transport (nmol/min/mg protein) in rat liver mitochondria. For details see Figure 2.

statistically significant effects. AEA and DMT inhibited respiration by about 89% and 40%, respectively.

The effects of these compounds on respiration in phosphorylating mitochondria are shown in Figure 4a. In this case the respiration rate is controlled jointly by the electron transport chain and by the phosphorylating system. Therefore, a decrease in respiration rate in this condition might result from an effect on one or both of these systems. Taurine stimulated the respiration rate, increasing it 14% above the control value, whereas AEA inhibited State 3 respiration by 45%. None of the other compounds tested had any significant effect.

DISCUSSION

Taurine stimulated Ca^{2+} uptake in a dose-dependent and saturable process that tended to a value of about twice that determined in its absence at 10 mM taurine. Among the taurine analogues tested, 10 mM DMT also stimulated Ca^{2+} uptake to an extent that was not significantly different from that given by the same concentration of taurine. However, APS and ISE had either no or only a very small effect, respectively. Substitution of the sulfonic acid group of taurine with an arsonic acid group transforms taurine into the 2-aminoethylarsonic acid (AEA), a compound that caused significant inhibition of Ca^{2+} uptake. Although the mechanism of this inhibition remains to be elucidated, our preliminary data indicate that AEA competitively antagonizes taurine-induced potentiation of Ca^{2+} transport.

In addition to the effects on Ca^{2+} uptake, these compounds also had different effects on the rate of mitochondrial respiration. Taurine activated the coupled respiration of the mitochondria both in the absence and presence of ADP (States 4 and 3 respiration, respectively) but had no effect on the rate of uncoupled respiration. AEA inhibited the respiration rate under all the three conditions whereas ISE and APS have no significant effects.

These results indicate that effects of taurine and its analogues on Ca^{2+} transport and respiration rate in mitochondria parallel one another and thus are consistent with the hypothesis that these two functions are intimately inter-dependent under physiological conditions[8]. Coupled respiration is largely controlled by the proton leak through the inner

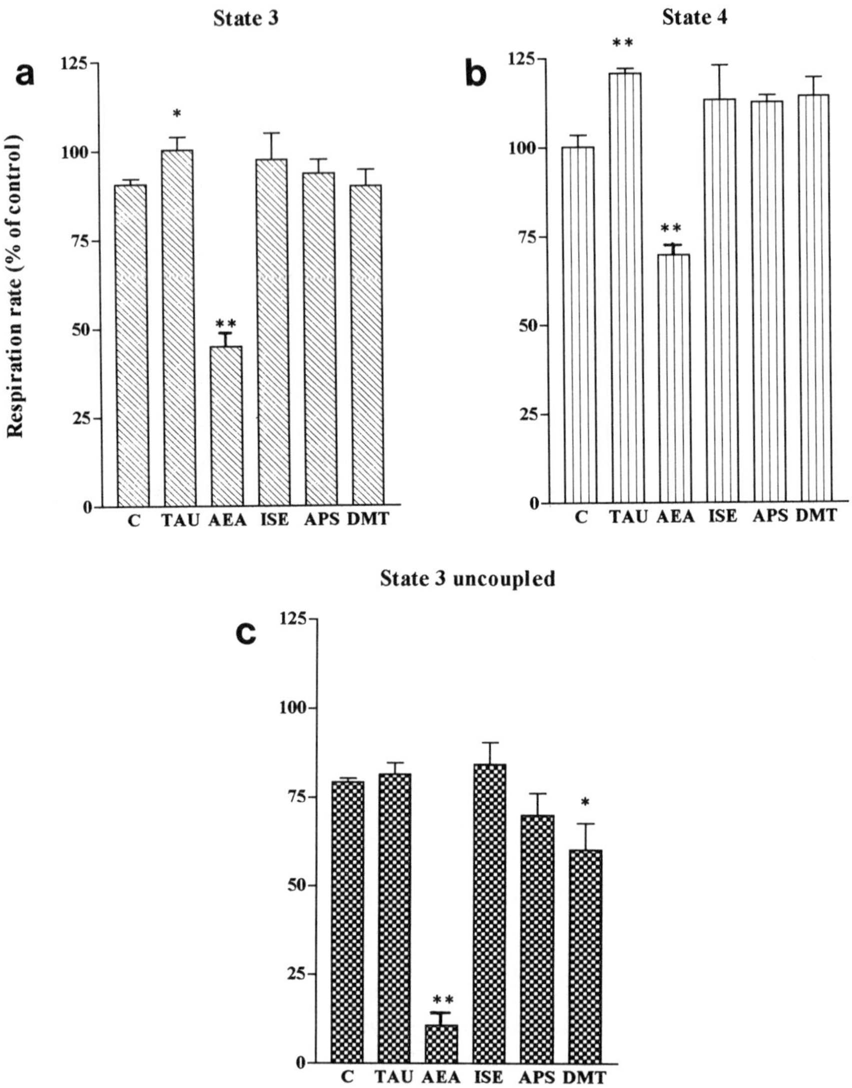

Figure 4. Effects of a single dose (10 mM) of taurine (TAU) aminoethylarsonic acid (AEA), isethionic acid (ISE), aminopropanesulphonic acid (APS) and *N,N*-dimethyltaurine (DMT) on mithocondrial oxidative phosphorylation in rat liver mitochondria in the presence of 8 μM Ca^{2+}. The respiration rate (nmol/min/mg of mitochondrial protein) was measured in conditions of coupled respiration (state 4), respiration in phosphorylating mitochondria (state 3) and uncoupled respiration (state 3 uncoupled). Values (means ± SEM of duplicate determinations from 3-5 separate experiments) represent the percentage between controls and treatments. 100% values are controls in Ca^{2+}-free medium. Student's t test was used for statistical significance between controls and treatments.

membrane of the mitochondrion and therefore compounds which increase this leak will stimulate the coupled respiration by acting as uncouplers. In addition to perturbation of membrane permeability or to a mechanism involving increased shuttling of protons across the membrane[17], an enhancement of proton leak may result from a modification of Ca^{2+} cycling[1,6,9,10]. Such a process can lead to partial dissipation of the electrochemical H^+ gradient and, in turn, to stimulation of the rate of outward H^+ pumping by the electron transport chain in order to re-establish the electrochemical gradient. Inhibition of the Ca^{2+} cycling system has, in fact, been shown to inhibit respiration in the state 4 condition, indicating that Ca^{2+} itself behaves as an uncoupler and that some part of coupled respiration is used to drive Ca^{2+} cycling. Thus the observed effects of taurine on coupled respiration may be attributed to its effect on Ca^{2+} cycling, since the data presented here shows that this amino acid does, indeed, stimulate the Ca^{2+} uptake process. Further support to this hypothesis comes from our preliminary determinations (data not shown) of mitochondrial respiration in Ca^{2+}-free medium, obtained by the addition of excess of EGTA. Under these conditions taurine failed to stimulate respiration rate in any of the three conditions (states 3 and 4 and state 3 uncoupled). The observation that, in the presence of Ca^{2+}, taurine stimulated state 3 but not state 3 uncoupled respiration might reflect a difference in rate-determining steps for these two processes. It is also possible that, in addition to the reported Ca^{2+}-mediated effects reported here, taurine also possesses a Ca^{2+}-independent action, possibly involving one or both the components of the phosphorylating system (adenine nucleotide translocase and the F_1F_o-ATP synthase).

DMT caused a modest degree of inhibition of the FCCP-uncoupled respiration rate, suggesting that this compound might have a direct inhibitory effect on the respiratory chain itself. Such an effect might explain why, in contrast to the behaviour of taurine, DMT activated Ca^{2+} transport but not coupled (state 3) respiration. Since the energy source for enhancing H^+ pumping in leaking mitochondria is provided by the electron transport chain, it would be reasonable to suppose that the proton leak across the inner mitochondrial membrane induced by DMT does not result in any significant stimulation of coupled respiration as the capacity of this process to increase in rate is limited by the mild inhibitory action of this compound. Further studies will be necessary to determine the site at which DMT exerts it inhibitory action. Since taurine itself does not directly affect the electron transport chain to any significant extent, the stimulation of Ca^{2+} transport is reflected in a stimulation of states 3 and 4 respiration.

In conclusion, the present data indicate that: (i) taurine may have an unique role in modulating mitochondrial Ca^{2+} homeostasis; (ii) the process whereby taurine influences Ca^{2+} transport is concentration-dependent and saturable and appears to be quite specific since some of the other analogues studied were without effect; and (iii) the effects of taurine on mitochondrial Ca^{2+} transport also account for its stimulatory effects on coupled respiration. Overall these data suggest a specific interaction between taurine and an, as yet unidentified site in mitochondrial membranes. Further studies on the effects of taurine on Ca^{2+} efflux from mitochondria are in progress.

ACKNOWLEDGMENTS

This work was partially supported by a grant from the European Community (BIOMED I, BHMT1 CT-94-1402). The authors thank B. Gorelli and A. Benocci for technical assistance.

REFERENCES

1. Brand, M.D. and Murphy M.P. 1987, Control of electron flux through the respiratory chain in mitochondria and cells. *Biol. Rev.* 62: 141-193.
2. Denton, R.M. and McCormack, J.G. 1985, Ca^{2+} transport by mammalian mitochondria and its role in hormone action. *Am. J. Physiol.* 249 (*Endocrinol. Metab.* 12): E 543-E 554.

3. Chappell J.B. and Hansford, R.G. 1972, Preparation of mitochondria from animal tissues and yeast, in: "Subcellular Components: Preparation and Fractionation", 2nd Ed., Birmie G.D. ed., London: Butterworth, pp. 77-91 (1972).
4. Fusi, F. Valoti, M., Sgaragli G.P. and Murphy, M.P. 1991, The interactions of antioxidants and structurally related compounds with mitochondrial oxidative phosphorylation. *Meth. Exp. Pharmacol.* 13: 599-603.
5. Gornall, A.G., Bardawill C.J. and David, M.M. 1949, Determination of serum protein by means of the biuret reaction. *J. Biol. Chem.* 177: 751-766.
6. Gunter, T.E., Gunter, K.K., Shey-Shing, S. and Gavin, C.E. 1994, Mitochondrial calcium transport: physiological and pathological relevance. *Am. J. Physiol.* 267 (*Cell Physiol* 36): C313-C339.
7. Gunter T.E. and Pfeiffer, D.R. 1990, Mechanism by which mitochondria transport calcium. *Am J. Physiol.*, 258 (*Cell Physiol.* 27): C755-C786.
8. Hansford, R.G. 1985, Relation between mitochondrial calcium transport and control of energy metabolism. *Rev. Physiol. Biochem. Pharmacol.* 102: 1-72.
9. Hansford, R.G. 1991, Dehydrogenase activation by Ca^{2+} in cells and tissues. *J. Bioenerg. Biomembr.* 23: 823-854.
10. Hansford, R.G. 1994, Physiological role of mitochondrial Ca^{2+} transport. *J. Bioenerg. Biomembr. 26: 495-508.*
11. Huxtable, R.J. 1987, From heart to hypothesis: a mechanism for the calcium modulatory actions of taurine, *in*: "The Biology of Taurine: Methods and Mechanisms", R.J. Huxtable, F. Franconi and A. Giotti eds., New York: Plenum Press, pp. 371-388.
12. Izumi, K., Butterworth R.F. and Barbeau, A. 1977, Effect of taurine on calcium binding to microsomes isolated from rat cerebral cortex. *Life Sci.* 20: 943-950.
13. Kuo, C.H. and Miki, N. 1980, Stimulatory effect of taurine on Ca^{2+} uptake by disc membranes from photoreceptors cell outer segments *Biochem. Biophys. Res. Commun.* 94: 646-651.
14. Kuriyama, K., Muratmatsu, M., Nakagawa K. and Kakita, K. 1978, Modulating role of taurine on release of neurotransmitters and calcium transport in excitable tissues, in: "Taurine and Neurological Disorders", A. Barbeau and R.J Huxtable, eds, New York: Raven Press, pp. 201-216.
15. Lombardini, J.B. 1988, Effects of taurine and mitochondrial metabolic inhibitors on ATP-dependent Ca^{2+} uptake in synaptosomal and mitochondrial subcellular fractions of rat retina. *J. Neurochem.* 51: 200-205.
16. McCormack, J.G., Halestrap A.P. and Denton, R.M. 1990, Role of calcium ions in the regulation of mammalian intramitochondrial metabolism. *Physiol. Rev.* 70: 391-425.
17. Murphy, M.P. 1989, Slip and leak in mitochondrial oxidative phosphorylation. *Biochem. Biophys. Acta*, 977: 123-141.
18. Palade, P. 1986, Drug-induced Ca^{2+} release from isolated sarcoplasmic reticulum by use of pyrophosphate to study caffeine-induced Ca^{2+} release. *J. Biol. Chem.*, 282: 6135-6141.
19. Palade, P. and Vergara, J. 1982, Arsenazo III and Antipyrylazo III Calcium transients in single skeletal muscle fibers. *J. Gen. Physiol.* 79: 679-707.
20. Pasantes-Morales, H. 1982, Taurine-calcium interactions in frog rod outer segments: taurine effects on an ATP-dependent calcium translocation process. *Vision Res.* 22: 1487-1493.
21. Reynafarje, B., Costa L.E. and Lehninger, A.L. 1985, O_2 solubility in aqueous media determined by a kinetic method. *Anal. Biochem.* 145: 406-418.
22. Sebring, L. and Huxtable, R.J. 1985, Taurine modulation of calcium binding to cardiac sarcolemma. *J. Pharmacol. Exp. Ther.* 232: 445-451.

15

CHANGES IN TAURINE AS AN INDICATOR OF HEPATIC DYSFUNCTION AND BIOCHEMICAL PERTURBATIONS

Studies *in Vivo* and *in Vitro*

John A. Timbrell and Catherine J. Waterfield

Toxicology Department
School of Pharmacy
Brunswick Square
London WC1N 1AX, United Kingdom

ABSTRACT

We have shown that urinary taurine level may be used as a biomarker of pathological and biochemical lesions. Detection of changes in the urinary concentration of this low molecular weight metabolite indicates biochemical lesions which may also be associated with pathological damage.

Hepatotoxic compounds such as CCl_4, galactosamine and thioacetamide that cause hepatic necrosis and compounds such as hydrazine and ethionine that cause fatty liver all result in elevated urinary taurine levels in rats. However compounds which do not cause liver damage, such as cycloheximide, also raise urinary taurine levels. All of these substances are known to or are believed to inhibit protein synthesis. Conversely, compounds which increase protein synthesis, such as phenobarbital and clenbuterol, significantly decrease urinary taurine levels. Compounds which interfere with hepatic GSH synthesis will also change urinary taurine levels. Thus, depletion of GSH with diethyl maleate or phorone decreases urinary taurine whereas inhibition of GSH synthesis with compounds such as buthionine sulphoximine increases urinary taurine levels.

In isolated hepatocytes *in vitro*, leakage of taurine occurs in response to cytotoxic compounds such as hydrazine and allyl alcohol. However, total taurine levels were increased by the hepatotoxicant CCl_4. Taurine synthesis is decreased by depletion of GSH with allyl alcohol in isolated hepatocytes.

Therefore taurine levels are an important potential biomarker for biochemical lesions induced by chemicals both *in vivo* and *in vitro*, in particular changes in protein and GSH synthesis.

Taurine 2, edited by Huxtable et al.
Plenum Press, New York, 1996

INTRODUCTION

Detection of damage to organs such as the liver normally involves techniques such as histopathology or serum analysis[16] which necessitate invasive procedures. Therefore urinary markers for such pathological damage that are noninvasive are of particular interest. Furthermore, some types of biochemical lesion such as fatty liver are difficult to detect by serum measurements[8]. Initial studies using proton NMR for analysis of urine revealed that taurine was raised in a dose-dependent manner by the hepatotoxicant hydrazine and other hepatotoxicants[7]. Previous studies have shown that urinary taurine may be influenced by exposure to xenobiotics[6,17].

We have explored these original findings in rats extensively and confirmed that changes in urinary taurine occur after exposure to some xenobiotics, especially compounds which are toxic to the liver or cause perturbations in protein and GSH synthesis[8,10,11,12]. This low molecular weight metabolite is readily measured in urine at the concentrations found.

Evaluation of changes in urinary taurine can allow a diagnosis of a variety of pathological and biochemical lesions. Indeed this marker often correlates well with other markers of damage[10,11].

The aims of the studies described were: (i) to determine the utility of urinary taurine as a noninvasive biomarker of organ dysfunction; and (ii) to compare with other biomarkers.

This paper represents an amalgamation of our data, some of which have been previously published[2,8,10,11,12].

METHODS

In vivo studies were carried out in male Sprague Dawley rats (150-250 g, supplied by Glaxo Group Research, Ware, U.K.) and housed in individual metabolism cages under controlled lighting and temperature conditions (21 ± 2°C, 12 h light, 12 h dark). Urine was collected over ice both before and after dosing for at least 48 h. Animals were dosed either orally or i.p. with various compounds. Controls received vehicle by the same route. At the end of the studies animals were anesthetized (diethyl ether), blood was removed from the abdominal aorta for preparation of serum, and liver and other tissues were removed for histopathology and biochemical measurements. Serum was collected for clinical chemical analysis (ALT, AST, ALP, bilirubin, urea, creatinine, total protein, albumin, triglycerides). Standard clinical chemical parameters in serum were measured in an automatic analyzer (BM Hitachi System 795). Fatty liver was detected histologically and total nonprotein sulphydryl levels in liver (predominantly reduced GSH) were determined as previously described[12].

For *in vitro* studies again male Sprague-Dawley rats (180-250 g) were used. Hepatocytes were prepared from liver by a recirculating two-step perfusion method, as previously described[13].

Hepatocytes were used as suspensions in Krebs-Henseleit buffer in rotating flasks gassed with CO_2/O_2 (95%:5%). Taurine precursors or toxic compounds were placed in the incubation buffer 1 h after preincubation.

Taurine in urine, liver tissue, hepatocytes or incubation buffer was measured by HPLC as previously described[14].

Lactate dehydrogenase was measured by the method of Bergemeyer *et al.*[1]

Reduced GSH was measured by the method of Hissin and Hilf[5] *in vitro* or the method of Ellman[4] *in vivo*.

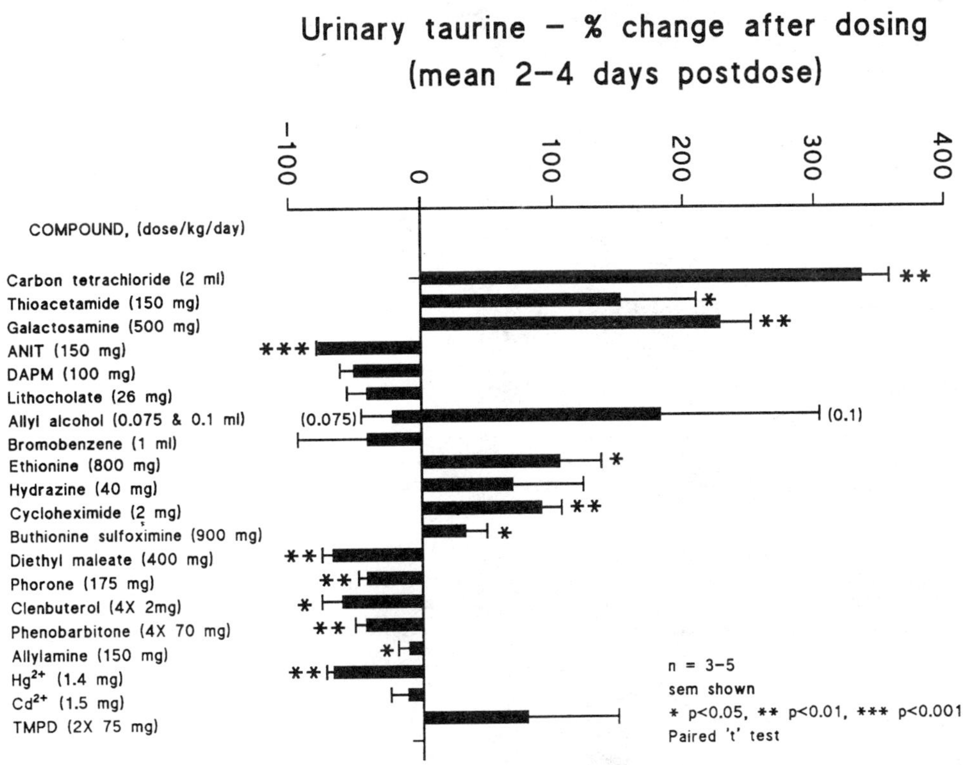

Figure 1. Percentage change in urinary taurine levels caused by various compounds. Results are means ± SEM, numbers of animals 3-8. $^*p < 0.05$; $^{**}p < 0.01$; $^{***}p < 0.001$. ANIT: α-naphthylisothiocyanate; DAPM: diaminodiphenyl methane; TMPD: 2,3,5,6-tetramethylphenylene diamine.

RESULTS

Various changes in urinary taurine were detected after exposure to different toxicants and xenobiotics (Fig. 1). These were compared with various other indicators of liver damage where appropriate (Tables 1a-1d).

Table 1a. Biochemical markers for hepatic damage

Treatment	Lesion	Urinary taurine[a] % Change	Serum ALT iu/L	Serum AST iu/L	Serum total protein g/L	Serum total bilirubin μmol/L	Serum albumin g/L
Control (Range)	–	–	35-89	73-99	55-59	1-11	33
CCl_4 (2 ml/kg)	Necrosis	↑337**(2)	344	1219*↑	57	9	35
Galactosamine (500 mg/kg)	Necrosis	↑229**(2)	4295*↑	6479*↑	48*↓	134*↑	31
Thioacetamide (150 mg/kg)	Necrosis	↑153*(2)	442*↑	1735*↑	49*↓	9*↑	28*↓

[a]Taurine values are mean % change per 24 h averaged over (n) days after dosing. $^*p < 0.05$; $^{**}p < 0.01$; significantly different from control by t-test. Data from Waterfield *et al.*[11]

Table 1b. Biochemical markers for compounds which affect GSH levels

Treatment	Lesion	Urinary Taurine % Change	Serum ALT iu/L	Serum AST iu/L	Serum Total Protein g/L	Serum Total Bilirubin μmol/L	Serum Albumin g/L
Diethyl maleate (400 mg/kg)	Depletes GSH	↓-69**(2)	67	104	57	4	34
Phorone (175 mg/kg)	Depletes GSH	↓-62**(2)	70	85	59	4	34
Allyl alcohol (0.075 ml/kg)	Periportal necrosis	↓-19(2)	100	138	57	4	32
Bromobenzene (1 ml/kg)	Centrilobular necrosis	↓-42(2)	173*↑	1185*↑	51*↑	10	31
Buthionine Sulphoximine (900 mg/kg)	Blocks GSH synthesis	↑33*(2)	67	85	55	5	33

[a]Taurine values are mean % change per 24 h averaged over (n) days after dosing. *p < 0.05; **p < 0.01; significantly different from control by t-test. Data from Waterfield *et al.*[11,12]

Table 1c. Biochemical markers for compounds which affect the biliary system

nTreatment	Lesion	Urinary Taurine % Change	Serum ALT iu/L	Serum AST iu/L	Serum Total Protein g/L	Serum Total Bilirubin μmol/L	Serum Albumin g/L
ANIT (150 mg/kg)	Bile duct damage	↓-78***(2)	723*↑	1501*↑	63*↑	177*↑	32
DAPM (100 mg/kg)	Bile duct damage	↓-52(2)	133*↑	197*↑	n.d.	51*↑	n.d
Lithocholate (26 mg/kg)	Cholestasis	↓-42(2)	32*↓	64	57	2	32

[a]Taurine values are mean % change per 24 h averaged over (n) days after dosing. *p < 0.05; **p < 0.001, significantly different from control by t-test. Data from Waterfield *et al.*[11] and Seabra and Timbrell, unpublished observations. ANIT: α-naphthylisothiocyanate; DAPM: diaminophenylmethane.

Table 1d. Biochemical markers for compounds which affect protein synthesis

Treatment	Lesion	Urinary Taurine % Change	Serum ALT iu/L	Serum AST iu/L	Serum Total Protein g/L	Serum Total Bilirubin μmol/L	Serum Albumin g/L
Hydrazine (40 mg/kg)	Fatty Liver	↑69(2)	24*↓	86	56	1	32
Cycloheximide (2 mg/kg)	—	↑-91**(2)	49*↓	90	46*↓	10	28*↓
Phenobarbital (4x70 mg/kg)	—	↓-44**(4)	33	72	58	1	33
Clenbuterol (4x 2 mg/kg)	CK-MM raised (63*v19 iu/L)	↓-62*(4)	64	108	52	6	29
Ethionine (800 mg/kg)	Fatty Liver	↑-105*(3)	43	46	54	7	33

[a]Taurine values are mean % change per 24 h averaged over (n) days after dosing. *p < 0.05; **p < 0.01; significantly different from control by t-test. Data from Waterfield *et al.*[11,12,15].

The changes in urinary taurine were found to reflect the type of damage induced by various xenobiotics and may indicate the biochemical perturbations underlying them (Fig. 1; Tables 1a-1d). Thus the hepatotoxicants which caused hepatic necrosis, as indicated by histology and other markers in serum (CCl_4, thioacetamide, galactosamine) increased taurine (Table 1a), whereas the hepatotoxicants which caused necrosis but also depleted GSH, (allyl alcohol, bromobenzene), resulted in reduced urinary taurine (Table 1b). Bile duct toxicants, as indicated by biochemical changes in serum (α-naphthyl-isothiocyanate, diphenylaminomethane, lithocholate) also decreased urinary taurine (Table 1c). The compounds which deplete GSH without causing liver damage (diethyl maleate, phorone) also decreased urinary taurine (Table 1b). In contrast, inhibition of GSH synthesis with buthionine sulphoximine (BSO) increased urinary taurine (Table 1b). These changes in GSH level do not cause changes in the other markers measured (Table 1b). The compounds which cause fatty liver, (ethionine, hydrazine), or inhibit protein synthesis (cycloheximide), also raised urinary

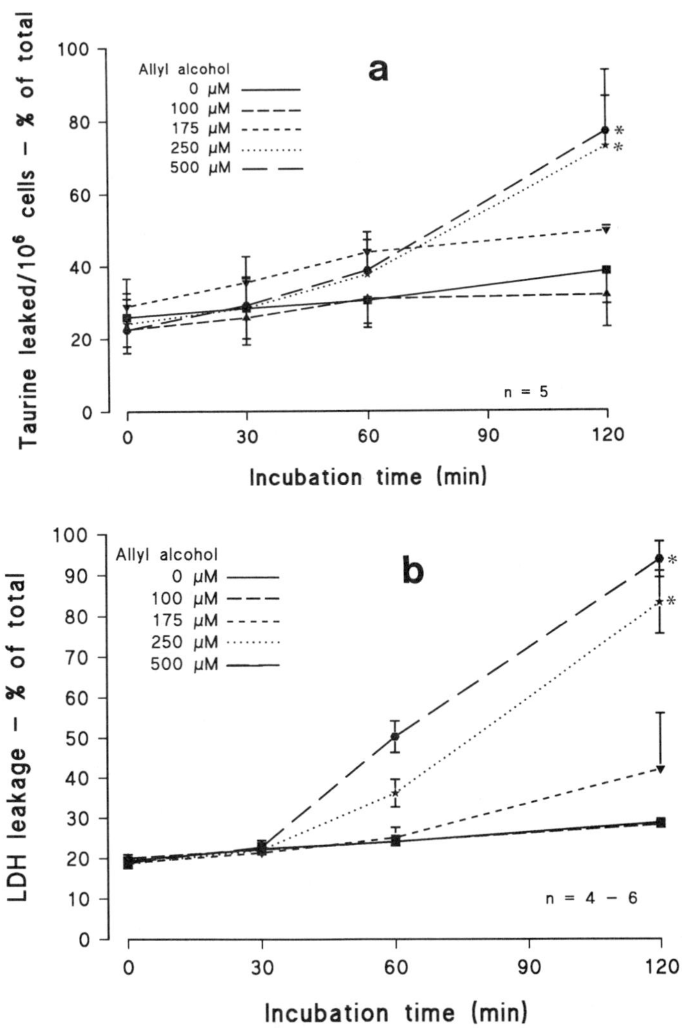

Figure 2. Leakage of (a) taurine and (b) LDH from rat hepatocytes exposed to various concentrations of allyl alcohol. Results are means ± SEM. n=4-6. *$p < 0.05$; **$p < 0.01$.

Table 2. Biochemical markers for miscellaneous toxic compounds

Treatment	Lesion	Urinary taurine[a] % change	Serum ALT iu/L	Serum AST iu/L	Urinary creatine μmol/kg/ day[b]	Serum testosterone pg/ml	Serum LDHC4 iu/L	Serum iu/L	Serum Albumin g/L
Control	—	—	35-89	73-99	5-10	750	19	265-383	33
TMPD (2x75 mg/kg)	Muscle necrosis; Testicular damage	80(2)	1117*	7174*	600*↑	77	43*	1255*	26*
Allyl amine (150 mg/kg)	Cardiac necrosis	↓-12*(2)	44*	132	20	n.d.	n.d.	461	35
Mercuric chloride (1.4 mg/kg)	Proximal tubule necrosis	↓69**(2)	51*	164*	81*↑	n.d.	n.d.	n.d.	29

[a]Taurine values are mean % change per 24 h averaged over (n) days after dosing. [b]Creatine values are total excretion 24-48 h postdosing. *$p < 0.05$; **$p < 0.01$; significantly different from control by t-test. TMPD: 1,2,5,6-tetramethylphenylenediamine. Data from Waterfield *et al.*[12,14].

taurine (Table 1d). However other markers in serum were either unchanged or only slightly decreased (Table 1d). Compounds which increase protein synthesis (phenobarbital, clenbuterol) decreased urinary taurine (Table 1d) but did not affect other markers in general. Damage to heart muscle by allylamine and renal tubular damage caused by $HgCl_2$ both significantly decreased urinary taurine (Fig. 1; Table 2) but testicular damage caused by $CdCl_2$ did not significantly affect urinary taurine (Fig. 1). Of particular interest is the small but nonsignificant rise in urinary taurine following extensive skeletal muscle damage caused by 2,3,5,6-tetramethyl *p*-phenylenediamine (TMPD) (Fig. 1; Table 2).

Overall, therefore, urinary taurine was changed in cases when other biomarkers were unaltered despite the presence of a biochemical or pathological lesion.

Damage to isolated hepatocytes *in vitro* leads to the release of taurine into the surrounding medium. This occurred with the three compounds studied; allyl alcohol, CCl_4

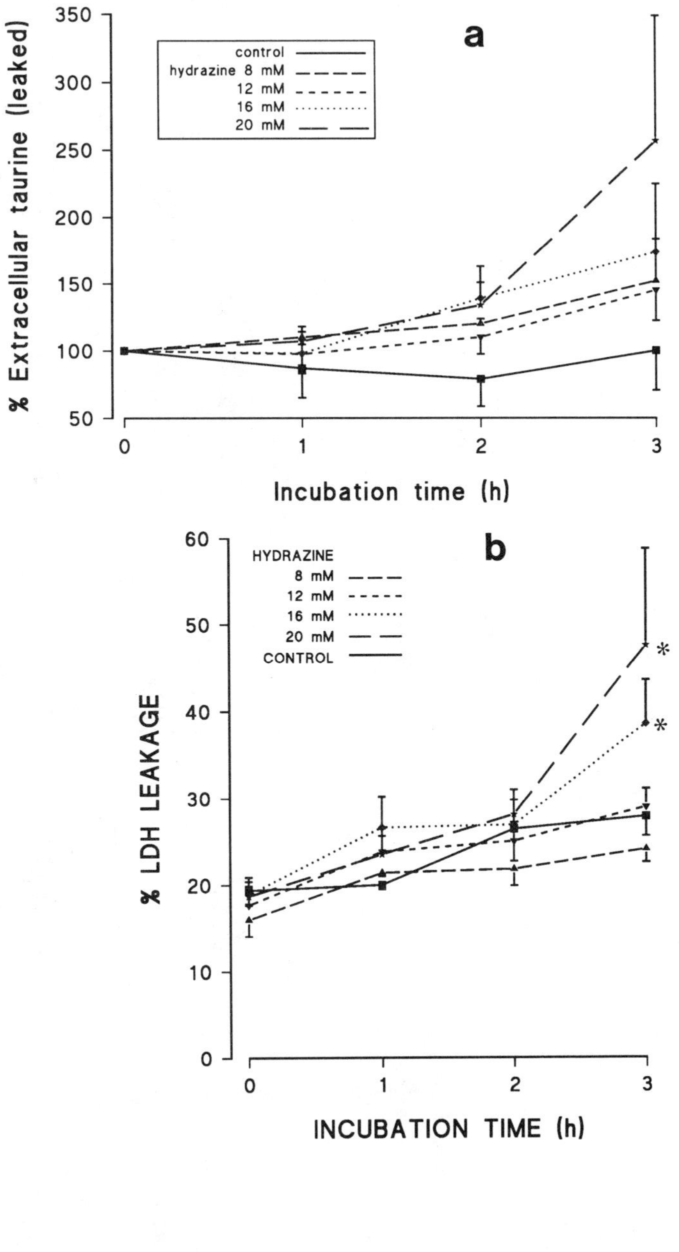

Figure 3. Leakage of (a) taurine and (b) LDH from rat hepatocytes exposed to various concentrations of hydrazine. Results are means ± SEM. n=4-6. *$p < 0.05$; **$p < 0.01$.

and hydrazine. In the case of allyl alcohol and hydrazine, the leakage was similar to leakage of LDH (Figs. 2,3). With CCl_4, however, there was a significant increase in taurine leakage without significant leakage of LDH (Fig. 4). Total taurine was also increased in CCl_4-treated cells, indicating that synthesis was increased (Fig. 4). This was not the case with hydrazine or allyl alcohol.

DISCUSSION

Our studies indicate that changes in urinary taurine, a noninvasive biomarker, may give an indication of organ dysfunction or metabolic perturbations. The explanations for the changes in urinary taurine observed are currently being investigated. However, our results suggest that simple leakage from damaged tissue is not the sole cause of the increased levels as changes in tissue levels do not account for the changes. Indeed if this were the case, TMPD which caused extensive skeletal muscle damage would have been expected to elevate urinary

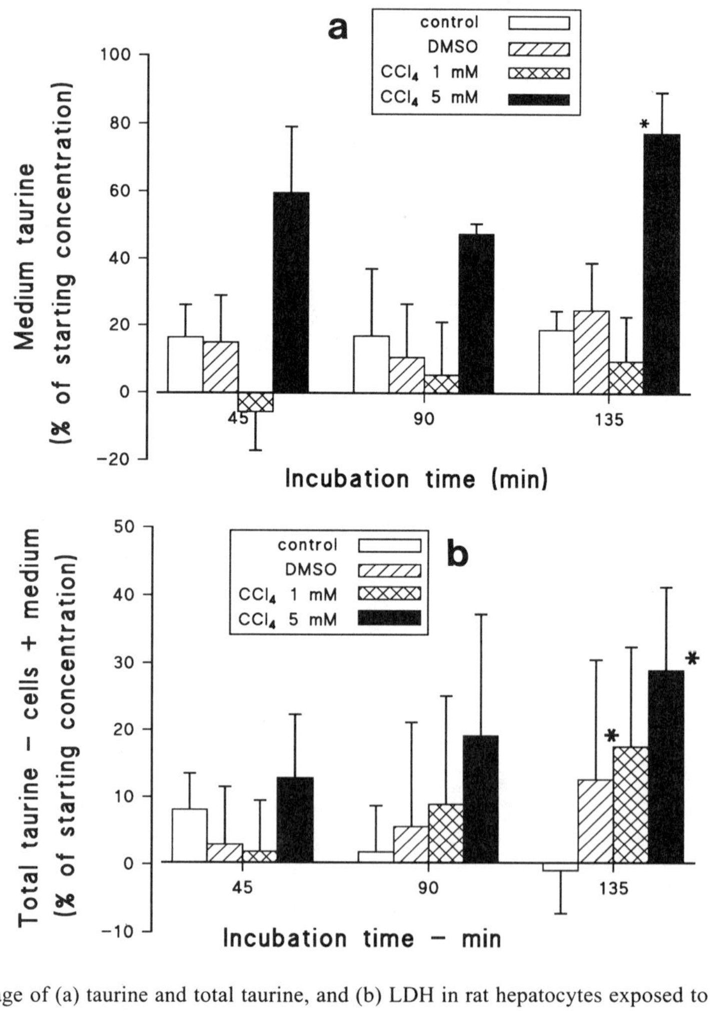

Figure 4. Leakage of (a) taurine and total taurine, and (b) LDH in rat hepatocytes exposed to CCl_4. Results are means ± SEM. n=4. *$p < 0.05$.

taurine significantly, but it did not (Fig. 1; Table 2). We believe the changes in taurine are due to changes in the precursor amino acid pool of cysteine in most cases. Thus, inhibition of protein synthesis by cycloheximide or compounds causing hepatic necrosis or fatty liver will increase cysteine levels and therefore taurine. Depletion of GSH is followed by increased synthesis of this tripeptide with a concomitant reduction in cysteine and therefore decreased taurine synthesis. Inhibition of GSH synthesis causes a decrease in the utilization of cysteine and hence urinary taurine is elevated. Interference with the biliary system resulting in cholestasis may reduce the enterohepatic cycling of taurine leading to reduced export of taurine and decreased urinary taurine but an increase in liver taurine, which was found in ANIT dosed animals and is consistent with this. Compounds which damage the heart or muscle, (allylamine, TMPD), tissues containing high levels of taurine but which do not synthesize it, do not raise urinary taurine. Damage to the kidney significantly reduces urinary taurine, perhaps by decreasing filtration.

Preliminary studies in industrial workers exposed to solvents have revealed significant increases in urinary taurine which was significantly correlated with other markers of liver dysfunction (3).

In conclusion, urinary taurine is a valuable noninvasive *in vivo* biomarker of liver dysfunction caused by chemicals. This biomarker may help in the diagnosis of a range of metabolic and pathological lesions without the need to take biopsies or blood samples. Furthermore with some of the toxicants studied serum markers were not altered, despite the presence of a biochemical or pathological lesion such as fatty liver and therefore dysfunction was not detected by the normal clinical chemical measurements.

Measurement of taurine leakage and concentration *in vitro* in isolated hepatocytes may also be a useful marker of cellular damage or biochemical dysfunction.

ACKNOWLEDGMENTS

The authors are grateful to Glaxo Research and Development for funding.

REFERENCES

1. Bergmeyer, M.U., Bernt, E. and Hess, B. 1965, Lactate dehydrogenase (II 2a) In: Methods of Enzymatic Analysis (Ed. Bergmeyer, H.) Academic Press, New York, pp 736-743.
2. Draper, R.P., Waterfield, C.J., York, M.J. and Timbrell, J.A. 1994, The effect of the muscle toxicant 2,3,5,6-Tetramethyl p-phenylenediamine (TMPD) on urinary taurine and creatine. *Arch. Toxicol.* 69:111-117.
3. Draper, R.P., Waterfield, C.J., Mason, H., Wheeler, J. and Timbrell, J.A. 1994, Study of the urinary excretion of taurine and creatine in workers exposed to solvents. *Human Exp. Toxicol.* 13:643.
4. Ellman, G.L. 1959, Tissue sulfhydryl groups. *Arch. Biochem. Biophys.* 82:70-77.
5. Hissin, P.J. and Hilf, R. 1976, A fluorometric method for determination of oxidized and reduced glutathione in tissues. *Anal. Biochem.* 74:214-226.
6. Kocsis, J.J., Harkway, S. and Snyder, R. 1975, Biological effects of the metabolites of dimethyl sulphoxide. *Ann. N.Y. Acad. Sci.* 243:104-109.
7. Sanins, S.M., Nicholson, J.K., Elcombe, C. and Timbrell, J.A. 1990, Hepatotoxin-induced hypertaurinuria: a proton NMR study. *Arch. Toxicol.* 64:401-411.
8. Timbrell, J.A., Draper, R.P. and Waterfield, C.J. 1994, Biomarkers in Toxicology. New uses for old molecules. *Toxicology and Ecotoxicology News* 1:1-11.
9. Waterfield, C.J., Turton, J.A., Scales, M.D.C. and Timbrell, J.A. 1990, Taurine synthesis in isolated rat hepatocytes in suspension exposed to carbon tetrachloride. *Biochem. Soc. Trans.* 18:1218-1219.
10. Waterfield, C.J., Turton, J.A., Scales, D.C., and Timbrell, J.A. 1991, Taurine, a possible urinary marker of liver damage: a study of taurine excretion in carbon tetrachloride treated rats. *Arch. Toxicol.* 65:548-555.

11. Waterfield, C.J., Turton, J.A., Scales, D.C. and Timbrell, J.A. 1993, Investigations into the effects of various hepatotoxic compounds on urinary and liver taurine levels in rats. *Arch. Toxicol.* 67:244-254.
12. Waterfield, C.J., Turton, J.A., Scales, D.C. and Timbrell, J.A. 1993, Effect of various non-hepatotoxic compounds on urinary and liver taurine levels in rats. *Arch. Toxicol.* 67:538-546.
13. Waterfield, C.J., Mesquita, M., Parnham, P. and Timbrell, J.A. 1993, Taurine protects against the cytotoxicity of hydrazine, 1,4-naphthoquinone and carbon tetrachloride in isolated rat hepatocytes. *Biochem. Pharmacol.* 46:589-595.
14. Waterfield, C.J. 1994, Determination of taurine in biological samples and isolated hepatocytes by high-performance liquid chromatography with fluorimetric detection. *J. Chromatog.* 657:37-45.
15. Waterfield, C.J., Jairath, M., Asker, D. and Timbrell, J.A. 1995, The biochemical effects of clenbuterol: With particular reference to taurine and muscle damage. *Europ. J. Pharmacol.* 293:141-149.
16. Woodman, D.D. 1988, Assessment of hepatic function and damage in animal species. *J. Appl. Toxicol.* 8(4), 249-254.
17. Yoshida, M. and Hara, I. 1985, Composition of urinary metabolites and variation of urinary taurine levels in rats injected with chlorobenzene. *Ind. Health* 23:239-243.

16

EFFECT OF THE PYRROLIZIDINE ALKALOID MONOCROTALINE ON TAURINE AND SULFUR AMINO ACID METABOLISM IN THE RAT LIVER

Chong Chao Yan and Ryan J. Huxtable

Department of Pharmacology
College of Medicine
University of Arizona
Tucson, Arizona 85724

INTRODUCTION

Pyrrolizidine alkaloids (PAs) comprise a large group of plant toxins, characteristically producing veno-occlusive disease of the liver as a result of hepatic bioactivation to pyrrolic dehydroalkaloids[4,5] (Fig. 1). These pyrrolic metabolites are highly reactive alkylating agents, with aqueous half-lives of only a few seconds[1]. Dehydroalkaloids are detoxified in the liver by conjugation with glutathione (GSH) to yield 7-glutathionyl-6,7-dihydro-1-hydroxymethyl-5H-pyrrolizine (GSDHP; Fig. 1). Following exposure *in vivo* to a PA, or perfusion of an isolated liver *in vitro*, high levels of GSDHP are released into the bile[11,13].

We have reported that metabolism and toxicity of PAs are markedly influenced by sulfur amino acid metabolites such as GSH or taurine[13,18]. Using monocrotaline (Fig. 1) as a prototypical PA, we now report that these alkaloids, in turn, markedly influence sulfur amino acid metabolism.

METHODS

Monocrotaline was isolated from the seeds of *Crotalaria spectabilis*, purified by counter-current distribution and identified by mass spectrometry. Bovine serum albumin (fraction V) was purchased from Sigma Chemical Co. (St. Louis, MO). Ehrlich reagent (0.5 g p-dimethylaminobenzaldehyde in 25 ml absolute ethanol acidified with 0.35 ml 37% perchloric acid) was prepared daily. Sprague-Dawley rats (body weight 240-270 g) were obtained from the University of Arizona Division of Animal Resources and were housed 2-3

Taurine 2, edited by Huxtable et al.
Plenum Press, New York, 1996

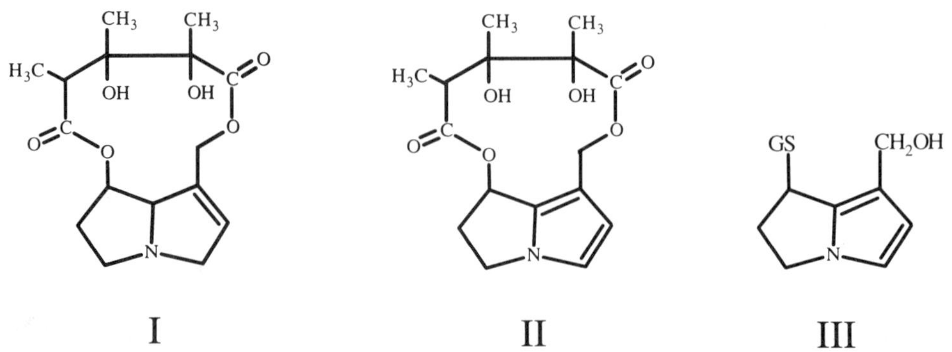

Figure 1. Structure of **I:** the pyrrolizidine alkaloid, monocrotaline; **II:** its toxic hepatic metabolite, dehydromonocrotaline; and **III:** GSDHP: the less toxic secondary metabolite formed by conjugation of GSH with dehydromonocrotaline (GS = Glutathionyl).

to a cage in a room kept on 12 hr cycles of light and dark. Rats were allowed food and tap water *ad libitum*, and were allowed to adapt for at least 3 days before being used.

Liver Perfusion

Rat liver was surgically removed, the bile duct cannulated and the liver then perfused as described by Yan and Huxtable[12]. Monocrotaline (0.5 mM final concentration) was dissolved in the buffer reservoir at the beginning of the experiment. The duration of perfusion was 60 min. At 15 min intervals, bile was collected for analysis. To prevent the oxidation of biliary GSH, the bile was collected in a cold microfuge placed on ice. Bile volume was determined gravimetrically, assuming a density of 1.0 gm/ml.

In Vivo Administration of Monocrotaline

Rats (6 animals per group) were injected i.p. either with 65 mg/kg monocrotaline or an equal volume of 0.9% saline. To avoid the effect of circadian variation in enzyme activities or metabolite levels, all rats were injected between 9:00 a.m. and 9:30 a.m. After 24 h, rats were anesthetized with sodium pentobarbital (50 mg/kg). Livers were removed, and cytosol and microsomes isolated for enzyme measurements as previously described[18].

Pair-Feeding Study

Twelve Male Sprague-Dawley rats (body weight 68-87 g) were obtained from the University of Arizona Division of Animal Resources and housed 2 a cage in a room maintaining 12 h cycles of light and dark. After 3 days for adaptation, the animals were divided randomly into 2 groups. One group was given monocrotaline (i.p., 65 mg/kg body weight) and the other group 0.85% NaCl i.p. in the same volume. The food intake of the monocrotaline group was recorded daily for each cage thereafter. The control group was given the weight of food consumed by the monocrotaline group on the previous day. Both groups were allowed tap water ad libitum. After two weeks, the study was terminated. To avoid differences due to diurnal variation in enzyme activities, animals were killed by decapitation between 9:00-10:00 a.m. Liver fractions were prepared for enzyme determinations as previously described[18].

Ehrlich Analysis for GSDHP in Bile

One ml Ehrlich reagent was added to 10 μl bile and the mixture was heated at 55°C for 5 min. The resulting color was quantitated at 566 nm.

Determination of Amino Acids, Cysteine, Cysteinylglycine and GSH in Bile

Analyses were performed by HPLC as previously described[14].

GSH Synthetase

This enzyme was assayed by following the formation of GSH in incubations of liver cytosol[18]. The incubation medium consisted of 10 mM glycine, 5 mM γ-glutamylcysteine, 10 mM ATP, 0.1 M Tris HCl (pH 8.2), 0.15 M KCl, 2 mM EDTA and 20 mM $MgCl_2$ in a 1.5 ml microfuge tube. After preincubation at 37°C for 3 min, cytosolic protein (≈1 mg) was added to initiate the enzyme reaction. The total volume was 1 ml, and the incubation time was 20 min. The formation of GSH was detected as previously described[14].

Other Enzyme Measurements

Determination of other enzyme activities are described in Yan and Huxtable.[19]

Statistics

Statistical comparisons were performed using the Student unpaired 't' test.

RESULTS

Monocrotaline on the Isolated Perfused Liver

Over the course of a 1 h period, monocrotaline perfusion (0.5 mM) resulted in an increased release of methionine and GSH into the perfusate of the isolated rat liver, but a decreased release of taurine (Fig. 2). The release of leucine, lysine and alanine was also increased (Table 1).

Monocrotaline perfusion had an insignificant effect on bile flow[14]. However, monocrotaline produced a large increase in the biliary excretion of GSH and its precursors, cysteine and cysteinylglycine (Fig. 3). GSH release was almost 30-fold greater than release from control livers, the latter being around 1000 nmol/h/g liver. Biliary concentrations in excess of 20 mM were achieved following perfusion with monocrotaline. GSH was additionally lost from the liver in the form of the conjugate, GSDHP (Fig. 4). Approximately equal amounts were lost into the perfusate and into bile. The concentration in bile, however, was very much higher, reaching 3.54 mM at 60 min compared to 14.1 ± 1.1 μM in the perfusate. Release of taurine into bile was much lower and was unaffected by monocrotaline (monocrotaline-perfused livers 46.7 ± 9.7 nmol/h/g liver; control livers 44.6 ± 5.6 nmol/h/g liver). Significant increases also occurred in the biliary excretion of aspartate and glycine (data not shown).

No significant difference was seen in hepatic taurine levels at the termination of perfusion (monocrotaline-perfused 0.77 ± 0.11 μmol/g protein; control 0.74 ± 0.17 μmol/g

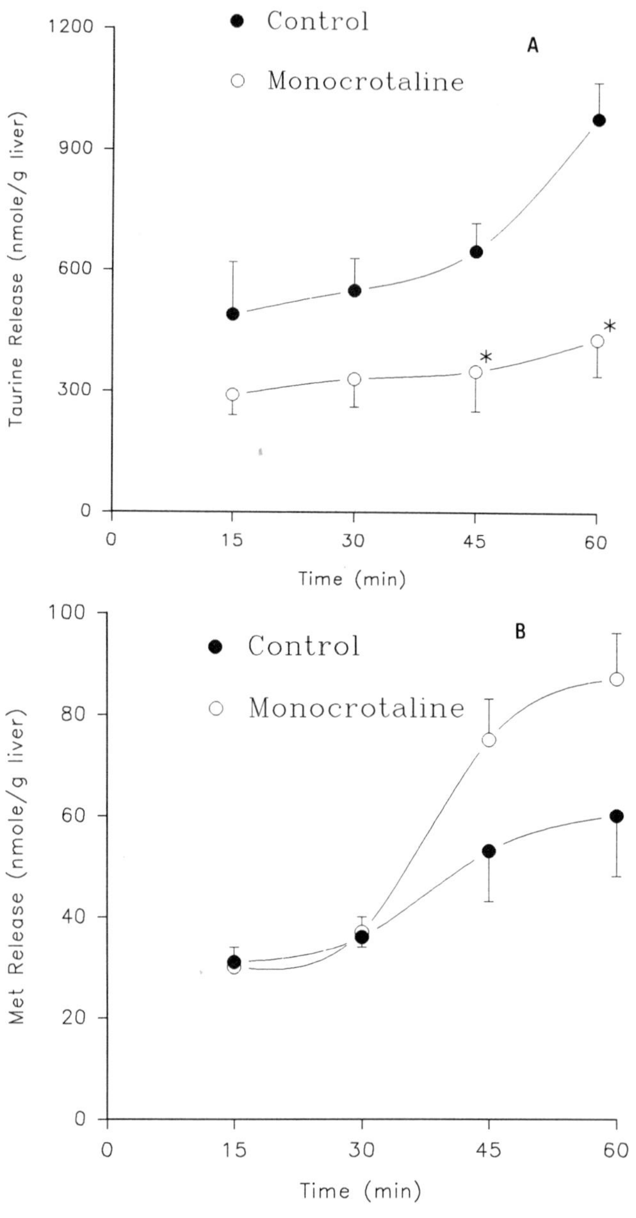

Figure 2. Cumulative release of sulfur amino acids into perfusate of control (•) and monocrotaline-perfused rat liver (○). Livers were perfused in a recirculating system with an initial concentration of 0.5 mM monocrotaline for a period of 1 h. Data are means ± SD for 5 livers. **A:** Taurine; **B:** Methionine; and **C:** GSH.

protein). Glutamate levels were, however, increased, from 1.69 ± 0.19 μmol/g protein to 2.23 μmol/g protein. Small, but significant, increases in GSH transferase activities were found in liver after perfusion (Table 2).

Effects of Monocrotaline *in Vivo* After 24 Hours

Monocrotaline (65 mg/kg; 60% of an LD_{50} dose) was given by i.p. injection, and rats killed 24 h later. At this time, a marked increase in GSH levels was seen in the liver, and hepatic activities of a number of enzymes involved in GSH synthesis and turnover were

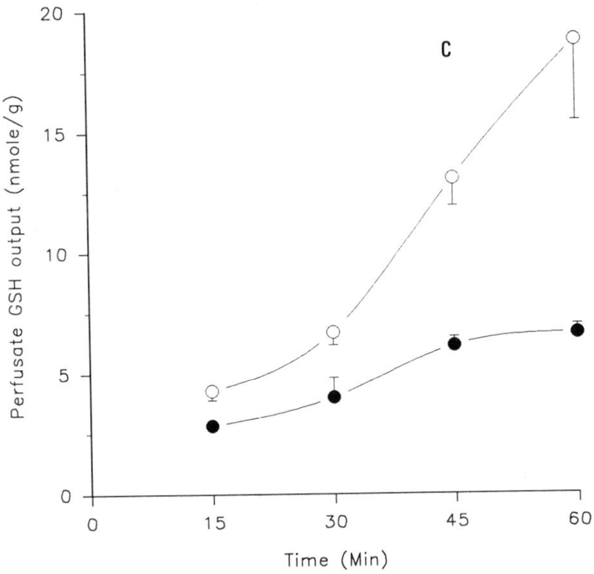

Figure 2C.

increased (Table 2). Taurine, methionine and arginine levels in the liver were significantly depressed (Table 3). Aspartate, glutamate and serine levels where significantly elevated (Table 3).

Effects of Monocrotaline *in Vivo* in Pair-Fed Rats After 14 Days

Rats were injected i.p. with either monocrotaline (65 mg/kg) or an equivalent volume of saline. The two groups were pair-fed as described in the Methods. Weight gains remained invariant between the two groups over the 14 day period (Fig. 5). At this time, monocro-

Table 1. Total release of amino acids in the perfusate of isolated rat liver perfused with monocrotaline (0.5 mM) for 1 h (nmol/g liver/h)

Amino Acid	Control	Monocrotaline
Asp	309 ± 22	294 ± 20
Glu	323 ± 67	289 ± 59
Ser	230 ± 37	226 ± 41
Gly	600 ± 48	619 ± 29
Tau	98 ± 9	43 ± 9*
Ala	389 ± 30	558 ± 58*
Met	60 ± 12	87 ± 9*
Val	377 ± 40	460 ± 45
Phe	723 ± 115	977 ± 155
Ile	601 ± 65	750 ± 106
Leu	947 ± 55	1234 ± 83*
Lys	209 ± 15	480 ± 76*

Data are means ± SD for 5 livers. *$p < 0.05$ compared to control perfusions.

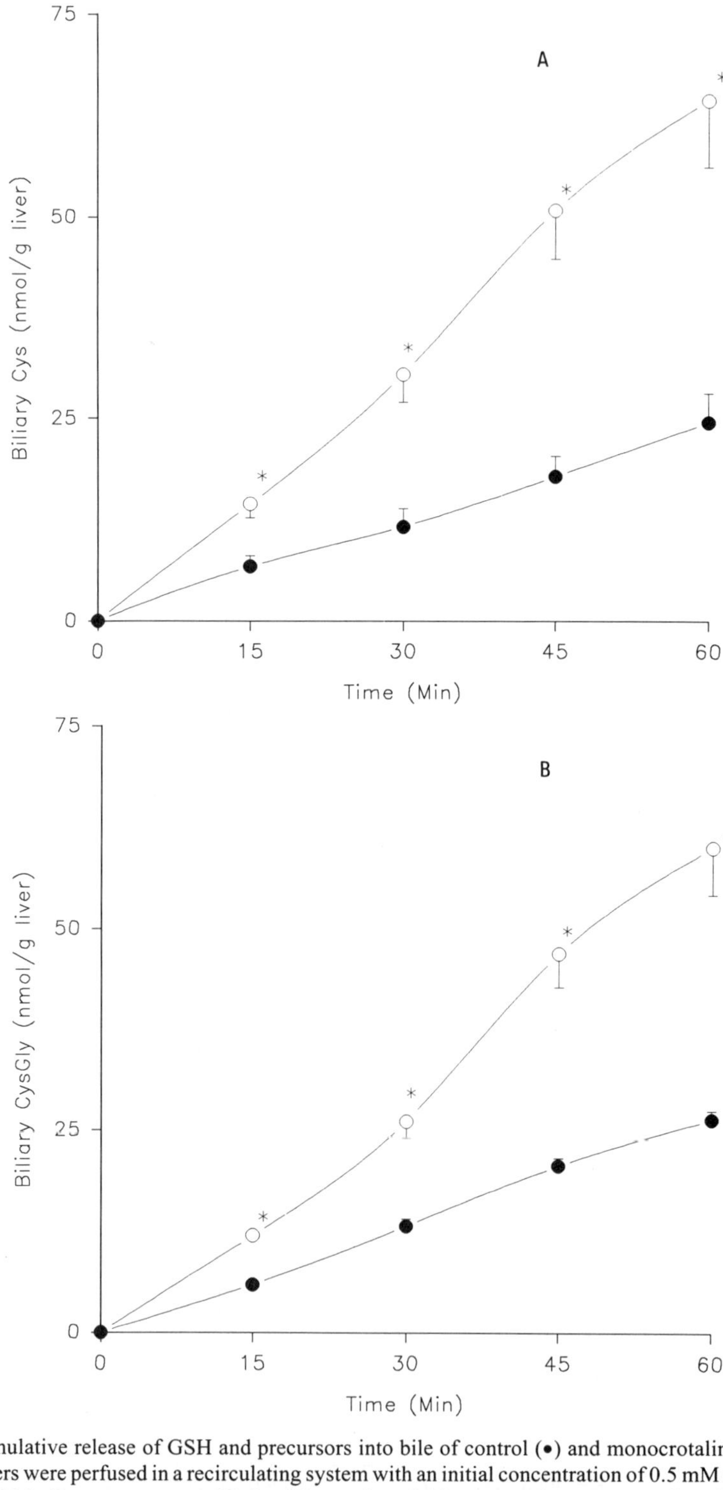

Figure 3. Cumulative release of GSH and precursors into bile of control (•) and monocrotaline-perfused rat liver (○). Livers were perfused in a recirculating system with an initial concentration of 0.5 mM monocrotaline for a period of 1 h. Data are means ± SD for 5 livers. *$p < 0.05$ compared to corresponding time for control livers. **A:** Cysteine; **B:** Cysteinylglycine; and **C:** GSH.

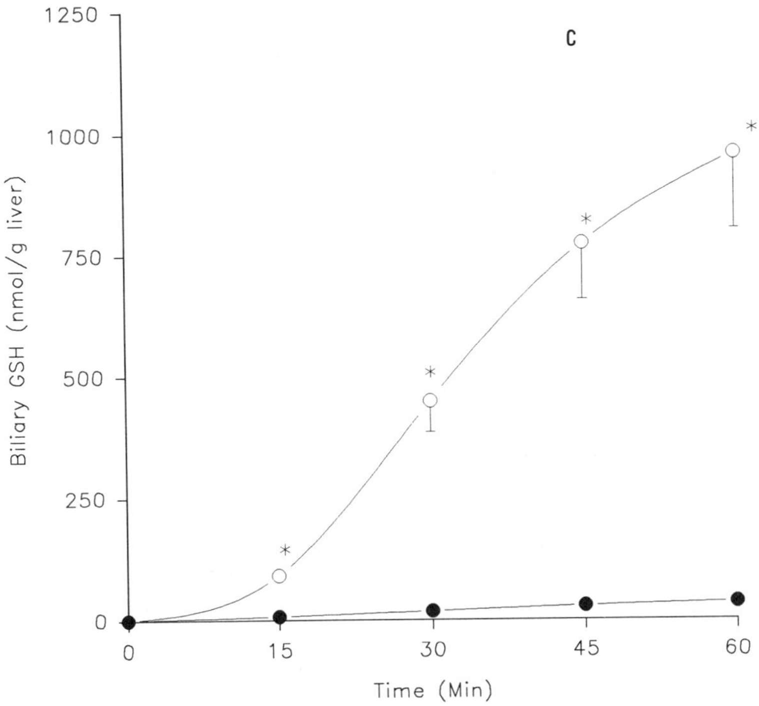

Figure 3C.

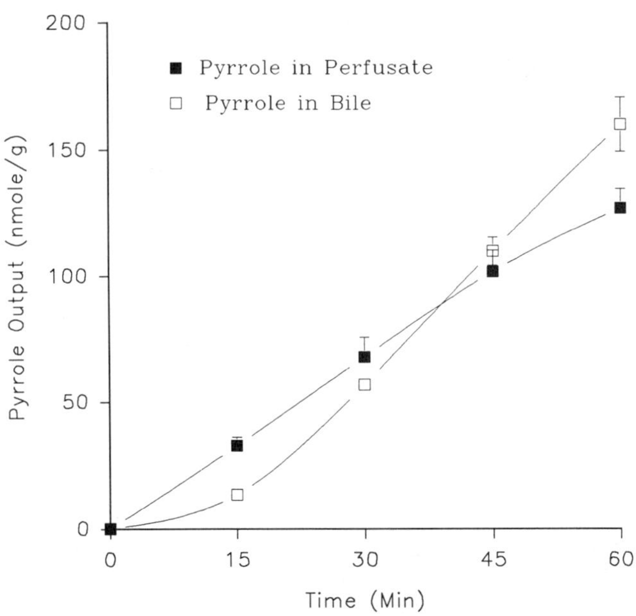

Figure 4. Cumulative release of GSDHP into bile (□) and perfusate (■) of monocrotaline-perfused rat liver. Livers were perfused in a recirculating system with an initial concentration of 0.5 mM monocrotaline for a period of 1 h. Data are means ± SD for 5 livers.

Table 2. Enzyme activities[a] related to GSH metabolism in rat isolated liver after 1 h perfusion with 0.5 mM monocrotaline or *in vivo* 24 h after injection of monocrotaline (65 mg/kg; i.p.)

	In Vitro		*In Vivo*	
	Control	Monocrotaline (0.5 mM perfusion)	Control	Monocrotaline (24 h after 60 mg/kg; i.p.
GSH (μmol/g protein)	87.1 ± 8.1	87.2 ± 6.6	97.4 ± 11.6	161.8 ± 18.3*
γ-glutamylcysteinyl synthetase	8.07 ± 0.45	8.03 ± 0.72	7.0 ± 0.8	9.3 ± 1.3
GSH synthetase	nd	nd	4.63 ± 0.38	5.86 ± 0.51*
Cytosolic GSH transferase	170 ± 4	188 ± 5*	388 ± 36	491 ± 107*
Microsomal GSH transferase	146 ± 3	189 ± 35*	194 ± 25	259 ± 22*
GSH reductase	16.5 ± 1.1	16.6 ± 1.4	nd	nd
GSH peroxidase	9.46 ± 0.37	9.27 ± 0.33	nd	nd
γ-Glutamyltranspeptidase	2.86 ± 0.40	2.81 ± 0.43	0.90 ± 0.08	1.08 ± 0.09*
Dipeptidase	0.77 ± 0.08	0.79 ± 0.08	0.53 ± 0.05	0.93 ± 0.12*

[a]nmol/min/mg protein; nd: not determined; *$p < 0.05$ compared to control

taline-injected animals had developed right ventricular hypertrophy and pulmonary hyperplasia. Hepatic GSH concentrations were significantly increased (Fig. 6). Hepatic aspartate, serine, and threonine levels were also increased (data not shown). Moderate, but significant, increases were observed in the activities of a number of enzymes involved in GSH synthesis and turnover (Table 4).

DISCUSSION

Monocrotaline is present in numerous species of *Crotalaria*, a genus in the legume family. In humans, livestock and experimental animals, monocrotaline produces veno-oc-

Table 3. GSH and amino acid levels in rat liver 24 h after administration of monocrotaline (65 mg/kg; i.p.) (nmol/g liver)

Amino Acid	Control	Monocrotaline
Asp	8.83 ± 1.05	12.05 ± 2.28*
Glu	7.35 ± 1.73	9.51 ± 1.85*
Asn	3.44 ± 0.70	4.18 ± 0.51
Ser	1.94 ± 0.20	3.37 ⊥ 0.40*
Gln	11.63 ± 1.68	14.06 ± 2.84
Gly	6.04 ± 0.73	5.89 ± 0.22
Arg	3.98 ± 0.64	1.88 ± 0.21*
Tau	6.63 ± 2.14	1.57 ± 0.19*
Ala	2.41 ± 0.52	2.04 ± 0.30
Met	0.46 ± 0.05	0.32 ± 0.06*
Val	0.73 ± 0.11	0.60 ± 0.16
Phe	0.76 ± 0.10	0.58 ± 0.18
Ile	1.08 ± 0.23	0.86 ± 0.34
Leu	2.05 ± 0.54	1.44 ± 0.56
Lys	0.58 ± 0.08	0.51 ± 0.07
GSH	5620 ± 330	8040 ± 450*

Data are means ± SD for 6 livers. *$p < 0.05$ compared to control value.

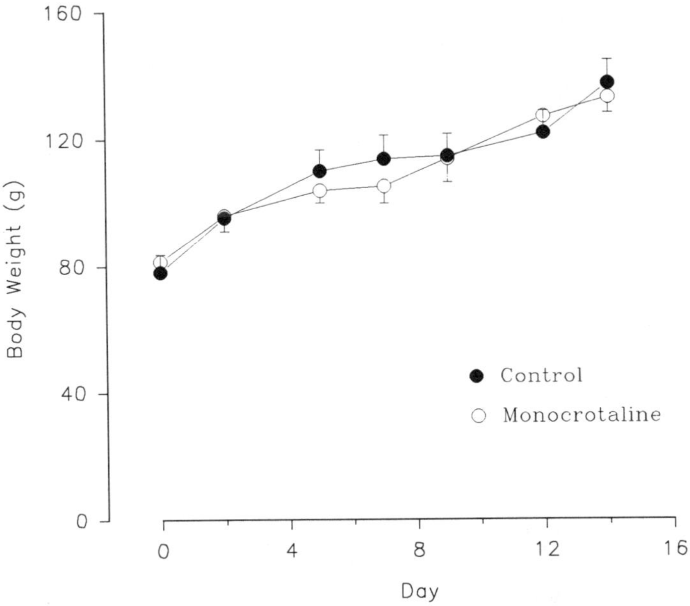

Figure 5. Body weight gain in monocrotaline-injected (65 mg/kg, i.p. on day 0) (○) and pair-fed control rats (•). Data are means ± SD for 6 animals per group.

clusive disease within a period of days or weeks of exposure. This alkaloid has caused numerous epidemics and individual cases of human poisoning[5,6]. Epidemics of PA poisoning typically occur following contamination of food grains with the seeds of PA-containing plants. Sizeable populations may be exposed for many months. Monocrotaline also causes pulmonary arterial hypertension and right ventricular hypertrophy in rats, dogs, nonhuman primates and other species[9].

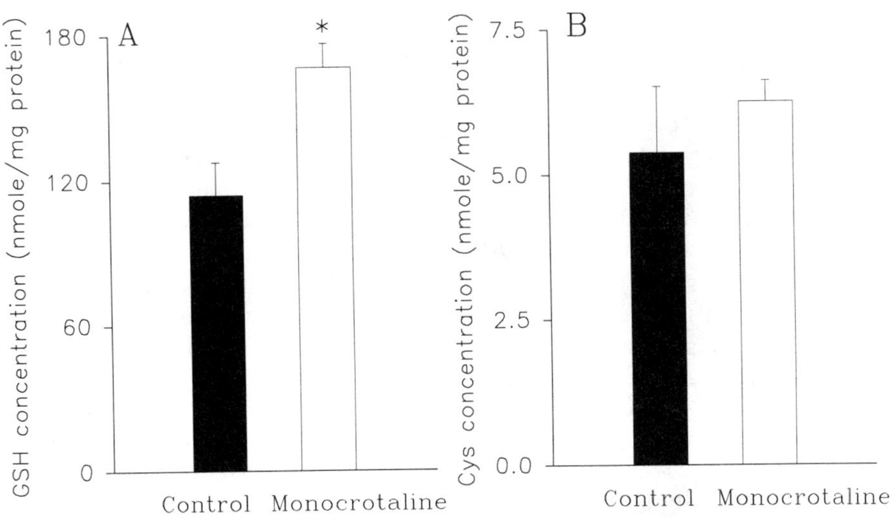

Figure 6. Hepatic GSH and cysteine concentrations in rats given monocrotaline (65 mg/kg, i.p.) two weeks earlier (open bars) and in pair-fed control rats (filled bars). *$p < 0.05$ compared to control levels.

Table 4. Enzyme activities related to GSH metabolism in livers of rats injected with monocrotaline (65 mg/kg; i.p.) and pair-fed for 14 days thereafter (nmol/min/mg protein)

	Control	Monocrotaline
γ-Glutamylcysteinyl Synthetase	13.19 ± 0.99	16.24 ± 1.46*
GSH S-Transferase (cytosolic)	794 ± 30	947 ± 54*
GSH S-Transferase (microsomal)	275 ± 12	343 ± 23*
GSH Reductase	14.39 ± 0.81	16.86 ± 0.46*
GSH Peroxidase	10.96 ± 1.33	10.19 ± 0.92
γ-Glutamyltranspeptidase	1.34 ± 0.29	1.44 ± 0.39
Dipeptidase	2.11 ± 0.25	1.98 ± 0.18

*$p < 0.05$ versus control.

In confirmation of the 1968 hypothesis of Mattocks[8], we have established that hepatic bioactivation of monocrotaline occurs via the formation of the dehydroalkaloid, dehydromonocrotaline (Fig. 1)[3]. This is a highly reactive electrophile, its half-life in water being only 3.39 s (Ref. 1). It is detoxified by conjugation with GSH to form GSDHP (Fig. 1), this being released in high concentrations into the bile. The proportion of dehydromonocrotaline which is detoxified by conjugation is a function of the GSH concentration in the liver[10,15].

The work reported herein establishes that monocrotaline, in turn, has a marked effect on hepatic sulfur metabolism. Perfusion of the isolated rat liver with 0.5 mM monocrotaline leads to a decreased release of taurine into the perfusate (Table 1), but an increased release of methionine and GSH and a marked biliary release of GSH and its precursors (Figs. 2 and 3). An additional drain on hepatic GSH comes from the release of GSH conjugates of monocrotaline metabolites (Fig. 4). *In vivo*, one day after administration of 65 mg/kg monocrotaline (representing 60% of an LD_{50} dose), GSH levels in liver are increased, while taurine and methionine levels are decreased (Table 3). The levels of glutamate, a precursor of GSH, are also elevated. In general, there are small but significant increases in the enzymes involved in GSH synthesis and turnover (Table 2). The elevations in GSH levels and in GSH-related enzymes persists for at least 14 days following exposure to monocrotaline (Table 4; Fig. 6). Because subacute exposure to monocrotaline leads to decreased weight gain due to lowered food consumption, control animals were pair-fed to control for this.

Similar changes in hepatic GSH levels have also been observed following *in vivo* exposure to the PAs retrorsine[16] and trichodesmine[17]. Retrorsine, which is a more toxic PA than monocrotaline, produced a profound fall in hepatic taurine levels that persisted through at least 48 h following injection of the PA[16].

Our data suggest that monocrotaline and other PAs modify sulfur amino acid fluxes in the liver, diverting more cysteine to GSH synthesis and away from taurine synthesis. Although GSH protects against PA toxicity, this modification of sulfur metabolism may not be altogether benign. Taurine has many protective effects on the liver[2,7]. In addition, taurine supplementation lowers the toxicity and lethality of monocrotaline[19]. The profound fall in hepatic taurine levels produced within hours of exposure to PAs such as monocrotaline or retrorsine may, therefore, be a contributory factor to PA-induced hepatotoxicity. Epidemics of PA poisoning typically occur in populations with low dietary intakes of taurine, which may also be an exacerbating factor predisposing such populations to PA-induced disease. Taurine supplementation as a prophylactic measure should be considered for such populations.

ACKNOWLEDGMENTS

These studies were partially supported by USPHS HL 25258 and the Taisho Pharmaceutical Co.

REFERENCES

1. Cooper, R.A. and Huxtable, R.J. 1996, A simple procedure for determining the aqueous half-lives of pyrrolic metabolites of pyrrolizidine alkaloids, *Toxicon*, in press.
2. Dokshina, G.A., Yartsev, E.I., Kolesnikov, Y.A., and Kostesha, N.Y. 1974, Influence of taurine on the recovery of tissue respiration and oxidative phosphorylation in liver mitochondria in irradiated rats, *Radiology*, 14:44-48.
3. Glowaz, S.L., Michnika, M., and Huxtable, R.J. 1992, Detection of a reactive pyrrole in the hepatic metabolism of the pyrrolizidine alkaloid, monocrotaline, *Toxicol.Appl.Pharmacol.* 115:168-173.
4. Huxtable, R.J. 1979, New aspects of the toxicology and pharmacology of pyrrolizidine alkaloids, *General Pharmacology*, 10:159-167.
5. Huxtable, R.J. 1989, Human health implications of pyrrolizidine alkaloids and herbs containing them, in: "Toxicants of Plant Origin, Vol I: Alkaloids", Cheeke, P.R. ed., CRC Press, Boca Raton, Florida, pp. 41-86.
6. Huxtable, R.J. 1992, The toxicology of alkaloids in foods and herbs, in: "Handbook of Natural Toxins: Vol. 7 Food Poisoning", Tu, A.T. ed., Marcel Dekker, New York, pp. 237-262.
7. Kupryszewski, G., Ragnarsson, U., Rolka, K., and Wilusz, T. 1985, Lowering of liver acetaldehyde but not ethanol concentrations by pretreatment with taurine in ethanol-loaded rats, *Experientia*, 41:1421.
8. Mattocks, A.R. 1968, Toxicity of pyrrolizidine alkaloids, *Nature (London)*, 217:723-728.
9. McLean, E.K. 1970, The toxic actions of pyrrolizidine (*Senecio*) alkaloids, *Pharm.Rev.* 22:429-483.
10. Nigra, L. and Huxtable, R.J. 1992, Hepatic glutathione concentrations and the release of pyrrolic metabolites of the pyrrolizidine alkaloid, monocrotaline, from the isolated perfused liver, *Toxicon*, 30:1195-1202.
11. Yan, C.C., Cooper, R.A., and Huxtable, R.J. 1995, The comparative metabolism of the four pyrrolizidine alkaloids, seneciphylline, retrorsine, monocrotaline, and trichodesmine in the isolated, perfused rat liver, *Toxicol.Appl.Pharmacol.* 133:277-284.
12. Yan, C.C. and Huxtable, R.J. 1994, Quantitation of the hepatic release of metabolites of the pyrrolizidine alkaloid, monocrotaline, *Toxicol.Appl.Pharmacol.* 127:58-63.
13. Yan, C.C. and Huxtable, R.J. 1995, Relationship between glutathione concentration and metabolism of the pyrrolizidine alkaloid, monocrotaline, in the isolated, perfused liver, *Toxicol.Appl.Pharmacol.* 130:132-139.
14. Yan, C.C. and Huxtable, R.J. 1995, Effect of the pyrrolizidine alkaloid, monocrotaline, on bile composition of the isolated, perfused rat liver, *Life Sci.* 57:617-626.
15. Yan, C.C. and Huxtable, R.J. 1995, Relationship between glutathione concentration and metabolism of the pyrrolizidine alkaloid, monocrotaline, in the isolated, perfused liver, *Toxicol.Appl.Pharmacol.* 130:132-139.
16. Yan, C.C. and Huxtable, R.J. 1995, Effects of the pyrrolizidine alkaloid, retrorsine, on sulfur metabolism in the liver, *Proc.Western Pharmacol.Soc.* 38:37-40.
17. Yan, C.C. and Huxtable, R.J. 1995, The effect of the pyrrolizidine alkaloids, monocrotaline and trichodesmine, on tissue pyrrole binding and glutathione metabolism in the rat, *Toxicon*, 33:627-634.
18. Yan, C.C. and Huxtable, R.J. 1996, Effects of taurine and guanidinoethane sulfonate on right ventricular hypertrophy induced by the pyrrolizidine alkaloid, monocrotaline, *Biochem.Pharmacol.* 51:321-329.
19. Yan, C.C. and Huxtable, R.J. 1996, Effect of taurine on toxicity of the pyrrolizidine alkaloid, monocrotaline, in rats, in: "Taurine: Basic and Clinical Aspects", Huxtable, R.J., Azuma, J., Nakagawa, M., Kuriyama, K. and Baba, A. eds., Plenum, New York.

17

PROTECTIVE EFFECT OF TAURINE ON INDOMETHACIN-INDUCED GASTRIC MUCOSAL INJURY

Miwon Son,[1] Hee Kee Kim,[1] Won Bae Kim,[1] Junnick Yang,[1] and Byong Kak Kim[2]

[1] Research Laboratories of Dong-A Pharmaceutical Co., Ltd.
Yongin-kun, Kyungki-do 449-900, Korea
[2] College of Pharmacy
Seoul National University
Seoul 151-742, Korea

ABSTRACT

It has been suggested that oxygen-derived free radicals play an important role in the pathophysiology of acute gastric ulceration induced by NSAIDs and ischemia-reperfusion. Taurine is hypothesized to exert its protective effect on NSAIDs-induced gastric injury by its antioxidant properties. The protective effect of taurine on indomethacin-induced gastric mucosal lesion and its protective mechanism were investigated. Intragastric administration of 25 mg/kg of indomethacin induced hemorrhagic lesions on the glandular stomach in rats. Pretreatment with 0.25 or 0.5 g/kg of taurine one day before or for 3 days significantly reduced gastric lesion formation and inhibited the elevation of lipid peroxide level in gastric mucosa. Both resting and FMLP-induced luminol-dependent chemiluminescence of rat peritoneal neutrophils increased immediately after treatment with indomethacin. Taurine (5-20 mM) inhibited chemiluminescence of neutrophils activated by FMLP. Human neutrophils (polymorphonuclear leukocytes) adhered to the confluent monolayer of human umbilical vein endothelial cells (HUVEC) after coincubation with indomethacin. This neutrophil adhesion induced by indomethacin to HUVEC was prevented by taurine in a dose-dependent manner.

These results indicate that the protective effect of taurine against NSAIDs-induced gastric mucosal injury is due to its antioxidant effect, which inhibits lipid peroxidation and neutrophil activation.

INTRODUCTION

It is well documented that nonsteroidal anti-inflammatory drugs (NSAIDs) induced ulceration in the upper gastrointestinal tract[18]. Although the underlying mechanisms respon-

Taurine 2, edited by Huxtable et al.
Plenum Press, New York, 1996

sible for this action are still unclear, the ability of these agents to inhibit gastric prostaglandin synthesis is an important contributing factor[13]. In recent years, a number of studies has highlighted the importance of alterations in mucosal blood flow after NSAIDs administration in the pathogenesis of ulceration seen with several experimental models. It was demonstrated that NSAIDs caused a reduction in gastric mucosal blood flow[2] and it was further noted that the reduction of mucosal blood flow induced by topical application of aspirin to the rat stomach was preceded by the adherence of white thrombi to vessel walls in the mucosal microcirculation and a reduction in gastric mucosal blood flow[11]. Hemorrhagic lesions subsequently formed in the regions of reduced perfusion. These observation suggested a role for circulating leukocytes in the pathogenesis of NSAIDs-induced ulceration[19].

The activation of neutrophils is accompanied by the release of oxygen-derived free radicals. Free radicals have been shown to play a role in various models of gastrointestinal injury such as ischemia-reperfusion injury[3,16,17,22] and some radical scavengers show a protective effect against the mucosal injury induced by active oxygen species resulted from ischemia-reperfusion and NSAIDs treatment[10].

The present study was undertaken to investigate the effect of taurine on indomethacin-induced gastric mucosal injury and to examine whether taurine was acting through its antioxidant effects such as inhibition of lipid peroxidation and neutrophil inactivation.

MATERIALS AND METHODS

Indomethacin-Induced Gastric Mucosal Injury

Male Sprague-Dawley rats (180 to 220 g) obtained from Charles River Japan, were fasted 18 h before the experiments but were allowed free access to water. Gastric mucosal damage was induced by the oral administration of 25 mg/kg of indomethacin suspended in 0.5% carboxymethyl cellulose (CMC) solution with a few drops of Tween 80 in a volume of 0.5 ml/100 g body weight. Taurine (Sigma Chemical Co., St. Louis. MO) was suspended in saline and administered i.p. 1 day before indomethacin (Sigma, Chemical Co., St. Louis. MO) treatment in a dose of 0.25 or 0.5 g/kg. An equivalent volume of vehicle was administered to the control rats. A respective treatment was done in a randomized order.

Rats were killed by exsanguination via the abdominal aorta under inhaled ether anesthesia 6 h after indomethacin administration. Their stomachs were removed and opened along the greater curvature, and carefully examined macroscopically and microscopically. The extent of gastric mucosal lesions was expressed as the total area of erosions observed with a dissecting microscope[17].

The levels of thiobarbituric acid (TBA)-reactive substances in gastric mucosal homogenates were measured and used as an index of lipid peroxidation after indomethacin administration. The level of TBA-reactive substances was expressed in terms of nmol malondialdehyde (Sigma Chemical Co., St. Louis. MO).

FMLP-Elicited Chemiluminescence from Neutrophils

The rats were injected i.p. with 20 ml of sterile 12% (w/v) Na^+ caseinate in iso-osmotic (0.9%) NaCl. The animals were killed 24 h later by ether asphyxiation, and the peritoneal exudate was collected. After being filtered through three layers of surgical gauze, the exudate was centrifuged at 200 g for 5 min, and the pellet was washed twice with Dulbecco's phosphate-buffered saline (DPBS) (Gibco, Life Technologies Inc., N.Y. USA). The pellet was suspended in 1 ml of iso-osmotic NaCl, 10 ml of distilled water, and 10 ml of 1.8% NaCl to cause hypotonic lysis of erythrocyte contaminants. The cell suspension was centrifuged and the pellet was resuspended in Hank's balanced salt solution (HBSS) (Sigma Chemical Co., St. Louis. MO.) with 1% BSA. The specificity of the cell population was determined by differential counting of smears stained with Wright's stain. Viability of the neutrophils was measured with the trypan blue exclusion technique.

Chemiluminescence was measured by using luminol with a microplate scintillation counter, Topcount (PACKARD, Canberra Co.). Reaction mixtures contained 1.5 x 10^6 granulocytes, 1 μM FMLP (<0.01% DMSO), 0-20 mM taurine, and 0.07 mM luminol (0.05 M Tris HCl, pH 7.4) in 200 μl of HBSS. Luminol was dissolved in 50 mM Tris solution and then adjusted to pH 7.4 with 0.1 N HCl. Taurine was diluted with HBSS. Control studies were performed by the addition of saline alone. The results were expressed in terms of the maximal counts per min[6].

Human Neutrophils Adherence to HUVEC Cells

Human umbilical vein endothelial cells (HUVEC) were plated in Medium 199 (Gibco, Life Technologies Inc. N.Y. USA) supplemented with 10% heat-inactivated fetal calf serum, thymidine (2.4 mg/L), glutamine (230 mg/L), heparin sodium (10 IU/ml), antibiotics (100 IU/ml penicillin, 100 μg/ml streptomycin, and 0.125 μg amphotericin B), and endothelial cell growth factor (80 μg/ml). The cell cultures were incubated at 37°C in a humidified atmosphere with 5% CO_2 and expanded by brief trypsinization (0.25% trypsin in PBS containing 0.02% EDTA). Cells were seeded into gelatin (0.1%) and fibronectin-coated (25 μg/ml) 11 mm, 96-well tissue culture plates and used when confluent.

Human neutrophilic polymorphonuclear leukocytes were isolated from the venous blood of healthy adults using Ficoll-plaque. Isolated neutrophils were suspended in PBS and radiolabeled by incubating the cells at 2 x 10^7 cells/ml with 30 μCi $^{51}NaCrO_4$/ml (Du Pont NEN, Massachusetts, USA) at 37°C for 60 min. The cells were then washed twice with cold PBS at 250 g for 8 min to remove unincorporated radioactivity and were resuspended in plasma-free HBSS. Labeled neutrophils were added to HUVEC monolayers at a neutrophil-to-endothelial cell ratio of 10:1 with various concentrations of indomethacin. After coincubation (30 min), the percent of added neutrophils that adhered to the HUVEC monolayers was quantitated. After a 30 min incubation HUVEC were washed to remove the drug, the labelled neutrophils were added to the monolayers, and neutrophil adhesion was assessed[12,21].

RESULTS

Effect of Taurine on Indomethacin-Induced Gastric Mucosal Injury

Multiple erosions and bleeding were present in the stomach 6 h after administration. The total area of gastric erosions was significantly larger in the indomethacin-treated rats than the normal rats (Fig. 1a,b). Histology showed that the indomethacin-induced lesions in the control rats penetrated deeply into mucosa. Hemorrhagic necrosis and diffused necrosis of villous ridges were present, and the upper of the mucosa was completely destroyed. Treatment with taurine (500 mg/kg) reduced the depth and severity of the lesions (Fig. 1c). Mild to moderate villous atrophy was noted and severe necrotic changes of mucus ridges were not seen.

The lipid peroxide level in the antral mucosa of the untreated rats was 85.1 ± 5.2 nmol/mg wet weight. TBA-reactive substances significantly increased 6 h after indomethacin administration compared to the value in normal rats. Taurine, 250 or 500 mg/kg,

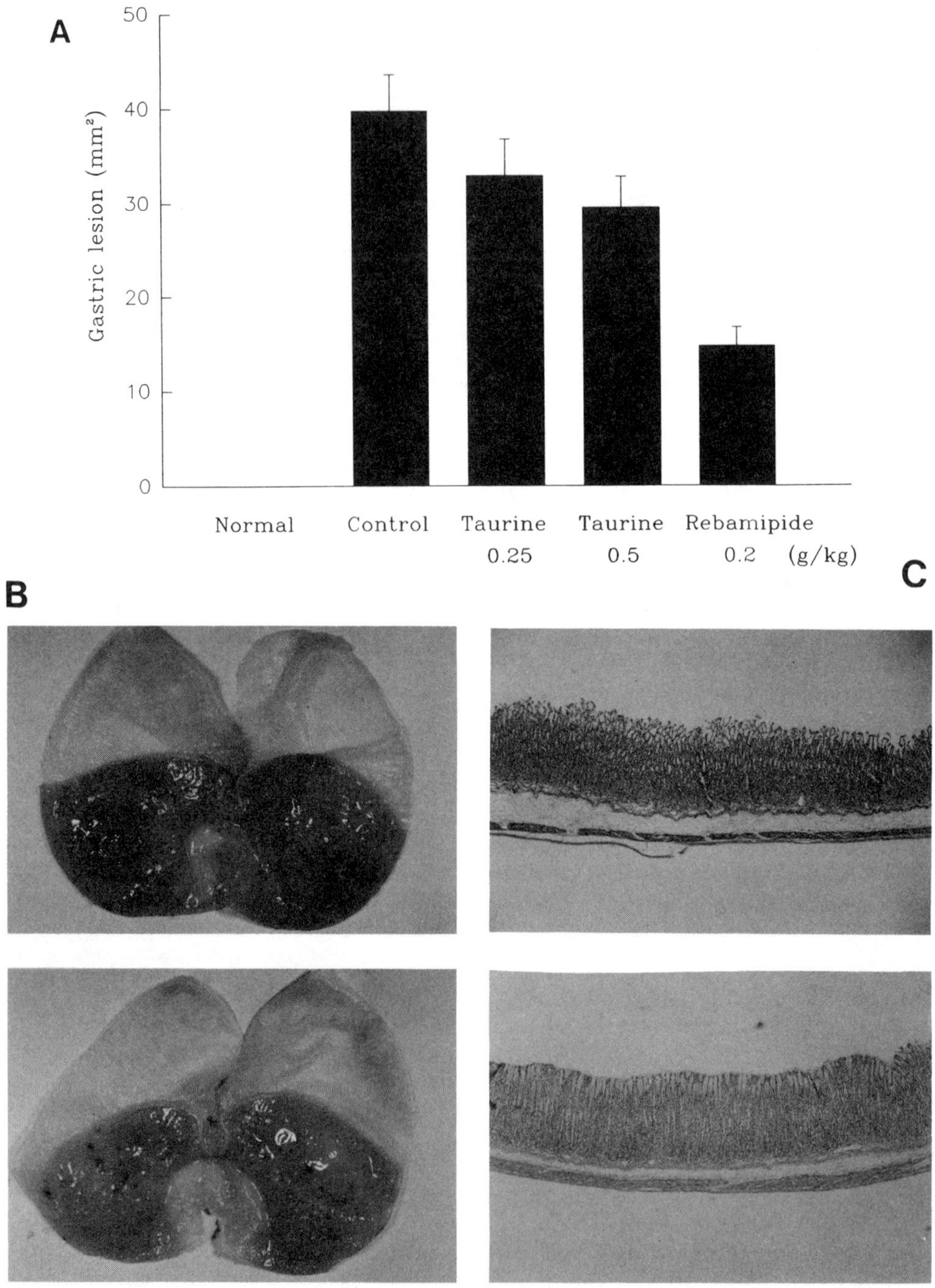

Figure 1. (a) Effect of taurine on the formation of gastric mucosal lesion 6 h after oral administration of indomethacin, 25 mg/kg to rats. Each value indicates mean ± SEM of 3 separate experiments (n=10). (b) Gross appearance of indomethacin-induced gastric mucosal injury in rats. Hemorrhagic erosions were observed linearly along the mucosal foldings in control rats (above) 6 h after indomethacin administration (25mg/kg). Injury was inhibited by treatment with taurine (below; 0.5 g/kg). (c) Histology of gastric mucosal lesion in rats (Hematoxylin and eosin stain x 40) Above: Control; below:taurine-treated rat.

Table 1. Effect of taurine on the lipid peroxide concentration in the antral mucosa. The gastric lesion was induced by intragastric administration of indomethacin, 25 mg/kg to rats. The mucosa was seperated from the antrum of stomach 6 h after administration of indomethacin. Lipid peroxide level is shown as the amount of malondialdehyde (MDA) determined

Treatment	TBA-Reactive Materials (nmol/g wet weight)	% Inhibition
Normal	85.1 ± 5.2	
Indomethacin		
± vehicle	127 ± 16	
± taurine 250 mg/kg	106 ± 9	50.1
± taurine 500 mg/kg	100 ± 18	64.4
± rabamipide 50 mg/kg	87.1 ± 14.3	95.5

Data represent mean ± SEM

significantly inhibited the increase in TBA-reactive substances (Table 1). But taurine had no inhibitive effect on NADPH-dependent microsomal lipid peroxidation in the rat liver.

Effect of Taurine on FMLP-Elicited Chemiluminescence from Neutrophils

When rat peritoneal neutrophils were exposed to FMLP a chemiluminescence response was obtained. The peak appeared at around 1 min after addition of FMLP. Indomethacin, like FMLP, induced luminol-dependent chemiluminescence immediately after treatment (Fig. 2). Taurine at 5-20 mM dose-dependently inhibited chemiluminescence of neutrophils activated by FMLP. The concentration for 50% inhibition (IC_{50}) was 5.52 mg/ml (Fig. 3). Hypochlorous acid produced as a result of the peroxidase-H_2O_2-halide reaction has been proposed as being involved in the luminol-dependent chemiluminescence. Taurine is a HOCl scavenger, and should thus inhibit the chemiluminescence activity dependent on HOCl. Taurine had no effect on neutrophil viability in this study.

Effect of Taurine on Adherence of Human Neutrophils to HUVEC Cells

Human neutrophils adhered to confluent monolayer of HUVEC. The basal adhesion (in the absence of drugs) of the neutrophils to the monolayers ranged between 14.6-24.8% adherence.

As shown in Fig. 4, indomethacin induced a dose-dependent increase in neutrophil adherence to HUVEC monolayers. The effect of taurine on the neutrophil adherence to the monolayer is seen in Fig. 5. Taurine inhibited neutrophil adhesion induced by indomethacin. The inhibition was below basal adhesion. Indomethacin altered the arachidonic acid metabolism pathway and was led to production of leukotrienes. Taurine reduced reactive oxygen and LTB_4 levels by manipulation of the MPO-H_2O_2-halide system, and indomethacin-induced neutrophil adhesion to endothelial cells was reduced.

DISCUSSION

Taurine was found to inhibit the increase in gastric mucosal lesions induced by indomethacin. It significantly inhibited the increase in TBA-reactive substances in the gastric

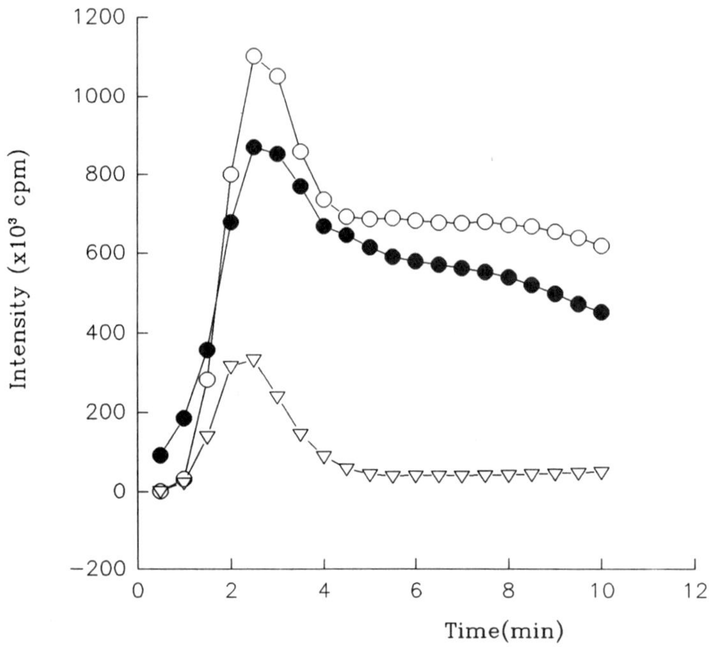

Figure 2. Luminol-dependent chemiluminescence of rat peritoneal neutrophils induced by 10 μM FMLP (O), 0.5 mM indomethacin (•) and HBSS (∇).

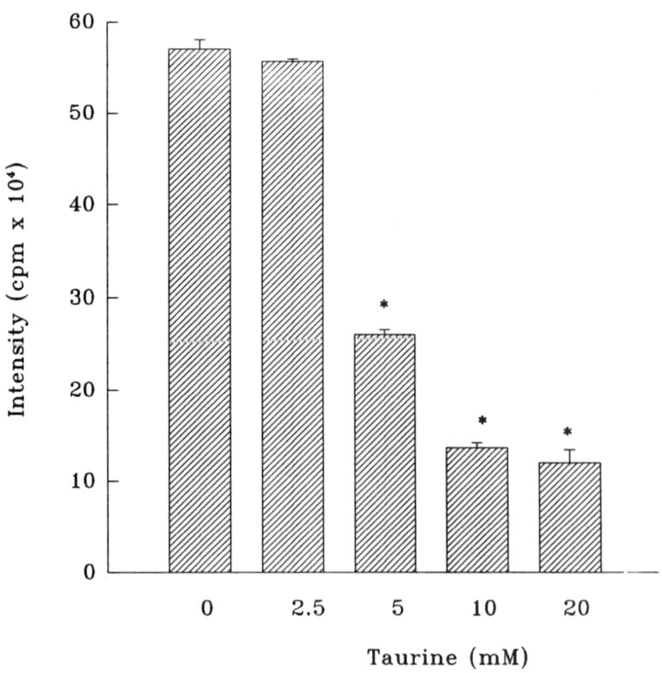

Figure 3. Effect of taurine on luminol-dependent chemiluminescence in FMLP-stimulated neutrophils. * $p < 0.05$ compared with HBSS.

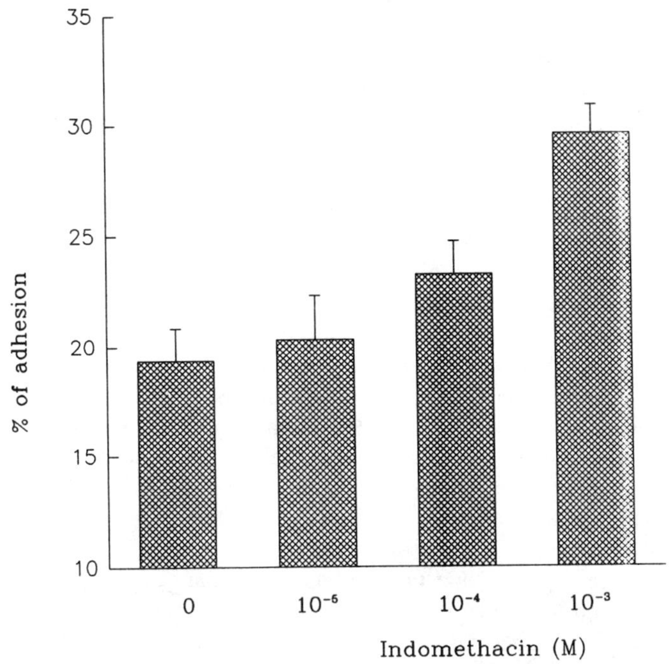

Figure 4. Stimulation of neutrophil adhesion to HUVEC monolayer by indomethacin. ^{51}Cr-labelled neutrophils were added to HUVEC monolayer in the presence of various concentrations of indomethacin.

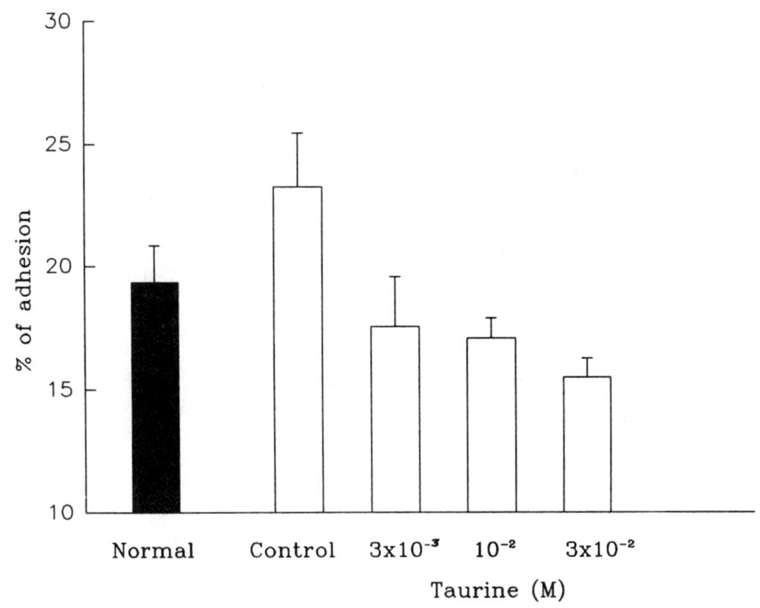

Figure 5. Inhibition of indomethacin-stimulated neutrophil adhesion to HUVEC monolayer by taurine. ^{51}Cr-labelled neutrophils were stimulated by indomethacin (100 μM) and added to HUVEC monolayer in the presence of various concentrations of taurine.

mucosa 6 h after indomethacin administration. Many studies have focused on the role of lipid peroxidation mediated by oxygen radicals in the development of gastric mucosal injury[3,16,17,22]. It was found that the gastric mucosal injury and the level of TBA-reactive substances in the gastric mucosa were significantly inhibited by pretreatment with superoxide dismutase, a scavenger of superoxide, and by dimethylsulfoxide, a hydroxy radical scavenger, indicating that oxygen radicals induced by lipid peroxidation play an important role in the pathogenesis of indomethacin-induced gastric mucosal damage as well as in gastric injury produced by ischemia-reperfusion[22]. Our study showed that taurine itself did not act as superoxide radical scavenger. The exact mechanism of the *in vivo* pharmacological effect of taurine remains to be determined.

In the present study taurine inhibited the neutrophil activation and the neutrophil adherence to endothelial cells. The source of active oxygen species in the indomethacin-induced gastric lesion model is known to be neutrophils which became adhered to endothelial cells and activated[18]. Depletion of circulating neutrophils, by treatment with antineutrophil serum or methotrexate, significantly reduce the damage induced by indomethacin[19]. Prevention of leukocyte adherence to the vascular endothelium through pretreatment with monoclonal antibody directed against the common subunit of the granulocyte adherence glycoprotein (CD11/CD18) resulted in significant reduction in the extent of indomethacin-induced gastric ulceration[7,17].

Much evidence suggests that taurine has antioxidant effect *in vivo* and *in vitro*[1,9]. Taurine has been found to scavenge hypochlorous acid generated by polymorphonuclear leukocytes[5] and inhibit lipid peroxidation of various membranes, such as rat mesangial cells and rod outer segments[14,15,20]. Taurine also protects against pulmonary fibrosis induced by chemicals and ozone[4,8,20]. These findings suggest that taurine acts as an antioxidant *in vivo*, inhibiting lipid peroxidation and neutrophil activation and that this action may be partially responsible for the mucus-protecting effect.

REFERENCES

1. Aruoma, O.I., Halliwell, B., Hoey, B.M., and Butler, J. 1988, The antioxidant action of taurine, hypotaurine and their metabolic precursors, *Biochem.J.* 256:251-255.
2. Ashey, S.W., Sonnenschein, L.A., and Cheung, L.Y. 1985, Focal gastric mucosal blood flow at the site of aspirin-induced ulceration, *Am.J.Surg.* 149:53-59.
3. Ayene, I.S., Dodia, C., and Fisher, A.B. 1992, Role of oxygen in oxidation of lipid and protein during ischemia/reperfusion in isolated perfused rat lung, *Arch.Biochem.Biophys.* 296:183-189.
4. Banks, M.A., Porter, D.W., Martin, W.G., and Castranova, V. 1990, Effects of *in vitro* ozone exposure on peroxidative damage, membrane leakage, and taurine content of rat alveolar macrophages, *Toxicol.Appl.Pharmacol.* 105:55-65.
5. Cantin, A.M. 1994, Taurine modulation of hypochlorous acid-induced lung epithelial cell injury in vitro. Role of anion transport, *J.Clin.Invest.* 93:606-614.
6. Dahlgren, C. 1991, Effect of different inhibitors on the intracellularly and extracellularly generated chemiluminescence induced by formylmethionyl-leucyl-phenylalanine in polymorphonuclear leukocytes. Cellular response in the presence of mannitol, benzoate, taurine, indomethacin and NDGA, *J.Biolum.Chemilum.* 6:29-34.
7. Gimbrone, M.A., Jr., Brock, A.F., and Schafer, A. 1984, Leukotriene B_4 stimulates polymorphonuclear leukocyte adhesion to cultured vascular endothelial cells, *J.Clin.Invest.* 74:1552-1555.
8. Giri, S.N. and Wang, Q. 1992, Taurine and niacin offer a novel therapeutic modality in prevention of chemically-induced pulmonary fibrosis in hamsters, *Adv.Exp.Med.Biol.* 315:329-340.
9. Green, T.R., Fellman, J.H., Eicher, A.L., and Pratt, K.L. 1991, Antioxidant role and subcellular location of hypotaurine and taurine in human neutrophils, *Biochim.Biophys.Acta Gen.Subj.* 1073:91-97.
10. Hirota, M., Inoue, M., Ando, Y., and Morino, Y. 1990, Inhibition of stress-induced gastric mucosal injury by a long acting superoxide dismutase that circulates bound to albumin, *Arch.Biochem.Biophys.* 280:269-273.

11. Kitahora, T. and Guth, P.H. 1987, Effect of aspirin plus hydrochloric acid on the gastric mucosal microcirculation, *Gastroenterology*, 93:810-817.
12. McGregor, P.E., Agrawal, D.K., and Edwards, J.D. 1994, Technique for assessment of leukocyte adherence to human umbilical vein endothelial cell monolayers, *J.Pharm.Mol.* 32:73-77.
13. Naito, Y., Yoshikawa, T., Kaneko, T., Iinuma, S., Nishimura, S., Takahashi, S., and Kondo, M. 1993, Role of oxygen radicals in indomethacin-induced gastric mucosal microvascular injury in rats, *J.Clin.Gastroenterol.* 17:S99-S103.
14. Pasantes-Morales, H. and Cruz, C. 1984, Protective effect of taurine and zinc on peroxidation-induced damage in photoreceptor outer segments, *J.Neurosci.Res.* 11:303-311.
15. Pasantes-Morales, H. and Cruz, C. 1985, Taurine and hypotaurine inhibit light-induced lipid peroxidation and protect rod outer segment structures, *Brain Res.* 330:154-157.
16. Takeuchi, K., Ueshima, K., Hironaka, Y., Fujioka, Y., Matsumoto, J., and Okabe, S. 1991, Oxygen free radicals and lipid peroxidation in the pathogenesis of gastric mucosal lesions induced by indomethacin in rats, *Digestion*, 49:175-184.
17. Vaananen, P.M., Meddings, J.B., and Wallace, J.L. 1991, Role of oxygen-derived free radicals in indomethacin-induced gastric injury, *Am.J.Physiol.* 261:G470-G475.
18. Wallace, J.L. and Granger, D.N. 1992, Pathogenesis of NSAID gastropathy: are neutrophils the culprits? *Trends Pharmacol.Sci.* 13:129-131.
19. Wallace, J.L., Keenan, C.M., and Granger, D.N. 1990, Gastric ulceration induced by nonsteroidal anti-inflammatory drugs is a neutrophil-dependent process, *J.Physiol.* 259:G462-G467.
20. Wang, Q., Hyde, D.M., and Giri, S.N. 1992, Abatement of bleomycin-induced increases in vascular permeability, inflammatory cell infiltration, and fibrotic lesions in hamster lungs by combined treatment with taurine and niacin, *Lab.Invest.* 67:234-242.
21. Yoshida, N., Takemura, T., Neil Granger, D., Anderson, D.C., Wolf, R.E., McIntire, L.V., and Kvietys, P.R. 1993, Molecular determinants of aspirin-induced neutrophil adherence to endothelial cells, *Gastroenterology*, 105:715-724.
22. Yoshikawa, T., Ueda, S., Naita, Y., Takahashi, S., Oyamada, H., Morita, Y., Yoneta, T., and Kondo, M. 1989, Role of oxygen-derived free radicals in gastric mucosal injury induced by ischemia or ischemia-reperfusion in rats, *Free Rad.Res.Comms.* 7:285-291.

18

TAURINE REDUCES EXPERIMENTAL LIVER INJURY AFTER COLD ISCHEMIC PRESERVATION AND A PERIOD OF REWARMING PRIOR TO REPERFUSION

Thomas Minor,[1] Tetsuya Yamaguchi,[1] Hiltrud Klauke,[1] Peter Wingenfeld,[2] Dietrich Michalk,[2] and Wolf Isselhard[1]

[1] Institute for Experimental Medicine
University of Cologne Robert Koch Str. 10
50931 Cologne, Germany
[2] Department of Pediatrics
University of Cologne
Cologne, Germany

ABSTRACT

Livers of male Wistar rats (250-300 g) were isolated and flushed with 10 ml of Ringer's solution and 10 ml of UW preservation solution. Then the organs were stored for 24 h at 4°C in UW solution. Livers of Group 1 were rinsed with 10 ml of Ringer's solution and reperfused after hypothermic storage with oxygenated Krebs-Henseleit solution (95% O_2; 5% CO_2) in a nonrecirculating system at constant pressure (10 mmHg) and 37°C.

Livers of Group 2 were incubated for 30 min at 37°C prior to reperfusion, in order to simulate rewarming of the organ upon surgical implantation. Livers of Group 3 were treated like Group 2, but taurine was admixed to the UW solution (1 mM).

Livers of Group 1 showed little signs of a preservation/reperfusion injury, with low enzyme activities of the parenchymal ALT and endothelial purine nucleoside phosphorylase (PNP) in the postischemic rinse solution (ALT: 19.9 ± 12. 4; PNP: 3.3 ± 0.4 U/liter), adequate portal flow values about 3 ml/g/min and high O_2 uptake at the end of the experiment (VO_2: 3.2 ± 0.4 ml/l00g/min). Livers of Group 2 exhibited nearly tenfold higher enzyme activities in the rinse solution (ALT: 247.0 ± 94.7*; PNP: 29.5 ± 17.0* U/l) and disturbed tissue perfusion with significantly reduced flow values of about 2 ml/g/min during the first 10 min of reperfusion. As a result, the recovery of O_2 uptake was only 2.2 ± 0.3 ml/100 g/min*.

Addition of taurine (Group 3) resulted in a significant reduction of the enzyme loss (ALT: 96.2 ± 50.0#; PNP:12.4 ± 7.0# U/liter) and improved portal flow values and O_2 uptake at the end of reperfusion (2.7 ± 0.3 ml/l00 g/min#). The results give evidence for the importance of the rewarming period after hypothermic storage, which is inevitable during implantation of the organ *in vivo*. Taurine seems to exert a protective effect, affecting both the vascular endothelium and parenchymal tissue (*$p< 0.05$ vs Group 1; # $p< 0.05$ vs Group 2).

INTRODUCTION

The recovery of the liver after global ischemia is often compromised by circulatory disturbances during the initial phase of reperfusion[2,3,7]. In this context, the vascular endothelium seems to play a pivotal role determining postischemic graft viability. The aim of the present study was to investigate the respective effects of cold preservation and a short period of subsequent rewarming on morphological alterations to the vascular endothelium and reflow characteristics upon reperfusion.

MATERIAL AND METHODS

Livers of male Wistar rats were excised under ether anesthesia and flushed via the portal vein with 10 ml of Ringer's solution and 10 ml of UW solution. The livers were then stored at 4°C in UW solution for 24 h. During this time the infrahepatic caval vein was ligated and a polyethylene catheter was telescoped into the suprahepatic caval vein and secured.

There were three experimental groups in this study: Livers of Group 1 were rinsed with 10 ml of Ringer's solution immediately after 24 h of cold preservation in UW solution and then reperfused. In Group 2, livers were incubated for 30 min in 37°C saline solution subsequent to cold storage and prior to rinsing and reperfusion. The slow rewarming any organ would undergo during surgical implantation in vivo, should thus be simulated. In Group 3, livers were subjected to the same protocol as in Group 2, but 1 mM taurine (Sigma Chemicals, St. Louis Mo. USA) was admixed to the UW solution during flush and preservation.

The post-ischemic rinse of the livers was performed under standardized conditions with 10 ml of room-temperature Ringer's solution and the primary effluent was collected for the determination of enzyme activities.

Reperfusion was carried out in a nonrecirculating system for 45 min at a constant perfusion pressure via the portal vein of 10 mmHg, regulated by overflow of the perfusate (Fig. 1). The liver was placed in a bath of Krebs-Henseleit solution in order to achieve a homogenous perfusion of the isolated organ *in vitro*. The perfusate consisted of O_2-saturated Krebs-Henseleit solution.

Portal venous flow was measured by collecting the effluent at given time intervals.

Effluent samples were taken during reperfusion for determination of hepatic O_2 consumption (acid-base laboratory ABL-2, Radiometer Copenhagen) and cellular enzyme loss (standard kits, Boehringer, Mannheim). Purine nucleoside phosphorylase was measured using the method of Dwenger and Trautschold[1].

Statistical analysis was performed by one way analysis of variance with the Bonferroni t-test being used for multiple comparison. Differences were considered to be significant if $p < 0.05$.

RESULTS AND DISCUSSION

The portal venous flow during reperfusion is depicted in Fig. 2. In Group 1, i.e. after hypothermic preservation alone, the transhepatic flow recovered rapidly to normal values of

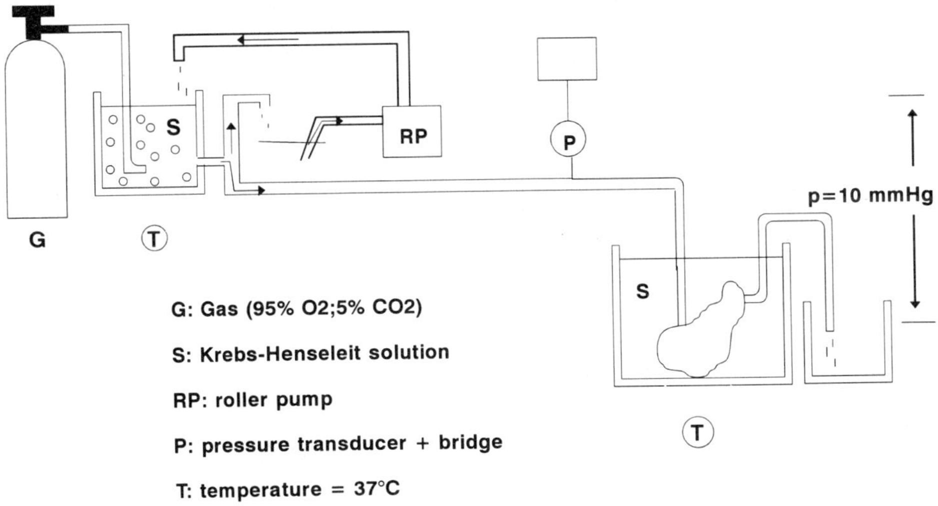

Figure 1. Schematic representation of the *in vitro* perfusion device

3-4 ml/g liver/min. However, if the livers were subjected to an additional 30 min of ischemic rewarming (Group 2), portal flow was significantly impaired during the first 10 min of reperfusion, and recovery to normal flow conditions was seriously postponed.

These circulatory disturbances could be effectively attenuated by the admixture of taurine to the preservation solution (Group 3).

Enzyme leakage into the effluent also revealed significant differences between the groups (Table 1).

Although after hypothermic storage alone only little activity of Alanine aminotransferase (ALT) was seen in the effluent, the additional rewarming led to a remarkable rise in ALT leakage during the initial rinse to up to 10-fold the values observed in Group 1. At the end of the experiment, the ALT release into the perfusate was also significantly elevated.

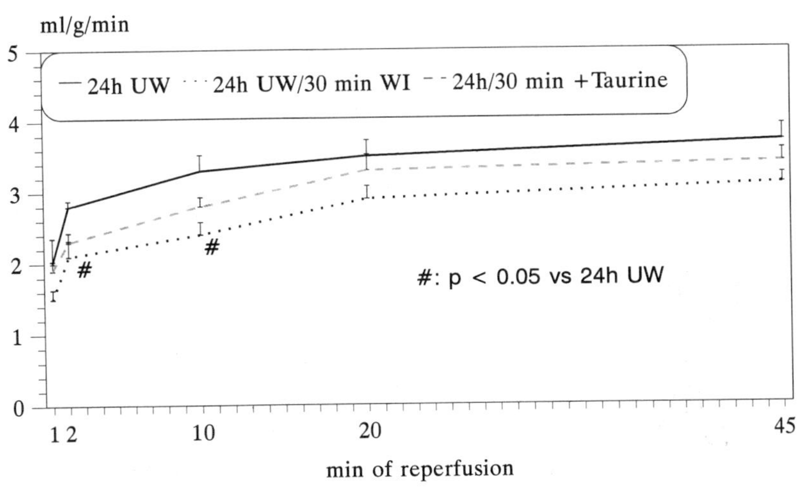

Figure 2. Portal venous flow upon post-ischemic reperfusion *in vitro*.

Table 1. Enzyme release after post-ischemic rinse and after 45 min of reperfusion *in vitro*

Rinse	ALT (IU/liter)	GLDH (IU/liter)	PNP (U/liter)
Group 1	19.9 ± 12.4	0.7 ± 0.7	3.3 ± 0.4
Group 2	247.0 ± 94.7*	3.8 ± 1.6*	29.5 ± 17.0*
Group 3	96.2 ± 50.0#*	1.8 ± 0.4#	12.4 ± 7.0*
Reperfusion (45 min)	**ALT (IU/g/min)**	**GLDH (IU/g/min)**	**PNP (U/g/min)**
Group 1	9.5 ± 4.4	0.7 ± 0.7	6.3 ± 1.8
Group 2	17.3 ± 3.8*	3.8 ± 1.6*	5.9 ± 4.3
Group 3	6.0 ± 1.5#	1.8 ± 0.4#	6.8 ± 1.5

Values are means ± SD; *$p < 0.05$ versus Group 1; #$p < 0.05$ versus Group 2

Treatment with taurine resulted in a significant reduction of the initial ALT release during post-ischemic rinse and minimized the perfusate activities after 45 min of reperfusion. The data on the intramitochondrial enzyme, glutamate dehydrogenase (GLDH), were comparable to those of the cytoplasmatic enzyme, ALT. Both groups with ischemic rewarming showed higher enzyme leakage upon post-ischemic rinse and during reperfusion, but the values were significantly reduced in Group 3 as compared to Group 2.

Purine nucleoside phosphorylase (PNP) has been reported as a marker enzyme for nonparenchymal liver cells and especially for vascular endothelial cells[4-6]. Livers of Group 1 showed only a few signs of preservation injury to endothelial cells, as judged from the release of PNP in the post-ischemic rinse solution, whereas livers of Group 2 exhibited up to tenfold higher activities of PNP at this time. It can thus be concluded that the short period of rewarming after hypothermic storage has an influence on the morphological integrity of the endothelial cells. Taurine led to a significant reduction of PNP liberation upon post-ischemic rinse as compared to Group 2, but yielding still significantly higher values with reference to Group 1.

Except for the initial minutes, the release of PNP was comparable in all three Groups during the 45 min of reperfusion.

In order to evaluate the functional recovery of the livers, the O_2 consumption at the end of the experiment and the cumulative bile production during reperfusion were investi-

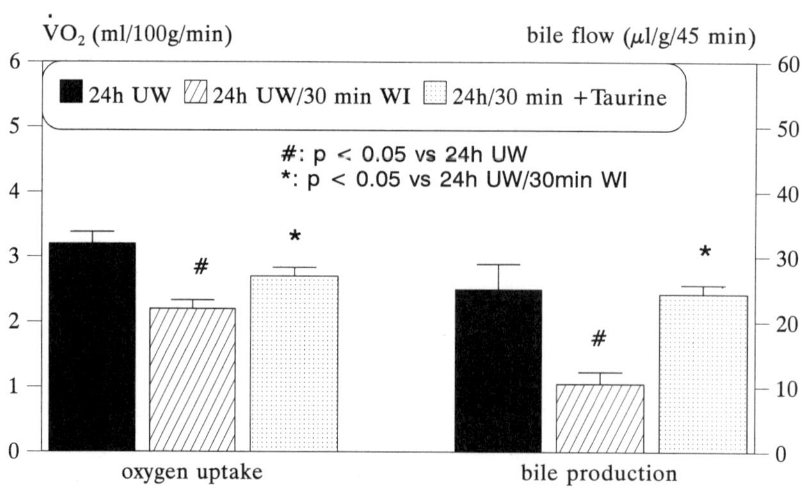

Figure 3. Functional recovery upon postischemic reperfusion *in vitro*

gated (Fig. 3). In accordance with the results on cellular enzyme loss, livers of Group 2 showed significantly reduced functional recovery concerning bile flow as well as O_2 utilization as compared to livers of Group 1. In both parameters, the addition of taurine led to a significant amelioration by comparison to Group 2.

In conclusion, our data indicate that a short period of ischemic rewarming may lead to circulatory disturbances, and substantially affect organ viability upon reperfusion. Addition of taurine to the UW preservation solution may be helpful in improving the recovery of cold stored livers upon post-ischemic reperfusion.

REFERENCES

1. Dwenger, A. and Trautschold, I. 1983, Purine-nucleoside phosphorylase, in: "Methods of Enzymatic Analysis," Bergmeyer, H.U. ed., Verlag Chemie, Weinheim, pp. 382.
2. Menger, M.D., Lehr, H.-A., and Messmer, K. 1991, Role of oxygen radicals in the microcirculatory manifestations of postischemic injury, *Klin.Wochenschr.* 69:1050-1055.
3. Minor, T. and Isselhard, W. 1993, Role of the hepatovasculature in free radical mediated reperfusion damage of the liver, *Eur.Surg.Res.* 25:287-293.
4. Minor, T., Osswald, B., Krauss, T.W., July, N., Isselhard, W., and Klar, E. 1995, Determination of plasma activities of purine nucleoside phosphorylase by high performance liquid chromatography: estimates of nonparenchymal cell injury after porcine liver transplantation, *J.Chromatogr.B.* 670:332-336.
5. Rao, P.N., Walsh, T.R., Makowka, L., Rubin, R.S., Weber, T., Snyder, J.T., and Starzl, T.E. 1990, Purine nucleoside phosphorylase: A new marker for free oxygen radical injury to the endothelial cell, *Hepatology*, 11:193-198.
6. Reinders, M.E., Vangulik, T.M., Boers, W., Chamuleau, R.A.P.M., and Klopper, P.J. 1993, Purine nucleoside phosphorylase (PNP) as a marker of endothelial cell injury in cold preserved dog livers, *Eur.Surg.Res.* 25:42-43.(Abstract)
7. Vollmar, B., Glasz, J., Leiderer, R., Post, S., and Menger, M.D. 1994, Hepatic microcirculatory perfusion failure is a determinant of liver dysfunction in warm ischemia-reperfusion, *Am.J.Pathol.* 145:1421-1431.

Part 3

TAURINE AND THE KIDNEY

19

TAURINE AND WATER CHANNELS ARE CO-LOCALIZED IN RENAL TUBULE CELLS AND OTHER TISSUES

Immunocytochemical Studies in Rats

M. Amiry-Moghaddam,[1] E. A. Nagelhus,[1] P. Agre,[2] S. Nielsen,[3] and O. P. Ottersen[1]

[1] Department of Anatomy
Institute of Basic Medical Sciences
University of Oslo, Oslo, Norway
[2] Johns Hopkins University School of Medicine
Baltimore, Maryland
[3] Department of Cell Biology
Institute of Anatomy
University of Aarhus, Aarhus, Denmark

INTRODUCTION

One of the best established roles of taurine is that of an osmoregulator[7]. Many cell types have been shown to sustain a loss of taurine upon exposure to hypoosmotic stress[11] while an increased uptake of this amino acid can be induced by hyperosmotic conditions[12]. The latter effect has been demonstrated in renal cells in culture, thus supporting the notion that taurine acts as an organic osmolyte in the kidney along with polyhydric alcohols and methylamines[5].

If taurine simply serves to balance the extracellular osmolarity one would expect a gradual increase in the intracellular contents of this amino acid when moving from the renal cortex to the tip of the medullary pyramids. We show in the present report that the cellular distribution of taurine does not conform to such a pattern. On the contrary, data obtained by light microscopic and quantitative electron microscopic immunocytochemistry point to a heterogeneous distribution of taurine among kidney tubule cells. This suggests that the functional role of taurine in kidney cells extends beyond that of serving as an organic osmolyte. The striking co-localization of taurine and water channels (aquaporins) demonstrated in the present report raises the possibility that taurine directly or indirectly regulates the expression or activity of these proteins.

Taurine 2, edited by Huxtable et al.
Plenum Press, New York, 1996

MATERIAL AND METHODS

Rats were perfusion-fixed through the heart with a mixture of glutaraldehyde and formaldehyde. For the first series of animals we used 2.5% glutaraldehyde and 1% formaldehyde (fixative #1) while a second series of animals was perfused with 0.1% glutaraldehyde and 4% formaldehyde (fixative #2). Kidney tissue specimens from the first series were treated with OsO_4, dehydrated, and embedded in an epoxy resin (Durcupan). Specimens from the second series were cryoprotected with glycerol, frozen in liquid propane, and subjected to a freeze substitution procedure as previously described[6]. Embedding was performed in a metachrylate resin (Lowicryl HM20) which was polymerized by UV light at low temperature.

Ultrathin sections from the Lowicryl-embedded material and adjacent ultrathin and semithin sections from the Durcupan-embedded material were incubated with antisera raised against glutaraldehyde-conjugated taurine. These antisera (#20 and #286) have been shown previously to react selectively with taurine, except for minor cross-reactivity with hypotaurine[2]. Bound antibodies were visualized by the peroxidase-antiperoxidase procedure for light microscopy, and by use of secondary gold-conjugated antibodies for electron microscopy.

Some ultrathin sections from the Lowicryl-embedded material were double-labelled with antisera to taurine and to aquaporin-1[9]. The two antigens were distinguished by use of different gold particle sizes (30 nm for aquaporin-1 and 10 nm for taurine). Interference between the two sequential immunoincubations was avoided by treating the sections with formaldehyde vapour after the first incubation[14,10].

In some rats, tissue from other organs was removed and processed similarly to the kidney tissue.

The density of gold particles over different tissue compartments was estimated by use of a tailor-made computer programme (Morforel[3]). Model sections containing a series of known antigen concentrations were incubated together with the tissue sections. This permitted the gold particle densities to be translated into approximate concentrations of fixed taurine. There was a close to linear relationship between the concentration of fixed taurine and the density of gold particles ($r=0.98$)[2].

RESULTS

The light and electron microscopic preparations revealed a differentiated distribution of taurine among kidney tubule cells. High levels of immunoreactivity occurred, inter alia, in type 2 epithelium in the descending thin limb of Henle, in proximal straight tubule cells, and in collecting duct cells (the latter known to express aquaporin-2), while very weak immunolabelling was found in cells of the thick and thin ascending limbs (Figs. 1 and 2). The intercalated cells of the collecting ducts showed considerably less labelling than the collecting duct cells. Judged by the calibration system, the gold particle densities corresponded to intracellular concentrations of fixed taurine ranging between 11 mM (for the type 2 cells) to close to zero (for cells of the thick ascending limb; see Fig. 4).

In double-labelled sections, the large gold particles (signalling aquaporin-1) were selectively associated with cells that were also heavily labelled with small particles (signalling taurine). Thus high levels of both immunoreactivities occurred in type 2 cells of the thin descending limb of Henle and in proximal straight tubule cells. No instance was found of large gold particles decorating cells that contained little or no immunoreactivity for taurine.

Within the individual double-labelled cells the two immunoreactivities displayed distinct distributions. While the taurine labelling was found throughout the cytoplasm and nucleus, the large gold particles representing aquaporin-1 immunoreactivity were primarily

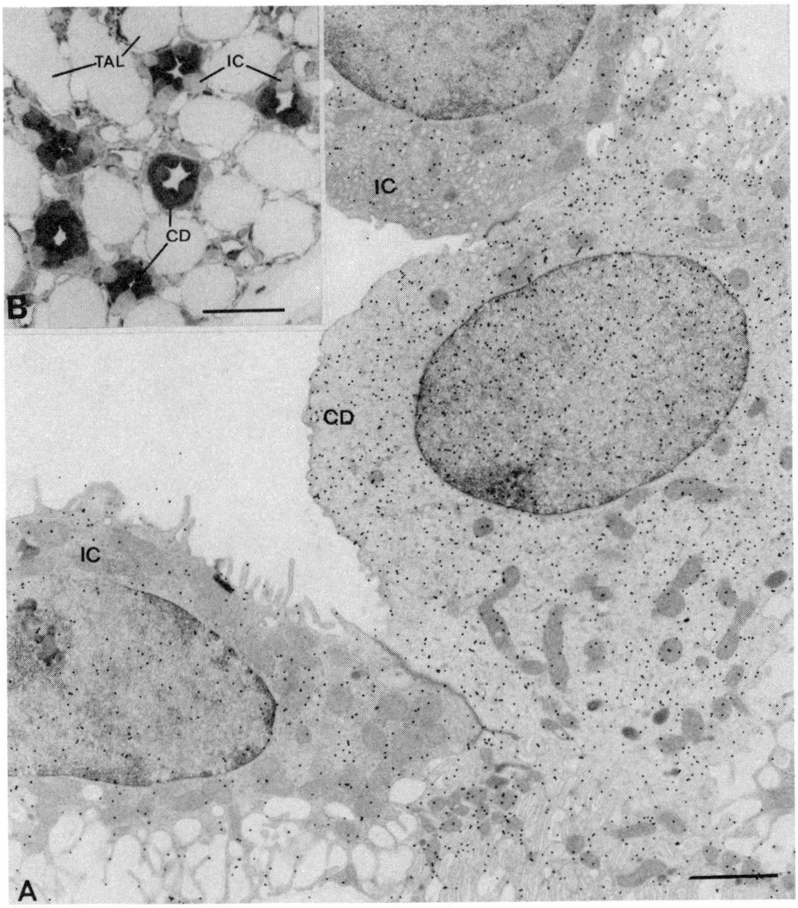

Figure 1. Electron microscopic (A) and light microscopic (B) distribution of taurine immunoreactivity in rat kidney tubules. The collecting duct cells (CD) are strongly labelled compared to the intercalated cells (IC) and cells of the thick ascending limb (TAL; B). Bars =1.7 μm (A); 40 μm (B).

associated with the apical plasma membrane, although significant labelling was also encountered over the basolateral membrane and over a subpopulation of vesicles (Fig. 3A; also see Nielsen *et al.*[9]).

Because our data from kidney indicated a close correspondence between the cellular distributions of taurine and water channels, we investigated several extrarenal tissues known to be equipped with different types of such channels[1], including red blood cells, excretory duct epithelium in the pancreas, biliary duct epithelium in liver, corneal endothelium and epithelium, capillary endothelium of several organs, and magnocellulary neurons of the hypothalamus. High levels of taurine immunoreactivity were detected in all of these cell categories.

DISCUSSION

The present immunocytochemical observations suggest that the level of taurine varies by a factor of more than 50 among different cells of rat kidney tubules. This variation is not

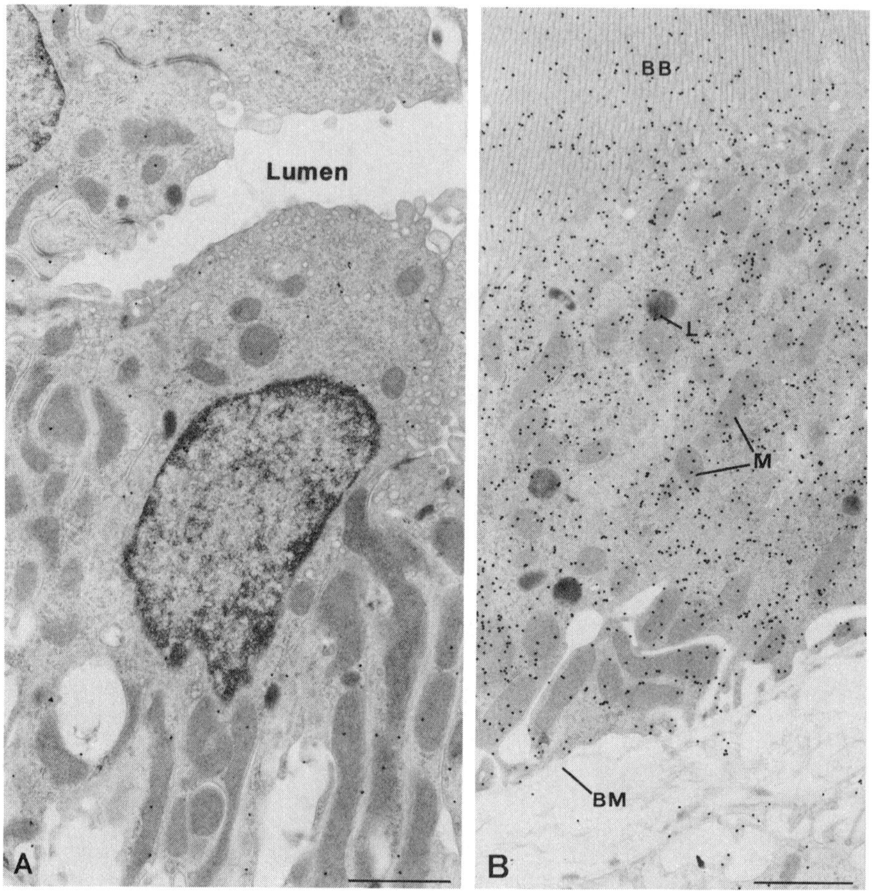

Figure 2. Electron micrographs showing taurine immunolabelling of thick ascending limb (A) and proximal straight tubule (B) from the same section. M, mitochondria; L, lysosome; BB, brush border; BM; basal membrane. Note the comparatively weak labelling of the thick ascending limb (compare with Figure 1B). Bars = 1.5 μm.

a function of the external osmolarity as the labelling intensity appeared to be independent of the cells' distance from the medullary tip. Rather, the labelling density was strongly correlated to cell type. This was particularly evident in the loop of Henle where type 2 cells of the descending limb were heavily labelled while cells in the ascending limb were almost devoid of immunoreactivity.

What do the strongly taurine immunoreactive cells have in common? Both in Henle's loop and in the remainder of the tubule system there is a close correspondence between taurine enrichment and water fluxing capacity[4,13]. Thus a high water fluxing capacity is a well known feature of part of the descending limb of Henle, and is also characteristic of the proximal tubule. In recent years it has been established that high water permeability in kidney cells as well as in other tissues is mediated by a special class of protein, the aquaporins. To corroborate our assumption of a correlation between taurine levels and water fluxing capacity we performed double-labelling with antisera to taurine and aquaporin-1. The double-labelled preparations clearly showed that cells equipped with aquaporin-1 invariably contained high levels of taurine. Although not demonstrated directly in this study, this correlation also holds

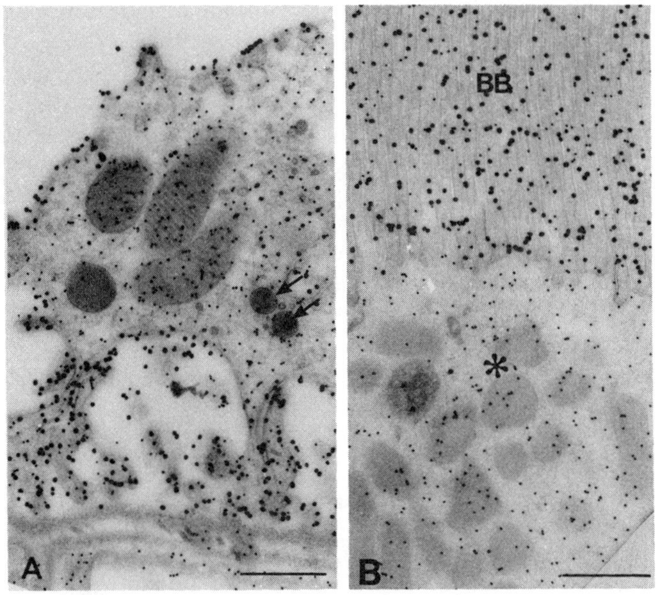

Figure 3. Double immunolabelling with antiserum to taurine (small particles) and aquaporin-1 (large particles). Type 2 cells of the descending thin limb of Henle (A) show intense taurine immunolabelling throughout the cytoplasm and aquaporin immunolabelling restricted to the apical and basal plasma membranes and to a subpopulation of vesicles (arrows). A co-localization of the two immunoreactivities is also found in proximal straight tubule cells (B). Note that aquaporin 1 immunolabelling is selectively present in the brush border (BB). The asterisk shows cytoplasm in the proximal tubule cell. Bars = 1.5 μm (A); 2.0 μm (B).

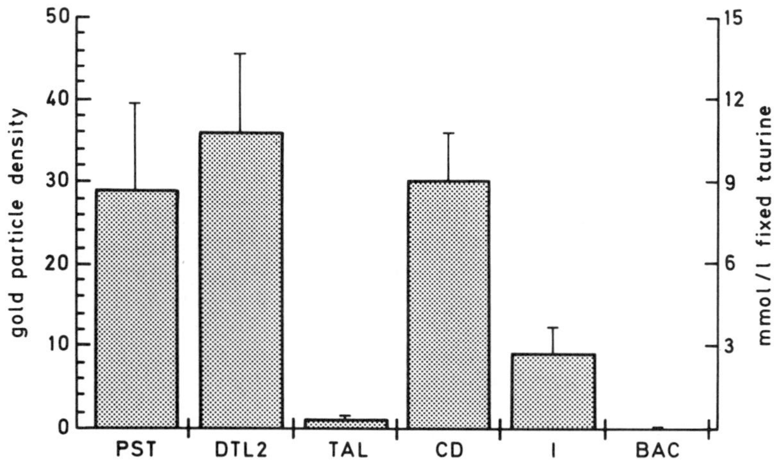

Figure 4. Diagram showing quantitative distribution of taurine immunolabelling in different tubule cells from perfusion fixed rat kidney. Gold particle densities (left axis) were translated into concentration of fixed taurine (right axis), using model sections incubated together with the tissue sections. The model sections contained a series of known concentrations of taurine[2]. Note that the mM values represent an underestimate of the *in vivo* concentrations of free taurine since the fixation efficiency is less than 100%[8]. PST, proximal straight tubules; DTL2, type 2 cells of the descending thin limb; TAL, thick ascending limb; CD, collecting duct cells; I, intercalated cells; BAC, background (empty resin).

for aquaporins 2, 3, and 4. The latter proteins are concentrated in collecting duct cells which were strongly taurine-immunoreactive in the present material.

To investigate whether the association between taurine and aquaporins is restricted to the kidney or whether it is a more general phenomenon we examined numerous extrarenal tissues known to harbour aquaporins. All aquaporin containing cells that have been investigated so far were found to be strongly labelled for taurine. Only a few mismatches were found. One notable example is the choroid plexus cells: these cells contain aquaporins but exhibited little taurine. Vice versa, although the parietal cells of Bowman's capsule in rats were strongly taurine-immunoreactive they have not been shown to contain any of the known aquaporins. However, human capsule cells do contain water channels (Nielsen, unpublished observations), raising the possibility that their rat counterparts are equipped with a hitherto unrecognized isoform of this protein.

In conclusion, it seems that a high taurine content is a hallmark of aquaporin-containing cells. Our findings suggest that taurine has an important function linked to the activity of aquaporins, but do not exclude any of the other roles that have been proposed for this amino acid. Of course, a spatial relationship between two molecules, no matter how close, is not necessarily indicative of a physiological interaction. Functional studies are now being conducted to clarify this issue.

ACKNOWLEDGMENTS

This work was supported by the Norwegian Research Council.

REFERENCES

1. Agre, P., Brown, D. and Nielsen, S. 1995, Aquaporin water channels: unanswered questions and unresolved controversies, *Curr. Opin. Cell Biol.*, in press.
2. Amiry-Moghaddam, M., Nagelhus, E.A. and Ottersen, O.P. 1994, Light - and electronmicroscopic distribution of taurine, an organic osmolyte, in rat renal tubule cells, *Kidney Int.* 45:10-22.
3. Blackstad, T.W., Karagülle, T. and Ottersen, O.P. 1990, MORFOREL a computer program for two-dimensional analysis of micrographs of biological specimens, with emphasis on immunogold preparations, *Comput. Biol. Med.* 20:15-34.
4. Chou, C-L. and Knepper, M.A. 1992, In vitro perfusion of chinchilla thin limb segments: Segmentation and osmotic water permeability. *Am. J. Physiol.* 263: F417-F426.
5. Garcia-Perez, A. and Burg, M.B. 1991, Renal medullary organic osmolytes, *Physiol. Rev.* 71:1081-1115.
6. Hjelle, O.P., Chaudhry, F.A. and Ottersen, O.P. 1994, Antisera to glutathione: characterization and immunocytochemical application to the rat cerebellum, *Eur. J. Neurosci.* 6:793-804.
7. Huxtable, R.J. 1992, The physiological action of taurine, *Physiol. Rev.* 72:101-163.
8. Madsen, S., Ottersen, O.P., Storm-Mathisen, J. and Sturman, J.A. 1990, Immunocytochemical localization of taurine: Methodological aspects, in: "Taurine: Functional Neurochemistry, Physiology, and Cardiology," Pasantes-Morales, H., Martin, D.L., Shain, W. and Martin Del Rio, R. eds., Wiley-Liss, New York, pp. 37-44.
9. Nielsen, S., Smith, B.L., Christensen, E.L., Knepper, M.A. and Agre, P. 1993, CHIP28 water channels are localized in constitutively water-permeable segments of the nephron, *J. Cell Biol.* 120:371-383.
10. Ottersen, O.P., Zhang, N. and Walberg, F. 1992, Metabolic compartmentation of glutamate and glutamine: Morphological evidence obtained by quantitative immunocytochemistry in rat cerebellum, *Neuroscience* 46:519-534.
11. Pasantes-Morales, H. and Schousboe, A. 1988, Volume regulation in astrocytes: A role for taurine as an osmoeffector, *J. Neurosci. Res.* 20:505-509.
12. Uchida, S., Nakanishi, T., Kwon, H.M., Preston, A.S. and Handler, J.S. 1991, Taurine behaves as an osmolyte in Madine-Darby Canine Kidney cells. Protection by polarized, regulated transport of taurine, *J. Clin. Invest.* 88:656-662.

13. Wade, J.B., Stanton, B.A. and Brown, D. 1992, Structural correlates of transport in distale tubule and collecting duct segments, in: "Handbook of Physiology: Renal Physiology," Windhager, E.E. ed., Oxford University Press, New York, pp. 169-226.
14. Wang, B.L. and Larsson, L.I. 1985, Simultaneous demonstration of multiple antigens by indirect immunofluorescence or immunogold staining. Novel light and electron microscopic double and triple staining method employing primary antibodies from the same species, *Histochemistry* 83:47-56.

20

REGULATION OF EXPRESSION OF TAURINE TRANSPORT IN TWO CONTINUOUS RENAL EPITHELIAL CELL LINES AND INHIBITION OF TAURINE TRANSPORTER BY A SITE-DIRECTED ANTIBODY

Xiaobin Han, Russell W. Chesney, Andrea M. Budreau, and Deborah P. Jones

Department of Pediatrics
University of Tennessee
Memphis, Tennessee

ABSTRACT

The renal tubular epithelium adapts to changes in the sulfur amino acid composition of the diet, particularly in terms of reabsorption of taurine. The adaptive response is expressed by enhanced or decreased NaCl-dependent taurine transport by rat renal brush border membrane vesicles (BBMV). Taurine transport activity in two cultured renal epithelial cell lines (MDCK and LLC-PK1) is up- or down-regulated by extracellular taurine concentration as the result of reciprocal changes in the V_{max} of the transporter. In MDCK cells, abundance of taurine transporter mRNA (pNCT mRNA) was up- or down-regulated after incubation in media containing 0, 50, or 500 μM taurine. Decreased mRNA was observed in both cell lines after 12 h, and it was appreciably reduced after 72 h exposure to 500 μM taurine. Northern blot analysis of mRNA from LLC-PK1 cells using pNCT cDNA as a riboprobe showed that two transcripts, 9.6 kb and 7.2 kb, were expressed; the abundance of mRNA was increased or decreased after incubation in taurine-free or high taurine medium, respectively. Down-regulation was observed primarily in the 7.2 kb transcript after 24 h incubation. Rapid up-regulation occurred in the 9.6 kb transcript within 12 h of transfer from high to low taurine. Nuclear run-off assays showed that the gene for pNCT is induced at the transcriptional level by taurine. Regulation of expression of the taurine transporter was also studied by injection of pNCT cRNA into *Xenopus laevis* oocytes. Expression of transport activity was significantly reduced (64%) when oocytes were incubated in 50 μM taurine as compared to 0 μM taurine. Transport activity was totally blocked when pNCT cRNA-injected oocytes were exposed to an active phorbol ester, PMA (10^{-6} M). Inhibition of uptake was reversed by staurosporine, an inhibitor of protein kinase C activity. An inactive phorbol

Taurine 2, edited by Huxtable et al.
Plenum Press, New York, 1996

ester, 4α-phorbol, had no effect on taurine transport. A polyclonal antibody directed against a highly conserved intracellular segment between homologous transmembrane domains VI and VII inhibited taurine transport activity in both pNCT cRNA-injected oocytes and BBMV. Incubation of oocytes with 10 μg/ml antibody (Ab) reduced taurine uptake to 46% of control, and 20-80 μg/ml Ab reduced uptake to 20% of control. In BBMV, active taurine uptake (10 μM) was inhibited ≈30% by 10 pg Ab/mg protein, whereas nonspecific IgG had no significant effect. Proline uptake (20 μM) by BBMV was not inhibited by the Ab, nor was GABA uptake (50 μM). Two pNCT proteins, ≈70 kD and ≈30 kD, were detected by Western blot, and the abundance of both was regulated by medium taurine. In conclusion: (i) regulation of taurine transport activity in LLC-PK1 cells by medium taurine occurs at the level of mRNA transcription; (ii) Regulation of pNCT occurs at both transcriptional and translational levels; (iii) pNCT expression is regulated by protein kinase C-dependent phosphorylation; and (iv) The intracellular segment between domains VI and VII may be required for activation of the taurine transporter; this segment may function as a gate in taurine transport.

INTRODUCTION

The renal tubular epithelium is capable of adapting to alterations in the sulfur amino acid composition of the diet, particularly in terms of the reabsorption of taurine[8-10,17]. This renal adaptive response is evidenced by augmented NaCl-dependent taurine transport by brush border membrane vesicles (BBMV) after a low taurine diet (LTD) as compared to a normal taurine diet (NTD). These changes in tubule and BBMV accumulation after dietary manipulation are associated with a change in the initial rate of taurine uptake (V_{max}) rather than the affinity of the transporter for taurine (K_m)[10,16,24]. This adaptive regulation has also been clearly demonstrated in two continuous renal epithelial cell lines: LLC-PK1 (a cell line of porcine proximal tubular origin) and MDCK (Madin-Darby canine kidney cell line, presumably of distal origin). Incubation of cell monolayers in taurine-free medium rapidly induces an increase in NaCl-dependent taurine uptake as compared to cells exposed to standard medium (50 μM taurine). Incubation of cells in high taurine medium (500 μM taurine) results in reduced uptake as compared to uptake by control cells. Again, changes in the apparent transport maximum (V_{max}) rather than in the apparent K_m account for this phenomenon[8,10,17,20].

The purpose of these studies is to define the factors that regulate this renal adaptive response to dietary sulfur amino acid and taurine intake. The continuous renal cell method allows examination of active transport as well as performance of the molecular biologic experiments necessary to define regulation. Two types of adaptive regulation, short-term and long-term, have been described previously[20,30]. In rats the long-term adaptive response occurs at the level of mRNA, and it is blocked by actinomycin D. In MDCK or LLC-PK1 cells, the short-term adaptive response is unaffected by actinomycin D or cyclohexamide. In *Xenopus laevis* oocytes injected with cRNA from pNCT, a taurine transporter cloned from MDCK cells[27], the expressed taurine transport activity is regulated by medium taurine. The mechanism of the adaptive response to alterations in diet or medium content is unknown. Based on the knowledge described above, the question was raised whether the renal adaptive response occurred at the level of transcription, the level of translation, or both. We also asked whether post-translational modification, particularly at the level of protein kinase C-mediated phosphorylation, might regulate the activity of the taurine transporter. We have used an antibody produced against a conserved peptide sequence of the taurine transporter to detect proteins localized to oocyte or BB vesicle membranes and to examine their relative density after media or dietary manipulation.

METHODS

Animals

Male Sprague-Dawley rats (Harlan Sprague-Dawley Inc., Indianapolis, Indiana), aged 56-60 days and weighing 250-300 g each, were fed low sulfur amino acid diet (LTD), normal taurine diet (NTD), or high taurine diet (HTD) for 28 days prior to sacrifice. This time interval permits full expression of the prolonged adaptive response[17]. HTD is supplemented with 3% taurine and 0.5% methionine (w/w), NTD is supplemented with 0.5% methionine, and LTD is supplemented with neither amino acid. The composition of each diet has previously been described in detail[11], as has its effects on tissue taurine values[8,10,11]. Adult female frogs (Xenopus I Co.) were kept in 14°C water and fed with chicken liver twice a week. New Zealand White rabbits were from Myrtle's Rabbitry, Thompson Station, TN.

Membrane Vesicle Preparation

Renal cortex BBMV were prepared by a modification of the method of Booth and Kenny[2]. Rats were decapitated and exsanguinated, the kidneys removed, decapsulated, and placed in a solution of 0.05 mM mannitol, 1 mM $MgSO_4$, 2 mM Tris, and 2 mM (HEPES, pH 7.1 (THM·$MgSO_4$). Renal cortex samples weighing 1.5 to 6.0 g each (wet weight) were homogenized in 20 volumes of THM·$MgS0_4$ for 1 min with a Polytron (Kinematica, Switzerland) at setting no. 4 for 45 s and setting no. 7 for 15 s. $MgCl_2$ was added to the homogenate to a final concentration of 10 mM and stirred on ice for 15 min to aggregate the intracellular and basolateral membranes. Repeated centrifugations and an additional $MgCl_2$ precipitation step were performed, as described elsewhere[12]. The final membrane preparation was suspended in 396 mM mannitol, 1 mM $MgS0_4$, 2 mM HEPES-Tris (pH 7.35), and contained approximately 10 mg protein/ml. The last two centrifugations and final suspension were done in the presence of 10 pg antibody (described below) or non-specific IgG (purified from sera of the rabbits before antigenic challenge) per mg protein to insure access of the antibodies to the intravesicular surface. In all experiments membrane vesicles were used for uptake studies immediately after preparation. Protein concentration was determined by the method of Bradford using the BioRad Coomassie Brilliant Blue Dye reagent[3].

RNA Preparation

Rats were anesthetized and exsanguinated; the kidneys were removed, decapsulated and total RNA was prepared as previously described[25]. Briefly, guanidinium thiocyanate was used to disrupt rat kidney cortex tissues, and the resulting homogenate was layered on a cushion of a dense solution of CsCl (5.7 M, 0.01 M EDTA, pH 7.5). Total RNA was pelleted from the lysate after centrifugation at 32,000 x g for 18 h at 25°C. Poly(A)+ RNA was isolated by oligo(dT)-cellulose chromatography, reprecipitated in ethanol, and then dissolved in DEPC (diethyl pyrocarbonate)-treated water to a final concentration of 1.0 μg/μl. Four rats on each diet were used separately for preparation of mRNA.

Microinjection of Oocytes

Ovarian lobes were dissected from anesthetized frogs (*Xenopus laevis*) and oocytes separated by incubation of ovarian fragments for 40 min with 2 mg/ml collagenase type II in Ca^{2+}-free buffer at room temperature. Defolliculated stage V-VI oocytes were selected and incubated overnight at 18°C in media containing 50% L-15 (Leibovitz, Sigma), 1 mM

L-glutamine, 15 mM HEPES, pH 7.6, and 100 μg/ml gentamicin sulfate) before injection[14,28,29]. 40 nl of water containing 0-40 ng of poly(A)+ RNA were injected into each oocyte. Injected oocytes were maintained at 18°C in the above media for 1-3 days.

BBMV Uptake Studies

Uptake of radiolabelled taurine, proline, and GABA by vesicles was assayed by a Millipore rapid filtration technique. In general, incubation of the membranes was initiated by adding freshly prepared vesicles to media containing NaCl and known amounts of labelled and unlabelled amino acid to achieve the desired final concentration. Experimental uptake media contained [^{3}H]taurine, 2 mM HEPES-Tris (pH 7.35), 1 mM $MgSO_4$, 100 mM NaCl, and mannitol to balance osmolarity with that of the intravesicular medium (300 mOsm). At specific time intervals, 50 μl samples of the incubation mixture were placed onto pre-wetted 0.45 μm Millipore filters (HAWP) and washed with 3 volumes (3.0 ml each) of an ice-cold "stop" solution that contained 2 mM HEPES-Tris, 1 mM $MgSO_4$, and 150 mM NaCl. All incubations were performed in triplicate and each experiment was performed a minimum of three times. Data are expressed as pmol amino acid/mg protein retained by the vesicles on the filters. Filters were dried overnight in open scintillation vials, then dissolved in a toluene-based cocktail and measured for radioactivity in a liquid scintillation counter (Packard Tri-Carb 2000CA). Nonspecific retention of radioactivity to detergent-lysed (0.1% SDS) membranes never exceeded 1-1.5% of the total amount found.

Oocyte Uptake Studies

Oocytes were transferred to Na^+-containing uptake solution (2 mM KCl, 1 mM $MgCl_2$, 100 mM NaCl) or Na^+-free uptake solution containing 10 μM unlabelled taurine and 0.5 μCi/ml [^{14}C]taurine. After incubation for an appropriate time at room temperature, oocytes were transferred to a 24-well cell culture plate in which they were washed rapidly 5 times with 1 ml of ice-cold Na^+-free uptake solution. Finally, individual oocytes were transferred to miniscintillation vials, solubilized in 100 μl of 10% sodium dodecyl sulfate (SDS), and counted in 2 ml of scintillation cocktail.

Reverse Transcription/Polymerase Chain Reaction (RT-PCR)

To determine if the renal adaptive response occurs at the level of taurine transporter mRNA, two primers were designed such that a 902-base pair fragment of rB16a could be amplified from cDNA and detected by Southern blot analysis. The reverse transcription reactions were performed individually with each mRNA in 20 μl of 1x reverse transcription buffer (10 mM Tris-HCl, pH 8.8, 50 mM KCl, 0.1% Triton X-100, 5 mM $MgCl_2$) containing 1 mM dNTP, RNasin ribonuclease inhibitor at 1 unit/μl, 1 μg oligo $(dT)_{15}$ primer, 1 μg mRNA and 30 units of AMV reverse transcriptase (H.C., Promega), at 42°C for 30 min. After cDNA synthesis, the reactions were stopped by heating at 95°C for 5 min, then the tubes were quickly chilled on ice. PCR reactions were carried out in 50 μl of 1x PCR buffer (10 mM Tris-HCl, pH 8.3, 50 mM KCl), 200 μM of each dNTP, 1.5 mM $MgCl_2$, 0.001% gelatin, 0.4 μM of each primer, 2.5 units of Taq polymerase and either cDNA or water using the following temperature profile: denaturation, 45 s at 94°C; annealing, 45 s at 66°C (58°C for β-actin); polymerization, 2 min (in the last cycle, 6 min) at 72°C for 30 cycles. The sequences of the sense and antisense primers were derived from amino acids 40-47 (5'-TCAGAGGGAGAAGTGGTCCAGCAAG-3')[26], and 335-341 (5'-AGCCAGACA-CAAAACTGGTACCA-3'), respectively. Primers were also designed to amplify the cDNA encoding β-actin (sense, 5'-ATGGATGACGATATCGCTG-3', antisense, 5'-

ATGGTAGTCTGTCAGGT-3', which generate a product of 568 bp)[23], as control. To detect amplified products, an oligonucleotide probe was synthesized (corresponding to amino acids 249-271) that is specific for rB16a[26], and an internal oligo was used (5'-AGCAAGA-GAGGTATTCCT-3') for the β-actin gene.

Northern Blot Analysis

Ten μg of poly(A)+ RNA isolated from kidney cortex of rats on each diet were separated in a 1.2% agarose gel containing 6.5% formaldehyde and transferred to a nylon membrane by overnight capillary blotting in l0x SSC (1.5 M NaCl, 1.6 M sodium citrate buffer). Before hybridization the membrane was incubated for 1 h at 42°C in a prehybridization solution containing 50% formamide, 1 M NaCl, 5% dextran sulfate, and 1% SDS. The Northern blot was hybridized overnight at 42°C with ^{32}P-labelled rB16a subclone (in pGEM-T vector, Promega) riboprobe in the hybridization solution. The blot was washed successively in 2x SSC/0.1% SDS, 1x SSC/0.1% SDS, and O.1x SSC/0.1% SDS at 65°C for 30 min respectively, and exposed to Kodak film with one intensifying screen at -80°C for three weeks.

Polyclonal Antibody Preparation

A 14 mer synthetic conserved peptide (GSYNKYKYNSYRDC, 313-326 residues) sequence from the cytoplasmic domain of predicted taurine transporter was coupled to keyhole limpet hemocyanin (KLH) using a kit (Imject SuperCarrier EDC System for Peptides, PIERCE). Rabbits were immunized with approximately 1000 μg of immunogen and boosted with an equivalent amount of antigen 2 weeks later. The animals were bled through the ear vein on days 21 and 36 to measure the antibody response. Antisera were collected on day 49 and purified by affinity chromatography.

Western Blot Analysis

Equal amounts of BBMV protein from rats fed different diets were separated by electrophoresis on a 12% SDS-polyacrylamide gel and transferred to a nitrocellulose membrane. Membranes were incubated with a 1:1000 dilution of antibody against the conserved peptide sequence of the taurine transporter. The blots were detected using an immunoblotting systems kit.

Nuclear Run-Off Assays

To determine the transcriptional regulation of pNCT mRNA by taurine, nuclear run-off assays were carried out according to methods described previously[16]. Medium was removed from five 150 x 20 mm tissue culture dishs, the cells placed on ice, and each dish washed twice with 5 ml ice-cold PBS. Cells were gently dislodged by scraping with a rubber policeman in 5.0 ml of fresh PBS, then centrifuged for 5 min at 500 x g at 4°C. The supernatant was removed and 4 ml of PNP-40 lysis buffer (10 mM Tris pH 7.4, 10 mM NaCl, 3 mM $MgCl_2$, 0.5% NP-40) were added. Cells were vortexed at half maximal speed for 10 s and incubated on ice for 5 min. After centrifugation at 500 x g for 5 min, the supernatant was discarded and the nuclear pellet washed with 4 ml NP-40 lysis buffer once, then resuspended in 200 μl of glycerol storage buffer (50 mM Tris, pH 8.3, 40% glycerol, 5 mM $MgCl_2$, 0.1 mM EDTA). Nuclei were stored at -80°C. For transcription, 200 μl of nuclei in glycerol storage buffer were incubated in a 400 μl reaction mixture containing 200 μl of 2x reaction buffer (for 1 ml, 10 μl each of 100 mM ATP, 100 mM GTP, 100 mM CTP, 5 μl 1M

DTT, dissolved in 0.5 mM EDTA, pH 8.0) plus 10 μl (containing 250 μCi) [α-^{32}P]UTP for 30 min at 30°C with shaking. Transcripts were isolated by standard phenol extraction and ethanol precipitation after DNase and proteinase K treatments of the nuclei. To quantify run-off transcripts of the taurine transporter and β-actin gene, plasmids carrying the corresponding cDNAs were immobilized on nitrocellulose membranes and hybridized transcripts were quantified by a Phosphor-Imager (Sony). Hybridizations were performed in 50% formamide, 5x SSC, 0.1% SDS, 20 mM Tris, pH 8.0, 5x Denhardt's solution, and 100 μg/ml salmon sperm DNA, and washed three times at 65°C in O.1x SSC/0.1% SDS.

Data Analysis

All experiments were performed at least twice; most were done in triplicate, including BBMV experiments. Student's t test was applied to determine significant differences in the means.

Materials

[^{14}C]Taurine (92.1 mCi/mmol), [^{3}H]taurine (21.9 Ci/mmol), [^{3}H]proline (60.0 Ci/mmol), [^{3}H]γ-aminobutyric acid (GABA) (33.05 Ci/mmol), [^{32}P]UTP (3000 Ci/mmol), and [^{32}P]dATP (3000 Ci/mmol) were purchased from Du Pont, NEN. pNCT was a generous gift from Dr. S. Uchida of the University of Tokyo.

RESULTS

Transcriptional Regulation of Expression of Taurine Transporter by Taurine

Figure 1A, lane 1, shows that the abundance of pNCT (MDCK cell taurine transporter) mRNA is increased in cells incubated in taurine-free medium. When cells were incubated in medium containing high taurine (500 μM), the abundance of pNCT mRNA was decreased (lane 3) as compared to control (lane 2, 50 μM taurine) after 72 h incubation. There was no influence of medium taurine content on the abundance of β-actin, which served as a control for lane loading.

The time course of down-regulation of pNCT mRNA by high medium taurine (500 μM) was also studied in MDCK cells. As shown in Fig. 2, the down-regulation of pNCT was observed after 12 h exposure and message expression was highly significantly reduced after 72 h incubation. Again, β-actin expression was unchanged.

The stability of pNCT mRNA was determined in cells exposed to 0 μM, 50 μM, or 500 μM taurine after treatment with 10 μg/ml actinomycin D to inhibit transcription. Equal quantities of total cellular RNA were loaded per track of Northern blot. The half-life of pNCT mRNA was greater than 24 h in all experimental groups (Fig. 3). These results indicate that medium taurine content has no effect on the stability of pNCT mRNA within a 24 h incubation period. The expression and stability of β-actin were also unchanged by taurine exposure.

To investigate whether transcription of the pNCT gene was decreased by the presence of taurine, nuclear run-off asssays were performed on nuclei isolated from MDCK cells after 72 h exposure to taurine-free or high taurine media. As shown in Fig. 4, the transcription rate of the gene for the taurine transporter was down-regulated after cells were incubated in medium containing 500 μM taurine for 72 h, compared to cells incubated in taurine-free

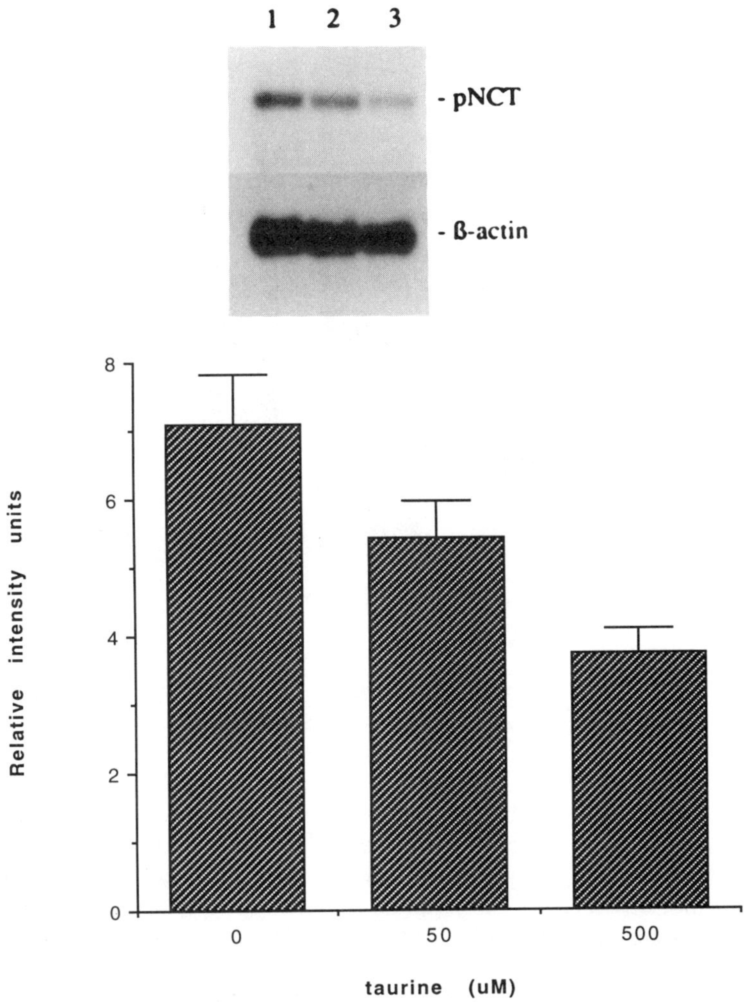

Figure 1. Adaptive regulation of expression of taurine transporter mRNA by taurine. MDCK cells were cultured in medium containing 0, 50, or 500 μM taurine for 24 h, then cells were harvested for preparation of total cellular RNA. Northern blot analysis was carried out as described in Methods. A: Lane 1, 0 μM taurine; Lane 2, 50 μM taurine; Lane 3, 500 μM taurine. β-actin was used as an internal control for loading. B: The relative levels of intensity of each band were assayed and quantitated using a Phosphor-Imager and Image Quant software (Molecular Dynamics). Values were expressed as mean ± SEM of three independent experiments.

medium ($p < 0.001$). These data indicate that this gene is a primary target for adaptive regulation by taurine.

Further Evidence of Transcriptional Regulation

Studies have shown that rB16a, a rat brain taurine transporter, is expressed in various tissues, including the kidney[26]. Southern blot analysis of PCR products showed that expression of rB16a in rat kidney was up-regulated by feeding rats LTD for two weeks and down-regulated by two weeks of HTD, as compared to NTD. Expression of β-actin was not

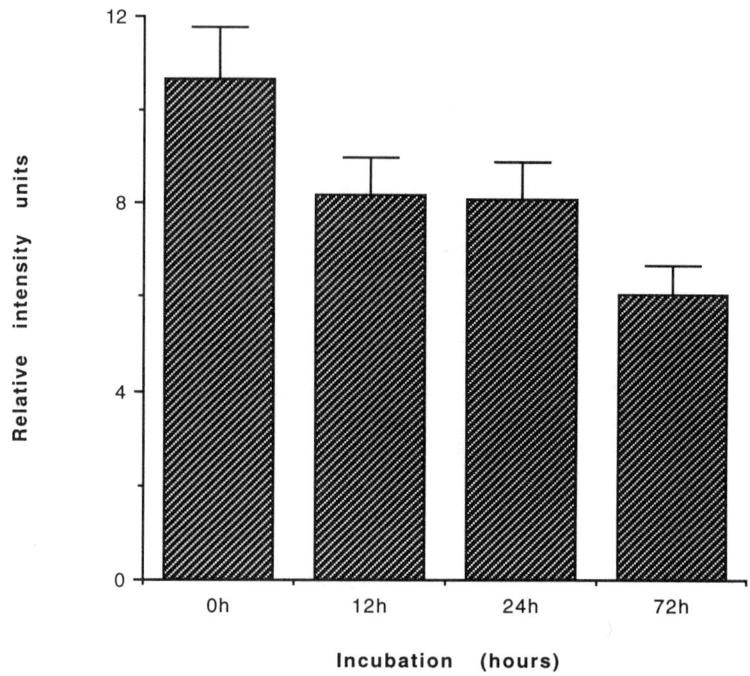

Figure 2. Down-regulation of taurine transporter mRNA by taurine. MDCK cells were exposed to 500 μM taurine for the periods of time indicated. Values are mean ± SEM of three independent experiments.

affected by diet (Fig. 5a). Northern blot analysis showed that a 2.4 kb transcript corresponding to the rB16a subclone probe was detected in mRNA isolated from the kidney cortex of each group of rats. β-actin was used as an internal control for gel loading. As shown in Fig. 5b, the abundance of the 2.4 kb transcripts was increased in renal cortex from rats fed LTD and reduced in tissue from rats fed HTD as compared to rats fed NTD. Diet had no influence on β-actin abundance.

Northern blot analysis of rB16a hybridized to mRNA from cat brain and kidney (kindly provided by Dr. J. A. Sturman) showed that prolonged dietary deprivation led to increased expression of cat kidney taurine transporter message, but that cat brain mRNA was identically expressed whether the cat was deprived or fed sufficient dietary taurine (Fig. 6).

A 900 base pair subclone product was amplified by PCR using primers that were specific for the cloned taurine transporter. Southern blot showed that several tissues hybridize to this subclone product of rB16a, including cat kidney and brain, rat kidney, MDCK

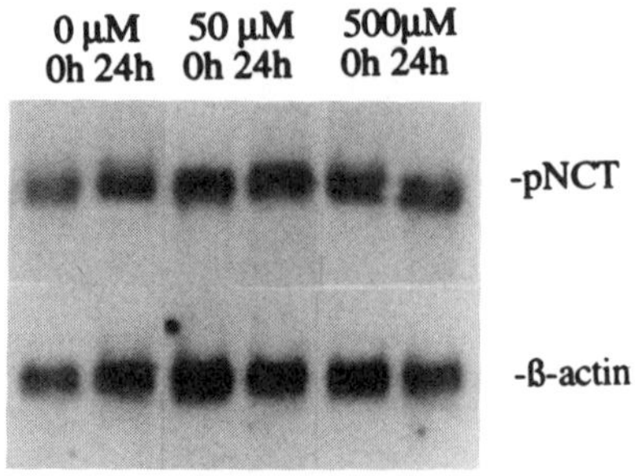

Figure 3. Stability of taurine transporter mRNA in the presence or absence of taurine. Actinomycin D (10 μg/ml) was added 20 min before taurine was added. Cells were continuously cultured for the time indicated. Northern blot analysis was done as for Fig. 1. Values are means ± SEM of three independent experiments.

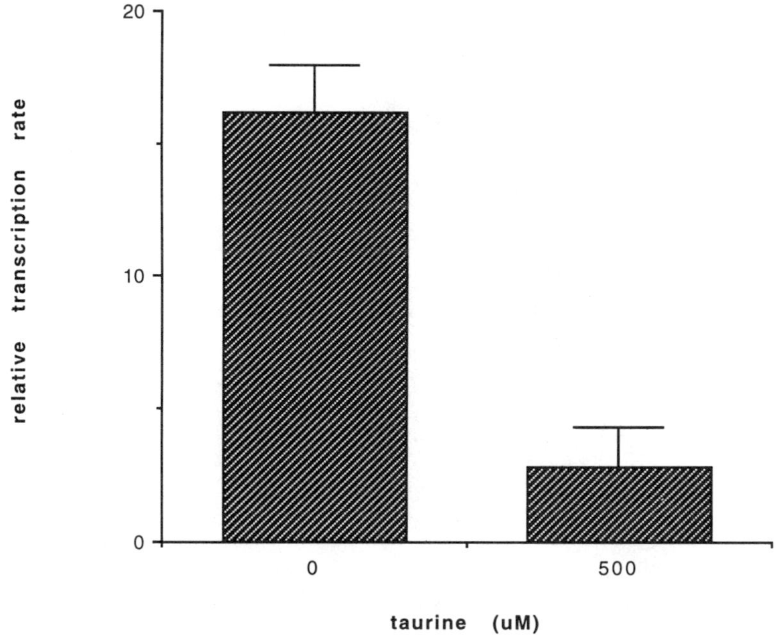

Figure 4. Nuclear run-off assay. pNCT cDNA (5 μg) was immobilized on nitrocellulose, and probed with ^{32}P-labelled nuclear transcripts isolated from nuclei which had been prepared from MDCK cells after 72 h incubation with (500 μM) or without taurine. pSPORT I DNA vector was used to control for nonspecific binding. Hybridization to a portion of cDNA encoding the β-actin gene was used as an internal control. The relative levels of intensity of each blot were assayed and quantitated as for Fig. 1. Values are expressed as means ± SEM of three independent experiments.

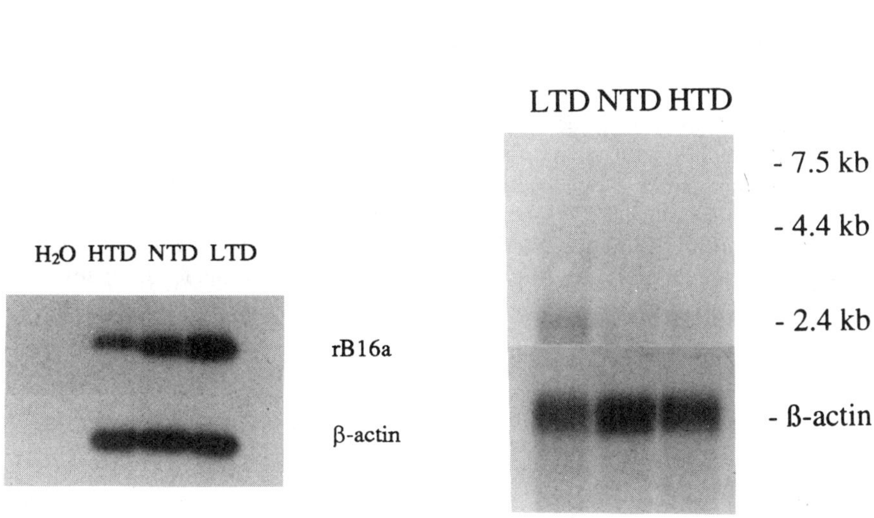

Figure 5. (a) Southern blot of PCR products shows up- and down-regulation of rB16a by dietary manipulation. (b) Northern blot showing 2.4 kb transcript corresponding to rB16a subclone probe detected in mRNA isolated from kidney cortex from rats on LTD, NTD, and HTD.

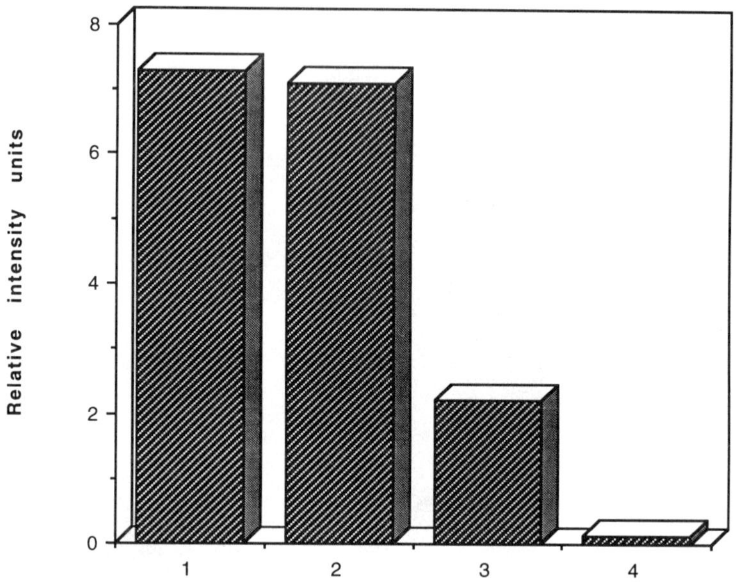

Figure 6. Relative levels of intensity of bands in Northern blot of rB16a hybridized to mRNA from cat brain and kidney. Lane 1: Cat brain mRNA following prolonged dietary taurine deprivation or (lane 2) availability. Lane 3: Cat kidney mRNA following prolonged dietary taurine deprivation or (lane 4) availability.

cells, and LLC-PK1 cells. In addition, PCR products derived from each of these tissues were detected by a ^{32}P-labelled rB16a riboprobe as well as a pNCT cDNA riboprobe. These results indicate that taurine transporters in kidney and/or brain of rat, cat, dog, and pig share a high degree of homology.

Translational Regulation

To investigate if up- or down-regulation also occurs at the translational level, *Xenopus laevis* oocytes were injected with 30 ng of pNCT cRNA and incubated in media containing 0, 50, or 500 μM taurine for 72 h, then taurine uptake was measured (Fig. 7). Results indicate that the expressed taurine transport activity was significantly reduced by the presence of 50 μM taurine (50% reduction, $p < 0.01$) or 500 μM taurine (64% reduction, $p < 0.01$) in the medium as compared to 0 μM taurine. Because the half-life of pNCT mRNA in actinomycin D-treated cells was greater than 24 h and was unaffected by medium taurine within the 24 h incubation period, this phenomenon relates to regulation at the level of translation.

Studies in LLC-PK1 Cells

At least two isoforms of mRNA that hybridize to pNCT are found in LLC-PK1 cells, a cell line of proximal origin. Expression of a 7.2 kb product is reduced by the presence of increasing amounts of taurine, whereas a 9.6 kb product is unaltered by diet (Fig. 8a). The time course of regulation is indicated in Figure 8b; expression of the 7.2 kb product declined progressively over 72 h incubation in 500 μM taurine, but the 9.6 kb product showed a decline only between 48 and 72 h. Incubation of LLC-PK1 cells in taurine-free medium for

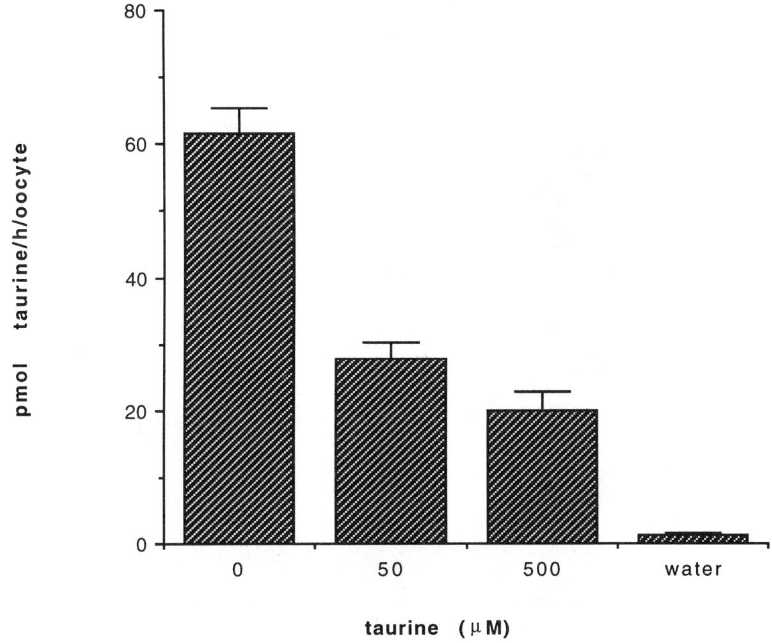

Figure 7. Post-transcriptional regulation of taurine uptake of pNCT cRNA-injected oocytes. pNCT cRNA (30 ng) was injected into each oocyte, and injected oocytes were incubated in media containing 0, 50, or 500 μM taurine for 3-4 days, then taurine uptake was measured.

12 h shows rapid up-regulation of the 9.6 kb fragment in particular (Fig. 8c). The 7.2 fragment was also up-regulated but to a lesser degree.

Post-Translational Modification

Our group has indicated that activation of protein kinase C may modify taurine accumulation by renal cell lines[22]. We therefore explored the regulation of taurine transport by protein kinase C by examining transport activity expressed in oocytes. pNCT cRNA (30 ng) was injected into each oocyte and the injected oocytes were incubated in taurine-free medium for 3 days. PMA and 4α-phorbol were added to the medium and incubated with oocytes for 2 h, then taurine uptake was measured. Taurine transport activity was significantly reduced in a dose-dependent manner in oocytes incubated with the active phorbol ester, PMA (Fig. 9). Taurine uptake was 30% of control when injected oocytes were exposed to 10^{-9} M of PMA, and was totally blocked at a concentration of 10^{-6} M. By contrast, 4α-phorbol, the inactive metabolite, had a slight effect on taurine uptake but no dose-dependence was evident.

To test further whether the effect of PMA on taurine uptake by injected oocytes is the result of activation of protein kinase C, PMA (10^{-6} M) and/or staurosporine (10^{-7} M) were added to the medium and incubated for 2 h as above (Fig. 10a,b). Control oocytes received no additions to the medium. Transport activity was significantly expressed ($p < 0.001$) compared to water-injected controls when pNCT cRNA was injected. PMA significantly reduced taurine uptake, but coincubation of PMA and staurosporine with oocytes or incubation of staurosporine alone led to restored transport activity.

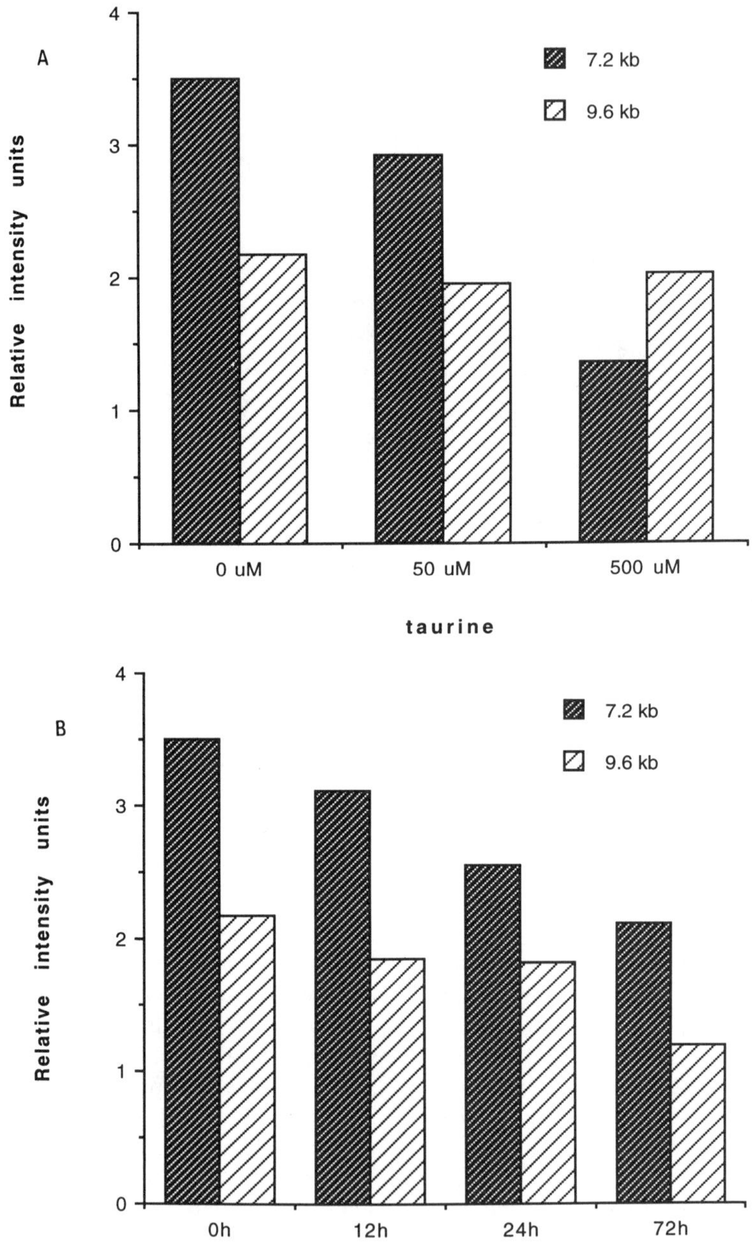

Figure 8. Adaptive regulation of expression of taurine transporter by taurine. LLC-PK1 cells were cultured in medium containing 0, 50, or 500 μM taurine for the indicated time, then total RNA was prepared and Northern blots done as described in the Methods. (a) LLC-PK1 cells were cultured in 0, 50, or 500 μM taurine for 24 h; (b) LLC-PK1 cells were cultured in 500 μM taurine for 72 h; (c) LLC-PK1 cells were cultured in 500 μM taurine for 72 h, then switched into taurine-free medium for the time period indicated.

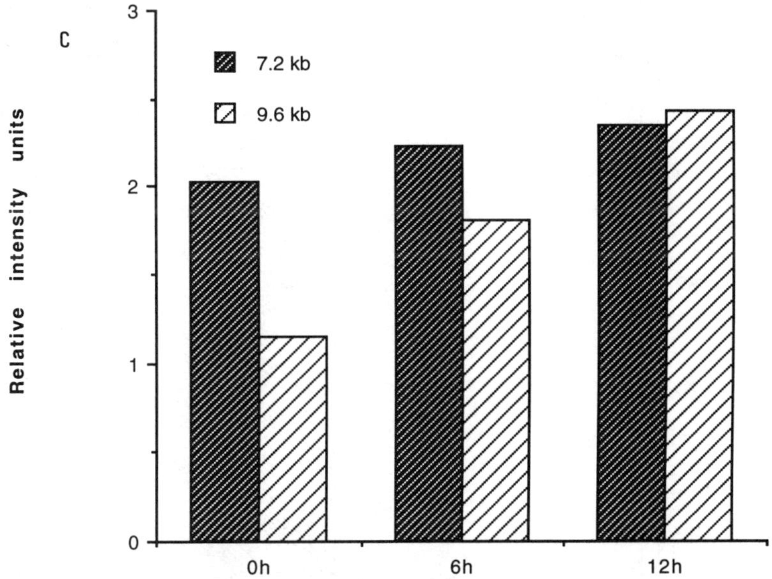

Figure 8C.

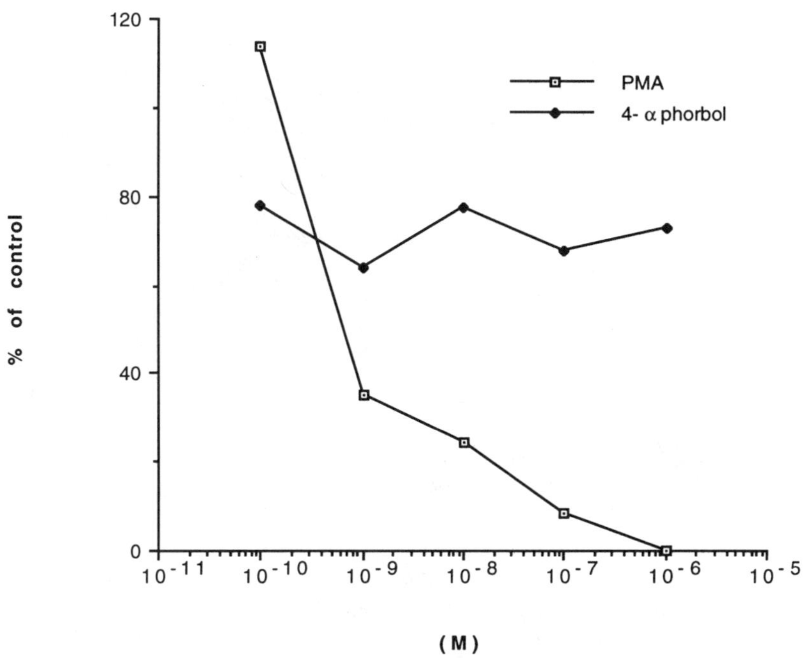

Figure 9. Effect of PMA and 4α-phorbol on taurine transport activity of pNCT cRNA-injected oocytes. 30 ng of pNCT cRNA were injected into each oocyte, and injected oocytes were incubated in taurine-free medium for 3 days. PMA and 4α-phorbol were added to the medium 2 h before taurine uptake was measured. Each data point represents the mean of six oocytes ± SEM of three independent experiments.

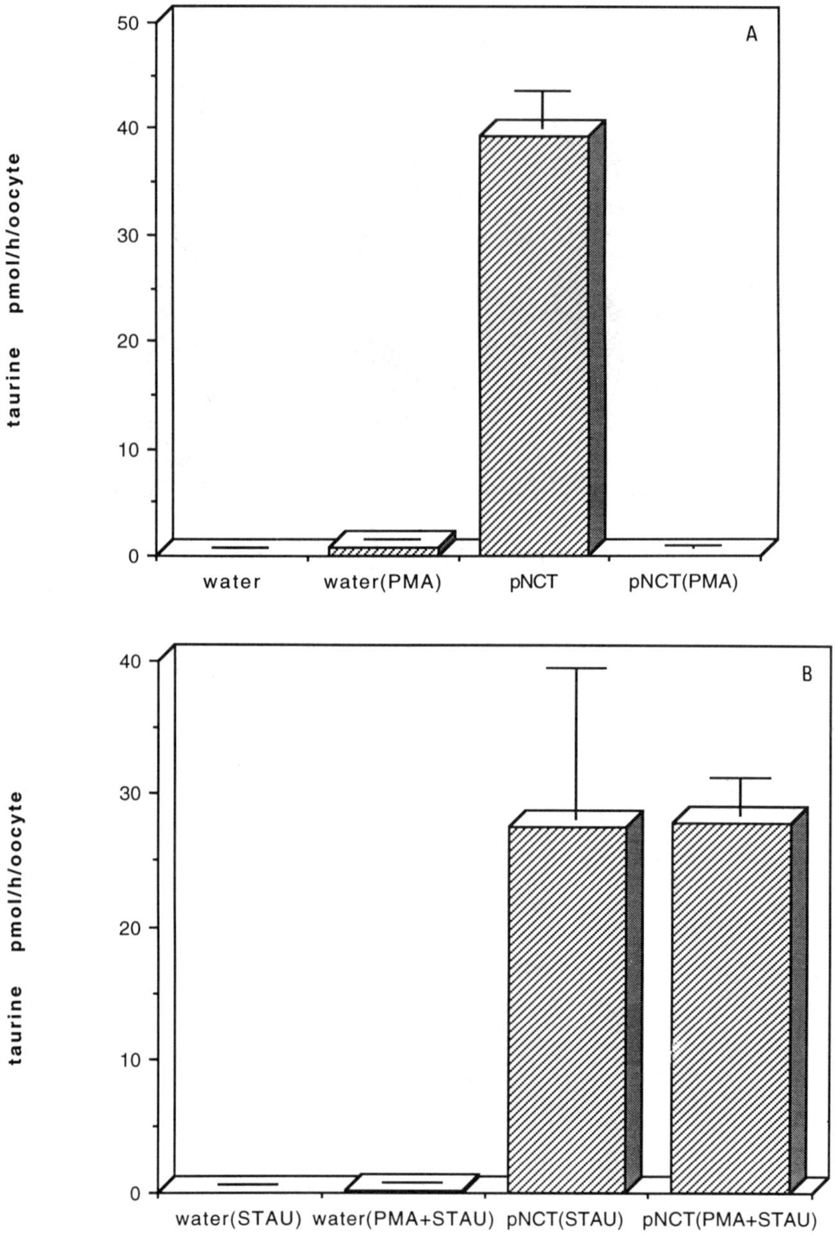

Figure 10. Effect of staurosporine on taurine transport activity expressed in oocytes. pNCT cRNA (30 ng) were injected into each oocyte, and injected oocytes were incubated in taurine-free medium for 3 days. A: PMA (10^{-6} M) or B: PMA (10^{-6} M) plus staurosporine (10^{-7} M) were added 2 h before taurine uptake was measured. Each data point represents the mean of six oocytes ± SEM of three independent experiments.

Site-Directed Antibodies Against the Taurine Transporter

To investigate if the open reading frame that defines pNCT encodes a protein expressed in oocytes or is similar to a protein found in rat kidney brush border membranes, a conserved 14 mer peptide sequence of the cloned MDCK taurine transporter was used to develop a polyclonal antibody to be used as a probe to search for novel proteins that might function as taurine transporters. Three days after injection of pNCT cRNA, oocytes were incubated with affinity-purified antibody for 1 h, then taurine uptake was determined. Uptake was significantly blocked (46.6% of control) by the antibody at 10 μg/ml and was reduced to 19.7% of control when the concentration of antibody exceeded 20 μg/ml. Non-specific IgG had no effect on taurine uptake by oocytes.

Preliminary BBMV experiments were performed to optimize the concentration of antibodies (specific and nonspecific control) such that maximal inhibition of 10 μM taurine uptake was achieved. At 10 pg/mg protein the polyclonal antibody to the taurine tranporter (Ab_{TBS1}) significantly inhibited taurine uptake in BBMV. Non-specific IgG, purified from rabbit serum before the animals were exposed to antigen had a slight inhibitory effect, but inhibition of taurine uptake by Ab_{TBS1} was significantly greater than inhibition by IgG ($p < 0.02$) (Fig. 11). Importantly, uptake of 50 μM GABA, which has been speculated to share a transporter with taurine[13], was not affected by the antibody. Uptake of an α-amino acid, proline (20 μM), was also unaffected by Ab_{TBS1} or IgG.

DISCUSSION

The renal adaptive response to altered sulfur amino acid intake has now been described in man, cat, mouse, rat, dog, and pig[12]. The observed phenomenon involves increased or decreased initial rate activity of the NaCl-dependent taurine transporter at the brush border membrane surface of the proximal tubule[8-10,17]. This transporter activity not only requires Na^+ and Cl^- for optimal activity but is inhibited by other β-amino acids. Enhanced taurine accumulation following dietary restriction of methionine and/or taurine results from a ten-fold increase in the renal reabsorption of taurine and reduced taurine excretion (0.9% of filtered load versus 9.0% under NTD conditions). After a high taurine diet there is reduced NaCl-taurine symport across the renal brush border surface coupled with an 18-fold increase in urinary taurine excretion as compared to controls. This adaptive response does not extend to methionine or cystine uptake or urinary excretion[7,12]. The phenomenon probably permits conservation of taurine in periods of dietary restriction and hyperexcretion in periods of dietary surfeit. This renal adaptive response is a biologic process that permits maintenance of the taurine body pool size and helps maintain tissue stores of this important osmolyte and neuromodulator.

The renal adaptive response to dietary taurine intake has been localized in part to the proximal tubule brush border membrane surface[10]. However, alterations in the extracellular concentration of taurine in a cell culture system involving renal cells of distal tubular origin demonstrates this same adaptive response with increased uptake at the basolateral surface of MDCK cells[20]. The intracellular physiologic signal for this renal adaptive response is only partially understood. Although plasma taurine concentrations correlate strongly with the fractional excretion of taurine[8], an even stronger correlation exists between renal cortex or cultured cell taurine content and both plasma (or medium) and urinary values for taurine[6,17]. Notably, increased NaCl-dependent taurine symporter activity was found when tissue slices were further depleted of taurine by either fasting or the presence of β-alanine despite the absence of a significant change in plasma concentrations[5,8]. Studies performed in continuous renal cell lines grown in culture, where intracellular taurine can be manipulated, mimic the

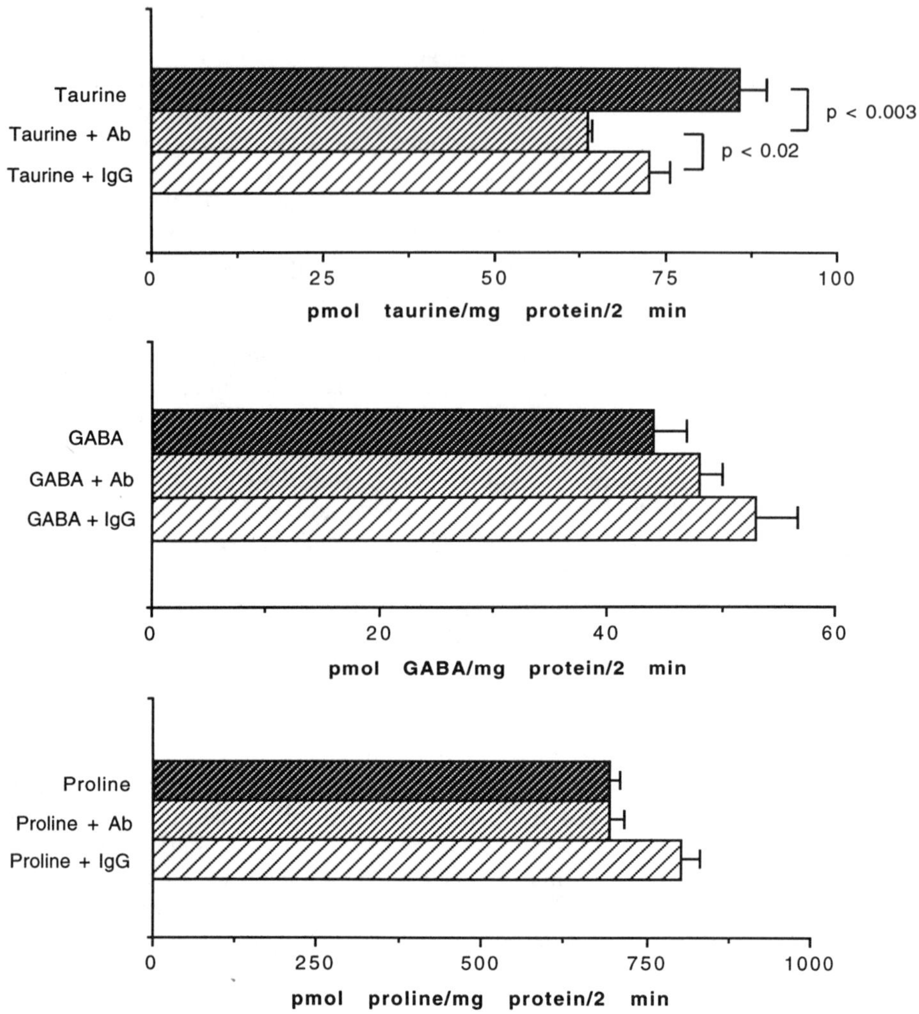

Figure 11. Uptake of 10 μM taurine, 50 μM GABA, and 20 μM proline by BBMV in the presence or absence of Ab_{TBS1} or nonspecific IgG at a final concentration of 10 pg/mg protein. Data are expressed as mean pmol amino acid/mg protein/2 min ± SEM for three filters.

actions observed in the brush border transporter. To wit, down-regulation of uptake in the presence of a 10-fold excess (500 μM) of medium taurine and up-regulation in the absence of taurine[21]. These findings support a role for intracellular taurine as the effector of renal adaptation rather than the plasma (or extracellular) concentration. The present study examines the role of NaCl-dependent taurine transporter gene transcription and translation as regulators of taurine transport and, hence, of the renal adaptive response.

Taurine uptake by oocytes injected with an equal quantity of pNCT cRNA is a function of the external taurine concentration. Expression of MDCK cell taurine transporter (pNCT) mRNA by oocytes was reduced following incubation for 24 h in medium containing high taurine and was increased following exposure to taurine-free medium, as compared to control. These results are consistent with observations made in rats fed diets containing defined quantities of sulfur amino acids, including taurine. When poly A+ RNA obtained from rat kidney is injected into oocytes, the uptake of taurine (taurine transporter activity) is greatest when the source of RNA is from rats fed a taurine-deficient diet and least when the RNA is from rats fed a high taurine diet[19].

Nuclear run-off transcription assays show that the pNCT expression gene was induced in MDCK cells incubated in taurine-free medium. Thus, changes in the external concentration of taurine modify the abundance and transcription rate of pNCT mRNA. The nature of the intracellular stimulus for increased transcription is not understood, but possibly intracellular taurine concentration, perhaps of a subcellular pool, regulates gene transcription or modifies a regulatory protein.

Another mechanism for adaptive regulation of taurine transport activity occurs at the level of protein synthesis. Two proteins, ≈30 kDa and ≈70 kDa, were identified by Western blot analysis using the antibody described above. The abundance of both proteins was reduced following incubation of injected oocytes in high medium taurine (in a manner similar to that of mRNA) while the abundance of the 30 kDa protein was increased after incubation in taurine-free medium. The molecular weight of the larger protein detected in this system is similar to the transporter protein detected in MDCK cells[27]. The 30 kDa protein may be the product of post-translational modification of the larger transporter protein. Consistent with this finding, SDS-gel electrophoresis analysis of membrane fractions derived from pNCT cRNA-injected oocytes gave rise to two [^{35}S]methionine-labelled translation products that corresponded in size (≈30 kDa and ≈70 kDa) to those identified by Ab_{TBS1} (data not shown). Therefore, in addition to an effect upon transcription, medium taurine concentration modifies a post-transcriptional event, either translation, post-translational modification, or cell membrane recycling, because incubation of oocytes injected with equivalent quantities of pNCT was associated with the anticipated reciprocal changes in transport activity as well as abundance of transporter protein.

The polyclonal antibody directed against a highly conserved region of the MDCK taurine transporter inhibited taurine transport by pNCT cRNA-injected oocytes. When oocytes were incubated with 10 µg Ab/ml, taurine uptake was significantly blocked, and it was reduced to about 20% of control when the concentration of Ab_{TBS1} exceeded 20 µg. Injection of 100 ng of Ab_{TBS1} into oocytes blocked approximately 60% of taurine transport, as compared to control. Assuming an intracellular volume of 1 µl[4], this corresponds to an intracellular antibody concentration of 100 µg/ml. Thus, the antibody effectively blocked taurine transport from either external or cytoplasmic surfaces, which is consistent with the proposed intramembranous location of the peptide from which it was raised. Indeed, the peptide chosen might be involved in activation of the taurine transporter, or function as a gate of the transporter. Failure of complete inhibition of taurine uptake might be due to the presence of more than one binding site for taurine on the intracellular domains of the transporter protein(s).

Protein kinase C (PKC) is involved in the control of many cellular processes, including membrane transport, by mechanisms such as transcriptional regulation and post-translational modification[1,15]. Activation of PKC by phorbol 12-myristate 13-acetate (PMA), an active phorbol ester known to stimulate protein kinase C, reduced Na^+-dependent taurine uptake into LLC-PK1 cells by a mechanism not blocked by treatment with colchicine, cytochalasin B, actinomycin D, or cycloheximide[4,22]. Several possible PKC phosphorylation sites are present along the amino acid sequence encoded by pNCT. Taurine transport activity was totally abolished when pNCT cRNA-injected oocytes were exposed to PMA. That inhibition of taurine transport by PMA is mediated through a PKC-dependent pathway is evident because staurosporine, a inhibitor of PKC, prevented the decrease in taurine transport by cells after exposure to PMA[4,21]. The inactive phorbol ester, 4α-phorbol 12,13-didecanoate, had a small effect on taurine transport by pNCT cRNA-injected oocytes, but no dose-dependence was observed.

Kinetic data supports the existence of at least two distinct systems for taurine transport in the renal epithelial cell lines. In addition, distinct systems have been postulated for proximal versus distal renal tubular cells. Differences in proximal and distal tubular taurine transport might be related to differences in the polarity of the taurine transporter rather than to structurally distinct transporter proteins; this remains to be determined. Incubation of pNCT-injected oocytes in taurine-free medium resulted in increased abundance of a 30 kDa protein immunoreactive with antibody which neutralized taurine uptake in the same system compared to control oocytes incubated in 50 μM taurine. Those oocytes exposed to high external taurine (500 μM) exhibited reduced expression of both the 30 kDa and a 70 kDa protein. The transcription of the gene(s) encoding the transporter protein is under the regulatory influence of external taurine concentration as is the transcription rate of the mRNA.

ACKNOWLEDGMENTS

We greatly appreciate the gift of the pNCT cDNA from Dr. S. Uchida. We are grateful to Dr. L. Albritton for technical advice. This study was supported in part by the LeBonheur Chair of Excellence of the Department of Pediatrics of the University of Tennessee, the Crippled Children's Foundation Research Center, and NIH grant DK37221-08.

REFERENCES

1. Bargon, J., Trapnell, B.C., Yoshimura, K., Dalemans, W., Pavirani, A., Lecocq, J.P., and Crystal, R.G. 1992, Expression of the cystic fibrosis transmembrane conductance regulator gene can be regulated by protein kinase C, *J.Biol.Chem.* 267:16056-16060.
2. Booth, A.G. and Kenny, A.J. 1974, A rapid method for the preparation of microvilli from rabbit kidney, *Biochem.J.* 142:575-581.
3. Bradford, M. 1976, A rapid and sensitive method for the quantitation of microgram quantities of protein utilizing the principle of protein-dye binding, *Anal.Biochem.* 72:248-254.
4. Brandsch, M., Miyamoto, Y., Ganapathy, V., and Leibach, F.H. 1993, Regulation of taurine transport in human colon carcinoma cell lines (HT-29 and Caco-2) by protein kinase C, *Am.J.Physiol.Gastrointest.Liver Physiol.* 264:G939-G946.
5. Chesney, R.W. 1985, Taurine: Its biologic role and clinical implications, *Yearbook, Chicago.*
6. Chesney, R.W., Friedman, A.L., Albright, P.W., and Gusowski, N. 1982, Fasting reverses the renal adaptation to altered dietary sulfur amino acid intake, *Proc.Soc.Exp.Biol.Med.* 170:493-501.
7. Chesney, R.W., Gusowski, N., and Theissen, M. 1984, Developmental aspects of renal β-amino acid transport. IV. Brush border membrane response to altered intake of sulfur amino acids, *Pediatr.Res.* 18:611-618.
8. Chesney, R.W., Gusowski, N., and Dabbagh, S. 1985, Renal cortex taurine content regulates renal adaptive response to altered dietary intake of sulfur amino acids, *J.Clin.Invest.* 76:2213-2221.
9. Chesney, R.W., Gusowski, N., and Dabbagh, S. 1987, Studies on renal adaptation to altered dietary amino acid intake: Reduced renal cortex taurine content increases the V_{max} of taurine uptake by brush border membrane vesicles, *Pediatr.Nephrol.* 1:9-15.
10. Chesney, R.W., Gusowski, N., and Freidman, A.L. 1983, Renal adaptation to altered amino acid intake occurs at the luminal brush border membrane, *Kidney Int.* 24:588-594.
11. Chesney, R.W., Lippincott, S., Gusowski, N., Padilla, M., and Zelikovic, I. 1986, Studies on renal adaptation to altered dietary amino acid intake: Tissue taurine responses in nursing and adult rats, *J.Nutr.* 116:1965-1976.
12. Chesney, R.W., Zelikovic, I., Jones, D.P., Budreau, A., and Jolly, K. 1990, The renal transport of taurine and regulation of renal sodium-chloride-dependent transporter activity, *Pediatr.Nephrol.* 4:399-407.
13. Dantzler, W.H. and Silbernagl, S. 1976, Renal tubular reabsorption of taurine, GABA, and β-alanine studied by continuous microperfusion, *Pflügers Arch.* 367:123-128.

14. Dumont, J.N. 1972, Oogenesis in *Xenopus laevis* (Daudin) I. Stages of oocyte development in laboratory maintained animals, *J.Morphol.* 136:153-180.
15. Fournier, F., Charnet, P., Bourinet, E., Vibert, C., Matifat, F. Charpentier, G., Navarre, P., Brule, G. and Marlot, D. 1993, Regulation by protein kinase C of putative P-type Ca channels expressed in *Xenopus* oocytes from cerebellar mRNA, *FEBS* 317:118-124.
16. Friedman, A., Albright, P.W., and Chesney, R.W. 1981, Dietary adaptation of taurine transport by rat renal epithelium, *Life Sci.* 29:2415-2420.
17. Friedman, A.L., Albright, P.W., Gusowski, N., Padilla, M., and Chesney, R.W. 1983, Renal adaptation to alteration in dietary amino acid intake, *Am.J.Physiol.* 245:F159-F166.
18. Frohman, M.A., Dush, M.K., and Mattinm, G.R. 1988, Rapid production of full-length cDNAs from rare transcripts using a single gene-specific oligonucleotide primer, *Proc.Natl.Acad.Sci USA*, 85:8998-9002.
19. Han, X. 1994, Expression of taurine transporter and its regulation by diet in *Xenopus laevis* oocytes following injection of rat kidney cortex mRNA, in: "Taurine in Health and Disease", Huxtable, R.J. and Michalk, D. eds., Plenum Press, New York, pp. 121-130.
20. Jones, D.P., Miller, L.A., and Chesney, R.W. 1990, Adaptive regulation of taurine transport in two continuous renal epithelial cell lines, *Kidney Int.* 38:219-226.
21. Jones, D.P., Miller, L.A., and Chesney, R.W. 1993, Polarity of taurine transport in cultured renal epithelial cell lines: LLC-PK_1 and MDCK, *Am.J.Physiol.Renal,Fluid Electrolyte Physiol.* 265:F137-F145.
22. Jones, D.P., Miller, L.A., Dowling, C., and Chesney, R.W. 1991, Regulation of taurine transporter activity in LLC-PK1 cells: role of protein synthesis and protein kinase C activation, *J.Am.Soc.Nephrol.* 2:1021-1029.
23. Rickwood, D., Hames, B.D., McPherson, M.J., Ouirke, P., and Taylor, G.R. 1991, PCR: A practical approach, in: "The Practical Approach Series", Anonymous Oxford University Press, New York,
24. Rozen, R. and Scriver, C.R. 1982, Renal transport of taurine adapts to perturbed taurine homeostasis, *Proc.Natl.Acad.Sci.USA*, 79:2101-2105.
25. Sambrook, J., Fritsch, E.F., and Maniatis, T. 1989, " Molecular cloning: a laboratory manual", Cold Spring Harbor Laboratory Press, Cold Spring Harbor.
26. Smith, K.E., Borden, L.A., Wang, C.D., Hartig, P.R., Branchek, T.A., and Weinshank, R.L. 1992, Cloning and expression of a high affinity taurine transporter from rat brain, *Mol.Pharmacol.* 42:563-569.
27. Uchida, S., Kwon, H.M., Preston, A.S., and Handler, J.S. 1991, Expression of Madin-Darby canine kidney cell Na^+- and Cl^--dependent taurine transporter in *Xenopus laevis* oocytes, *J.Biol.Chem.* 266:9605-9609.
28. Wallace, R.A., Jared, D.W., Dumont, J.N., and Sega, M.W. 1973, Protein incorporation by isolated amphibian oocytes. III. optimum incubation conditions, *J.Exp.Zool.* 184:321-334.
29. Wallace, R.A. and Misulovin, Z. 1978, Long-term growth and differentiation of *Xenopus* oocytes in a defined growth medium, *Proc.Natl.Acad.Sci.USA*, 75:5534-5538.
30. Zelikovic, I. and Chesney, R.W. 1989, Ionic requirements for amino acid transport, *Am.J.Kidney Dis.* 14:313-316.

21

RENAL CONCENTRATING DEFECT AND ORGANIC OSMOLYTES

Paradoxical Changes of Renal Medullary Taurine Contents in Potassium-Depleted Rats

Takeshi Nakanishi and Yoshihiro Takamitsu

Department of Kidney and Dialysis
Hyogo College of Medicine
1-1 Mukogawacho, Nishinomiya 663, Japan

INTRODUCTION

Under long-term hyperosmotic conditions, renal cells accumulate organic osmolytes preferentially over inorganic ions, which maintain osmotic balance between the inside and the outside of the cells[6,7]. Unlike "perturbing" solutes such as NaCl, KCl and urea, the organic osmolytes have relatively weak inhibitory action or even protective effects on enzymes and various cellular processes[15]. Organic osmolytes play an important role in the osmoregulation of the kidney. However, the relationship between renal concentrating defect and organic osmolytes has not been well clarified. In kidney in situ, medullary hypertonicity is constructed by medullary cells themselves. So, delayed or decreased accumulation of osmolytes in the renal medulla may be the cause of renal concentrating defects. In this study, we chose K^+-depleted rats as a model of renal concentrating defect. In K^+ depletion, a renal concentrating defect is a prominent and consistent finding. Several defects associated with urinary concentrating ability have been reported, including prostaglandin overproduction[9], blunted responsiveness to antidiuretic hormone[2], reduced papillary solute content[4], abnormal medullary oxidative metabolism[3,13], altered ADH release[10], and primary polydipsia[1]. Decreased accumulation of organic osmolytes in renal medulla could explain some of these abnormalities associated with the concentrating defect.

Our object was to clarify the relationship between inner medullary osmolytes and urine concentrating ability or to determine whether decreased medullary organic osmolytes cause a urine concentration defect associated with K^+ depletion. Two significant points should be taken into consideration in determining the role of organic osmolytes in urinary concentration. Firstly, polydipsia and polyuria are common complications in K^+-depleted animals[5]. In order to discriminate the effects of polyuria and polydipsia from K^+ depletion, "water diuresis" as well as "K^+ depletion" were assigned to the animals as the pretreatment. Secondly, decreased medullary tonicity could reduce the need for medullary organic os-

Taurine 2, edited by Huxtable et al.
Plenum Press, New York, 1996

molytes during maximal urine concentration. As for the protocol of hyperosmolality, we used a protocol of water deprivation to study the accumulation of osmolytes at the maximal urinary concentration. Sodium infusion was used to induce hyperosmotic stress to the equivalent level in all experimental groups.

METHODS

Male Wistar rats (260 to 350 g) were separated into three groups: untreated (control), water diuresis, and K^+ depletion: (i) Untreated rats had free access to regular laboratory rat chow and water for 21 days; (ii) Rats in the water diuresis group had free access to regular laboratory rat chow and 3% sucrose solution for 21 days; and (iii) K^+-depleted animals also had free access to food and water for 21 days, but their diet contained less than 0.5 mEq of K^+ per 100 g.

After the 3 weeks of dietary pretreatment, rats in each group were subdivided into three subgroups for hyperosmolality studies; no hyperosmosis, dehydration, or Na^+ infusion: (i) Rats assigned to the protocol of no hyperosmotic stress underwent a sham operation of femoral vein cannulation and were given free access to water for 4 days: (ii) In the dehydration protocol, rats underwent a sham operation of femoral vein cannulation and were depleted of water for 4 days; and (iii) Rats in the third group were infused continuously with hypertonic salt solution (4.08% NaCl solution; 18 ml/day) for 72 h before death.

Urine samples were collected during the 3 h period prior to sacrifice (during the last 12 h in the case of dehydrated rats). Na^+ and K^+ concentrations were measured using an electrolyte autoanalyzer (Hitachi, TYPE-710, Tokyo, Japan). Urea was measured with an autoanalyzer (Hitachi, TYPE-736, Tokyo, Japan). Osmolality was measured with a freezing point osmometer (Kyoto-Daiichi-Kagaku, OM-6010, Kyoto, Japan).

Inner medullary tissues were dissected from kidney slices, lyophilized for 24 h, and weighed to determine dry weight. The dried sections were homogenized with a glass homogenizer in perchloric acid. The homogenates were centrifuged for 20 min at 4°C. The pellets were then dissolved in 0.25 N NaOH and tested for protein content with use of the BioRad protein assay (Bio-Rad Labs, Richmond, California). Bovine γ-globulin was treated in an identical fashion and used as a protein standard.

Betaine, glycerophosphorylcholine (GPC), sorbitol, and myo-inositol were measured in the processed extracts by HPLC with the use of a Sugar-Pak Ca column (Nippon Waters, Tokyo, Japan)[14]. Taurine was measured with a high-speed amino acid analyzer (Hitachi L8500, Tokyo, Japan) in the same samples. Tissue Na^+ and K^+ were determined in the supernatants of perchloric acid extracts using a two-wavelength atomic absorption spectrophotometer (Hitachi, TYPE-308, Tokyo, Japan).

RESULTS

Fig. 1A shows the urine Na^+, K^+, and urea concentrations and osmolality after the hyperosmolality protocol of water deprivation. After 3 weeks pretreatment of control, hydration, and K^+ deprivation, rats were subjected to conditions of hyperosmolality, no hyperosmosis or water deprivation. Compared with the other protocol, the dehydration protocol increased all urine parameters except K^+ in K^+-deprived rats. After the dehydration protocol, urine Na^+ and urea concentrations did not differ among three pretreatment groups. Urine osmolality and K^+ concentration were significantly less in K^+-deprived rats but not in water diuresis rats.

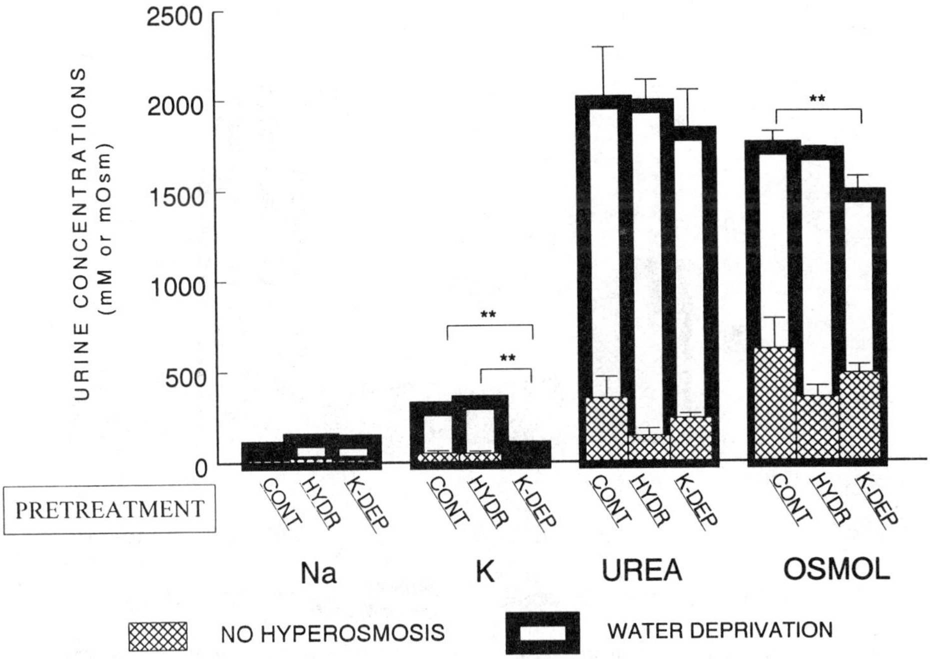

Figure 1A. Effect of water deprivation protocol on urine parameters in rats previously subjected to water diuresis or K^+ deprivation.

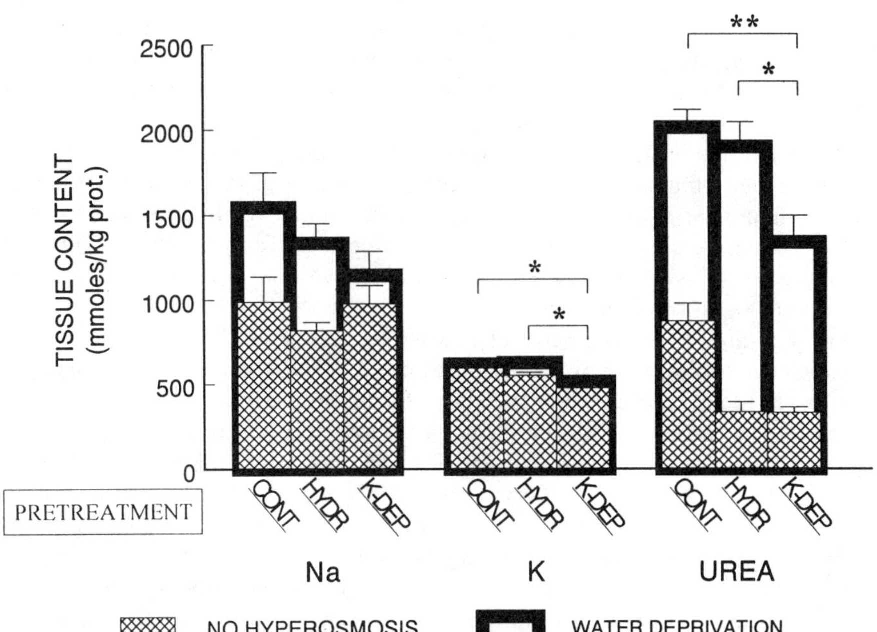

Figure 1B. Effect of water deprivation protocol on medullary contents of Na^+, K^+, and urea in rats previously subjected to water diuresis or K^+ deprivation.

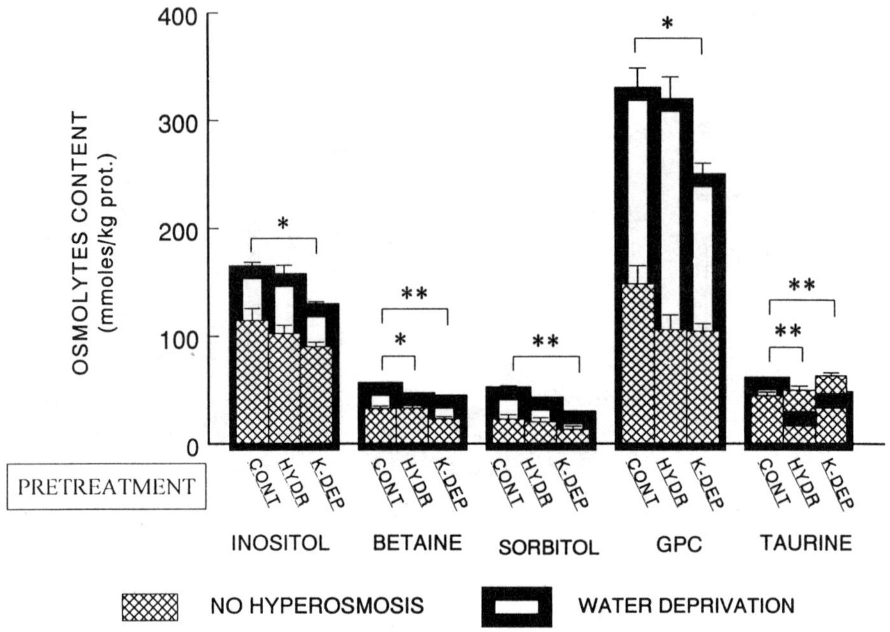

Figure 1C. Effect of water deprivation protocol on medullary contents of organic osmolytes in rats previously subjected to water diuresis or K^+ deprivation. CONT: control; HYDR, water diuresis; K-DEP, K^+ deprivation. Results are means ± SEM. *$p < 0.05$; **$p < 0.01$.

Water deprivation increased tissue Na^+ and urea contents in all three pretreatment groups (Fig. 1B). After the dehydration protocol, the pretreatment of water diuresis or K^+ deprivation did not significantly affect inner medullary Na^+ content, although it was apparently smaller in K^+-depleted rats. K^+ and urea contents were significantly lower in K^+-deprived rats.

Water deprivation increased myo-inositol, betaine, sorbitol and GPC in all three pretreatment groups, but taurine only in the control group (Fig. 1C). In the no hyperosmosis study, the pretreatment of water diuresis and K^+ deprivation even increased taurine content. After dehydration, betaine and taurine contents were significantly lower in the rats previously subjected to water diuresis. In K^+-depleted rats, on the other hand, all organic osmolytes were significantly decreased.

Fig. 2A shows the urine parameters after Na^+ infusion. Compared between the no hyperosmosis and Na^+ infusion protocols, the infusion of NaCl increased all urine parameters except K^+ in K^+-deprived rats. After the Na^+ infusion protocol, neither urine Na^+ nor urea concentrations, nor osmolality, differed among the three pretreatment groups.

The infusion of Na^+ increased tissue Na^+ and urea contents in all three pretreatment groups (Fig. 2B). After the Na^+ infusion protocol, water diuresis or K^+ deprivation did not significantly affect inner medullary Na^+ or urea contents. K^+ content was significantly smaller in K^+-deprived rats than in the hydrated group.

Na^+ infusion increased betaine, sorbitol and GPC in all three pretreatment groups, myo-inositol in hydrated rats and taurine in the control group (Fig. 2C). After the Na^+ infusion protocol, taurine content was significantly lower and myo-inositol content was higher in hydrated rats. In K^+-depleted rats followed by NaCl infusion, betaine, sorbitol and taurine were significantly decreased.

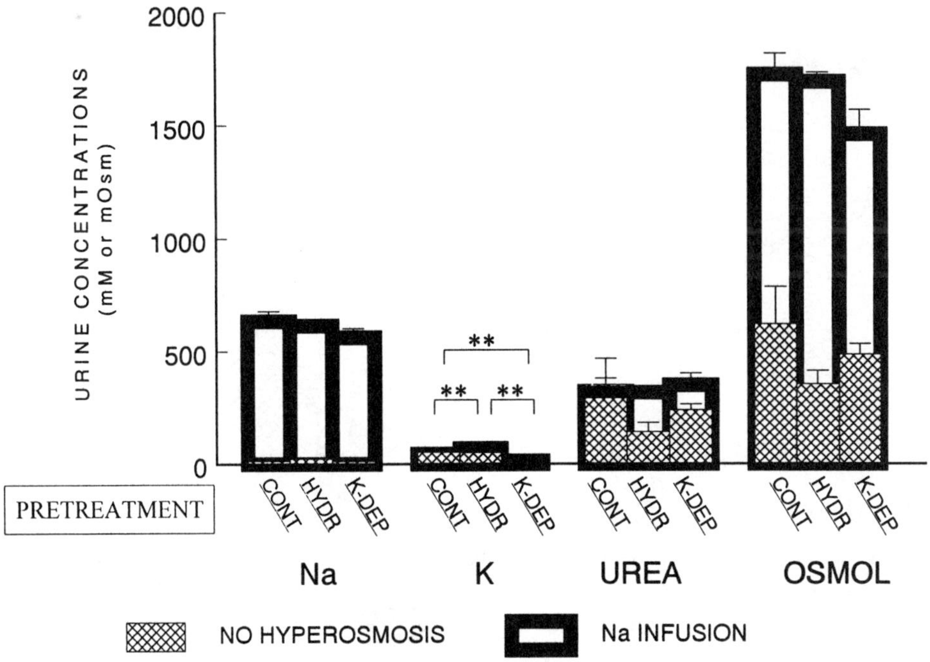

Figure 2A. Effect of NaCl infusion on urinary parameters in rats previously subjected to water diuresis or K^+-deprivation.

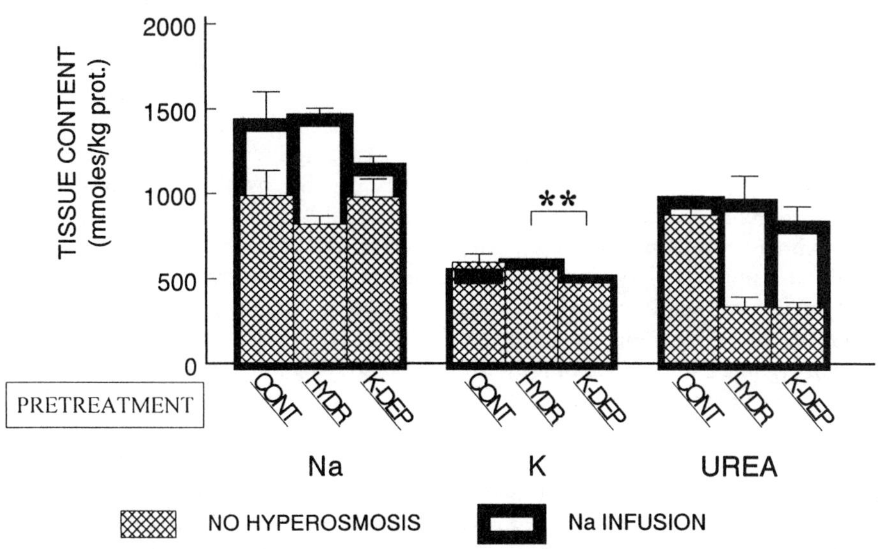

Figure 2B. Effect of NaCl infusion protocol on medullary contents of Na^+, K^+, and urea in rats previously subjected to water diuresis or K^+ deprivation.

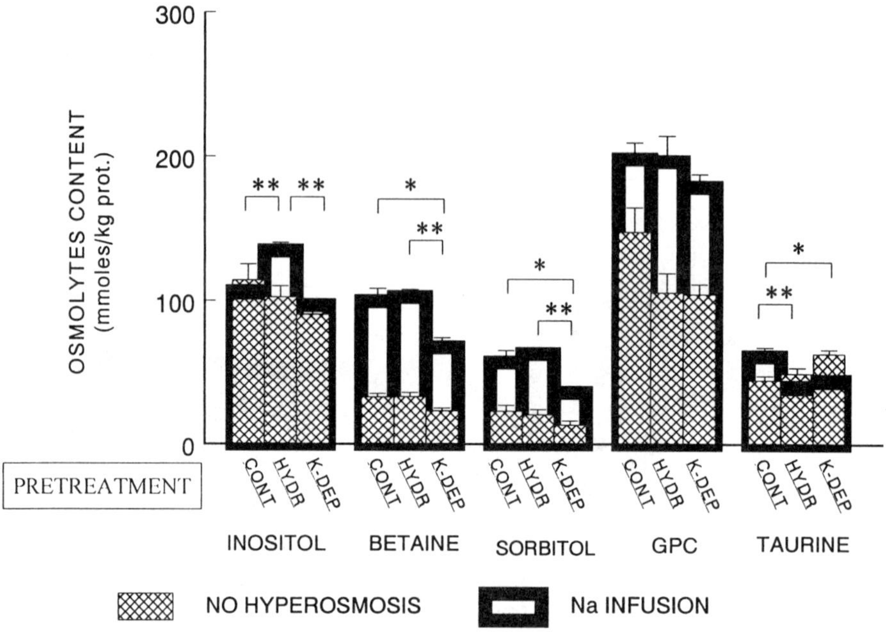

Figure 2C. Effect of NaCl infusion on medullary contents of organic osmolytes in rats previously subjected to water diuresis or K^+ deprivation. CONT: control; HYDR: water diuresis; K-DEP, K^+ deprivation. Results are means ± SEM. *p < 0.05; **p < 0.01.

DISCUSSION

We examined the relationship between urinary concentrating ability and medullary organic osmolytes, using rats subjected to water diuresis or K^+ deprivation for 3 weeks. In the 4 days dehydration protocol, which induced the maximal urine concentration, urinary concentrating defect or decreased urine osmolality was demonstrated in K^+-depleted rats, but not in animals with water diuresis. In K^+-deprived rats, urine and tissue K^+ and urine urea decreased significantly, but urine and tissue Na^+ did not differ among the three pretreatment groups. Urinary concentrating defect associated with K^+ depletion was accompanied by decreased accumulation of all medullary organic osmolytes. Medullary taurine and betaine contents decreased in water-diuresed animals.

Maximal urine concentrating ability by water deprivation was deranged in K^+-depleted animals, and was associated with decreased medullary organic osmolytes. However, we could not determine directly whether the reduction in organic osmolytes deranged urine concentration, or whether decreased urine concentration or medullary tonicity cause the reduction in organic osmolytes. For the purpose of clarifying the point, we chose the protocol of Na^+ infusion to impose an equivalent level of tonicity on medulla in all pretreatment groups.

In the Na^+ infusion protocol, urinary Na^+ and urea concentrations and osmolality, and tissue Na^+, K^+, and urea contents in water diuresis and K^+-depleted animals were comparable to those in control animals. In K^+-depleted animals, the medullary contents of taurine, betaine, and sorbitol did not restore to the untreated levels, although medullary myo-inositol and GPC contents increased. In water diuresis animals, medullary taurine content decreased significantly. From the urine and tissue parameters, Na^+ infusion could produce the same level of tonicity in renal medulla in all pretreatment groups. However, the contents of taurine, betaine and sorbitol were significantly decreased. These data suggested that the process of accumulating these organic osmolytes could be deranged in K^+ depletion. Therefore, we concluded that decreased accumulation of medullary organic osmolytes might impair the renal concentrating ability.

In the bacteria, *Escherichia coli* and *Salmonella typhimurium*, it was demonstrated that the ionic composition of the intracellular milieu, especially the K^+ concentration can regulate gene expression of proU gene which is related to betaine transport for osmoprotection[12]. In cultured cells, Smardo *et al.*[11] proposed that the hypertonicity-induced increase in the transcription of the genes coding for aldose reductase for sorbitol accumulation probably depends on increased intracellular Na^+ plus K^+. So it is reasonable to presume that the gene regulation of these osmoregulatory proteins might be modulated by intracellular K^+, although we had not yet determined this.

The patterns of accumulation of betaine, myo-inositol, GPC or sorbitol in the renal inner medulla were similar. However, taurine was strikingly different from other organic osmolytes (Fig. 3). The sum of all organic osmolytes except taurine increased in the animals subjected to water deprivation or Na^+ infusion, irrespective of the pretreatments. The pattern of taurine content is complementary to the sum of other organic osmolytes. In the control

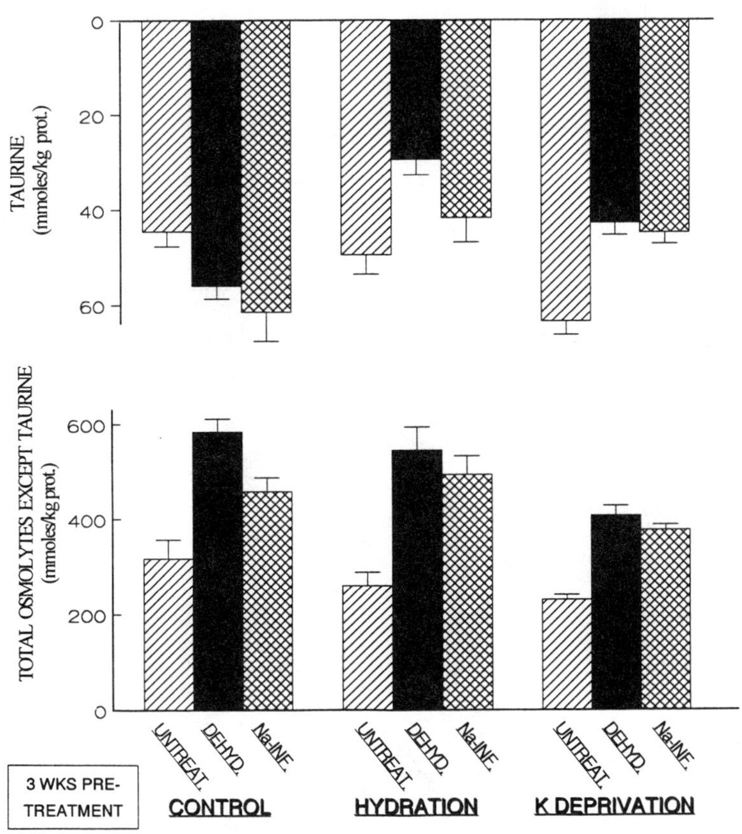

Figure 3. Inner medullary contents of taurine and the sum of organic osmolytes except taurine after water deprivation and NaCl infusion protocols in rats previously subjected to water diuresis and K^+ deprivation. UNTREAT: no hyperosmosis; DEHYD, water deprivation; Na-INF: NaCl infusion.

rats, taurine content increased significantly after water deprivation or Na^+ infusion. In the rats subjected to water diuresis or K^+ deprivation as the pretreatment, taurine content decreased significantly after water deprivation or Na^+ infusion. Medullary taurine content was dependent on extracellular hypertonicity in control rats, but not after water diuresis with or without K^+ depletion. We suppose that organic osmolytes other than taurine accumulated first, or to the limited level, and then, taurine accumulated to compensate for the remaining difference between intra- and extracellular osmolality.

The mechanism of "the paradoxical change of taurine accumulation" has not been clarified. But we have another observation of a paradoxical change in taurine contents[8]. The administration of vasopressin V2 receptor antagonists to dehydrated rats decreased sorbitol, betaine, GPC as well as urine osmolality and tissue urea content, but increased taurine contents. Thus, the vasopressin V2 receptor also might be related to the paradoxical taurine change.

A possible explanation is that the regulation of organic osmolytes in the medullary cells for balancing intracellular osmolality with extracellular is deranged after the V2 receptor is blocked by water diuresis or V2 antagonist. This might be consistent with the paradoxical increase in taurine content in water diuresis and K^+ depletion. When renal medullary cells are made hypotonic by water diuresis, intracellular organic osmolytes should decrease. All organic osmolytes other than taurine leak out of these cells excessively. The sum of these organic osmolytes then becomes too low to balance the extracellular osmolality. Taurine accumulates to compensate the osmolality imbalance, which might be over the control level.

This could also explain the paradoxical decrease in animals subjected to water diuresis followed by water deprivation and Na^+ infusion. When switching from hypotonic to hypertonic condition, the accumulation of organic osmolytes in the medulla also exceed the balance of extracellular osmolality, and decrease the need for taurine accumulation. However, further investigation is necessary for clarifying the complementary relationship between taurine and other organic osmolytes.

In conclusion, the renal concentrating defect due to K^+ deprivation was associated with decreased accumulation of all the organic osmolytes. A decrease in medullary osmolytes might precede and impair the urine concentration. The pattern of taurine accumulation in renal inner medulla was different from other organic osmolytes and was not dependent on extracellular tonicity after water diuresis with or without K^+ depletion.

REFERENCES

1. Berl, T., Linas, S.L., Aisenbrey, G.A. and Anderson, R.J. 1977, On the mechanism of polyuria in potassium depletion: The role of polydipsia, *J. Clin. Invest.,* 60:620-625.
2. Gullner, H.G. and Bartter, F.C. 1980, The role of urinary prostaglandin E and cyclic AMP in the polyuria of hypokalemia in rats, *Prostaglandins Med.*, 4:13-19.
3. Kannegiesser, H., and Lee, J.B. 1971, Role of outer renal medullary metabolism in the concentrating defect of K depletion, Am. J. Physiol. 220:1701-1707.
4. Manitius, A., Levitin, H., Beck, D. and Epstein, F.H. 1960, On the mechanism of impairment of renal concentrating ability in potassium deficiency, *J. Clin. Invest.,* 39:684-692.
5. Nakanishi, T., Takamitsu, Y., Nakahama, H. and Sugita, M. 1994, Impairment of renal medullary osmolyte accumulation in potassium-depleted rats, *Am. J. Physiol.,* 267:F139-F145.
6. Nakanishi, T., Uyama, O. and Sugita, M. 1991, Osmotically regulated taurine content in rat renal inner medulla, *Am. J. Physiol.*, 261:F957-F962.
7. Nakanishi, T., Uyama, O. and Sugita, M. 1992, Amino acids as well as polyols and methylamines accumulated in rat kidney during dehydration, *Amino Acids*, 3:131-138.

8. Nakanishi, T., Yamauchi, A., Nakahama, H., Yamamura, Y., Yamada, Y., Orita, Y., Fujiwara, Y., Uyeda, N., Takamitsu, Y. and Sugita, M. 1994, Organic osmolytes in rat renal inner medulla are modulated by vasopressin V1 and/or V2 antagonists, *Am. J. Physiol.,* 267:F146-F152.
9. Raymond, K.H., Lifschitz, M.D. and McKinney, T.D. 1987, Prostaglandins and the urinary concentrating defect in potassium-depleted rabbits, *Am. J. Physiol.*, 253:F1113-F1119.
10. Rutecki, G.W., Cox, J.W., Robertson, G.W., Francisco, L.L. and Ferris, T.F. 1982, Urinary concentrating ability and antidiuretic hormone responsiveness in the potassium-depleted dog, *J. Lab. Clin. Med.* 100:53-60.
11. Smardo, F.L.Jr., Burg, M.B. and Garcia-Perez, A. 1992, Kidney aldose reductase gene transcription is osmotically regulated, *Am. J. Physiol.,* 262:C776-782.
12. Sutherland, L., Cairney, J., Elmore, M.J., Booth, I.R. and Higgins, C.F. 1986, Osmotic regulation of transcription: Induction of proU betaine transport gene is dependent on accumulation of intracellular potassium, *J. Bacteriol.* 168:805-814.
13. Weiner, M.W., Sauer, L.A., Torretti, J. and Epstein, F.H. 1971, Renal mitochondrial enzymes in potassium depletion, *Am. J. Physiol.,* 221:613-617.
14. Wolff, S.D., Yancey, P.H., Stanton, T.S. and Balaban, R.S. 1989, A simple HPLC method for quantitating major organic solutes of renal medulla, *Am. J. Physiol.,* 256:F954-F956.
15. Yancey, P.H., Clark, M.E., Hand, S.C., Bowlus, R.D. and Somero, G.N. 1982, Living with water stress: Evolution of osmolyte system, *Science,* 217:1214-1222.

22

LONG-LASTING HYPOXIC PRESERVATION OF PORCINE KIDNEY CELLS

Beneficial Effect of Taurine on Viability and Metabolism in a Simplified Transplantation Model

Peter Wingenfeld,[1] Ute Gehrmann,[1] Stefan Strübind,[1] Thomas Minor,[2] Wolf Isselhard,[2] and Dietrich Volker Michalk[1]

[1] Department of Pediatrics
[2] Institute for Experimental Medicine
University of Cologne
Cologne, Federal Republic of Germany

INTRODUCTION

In the field of transplantation medicine one of the determining factors for the outcome of the transplanted organ is the cellular deterioration due to hypoxia and reoxygenation and the protection against it. Frequently the hypoxic storage of the removed organ limits the possibility of successful transplantation. The hypoxic cell damage due to insufficient organ preservation is a major cause for the loss or the dysfunction of transplanted organs like hearts and kidneys[2,5,6,15,17].

Some of the factors determining the extent of hypoxic and post-hypoxic deterioration are the energy deficiency of the organ[3], intracellular Ca^{2+} overload[22], osmoregulatory dysfunction[18,19], the O_2 paradox and membrane oxidation induced by toxic metabolites and free peroxide radicals[3].

The methods of preservation and suitable solutions for the adequate preservation of the graft are still under investigation.[2,5,6,11-13,15,17] In 1988, Belzer and-co-workers introduced the University of Wisconsin solution (UW) for hypothermic preservation of solid organs[2]. In the following decade this solution was claimed to be the superior preservation solution for the cold storage of the human kidney during transplantation.

Because of the unique functions of the phylogenetically ancient amino acid taurine as a modulator of transmembraneous Ca^{2+} transport[9], as osmoregulator[18,19] and as free radical scavenger[14], this amino acid seems to be a suitable protective agent against the cellular damage due to hypoxia and reoxygenation.

In previous studies we could demonstrate that supplementation by taurine of a conventionally used kidney preservation solution was able to increase viability rates significantly, and that the addition leads to a markedly improved metabolic status after hypoxia

Taurine 2, edited by Huxtable et al.
Plenum Press, New York, 1996

and reoxygenation[23,24]. The purpose of this study was to investigate in a cell culture model whether the University of Wisconsin solution, the present "gold standard" preservation solution, can be improved by supplementation of taurine and, furthermore, if the supplementation by taurine during hypoxic preservation is able to improve the post-hypoxic outcome.

MATERIALS AND METHODS

The experiment was divided into three time periods: aerobic culturing, hypoxic preservation and reoxygenation (Fig. 1).

Aerobic Culturing

Proximal porcine renal tubular cells from the cell strain LLC PK 1 (Kasza pig kidney; ICN Biomedical, Meckenheim, FRG) were cultured as monolayers in flasks aerobically for 48 h in M 199 medium with an admixture of fetal calf serum (50 ml/liter, 20%) and penicillin as well as streptomycin and amphotericin B. The cultures were cultivated in an atmosphere containing 21% O_2 and 5% CO_2. The temperature was kept constantly at 37°C and the O_2 partial pressure was about 140 mmHg. The cultures were fed three times a week. After 48 h of subculturing each confluent monolayer contained between 1 and 2 million cells.

Hypoxic Preservation

Subsequent to aerobiosis, the culture medium was removed and replaced by 5 ml University of Wisconsin solution (Fig. 1). The preservation solution was rendered hypoxic

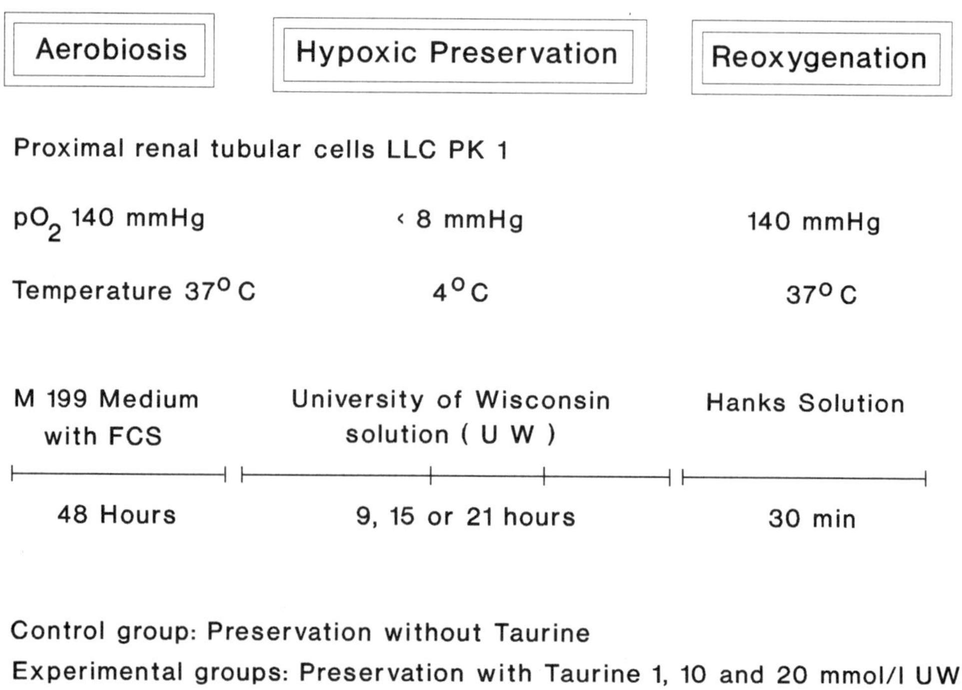

Figure 1. Experimental course and groups.

before by equilibration with purified N_2 resulting in O_2 partial pressures below 8 mmHg, as determined by means of a acid base laboratory (ABL 300, Radiometer Copenhagen).

With a new and simple method the flasks (Easy access, 25 cm^2, canted neck; Falcon #3813) were plugged with a seal screw cap, described elsewhere[23], which allowed persufflation with gaseous N_2 and the realization of O_2 partial pressures below 8 mmHg in the supernatant of the culture.

With constant low flow gas persufflation and defined supernatant volume the flasks were placed in a water bath at 4°C for different time courses.

Reoxygenation

After 9, 15 or 21 h, the hypoxic preservation medium was removed and replaced by a crystalloid perfusion solution without taurine admixture (Hank's balanced salt solution without Ca^{2+}, Sigma Chemicals, St. Louis, USA). Thereafter the cell cultures were stored aerobically again for 30 min at a temperature of 37°C and an O_2 partial pressure of 140 mmHg.

During hypoxic storage, taurine (Sigma Chemicals, St. Louis USA) was added to the preservation solution resulting in final concentrations of 1, 10 or 20 mM, ranging from physiological up to pharmacological values. Controls were preserved hypoxically without taurine additives.

Measurements

After each experimental period, 1 ml of the supernatant was removed for the determination of enzyme leakage during experimental course and the cells were rinsed from the bottom of the flask by treatment with Trypsine-EDTA solution (2.5% Trypsine - 0.025% EDTA, 1 ml, Fa. Gibco) added to the supernatant. After centrifugation (300 g, 5 min and 30°C) the supernatant was removed and the cells were resuspended. In an small aliquot we determined the % of remaining cells by cell counting with a Neubauer chamber and measured the viability of these cells by the trypan blue exclusion test.

The product of both parameters, regarded as survival index (SI), was calculated (SI = [remaining cells/cells at the beginning] x 100 x viability of the remaining cells [% x %]) and served as an index for the survival of the cultures. This parameter reflects the amount of cells which remain after the experimental period and which are still viable, and it enables the detection of differences in the number of remaining cells after experimental treatment or effects like cell lysis or detachment from the monolayer surface during the experimental course[9,22]. The measurement of viability alone is not sufficient to consider these effects adequately.

Additionally, the leakage of lactate dehydrogenase was measured with a conventional biochemical assay after each experimental period, in an aliquot of the supernatant of the cultures.

As parameters of the metabolic status of the whole culture after reoxygenation, the content of high energy phosphates and adenine nucleotides was measured by HPLC of neutralized perchloric acid extracts. Furthermore, the energy charge potential (ECP) was calculated: $ECP = ([ATP] + 0.5[ADP])/([ATP] + [ADP] + [AMP])$[1].

Before denaturation of proteins with perchloric acid (ice bath, 0.25 ml 1 M $HClO_4$) for the HPLC, the protein content of each culture was measured in an aliquot of the rinsed culture with the method described by Bradford[4].

Analyses of 5-9 cell cultures per group were used for the determination of each parameter.

Table 1. Percentage of remaining cells, viability rates and survival index after different periods of hypoxic preservation with UW solution

Preservation Time (h)	Taurine (mM)	Remaining Cells (%)	Viability (%)	SI [% x %] x 10^3
9	0	68 ± 19	89 ± 3	5.99 ± 1.69
	1	82 ± 18	92 ± 2	7.45 ± 1.47
	10	98 ± 13[b]	89 ± 4	8.78 ± 1.41[b]
	20	100 ± 13[b]	91 ± 2	9.04 ± 1.28[b]
15	0	67 ± 20	88 ± 2	5.88 ± 1.76
	1	60 ± 16	91 ± 4[a]	3.82 ± 0.89
	10	98 ± 19[a]	84 ± 4	8.29 ± 1.59[a]
	20	72 ± 8	93 ± 2[c]	6.70 ± 0.82
21	0	69 ± 14	84 ± 5	5.76 ± 1.03
	1	55 ± 9	90 ± 3[b]	4.94 ± 0.74
	10	87 ± 11[a]	91 ± 2[c]	7.96 ± 1.01[b]
	20	69 ± 15	84 ± 1	5.83 ± 1.24

n = 5 cultures per experiment, mean ± SD; Student's t-test vs. control: $^{a}p < 0.05$; $^{b}p < 0.01$; $^{c}p < 0.001$.

Statistical analyses were calculated parametrically with Student's t-test in comparison to the controls. Different levels of significance were accepted as significant ($^{a}p < 0.05$; $^{b}p < 0.01$; $^{c}p < 0.001$). Data are presented as mean values ± SD.

RESULTS

Hypoxic Preservation with UW Solution

After 9 h of hypoxic preservation with UW solution, taurine supplementation in concentrations of 10 and 20 mM significantly increased the amount of remaining cells and the survival index (Table 1).

After 15 h of cold storage, comparable effects occurred. Remaining cells, viability and survival index were markedly improved in the experimental groups treated with taurine in nearly physiological doses and pharmacological concentrations (Table 1).

Subsequent to 21 h of hypoxic storage, groups preserved with taurine in a concentration of 1 or 10 mM, respectively, showed markedly higher values of remaining cells, significantly increased viability rates and markedly higher survival indices (Table 1). With longer preservation time, higher concentrations of taurine were needed for the same protective effect. A possible explanation for the dose- and preservation time-dependent action of taurine is an osmotic swelling of the cells during longer preservation periods, which may be counteracted by the osmoregulatory function of higher taurine doses[10].

During hypoxic preservation, considerable amounts of lactate dehydrogenase were released (Table 2). As expected, higher amounts of lactate dehydrogenase were released into the supernatant of the cultures after longer preservation times, indicating a major cellular deterioration during long-lasting hypoxic storage.

The pattern of LDH leakage was inverse to the measured viability rates and survival indices. Significantly higher values of remaining cells or increased viability rates as well as survival indices were associated with reduced enzyme release during hypoxic preservation in the cultures preserved with taurine in concentrations between 1 and 20 mM.

The additional LDH leakage after reoxygenation compared to that during hypoxic preservation indicates further cell damage due to the metabolic events of reoxygenation. The

Table 2. Leakage of lactatedehydrogenase (LDH) before and during aerobiosis, hypoxic preservation with UK solution and during 30 min reoxygenation

Taurine (mM)	LDH leakage [units LDH / liter supernatant]	
	LDH leakage during aerobical culturing	
	24 h	48 h
0	0.8 ± 0.3	1.2 ± 0.4
	LDH leakage during hypoxia and reoxygenation	
	9 h	30 min
0	10.3 ± 2.6	9.6 ± 2.4
1	10.0 ± 2.1	7.6 ± 1.7
10	9.0 ± 1.0	10.2 ± 1.5
20	6.2 ± 1.5[a]	7.8 ± 2.6
	15 h	30 min
0	8.4 ± 0.9	11.4 ± 2.5
1	6.8 ± 0.9[a]	8.5 ± 1.3[a]
10	8.0 ± 2.1	10.8 ± 1.5
20	16.4 ± 1.6[a]	16.4 ± 1.6
	21 h	30 min
0	13.2 ± 1.4	14.5 ± 2.7
1	14.4 ± 1.9	9.7 ± 1.2[b]
10	10.5 ± 1.1[b]	12.5 ± 1.5
20	3.9 ± 1.1[c]	6.7 ± 1.3[c]

n = 5 cultures per experiment, mean ± SD; Student's t-test vs. control: [a]$p < 0.05$; [b]$p < 0.01$; [c]$p < 0.001$

total cell damage indicated by the overall enzyme release was caused by deteriorating effects induced by hypoxic preservation as well as by reoxygenation. Taurine supplementation reduced the general enzyme leakage markedly, especially after longer preservation times (Fig. 2).

Hypoxic Preservation with UW solution and Reoxygenation

The viability rates of the remaining cells after reoxygenation were markedly higher in comparison to the post-hypoxic values (Fig. 3). The number of remaining cells, and the survival indices after hypoxia and reoxygenation, indicated a further growing of the cell cultures after O_2 deficiency in spite of a deterioration of the cultures due to reoxygenation (Fig. 3, Table 3).

In some taurine-treated groups, slightly higher viability rates were found in comparison to the controls after long preservation periods and subsequent to 30 min reoxygenation (Fig. 3).

Whereas viability was only slightly improved, significant differences were seen in the number of remaining cells, and consequently in survival rates, when 10 mM taurine was added to the preservation solution. In this group, significantly higher survival rates and an increased numbers of remaining cells were determined 30 min after reoxygenation (Table 3). The same dose of taurine, which improved post-hypoxic viability and survival indices, exerted beneficial effects during hypoxia and reoxygenation.

The data show that taurine administered only during hypoxia seems to be able to protect the cells against the hypoxic cell damage and to improve the post-hypoxic recovery.

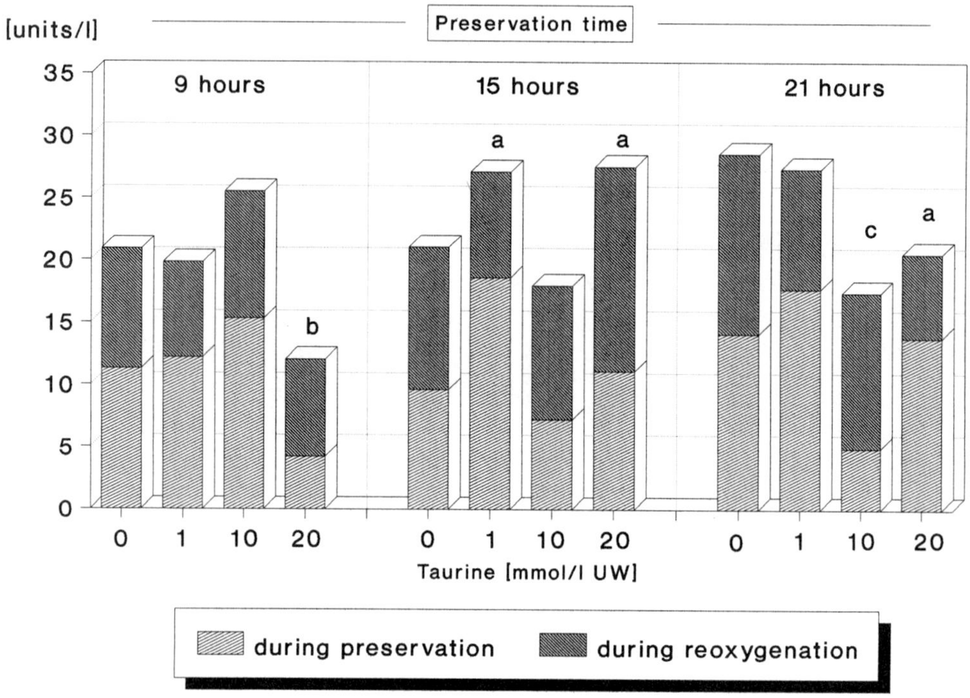

Figure 2. Leakage of lactate dehydrogenase (LDH) into the supernatant of the cultures during hypoxic preservation and subsequent to reoxygenation (compared to control group: [a]$p < 0.05$; [b]$p < 0.01$; [c]$p < 0.001$).

Table 3. Remaining cells and survival indices after different times of hypoxic preservation with UW solution and 30 min reoxygenation

Time (h)	Taurine (mM)	Remaining Cells (%)	survival index [% x %] x 10^3
9 + 0.5	0	93.9 ± 23.4	8.95 ± 2.11
	1	111.6 ± 19.3	10.68 ± 2.01
	10	146.9 ± 23.9[b]	14.25 ± 2.34[b]
	20	101.7 ± 19.2	9.87 ± 1.96
15 + 0.5	0	83.3 ± 12.2	8.00 ± 1.06
	1	95.4 ± 16.7	9.14 ± 1.61
	10	106.9 ± 8.3[a]	10.15 ± 0.87[a]
	20	90.4 ± 6.7	8.41 ± 0.74
21 + 0.5	0	85.1 ± 14.9	8.06 ± 1.39
	1	74.9 ± 12.4	7.10 ± 1.15
	10	93.7 ± 17.2	8.99 ± 1.68[a]
	20	80.6 ± 11.6	7.61 ± 1.22

n = 5 cultures per experiment, mean ± SD; Student's t-test vs. control: [a]$p < 0.05$; [b]$p < 0.01$; [c]$p < 0.001$

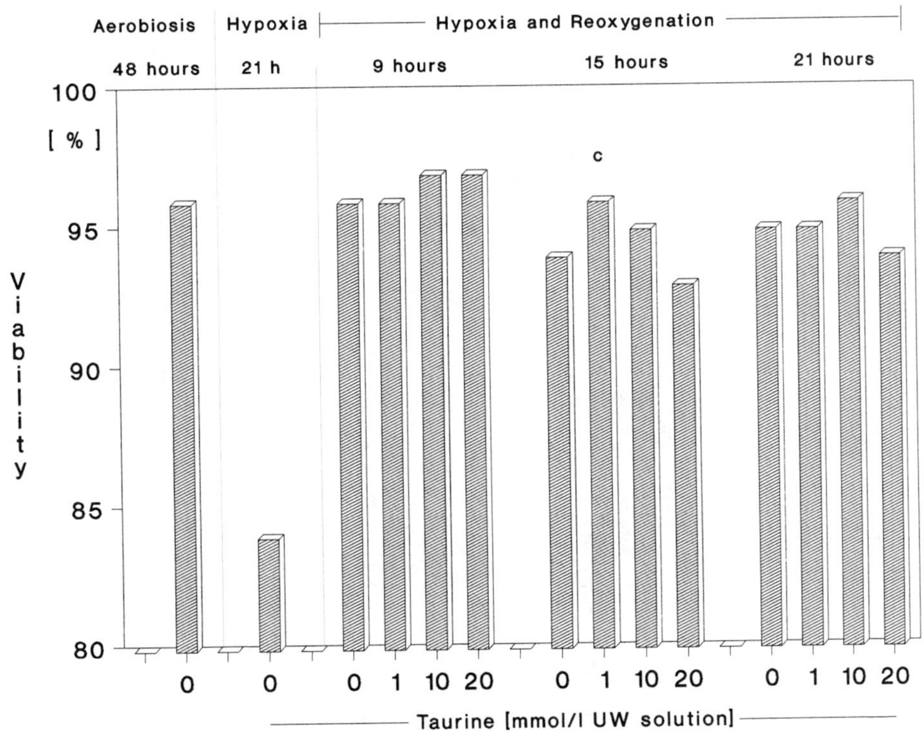

Figure 3. Viability of the cultures after aerobic culturing, 21 h hypoxic preservation and different times of hypoxia and reoxygenation (compared to control group: $^{c}p < 0.001$).

The measurements of metabolic parameters after cold hypoxic storage and reoxygenation confirmed the post-hypoxic recovery of the cells (Table 4). The content of high energy phosphates as well as the calculated energy charge potentials argue for viable cell cultures. The measured metabolite concentrations of adenine nucleotides were slightly higher than described by others[7,16,21]. The differences in metabolite concentrations between the several groups may be due to the variable content of high energy phosphates of the cell cultures[8] and different methods of determination[7,16,21].

In comparison to the prehypoxic values of the control group, there was a significant increase in the cellular content of high energy phosphates and the sum of adenine nucleotides when 1-10 mM taurine was added to the preservation solution. Apparently, the cells were stimulated to improved *de novo* synthesis of high energy phosphates, induced by hypoxia and following reoxygenation, and this effect was most pronounced in the presence of taurine. Furthermore taurine, administered during hypoxia, seemed to be able to reduce the loss of adenine nucleotides markedly.

DISCUSSION

One factor in the beneficial effect of taurine may be its osmoregulatory effect, which counteracted the osmotic cell swelling during hypoxia[18,19]. Furthermore an improvement of Ca^{2+} homeostasis and a protection of cellular biomembranes against free radical-induced

Table 4. High energy phosphate content and sum of adenine nucleotides (SAN) after 48 h aerobical culturing in M 199 medium and subsequent to hypoxic preservation with UK solution and 30 min reoxygenation with Hanke solution (nmol/mg protein)

Taurine (mM)	ATP	ADP	AMP	SAN
	Aerobical culturing with M 199			
0	36 ± 3.6	12 ± 9.4	30 ± 4.9	77 ± 16.5
	After hypoxic preservation and reoxygenation			
		9 h + 30 min		
0	20 ± 7.5	5 ± 1.1	—	25 ± 7.9
1	39 ± 14.1	6 ± 1.5	—	45 ± 14.9
10	50 ± 13.9[a]	9 ± 2.4[b]	4 ± 2.2[c]	63 ± 11.0[b]
20	98 ± 13.6[c]	12 ± 0.8[a]	—	110 ± 21.5[b]
		15 h + 30 min		
0	35 ± 3.4	5 ± 0.5	—	40 ± 4.4
1	20 ± 4.1[c]	6 ± 1.1	—	26 ± 4.8[b]
10	39 ± 4.0[a]	10 ± 2.8[b]	—	50 ± 5.0[a]
20	34 ± 4.9	11 ± 1.0[b]	—	43 ± 7.0
		-21 h + 30 min		
0	34 ± 3.8	6 ± 0.8	—	40 ± 3.6
1	60 ± 15.9[a]	6 ± 1.9	—	66 ± 17.5[a]
10	60 ± 14.3[a]	9 ± 3.2	3.6 ± 1.8 c	70 ± 19.4[a]
20	26 ± 5.7[b]	6 ± 1.7	—	32 ± 6.0[a]

n = 5 cultures per experiment, mean ± SD; No value indicates nucleotide was not detectable. Student's t-test vs. control: $^{a}p < 0.05$; $^{b}p < 0.01$; $^{c}p < 0.001$

peroxidative damage could be responsible for the reduced cellular deterioration during hypoxia and reoxygenation[3,14,20].

Viability and survival rates of the cell cultures suggest a taurine-induced stimulation of protein synthesis and cell growth in spite of O_2 deprivation and considerable deterioration of cellular integrity during hypoxia. In face of the averaged generation time of the cell strain LLC PK 1 of 30 h and generation times during the growth period of 7.5 to 9 h[7] accelerated growth processes have to take place during the observed short reoxygenation period. It is not yet evident if taurine administered during hypoxia accelerated cell generation by itself or if different mediators are involved.

SUMMARY

Taurine administered during hypoxia reduced the cell damage due to O_2 deficiency markedly. The beneficial effect outlasted the period of reoxygenation. The mechanisms for the improved survival rates are postulated in a reduced osmoregulatory disturbance of cellular integrity, improved Ca^{2+} homeostasis and induction of accelerated cellular growth processes.

We conclude that taurine supplementation of the conventionally used kidney preservation solution (UW) improves this "gold standard" kidney preservation solution markedly.

ACKNOWLEDGMENTS

We gratefully acknowledge the skilful technical assistance of Mrs. Anke Gradehand, Mr. Hans Günter Herb and Mr. Norbert July.

REFERENCES

1. Atkinson, D.E. 1968, Energy charge of adenylate pool as a regulatory parameter. Interaction with feedback modifiers, *Biochemistry*, 7:4030-4034.
2. Belzer, F.O. and Southhard, J.H. 1988, Principles of solid-organ preservation by cold storage, *Transplantation*, 45:673-676.
3. Bonventre, J.V. 1993, Mechanisms of ischemic acute renal failure, *Kidney Int.* 43:1160-1178.
4. Bradford, M. 1976, A rapid and sensitive method for the quantitation of microgram quantities of protein utilizing the priciple of protein-dye binding, *Anal.Biochem.* 72:248-254.
5. Collins, G.M., Bravo-Shugarman, M., and Terasaki, P.I. 1969, Kidney preservation for transportation, *Lancet*, 2:1219-1222.
6. Collins, G.M. and Wicomb, W.N. 1992, New organ preservation solutions, *Kidney Int.* 42 (Suppl. 38):197-202.
7. Gstraunthaler, G., Pfaller, W., and Kotanko, P. 1985, Biochemical characterization of renal epithelial cell cultures (LLC-PK1 and MDCK), *Am.J.Physiol.* 248:F536-F544.
8. Hull, R.N., Cherry, W.R., and Weaver, G.W. 1976, The origin and characteristics of a pig kidney cell strain, LLC PK1, *In Vitro*, 12:670-677.
9. Huxtable, R.J. 1992, Physiological actions of taurine, *Physiol.Rev.* 72:101-163.
10. Jauregul, H.O., Hayner, N.T., and Driscoll, J.L. 1981, Trypan blue dye uptake and lactate dehydrogenase in adult rat hepatocytes - freshly isolated cells, cell suspensions, and primary monolayer cultures, *In Vitro*, 17:1100-1110.
11. Lindell, S.L., Gandolph, D., Southard, J.H., and Belzer, F.O. 1991, Comparison of PBS, HTK, and UW solutions for kidney preservation, *Transplantation Proc.* 23:2399-2401.
12. Marshall, V.C., Biguzas, M., Jablonski, P., Scott, D.F., Howden, B.O., Thomas, A.C., Cham, C.W., and Walls, K. 1990, UW solution for kidney preservation, *Transplantation Proc.* 22:496-497.
13. Marshall, V.C., Jablonski, P., Biguzas, M., Howden, B.O., and Walls, K. 1990, Kidney preservation with UW solution: the nature of the impermeant, *Transplantation Proc.* 22:2131-2132.
14. Nakamori, K., Koyama, I., Yoshida, T., Umeda, M., and Inoue, K. 1990, Effectivness of taurine in protecting biomembrane against oxidant, *Chem.Pharm.Bull.* 38:3116-3119.
15. Ploeg, R.J., van Bockel, J.H., Langendijk, P.T.H., Groenewegen, M., van der Woude, F.J., Persijn, G.G., Thoregood, J., and Hermans, J. 1992, Effect of preservation solution on results of cadaveric kidney transplantation, *Lancet*, 340:129-136.
16. Sahai, A., Cole, L.A., Clarke, D.L., and Tannen, R.L. 1989, Rocking promotes differentiated properties in LLC-PK1 cells by improved oxygenation, *Am.J.Physiol.* 315:C1064-C1069.
17. Torras, J., Bordalba, J.R., Seron, D., Carrera, M., Castelao, A.M., Poveda, R., Alsina, J., and Grino, J. 1992, An experimental comparison between isotonic saline solution, Euro-Collins, and a flush solution with mannitol in the prevention of renal damage due to warm ischemia, *Transplantation Proc.* 24:54-55.
18. Trachtman, H. 1991, Cell volume regulation: a review of cerebral adaptive mechanisms and implications for clinical treatment of osmolal disturbances: I, *Pediatr.Nephrol.* 5:743-750.
19. Trachtman, H. 1992, Cell volume regulation: a review of cerebral adaptive mechanisms and implications for clinical treatment of osmolal disturbances: II, *Pediatr.Nephrol.* 6:104-112.
20. Troyer, D.A., Kreisberg, J.I., and Venkatachalam, M.A. 1986, Lipid alterations in LLC-PK1 cells exposed to mercuric chloride, *Kidney Int.* 29:530-538.
21. Venkatachalam, M.A., Patel, Y.J., Kreisberg, J.I., and Weinberg, J.M. 1988, Energy thresholds that determine membrane integrity and injury in a renal epithelial cell line (LLC-PK1), *J.Clin.Invest.* 81:745-758.
22. Weinberg, J.M., Davis, J.A., Roeser, N.F., and Venkatachalam, M.A. 1990, Role of increased cytosolic free calcium in the pathogenesis of rabbit proximal cell injury and protection by glycine or acidosis, *J.Clin.Invest.* 87:581-590.
23. Wingenfeld, P., Minor, T., Gehrmann, U., Strübind, S., Isselhard, W., and Michalk, D. 1995, Hypoxic cellular deterioration and its prevention by the amino acid taurine in a transplantation model with renal tubular cells (LLC PK1), *In Vitro Cell Dev.Biol.* 31:483-486.

24. Wingenfeld, P., Strübind, S., Gehrmann, U., Minor, T., Isselhard, W., and Michalk, D. 1994, Protecting effects of taurine against hypoxic cell damage in renal tubular cells cultured in different preservation solutions, in: "Taurine in Health and Disease", Huxtable, R.J. and Michalk, D.V. eds., Plenum Press, New York, pp. 159-169.

23

PROTECTIVE EFFECT OF TAURINE ON HYPOXIA AND REOXYGENATION-INDUCED DAMAGE OF HUMAN COLON CELLS (HT 29)

Peter Wingenfeld,[1] Dietrich Volker Michalk,[1] Angela Sonntag,[1]
Stefan Paas,[1] Thomas Minor,[2] and Wolf Isselhard[2]

[1] Department of Pediatrics
[2] Institute for Experimental Medicine
University of Cologne
Cologne, Federal Republic of Germany

INTRODUCTION

Transplantation of the small and large bowel may eventually become a curative treatment for pediatric and adult patients with intestinal failure and short gut syndrome. Since the first description of human allogenic transplantation of intestinal organs such as small and large bowel[13], the problem of the suitable preservation of the explanted organ is one of the determining factors for sucessful transplantation. With the recent sucess in transplantation of intestinal organs it has become important to develop better techniques for the preservation of the grafts[1,3-10,21]. Similar to the proceedings of other allotransplantations, the outcome highly depends on the quality of the preservation of the allograft[14,18,22,24,25,27,28]. Frequently the long lasting period of cold (hypoxic) storage of the organ and the induced cellular deterioration due to hypoxia limits the possibility of sucessful transplantation. Furthermore the pathophysiological events of the period of reoxygenation subsequent to hypoxia were limiting factors in the transplantation of human organs. Some of these limiting factors are the posthypoxic energy deficiency of an organ[2,26], e.g. a human kidney, the intracellular Ca^{2+} overload[6], osmoregulatory dysfunction induced by deterioration of membrane integrity[23,29], the O_2 paradox and membrane oxidation induced by toxic metabolites and free peroxide radicals[2,19].

Because of the unique functions of taurine as modulator of transmembraneous Ca^{2+} transport[11], as osmoregulator[29] as membrane stabilizing factor and free radical scavenger[20], this endogenous amino acid seems to be a suitable protective agent against the cellular damage due to hypoxia and reoxygenation. In previous studies with porcine renal tubular cells, it has been shown that taurine in a simplified transplantation model in adequate dosage and mode of application was able to reduce markedly the cellular damage due to hypoxia and reoxygenation[30]. Supplementation by taurine of a conventionally used kidney preservation solution was able to increase viability rates significantly. Furthermore, taurine leads to a markedly improved metabolic status after hypoxia and reoxygenation[31].

The purpose of this study was to investigate, in a cell culture model with human colon cells, if the protective effect of taurine during and subsequent to hypoxia was also present in an intestinal cell culture model. Additionally we wanted to compare the effectiveness of different preservation solutions for the preservation of intestinal cells in a simplified transplantation model. Preservation solutions compared include kidney preservation solutions like Euro Collins solution or University of Wisconsin solution[1,14,16,17] as well as the HTK preservation solution (Bretschneider solution) used in the transplantations of kidneys, hearts and lungs[15].

Another purpose of this study was to investigate if the most suitable preservation solution could be improved by supplementation with taurine.

MATERIALS AND METHODS

The experiment was divided in three time periods: aerobiosis, hypoxia and reoxygenation (Fig. 1).

Aerobiosis

Human colon cells of a carcinoma cell line from the available cell strain HT 29 (American type culture collection, Rockville Maryland, USA) were cultured aerobically as

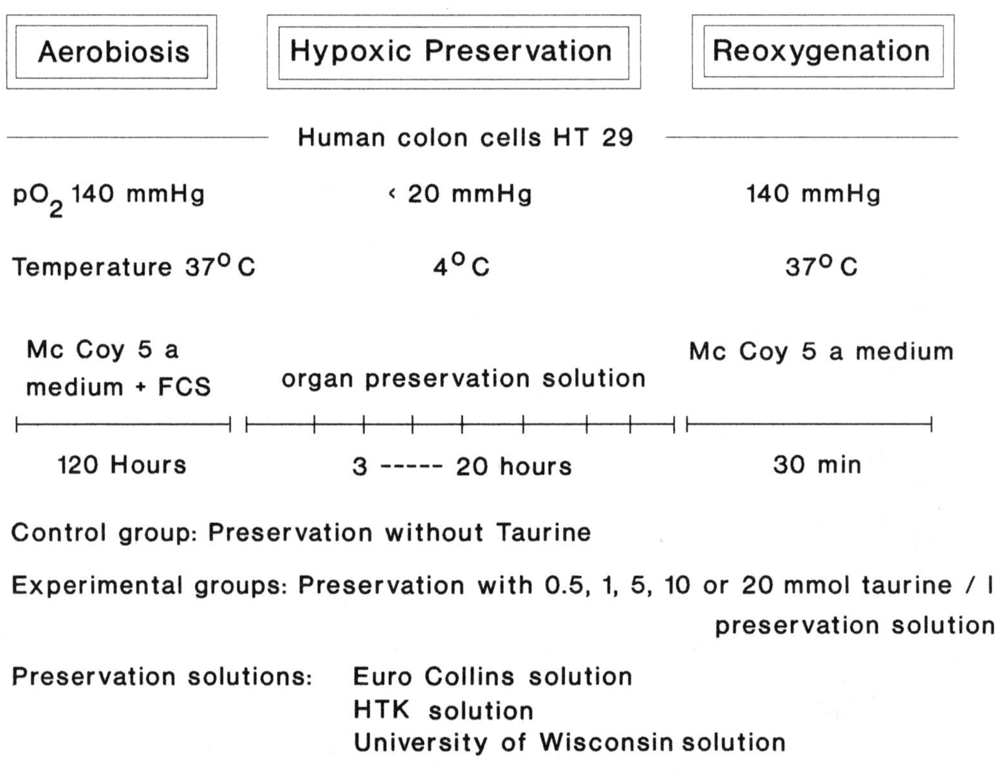

Figure 1. Experimental course and groups.

monolayers in flasks (Easy access, 25 cm^2, canted neck; Falcon # 3813) for 120 h in McCoy 5, a medium with an admixture of fetal calf serum (100 ml serum/liter, serum diluted 1:4) and penicillin (100 IU/ml) as well as streptomycin (100 μg/ml) and amphotericin B (250 μg/liter). The cultures were incubated with 21% O_2 and 5% CO_2. The temperature was kept constantly at 37°C and the O_2 partial pressure was about 140 mmHg. The cultures were fed 3-4 times a week. After 120 h of subculturing, each monolayer contained between 3 and 5 1/2 million cells.

Hypoxia

At the end of this period the culture medium was removed and replaced by 5 ml preservation solution (either Euro Collins or University of Wisconsin solution or HTK solution, also known as Bretschneider solution) (Fig. 1). The renewed solution were equilibrated beforehand with pure N_2 (UW and HTK) or N_2 plus an admixture of 5% CO_2 (EC) resulting in O_2 partial pressures below 20 mmHg, determined by means of a acid-base laboratory (ABL 300, Radiometer Copenhagen).

The flasks were plugged with a seal screw cap, described previously[30], which allowed the filling and persufflation with a nearly O_2-free atmosphere and to realize O_2 partial pressures below 20 mmHg in the supernatant of the culture.

With constant low flow gas persufflation and defined supernatant volume, the flasks were placed in a water bath at 4°C for different time courses. The gas persufflation was performed with gaseous N_2 (UW, HTK) or N_2 plus CO_2 (EC).

Reoxygenation

After different periods of hypoxic storage the preservation medium was removed and replaced by the cultivation medium used during aerobical culture. During this period McCoy medium contained no fetal calf serum supplementation. Thereafter the cell cultures were stored aerobically again for 30 min at a temperature of 37°C and an O_2 partial pressure of about 140 mmHg.

The experimental course imitated the conditions of bowel transplantation, where the intestinal organ is removed from a well oxygenized organism, then stored under either normoxic or partly hypoxic or ischemic conditions and reoxygenized after transplantation.

Experimental Groups

During hypoxic storage, taurine (Sigma chemicals, St. Louis USA) was added to the preservation solution resulting in final concentrations of 0.5, 1, 5, 10 or 20 mM, ranging from physiological up to pharmacological values. Controls were preserved hypoxically without taurine additives.

Subsequent to hypoxia all cell cultures were reoxygenated without additives to the McCoy medium.

Measurements

Before each hypoxic preservation and reoxygenation, 2 monolayer cultures were rinsed with Trypsine-EDTA (0.5% Trypsine - 0.025% EDTA, Fa. Gibco) and the cells were counted in a Neubauer chamber to get basic values of cell numbers. Additionally, viability rates were determined by trypan blue exclusion test[12].

Subsequent to hypoxia and reoxygenation the cells were rinsed from the bottom of the flask by treatment with Trypsine-EDTA solution. After centrifugation (300 g, 5 min and

Table 1. Percentage of remaining cells, viability rates and survival index after different periods of hypoxic preservation and reoxygenation with different preservation solutions without taurine supplementation

Preservation solution	Time [h]	Remaining cells [%]	Viability [%]	SI [% x %] x 10^3
EC	1	58 ± 8	91 ± 5	4.96 ± 0.82
	3	72 ± 24	64 ± 10	4.52 ± 1.64
	5	70 ± 20	53 ± 9	3.77 ± 1.51
	7	85 ± 19	47 ± 9	4.11 ± 1.53
	10	74 ± 6	33 ± 7	2.48 ± 0.64
HTK	9	92 ± 37	48 ± 9	4.58 ± 2.07
	15	56 ± 14	58 ± 9	3.21 ± 0.84
	17	72 ± 7	45 ± 14	3.26 ± 1.18
	20	47 ± 6	55 ± 12	2.57 ± 0.64
UW	9	90 ± 15	75 ± 5	6.64 ± 0.64
	15	86 ± 9	82 ± 5	6.97 ± 0.91
	20	63 ± 12	66 ± 9	4.23 ± 1.36

n = 5 - 8 cultures per experiment, mean values ± SD

30°C) the supernatant was removed and the cells were resuspended. In an small aliquot we determined the percentage of remaining cells and viability. The product of both parameters, regarded as survival index (SI), was calculated (SI = [remaining cells/cells at the beginning] x 100 x viability of the remaining cells [% x %]) and served as an index for the survival of the cultures. This parameter enabled a better comparison between different preservation periods and preservation solutions.

Additionally the leakage of lactate dehydrogenase (LDH) was measured with standard biochemical assays (LDH Monotest, Boehringer Mannheim) in an aliquot of the supernatant of the cultures subsequent to cold hypoxic storage.

Analyses of 5 to 9 cell cultures per group were used for the determination of each parameter. Statistical analyses were done parametrically with Student's t-test in comparison to the controls (levels of significance: $^a p< 0.05$; $^b p< 0.01$; $^c p< 0.001$). Data represent means ± SD.

RESULTS AND DISCUSSION

Hypoxic Preservation with Euro Collins Solution and Reoxygenation

Euro Collins solution enabled only short periods of sufficient preservation of the colon cell cultures (Table 1). Subsequent to 7 h and 10 h of cold hypoxic preservation and following reoxygenation viability rates and the amount of remaining cells argued for a markedly deterioration of the cultures.

Taurine supplementation during hypoxia led to significantly increased viability rates subsequent to the different periods of hypoxia and reoxygenation. The numbers of remaining cells and the values of the survival index were markedly higher in the groups preserved with 0.5 mM taurine. In the groups preserved 7 h with EC solution, nearly two-fold increased viability rates were measured when taurine was added to the preservation solution in concentrations of 0.5, 5 or 10 mM. With longer preservation periods higher concentrations of taurine seemed to be most effective in exerting beneficial effects (Table 2). A possible explanation for the dose- and preservation time-dependent action of taurine is an osmotic

Table 2. Percentage of remaining cells, viability rates and survival index after different periods of hypoxic preservation and reoxygenation with EC solution with or without taurine supplemtation

Preservation time [h]	Taurine [mmol]	Remaining cells [%]	Viability [%]	SI [% x %] x 10^3
3	0	72 ± 24	64 ± 10	4.52 ± 1.64
	0.5	121 ± 23[b]	86 ± 6 [b]	10.37 ± 1.90[b]
	1	97 ± 19	82 ± 5[b]	8.00 ± 1.55[a]
	5	58 ± 10	85 ± 7[b]	4.86 ± 0.50
	10	142 ± 36[b]	77 ± 16	11.20 ± 4.17[a]
	20	98 ± 15	86 ± 3[b]	8.41 ± 1.11[b]
5	0	69 ± 20	54 ± 9	3.77 ± 1.51
	0.5	95 ± 12[a]	90 ± 6[c]	8.60 ± 1.56[c]
	1	77 ± 15	77 ± 13[a]	5.84 ± 1.23[a]
	5	92 ± 26	73 ± 16[a]	5.29 ± 2.95
	10	89 ± 24	75 ± 12[b]	6.64 ± 1.97[a]
	20	97 ± 25	84 ± 9[c]	7.62 ± 2.83[a]
7	0	85 ± 19	48 ± 10	4.12 ± 1.53
	0.5	93 ± 2	76 ± 12[a]	7.19 ± 1.12[b]
	1	65 ± 32	58 ± 12	3.62 ± 1.58
	5	73 ± 8	84 ± 7[c]	6.14 ± 0.82[a]
	10	84 ± 19	74 ± 7[c]	6.30 ± 1.40
	20	86 ± 30	70 ± 11[b]	6.17 ± 2.60

n = 5 - 8 cultures per experiment, mean values ± SD
Students t-test vs. control: [a]$p<0.05$, [b]$p<0.01$,[c]$p<0.001$

swelling of the cells during longer preservation periods, which may be counteracted by the osmoregulatory function of higher taurine doses[29].

Hypoxic Preservation with HTK Solution and Reoxygenation

In the transplantation of human kidneys the HTK solution allows for longer ischemia times than EC solution. Therefore the cultures of this experimental group were preserved 2-3 fold longer than cells preserved with EC solution (Table 1).

Comparable to the experiments with EC solution, the taurine supplementation led to markedly increased viability rates and significantly higher numbers of remaining cells. In the group preserved 15 h with HTK solution the highest values of viability, remaining cells and survival rates were determined when taurine was added in a concentration of 1 and 5 mM HTK solution, respectively (Fig. 2).

Subsequent to a preservation period of 20 h, improved viability and survival rates were measured in the presence of 1-20 mM taurine during hypoxia in HTK solution (Fig. 2). Similar to the experiments with EC solution, higher concentrations of taurine were needed at longer preservation periods. The increased rates of remeaining cells which were determined after hypoxia and reoxygenation argued for apparently stimulated growth processes which have to take place during hypoxia and reoxygenation. Furthermore taurine seemed to improve the osmoregulation of the cultures, especially subsequent to long lasting preservation periods.

In this experimental model the HTK solution seemed to be more suitable than EC solution for the cold preservation of human colon cells.

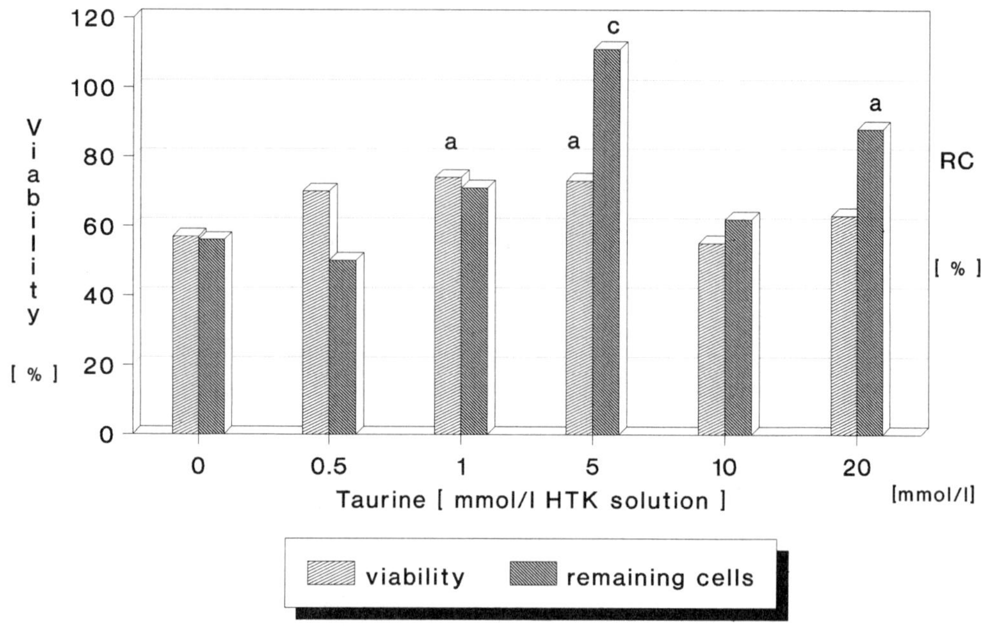

Figure 2. Viability rates and remaining cells subsequent to 15 h of hypoxic preservation with HTK solution and 30 min of reoxygenation in the presence or absence of taurine during hypoxia (vs. control group: $^{a}p < 0.05$; $^{c}p < 0.001$).

Hypoxic Preservation with UW Solution and Reoxygenation

Preservation with UW solution enabled the longest preservation periods. The numbers of remaining cells and the values of viability and survival indices were markedly increased in comparison to the other preservation fluids, especially after longer periods of O_2 deficiency (Table 1).

In spite of the higher effectivity of the UW solution for the preservation of the cultures, taurine supplementation resulted in further improvement of the solution. After 9 h of hypoxic preservation with UW solution, 1 mM taurine supplementation slightly increased the viability of the cells as well as the number of remaining cells and consecutively the survival index (Table 3).

After 15 h of cold storage, comparable effects occured. Remaining cells and survival index were markedly improved in the experimental group treated with 1 mM taurine mmol/l (Table 1). In contrast to this group, higher concentrations of taurine slightly reduced the measured viability rates and numbers of remaining cells. This effect disappeared with longer preservation times.

Subsequent to 20 h of hypoxic storage, groups preserved with 5-10 mM taurine showed markedly higher values of remaining cells, slightly or markedly improved viability rates and markedly higher survival indices (Table 3). With longer preservation time higher concentrations of taurine were again needed to exert the same protective effect; the improved osmoregulation due to the addition of taurine might be also a possible explanation for the beneficial effect in this experimental group. This appeared more surprising in face of the different ingredients of the conventionally used UW solution, which are added for edema reduction during hypoxia. If the improved osmoregulation due to taurine supplementation

Table 3. Remaining cells, viability rates and survival index after different times of hypoxic preservation with UW solution and 30 min reoxygenation with or without taurine supplementation

Preservation time [h]	Taurine [mmol/l]	Remaining cells [%]	Viability [%]	SI [% x %] x 10^3
9	0	90 ± 15	75 ± 5	6.64 ± 0.67
	0.5	94 ± 30	79 ± 9	7.71 ± 2.81
	1	94 ± 37	80 ± 6	7.52 ± 3.40
	5	73 ± 18	80 ± 13	6.34 ± 1.35
	10	88 ± 13	65 ± 9	5.78 ± 1.55
	20	71 ± 10	78 ± 12	5.58 ± 1.07
15	0	86 ± 9	81 ± 5	6.97 ± 0.91
	0.5	91 ± 19	71 ± 14	6.32 ± 1.50
	1	156 ± 40[b]	57 ± 8[c]	8.62 ± 1.61
	5	89 ± 27	78 ± 9	6.76 ± 1.68
	10	69 ± 16	72 ± 12	4.81 ± 0.47[b]
	20	68 ± 30	52 ± 8[a]	3.39 ± 1.32[c]
20	0	63 ± 12	66 ± 9	4.23 ± 1.36
	0.5	73 ± 19	68 ± 8	5.00 ± 1.73
	1	58 ± 8	69 ± 8	4.00 ± 0.85
	5	112 ± 22[b]	64 ± 11	6.99 ± 0.91[b]
	10	92 ± 5b	84 ± 6a	7.72 ± 0.96[b]
	20	68 ± 17	59 ± 13	4.17 ± 2.00

n = 5 cultures per experiment, mean values ± SD;
Students t-test vs. control: a $p<0.05$, b $p<0.01$, c $p<0.001$

was partly responsible for the beneficial effect, taurine have had to be more effective than the conventionally additives in the preservation fluid.

Comparable to the results of the preservation with HTK solution, one of the mainly responsible mechanisms for the taurine-induced increase in remaining cells after hypoxia and reoxygenation seemed to be the induction of cellular growth processes.

During hypoxic preservation, considerable amounts of lactate dehydrogenase were released. As expected, higher amounts of lactate dehydrogenase were released into the supernatant of the cultures after longer preservation times (Fig. 2), indicating a major cellular deterioration during long-lasting hypoxic storage.

During prolonged preservation with UW solution, the pattern of LDH leakage was inverse to the measured viability rates and survival indices. Significantly higher numbers of remaining cells or increased viability rates as well as survival indices were associated with reduced enzyme release during hypoxic preservation in the cultures preserved with 10-20 mM taurine. Cell cultures preserved with HTK solution showed similar results, the monolayers with higher rates of viable cells and increased survival rates releasing slightly or markedly reduced enzyme activity, indicating a reduced cellular damage during the period of O_2 deficiency (Fig. 3).

Surprisingly, the cells which were preserved with UW solution have lost higher amounts of enzyme activity in comparison to the cells preserved with HTK solution. The more effective preservation with UW indicated by higher posthypoxic values of remaining and trypan blue-excluding cells was associated by with increased cellular deterioration during hypoxic storage. Obviously the period of reoxygenation was influenced by the previous preservation. In spite of increased extent of cellular deterioration during hypoxia of the cells preserved with UW solution, these cells have been protected more effectively against the deteriorating effects of renewed oxygenation.

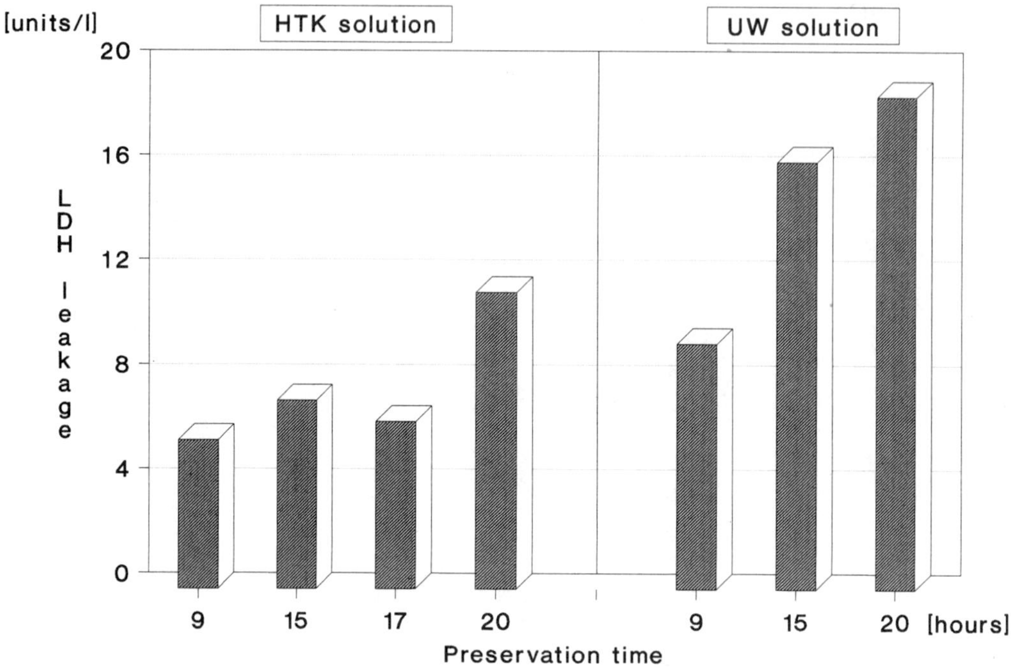

Figure 3. Leakage of lactate dehydrogenase (LDH) into the supernatant of the cultures during hypoxic preservation with HTK solution and UW solution.

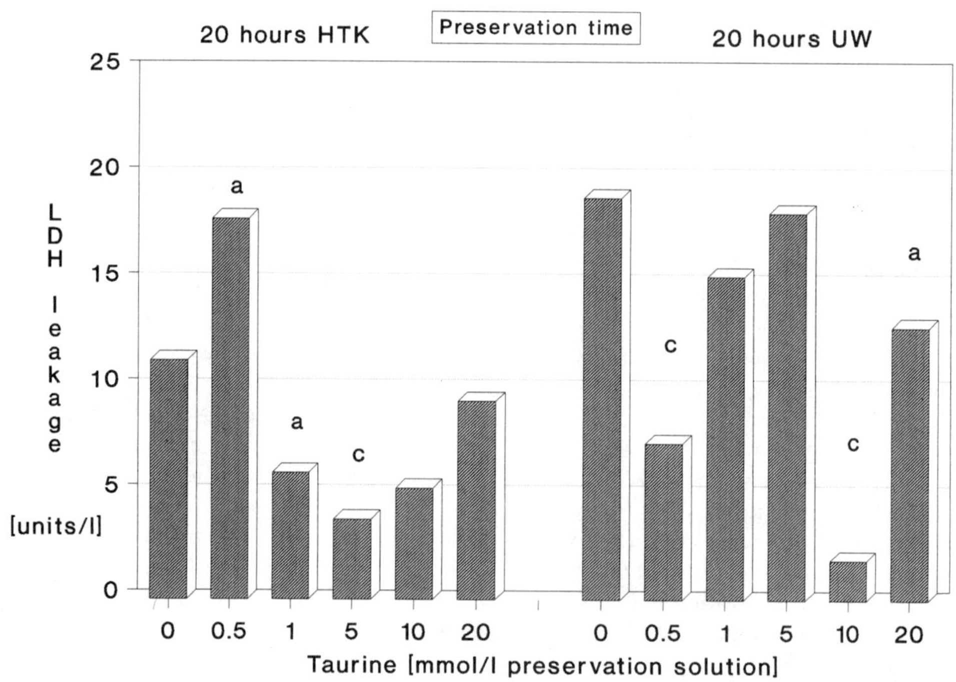

Figure 4. Leakage of lactate dehydrogenase (LDH) into the supernatant of the cultures preserved 20 h with HTK or UW solution with or without taurine supplementation (vs. control: $^{a}p < 0.05$; $^{c}p < 0.001$).

This experimental procedure imitated the conditions of common organ transplantation, where the organ is removed from a well-oxygenized organism, then stored under hypoxic conditions and reoxygenized after transplantation.

One factor in the beneficial effect of taurine may be its osmoregulatory effect, which counteracted the osmotic cell swelling during hypoxia[29]. It is not evident if the reduced deterioration of Ca^{2+} homeostasis could be responsible for the reduced cellular deterioration during hypoxia and reoxygenation.

Further studies are needed to determine, if a reduction of oxygen free radicals after reoxygenation could be responsible for the cell protective effect of taurine[19].

The changes of viability and survival rates of the cell cultures during and after hypoxic preservation with the different solutions suggested a taurine-induced stimulation of protein synthesis and cell growth in spite of O_2 deprivation and considerable deterioration of cellular integrity during hypoxia. Taurine, administered during hypoxia, seemed to be a endogenous growth stimulating factor, accelerating cell generation by itself or by different mediators.

SUMMARY

In this experimental model, taurine administered during hypoxia markedly reduced the cell damage due to O_2 deficiency, and the beneficial effect outlasted the period of reoxygenation. The mechanisms for the improved survival rates are postulated to be a reduced osmoregulatory disturbance of cellular integrity, improved Ca^{2+} homeostasis and induction of accelerated cellular growth processes.

In our simplified cell culture model the UW solution seems to be the most appropriate solution for the cold (hypoxic) preservation of human colon cells.

We conclude, that within this experimental model and under these experimental conditions, taurine supplementation of the conventionally used preservation solutions improved the solutions markedly. Considering our previous studies, taurine seems to be a potent endogenous protective agent against cellular deterioration due to hypoxia and reoxygenation.

ACKNOWLEDGMENTS

We gratefully acknowledge the skilful technical assistance of Mrs. Anke Geradehand, Mr. Hans Günter Herb and Mr. Norbert July.

REFERENCES

1. Belzer, F.O. and Southhard, J.H. 1988, Principles of solid-organ preservation by cold storage, *Transplantation*, 45:673-676.
2. Bonventre, J.V. 1993, Mechanisms of ischemic acute renal failure, *Kidney Int.* 43:1160-1178.
3. Collins, G.M. and Wicomb, W.N. 1992, New organ preservation solutions, *Kidney Int.* 42 (Suppl. 38):197-202.
4. de Bruin, R.W., Heinemann, E., and Marquet, R.L. 1994, Small bowel transplantation: an overview, *Transpl. Int.*, 7:47-61.
5. Deltz, E., Mengel, W., and Hamelmann, H. 1990, Small bowel tansplantation: report of a clinical case, *Prog. Pediatr. Surg.*, 25:91-96.
6. Filez, L., Pennickx, F., Stalmans, W., Kerremans, R., and Geboes, K. 1994, Prevention of mucosal reperfusion damage after orthotopic small bowel autotransplantation in cats, *Transplant Proc.*, 26:1485-1488.

7. Fujiwara, H., Raju, S., Grogan, J.B., and Johnson, W. 1990, Organ preservation in small bowel transplantation, *J. Invest. Surg.*, 3:
8. Goulet, O.J., Revillon, Y., and Cerf-Bensussan, N. 1988, Small intestinal transplantation in a child using cyclosporine, *Transplant Proc.*, 20:288-296.
9. Grant, D., Sommeraver, J., and Mimeault, R. 1989, Treatment with continous high dose cyclosporine following clinical intestinal transplantation, *Transplantation*, 48:151-153.
10. Grant, D., Wall, W., Mimeault, R., Zhong, R., Garcia, B., Stiller, C., and Duff, J. 1990, Successful small bowel/liver transplantation, *Lancet*, 335:181-184.
11. Huxtable, R.J. 1992, Physiological actions of taurine, *Physiol.Rev.* 72:101-163.
12. Jauregul, H.O., Hayner, N.T., and Driscoll, J.L. 1981, Trypan blue dye uptake and lactate dehydrogenase in adult rat hepatocytes - freshly isolated cells, cell suspensions, and primary monolayer cultures, *In Vitro*, 17:1100-1110.
13. Johnson, C. 1989, Canadian surgeons lay claim to world's first successful bowel transplantation, *Med. Assoc. J.*, 141, 156
14. Kukudo, Y., Furuya, T., Takeyoshi, I., Nakamura, K., Zhang, S., Murase, N., and Todo, S. 1994, Comparison of University of Wisconsin solution, Euro-Collins, and lactated Ringer's solution in rat small bowel preservation for orthotopic small bowel transplantation, *Transplant Proc.*, 26:1492-1493.
15. Lindell, S.L., Gandolph, D., Southard, J.H., and Belzer, F.O. 1991, Comparison of PBS, HTK, and UW solutions for kidney preservation, *Transplantation Proc.* 23:2399-2401.
16. Marshall, V.C., Jablonski, P., Biguzas, M., Howden, B.O., and Walls, K. 1990, Kidney preservation with UW solution: the nature of the impermeant, *Transplantation Proc.* 22:2131-2132.
17. McAnulty, J.F., Vreugdenhil, P.K., Southard, J.H., and Belzer, F.O. 1990, Use of UW cold storage solution for machine perfusion of kidneys, *Transplantation Proc.* 22:458-459.
18. Müller, A.R., Nalesnik, M., Platz, K.P., Langrehr, J.M., Hoffman, R.A., and Schraut, W.H. 1994, Evaluation of preservation conditions and various solutions for small bowel preservation, *Transplantation*, 57:649-655.
19. Müller, A.R., Platz, K.P., Langrehr, J.M., Hoffman, R.A., Nussler, A.K., Neuhaus, P., and Schraut, W.H. 1994, Function of nitric oxide during reperfudion of cold preserved small bowel grafts, *Transplant Proc.*, 26:1501-1503.
20. Nakamori, K., Koyama, I., Yoshida, T., Umeda, M., and Inoue, K. 1990, Effectivness of taurine in protecting biomembrane against oxidant, *Chem.Pharm.Bull.* 38:3116-3119.
21. Robinson, J.D. 1990, Evolution of transport ATPases, *Trends Biochem.Sci.* 15:180
22. Rodriguez, F.J., Toledo-Pereyra, L.H., and Suzuki, S. 1993, Total heterotopic transplant of the small intestine in rats. Technical considerations on organ acquiring, preservation and transplantation, *Gac. Med. Mex.*, 129:131-137.
23. Slocum, M.M. and Granger, D.N. 1993, Early mucosal and microvascular changes in feline intestinal transplants, *Gastroenterology*, 105:1761-1768.
24. Sonino, R.E., Pritchard, T., and Riddle, J.M. 1993, Limited survival of rat small bowel transplants preserved in University of Wisconsin solution for 48 hours, *J. Invest. Surg.*, 6:185-199.
25. Takara, T., Guttman, E., Zorychta, E., Lubin, A., Mayer, S.K., Rosenmann, E., and Richardson, J. 1993, Rat small bowel preservation for 24 and 48 hours, *Transplant Proc.*, 25:2999-3000.
26. Takeyoshi, I., Kokudo, Y., Zhang, S., Nakamura, K., Ikoma, A., Suzuki, T., Zhu, Y., and Todo, S. 1994, Susceptibility to ischemia: the large bowel versus the small bowel, *Transplant Proc.*, 26:1491
27. Thaler, W., Oberhuber, G., Klima, G., Margreiter, R., and Schmid, T. 1993, Preservation of small bowel grafts-a comparison with two standard solutions, *Eur. Surg. Res.*, 25:181-186.
28. Todo, S., Tzakis, A., Abu Elamagd, K., Reyers, J., Fung, J., Casavilla, A., Nakamura, K., Yagihashi, A., Jain, A., Murase, N., Iwaki, Y., Demetris, A., van Thiel, D., and Starzl, T. 1992, Cadaveric small bowel and small bowel-liver transplantation in humans, *Transplantation*, 53:369-376.
29. Trachtman, H. 1992, Cell volume regulation: a review of cerebral adaptive mechanisms and implications for clinical treatment of osmolal disturbances: II, *Pediatr.Nephrol.* 6:104-112.
30. Wingenfeld, P., Minor, T., Gehrmann, U., Strübind, S., Isselhard, W., and Michalk, D. 1995, Hypoxic cellular deterioration and its prevention by the amino acid taurine in a transplantation model with renal tubular cells (LLC PK1), *In Vitro Cell Dev.Biol.* in press
31. Wingenfeld, P., Strübind, S., Gehrmann, U., Minor, T., Isselhard, W., and Michalk, D. 1994, Protecting effects of taurine against hypoxic cell damage in renal tubular cells cultured in different preservation solutions, in: "Taurine in Health and Disease", Huxtable, R.J. and Michalk, D.V. eds., Plenum Press, New York, pp. 159-169.

24

THE MECHANISMS OF TAURINE MEDIATED PROTECTION AGAINST CELL DAMAGE INDUCED BY HYPOXIA AND REOXYGENATION

Dietrich Volker Michalk, Peter Wingenfeld, Christoph Licht, Taner Ugur, and Lilly Farrokh Siar

Department of Pediatrics
University of Cologne
Cologne, Federal Republic of Germany

INTRODUCTION

Recent studies from our laboratory with renal porcine tubular cell cultures gave evidence that taurine in an appropriate dosage and mode of application could markedly reduce the cellular damage due to hypoxia and reoxygenation[20,21]. Furthermore, the taurine supplementation led to significantly improved metabolic conditions and stimulated cellular growth processes. Because of the well known pharmacological actions of taurine as modulator of Ca^{2+} homeostasis[6,10,12,13], free radical inhibitor[9,14], osmoregulator[6,16,17] and the possible role as growth modulating factor, one aim of this study was to elucidate mechanisms which could have been responsible for the protection of the cell cultures during and subsequent to hypoxia. Since many cellular processes have now been recognized to be mediated by changes in cytosolic free Ca^{2+} and since the free cytosolic free Ca^{2+} concentration is involved in different cell functions[1-3,5,8,11,15,18,19] the influence of taurine on intracellular Ca^{2+} homeostasis and transmembraneous Ca^{2+} currents was also investigated.

Furthermore, we wanted to know if the protective effect of taurine was related to an active uptake of the amino acid either during hypoxia or during reoxygenation.

MATERIALS AND METHODS

The experimental course was divided in three time periods, aerobiosis, hypoxia, and reoxygenation (Fig. 1), unless otherwise indicated in the text.

Taurine 2, edited by Huxtable et al.
Plenum Press, New York, 1996

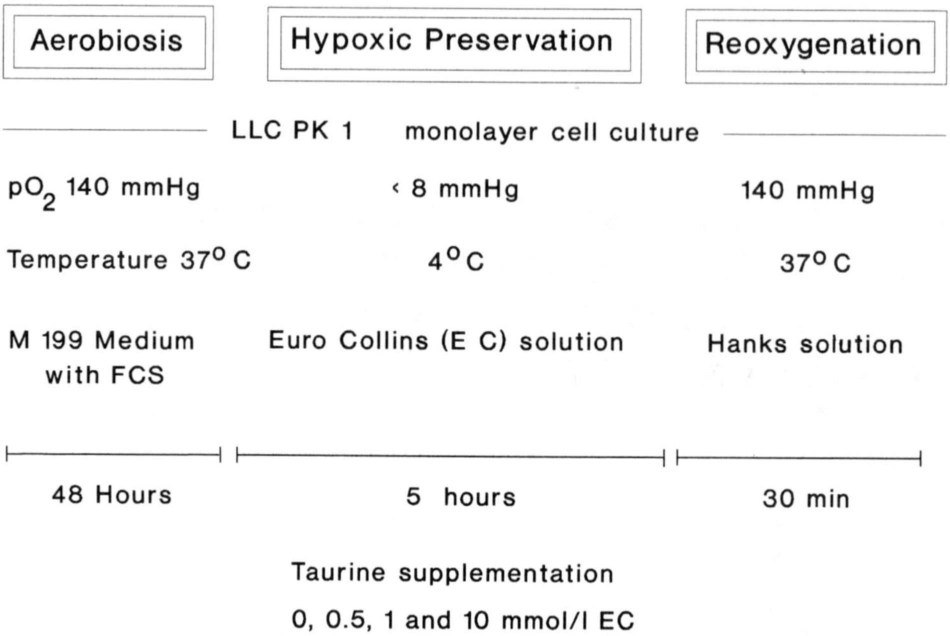

Figure 1. Experimental course and groups.

Aerobiosis

Proximal porcine renal tubular cells from the cell strain LLC PK 1 (ICN Biomedical, Meckenheim, FRG) were aerobically cultured as monolayers in flasks for 48 h in M 199 medium with an admixture of fetal calf serum (50 ml/l, 20%) and penicillin (100 I.U./ml) as well as streptomycin (100 μg/ml) and amphotericin B (250 μg/l). The cultures were cultivated in an atmosphere containing 21% O_2 and 5% CO_2. The temperature was maintained at 37°C with an O_2 partial pressure of about 140 mm Hg. After 48 h of subculturing, each confluent monolayer contained between 1 and 2 million cells.

Hypoxia

Subsequent to the period of aerobiosis the culture medium was removed and replaced by 5 ml Euro Collins solution (glucose 39.27 g/l, KH_2PO_4 2.092 g/l, $KHPO_4^-$ 7.548 g/l, KCl 1.14 g/l, $NaHCO_3$ 0.86 g/l). The preservation solution was rendered hypoxic by prior equilibration with purified N_2 resulting in O_2 partial pressures below 8 mm Hg (Fig. 1).

The flasks (Easy access, 25 ml, canted neck; Falcon #3813) were plugged with a seal screw cap, described previously[20], which allowed the persufflation with gaseous N_2 and the realization of O_2 partial pressures below 8 mm Hg in the supernatant of the culture.

With constant low flow gas persufflation (0.5 l/min) and defined supernatant volume, the flasks were placed in a water bath kept at 4°C.

Reoxygenation

After 5 h, the hypoxic preservation medium was removed and replaced by a crystalloid perfusion solution (Hank's balanced salt solution without Ca^{2+}, containing glucose

600 mg/l, NaCl 80 g/l, KCl 4 g/l, KH_2PO_4 10 g/l, Na_2HPO_4 475 mg/l, [Sigma chemicals, USA]). Thereafter the cell cultures were stored aerobically again for 30 min at of 37°C under O_2 partial pressure of 140 mm Hg.

During hypoxic storage taurine (Sigma chemicals, St. Louis, USA) was added to the preservation solution to give final concentrations of 0.5, 1 or 10 mM.

Controls were preserved hypoxically without addition of taurine. Whenever the experimental course was different to that in the schematic diagram (Fig. 1) it is mentioned in the text.

Measurements

Before hypoxia or reoxygenation, 2 cultures were rinsed from the monolayer surface with Trypsin-EDTA (0.5% Trypsin - 0.025% EDTA, Fa. Gibco), the cells were counted in a Neubauer chamber and their viability rates were determined by trypan blue exclusion test[7].

The product of both parameters, regarded as survival index (SI), was calculated (SI = (remaining cells/cells at the beginning) x 100 x viability of the remaining cells [% x %]) and served as an index for the survival of each culture.

Measurement of Mean Cellular Volume and Osmoregulation

For the determination of the mean cellular volume of the cells subsequent to aerobiosis, hypoxia, and reoxygenation, we used a blood cell counter with computerized volume analysis (Cobas Minos STX, Hoffmann La Roche, Switzerland). The data represent the mean cellular volume of 1.2 to 2.5 x 10^6 cells.

Measurements on Cellular Calcium Homeostasis

In order to measure overall intracellular Ca^{2+} we used modified flame photometry (EFOX 5053, Eppendorf, Germany). To eliminate extracellular Ca^{2+} and to prevent contamination of the extracts all solutions used were pretreated with Tritriplex VI (EGTA, Merck, Germany). The cells were concentrated by centrifugation and cellular membranes were disintegrated by ultrasonic treatment. Data represent mean overall Ca^{2+} values of 2.5 to 7 x 10^6 cells.

Inhibition of Active Intracellular Taurine Uptake

Experiments were made with an supplementation of phorbol ester (30 nM) to the solution used during hypoxic preservation and with an addition of both phorbol ester and cyclosporin (100 ng/ml) during hypoxia. Both substances prevent the active cellular uptake of taurine during hypoxic preservation as well as during reoxygenation[4].

Cellular Growth Measurement

Cellular growth and DNA synthesis was determined by the use of the 5-Bromo-2-desoxy-uridine test (5-BrdU Labelling and Detection Kit III, Boehringer Diagnostics, Germany), a nonradioactive alternative for the [^{3}H]thymidine incorporation assay.

Cell cultures were incubated with 5-BrdU during 7 h of hypoxia in the presence or absence of 10 mM taurine in the preservation solution. The reoxygenation period was 1 h.

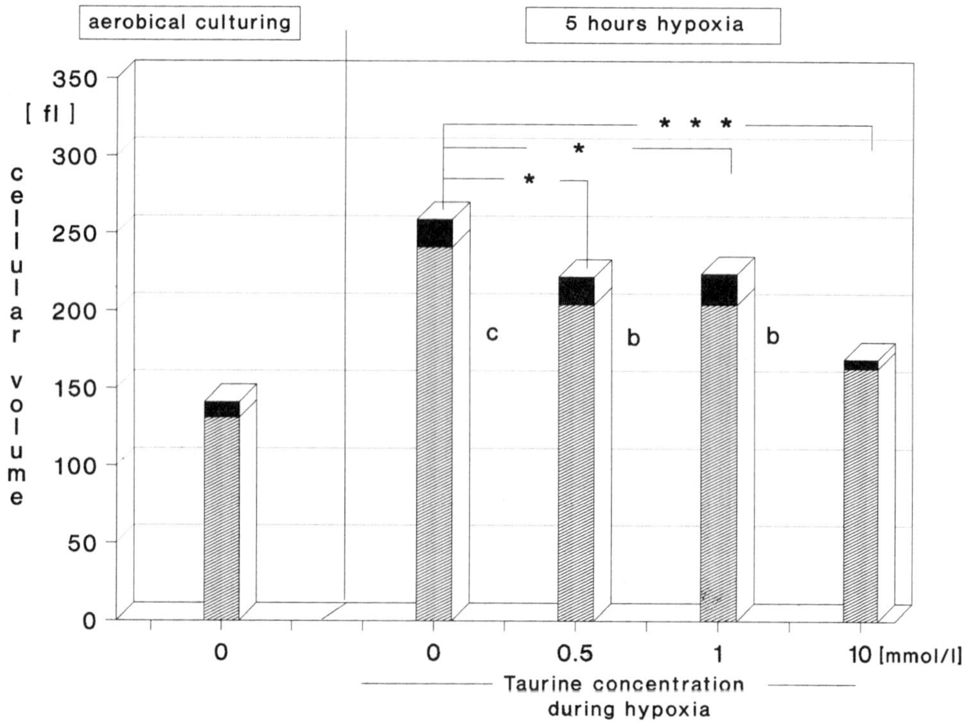

Figure 2. Mean cellular volume after 5 h hypoxic preservation (Students t-test vs. aerobic control: b $p < 0.01$, c $p < 0.001$; vs. treated group: *$p < 0.05$, ***$p < 0.001$; solid bars represent SD).

Statistics

Statistical analyses were calculated parametrically with Student's t-test in comparison to the controls. Different limits of significance were accepted (a: $p < 0.05$; b: $p < 0.01$; c: $p < 0.001$). Data are presented as means ± SD.

RESULTS

Osmoregulation

In comparison to the aerobic controls the cellular volume increased during the period of O_2 deficiency. The addition of taurine showed a dose dependent reduction of cellular swelling. Surprisingly, taurine in concentrations as low as 0.5 mM reduced the mean cellular volume. Higher concentrations of taurine reduced the osmotic swelling more markedly (Fig. 4). At a taurine concentration of 10 mM, the osmoregulatory deterioration due to hypoxia was completely prevented.

Calcium Metabolism

During hypoxic preservation and reoxygenation with Ca^{2+}-free solutions, intracellular Ca^{2+} decreased to 25% of the aerobic control values. In spite of the short period of

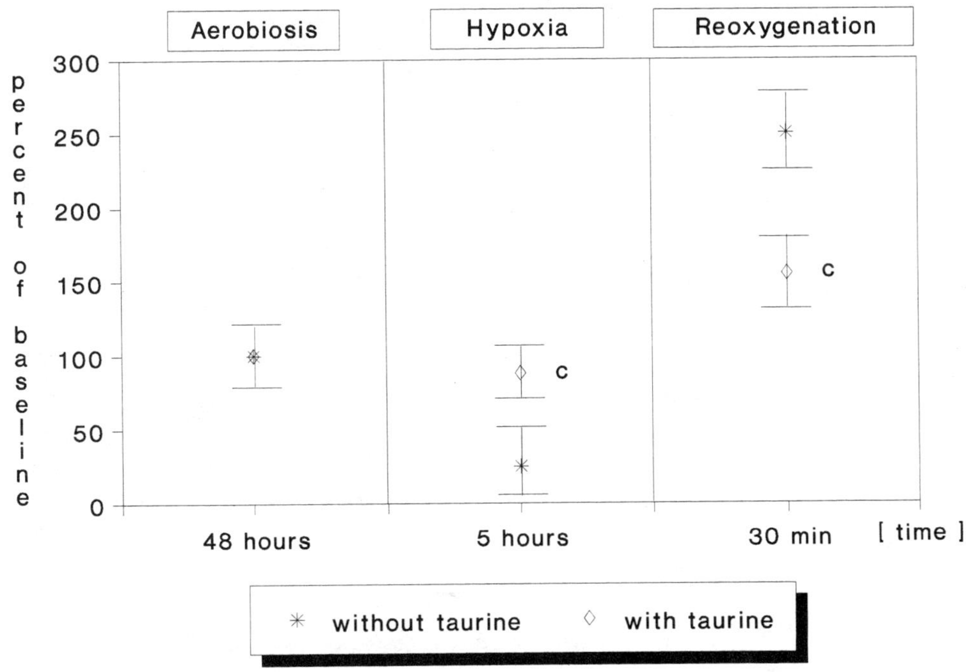

Figure 3. Effect of 0.5 mM taurine supplementation during hypoxia on cellular Ca^{2+} content after hypoxia and subsequent to reoxygenation (Students t-test vs. aerobical cultured cells: c $p < 0.001$; Ca^{2+} content expressed as % of baseline = values after aerobical culturing).

reoxygenation, an massive increase in cellular Ca^{2+} content occurred (Ca^{2+} overload). Taurine (0.5 mM) added during hypoxia reduced the loss of Ca^{2+} during hypoxia and the taurine-induced effect outlasted the period of hypoxia and protected the cells from Ca^{2+} overload during reoxygenation (Fig. 3).

During the period of hypoxia, physiological and supraphysiological Ca^{2+} concentrations in the supernatant of the cultures showed no deteriorating effect on the number of remaining cells or the viability of these cells. During reoxygenation, in contrast, higher concentrations of Ca^{2+} led to reduced rates of rates of remaining cells without a change of the viability rates (Table 1).

Table 1. Viability and percentage of remaining cells subsequent to hypoxia and reoxygenation in the presence of different extracellular Ca^{2+} concentrations

Experimental Period	Ca^{2+} (mM)	Remaining Cells (%)	Viability (%)
Hypoxia			
	0	85.5 ± 19.7	91.2 ± 8.1
	2.5	84.4 ± 13.3	86.6 ± 2.8
	5	84.8 ± 16.4	90.6 ± 2.5
	10	84.1 ± 17.5	92.3 ± 3.2
Reoxygenation			
	0	102.9 ± 26.7	93.0 ± 3.6
	2.5	83.1 ± 26.8	93.4 ± 3.5
	5	83.0 ± 14.4	92.3 ± 6.4

n = 5 cultures per experiment, mean values ± SD

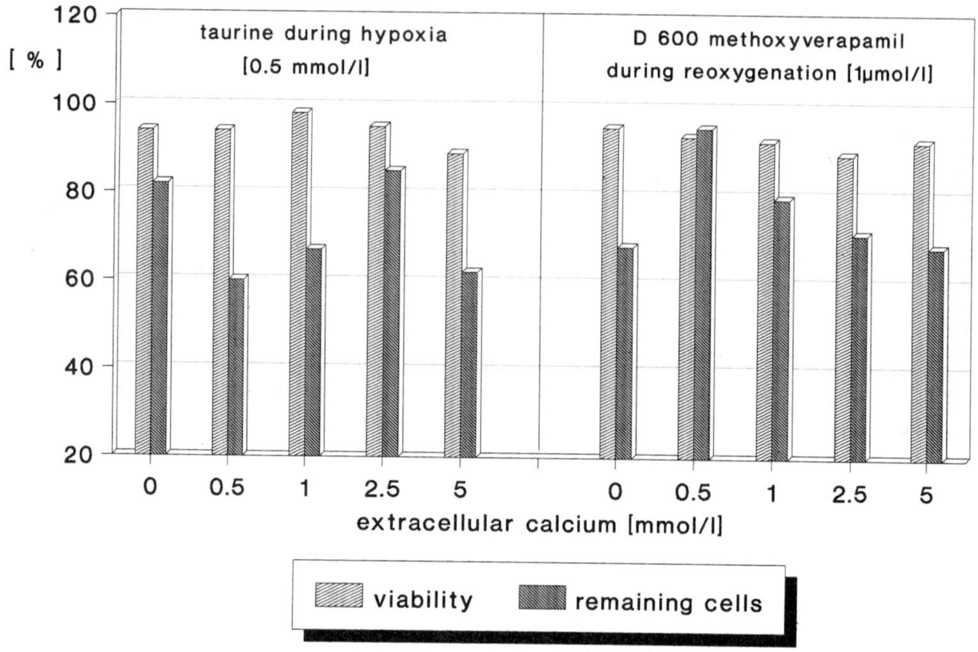

Figure 4. Effect of 0.5 mM taurine supplementation during hypoxia and supplementation of methoxyverapamil (100 µg/l) during reoxygenation on viability and remaining cells in the presence or absence of different Ca^{2+} concentrations during reoxygenation.

When taurine was added to the preservation fluids, the viability was slightly improved, but surprisingly the number of the remaining cells were even more depressed than with Ca^{2+} alone, with one exception: at a Ca^{2+} concentration of 2.5 mM, the human physiologic Ca^{2+} concentration, this negative effect of Ca^{2+} was compensated by the addition of taurine.

Almost comparable results were obtained by the addition of the Ca^{2+} antagonist methoxyverapamil (1 µM) (Fig. 4).

Inhibition of Taurine Uptake

As demonstrated in our previous experiments, taurine improved the number of remaining cells as well as the viability of these cells. This effect seemed to be unrelated to an active uptake of taurine during hypoxia since it was not impaired by the addition of phorbol ester and cyclosporin (Fig. 5).

In contrast, there was a prevention of the taurine-induced regeneration and growth of the cells by the transport inhibitors during the period of reoxygenation, suggesting that taurine must be actively taken up by the cells to exert its beneficial effect against the reoxygenation cell damage (Fig. 6).

Cellular Growth

BrdU uptake during 7 h of hypoxia and 1 h reoxygenation was significantly increased when taurine was added during anaerobiosis (Fig. 7). Obviously taurine stimulated cellular growth processes during hypoxia *and* reoxygenation.

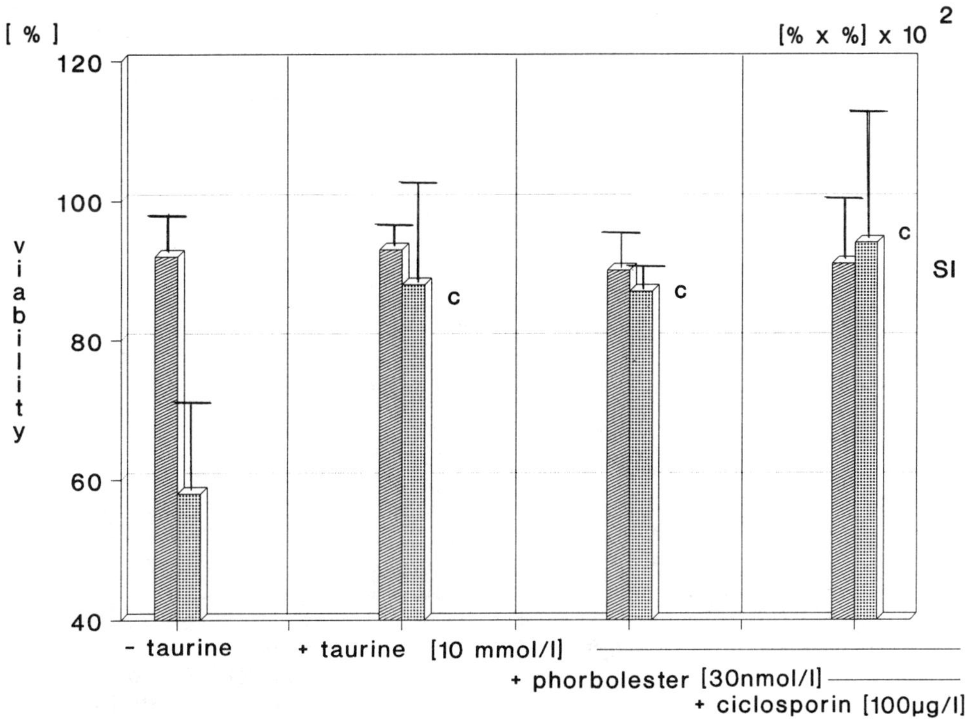

Figure 5. Inhibition of active cellular taurine uptake during hypoxia with cyclosporin and/or phorbol ester and its effect on viability rates and survival index (SI) (Student's t-test vs. control: c $p < 0.001$. Data are means ± SD).

DISCUSSION

One factor in the taurine-induced cellular protection against hypoxia and reoxygenation was the improved osmoregulation. Taurine reduced the cellular swelling during hypoxia and reoxygenation in a dose-dependent manner. It was not evident whether the effect of taurine at a lower concentration was a direct one or rather mediated by an influence on the activity of Na^+ or Cl^- transporter mechanisms[12,13]. Furthermore other electrolyte transporting transmembraneous systems like the Na^+/K^+ or Na^+/H^+ transporter could be influenced by taurine during hypoxia. Higher concentrations of taurine were presumed to reduce the osmotic swelling of the cells more directly by the increase of the osmolality of the culture medium[6,16,17].

In the absence of extracellular Ca^{2+}, taurine reduced the cellular Ca^{2+} efflux during hypoxia markedly and impaired the Ca^{2+} influx or Ca^{2+} overload during reoxygenation. Taurine acted as modulator of transmembranous Ca^{2+} currents. But it remains unclear if alterations of the cellular membrane; e.g. an increased transmembraneous Ca^{2+} flux because of osmotic dysregulation, may have played a role in the alterations of Ca^{2+} fluxes. Nor was it evident if the measured overall Ca^{2+} reflected changes in cytosolic, mitochondrial, membrane-associated or nuclear Ca^{2+} and which intracellular Ca^{2+} compartment was mainly affected[15]. Furthermore the taurine induced modulation of Na^+ or K^+ channels and the induced electrolytic imbalances perhaps could have been responsible for the present alterations of the overall Ca^{2+} content.

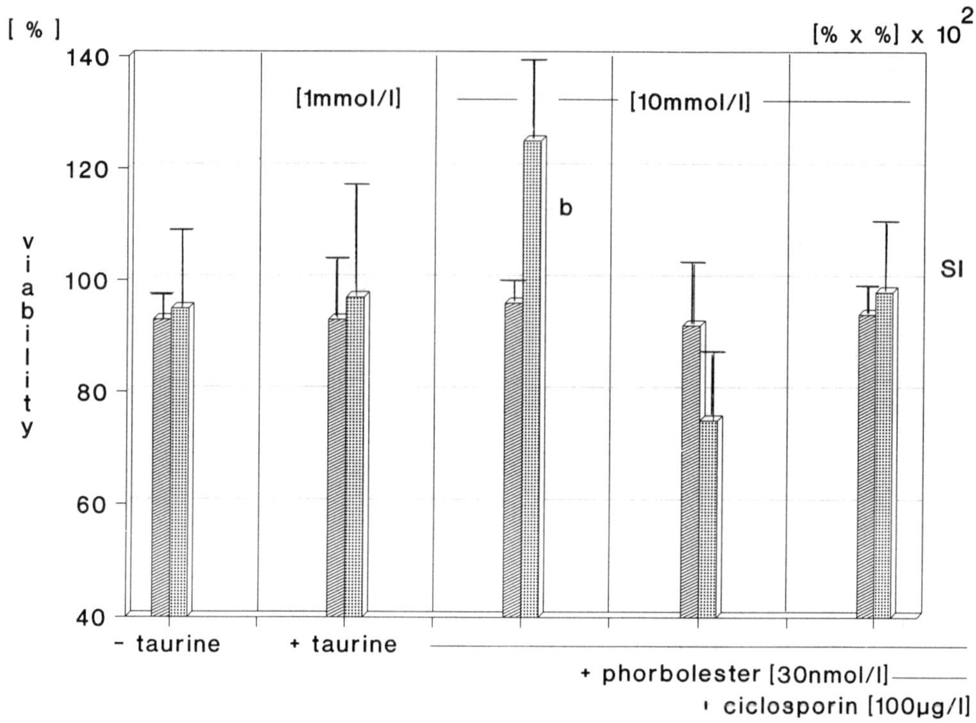

Figure 6. Inhibition of active cellular taurine uptake during reoxygenation after taurine supplementation during hypoxia and its effect on viability rates and survival index (SI) (Student's t-test vs. control group: b $p < 0.01$; data are means ± SD).

Obviously the taurine supplementation during hypoxia exerted effects outlasting the period of hypoxia, perhaps due to taurine uptake during hypoxia and release during reoxygenation. This released taurine could reduce the Ca^{2+} overload during reoxygenation by inhibition of Ca^{2+} uptake.

In this experimental model, the exposure of the cell cultures to physiological and supraphysiological Ca^{2+} concentrations during hypoxia showed no deleterious effect on viability or remaining cells. In contrast, after reoxygenation the cell cultures showed slightly reduced values of viability and remaining cells. A Ca^{2+}-induced detachment of cells during reoxygenation could be a possible explanation for this alteration. in face of the short period of reoxygenation and the measured viability rates.

The cytoprotective effect of taurine seems to be partly related to its action as a Ca^{2+} channel blocker since the taurine-induced changes in viability rates and amounts of remaining cells were comparable to that of methoxyverapamil. Whereas the effect was not detectable in the presence of lower or supraphysiological Ca^{2+} concentrations it could be detected in the presence of physiological Ca^{2+} concentrations.

The analyses of BrdU uptake during hypoxia and reoxygenation suggests a taurine-induced stimulation of protein synthesis and cellular growth in spite of O_2 deprivation and considerable deterioration of cellular integrity. It is not yet evident if taurine administered during hypoxia accelerated cell generation by itself or if different mediators are involved, but the induction of cellular growth seemed to be one of the mechanisms mainly responsible for the protection of the cultures during and after O_2 deficiency.

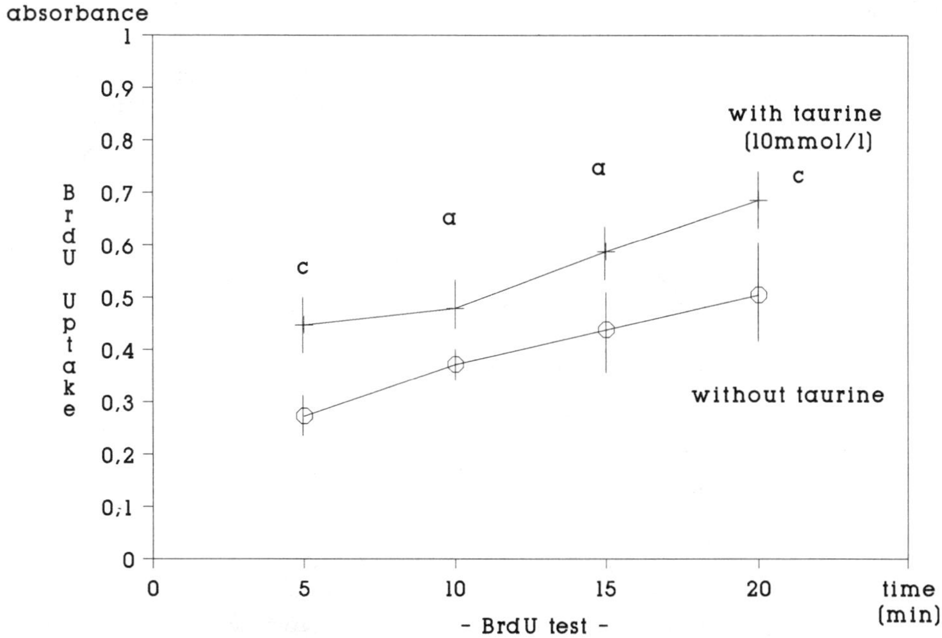

Figure 7. BrdU Uptake during 7 h of hypoxia and 1 h of reoxygenation in the presence or absence of 10 mM taurine during hypoxic preservation (Student's t-test vs. control group: a $p < 0.05$, c $p < 0.001$; data are means ± SD).

SUMMARY

Taurine administered during hypoxia markedly reduced the cell damage due to O_2 deficiency and reoxygenation. Different mechanisms are responsible for the improved survival of the renal cell cultures.

Taurine markedly reduces the osmoregulatory deterioration during hypoxia and reoxygenation. Calcium homeostasis was markedly improved. Ca^{2+} efflux during hypoxia as well as Ca^{2+} overload during reoxygenation was significantly reduced by the amino acid. The effect of taurine was partly comparable to the effect induced by Ca^{2+} channel blockers.

One of the effects mainly responsible for cellular protection seems to be the taurine-induced acceleration of cellular growth processes in spite of hypoxia and reoxygenation.

The spectrum of cytoprotective effects of taurine predisposes this substance to be a physiological protective agent responsible for cellular homeostasis or enantiostasis.

ACKNOWLEDGMENTS

We gratefully acknowledge the assistance of Mrs. Anke Gradehand from the Department of Pediatrics for maintenance of the cell cultures and Mr. Hans Günter Herb and Mr. Norbert July from the Institute of Experimental Medicine, Cologne, for excellent technical assistance.

REFERENCES

1. Almeida, A.R.P., Bunnachak, D., Burnier, M., Wetzel, J.F.M., Burke, T.J., and Schrier, R.W. 1992, Time dependent protective effects of calcium channel blockers on anoxia and hypoxia induced proximal tubule injury, *J.Pharm.Exp.Ther.* 260:526-532.
2. Bonventre, J.V. 1993, Mechanisms of ischemic acute renal failure, *Kidney Int.* 43:1160-1178.
3. Cheung, J.Y., Bonventre, J.V., Malis, C.D., and Leaf, A. 1986, Calcium and ischemic injury, *New Eng.J.Med.* 314:1670-1676.
4. Ganapathy, V. and Leibach, F.H. 1994, Expression and regulation of the taurine transporter in cultured cell lines of human origin, in: "Taurine in health and disease," Huxtable, R.J. and Michalk, D.V. eds., Plenum Press, New York, pp. 51-57.
5. Greene, E.L. and Paller, M.S. 1994, Calcium and free radicals in hypoxia/reoxygenation injury of renal epithelial cells, *Am.J.Physiol.* 266:13-20.
6. Huxtable, R.J. 1992, Physiological actions of taurine, *Physiol.Rev.* 72:101-163.
7. Jauregul, H.O., Hayner, N.T., and Driscoll, J.L. 1981, Trypan blue dye uptake and lactate dehydrogenase in adult rat hepatocytes - freshly isolated cells, cell suspensions, and primary monolayer cultures, *In Vitro*, 17:1100-1110.
8. Kribben, A., Wieder, E.D., Wetzels, J.F., Yu, L., Gengaro, P.E., Burke, T.J., and Schrier, R.W. 1994, Evidence for role of cytosolic free calcium in hypoxia-induced proximal tubule injury, *J.Clin.Invest.* 93:1922-1929.
9. Nakamori, K., Koyama, I., Yoshida, T., Umeda, M., and Inoue, K. 1990, Effectivness of taurine in protecting biomembrane against oxidant, *Chem.Pharm.Bull.* 38:3116-3119.
10. Nakashima, T., Seto, Y., Nakajima, T., Shima, T., Sakamoto, Y., Cho, N., Sano, A., Iwai, M., Kagawa, K., Okanoue, T., and Kashima, K. 1990, Calcium-associated cytoprotective effects of taurine on the calcium and oxygen paradoxes in isolated rat hepatocytes, *Liver*, 10:167-172.
11. Piper, H.M. 1989, Energy deficiency, calcium overload and oxidative stress: possible causes of irreversible ischemic myocardial injury, *Klin.Wochenschr.* 67:465-476.
12. Satoh, H. 1994, Cardioprotective actions of taurine against intracellular and extracellular calcium-induced effects, in: "Taurine in Health and Disease," Huxtable, R.J. and Michalk, D. eds., Plenum Press, New York, pp. 181-196.
13. Schaffer, S.W., Ballard, C., and Azuma, J. 1994, Mechanisms underlying physiological and pharmacological actions of taurine on myocardial calcium transport, in: "Taurine in health and disease," Huxtable, R.J. and Michalk, D.V. eds., Plenum Press, New York, pp. 171-176.
14. Schuller-Levis, G., Quinn, M.R., Right, C., and Park, E. 1994, Taurine protects against oxidant induced lung injury: possible mechanisms of action, in: "Taurine in health and disease," Huxtable, R.J. and Michak, D.V. eds., Plenum Press, New York, pp. 159-169.
15. Silverman, H.S. 1993, Mitochondrial free calcium regulation in hypoxia and reoxygenation: relation to cellular injury, *Basic Res.Cardiol.* 88:483-494.
16. Trachtman, H. 1991, Cell volume regulation: a review of cerebral adaptive mechanisms and implications for clinical treatment of osmolal disturbances: I, *Pediatr.Nephrol.* 5:743-750.
17. Trachtman, H. 1992, Cell volume regulation: a review of cerebral adaptive mechanisms and implications for clinical treatment of osmolal disturbances: II, *Pediatr.Nephrol.* 6:104-112.
18. Weinberg, J.M., Davis, J.A., Roeser, N.F., and Venkatachalam, M.A. 1990, Role of increased cytosolic free calcium in the pathogenesis of rabbit proximal cell injury and protection by glycine or acidosis, *J.Clin.Invest.* 87:581-590.
19. Windhager, E., Frindt, G., Yang, J.M., and Lee, C.O. 1986, Intracellular calcium ions as regulators of renal tubular sodium transport, *Klin.Wochenschr.* 64:847-852.
20. Wingenfeld, P., Minor, T., Gehrmann, U., Strübind, S., Isselhard, W., and Michalk, D. 1995, Hypoxic cellular deterioration and its prevention by the amino acid taurine in a transplantation model with renal tubular cells (LLC PK1), *In Vitro Cell.Dev.Biol.* 31:483-486.
21. Wingenfeld, P., Strübind, S., Gehrmann, U., Minor, T., Isselhard, W., and Michalk, D. 1994, Protecting effects of taurine against hypoxic cell damage in renal tubular cells cultured in different preservation solutions, in: "Taurine in Health and Disease," Huxtable, R.J. and Michalk, D.V. eds., Plenum Press, New York, pp. 159-169.

25

EFFECT OF TREATMENT WITH β-AGONISTS ON TISSUE AND URINARY TAURINE LEVELS IN RATS

Mechanism and Implications for Protection

Catherine J. Waterfield,[1] Felix Carvalho,[2] and John A. Timbrell[1]

[1] The Department of Toxicology
School of Pharmacy
Brunswick Square, London WC1N 1AX, United Kingdom
[2] Toxicology Laboratory
Faculdade de Farmacia
Univesidade do Porto
Rue Anibal Cunah, Porto, Portugal

ABSTRACT

Administration of clenbuterol to rats in the drinking water over a four day period increased incorporation of [^{3}H]leucine into muscle protein but did not result in an increase in body or muscle weight. However, both urinary and liver taurine were significantly reduced at the highest dose of clenbuterol (2 $mg.kg^{-1}.day^{-1}$). Salbutamol also reduced urinary levels of taurine in both rats and humans. The reduction in the body pool of taurine caused by β-agonists may be of concern as taurine has been shown to have protective properties.

INTRODUCTION

Previous work has shown that urinary taurine levels in rats are increased by compounds which inhibit protein synthesis (e.g. cycloheximide) and reduced by compounds which increase protein synthesis (e.g. phenobarbital)[20]. This is probably due to the respective increased and decreased pool of cysteine available for taurine synthesis. Urinary taurine levels appear to be a good indicator of the body pool of taurine, especially in the liver (Fig. 1)[21]. This suggests that compounds which reduce urinary taurine levels may also reduce liver levels of taurine. Thus, compounds which increase protein or GSH synthesis (or reduce protein degradation) may lower the body pool of taurine. These reduced levels may have implications in relation to the recognised protective role of taurine[1,22,24] especially if taurine levels are compromised by diet or medical treatment (e.g. parenteral nutrition).

Taurine 2, edited by Huxtable et al.
Plenum Press, New York, 1996

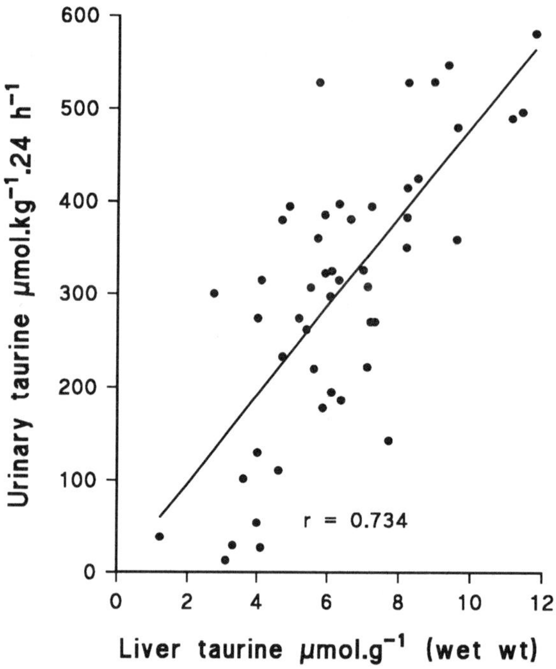

Figure 1. Correlation between the level of taurine in the urine and liver taurine concentrations measured after collection of final 24 h urine sample: r 0.734; $p < 0.001$; N = 50.

For this reason we have examined the effects on urinary and tissue levels of taurine of two β-agonists, clenbuterol and salbutamol, which are known to affect protein synthesis. Some of these data have been published[2,19].

Clenbuterol is a potent β-adrenergic agonist, which is used to treat respiratory diseases in cattle and horses and to delay parturition in cows[12]. It has well-known repartitioning effects, resulting in an increase in muscle growth and a decrease in fat deposition[14,23]. The clinical use of clenbuterol for this purpose in humans has been proposed as it is effective in reducing or restoring protein loss in animals with inherited muscular dystrophy, muscle wasting and endotoxeamia[10]. However, clenbuterol is already used illegally to enhance muscle growth in athletes and to improve the protein:fat ratio in live stock[23]. This has important implications in several areas of human health, both by the direct effects of the drug and also by possible side effects.

Salbutamol is a β-adrenergic agonist used for the treatment of asthma. It also has growth-promoting effects which have been utilised to increase the protein:fat ratio in livestock and may have clinical relevance in treating muscle wasting in patients.

The aim of the following study was to investigate the effect of repeated administration of clenbuterol and salbutamol in rats and to obtain preliminary data from human volunteers treated with slow release salbutamol.

METHODS

Clenbuterol in Rats

Male rats (Sprague-Dawley stock, Glaxo Research and Development), weighing 165-215 g were acclimatised for eight days in communal cages after delivery then housed in individual metabolism cages during the experiment[20]. The rats were allocated to one of four treatment groups (4 in each group), with similar mean body weights (189 g ± 4 g). Powdered diet (containing 2 μmol taurine.g^{-1} dry weight) and drinking water were allowed *ad libitum*. Lighting was controlled to give a regular 12 h light-dark cycle (7 am on - 7 pm off) and room temperature was maintained at 21 ± 2°C.

Animals were weighed and food and water intake were measured daily for 4 days. On day 5, three dosing solutions were made up in the drinking water, estimated to provide 2.0, 0.2 and 0.02 mg.kg^{-1} body weight clenbuterol per day (16.6, 1.66 and 0.166 mg.liter^{-1}). These doses were similar to those which had been shown to result in repartitioning of fat:protein in previous studies[11]. Solutions were prepared daily in tap water and HCl (0.1 M) was added to each drug solution and control to aid initial dissolution. The pH was then adjusted to 7.4 using NaOH (0.1 M). All drinking water solutions were protected from light by covering the water bottles with foil.

Urine (24 h) was collected over ice, made up to 25 ml with ultra high quality water and stored in 5 ml aliquots at -80°C for later analysis. Drinking water solution and food intake and body weights were also monitored daily. On day 3 of the study, approximately 1 μmol of [^{3}H]leucine, 0.25 μCi.g^{-1} body weight, was administered in the drinking water for 24 h before post-mortem. At the end of the study, the animals were anaesthetized and exsanguinated from the abdominal aorta. Tissues and serum were taken for histological and biochemical analysis, including taurine, total nonprotein sulphydryls, [^{3}H]leucine incorporation into acid-precipitable protein and serum enzymes.

Salbutamol in Rats

This investigation was carried out essentially in the same way as study 1. Rats were divided into four groups of 4 animals. Three groups were given salbutamol in their drinking water at concentrations of 12, 24 or 36 mg free base per liter of tap water. These concentrations were estimated to provide 1.5, 3 or 4.5 mg.kg^{-1}.day^{-1}. The remaining group was given tap water.

Two 24 h urine collections were made before administering salbutamol in the drinking water and then for a further 13 days during the treatment period. On day 10 of the study animals were removed from the metabolism cages for a 24 h period to allow for cleaning and to comply with U.K. Home Office requirements. Animals were not given [^{3}H]leucine in the drinking water in this study.

Serum Enzymes

Serum levels of aspartate aminotransferase, alanine aminotransferase, lactate dehydrogenase and other serum parameters were measured at 37°C using the appropriate kits and a IL Monarch 2000 analyser. Isoenzymes of creatine kinase (CK) were separated using electrophoresis and quantified by scanning densitometry using the (REP) automated electrophoresis system (Helena Laboratories (UK) Ltd).

Taurine and Total Nonprotein Sulphydryls

Taurine was extracted from diluted urine (25 μl) and serum using Dowex resins similar to the method previously described[18] modified from Larsen *et al.*[8]. Taurine and total nonprotein sulphydryls were measured in the supernatant prepared from liver and gastrocnemius muscle samples (300 mg) homogenised in sulphosalicylic acid (4 ml, 0.2 M, 4°C). Supernatant was applied to Dowex columns (25 μl) for taurine extraction or assayed for total nonprotein sulphydryls[4]. As reduced GSH constitutes >95% of liver and muscle total nonprotein sulphydryls, this was used as a measure of GSH[3,13]. Taurine was analysed in the eluates from the ion exchange columns used for a clean-up step by HPLC[18].

[^{3}H]Leucine Incorporation into Protein

[^{3}H]Leucine incorporation into liver, heart, gastrocnemius muscle and serum was measured in the sulphosalicylic acid-precipitated protein pellets prepared when measuring taurine and total nonprotein sulphydryls. The pellets were resuspended and washed with sulphosalicylic acid (4°C, 4x), then the pellets were dissolved in NaOH (1-10 ml, 1 M, 60°C) and sonicated to aid dissolution. Aliquots (200 μl) were mixed with Eco-lite scintillation fluid (3.5 ml) and counted for 15 min. Protein was measured by the method of Lowry[9] using bovine albumin as standard.

Urinary Creatinine

Urinary creatinine were measured in 1 to 20 dilutions of urine using the enzymatic assay of Siedel *et al.*[16]

Human Volunteers

A complete 24 h pretreatment urine sample was taken from each of 8 female volunteers (ages 35-50 yr). Four people were given slow-release salbutamol in tablet form (8 mg) twice a day (every 12 h) for 14 days. The remaining 4 volunteers received placebo. At the end of that time a second complete 24 h urine collection was made. Urinary taurine and creatinine were measured and urinary taurine was expressed as μmol taurine/mmol creatinine.

Statistical Analysis

Urinary levels of taurine and creatinine in samples taken from the same animal on different days were compared with predose values for each animal using a paired 't' test. Dunnett's test for multiple comparisons with a single control was used to determine significance between treatment groups and the control group for all other comparisons, unless otherwise stated. Data are reported as means ± SEM unless otherwise stated.

RESULTS

Clinical Effects of β-Agonists in Rats

Body weight gain was not significantly different between the groups. There was a significant reduction in food and water intake in those animals given the highest dose of clenbuterol, 0-48 h after the start of the treatment. However, both food and water intake returned to control levels after this time. There was a slightly higher body weight gain/food intake over the 4 day treatment period in animals given 0.2 $mg.kg^{-1}.day^{-1}$ clenbuterol, (0.32 ± 0.06 versus 0.28 ± 0.02) but the difference was not significant. None of these parameters were altered by salbutamol treatment (Table 1).

There was no effect of clenbuterol on kidney or soleus muscle weights, although there were increases in gastrocnemius muscle weight and heart (at the highest dose) but these increases were not significant. There was, however, a dose-related reduction in liver weights

Table 1. Food and water intake, urine output and body weight increase in control and salbutamol treated animals

	Dose Group			
Parameter	Control	Low Dose	Mid-dose	High Dose
Mean Daily Salbutamol Intake[a]	0	1.47 ± 0.04	3.05 ± 0.06	4.52 ± 0.9
Mean Daily Food Intake[b]	27.1 ± 0.6	26.8 ± 0.4	25.9 ± 0.4	25.6 ± 0.6
Mean Daily Body Weight Increase[b] (Range)	5.4 ± 0.5	5.3 ± 0.3	5.4 ± 0.6	5.2 ± 0.3
	(2.5-8.3)	(3.3-7.5)	(2.3-9.0)	(4.0-7.0)
Mean Daily Water Intake[b]	34.9 ± 0.9	33.8 ± 0.8	35.5 ± 0.3	34.3 ± 0.5
Mean Daily Urine Output[b]	10.5 ± 0.4	10.0 ± 0.5	10.8 ± 0.5	10.8 ± 0.7

Values are means of 12 daily averages ± SEM. [a]mg/kg/day; [b]g/day

which was significant in the highest dose group (Table 2). Salbutamol treated animals showed no changes in these parameters.

Sections taken from muscle of animals treated with both the intermediate and highest doses of clenbuterol showed some abnormal muscle fibres in the soleus muscle. Thus some muscle fibres showed segmental degeneration. The affected fibres were basophilic with peripheral proliferation and "rowing" of nuclei. Liver, heart, kidney and testicular tissue all appeared normal.

Biochemical Effects of β-Agonists in Rats

Serum alkaline phosphatase, alanine aminotransferase and aspartate aminotransferase levels were unaffected by treatment with clenbuterol or salbutamol. Serum albumin, total protein and urea were reduced by the two highest doses of clenbuterol, but not significantly and salbutamol had no effect. Serum glucose, however, was significantly reduced by the highest dose of clenbuterol. The percentage of serum creatine kinase attributed to the muscle isoenzyme (CK-MM) showed a dose-related increase and was significantly raised in animals given the highest dose of clenbuterol (Table 2). There was no muscle damage after salbutamol treatment and serum chemistry appeared normal (Table 3).

Taurine Levels

There was a dose-related decrease in taurine levels in liver, serum and muscle, an effect which was significant in the liver of animals given the highest dose of clenbuterol (Table 2), but there was no effect in animals treated with salbutamol (Table 4).

Urinary taurine levels were variable in all animals. However, there was a reduction in urinary taurine excretion in the animals receiving the highest dose of clenbuterol 2-4 days after the start of treatment (Fig. 2) which was significant.

The decrease in urinary taurine was significantly correlated with the cumulative dose of clenbuterol (Fig. 3). All doses of salbutamol reduced urinary taurine significantly after six days of treatment (Table 4) and the mean urinary taurine level throughout the treatment period was significantly lower at the highest dose. The mean urinary taurine level showed a significant correlation with the mean daily intake of salbutamol when plotted for each animal (Fig. 4).

Table 2. Effect of clenbuterol on biochemical and other parameters in tissues.

Clenbuterol mg.kg^{-1}day^{-1}[a]	Heart weight (% body wt)	Serum CK-MM[c] (% total CK[d])	Serum glucose (mM)	Serum taurine (μM)	Muscle [^{3}H]leucine incorporation (dpm.mg^{-1} protein (c10^2)	Muscle taurine (μmol.g^{-1} wet weight)	Muscle weight (g wet weight)	Liver taurine μmol.g^{-1} wet weight	Liver weight (% body weight)
0.0	0.396 ± 0.035	7.9 ± 0.4	9.5 ± 0.5	267 ± 16	13.6 ± 0.5	12.2 ± 1.5	1.14 ± 0.08	7.1 ± 0.6	4.5 ± 0.2
0.02	0.410 ± 0.004	10.6 ± 1.9[b]	9.1 ± 0.4[b]	295 ± 18	15.2 ± 2.2	11.9 ± 0.7	1.16 ± 0.02	4.9 ± 1.3	4.4 ± 0.1
0.2	0.404 ± 0.012	11.0 ± 0.3**	8.7 ± 0.3	258 ± 8	16.1 ± 0.5	11.6 ± 0.7	1.22 ± 0.04	4.3 ± 1.0	4.1 ± 0.2
2.0	0.439 ± 0.007	19.1 ± 4.5*	7.6 ± 0.4*	208 ± 24	22.6 ± 3.7*	10.5 ± 1.4	1.23 ± 0.08	3.3 ± 0.8*	3.9 ± 0.1*

All measurements were made on tissue taken at necropsy 4 days after start of treatment. [a]Approximate dose; Values are means ± sem; N= 4, except [b]where n = 3; [c]creatine kinase muscle isoenzyme; [d]creatine kinase; muscle analysed was gastrocnemius; *$p < 0.05$ compared to control (Dunnett's test for multiple comparisons with a single control); **$p < 0.001$ (t test).

Table 3. Serum biochemical parameters in control and salbutamol treated animals

	Dose Group			
Parameter	Control (3)	1.5 mg/kg/day (3)	3 mg/kg/day (4)	4.5 mg/kg/day (4)
Serum Total Protein (g/l)	57.3 ± 2.5	57.5 ± 0.7	60.3 ± 0.4	60.9 ± 0.5
Serum Albumin (g/l)	30.9 ± 1.1	31.8 ± 0.2	32.4 ± 0.3	31.9 ± 0.8
Serum urea (mM)	6.2 ± 0.6	6.0 ± 0.1	6.1 ± 0.3	5.5 ± 0.2
Serum creatinine (mM)	48.0 ± 3.5	56.3 ± 4.9	58.0 ± 3.9	47.5 ± 0.9

Results are means ± SEM, numbers of animals in brackets.

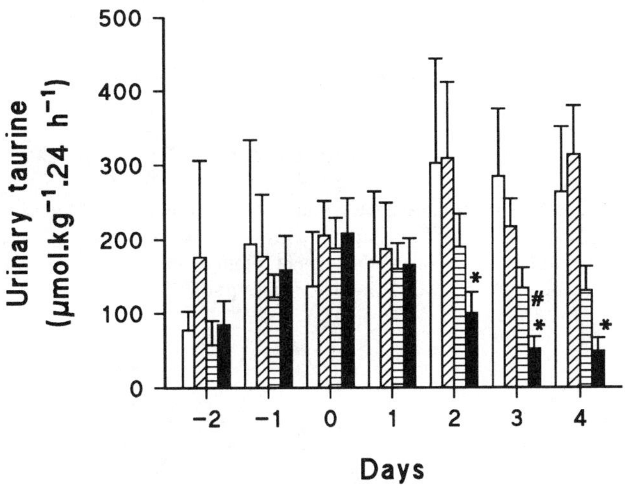

Figure 2. Urinary taurine excretion in rats treated with clenbuterol in the drinking water. Values are means ± SEM; N = 4; *p < 0.05, paired "t" test, predose values compared to postdose values; #p < 0.05 high dose group compared to control using Dunnett's test. Clenbuterol (mg/kg/day): ☐ 0; ▨ 0.02; ▤ 0.2; ■ 2.0.

Table 4. Liver taurine, glutathione and triglyceride levels in control and salbutamol treated animals

	Dose Group			
Parameter	Control (3)	1.5 mg/kg/day (3)	3 mg/kg/day (4)	4.5 mg/kg/day (4)
Urinary taurine[#] (μmol/kg/24h)	161.46 ± 9.3	117.0 ± 12.4*	68.9 ± 20.6*	86.87 ± 23.3*
Liver taurine (μmol/g liver)	7.7 ± 1.3	10.8 ± 1.5	10.8 ± 0.9	6.2 ± 0.9
Liver TNPSH (μmol/g liver)	7.7 ± 0.7	8.1 ± 0.5	7.2 ± 0.8	5.6 ± 0.8
Mean daily urinary creatinine (μmol/kg/24h)	301.8 ± 8.7	284.6 ± 7.2	272.8 ± 10.8	277.9 ± 12.2

[#]Day 6 of treatment. Results are means ± SEM, numbers of animals in brackets. [*] p<0.05, compared with control (Students t-test, unpaired).

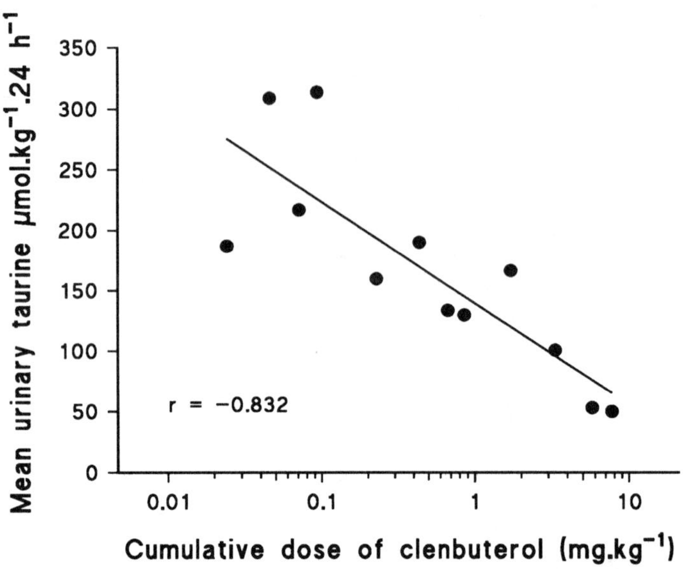

Figure 3. Correlation between urinary taurine excretion and the cumulative dose of clenbuterol received for the four days following start of treatment. Each point represents the mean urinary taurine for one treatment group for one 24 h period following the start of treatment, plotted against the total clenbuterol intake at the time of taurine measurement. N = 4 for each point; r = -0.832; p<0.001.

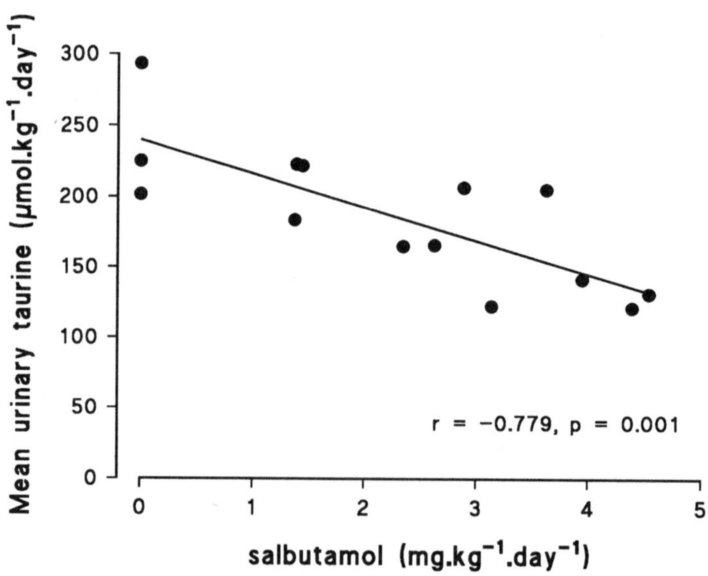

Figure 4. Correlation between mean urinary taurine excretion and the mean daily intake of salbutamol for individual animals. r=-0.779; p<0.001.

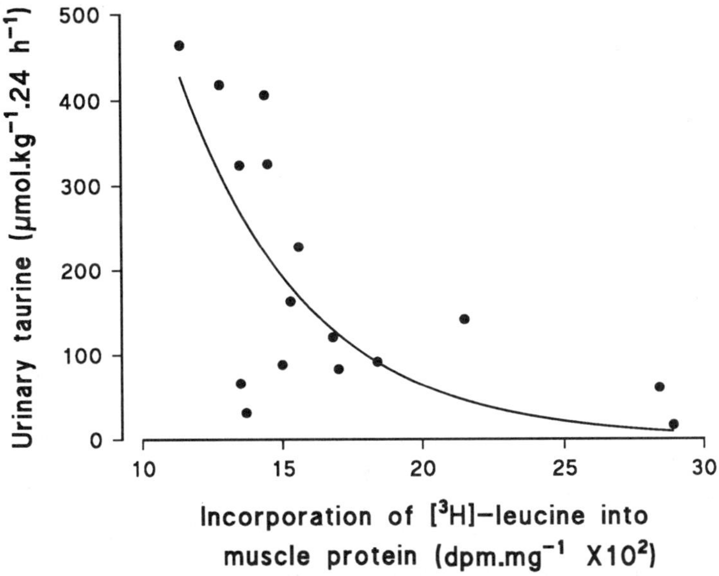

Figure 5. Mean urinary taurine excretion during 4 days of treatment with clenbuterol compared with the incorporation of [^{3}H]leucine into muscle protein taken after day 4 of treatment. Each point represents a value for a single animal.

Protein Synthesis

There was a significant increase in [^{3}H]leucine incorporation into acid precipitable protein from the gastrocnemius muscle of animals given 2 mg.kg^{-1} clenbuterol (Table 2) and a reduction in the urinary excretion of [^{3}H]leucine as a % of the amount administered, expressed as a % of the control values (low, 109%; middle, 79%; high, 68%). The level of urinary taurine showed a negative correlation with increased [^{3}H]leucine incorporation into muscle protein (Fig. 5). Although protein synthesis was not measured in animals treated with salbutamol there was a positive correlation between the mean salbutamol intake of each rat and the serum total protein at the time of post mortem (Fig. 6).

Total Nonprotein Sulphydryls

Levels of total nonprotein sulphydryls in liver, heart and muscle were not affected by treatment with clenbuterol (data not shown) or salbutamol (Table 4).

Urinary Taurine Levels in Human Volunteers

Although there was variation between values for individual volunteers for urinary taurine excretion, there was a significant reduction in urinary taurine expressed as a ratio to creatinine excretion after 14 days treatment with salbutamol. The values were also significantly lower when an individual's level was compared with his predose level of urinary taurine (Table 5). Creatinine values were not different after treatment.

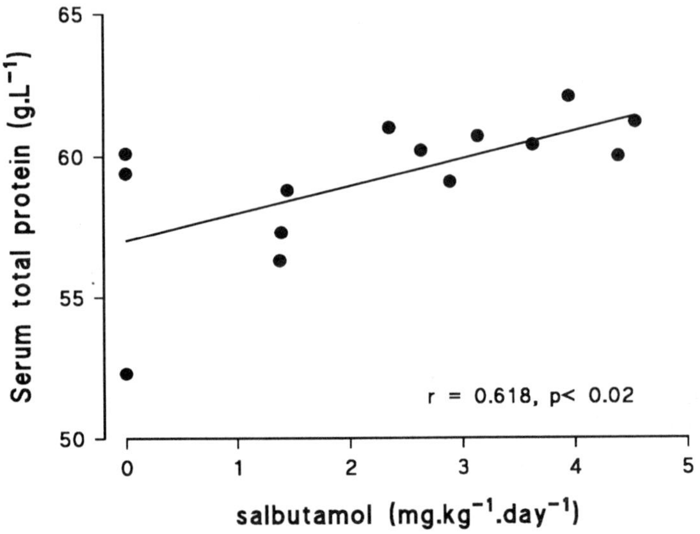

Figure 6. Correlation between serum protein levels after 13 days of salbutamol treatment with the mean intake of salbutamol for individual animals. r=0.618; p<0.02.

DISCUSSION

The aim of this investigation was to examine the effects of short term continuous administration of β-adrenoceptor agonists on tissue and urinary taurine levels in rats and to correlate any changes with muscle or body weight changes, protein synthesis and protein degradation. The effect of salbutamol on human volunteers was also investigated.

Animals given clenbuterol and salbutamol did not increase their body weight or muscle mass significantly above those of control animals. These results are similar to those reported by Rothwell and Stock[15] in rats receiving 1 $mg.kg^{-1}$ clenbuterol subcutaneously but are contrary to those found in a similar study carried out by Reeds *et al.*[14] who found a significant increase in heart, muscle and body weights in rats after 4 days of treatment with clenbuterol at a level of 2 $mg.kg^{-1}$ diet. This would correspond approximately to the middle dose in this study. McElligott *et al.*[11] also found an increase in body weight and muscle weight 14 days after the start of the treatment in rats given clenbuterol in the diet at a similar dose to the highest in the current study. The lack of effect found in this study could reflect the shorter administration time.

Table 5. Urinary taurine in female volunteers before and after treatment for 14 days with salbutamol (8 mg twice daily)

	Urinary taurine ($\mu mol.24\ h^{-1}$)		Taurine:creatinine (μmol:mM)	
Groups (N = 4)	predose	postdose	predose	postdose
Placebo	439 ± 458	314 ± 238	88 ± 28	109 ± 138
Salbutamol	279 ± 179	60 ± 12	34 ± 17	12 ± 8*

Values are means ± SD; *p < 0.05 compared with post-dose value ('t' test)

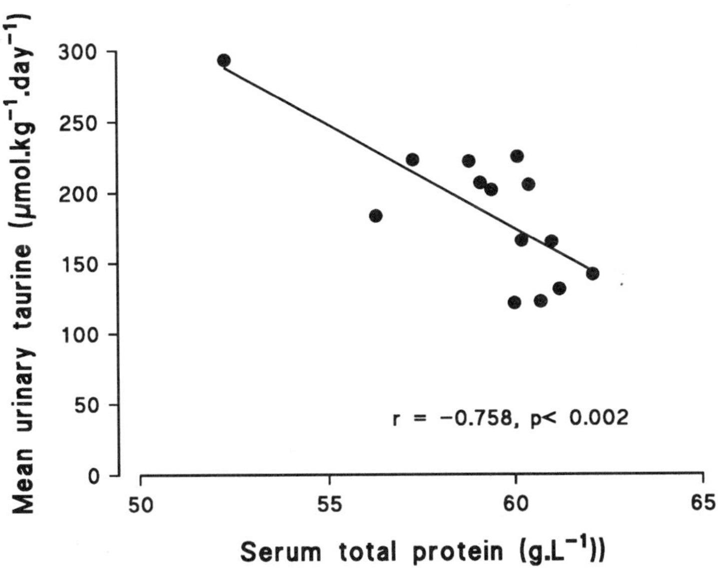

Figure 7. Correlation between mean urinary taurine excretion for individual animals over the 13 day treatment period and serum total protein at the time of post portem. Data from individual animals. r=-0.758; p<0.002.

The muscle fibre injury observed histopathologically and biochemically at both the top and intermediate doses of clenbuterol is of particular interest, especially as it was confined to type I muscle fibres in the soleus muscle.

The reduction in liver weights in clenbuterol-treated animals in this study was also observed by Reeds *et al.*[14], although no explanation was given for this observation. We have also been unable to determine the reason for this reduction in liver weight but found it was not due to a depletion of glycogen stores.

The increase in [^{3}H]leucine incorporation into muscle protein suggests that clenbuterol at the highest dose increased protein synthesis, inhibited degradation, or both. As the labelling of proteins was carried out by continuous administration of [^{3}H]leucine in the drinking water, incorporation would represent proteins with both a short and long half life[5]. The reduced rate of urinary radioactivity excreted would also suggest an increase in the utilisation of [^{3}H]leucine in protein synthesis. Although protein synthesis was not measured using [^{3}H]leucine incorporation in the salbutamol study, total serum protein was significantly correlated with dose of salbutamol in individual animals (Fig. 6). This is consistent with an increased synthesis of protein in the liver after salbutamol treatment. Furthermore the serum total protein was significantly negatively correlated with urinary taurine (Fig. 7). Thus increased synthesis of serum proteins correlates with decreased excretion of taurine.

More striking are the reduced levels of taurine in the urine during the treatment period, particularly after treatment with the highest dose of clenbuterol. This is unlikely to be due to the reduction in food intake as we have previously shown that dietary restriction must be 60% or more to affect urinary taurine significantly and urinary taurine decreased further after consumption had returned to control levels. The most likely reason for the reduced levels is the increased level of protein synthesis and concomitant accretion in skeletal muscle. This would have reduced the pool of amino acids available for taurine synthesis, in particular the precursors cysteine and methionine, and is consistent with our previous studies with cycloheximide and phenobarbital[20]. The reduced levels of liver taurine after clenbuterol treatment would also tend to suggest that there was a reduction in taurine synthesis as the liver is the primary site of taurine synthesis[24]. The observed reduction in serum and muscle taurine would be consistent within the dependence of muscle on circulating taurine to maintain tissue levels. The levels of glutathione were maintained in the liver, heart and muscle. The lack of effect of salbutamol on these parameters may reflect the lower potency of this β-agonist.

The effects of β-adrenoceptor agonists on protein metabolism are not always clear as there appears to be evidence that these compounds both reduce protein degradation and increase protein synthesis[6,14,23]. Our data are consistent with this.

The ease of measuring taurine, its relatively high concentration in urine and the rapid clearance of any excess from the kidney make it an attractive noninvasive biochemical marker of protein status. This could be particularly important in the clinical situation where β-adrenoceptor agonists may be used to treat muscle wasting, when factors such as oedema may mask weight changes.

The preliminary data suggest that patients receiving β-adrenoceptor agonists for the treatment of asthma may also have reduced levels of taurine. Furthermore, if a reduction in tissue taurine levels occurs during administration of these drugs, this could have important implications for the use of β-agonists. Thus, as taurine has an important role in protection of various organs and tissues, such as heart, lung and retina, against pathological damage[7,17,24], a reduction in the levels may reduce the level of protection afforded by high tissue taurine levels. The widespread use of β-agonists for the treatment of asthma and the misuse by athletes for increasing muscle mass could mean that many individuals may have reduced taurine levels and thereby increased susceptibility.

This study has shown that continuous administration of clenbuterol and salbutamol cause a dose-related decrease in urinary taurine. This effect occurred in the absence of significant changes in body weight.

ACKNOWLEDGMENTS

CJW is grateful to Glaxo Research and Development for financial support. We thank Daniel Asker and Minakshi Jairath for technical help, Malcolm York for analysis of serum clinical chemistry and Mary Fagg for help in the preparation of the manuscript. The human studies were carried out by Dr. John Clague, Department of Geriatric Medicine, Withington Hospital, University of Manchester.

REFERENCES

1. Azuma, J., Hamaguchi, T., Ohta, H., Takihara, K., Awata, N., Sawamura, A., Harada, H., Tanaka, Y., and Kishimoto, S. 1987, Calcium overload-induced myocardial damage caused by isoproterenol and by adriamycin: possible role of taurine in its prevention, in: "The Biology of Taurine: Methods and Mechanisms," Huxtable, R.J., Franconi, F. and Giotti, A. eds., Plenum Press, New york, pp. 167-179.
2. Carvalho, F., Waterfield, C.J., de Lourdes Bastos, M., Ferreira, M., and Timbrell, J.A. 1995, Effect of repeated administration of salbutamol on urinary and liver taurine levels in rats, *Pharmacol.Commun.* 5:171-180.
3. DeMaster, E.G. and Redfern, B. 1987, High-performance liquid chromatography of hepatic thiols with electrochemical detection, in: "Methods in Enzymology", Jakoby, W.B. and Griffith, O.W. eds., Academic Press, New York City, NY, pp. 110-114.
4. Ellman, G.L. 1959, Tissue sulfhydryl groups, *Arch.Biochem.Biophys.* 82:70-77.
5. Hasselgren, P.-O., Pedersen, P., Sax, H.C., Warner, B.W., and Fischer 1988, Current research review: Methods for studying protein synthesis and degradation in liver and skeletal muscle, *J.Surg.Res.* 45:389-415.

6. Hesketh, J.E., Campbell, G.P., Lobley, G.E., Maltin, C.A., Acamovic, F., and Palmer, R.M. 1992, Stimulation of actin and myosin synthesis in rat gastrocnemius muscle by clenbuterol; Evidence for translational control, *Comp.Biochem.Physiol.* 102C:23-27.
7. Huxtable, R.J. 1992, Physiological actions of taurine, *Physiol.Rev.* 72:101-163.
8. Larsen, B.R., Grasso, D.S., and Chang, S.Y. 1980, A rapid method for taurine quantitation using high performance liquid chromatography, *J.Chromatog.Sci.* 18:233-236.
9. Lowry, O.J., Rosebrough, N.J., Farr, A.L., and Randall, R.J. 1951, Protein measurement with folin phenol reagent, *J.Biol.Chem.* 193:265-275.
10. Martineau, L., Horan, M.A., Rothwell, N.J., and Little, R.A. 1992, Salbutamol, a β_2-adrenoceptor agonist, increases skeletal muscle strength in young men, *Clin.Sci.* 83:615-621.
11. McElligott, M.A., Barreto, A., Jr., and Chaung, L.-Y. 1989, Effect of continuous and intermittent clenbuterol feeding on rat growth and muscle, *Comp.Biochem.Physiol.* 92c:135-138.
12. Nakeo, T.K., Sato, T., Nakamura, T., Taguchi, K., Moriyoshi, M., and Kawaka, K. 1992, Use of a beta 2-adrenergic stimulant (clenbuterol) for eliminating night-time calving, *J.Vet.Med.Sci.* 54:19-22.
13. Potter, D.W. and Tran, T.-B. 1993, Apparent rates of glutathione turnover in rat tissues, *Toxicol.Appl.Pharmacol.* 120:186-192.
14. Reeds, P.J., Hay, S.M., Dorwood, P.M., and Palmer, R.H. 1986, Stimulation of muscle growth by clenbuterol: lack of effect on muscle protein biosynthesis, *Br.J.Nutr.* 56:249-258.
15. Rothwell, N.J. and Stock, M.J. 1987, Effect of a selective β_2-adrenergic agonist (clenbuterol) on energy balance and body composition in normal and protein deficient rats, *Biosci.Reps.* 7:933-940.
16. Siedel, J., Mullering, H., and Ziegenhorn, J. 1992, Sensitive color reagent for the enzymatic determination of creatinine, *Clin.Chem.* 30:968-969.
17. Sturman, J.A. 1993, Taurine in development, *Physiol.Rev.* 73:119-147.
18. Waterfield, C.J. 1994, Determination of taurine in biological samples and isolated hepatocytes by high-performance liquid chromatography with fluorimetric detection, *J.Chromatogr.B Biomed.Appl.* 657:37-45.
19. Waterfield, C.J., Jairath, M., Asker, D.S., and Timbrell, J.A. 1995, The biochemical effects of clenbuterol: With particular reference to taurine and muscle damage, *Eur.J.Pharmacol.Environ.Toxicol.Pharmacol.* 293:141-149.
20. Waterfield, C.J., Turton, J.A., Scales, M.D., and Timbrell, J.A. 1993, Effect of various non-hepatotoxic compounds on urinary and liver taurine levels in rats, *Arch.Toxicol.* 67:538-546.
21. Waterfield, C.J., Turton, J.A., Scales, M.D.C., and Timbrell, J.A. 1992, The correlation between urinary and liver taurine levels and between predose urinary taurine and liver damage, *Toxicology*, 77:1-5.
22. Waterfield, C.J., Turton, J.A., Scales, M.D.C., and Timbrell, J.A. 1992, Reduction of liver taurine in rats by β-alanine treatment increases carbon tetrachloride toxicity, *Toxicology*, 77:7-20.
23. Yang, Y.T. and McElligott, M.A. 1989, Multiple actions of β-adrenergic agonists on skeletal muscle and adipose tissue, *Biochem.J.* 261:1-10.
24. Zelikovic, I. and Chesney, R.W. 1989, Taurine, in: "New protective roles for selected nutrients", Spiller, G.A. and Scala, J. eds., Alan R. Liss, New York, pp. 253-297.

Part 4

TAURINE IN MUSCLE AND THE CARDIOVASCULAR SYSTEM

26

EFFECTS OF CHRONIC TAURINE TREATMENT ON THE ELECTRICAL AND CONTRACTILE PROPERTIES OF SKELETAL MUSCLE FIBERS OF AGED RATS

Sabata Pierno,[1] Annamaria De Luca,[1] Ryan J. Huxtable,[2] and Diana Conte Camerino[1]

[1] Unit of Pharmacology
Department of Pharmacobiology
Faculty of Pharmacy
University of Bari, Italy
[2] Department of Pharmacology
College of Medicine
University of Arizona, Tucson, Arizona

INTRODUCTION

Taurine has an important role in stabilizing mammalian skeletal muscle membrane. Indeed previous studies have shown that *in vitro* application of taurine reduces fiber excitability by a specific and concentration-dependent increase of resting Cl^- conductance (G_{Cl})[4, 20], this parameter largely contributing to the electrical stability of sarcolemma[21]. A better evaluation of the role of taurine in skeletal muscle derives from studies on taurine-depleted muscles. Indeed chronic treatment with guanidinoethane sulfonate (GES), an inhibitor of taurine transport[15], caused a fall in muscular taurine content and a marked decrease of G_{Cl} along with an increase of excitability of rat skeletal muscle fibers[9, 19]. Furthermore we found that the excitation-contraction coupling process of taurine-depleted muscles was significantly changed, the GES-treated fibers contracting at more negative potentials than normal controls[9].

Alterations of the electrical and contractile properties of muscle fibers, similar to those produced by taurine depletion, spontaneously occur in rats during aging. A reduction of G_{Cl} and a shift of the voltage threshold for contraction towards more negative potentials were observed in aged rat skeletal muscle fibers[6,10]. Interestingly, Dawson and Wallace[5] found a decrease of taurine content in aged rats both in plasma and in various tissues with respect to adults.

All these findings led us to hypothesize that the changes occurring in skeletal muscle function during aging could be related to a reduction of taurine content. To test this

Taurine 2, edited by Huxtable et al.
Plenum Press, New York, 1996

hypothesis the present study was aimed at evaluating the effects of an *in vivo* chronic treatment with taurine on the ionic conductances and excitability characteristics of extensor digitorum longus (EDL) muscle of aged rats. Also the effect of this treatment was evaluated on the excitation-contraction coupling process by measuring the mechanical threshold for contraction. Recent pharmacological studies with phorbol esters, specific activators of Ca^{2+}- and phospholipid-dependent protein kinase C (PKC) have suggested that Cl^- channels of mammalian skeletal muscle can be phosphorylated and closed upon PKC activation[10]. Moreover the low G_{Cl} found in aged muscles has been proposed to be due to PKC overactivity[10]. Since taurine has been demonstrated to inhibit protein phosphorylation by PKC in rat cortex[18], we evaluated the possible involvement of taurine in the PKC mediated phosphorylation of Cl^- channels by testing the sensitivity to phorbol esters of the taurine treated aged muscles.

METHODS

Aged (24 month old) and adult (6 month old) male Wistar rats were used for all the experiments. The aged animals used were 22 months old at the beginning of the treatment and were in good health, with no impairment of hind limb movements or locomotor activity[7] throughout the period of study. Taurine (Teofarma, Italy) was given to aged rats daily at the dose of 1 g/kg dissolved in drinking water, for 2-3 months.

The electrophysiological experiments were done *in vitro* on extensor digitorum longus (EDL) muscles, removed under urethane anaesthesia from taurine-treated and untreated rats, placed in a 20 ml muscle bath maintained at 30°C, and perfused with physiological solutions (8).

A standard two microelectrode technique was used for making intracellular measurements of membrane potential and cable properties from single muscle fibers[2]. The current pulse generation, the acquisition of the voltage records and the calculation of fiber cable constant, by standard cable analysis, were done in real-time under computer control as described in detail elsewhere[2]. The reciprocal of membrane resistance in normal physiological solution was assumed to be the total membrane conductance (G_m) and the same parameter measured in Cl^--free solution was considered largely the K^+ conductance (G_K). The mean Cl^- conductance (G_{Cl}) of each preparation was estimated as the mean G_m minus the mean G_K. The excitability characteristics were determined by recording the intracellular membrane potential response to depolarizing square-wave constant current pulses. In each fiber the membrane potential was set by a steady holding current to -80 mV, before passing the depolarizing pulses[8].

The mechanical threshold of EDL muscle fibers was determined by using a two microelectrode point voltage clamp method[6] in the presence of 3 μM tetrodotoxin (TTX). The holding current was set at -90 mV. To determine the rheobase mechanical threshold, depolarizing command pulses of 500 ms duration were given repetitively at the rate of 0.3 Hz, while the impaled fibers were viewed continuously with a stereomicroscope. The command voltage was increased until contractions were just visible. The threshold membrane potential at this point could be read from a digital sample-and-hold voltmeter (6).

The normal physiological solution had the following composition (mM): NaCl 148; KCl 4.5; $CaCl_2$ 2.0; $MgCl_2$ 1.0; $NaHCO_3$ 12.0; NaH_2PO_4 0.44 and glucose 5.55. The Cl^--free solution was prepared by equimolar substitution of methylsulfate salts for NaCl and KCl, and nitrate salts for $CaCl_2$ and $MgCl_2$. Both solutions were continuously gassed with 95% O_2 and 5% CO_2 and the pH was carefully maintained between 7.2-7.3 during each experiment[4].

Table 1. Effects of taurine treatment on membrane ionic conductances of extensor digitorum longus muscle fibers of aged rats

Experimental conditions	N fibers	Gm ($\mu S/cm^2$)	N fibers	Gk ($\mu S/cm^2$)	GCl ($\mu S/cm^2$)
Mean of 6 adult rats	74	3124 ± 80	30	373 ± 30	2751 ± 69
Mean of 7 aged rats	106	2787 ± 72[a]	66	407 ± 19	2380 ± 58[a]
Mean of 7 taurine-treated aged rats	112	3194 ± 65[b]	89	435 ± 17	2759 ± 50[b]

G_m: total membrane conductance; G_K: membrane K^+ conductance; G_{Cl}: membrane Cl^- conductance. [a]Significantly different with respect to adult rats by Student's t-test ($p < 0.001$); [b]Significantly different with respect to age-matched control rats by Student's t-test ($p < 0.001$).

Data are expressed as mean ± SE. The estimates for SE and number of fibers (N) of G_{Cl} were obtained from the variance of G_m and G_K, assuming no covariance and from the number of fibers sampled for G_m and G_K, respectively[12]. The estimates of SE and N of normalized G_{Cl} values have been obtained as described by Green and Margerison[13]. Statistical significance of differences between group means was calculated by unpaired Student's t-test.

RESULTS

Effects of Taurine Treatment on Ionic Conductances and Excitability Characteristics of Muscle Fibers of Aged Rats

In agreement with previous findings[10], the Cl^- conductance (G_{Cl}) of EDL muscle fibers was significantly lower in the untreated aged rats than in the adult rats (Table 1). Such a significant decrease of G_{Cl} was recorded in each of the seven muscles removed from 24 month old rats, even if a slight inter-individual variability was observed. The excitability parameters related to G_{Cl} were also changed[4]; in particular a prolongation of the latency (delay between the application of the depolarizing pulse and the onset of the spike, Lat) and the duration (Dur) of action potential were observed (Table 2). However the amplitude of the action potential (AP) and the firing capability (N spikes) were significantly reduced (Table 2) in spite of the decreased G_{Cl}[1,3], probably due to the involvement of age-dependent alterations of Na^+ channels, as already hypothesized[7]. The K^+ conductance (G_K) was slightly increased in the aged untreated muscles with respect to the adults (Table 1).

Two months of taurine treatment restored the values of G_{Cl} lowered by aging (Table 1). Indeed, the administration of this amino acid significantly increased G_{Cl} in six out of seven treated rats; only in one aged rat was the value of G_{Cl} not ameliorated by taurine treatment, being 2326 ± 152 $\mu S/cm^2$. Moreover, G_K was only slightly and not significantly modified by this treatment (Table 1). In taurine-treated aged rats, in parallel to the restoration of G_{Cl}, a significant improvement of membrane excitability was observed. Indeed the latency and the duration of action potential, as well as the firing capability of the fibers were significantly restored towards the adult values (Table 2).

Effects of Taurine Treatment on the Mechanical Threshold for Contraction of Muscle Fibers of Aged Rats

The threshold potential for contraction of mammalian skeletal muscle fibers depends on command pulse duration. It becomes more negative the longer the duration of the pulse.

Table 2. Effects of taurine treatment on the excitability parameters of extensor digitorum longus muscle fibers of aged rats.

Experimental conditions	N fibers	RP (-mV)	AP (mV)	Lat (ms)	Dur (ms)	N Spikes
Mean of 3 adult rats	17	67 ± 2.0	102 ± 7.0	5.3 ± 0.1	9.4 ± 0.4	3.4 ± 0.5
Mean of 5 aged rats	34	66 ± 1.0	88.5 ± 2.5[a]	9.0 ± 0.7[a]	12.7 ± 0.6[a]	1.9 ± 0.2[a]
Mean of 7 taurine-treated aged rats	35	68 ± 1.0	89.2 ± 2.3	6.5 ± 0.3[b]	10.9 ± 0.4[b]	2.5 ± 0.3

RP: membrane resting potential; AP: amplitude of action potential; Lat and Dur: latency and duration of action potential, respectively; N spikes: maximum number of action potentials elicited by raising the intensity of a long duration pulse. [a]Significantly different with respect to adult rats by Student's t-test ($p < 0.05$ or less); [b]Significantly different with respect to age-matched control rats by Student's t-test ($p < 0.05$ or less).

A constant rheobase value is almost fully reached at the longest pulses used. we evaluated, therefore, the rheobase voltage with pulses of 500 ms (6). As previously shown[6], the mechanical threshold of muscle fibers of aged rats was significantly different with respect to adults. Indeed the muscle fibers from aged rats needed significantly less depolarization to contract and the rheobase voltage was more negative with respect to that of the adult fibers (Table 3). Taurine treatment significantly restored the mechanical threshold for contraction changed by aging. The rheobase voltage for mechanical activation of the aged rats treated with taurine was similar to those of the adult animals (Table 3).

Effects of *in Vitro* Application of Phorbol Esters on Muscle Fibers of Taurine-Treated Aged Rats

The *in vitro* application of 4-β-phorbol 12,13-dibutyrate (4-β-PDB) in the range from 3 nM (90 min incubation) to 50 nM (30 min incubation) produced a clear concentration-dependent block of G_{Cl} in adult, aged and taurine-treated aged muscle fibers. However the potency in blocking G_{Cl} was much greater in EDL muscle fibers from aged rats than in those from adult controls (Figure 1).

Moreover, 50 nM 4-β-PDB produced a complete block of G_{Cl} (99%) in aged muscle fibers, whereas the same concentration produced a 76% block of G_{Cl} in the adult (not shown). Surprisingly the block of G_{Cl} produced by application of 4-β-PDB in taurine-treated aged rats was similar to that observed in the adult, particularly at the lower concentrations of 3 nM and 10 nM (Figure 1), being 64% and 89%, at 30 nM and 50 nM, respectively.

Table 3. Effects of taurine treatment on the mechanical threshold of extensor digitorum longus muscle fibers of aged rats

Experimental Conditions	N fibers	Rheobase voltage (mV)
Mean of 4 adult rats	34	-56.1 ± 0.9
Mean of 5 aged rats	41	-63.7 ± 0.4[a]
Mean of 5 taurine-treated aged rats	48	-57 ± 0.4[b]

N fibers: number of fibers sampled; Rheobase voltage expressed as mean ± SE.
[a]Significantly different with respect to adult rats by Student's t-test ($p < 0.001$);
[b]Significantly different with respect to age-matched control rats by Student's t-test ($p < 0.001$).

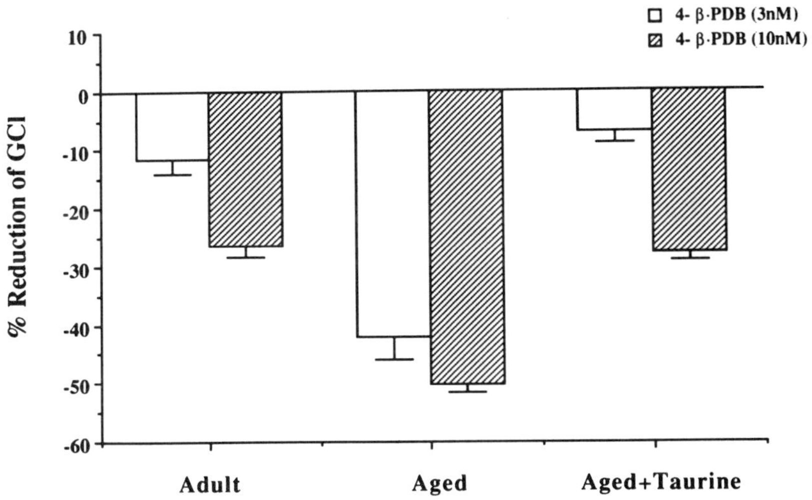

Figure 1. Normalized % reduction of G_{Cl} by *in vitro* application of different concentrations of 4-β-PDB in extensor digitorum longus muscle fibers from adult, aged and taurine treated aged rats. The values are mean ± SE from 10 to 25 fibers.

DISCUSSION

The aging process is characterized by a decline of muscle performance[17]. In aged rat skeletal muscle fibers, a decrease of resting ionic conductance to Cl^- ions (G_{Cl}) with consequent increase of membrane excitability has been reported[8]. A high G_{Cl} plays a fundamental role in stabilizing membrane potential and a malfunctional Cl^- channel is the genetic defect responsible for the low G_{Cl} and for the abnormal hyperexcitability observed in muscle of subjects affected by myotonic syndromes[21, 16]. It is known that in adult fibers the activation of Ca^{2+}- and phospholipid-dependent protein kinase C (PKC) decreases G_{Cl}, suggesting its involvement in the intracellular modulatory pathway controlling the channel state[10]. Our recent pharmacological studies led us to propose that the age-related decline of muscle G_{Cl} may be due to concurrent events, many of them mediated by Ca^{2+}, involving an overactivity of PKC[10]. Alterations in Ca^{2+} homeostasis were described to occur in skeletal muscle of aged rats as a consequence of changes in the sarcoplasmic reticulum protein composition and Ca^{2+} pumping ability[17]. In consequence of the resulting higher cytosolic Ca^{2+} concentration, the voltage threshold for mechanical activation in aged rat skeletal muscle was shifted towards more negative potentials[6]. Moreover we found that the *in vitro* application to muscle fibers of the Ca^{2+} ionophore, A23187, produced a reduction of G_{Cl} and that this effect was greater in aged muscles than in adults[10]. In conclusion, the higher cytosolic Ca^{2+} concentration of aged muscles might cause a direct or indirect (via phospholipase C) increased activity of PKC (11) as demonstrated by the higher potency of 4-β-PDB in reducing muscle G_{Cl} in aged more than in adult rats[10].

The present results show that taurine supplementation improves the electrical properties of skeletal muscle membrane which are altered by the aging process. In particular, the lower G_{Cl} found in the aged subjects was increased towards the adult values by taurine administration. Concomitantly also the mechanical threshold of aged muscle fibers was restored by taurine treatment. In this regard it is known that taurine is able to regulate intracellular levels of Ca^{2+}, by stimulating Ca^{2+} reuptake by the intracellular stores[14]. Recently it has been proposed that by this mechanism taurine can inhibit the activity of various protein kinases such as protein kinase C[18]. It is possible to hypothesize that in the aged muscles a reduction of taurine content may occur, as already proposed in other tissues[5], that can contribute to the increase of the cytosolic Ca^{2+} levels and to the alterations of kinase-mediated regulatory pathways. Thus the restorative effects on G_{Cl} and on the mechanical threshold observed after taurine administration can be due to normalized intracellular levels of the amino acid and consequently of the Ca^{2+}-dependent cellular functions. In support of this hypothesis the enhanced inhibitory effect of phorbol esters on G_{Cl}, observed in aged muscles, was counteracted by taurine administration, in that the block of G_{Cl} by 4-β-PDB in taurine-treated aged rats was similar to that of adults. Our results suggest that an alteration of taurine content can at least in part account for some of the features of aged muscles and are in favor of a potential benefit of taurine administration to aged subjects to improve muscle performance.

ACKNOWLEDGMENTS

This work is financially supported by the Italian CNR #93.227 and #94.2368 and P. F. Invecchiamento.

REFERENCES

1. Adrian, R.H. and Bryant, S.H. 1974, On the repetitive discharge in myotonic muscle fibres, *J. Physiol.* 240:505-515.
2. Bryant, S.H. and Conte Camerino, D. 1991, Chloride channel regulation in the skeletal muscle of normal and myotonic goats, *Pflügers Arch.* 417:605-610.
3. Camerino, D. and Bryant, S.H. 1976, Effects of denervation and colchicine treatment on the chloride conductance of rat skeletal muscle fibers, *J. Neurobiol.* 7:221-228.
4. Conte Camerino, D., Franconi, F., Mambrini, M., Bennardini, F., Failli, P., Bryant, S.H. and Giotti, A. 1987, The action of taurine on chloride conductance and excitability characteristics of rat striated muscle fibers, *Pharmacol. Res. Commun.* 19:685-701.
5. Dawson, R. and Wallace, D.R. 1992, Taurine content in tissues from aged Fischer 344 rats, *Age* 15:73-81.
6. De Luca, A. and Conte Camerino, D. 1992, Effects of aging on the mechanical threshold of rat skeletal muscle fibers, *Pflügers Arch.* 420:407-409.
7. De Luca, A., Mambrini, M. and Conte Camerino, D. 1990, Changes in membrane ionic conductances and excitability characteristics of rat skeletal muscle during aging, *Pflügers Arch.* 415:642-644.
8. De Luca, A., Pierno, S., Cocchi, D. and Conte Camerino, D. 1994, Growth hormone administration to aged rats improves membrane electrical properties of skeletal muscle fibers, *J. Pharmacol. Exp. Ther.* 269:948-953.
9. De Luca, A., Pierno, S., Huxtable, R.J., Failli, P., Franconi, F., Giotti, A. and Conte Camerino, D. 1992, Effects of taurine depletion on membrane electrical properties of rat skeletal muscle, in: "Taurine Nutritional Value and Mechanisms of Action," J.B. Lombardini, S.W. Schaffer and J. Azuma, eds., Plenum Press, New York, Vol 315 pp. 199-205.
10. De Luca, A., Tricarico, D., Pierno, S. and Conte Camerino, D. 1994, Aging and chloride channel regulation in rat fast-twitch muscle fibers, *Pflügers Arch.* 427:80-85.
11. Eberhard, D.A. and Holtz, R.W. 1988, Intracellular Ca^{2+} activates phospholipase C, *Trends Neurosci.* 11:517-520.
12. Eisenberg, R.S. and Gage, P.W. 1969, Ionic conductance of the surface and trasverse tubular membranes of frog sartorius fibers, *J. Gen. Physiol.* 53:279-297.
13. Green, J.R. and Margerison, D. 1978, in: "Statistical Treatment of Experimental Data," New York, Elsevier, pp. 86-88.
14. Huxtable, R.J. 1992, The physiological actions of taurine, *Physiol. Rev.* 72:101-163.
15. Huxtable, R.J., Laird, H.E. and Lippincott, S.E. 1979, The transport of taurine in the heart and the rapid depletion of tissue taurine content by guanidinoethyl sulfonate, *J. Pharmacol. Exp. Ther.* 211:465-471.

16. Koch, M.C., Steinmeyer, K., Lorenz, C., Ricker, K., Wolf, F., Otto, M., Zoll, B., Lehmann-Horn, F., Grzeschik, K.H. and Jentsch, T.J. 1992, The skeletal muscle chloride channel in dominant and recessive human myotonia, *Science* 257:797-800.
17. Larsson, L. and Salviati, G. 1989, Effects of age on calcium transport activity of sarcoplasmic reticulum in fast- and slow-twitch rat muscle fibers, *J. Physiol.* 419:253-264.
18. Li, Y.-P. and Lombardini, J.B. 1991, Inhibition by taurine of the phosphorylation of specific synaptosomal proteins in the rat cortex: effects of taurine on the stimulation of calcium uptake in mitochondria and inhibition of phosphoinositide turnover, *Brain Res.* 553:89-96.
19. Pierno, S., De Luca, A., Huxtable, R.J. and Conte Camerino, D. 1994, Dual effects of taurine on membrane ionic conductances of rat skeletal muscle fibers, in: "Taurine in Health and Desease," R.J. Huxtable and D.V. Michalk, eds., Plenum Press, New York, pp. 217-224.
20. Pierno, S., Tricarico, D., De Luca, A., Campagna, F., Carotti, A., Casini, G. and Conte Camerino, D. 1994, Effects of taurine analogues on chloride channel conductance of rat skeletal muscle fibers: a structure-activity relationship investigation, *Naunyn-Schmiedeberg's Arch. Pharmacol.* 349:416-421.
21. Rüdel, R. and Lehmann-Horn, F. 1985, Membrane change in cells from myotonia patients, *Physiol. Rev.* 65:310-356.

27

DIRECT INHIBITORY EFFECTS OF TAURINE ON NOREPINEPHRINE-INDUCED CONTRACTION IN MESENTERIC ARTERY OF STROKE-PRONE SPONTANEOUSLY HYPERTENSIVE RATS

Nanfang Li, Makoto Sawamura, Yasuo Nara, Katsumi Ikeda, and Yukio Yamori

Graduate School of Human and Environmental Studies
Kyoto University, Kyoto Japan

SUMMARY

The effect of taurine on vascular reactivity was investigated in the mesenteric artery of Wistar Kyoto (WKY), and stroke-prone spontaneously hypertensive rats (SHRSP). Administration of taurine significantly decreased blood pressure of SHRSP but not WKY. The mesenteric artery of taurine-treated SHRSP was excised to observe the contractile responses to transmural electrical stimulation and to several vasoactive substances. Vasocontraction induced by norepinephrine (NE) was significantly decreased by taurine. However, contraction elicited by the other substances tested was not affected. Consistently, when the vessels of SHRSP were pretreated with taurine *in vitro*, only the NE-induced vasoconstraction was specifically attenuated. No effect was observed in WKY vessels. These results indicate that the hypotensive effect of taurine is accounted for at least in part its direct action on blood vessels.

INTRODUCTION

Our world-wide epidemiological study on cardiovascular diseases and nutritional factors (CARDIAC study) demonstrated an inverse relationship between 24 h urinary taurine excretion and blood pressure[8]. In essential hypertensive patients, urinary excretion of taurine was decreased. The oral administration of taurine significantly decreased both systolic (SBP) and diastolic blood pressure (DBP)[4]. Moreover SHRSP, a genetic hypertensive rat strain, is characterized by high blood pressure, lower endogenous taurine content in the liver and lower serum taurine concentration in comparison to the WKY rat[7]. Several studies showed

Taurine 2, edited by Huxtable et al.
Plenum Press, New York, 1996

that taurine could decrease blood pressure (BP) and attenuate blood pressure increase in SHRSP[7,10].

Injection of taurine (icv) in rats decreased BP, heart rate and abdominal sympathetic nerve activity[3]. It was suggested that taurine decreased BP through decreasing sympathetic nerve activity in the CNS. However the effect of taurine on the peripheral cardiovascular system remains to be resolved.

In the present study, we used the mesenteric artery of SHRSP and WKY to investigate the direct effect of taurine on vascular reactivity in response to transmural electrical stimulation and to vasoactive substances such as NE, KCl, Angiotensin II (Ag 11) and acetylcholine (ACh).

METHODS

Animals used were 5 week-old male WKY/lzm and SHRSP/lzm, both from inbred original strains established at Kyoto University and Simane Medical University, and maintained at Funabashi farm under strict genetic control.

For *in vivo* experiments, 5 rats each were allocated either to taurine or control group. Rats in the taurine group received tap water containing 3% taurine, and control rats were given only tap water to drink. SBP was measured without anesthesia once a week by a tail cuff photoplethysmographic method.

When blood pressure was significantly decreased by taurine treatment for 5 weeks as compared with control rats, the mesenteric artery was dissected out under pentobarbital anesthesia. The fat and connective tissue around the mesenteric artery segment were neatly removed. The vessel segments were cut into rings 4 mm in length. Then each ring was suspended between two stainless steel wires mounted in an organ bath chamber filled with 5 ml of Krebs-Henseleit solution. The composition of this solution was as follows (mM); NaCl 20, KCl 4.7, $MgSO_4$ 1.2, KH_2PO_4 1.2, $CaCl_2$ 2.5, $NaHCO_3$ 25 and glucose 10 (pH 7.4). The perfusion fluid was gassed with 95% O_2 and 5% CO_2 and maintained at 37°C. The rings were maintained at a resting tension of 0.5 g and allowed to equilibrate for about 60 min. Before experiments began, the solution was replaced and the applied tension was readjusted every 10 min. The cumulative dose-response curves both for NE (10^{-9} - $3x10^{-3}$ M) and KCl (10 - 60 mM) were constructed step by adding progressively higher concentrations of the vasoconstrictor agents. The concentration was increased when a steady-state increase in vascular tone was achieved in response to the previous concentration. Ag II (3 μM) was also used to achieve vasoconstriction. For ACh-induced (10^{-9} - 10^{-5} M) relaxation, precontraction was evoked by 1 μM NE. For transmural electrical stimulation, the ring was placed between a pair of platinum wires (0.5 mm in diameter), 2 mm apart. As a stimulus, pulses of 0.3 ms in duration, 50 V in intensity, were given at 5-30 Hz frequency. Isometric tension was measured and recorded on a polygraph.

For *in vitro* experiments, mesenteric arteries were prepared from SHRSP/lzm and WKY/lzm (12-14 week-old males, n=8). After an equilibration period, the rings of mesenteric artery were exposed to 10 mM taurine for 15 min. Then NE, KCl, Ag II or ACh, at the concentrations described above, were used to evaluate vascular response, as described above.

Statistical Analysis

Data were analyzed by ANOVA with a significance level of $p < 0.05$. All results are expressed as mean ± SE.

RESULTS

In Vivo Experiments

The effect of taurine on SBP was assessed in SHRSP and WKY. As shown previously[7], SBP of taurine-treated SHRSP was significantly decreased as compared with the control rats. The hypotensive effect was observed following taurine administration for 2-5 weeks (taurine group: 167.6 ± 2.9 mmHg; control group: 182.4 ± 4.0 mmHg, $p < 0.05$, at the second week) (Fig. 1). However, taurine administration did not change SBP in WKY until 6 weeks (control group: 144.3 ± 3.8 mmHg; taurine group: 147.4 ± 4.0 mmHg).

Ex Vivo Experiments

The above result led us to investigate the vascular reactivity of resistance vessels to various vasoactive substances in SHRSP. SHRSP were given drinking water containing taurine for 5 weeks, and the mesenteric arteries were excised. As shown in Fig. 2a, NE-induced vasocontraction was significantly decreased as compared with the control rats. This effect of taurine was greater at the higher concentrations of NE ($3x10^{-7}$ - $3x10^{-5}$ M).

This effect of taurine was not observed in the contraction induced by KCl (10 - 60 mM) (Fig. 2b) or Ag II (3 μM) (data not shown). There was also no change observed in relaxation induced by ACh (10^{-9} - 10^{-5} M) (data not shown). Since NE is a neurotransmitter in blood vessels, the contractile response to transmural electrical stimulation was examined in the mesenteric artery of taurine-treated SHRSP. Vasocontraction elicited by electrical stimulation at 1O, 20 and 30 Hz tended to be decreased by taurine but the effect was not significant (Fig. 2c).

In Vitro Experiment

The above results indicate that taurine has an important effect on NE-induced vascular contraction *in vivo*. Therefore, the next set of experiments was carried out to examine the effect of taurine on vascular reactivity *in vitro*. Mesenteric arteries of untreated rats were excised and preincubated with taurine, followed by stimulation with the various agonists used in the *ex vivo* experiments. In SHRSP, taurine significantly decreased contraction induced by the lower concentration of NE as compared with the control (Fig. 3).

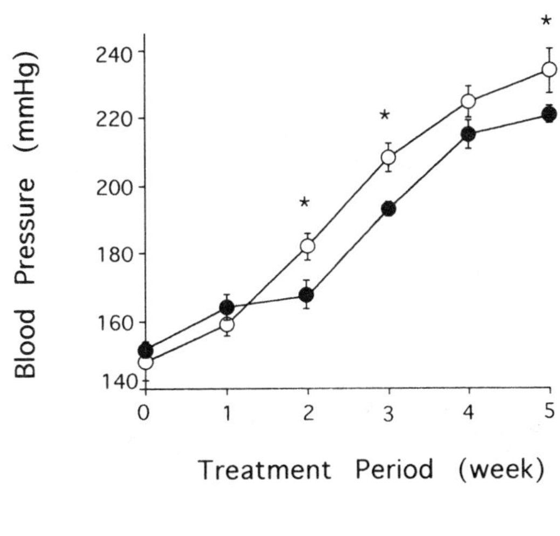

Figure 1. The effect of taurine on systolic blood pressure in SHRSP. Rats were given either 3% taurine in tap water (●) or tap water alone (O). *$p < 0.05$; n = 5.

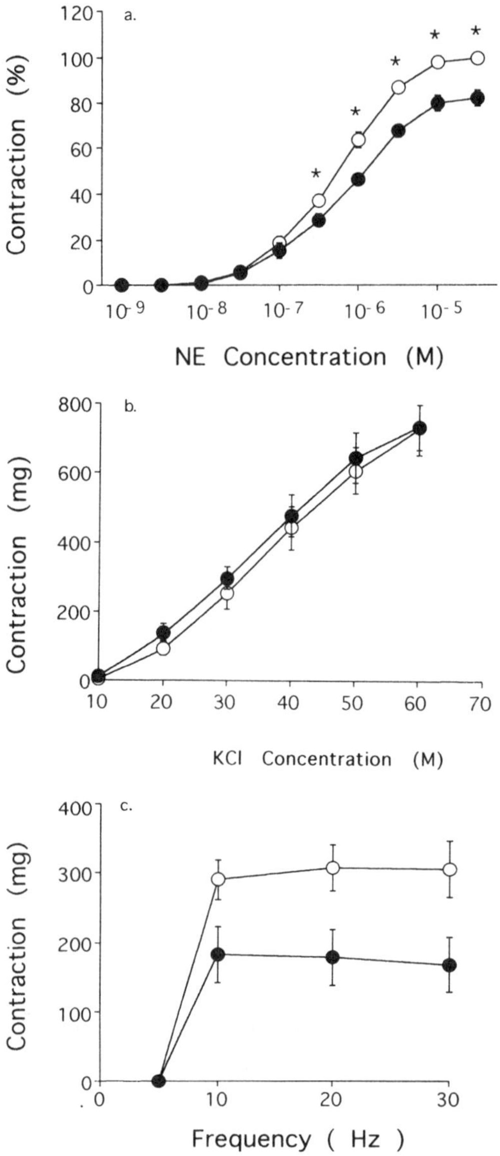

Figure 2. The effect of taurine on vascular reactivity in mesenteric artery of SHRSP given drinking water containing 3% taurine for 5 weeks. The mesenteric arteries of taurine-treated (●) or untreated (○) rats were stimulated by: a) NE (n = 5); b) KCl (n = 5); or c) transmural electrical stimulation (n = 4), and the isometric contractions were measured as described in Methods. In a), results were expressed as the % of the maximum contraction of the control. *$p < 0.05$.

The effect of taurine was again not observed in the contraction induced by KCl or Ag II, or on the relaxation induced by ACh (data not shown).

In WKY, taurine affected neither the contractions induced by NE, KCl or Ag II, nor the relaxation induced by ACh (data not shown).

DISCUSSION

The present studies showed that in the mesenteric artery of SHRSP, the vascular contraction induced by NE was decreased by taurine administration. This effect was specific

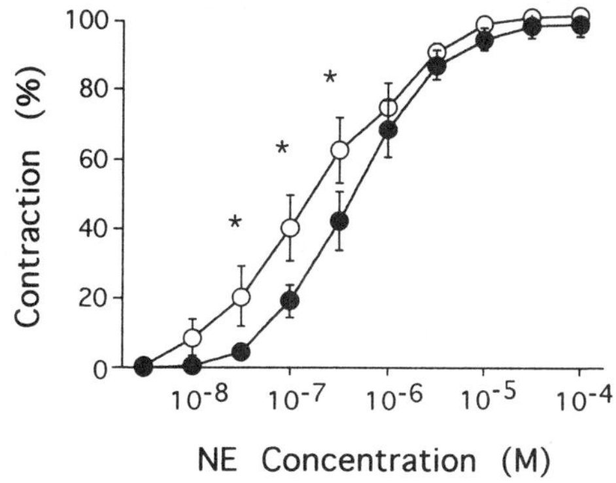

Figure 3. The effect of taurine preincubation on the vascular reactivity of SHRSP stimulated by NE *in vitro*. The mesenteric arteries of SHRSP were pretreated with 1 mM taurine (●) or vehicle (○) for 15 min, followed by stimulation with various amounts of NE. Results were expressed as the % of the maximum contraction of the control. *$p < 0.05$ (n = 7).

to NE and was not observed for KCl- or Ag II-induced contractions. ACh-induced relaxation was not affected by taurine.

Franconi *et al.* reported that the intraluminal administration of taurine induced concentration-dependent vasodilatation in rabbit ear artery preconstricted with KCl[2]. However, they used pharmacologically high concentrations of taurine (10-80 mM). Ristori *et al.* reported the taurine-induced vasorelaxation of KCl- or NE-preconstricted aortic rings of Wistar rats[9]. Since taurine is reported to reduce intracellular Ca^{2+} concentration in nerve tissues, these phenomena are possibly associated with Ca^{2+} availability in vascular smooth muscle cells. However, in our experiments the effect of taurine was only observed for NE-induced contraction. This discrepancy may be accounted for by a difference between the mesenteric artery and the aorta.

The specific effect of taurine on NE-induced vasocontraction suggests that taurine specifically desensitizes adrenoceptors on vascular smooth muscle. Receptor number, the affinity for NE, or signal transduction through the receptor should be examined in taurine-treated vessels.

Alternatively, taurine may enhance the catabolism of NE at vascular nerve terminals. In fact Chahine *et al.* reported the enhanced catabolism of NE by taurine in rabbit heart[1].

Our result shows that taurine attenuated vasocontraction induced by transmural electrical stimulation. This suggests the inhibition of NE release from nerve ending by taurine, consistent with earlier reports. For example, Chahine *et al.* reported [^{3}H]NE release induced by electrical stimulation was suppressed in the taurine-treated heart[1]. KCl-induced [^{3}H]NE release from the rat brain was also inhibited by taurine[6]. However, there was no effect of taurine on NE uptake by nerve terminals[6]. investigations are needed to examine the effect of taurine on electrically-evoked NE release and NE catabolism in mesenteric arteries.

in the *in vitro* experiment, the effect of taurine was observed at lower concentrations of NE, while in *ex vivo* studies, an effect was observed only at the higher concentrations of NE. This may be due to the difference in experimental conditions. Morphological or functional changes in blood vessels during the taurine treatment could also explain the difference.

In our study, the inhibitory effect of taurine was observed only in SHRSP. Previous studies reported that, in SHRSP, the content of taurine in the liver[10] and the brain[5], and the serum concentration[10] were markedly lower than those in WKY. It is possible that the tissues of WKY contain enough taurine to suppress neuromuscular transmission and vascular reactivity. Thus the exogenous taurine might fail to exert an additional effect.

In conclusion, we have demonstrated the direct inhibitory effect of taurine on the vascular contraction of mesenteric artery of SHRSP, which is specific for electrically released endogenous or exogenous NE-induced contraction. This effect contributes partially to the hypotensive effect of taurine in SHRSP.

ACKNOWLEDGMENT

We thank Dr. Kazuyoshi Kurahashi and Dr. Hachiro Usui of the Radioisotope Center of Kyoto University for their help in this work, and Dr. Yoshimi Ooda for his scholarship to Nanfang Li.

REFERENCES

1. Chahine, R., Hanna, H., Abou Khalil, K., Cheav, S.L., Hatala, R., Bouchi, N., and Mounayar, A. 1994, Taurine and myocardial noradrenaline, *Arzneimittelforschung*, 44:126-128.
2. Franconi, F., Giotti, A., Manzini, S., Martini, F., Stendardi, I., and Zilletti, L. 1982, The effect of taurine on high potassium- and noradrenaline-induced contraction in rabbit ear artery, *Brit.J.Pharmacol.* 75:605-612.
3. Inoue, A., Takahashi, H., Lee, L., Iyoda, I., Sasaki, S., Okajima, H., Takeda, K., Yoshimura, M., Nakagawa, M., and Ijichi, H. 1985, Centrally induced vasodepressor and sympathetic nerve responses to taurine, *Jpn.Circ.J.* 49:1180-1184.
4. Kohashi, N. and Katori, R. 1983, Decrease of urinary taurine in essential hypertension, *Jap.Heart J.* 24:91-102.
5. Kuriyama, K., Ida, S., and Ohkuma, S. 1984, Alteration of cerebral taurine in spontaneously hypertensive rats, *J.Neurochem.* 42:1600-1606.
6. Muramatsu, M., Kakita, K., and Kuriyama, K. 1978, A modulating role of taurine on release of acetyl choline and norepinephrine from neuronal tissue, *Jpn.J.Pharmacol.* 28:259-268.
7. Nara, Y., Yamori, Y., and Lovenberg, W. 1978, Effects of dietary taurine on blood pressure in spontaneously hypertensive rats, *Biochem.Pharmacol.* 27:2689-2692.
8. Nara, Y., Zhao, G.S., Huang, Z.D., Li, Y.H., Mizushima, S., Mano, M., Zhang, H.X., Sun, S.F., Sato, T., Horie, R., Zhang, M.X., He, B.S., Mori, C., Hatano, S., Liu, L.S., and Yamori, Y. 1990, Relationship between dietary factors and blood pressure in China, *J.Cardiovasc.Pharmacol.* 16 Suppl. 8:S40-S42.
9. Ristori, M.T. and Verdetti, J. 1991, Effects of taurine on rat aorta *in vitro*, *Fundam.Clin.Pharmacol.* 5:245-258.
10. Yamamoto, J., Akabane, S., Yoshimi, H., Nakai, M., and Ikeda, M. 1985, Effects of taurine on stress-evoked hemodynamic and plasma catecholamine changes in spontaneous hypertensive rats, *Hypertension*, 7:913-922.

28

MODULATION OF Ca^{2+} AND Na^{+} TRANSPORT BY TAURINE IN HEART AND VASCULAR SMOOTH MUSCLE

Ghassan Bkaily,[1] George Haddad,[1] Doris Jaalouk,[1] Nadine Gros-Louis,[1] Majda Taoudi Benchekroun,[1] Radha Naik,[1] Pierre Pothier,[1] Pedro D'Orléans-Juste,[2] Michel Bui,[1] Shimin Wang,[1] and Nicholas Sperelakis[3]

[1,2] MRCC Group in immuno-cardiovascular interaction
[1] Department of Physiology and Biophysics
[2] Department of Pharmacology
Faculty of Medicine
Université de Sherbrooke
Sherbrooke, Quebec, Canada, J1H 5N4
[3] Department of Physiology and Biophysics
College of Medicine, University of Cincinnati
Cincinnati, Ohio, 45267-0576

INTRODUCTION

Taurine has been reported to produce (i) a positive inotropic effect in heart muscle[15,20], (ii) beneficial effects against treatment of congestive heart failure[1], and (iii) protective effects against Ca^{2+} overload[27,28]. Compared to cardiac muscle, little is known on the action of taurine on vascular smooth muscle.

This first section comprises a brief review of data previously published from our laboratories using conventional intracellular microelectrode recording techniques, contraction recordings and recording of ionic currents using the whole-cell voltage clamp technique from heart cell types obtained from chick heart, guinea pig papillary muscle and single ventricular myocytes as well as rabbit aortic single cells.

In perfused 10 to 20 day-old embryonic chick heart, taurine exerted a positive inotropic effect without greatly affecting the slow Ca^{2+}-dependent action potentials (AP_s) accompanying the contraction. It has been suggested that the positive inotropic effect of taurine is not mediated through an increase in the slow inward Ca^{2+} current[23].

In addition, taurine did not induce Ca^{2+}-dependent slow APs[23], nor did it elevate the level of cyclic AMP[20]. Taurine (10 mM) also did not significantly increase the maximum rate of rise (max dV/dt), action potential (AP) amplitude (APA), or duration (at 50% repolarization, APD_{50}) of the slow APs induced by isoproterenol or histamine[23]. In contrast,

Taurine 2, edited by Huxtable et al.
Plenum Press, New York, 1996

Table 1. Summary of the effects of taurine on myocardial cell (chick, rat) ionic currents

Positive inotropic effect		
$[Ca^{2+}]_i$ normalized		
Protection against hypoxic decline of slow APs		
$I_{Ca}(L)$	PCa_i 7	depressed
	PCa_i 10	stimulated
	PCa_i 10	depressed (intracellular taurine)
	$[Ca]_o$ 3.6 mM	depressed
	$[Ca]_o$ 0.8 mM	stimulated
$I_{Ca}(T)$		stimulated
$I_{Na}(s)$ (fast component)		stimulated
$I_{Na}(F)$	PCa_i 10	stimulated (some cells, low dose)
		depressed
		stimulated (intracellular taurine)
$I_K(del)$	PCa_i 7	depressed
	PCa_i 10	stimulated

taurine exerted a significant positive inotropic effect (Table 1), on slow APs induced by tetraethylammonium (TEA, 10 mM) (to depress the outward K^+ current). Taurine also had no effect on max dV/dt and APA; however, the APD_{50} was slightly, but significantly, increased.

Taurine not only did not induce or stimulate slow APs, but actually exerted a transient depressant action on ongoing slow APs[23]. The transient depression of the slow APs occurred within 0.5 min and lasted for about 2-3 min; by 7 min, the response had spontaneously recovered and returned to control levels. There was an initial transient negative inotropic effect of taurine (3 min duration), followed by a positive inotropic effect (still not maximal at 10 min). The contractions were slightly depressed in parallel with taurine depression of the slow APs. By 7 min, a clear positive inotropic effect was evident. Therefore, these data suggested that the positive inotropic effect of taurine was not due to stimulation of the total inward Ca^{2+} slow currents (I_{Ca}). There were no major differences in the positive inotropy of taurine related to the agent used to induce the slow APs and contractions (e.g., histamine or isoproterenol).

The positive inotropic effect of taurine (10 mM) also occurred in low $[Ca]_o$ (0.4 mM) solutions. Taurine did not significantly affect max dV/dt, APA, nor APD_{50} of the fast APs and did not affect the resting potential.

Although taurine did not stimulate the slow APs under normal conditions, under hypoxic conditions taurine did exert a stimulant effect on guinea pig papillary muscle[23]. Hypoxia substantially depressed the slow APs within 5 min and induced a complete blockage by 20 min. Addition of 10 mM taurine partially restored the slow APs within 5 min. Furthermore, pretreatment with taurine was also able to protect against the depressant effect of hypoxia on the slow APs[22]. Since Ca^{2+} slow channel function is dependent on metabolism[25] and is regulated by cyclic nucleotides and phosphorylation[30,17,13], the protective/restorative effect of taurine was suggested to be mediated either indirectly by stimulation of metabolism or phosphorylation or by a direct effect on one or more types of ion channels.

Using the whole-cell voltage clamp technique on young 3-day-old embryonic chick heart cells, taurine activated the fast transient (ft) and blocked the slow transient and sustained components of the TTX and Mn^{2+}-insensitive slow Na^+ current. In older (10 to 20 day-old) embryonic chick heart as well as in adult heart cells, taurine stimulated the low threshold T-type Ca^{2+} current in all ventricular heart cells tested including human and inhibited the high threshold L-type Ca^{2+} current[8]. The effect of taurine on the T-type calcium current depends on the level of intracellular and extracellular free Ca^{2+} concentration[21,24]. The TTX-sensitive fast Na^+ current in old embryonic chick heart cells seemed to be stimulated by low concentrations of taurine (1 to 5 mM) or by intracellular application of taurine. In contrast, there was a marked depression of this type of current when high concentrations of taurine (10 to 20 mM) were applied extracellularly[18,24]. Table 1 summarizes the effect of taurine on ionic currents of myocardial cells. Thus, in normal heart cells it appears that the positive inotropic effect of taurine on heart muscle is due in part to stimulation of the T-type Ca^{2+} current by this amino acid. However, in situations in which intracellular free Ca^{2+} in the patch pipette solution is decreased from pCa 7 to pCa 10[16,29], extracellular application of taurine stimulated $I_{Ca}(L)$, TTX-sensitive fast I_{Na} and the delayed outward K^+ current (Table 1).

METHODS

Studies on Single Cells

Single ventricular cells of 10 day-old chick embryo heart muscle as well as single aortic cells of the rabbit were isolated using standard techniques described previously[7,9]. The techniques used for intracellular and intranuclear free Ca^{2+} and Na^+ measurements using ion sensitive dye and 2D and 3D imaging techniques have also been described elsewhere[6,9,13,10,12]. For more information concerning 2D and 3D Ca^{2+} imaging confocal techniques, please refer to Bkaily[6] and Bkaily *et al.*,[10,12]. Loading and measurement of intracellular ($[\,]_i$), cytosolic ($[\,]_c$) and intranuclear ($[\,]_n$) Na^+ using the Na^+ fluorescent probe sodium green was similar to that described for the Ca^{2+} probe, fluo-3[6,10,12].

Studies on Vascular Smooth Muscle Strips

Newborn albino New-Zealand rabbits of either sex were killed by stunning and exsanguination. The thoracic aortas were quickly removed, placed in buffer (PSS) and cut helicoidally. Segments 2 mm wide and 10 mm long were prepared and mounted in a 2.5 ml quartz organ bath chamber containing PSS. A cotton swab was passed over the luminal surface to remove the endothelium. The muscle strips were then loaded with the Ca^{2+} indicator dye, Fura-2, by incubation in medium containing 4 μM Fura-2/AM (an acetoxymethyl ester form of Fura-2) and 0.3% bovine serum albumin (BSA) for 90 to 120 min at room temperature. After loading, the strips were rinsed with normal oxygenated (95% O_2, 5% CO_2) PSS for at least 30 min at 37°C to remove any remaining dye in the extracellular space and to equilibrate the strips before initiating the measurements.

Contractile responses to taurine were recorded isometrically as changes in tension (g) with Grass FT03 isometric force transducers coupled to a Grass polygraph (Model 7D). Experiments were performed with a setup designed to measure both Fura-2 Ca^{2+} fluorescence and contractile tension simultaneously. Fluorescence measurements were made using double excitatory wavelength measurements (340/380 ratio) using a Deltascan microfluorometer (Photon Technology International Inc., PTi, Princeton, N.J.).

The physiological salt solution (PSS) had the following composition (mM): NaCl 136.9, KCl 5.4, $MgCl_2$ 1.0, $CaCl_2$ 1.5, D-Glucose 5.5, EGTA 0.01, Hepes 20.0.

Results are shown as mean values ± SEM for *n* experiments. All data were subjected to comparison and analysis by a one-way analysis of variance (ANOVA), followed by a Tukey-Kramer multiple comparisons tests to assess the statistical significance of the results. *P* values less than 0.05 were considered statistically significant.

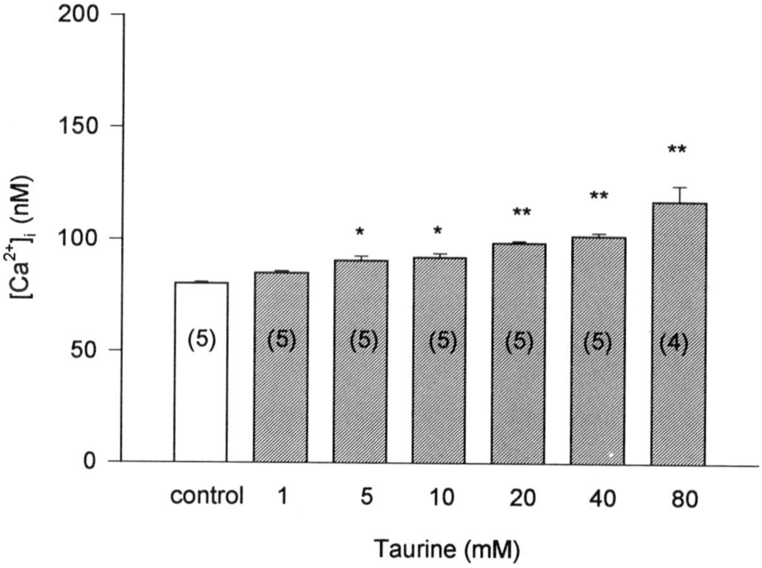

Figure 1. Short-term exposure to taurine induced a sustained increase of total intracellular free Ca^{2+} in 10-day-old chick heart single cells using the standard Fura-2 Ca^{2+} imaging technique. Addition of 1mM taurine slightly increased the sustained level of $[Ca^{2+}]_i$. Increasing the concentration of taurine from 5 to 80 mM increased the sustained $[Ca^{2+}]_i$ in a dose-dependent manner. Experiments were carried out using a PTI (Photon Technology International) microfluorometer. Vertical bars represent mean ± SEM; $*p < 0.05$, $**p < 0.01$.

RESULTS

Using Fura-2 Ca^{2+} two-dimensional imaging technique[6,9,13] we tested the effect of short term exposure (10 to 20 min) to taurine (1 to 80 mM) on the level of total intracellular free Ca^{2+} ($[Ca^{2+}]_i$) in single ventricular heart cells of 10 day-old embryonic chick. Short exposure to 1 mM taurine induced a slight sustained increase of $[Ca^{2+}]_i$. Increasing the concentration of taurine up to 5 and 10 mM significantly increased the sustained level of $[Ca^{2+}]_i$ in a concentration-dependent manner. Further increases of extracellular taurine up to 20, 40 and 80 mM induced, in a concentration-dependent manner, a further rise in the sustained level of $[Ca^{2+}]_i$ which was highly significant when compared to control levels. Fig. 1 summarizes the effect of short-term exposure to different concentrations of taurine on the total intracellular free Ca^{2+} in chick heart cells.

Three-dimensional cytosolic free concentration ($[\ ]_c$) of Ca^{2+} and Na^+ as well as intranuclear free concentration ($[\ ]_n$) of both Ca^{2+} and Na^+ ions in single ventricular cells from 10 day-old chick heart single cells were measured by confocal microscopy using the probes fluo-3 and sodium green, respectively[6,12]. As can be seen in Fig. 2, the resting basal levels of $[Ca]_n$ and $[Na]_n$ in heart cells were higher than that in the cytosol (Fig. 2A and F). These results with $[Ca]_n$ an $[Ca]_c$ are similar to those recently reported by our laboratory in the same heart cell preparations[10]. Prolonged exposure (12 to 24 h) to a low concentration of taurine (1 mM) significantly decreased both $[Ca]_n$ and $[Ca]_c$ (Fig. 2B) as well as $[Na]_n$, although not significant, without any apparent change in $[Na]_c$ (Fig. 2G).

In order to induce Ca^{2+} overload in heart cells, membranes of single cells were depolarized by increasing extracellular K^+ from 5 to 30 mM[9]. The cells were exposed to 30 mM $[K]_o$ for 12 to 24 h. As can be seen in Fig. 2D and I, long-term sustained depolarization of the cell membrane with high extracellular K^+ significantly increased $[Ca]_n$ without

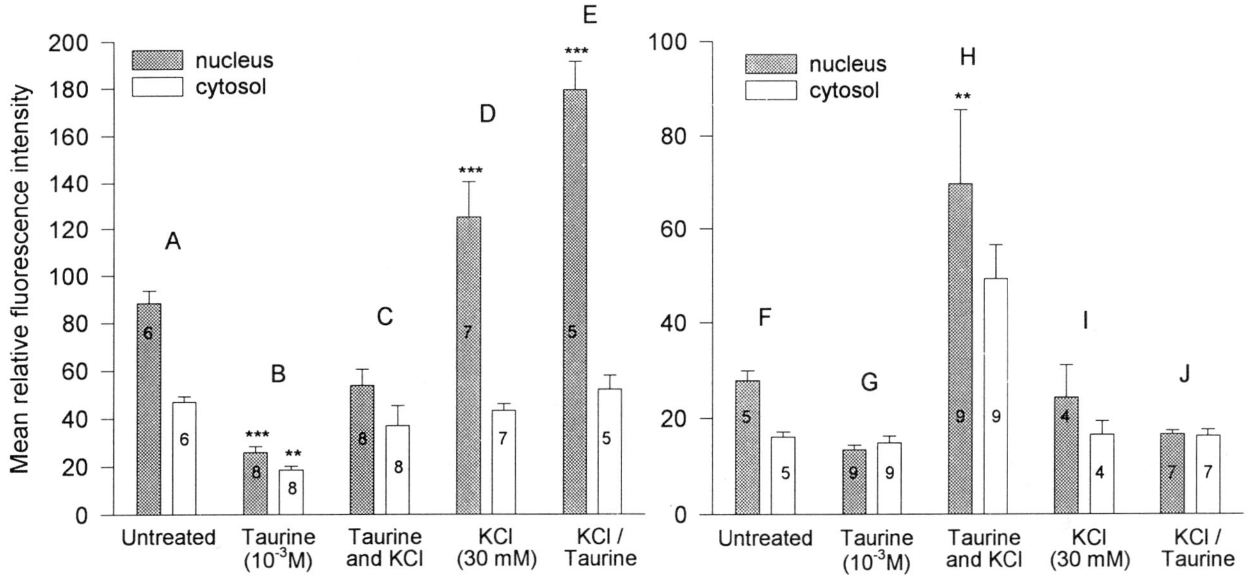

Figure 2. The effect of long-term exposure to taurine and high extracellular K^{+} on cytosolic and nuclear free Ca^{2+} and Na^{+} in 10-day-old chick embryonic heart cells using fluo-3 and Sodium Green dyes and 3-dimensional laser confocal microscopy (Molecular Dynamics). A and F: Resting basal fluo-3 Ca^{2+} fluorescence (A) and Sodium Green fluorescence (F) in the nucleus (■) and the cytosol (□). B and G: Effect of long-term exposure (12 to 24 h) to 2 mM taurine induced a significant sustained decrease of cytosolic and nuclear free Ca^{2+} (B) with an apparent decrease of nuclear free Na^{+} (G). D and I: Long-term (12 to 24 h) depolarization of the cell membrane with 30 mM $[K]_o$ induced only a sustained elevation of nuclear free Ca^{2+} (D) without affecting $[Na]_c$ and $[Na]_n$ (I). C and H: pretreatment with taurine for 12 h and then depolarization of the cell membrane for an additional 12 h (in the presence of taurine) prevents nuclear Ca^{2+} overload (C) induced by high $[K]_o$ (D) but increased both $[Na]_n$ and $[Na]_c$ (H). E and J: Depolarization of the cell membrane for 12 h and then addition of 1 mM of taurine (in the presence of high $[K]_o$) for an additional 12 h did not block the increase of nuclear free Ca^{2+} (E). Vertical bar represents mean ± SEM; $^{*}p < 0.05$, $^{**}p < 0.01$ and $^{***}p < 0.001$.

affecting the level of free cytosolic Ca^{2+} (Fig. 2D); however there were no changes in either $[Na]_n$ or $[Na]_c$ with permanent long-term depolarization of heart cells (Fig. 2I). After inducing nuclear overload with long-term permanent depolarization of the cell membrane of heart cells (12 h) followed by long-term exposure to 1 mM of taurine (in presence of high $[K]_o$) for an additional 12 h, there was a further increase of $[Ca]_n$ (Fig. 2E) with an apparent decrease of $[Na]_n$ (Fig. 2J), without any changes in either cytosolic levels of free Ca^{2+} or Na^+.

However, long-term pre-treatment with 1 mM taurine (12 to 24 h), followed by long term depolarization (12 h) of heart cell membranes with high extracellular K^+ (in the presence of taurine) prevented the increase of $[Ca]_n$ (Fig. 2C) induced by long-term depolarization alone (Fig. 2E), but increased both cytosolic and nuclear free level of Na^+ (Fig. 2H).

Effect of Short (10 to 20 min) and Long-Term (12 to 24 h) Exposure to Taurine on Total $[Ca^{2+}]_i$ of Human Aortic VSM Cells

Using the two-dimensional Fura-2 Ca^{2+} imaging technique[9], we tested the effect of different concentrations of taurine (1 to 80 mM) on the total intracellular free concentration of Ca^{2+} in single human aortic vascular smooth muscle (VSM) cells (Fig. 3A).

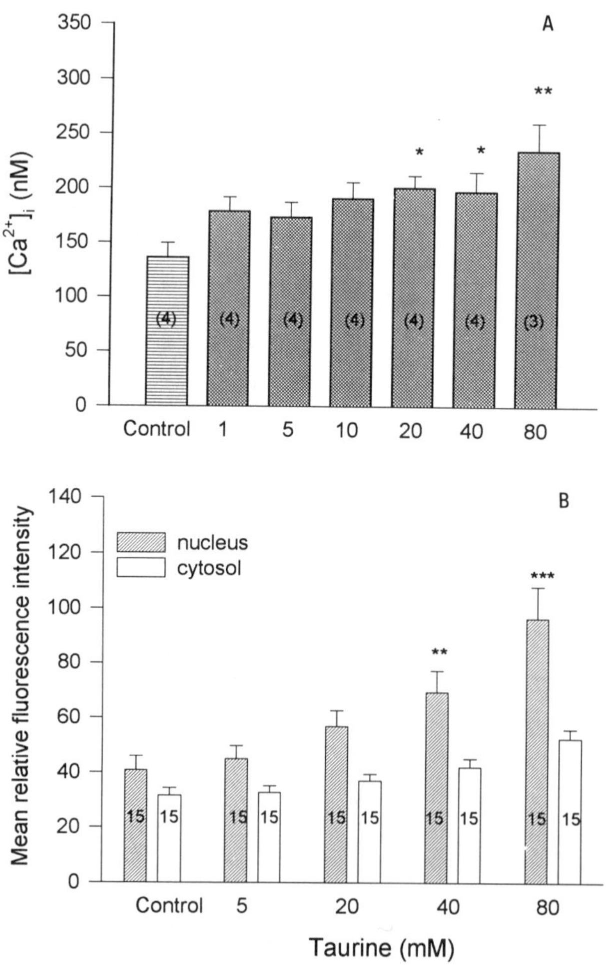

Figure 3. Effect of short-term exposure (5 to 10 min) to different concentrations of taurine (1-80 mM) on total intracellular (A) as well as nuclear (B■) and cytosolic (B□) free calcium in human aortic vascular smooth muscle (VSM) single cells. A: Using the Fura-2 Ca^{2+} imaging technique, 1 to 10 mM of taurine had no significant effect on total intracellular free Ca $[Ca^{2+}]_i$. However, a significant increase of $[Ca^{2+}]_i$ was observed at concentrations ≥20 mM. B: Using the 3D laser confocal microscopy technique, the level of free Ca^{2+} is higher in the nucleus than in the cytosol of VSM cells. Taurine from 5 to 20 mM had no effect $[Ca]_c$ and $[Ca]_n$ and significantly increased $[Ca]_n$ at concentrations ≥40 mM. Vertical bar represents mean ± SEM; $*p < 0.05$, $**p < 0.01$ and $***p < 0.001$.

The resting level of total $[Ca^{2+}]_i$ in these human single VSM cells (near 125 nM) was similar to those reported in rabbit aortic VSM single cells[9]. Short term exposure to low concentrations of taurine from 1 to 10 mM did not increase the total level of $[Ca^{2+}]_i$. At concentrations of 20 mM and up to 40 and 80 mM, taurine significantly increased the total level of $[Ca^{2+}]_i$. Using three-dimensional cytosolic and nuclear free Ca^{2+} measurements and the Ca^{2+} probe fluo-3, the nuclear free Ca^{2+} level of VSM cells was found to be higher than that in the cytosol (Fig. 3B). Short-term exposure to taurine concentrations from 5 to 20 mM apparently (but not significantly) increased both $[Ca]_n$ and $[Ca]_c$ in a concentration-dependent manner (Fig. 3B). Increasing the concentration of taurine up to 40 and 80 mM significantly increased $[Ca]_n$; however, $[Ca]_c$ was not significantly affected (Fig. 3B).

Effect of Short-Term Exposure to Different Concentrations of Taurine on Tension and $[Ca^{2+}]_i$ of Rabbit Aortic Muscle

The effect of different concentrations of taurine on simultaneous measurements of tension and intracellular free Ca^{2+} concentrations were studied in order to see if taurine, at

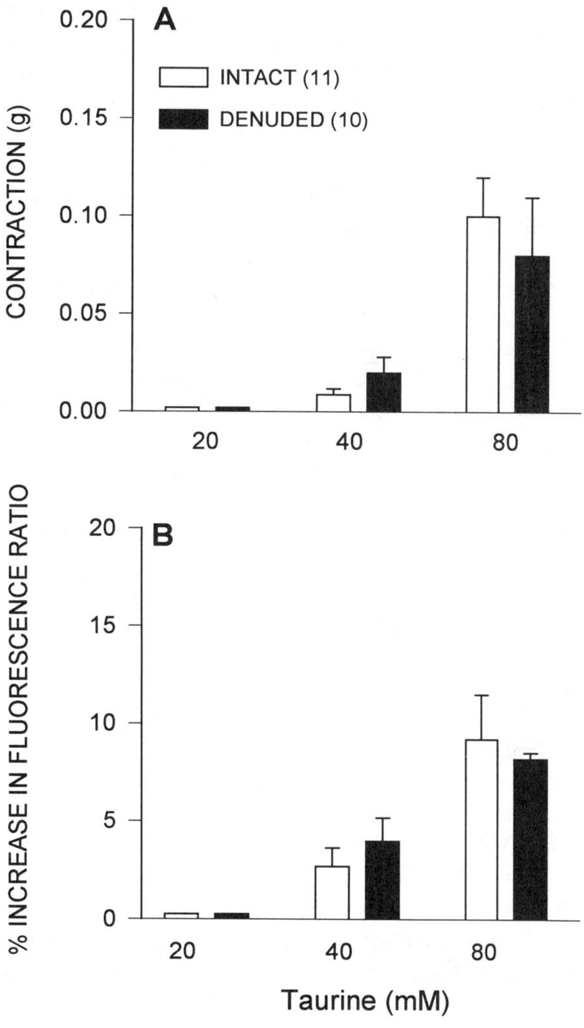

Figure 4. Effect of different concentrations of taurine on muscle tension of intact (□) or endothelium-denuded (■) rabbit aorta (A) and intracellular free Ca^{2+} (B). Each column represents the mean ± SEM of (n) different muscle preparations.

concentrations shown to significantly increase total $[Ca^{2+}]_i$ (Fig. 3A) and $[Ca^{2+}]_n$ (Fig. 3B) in single aortic cells (20 to 80 mM), induces both an increase in $[Ca^{2+}]_i$ and in contractile tension in rabbit aortic vascular smooth muscle strips. As can be seen in Fig. 4, in both intact and endothelium-denuded aortic muscle, 20 mM taurine had no effect on either resting $[Ca^{2+}]_i$ or associated tension (Fig. 4). Increasing the concentration of taurine up to 40 mM and 80 mM increased $[Ca^{2+}]_i$ and the associated tension in both intact and denuded aortic vascular smooth muscle (Fig. 4). This increase in $[Ca^{2+}]_i$ and tension induced by 40 mM taurine was apparently higher in endothelium-free muscle when compared to intact muscle (Fig. 4A and B). However, at high concentrations of taurine (80 mM), the increase in $[Ca^{2+}]_i$ and associated tension was slightly higher in intact compared to endothelium-free muscle (Fig. 4).

DISCUSSION

Earlier studies showed that in single 3-, 10-, and 17-day-old embryonic chick heart cells in culture, taurine (5 mM) activates a T-type Ca^{2+} channel and a TTX- and Mn^{2+}-insensitive fast transient slow Na^+ channel (Table 1). Also, taurine induced a fast transient slow Na^+ current in 3-day-old embryonic cells that did not exhibit this type of current in the absence of taurine. In addition, taurine accelerated the decay of the sustained slow Na^+ current. In single cells of 3-day-old and 10-day-old chick embryos which showed only a L-type component of Ca^{2+} current ($I_{Ca(L)}$), taurine blocked this current and induced a T-type component of I_{Ca} ($I_{Ca(T)}$) (Table 1).

The rapid (1 min) transient decrease of slow AP amplitude and max dV/dt by 1 mM taurine reported by Sawamura *et al.*[23] could be explained by the decrease of the slow $I_{Ca(L)}$. The spontaneous recovery of the AP amplitude and max dV/dt after a few minutes in the continued presence of taurine could be explained by the activation of the T-type Ca^{2+} current ($I_{Ca(T)}$). The blockade of $I_{Ca(L)}$ and the stimulation of $I_{Ca(T)}$ might explain the apparent lack of effect of taurine on the steady state, Ca^{2+}-dependent, slow APs in embryonic heart[23]. It is also possible that the sustained increase of total $[Ca^{2+}]_i$ by short-exposure to taurine could be due to stimulation of the R-type Ca^{2+} channel[9] by this amino acid. This sustained increase of $[Ca^{2+}]_i$ would accelerate the inactivation of the L-type Ca^{2+}, thus explaining in part the decrease of this type of current by taurine.

The increase or activation of a fast TTX-sensitive Na^+ current by taurine would increase $[Na]_i$ near the sarcolemma, which in turn may reverse the Na^+-Ca^{2+} exchange and thus allow Ca^{2+} to flow inside the cell by this pathway. This may suggest that if the positive inotropic effect of taurine is not mediated through an increase in the inward L-type Ca^{2+} current[23,26], it could be due, however, to taurine-stimulated Ca^{2+} influx through the T-type Ca^{2+} channel and the Na^+-Ca^{2+} exchanger, resulting from an increase in Na^+ influx via activation of the TTX-sensitive fast Na^+ current.

On the other hand, taurine exerts a dual effect on the TTX-sensitive fast I_{Na}, with inhibition or stimulation at lower concentrations (1 and 5 mM) and inhibition only at higher concentrations (10 and 20 mM) (Table 1). However, a dual action of taurine is not unique. For example, some local anesthetics have a dual action on max dV/dt: at low concentrations, max dV/dt is increased while at high concentrations, it is decreased[3,4]. The dual action of taurine on the TTX-sensitive I_{Na} may reflect two separate sites of action: (a) a direct action by stimulating the fast Na^+ channels, and (b) an indirect action by inhibiting the channels.

The effect of taurine on peak I_{Ca} was dependent on both $[Ca^{2+}]_o$ and $[Ca^{2+}]_i$. At low $[Ca^{2+}]_i$ or $[Ca]_o$, taurine stimulated the L-type Ca^{2+} current whereas taurine inhibited this type of current at high $[Ca^{2+}]_i$ or $[Ca]_o$. It is unknown whether taurine acts directly on Ca^{2+}

channels. One regulator of the L-type Ca^{2+} channels is the level of $[Ca^{2+}]_i$ with inhibition occurring at high $[Ca^{2+}]_i$ levels.

The effect of taurine on total outward K^+ current was also dependent on $[Ca^{2+}]_i$. Taurine stimulated I_K at low $[Ca^{2+}]_i$ and inhibited I_K at high $[Ca^{2+}]_i$. These results could explain the action of taurine on the APs: APD prolongation at high $[Ca]_o$ and APD shortening at low $[Ca^{2+}]_i$.

Short-term exposure to taurine actually elevated the total sustained $[Ca^{2+}]_i$, presumably resulting from the stimulation of the R-type Ca^{2+} channel, the slow I_{Na}, the fast I_{Na}, the $I_{Ca(T)}$ and the release of Ca^{2+} from the SR. Thus, taurine may exert a positive inotropic effect by such mechanisms.

Our previous and recent results showed that modulation of ion channels by extracellular application of taurine depends on the level of total intracellular free Ca^{2+} ($[Ca^{2+}]_i$). In normal situations, taurine seems to mainly stimulate the T-type slow Ca^{2+} channels in adult heart cells and the early embryonic TTX and Mn^{2+}-insensitive slow Na^+ channels without directly affecting the Na^+-Ca^{2+} exchange current. Under normal conditions, short-term exposure of heart cells to taurine will certainly increase total intracellular free Ca^{2+}: however, long-term exposure would decrease both cytosolic as well as nuclear free Ca^{2+} via activation of Ca^{2+} outflow through the Na^+-Ca^{2+} exchange. In situations where intracellular overload has already taken place in heart cells, taurine did not seem to block this phenomenon. However, our results show that taurine could prevent cytosolic and mainly intranuclear Ca^{2+} overload. The prevention of total intracellular Ca^{2+} overload by long-term exposure to taurine could be due to prestimulation of sarcolemmal and nuclear Ca^{2+} outflows and to Na^+ inflow through the Na^+-Ca^{2+} exchange by taurine as suggested by the data presented in Fig. 2.

In vascular smooth muscle and normal physiological conditions, taurine induced a rise in intracellular Ca^{2+} and tension that did not seem to be highly dependent of the presence of endothelial cells. This was only detected at a high concentrations of taurine (80 mM). Low concentrations of taurine (1 to 10 mM) were found to induce decreases in tension of aortic vascular smooth muscle. The vasorelaxant effect of low concentrations of taurine as well the failure of this amino acid to significantly increase tension at a concentration near 20 mM could be due in part to an increase in the kinetics of cytosolic free Ca^{2+} buffering by the nucleus[12] as suggested by Fig. 3B.

SUMMARY

Using the whole-cell voltage clamp technique, taurine was found to affect different types of various ionic currents including T and L-type Ca^{2+} currents, slow Na^+ and fast Na^+ currents as well as the delayed outward K^+ current. Also, in normal situations, taurine had no effect on the Na^+-Ca^{2+} exchange current. The effect of taurine on the different types of ionic currents appears to depend on $[Ca^{2+}]_o$ and $[Ca^{2+}]_i$ and may also vary according to the tissue or cell type studied. Using standard Ca^{2+} imaging techniques, short-term exposure (10 to 20 min) of single heart cells and aortic vascular smooth muscle cells was found to increase total intracellular free Ca^{2+} in a dose-dependent manner. However, using 3-dimensional Ca^{2+} and Na^+ imaging techniques, long-term exposure of heart and vascular smooth muscle cells to taurine was found to decrease both nuclear and cytosolic Ca^{2+} without significantly changing either nuclear or cytosolic Na^+ levels. Long- term exposure to taurine was found to prevent cytosolic and nuclear increases of Ca^{2+} induced by permanent depolarization of heart cells with high $[K^+]_o$. This preventive effect of taurine on nuclear Ca^{2+} overload was associated with an increase of both cytosolic and nuclear free Na^+. Thus, the effect of long-term exposure to taurine on intranuclear Ca^{2+} overload in heart cells seems to be mediated via stimulation of sarcolemma and nuclear Ca^{2+} outflow through the Na^+-Ca^{2+} exchanger.

ACKNOWLEDGMENTS

This study was supported by Medical Research Council grant no. PG-12350. Drs. Bkaily and D'Orléans-Juste are scholars of the Fonds de la Recherche en Santé du Québec. The authors thank Ms. Mireille Dussault for her secretarial assistance.

REFERENCES

1. Azuma, J., Taihara, K., Awata, N., Ohta, H., Sawamura, A., Harada, H., and Kishimoto, S., 1984, Beneficial effect of taurine on congestive heart failure induced by chronic aortic regurgitation in rabbits, *Res. Commun. Chem. Pathol. Pharmacol.* 45:261-270.
2. Baskin, S.I., and Finney, C.M., 1979, Effects of taurine and taurine analogue on the cardiovascular system, *Gamyu Aminosan* (Sulfur-containing Amino Acids) 2:1-18.
3. Bigger, J.T., and Mandel, W.J., 1970, Effect of lidocaine on the electrophysiological properties of ventricular muscle and Purkinje fibers, *J. Clin. Invest.* 46:63-77.
4. Bigger, J.T., Basset, A.L. and Hoffman, B.F., 1968, Electrophysiological effects of diphenylhydantoin on canine Purkinje fibers, *Circ. Res.* 22:221-236.
5. Bkaily, G., 1992, Single heart cells as model for studying cardiac toxicology. In: In vitro methods in toxicology, edited by G. Jolles and A. Cordier. Academic Press, London, pp. 289-334.
6. Bkaily, G., 1994, Regulation of R-type Ca^{2+} channels by insulin and ET-1 in VSM. In: Ionic channels in vascular smooth muscle, edited by G. Bkaily, Molecular Biol. Intelligence Unit. Austin, pp. 41-52.
7. Bkaily, G., Caillé, J.P., Payet, M.D., Peyrow, M., Sauvé, R., Renaud, J.F., Sperelakis, N., 1988, Bethanidine increases one type of K^+ current and relaxes aortic muscle. *Can. J. Pharmacol.* 66:731-736.
8. Bkaily, G., Chahine, M., Sperelakis, N., Yamamoto, T., 1988, Taurine increases one type of slow Na^+ and Ca^{2+} currents in embryonic heart. *J. Physiol.* (Lond.) 406:91
9. Bkaily, G., Economos, D., Potvin, L., Ardilouze, J.L., Mariott, C., Corcos, J., Bonneau, D., and Fong, C.N., 1992, Blockade of insulin steady-state R-type Ca^{2+} channel by PN200-110 in heart and vascular smooth muscle. *Mol. Cell. Biochem.* 117:93-106.
10. Bkaily, G., Gros-Louis, N., Naik, R., Jaalouk, D., Pothier, P., 1996, Implication of the nucleus in excitation contraction coupling of heart cells. *Mol. Cell Biochem.* (In press).
11. Bkaily, G., Jacques, D., Sculptoreanu, A., Yamamoto, T., Carrier, D., Vigneault, D., Sperelakis, N., 1991, Apamin, a highly potent blocker of the TTX- and Mn^{2+}-insensitive fast transient Na^+ current in young embryonic heart. *J. Mol. Cell Cardiol.* 23:25-39.
12. Bkaily, G., Naik, R., D'Orléans-Juste, P., Wang, S., and Fong, C.N., 1995, Endothelin-1 activates the R-type Ca^{2+} channel in vascular smooth muscle cells. *J. Cardiovas. Pharmacol.* 26:S303-S306.
13. Bkaily, G., Perron, N., Wang, S., Sculptoreanu, A., Jacques, D., Ménard, D., 1993, Atrial natriuretic factor blocks the high-threshold Ca^{2+} current and increased K^+ current in fetal single ventricular cells, *J. Mol. Cell Cardiol.* 25:1305-1316.
14. Bkaily, G., Peyrow, M., Yamamoto, T., Sculptoreanu, A., Jacques, D., Sperelakis, N., 1986, Macroscopic Ca^{2+}-Na^+ and K^+ currents in single heart and rabbit aortic cells. *Mol. Cell. Biochem.* 80:59-72.
15. Dietrich, J., and Diacono, J., 1971, Comparison between ouabain and taurine effects on isolated rat and guinea-pig hearts in low calcium medium, *Life Sci.* 10:499-507.
16. Fabiato, A., and Fabiato, F., 1979, Calculator programs for computing the composition of the solutions containing multiple metals and ligands used for experiments in skinned muscle cells, *J. Physiol.* (Lond.) 75:463-505.
17. Li, T., and Sperelakis, N., 1983, Stimulation of slow action potentials in guinea pig papillary muscle cells by intracellular injection of cAMP, Gpp(NH)p, and cholera toxin, *Circ. Res.* 52:111-117.
18. Satoh, H., and Sperelakis, N., 1991, Actions of taurine on fast Na^+ current (I_{Na}) in embryonic chick ventricular myocytes, *Faseb J.* 5(6):A1742.
19. Satoh, H., and Sperelakis, N., 1991, Identification of the hyperpolarization-activated inward current in young embryonic chick heart myocytes, *J. Develop. Physiol.* 15:247-252.

20. Sawamura, A., Azuma, J., Harada, H., Hasegawa, H., Ogura, K., Sperelakis, N., and Kishimoto, S., 1983, Protection by oral pretreatment with taurine against the negative inotropic effects of low calcium medium on isolated perfused chick heart, *Cardiovas. Res.* 17:620-626.
21. Sawamura, A., Sada., H., Azuma, J., Kishimoto, S., and Sperelakis, N., 1990, Taurine modulates ion influx through cardiac Ca^{2+} channels, *Cell Calcium* 11:251-259.
22. Sawamura, A., Sperelakis, N., and Azuma, J., 1986, Protective effect of taurine against decline of cardiac slow action potentials during hypoxia. *Eur. J. Pharm.* 120:235-239.
23. Sawamura, A., Sperelakis, N., Azuma, J., and Kishimoto, S., 1986, Effects of taurine on the electrical and mechanical activities of embryonic chick heart, *Can. J. Physiol. Pharmacol.* 64:649-655.
24. Sperelakis, N., Satoh, H., and Bkaily, G., 1991, Taurine effects on ionic currents in myocardial cells. *Adv. Exper. Med. Biol.* 359:9-17.
25. Sperelakis, N., and Schneider, J.A., 1976, A metabolic control mechanism for calcium ion influx that may protect the ventricular myocardial cell, *Am. J. Cardiol.* 37:1079-1085.
26. Sperelakis, N., Yamamoto, T., Bkaily, G., Sada, H., Sawamura, A., and Azuma, J., 1989, Taurine effects on action potentials and ionic currents in chick embryonic cells. In: Taurine and the Heart, edited by H. Iwata, J.B. Lombardini, and T. Segawa, Kluwer Academic Pub., Boston, pp. 1-20.
27. Takihara, K., Azuma, J., Awata, N., Ohta, H., Hamaguchi, T., Sawamura, A., Tanaka, Y., Kishimoto, S., and Sperelakis, N., 1986, Beneficial effect of taurine in rabbits with chronic congestive heart failure, *Am. Heart J.* 112:1278-1284.
28. Takihara, K., Azuma, J., Awata, N., Ohta, H., Sawamura, A., Kishimoto, S., and Sperelakis, N., 1985, Taurine's possible protective role in age-dependent response to calcium paradox, *Life Sci.* 37:1705-1710.
29. Tsien, R.Y., and Rink, T.J., 1980, Neutral carrier ion-selective microelectrodes for measurements of intracellular-free calcium, *Biochim. Biophys. Acta* 559:623-638.
30. Vogel, S., and Sperelakis, N., 1981, Induction of slow action potentials by microiontophoresis of cyclic AMP into heart cells. *J. Mol. Cell Cardiol.* 13:51-64.

29

SOME ACTIONS OF TAURINE ON IONIC CURRENTS OF MYOCARDIAL CELLS AND MYOMETRIAL CELLS*

N. Sperelakis, Y. Katsube, and M. Kusaka

Department of Molecular and Cellular Physiology
University of Cincinnati, College of Medicine
Cincinnati, Ohio 45267-0576

INTRODUCTION

The sulfur amino acid taurine has been reported: (i) to be a positive inotropic (cardiotonic) agent in failing hearts, in both animals and human; (ii) to be depleted in myocardial cells in ischemic heart disease; and (iii) to cause heart failure when deficient in diet of animals (e.g., cats) that cannot synthesize taurine. The taurine content of the heart is elevated in patients and animals with congestive heart failure[1]. Taurine has been used as a treatment for patients with congestive heart failure (e.g., 3 g per day) by Dr. Azuma *et al.* In Japan and Korea, soft drinks (e.g., soda pop and fruit drinks) that contain taurine (e.g., 1 g) as an additive are widely available.

Taurine exerts a number of cardioprotective actions. It protected the hearts of chicks and mice against the cardiotoxicity produced by the antineoplastic agent doxorubicin (Adriamycin)[5]. In addition, taurine was reported to protect the heart against damage under Ca^{2+} paradox conditions[18]. Taurine also protects cardiac muscle against hypoxic injury[12] and isoproterenol-induced cardiomyopathy. The area has been reviewed[1].

Our laboratory became involved in studying mechanism(s) of action of taurine when Dr. J. Azuma, and later his colleague, Dr. A. Sawamura, came to our laboratory. We published several papers on the effects of taurine on the electrical and contractile activities of chick cardiac muscle under various conditions[11,12,13,17]. We subsequently pursued the effects of taurine using whole-cell voltage clamp of single myocardial cells to measure the slow (L-type) Ca^{++} current ($I_{Ca(L)}$) with the collaboration of Sada, Bkaily, and Satoh[9,10,11,15,16].

The results of these studies were complex (Table 1). As shown, taurine (1-20 mM) added to the bath *depressed* $I_{Ca(L)}$ when $[Ca]_i$ was high ($pCa_i = 7$) or when $[Ca]_o$ was high (3.6 mM). In contrast, taurine *stimulated* $I_{Ca(L)}$ when $[Ca]_i$ was low ($pCa_i = 10$) or when $[Ca]_o$ was low (0.8 mM). Taurine also protected against the decline of the Ca^{++}-current-dependent

* Due to travel problems, Dr. Sperelakis was unable to present this paper.

Taurine 2, edited by Huxtable et al.
Plenum Press, New York, 1996

Table 1. Summary of taurine effects on ionic currents of myocardial cells (chick, rat)

$[Ca]_i$ normalized		
Protect against hypoxic decline of Ca^{++}-dependent slow APs		
$I_{Ca(L)}$	pCa_i 7	depressed
	pCa_i 10	stimulated
	pCa_i 10	depressed (intracell taur)
$[Ca]_o$	3.6 mM	depressed
$[Ca]_o$	0.8 mM	stimulated
$I_{Ca(T)}$		stimulated
$I_{Na(s)}$	(trans comp)	stimulated
$I_{Na(f)}$	pCa_i10	stimulated (some cells, low dose)
		depressed
		stimulated (intracell taur)
$I_{K(del)}$	pCa_i 7	depressed
	pCa_i 10	stimulated

slow action potentials produced by hypoxia[12,13]. Thus, it appears that taurine exerts some regulation of $I_{Ca(L)}$, trying to maintain it in a "normal" range, and therefore acting to stabilize (normalize) the level of $[Ca]_i$. In addition, taurine appeared to stimulate the fast (T-type) Ca^{++} current ($I_{Ca(T)}$) and a fast transient component of the TTX-insensitive Na^+ current ($I_{Na(s)}$). The TTX-sensitive fast Na^+ current ($I_{Na(f)}$) was stimulated in some cells by the lower concentrations of taurine (1 and 5 mM), but depressed in the other cells in a dose-dependent manner[9].

NEW RESULTS

Na/Ca Exchange Current in Myocardial Cells

In recent studies, we wanted to determine whether taurine had any direct effect on the Na^+/Ca^{++} exchange current ($I_{Na/Ca}$) of rat myocardial cells, as measured by the whole-cell voltage clamp and using the ramp pulse protocol of Junko Kimura (Fig. 1). We found no evidence for an effect of extracellularly-applied taurine (30 mM) on $I_{Na/Ca}$, either in the forward mode (3 Na_o:1 Ca_i) (Fig. 2) or reverse mode (1 Ca_o:3 Na_i) (Fig. 3) (Katsube & Sperelakis, unpublished). These results are summarized in Figure 4 and Table 2.

In contrast to our negative findings, it was reported by Earm *et al.*[4] that the $I_{Na/Ca}$ current of rabbit atrial cells was stimulated by taurine. This stimulation was presumably indirect, caused by release of Ca^{++} from the SR, thus elevating $[Ca]_i$. In our experiments, 10 mM EGTA was present in the patch pipette to prevent indirect effects on $I_{Na/Ca}$ due to possible Ca^{++} releae from the SR caused by taurine.

$I_{Ca(L)}$ Current in Myometrial Cells

In recent studies, we also studied the effect of extracellularly-applied taurine (20 mM) on $I_{Ca(L)}$ of rat (18-day pregnant) uterine smooth muscle cells (Kusaka & Sperelakis, unpublished). We found that taurine had no effect on $I_{Ca(L)}$ at pCa_i of 7, but slightly depressed the current at pCa_i of 10 (Fig. 5; Table 2). This result is nearly opposite to that summarized above for myocardial cells (see Table 1). We do not know the taurine content of normal uterine smooth muscle. The effects of taurine on $I_{Ca(L)}$ may depend on the specific isoform of the slow Ca^{++} channel.

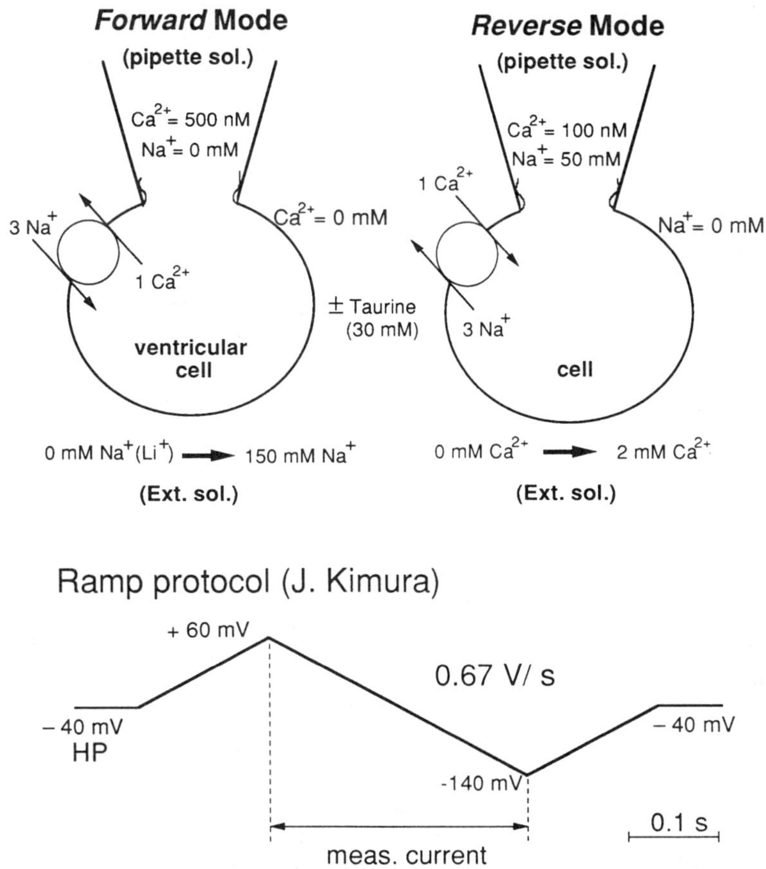

Figure 1. Diagram of method used to measure the Na/Ca exchange current in single cardiomyocytes (rat). The ramp pulse had three phases: an initial 100 mV depolarizing phase from the holding potential (HP) of -40 mV, a second hyperpolarizing phase of 200 mV, and a third phase returning to the HP. The slope of the ramp was ± 100 mV/0.3 s. The current/voltage (I/V) curve was measured during the second hyperpolarizing portion. *Forward Mode*: The pipette solution contained 500 nM Ca^{2+} (pCa 6.3). Changing the bathing solution from 150 mM Li^+ (Na^+-free) to 150 mM Na^+, the inward Na/Ca exchange current (carried by Na^+ influx) was started. *Reverse Mode*: The pipette solution contained 50 mM Na^+. Changing the bathing solution from Ca^{2+}-free to 2 mM Ca^{2+}, the outward Na/Ca exchange current (carried by Na^+ efflux) was initiated. Taurine (30 mM) was added in the external solution. The external solution contained ouabain (20 μM) to block the Na-K pump, $BaCl_2$ (1 mM) and CsCl (2 mM) to block K channels, and nifedipine (2 μM) to block slow Ca^{2+} channels. In the presence of these blockers, time-dependent ionic currents and Na-K pump current were mostly blocked, and the remaining conductance showed a nearly linear I/V relation.

$I_{Na(F)}$ Current in Uterine Leiomyosarcoma Cells

We previously demonstrated that TTX-sensitive fast Na^+ channels are gained by rat uterine smooth muscle cells during pregnancy and may be involved in parturition[6,8]. We had also found that a factor in serum induces fast Na^+ channels in human uterine leiomyosarcoma cells, and produces a very large $I_{Na(f)}$[7]. As summarized above, we had previously shown that $I_{Na(f)}$ of myocardial cells (chick, rat) is affected by taurine (see Table 1). Therefore, we tested the effect of taurine on $I_{Na(f)}$ in the leiomyosarcoma cells (Kusaka & Sperelakis, unpublished). We found that taurine (20 mM) had no effect on $I_{Na(f)}$ when pCa_i was 7.0 (Fig. 6; Table 2). Taurine was not tested at a pCa_i of 10 (Table 2).

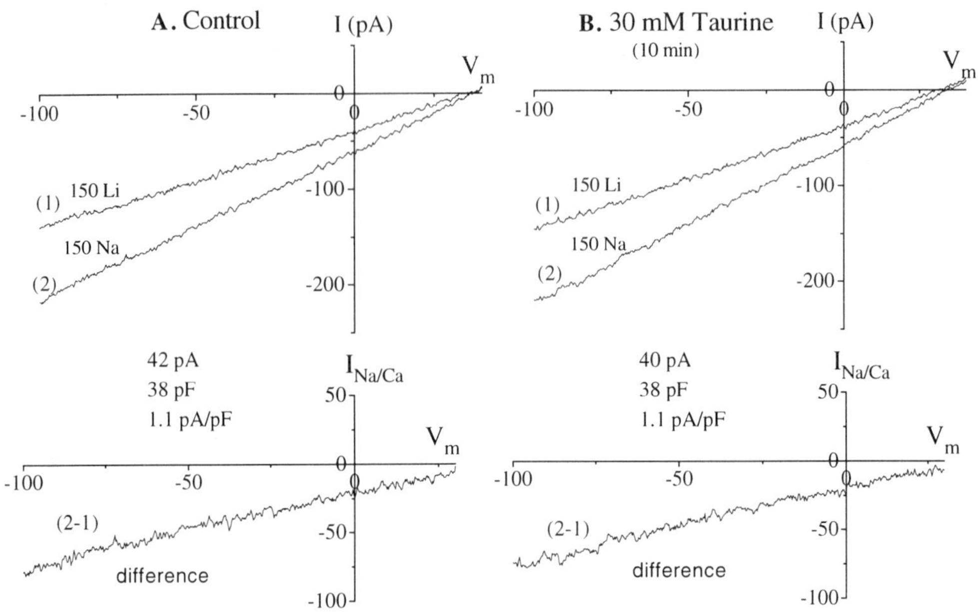

Figure 2. Effect of taurine on the *inward* Na/Ca exchange current *(forward mode)* on one cell (10-day-old rat ventricular cell). **A**. Control in absence of taurine. In the condition of 500 nM Ca^{2+} (pCa 6.3) in the pipette, exchanging the bathing solution from 150 mM Li^+ (1) to 150 mM Na^+ (2), the membrane current shifted inward (at a HP of -40 mV). The difference current (2 - 1) indicates the inward Na/Ca exchange current carried by Na^+ influx. Since the external solution did not contain added Ca^{2+}, the theoretical reversal potential (V_{rev}) should be at a very positive potential. Therefore, the exchange current ($I_{Na/Ca}$) is inward at all membrane potentials tested. **B.** Same cell in presence of taurine (30 mM) added to the bath. The Na/Ca exchange current was measured after 10 min. of incubation. The membrane current at -50 mV was 42 pA in **A** (C_m = 38 pF) and 40 pA in **B**. The current density was calculated to be 1.1 pA/pF, in both cases.

This result differs from those on myocardial cells, in which low doses of taurine stimulated $I_{Na(f)}$ in some cells and inhibited the current in the other cells in a dose-dependent manner (see Table 1). Therefore, the effects of taurine on $I_{Na(f)}$ must depend to a large extent on the specific isoform of the fast Na^+ channel present.

DISCUSSION

The present results show that taurine: (i) has no direct effect on the Na/Ca exchanger; (ii) does not affect the isoform of fast Na^+ channels in uterine smooth muscle cells; and (iii) produces slight inhibition of the slow Ca^{++} channels of myometrial cells at pCa_i of 10, but has no effect at pCa_i of 7. Thus, the results on $I_{Na(f)}$ and $I_{Ca(L)}$ of myometrial cells are substantially different from those on myocardial cells, perhaps reflecting different isoforms of the ion channels in heart versus uterus.

How can we explain the mechanism of the positive inotropic effect of taurine in normal and failing heart? The mechanisms whereby some positive inotropic drugs, such as cardiac glycosides, β-adrenergic agonists, and phosphodiesterase (PDE) inhibitors act are

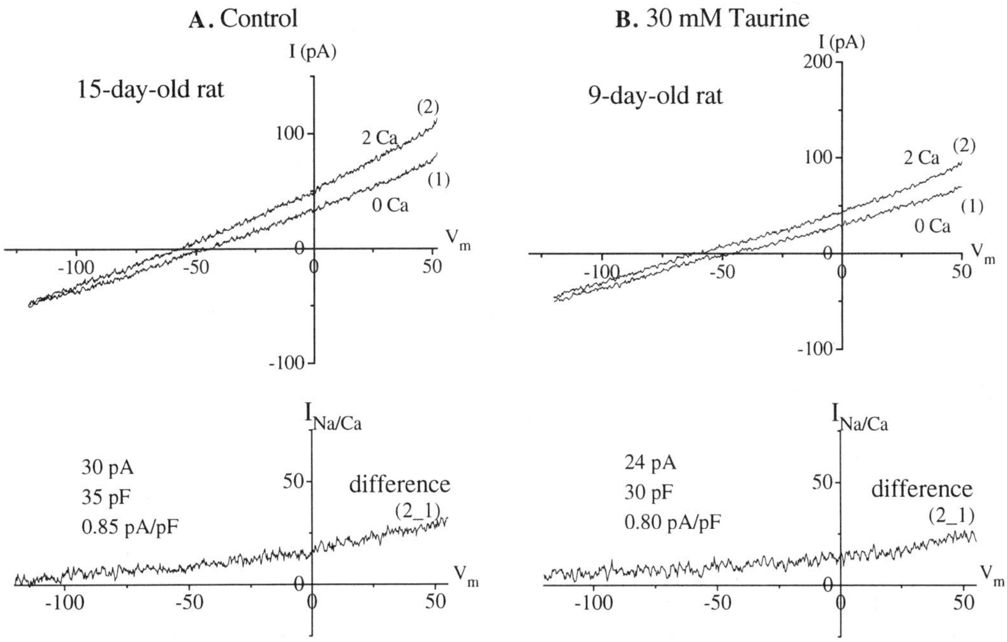

Figure 3. Lack of effect of taurine (30 mM) on *outward* Na/Ca exchange current *(reverse mode)* from 9-day-old rat ventricular cells (15-day and 9-day). **A** and **B** are from different cells. 50 mM Na^+ was in the pipette. Changing the bathing solution from Ca^{2+}-free (1) to 2 mM Ca^{2+} (2) caused the membrane current to shift outward (at a HP of -40 mV). The difference current (2 - 1) indicates the outward Na/Ca exchange current carried by Na^+ efflux. Since the external solution did not contain Na^+ (the theoretical V_{rev} should be at plus infinity), $I_{Na/Ca}$ is outward at all membrane potentials tested. The membrane current at +50 mV was 30 pA (C_m = 35 pF) in **A**, and 24 pA (C_m = 30 pF) in **B**. The current density was 0.85 and 0.80 pA/pF, respectively. Therefore, taurine did not stimulate $I_{Na/Ca}$.

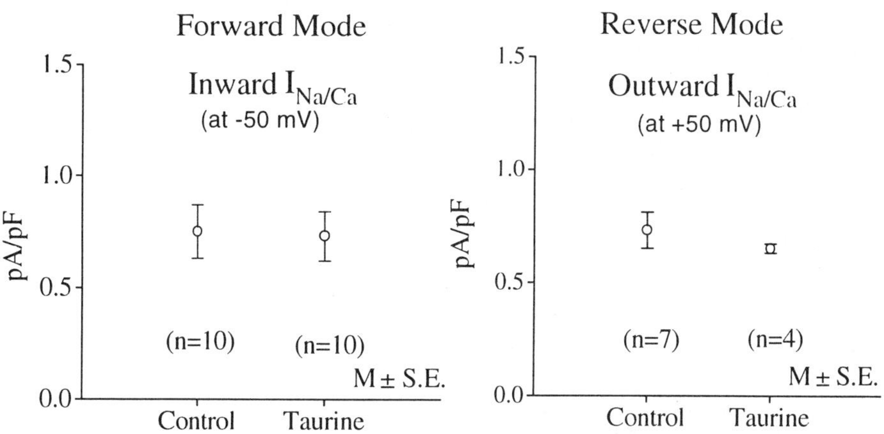

Figure 4. Summary of taurine effect on Na/Ca exchange current. There were no significant differences between control and 30 mM taurine on *inward* and *outward* Na/Ca exchange currents.

Table 2. Summary of taurine effects on ionic current

Myocardial cells (rat)		
$I_{Na/Ca}$	forward/reverse:	no effect
Myometrial smooth muscle cells (pregnant rats)		
$I_{Ca(L)}$	pCa_i 7:	no effect
	pCa_i 10:	depressed
Leiomyosarcoma cells (SK-UT-1B) (human uterine)		
$I_{Na(f)}$	pCa_i 7:	no effect
	pCa_i 10:	not done

depicted in Figure 7.* As shown (right side), the β-agonists and PDE inhibitors act via elevation of cAMP, activation of PK-A, and phosphorylation of the slow Ca^{++} channels, resulting in stimulation of $I_{Ca(L)}$.† The greater $I_{Ca(L)}$ causes more Ca^{++} to be released from the SR, due to Ca^{++}-induced Ca^{++} release from the SR (proportional release), and hence a greater force of contraction.

As shown in the left side of Figure 7, digitalis inhibits the Na^+/K^+-ATPase/pump in the sarcolemma, thereby leading to a rise in $[Na]_i$. Due to the Na/Ca exchange system, the rise in $[Na]_i$ leads to a concomitant rise in $[Ca]_i$, which allows the SR to become more loaded with Ca^{++}. Consequently, there can be greater release of Ca^{++} from the SR during excitation-contraction coupling, and hence a greater force of contraction.

It has been shown that there is a Na^+-taurine co-transport system in myocardial cells that is electrogenic.[2,3,14] ‡Therefore, adding taurine externally should cause $[Na]_i$ to rise to a moderate degree, and this, in turn, would cause $[Ca]_i$ to rise (due to the Na/Ca exchanger) and $[Ca]_{SR}$ to rise. Thus, there can be greater Ca^{++} release from the SR during excitation, and causing a greater contraction. This mechanism is depicted in Figure 8. Also diagrammed is the fact that the rise in taurine intracellularly may stimulate $I_{Ca(L)}$ (at low $[Ca]_o$ or $[Ca]_i$) and $I_{Ca(T)}$, both of which would elevate $[Ca]_{SR}$ and increase contraction. The decrease in $I_{K(del)}$ produced by taurine at the physiological level of $[Ca]_i$ (e.g., $pCa_i = 7$) would increase the duration of the action potential (APD_{50}), and thereby prolong the Ca^{++} influx, resulting in greater contractile force (Fig. 8). If $I_{Na(f)}$ and $I_{Na(s)}$ are stimulated in some cells at the lower taurine concentrations (bath), then the increase in Na^+ influx would act to elevate $[Na]_i$, and this would lead to elevated $[Ca]_i$ and stronger contraction (Fig. 8).

Taurine was also reported to (i) stimulate the inward background current activated by isoproterenol in heart cells (rabbit, guinea pig); (ii) inhibit the ATP-sensitive K^+ channel (K_{ATP} and $I_{K(ATP)}$) by directly interacting with the channel; and (iii) stimulate the inward Na/Ca exchange current ($I_{Na/Ca}$) in atrial cells (rabbit) that was Ca^{2+}-release dependent[3]. If taurine causes Ca^{++} release from the SR, the effect of this action on twitch contraction of cardiac muscle is difficult to predict.

---

* Not shown are Ca^{++} sensitizing drugs that act to increase the sensitivity of the contractile proteins (e.g., actin, myosin) and troponin-C to Ca^{++}, i.e., the force vs. pCa curve of permeabilized muscles is shifted to the left.

† Phospholamban protein in the SR membrane is also phosphorylated by PK-A, which removes the inhibition of the Ca^{2+}-ATPase/pump, thereby stimulating Ca^{++} uptake into the SR, thus making more Ca^{++} available for release during the twitch contractions of the heart.

‡ This co-transport operates in both directions, and it has been suggested that one important role played by taurine in myocardial cells is to serve to counter any severe elevation of $[Na]_i$ that would be produced under adverse conditions (e.g. hypoxia, ischemia); that is co-transport of taurine + Na^+ outward could act to lower an elevated $[Na]_i$.

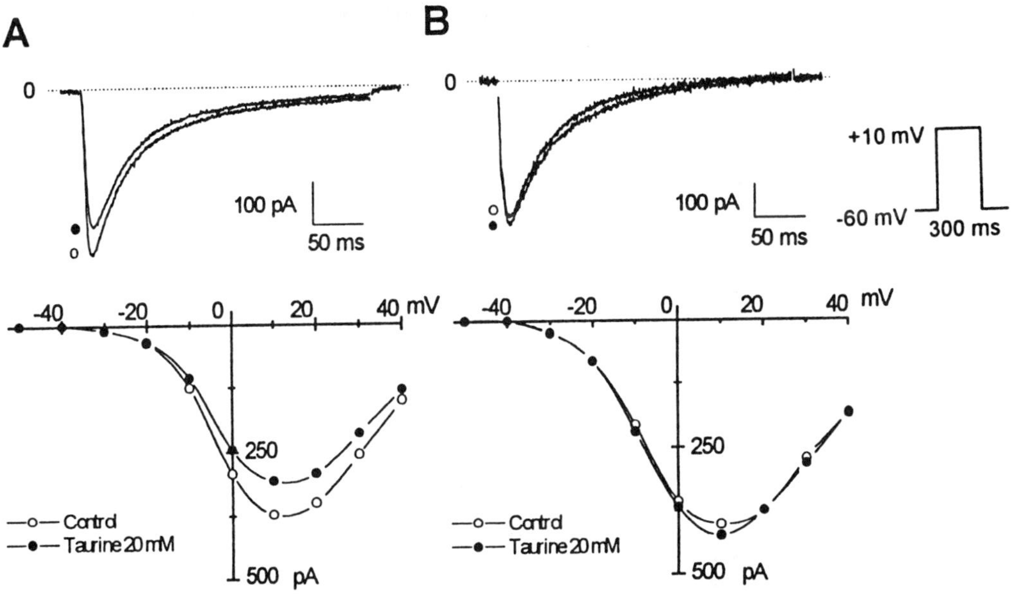

Figure 5. Effect of taurine (20 mM) on $I_{Ca(L)}$ in late pregnant rat (19-day) myometrial cells measured using the whole-cell voltage clamp. $I_{Ca(L)}$ was evoked by 300 ms depolarizing pulses from a HP of -60 mV every 20 s. *Upper panels*: Superimposed current traces before (open circles) and 10 min after (closed circles) taurine application into the bath. Test pulse was to +10 mV (*inset*). *Lower panels*: Current/voltage relationships of $I_{Ca(L)}$. Bath solution contained (in mM) 140 TEA, 2 $MgCl_2$, 10 glucose, 10 HEPES, and 3,4-aminopyridine (pH 7.35). Pipette solution contained (mM) 110 CsOH, 20 CsCl, 110 glutamic acid, 5 $MgCl_2$, 10 Hepes, 5 Na_2ATP, 10 EGTA (pH 7.2), $CaCl_2$ (sufficient to give pCa 10 or 7). **A.** Bath application of taurine slightly decreased $I_{Ca(L)}$ at pCa 10 in the pipette solution. **B.** At pCa 7, taurine did not affect the current.

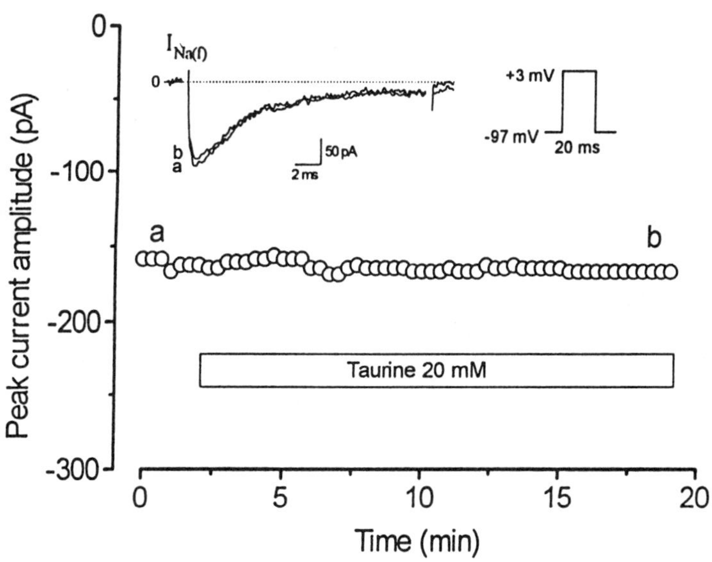

Figure 6. Lack of effect of taurine (20 mM) on $I_{Na(f)}$ of a human uterine leiomyosarcoma cell. $I_{Na(f)}$ was evoked by 20 ms depolarizing pulses to +3 mV from a HP of -97 mV (*inset*) every 20 s. Time-course plot of the peak $I_{Na(f)}$ and two superimposed current traces are shown (taken at time points **a** and **b**). Bath application of 20 mM taurine did not affect the current. Bath solution contained (in mM) 140 NaCl, 2 $NiCl_2$, 10 glucose, 10 HEPES, and 3, 4-aminopyridine (pH 7.35). Pipette solution contained (mM) 110 CsOH, 20 CsCl, 110 glutamic acid, 5 $MgCl_2$, 10 HEPES, 5 Na_2ATP, and 10 EGTA (pH 7.2) (pCa_i = 7).

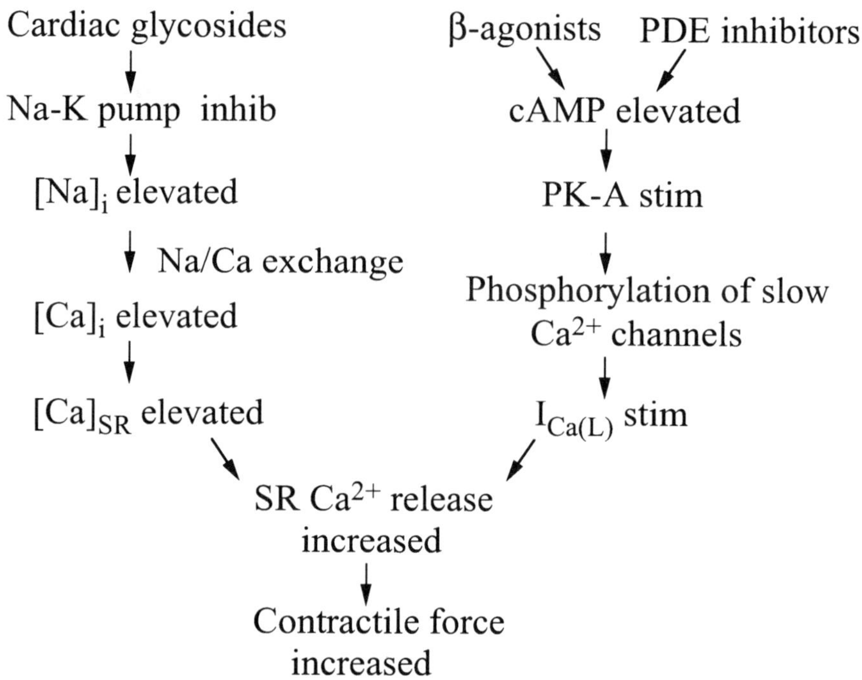

Figure 7. Positive inotropic drugs.

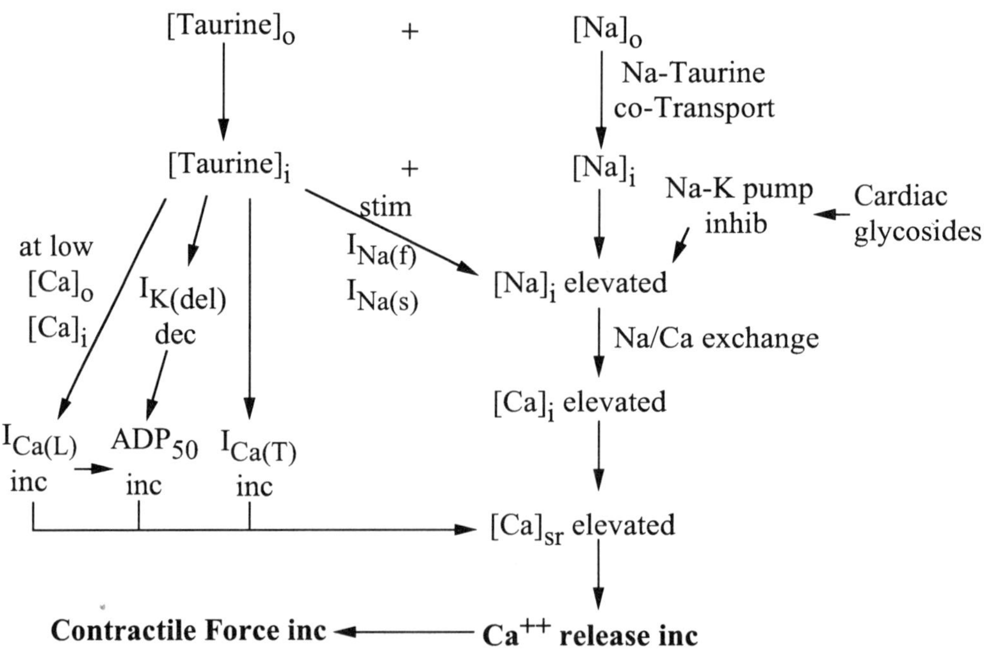

Figure 8. Possible positive inotropic mechanism of taurine.

Table 3. Effects of taurine (20 mM) on peak $I_{Ca(L)}$ and $I_{Na(f)}$ in uterine smooth muscle cells

	$[Ca]_i$	% Control
$I_{Ca(L)}$	pCa 7	106.3 ± 4.1 (6)
	PCa 10	83.1 ± 6.0* (6)
$I_{Na(f)}$	pCa 7	99.0 ± 3.2 (4)

$I_{Ca(L)}$ was evoked by C.P. to +10 mV from a H.P. of -60 mV (pregnant rat myometrial cells); $I_{Na(f)}$ was evoked by C.P. to +3 mV from a H.P. of -97 mV (human uterine leiomyosarcoma cells). Data are given as percent of control (M ± SE (n), where n = number of cells tested). *$p < 0.05$.

REFERENCES

1. Azuma, J. and Schaffer, S.W. 1993, Protective effect of taurine on the failing heart and its clinical application, in: "Ionic Channels and Effect of Taurine on the Heart", Noble, D and Earm, Y.E., eds., Kluwer, Boston, pp. 139-162, 1993.
2. Chapman, R.A., Suleiman, M.S. Rodrigo, G.C. Minazaki, K.K. Chatamra, K.R. Little, C.R. Mistry, D.K. AND Allen, T.J.A. 1993, Intracellular taurine, intracellular sodium and defense against cellular damage, in: "Ionic Channels and Effect of Taurine on the Heart", Noble, D. and Earm, Y.E. eds, Kluwer, Boston, pp. 73-91.
3. Earm, Y.E., Ho, W.K., So, I., Leem, C.H. and Han, J. 1993, Effect of taurine on the activation of background current in cardiac myocytes of the rabbit, in: "Ionic Channels and Effect of Taurine on the Heart", Noble, D. and Earm, Y.E., eds, Kluwer, Boston, pp. 119-138.
4. Earm, Y.E., Noble, D., Noble, S.J. and Spindler, A.J. 1990, Taurine activates a sodium-dependent inward current in isolated guinea-pig ventricular cells. *J. Physiol.* 417:56P.
5. Hamaguchi, T., Azuma, J. Yamauchi-Takihara, K. and Sperelakis, N. 1988, Reduction of doxorubicin-induced cardiotoxicity in mice by taurine. *Res. Commun. Chem. Pathol. & Pharmacol.* 59:21-30.
6. Inoue, Y. and Sperelakis, N. 1991, Gestational change in Na^+ and Ca^{2+} channel current densities in rat myometrial smooth muscle cells. *Am. J. Physiol.* 260:C658-C663.
7. Kusaka, M. and Sperelakis, N. 1994, Fast sodium currents induced by serum in human uterine leiomyosarcoma cells. *Am. J. Physiol./Cell.* 267:C1288-C1294.
8. Ohya, Y. and Sperelakis, N. 1989, Fast Na^+ and slow Ca^{2+} channels in single uterine muscle cells from pregnant rat. *Am. J. Physiol./Cell* 257: C408-C412.
9. Satoh, H. and Sperelakis, N. 1992, Taurine inhibition of fast Na^+ current in embryonic chick ventricular myocytes. *Eur. J. Pharmacol.* 218:83-89.
10. Satoh, H. and Sperelakis, N. 1993, Effects of taurine on Ca^{2+} currents in young embryonic chick cardiomyocytes. *Eur. J. Pharmacol.* 231:443-449.
11. Sawamura, A., Sada, H., Azuma, J., Kishimoto, S. and Sperelakis, N. 1990, Taurine modulates ion influx through cardiac Ca^{2+} channels. *Cell Calcium* 11:251-259.
12. Sawamura, A., Sperelakis, N. and Azuma, J. 1986Protective effect of taurine against decline of cardiac slow action potentials during hypoxia. *Eur. J. Pharmacol.* 120:235-239.
13. Sawamura, A., Sperelakis, N., Azuma, J. and Kishimoto, J. 1986, Effects of taurine on the electrical and mechanical activities of embryonic chick heart. *Can. J. Physiol. Pharmacol.* 64:649-655.
14. Schaffer, S.W., Kulakowski, E.C. and Kramer, J.H. 1982, Taurine transport by reconstituted membrane vesicles, in: "Taurine in Nutrition and Neurology," Huxtable, R.J. and Pasantes-Morales, H., eds. Plenum Press, New York, pp. 143-160.
15. Sperelakis, N. and Satoh, H. 1993, Taurine effects on ion channels of cardiac muscle, in: "Ionic Channels and Effect of Taurine on the Heart," Noble, D. and Earm, Y.E. eds. Kluwer Academic Publishers, pp. 93-118.
16. Sperelakis, N., Satoh, H. and Bkaily, G. 1992, Taurine effects on ionic currents in myocardial cells, in: "Taurine: Nutritional Value and Mechanisms of Action," Lombardini, B., Schaeffer S. and Azuma, J. eds. Plenum Press, New York, 129-143.

17. Sperelakis, N., Yamamoto, T., Bkaily, G., Sada, H., Sawamura, A. and Azuma, J. 1988, Taurine effects on action potentials and ionic currents in chick myocardial cells, in: "Taurine and the Heart," Iwata, H., Lombardini, J.B. and Segawa, T. eds. Kluwer Academic Publishers Chap. 1, pp. 1-19.
18. Yamauchi-Takihara, K., Azuma, J., Kishimoto, S., Onishi, S. and Sperelakis, N. 1988, Taurine prevention of calcium paradox-related damage in cardiac muscle: Its regulatory action on intracellular cation contents. *Biochem. Pharm.* 37:2651-2658.

30

ELECTROPHYSIOLOGICAL AND ELECTROPHARMACOLOGICAL ACTIONS OF TAURINE ON CARDIAC CELLS

Hiroyasu Satoh

Department of Pharmacology
Division of Molecular and Cellular Biology
Nara Medical University, Kashihara, Nara 634, Japan

INTRODUCTION

Myocardial levels of taurine fall in ischemia, hypoxia and cardiac failure, with the depletion correlated with the degree of mechanical dysfunction[17,19,21]. Taurine is abundant in myocardium (5.6 μmol/g wet weight in human). This suggests that taurine may be essential to maintain cardiac functions, and involved in cardioprotective functions.

Recently, many electrical and mechanical actions of taurine have been described[11,29,30,38,39]. Taurine produces a positive inotropic effect without affecting the resting potential, and modulates ionic currents across the cell membrane[14,25,40,43,44]. Taurine stimulates a TTX-insensitive Na^+ current, but depresses the fast (TTX-sensitive) Na^+ current[35,44]. In addition, Satoh and Sperelakis[36] and Satoh[28,30] have also reported that taurine exerts electrophysiological actions dependent on intracellular and extracellular Ca^{2+} concentrations ($[Ca]_i$ and $[Ca]_o$) in embryonic chick cardiomyocytes, guinea pig ventricular muscle, and rabbit sino-atrial (SA) nodal cells. Taurine inhibits the L-type Ca^{2+} current (I_{Ca}) and the delayed rectifier K^+ current (I_{Krec}) at pCa 7, whereas taurine enhanced both currents at pCa 10. Thus, the effects of taurine on cardiac cells are complex.

In addition, taurine has been demonstrated to abolish arrhythmias in guinea pig and rabbit hearts[26,29], and to attenuate the development of hypertension in SHR[22]. These results suggest that taurine possesses cardioprotective activity related to its ability to regulate Ca^{2+} movement in cytoplasm. In the present experiments, the modulations by taurine of the macroscopic and unitary L-type Ca^{2+} channel currents in adult cardiac cells were studied. Experiments were conducted using two $[Ca]_i$ (pCa 6 and 8) and $[Ca]_o$ (5.4 and 0.9 mM) levels.

Taurine 2, edited by Huxtable et al.
Plenum Press, New York, 1996

METHODS AND MATERIALS

Cell Preparation

Cells were prepared from tissue taken from the ventricle muscle of adult guinea pig hearts[23,35,36]. With animals under sodium pentobarbital anesthesia (30 mg/kg), the chest was opened and the aorta was cannulated in situ. The heart was dissected out and perfused with normal Tyrode solution on the Langendorff apparatus. After washout of blood, the heart was perfused with Ca^{2+}-free Tyrode solution. Spontaneous beating ceased. Then, the perfusate was switched to low-Ca^{2+} (30 to 60 μM) Tyrode solution containing 0.4 mg/ml collagenase (Type I, Sigma Chemical, St. Louis, MO, USA) for about 20 min. The heart was washed out by high-K^+ and low-Cl^- solution (KB solution), and was dissected with scissors. The temperature of all solutions was maintained at 36°C.

Whole-Cell Voltage-Clamp Experiments

Whole-cell voltage-clamp recordings were performed using a Axopatch patch-clamp amplifier (Axon Instruments, Burlingame, CA, USA) and standard techniques. Patch pipettes were fabricated using a two-stage puller, and had a resistance of 3-5 MΩ for whole-cell experiments. The series resistance error was less than 3-7 mV, and no compensation was used. Experiments were carried out at 36°C. The data were stored and analyzed on an IBM-AT microcomputer, using the PCLAMP analysis program (Axon Instruments). Current traces were filtered using a cut-off frequency of 1 KHz for plotting. All values are given as mean ± SEM. The differences of the mean values were analyzed by ANOVA for paired data, and $p < 0.05$ was considered significant.

The composition of Tyrode solution was (in mM): NaCl 137, KCl 5.4, $CaCl_2$ 1.8, $MgCl_2$ 1.0, NaH_2PO_4 0.3, glucose 5.0, and HEPES 5.0. The pH was adjusted to 7.4 with NaOH. To avoid the interference of other currents, 10 μM tetrodotoxin (TTX) was added to the external Tyrode solution to block the fast Na^+ current. Taurine (Sigma Chemical Co., St. Louis, MO, USA) was dissolved to the desired concentrations directly in the bath solution, and the solution was superfused. The pipette solution (intracellular) contained (in mM): L-aspartate 85; TEA-Cl 20, $MgCl_2$ 0.5, EGTA 10, Mg-ATP 10, Na_2-creatine phosphate 5, HEPES 10, and glucose 5.5 (pH 7.2). The concentration of Ca^{2+} in the pipette solution was determined according to the calculation of Fabiato and Fabiato[10] and the correction of Tsien and Rink[47].

Unitary Current Recording

Cell-attached patch-clamp was performed according to Hamil *et al.*[13]. The resistance of the patch electrode was 3-10 MΩ. The tip of the electrode was coated with Sylgard (KE106, Shin-etsu Chemical Co.). Unitary current recordings were carried out by PCLAMP system according to the whole-cell voltage-clamp experiments. Current traces were filtered with a cut-off frequency of 1 or 2 KHz for plotting (FV-625, NF, Tokyo).

The pipette solution contained (in mM): $BaCl_2$ 110, TEA-Cl 20, $MgCl_2$ 0.5, Mg-ATP 10, Na_2-creatine phosphate 5, BAPTA 10, and glucose 5.5 (pH 7.4). Bay K 8644 (100 nM) was added to pipette solution. All experiments were performed at 36°C.

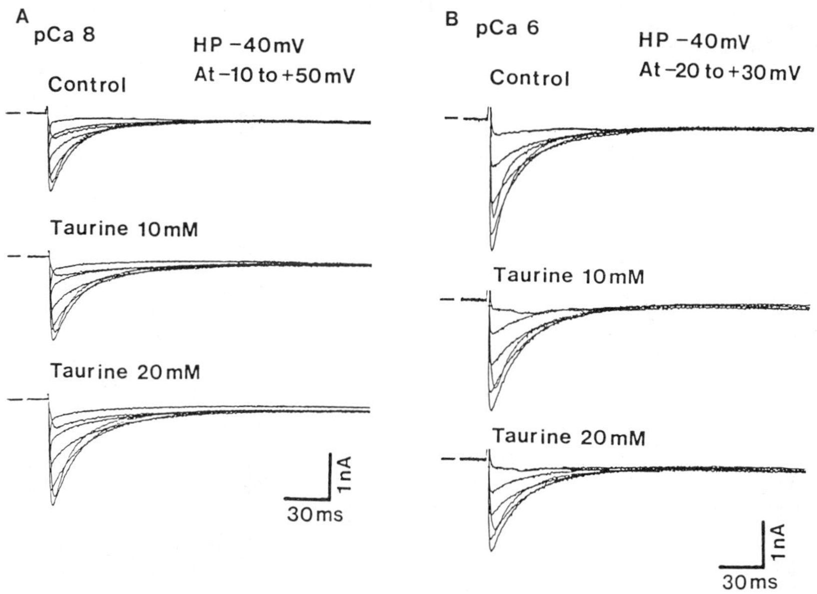

Figure 1. Modulation by taurine of the L-type Ca^{2+} current at different pCa levels in guinea pig ventricular myocytes. A: Current traces at pCa 8 in the absence and presence of 10 and 20 mM taurine. Test pulses for 200 ms were applied between -10 to +50 mV from a holding potential of -40 mV. B: Current traces at pCa 6 in the absence and presence of 10 and 20 mM taurine. Test pulses for 200 ms were applied between -20 to +30 mV from a holding potential of -40 mV. The short line at the left of the current records represents the zero current level.

RESULTS

Whole-Cell Voltage-Clamp Experiments

Effects of taurine on a macroscopic I_{Ca} current were examined using a whole-cell voltage-clamp mode. Taurine was cumulatively administered to the bath solution. The peak of I_{Ca} was modulated by application of taurine in a concentration-dependent manner, as shown in Fig. 1. At pCa 8, test pulses were applied between -10 to +50 mV from a holding potential of -40 mV. At pCa 8, taurine application increased I_{Ca} (Fig. 1A), whereas at pCa 6, taurine decreased I_{Ca} (Fig. 1B). After a washout, the current amplitude was irreversible.

Figure 2 shows the current-voltage relationships for I_{Ca}. The peak amplitudes of I_{Ca} at different pCa levels in normal Tyrode solution were different; -1.4 ± 0.2nA (at +10 mV, n=8) at pCa 8 and -2.7 ± 0.3 nA (at 0 mV, n=8) at pCa 6. When $[Ca]_i$ was pCa 8, I_{Ca} at +10 mV was enhanced by $10.1 \pm 3.2\%$ (n = 7, $p < 0.05$) at 10 mM taurine, and by $41.7 \pm 2.1\%$ (n=7, $p < 0.001$) at 20 mM taurine (Fig. 2A). In contrast, when $[Ca]_i$ was pCa 6, addition of 10 and 20 mM taurine reduced the Ca^{2+} current (I_{Ca}) at 0 mV by $14.4 \pm 2.0\%$ (n=8, $p < 0.01$) and $31.5 \pm 2.2\%$ (n = 8, $p < 0.001$), respectively (Fig. 2B).

Time constants of the inactivation phase of the I_{Ca} current are summarized in Table 1. The inactivation is composed of two exponentials. Fast and slow components (τ_f and τ_s) in normal Tyrode solution were 11.6 ± 1.1 ms and 23.3 ± 2.5 ms at pCa 8 (n=8), and 8.2 ± 2.1 ms and 25.3 ± 2.4 ms at pCa 6 (n=7), respectively. Taurine increased both components at low and high pCa. At pCa 8 taurine (10 and 20 mM) increased τ_f by $1.7 \pm 1.4\%$ (n=8, $p > 0.05$) and by $56.9 \pm 5.1\%$ (n=8, $p < 0.001$), and τ_s by $9.4 \pm 3.2\%$ (n=8, $p > 0.05$) and by 24.0

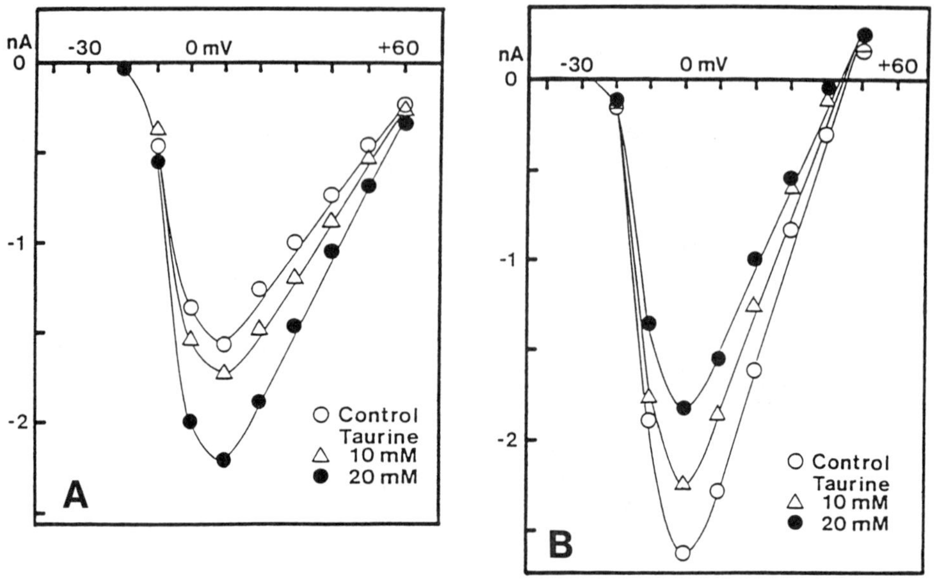

Figure 2. Current-voltage relationships for the Ca^{2+} current at different pCa levels. A: Current-voltage (I-V) curves at pCa 8. B: I-V curves at pCa 6. Symbols used are control (open circles), 10 mM taurine (triangles) and 20 mM taurine (filled circles).

± 3.4% (n=8, $p < 0.01$), respectively. At pCa 6, taurine (10 and 20mM) also increased τ_f by 140.2 ± 5.5% (n=7, $p < 0.001$) and by 204.8 ± 6.7% (n=7, $p < 0.001$), and τ_s by 41.1 ± 5.3% (n=7, $p < 0.01$) and by 72.3± 5.1% (n=7, $p < 0.001$), respectively.

Cell-Attached Patch-Clamp Experiments

Effects of taurine on a unitary Ba^{2+} channel current were examined using a cell-attached patch-clamp mode. To facilitate channel activity, Ca^{2+} in the pipette solution was replaced by Ba^{2+} (110 mM) and Bay K 8644 (100 nM) was added to the pipette solution. The channel conductance was 20.2 ± 1.7 pS (n=6) at 0.9 mM $[Ca]_o$, and 18.8 ± 1.3 pS (n=6) at 5.4 mM $[Ca]_o$. Taurine did not affect the channel conductances at either 0.9 or 5.4 mM $[Ca]_o$.

Table 1. Time constants of the inactivation phase of the L-type Ca^{2+} current in the absence and presence of taurine

	τ_f (ms)	τ_s (ms)
pCa 8[a]		
Control	11.6 ± 1.1	23.3 ± 2.5
Taurine 10 mM	11.8 ± 1.3	25.5 ± 2.0
Taurine 20 mM	18.2 ± 1.4***	28.9 ± 2.6**
pCa 6[b]		
Control	8.2 ± 2.1	25.3 ± 2.4
Taurine 10 mM	19.7 ± 2.2***	35.7 ± 2.7**
Taurine 20 mM	25.0 ± 2.1***	43.6 ± 2.5***

Data are means ± SEM of [a]8 or [b]7 experiments. **$p < 0.01$, ***$p < 0.001$, with respect to control value.

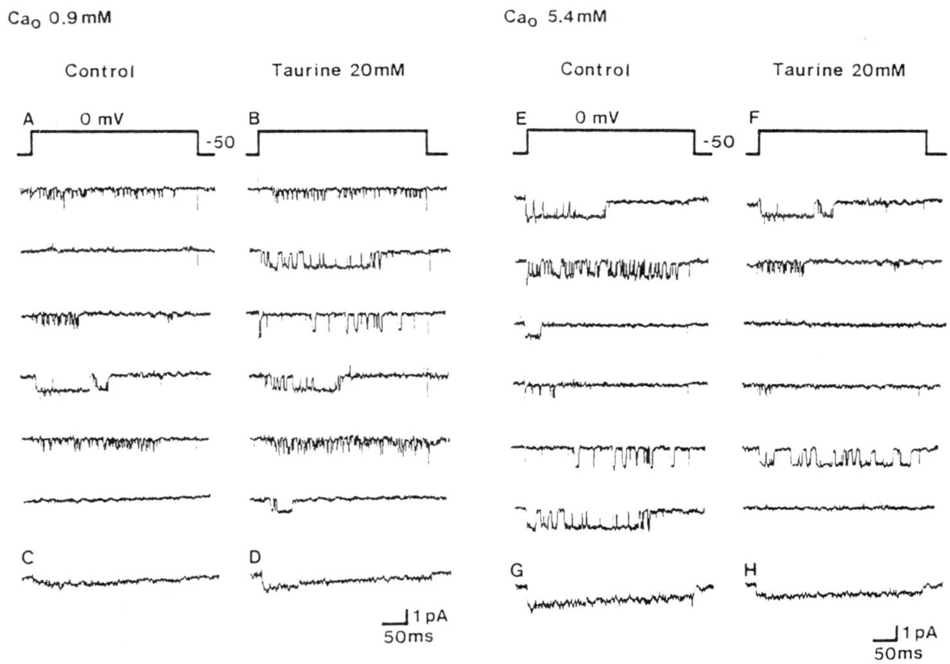

Figure 3. Effects of taurine on the unitary Ba^{2+} current at different $[Ca]_o$ levels. A test pulse for 300 ms was applied to 0 mV from a holding potential of -50 mV, using cell-attached mode. Ba^{2+} (110 mM), TEA (20 mM) and Bay K 8644 (100 nM) were added to intracellular pipette solution. A-B: Single channel currents in control and in 20 mM taurine at 0.9 mM $[Ca]_o$. C-D: Average current traces from 45 sweeps in control and from 40 sweeps in taurine 20mM. E-F: Single channel currents at 5.4 mM $[Ca]_o$ in control and in 20 mM taurine. G-H: Averaged current traces from 30 sweeps in control and from 35 sweeps in 20 mM taurine.

Taurine (20 mM) stimulated channel opening at 0.9 mM $[Ca]_o$ (Fig. 3A and B). Test pulse for 300 ms was applied to 0 mV from -50mV of a holding potential. The averaged current was increased by taurine, and the differences between the absence and presence of 20 mM taurine were remarkable, as shown in Fig. 3C and D. At 5.4 mM $[Ca]_o$, taurine (20 mM) inhibited the channel opening (Fig. 3E and F), and decreased the averaged current (Fig. 3G and H).

The open probability (P_o) of unitary Ba^{2+} current during each sweep at different $[Ca]_o$ levels in control and in 20 mM taurine is shown on Fig. 4. The averaged open probabilities in normal Tyrode solution were 0.21 ± 0.04 for 40 sweeps at 0.9 mM $[Ca]_o$ (Fig. 4A), and 0.63 ± 0.06 for 40 sweeps at 5.4 mM $[Ca]_o$ (Fig. 4C). The P_o was increased by 20 mM taurine at 0.9 mM $[Ca]_o$, and was decreased at 5.4 mM $[Ca]_o$ (Fig. 4B and D). Taurine at 10 and 20 mM increased it (for 40 to 45 sweeps) to 0.32 ± 0.06 (n=5, $p < 0.001$) and 0.48 ± 0.07 (n=4, $p < 0.01$) at 0.9 mM $[Ca]_o$, respectively. In contrast, taurine at 10 and 20 mM decreased the open probability (40 to 45 sweeps) to 0.49 ± 0.07 (n=5, $p < 0.01$) and 0.39 ± 0.09 (n=5, $p < 0.01$) at 5.4 mM $[Ca]_o$, respectively.

Figure 4 shows the histograms of dwell (open and closed) times at two $[Ca]_o$ levels. Both open and closed times were fitted by two exponential components. At high $[Ca]_o$, the time constants (τ_{of} and τ_{os}) of the open time were slightly increased, whereas those (τ_{cf} and τ_{cs}) of the closed time were decreased. In the absence of taurine, at 0.9 mM $[Ca]_o$ (n=4), τ_{of} and τ_{os} of the open time were 0.4 ± 0.1 ms and 7.0 ± 0.1 ms, and τ_{cf} and τ_{cs} of the closed time were 0.6 ± 0.1 ms and 22.3 ± 0.8 ms, respectively. At 5.4 mM $[Ca]_o$, τ_{of} and τ_{os} of the open time (n=5) were 0.5 ± 0.1 ms and 8.3 ± 0.5 ms, and τ_{cf} and τ_{cs} of the closed time (n=5) were 0.7 ± 0.2 ms and 16.4 ± 1.0 ms, respectively. Taurine (20 mM) increased τ_{of} and τ_{os} at

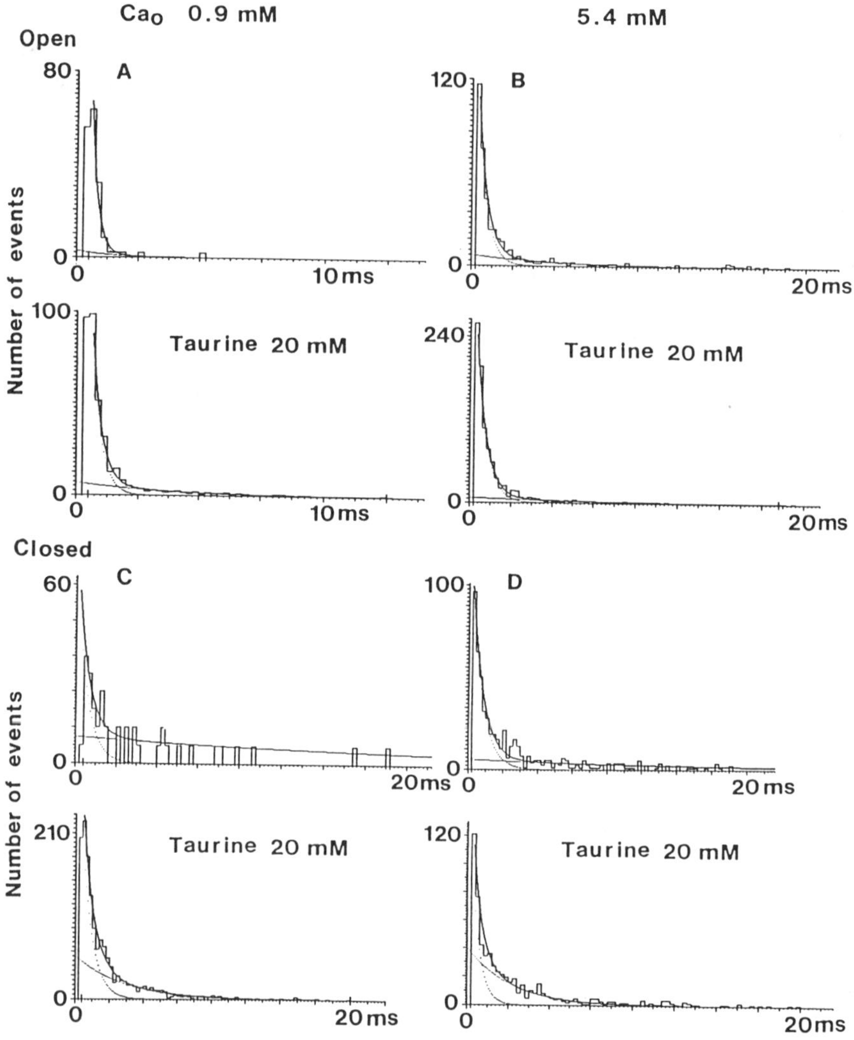

Figure 4. Open and closed (dwell) time histograms for the Ba^{2+} current at low and high $[Ca]_o$ levels. A: Open times at 0.9 mM $[Ca]_o$ in control and 20 mM taurine. B: Open times at 5.4 mM $[Ca]_o$ in control and 20 mM taurine. C-D: Closed times at 0.9 and 5.4 mM $[Ca]_o$ in control and 20 mM taurine. Two exponentials were fitted in both open and closed times for the Ba^{2+} current.

0.9 mM $[Ca]_o$, and did not affect τ_{of} but significantly slowed τ_{os} of the open time at 5.4 mM $[Ca]_o$ (Fig. 4A and B). On the other hand, taurine did not affect τ_{cf}, but significantly shortened only τ_{cs} of the closed time at both 0.9 and 5.4 mM $[Ca]_o$ (Fig. 4C and D). The effects of taurine on the time constant of dwell times for Ba^{2+} channel are summarized in Table 2.

DISCUSSION

Taurine, an amino acid largely obtained from the diet, exerts many important physiological functions[11,14,35,36,38-40,43]. The electrical and mechanical actions induced by

Table 2. Modulation by taurine of the time constants of dwell time histograms for the Ca^{2+} channel in guinea pig ventricular cells

Conditions		τ_{of} (ms)	τ_{os} (ms)
Open time $[Ca]_o$ 0.9 mM	Control	0.2 ± 0.1	1.5 ± 0.1
	Taurine	0.4 ± 0.1**	3.5 ± 0.2***
Open time $[Ca]_o$ 5.4 mM	Control	0.5 ± 0.1	4.8 ± 0.5
	Taurine	0.6 ± 0.1	6.9 ± 0.4**
		τ_{cf} (ms)	τ_{cs} (ms)
Closed time $[Ca]_o$ 0.9 mM	Control	0.6 ± 0.1	22.3 ± 0.8
	Taurine	0.7 ± 0.2	3.5 ± 0.2***
Closed time $[Ca]_o$ 5.4 mM	Control	0.7 ± 0.2	16.4 ± 1.0
	Taurine	0.9 ± 0.1*	6.9 ± 0.5***

Taurine concentration was 20 mM. Values represent mean ± SEM for 4-5 experiments. *$p < 0.05$, **$p < 0.01$, ***$p < 0.001$, with respect to control value.

taurine are independent of known regulators, such as cyclic AMP and cyclic GMP levels, Na^+/K^+-ATPase activity and calmodulin-dependent protein kinase action, although taurine may inhibit protein kinase C activity[20,41]. The present experiments were designed to examine the effects of taurine on the L-type Ca^{2+} channel current at low and high pCa and $[Ca]_o$ levels in single isolated guinea pig ventricular cardiomyocytes. Results were: (i) taurine stimulated I_{Ca} at low pCa, and depressed I_{Ca} at high pCa; (ii) taurine increased the time constants (τ_f and τ_s) of inactivation phase at both pCa 8 and 6; (iii) taurine decreased the open probability (P_o) of unitary Ba^{2+} current at high $[Ca]_o$, and increased it at low $[Ca]_o$, without affecting the channel conductance; and (iv) taurine at high $[Ca]_o$ slowed only the τ_{os} of open time and shortened τ_{cs} of the closed time. At low $[Ca]_o$, taurine increased both τ_{of} and τ_{os} of the open time and decreased τ_{cs} of the closed time. Thus, the effects induced by taurine in adult cardiomyocytes behaved in a Ca^{2+}-dependent manner, consistent with the actions on the I_{Ca} in embryonic chick cardiomyocytes[30,36] and in guinea pig cardiomyocytes by changing $[Ca]_o$ levels[38]. This is also similar to the findings that taurine stimulates I_{Krec} at pCa 10, whereas it depresses I_{Krec} at pCa 7[28]. However, the effects of taurine on the action potential duration (APD) are controversial. Taurine shortened APD at pCa 10, and had no effect at pCa 7 in embryonic chick cardiomyocytes[28]. In contrast, APD was prolonged at low $[Ca]_o$ and shortened at high $[Ca]_o$ in rabbit SA node cells and guinea pig ventricular muscles[29]. In our recent experiments using guinea pig ventricular cardiomyocytes (unpublished data), taurine shortened APD at pCa 8, whereas taurine prolonged APD at pCa 6. Thus, the actions of taurine on various ionic currents are not only complex and dependent on $[Ca]_i$ and $[Ca]_o$ levels, but also vary with different tissue isoforms.

Actions on the Inactivation Phase of I_{Ca}

The time decay of the inactivation phase for I_{Ca} is one of the regulatory factors of $[Ca]_i$ level. The inactivation is composed of two (fast and slow) exponential components (τ_f and τ_s)[15,18,36]. τ_f is $[Ca]_i$-dependent, and τ_s is voltage-dependent[5,9]. In the present experiments, τ_f and τ_s increased during exposure to taurine. This was consistent with results in embryonic chick cells[36]. Amino acids, such as taurine, alanine and glutamine, and Na^+ can cotransport across the membrane[4,45]. Taurine shifts the reversal potential of the fast Na^+ current (I_{Na}) in the hyperpolarizing direction and reduces cellular Na^+ content in embryonic chick cardiomyocytes[6,27,35]. This cotransport is considered to be the energy basis for a simple symport of 1 Na^+/1 taurine[7]. Change in cellular Na^+ content may alter $[Ca]_i$ level via Na^+-Ca^{2+} exchange. In the present experiments, however, $[Ca]_i$ was buffered with 10 mM of EGTA or BAPTA. No alteration of $[Ca]_i$ occurs, and thus $[Ca]_i$ would not act on the ionic channels, although Ca^{2+} may alter the permeation properties by interacting with the external mouth of the channel. Therefore, the slowing of the inactivation induced by taurine is due to a direct action.

Unitary Ca^{2+} Channel Current

Taurine did not affect the channel conductance, although the conductance was changed by different $[Ca]_o$ levels due to a shift of the driving force. During exposure to taurine, the P_o increased at low $[Ca]_o$, but decreased at high $[Ca]_o$.

Not only the closed time but also the open time were fitted with two exponentials. Both τ_{of} and τ_{os} for the open time were increased, and τ_{cs} of the closed time was decreased, by taurine application at low $[Ca]_o$. The enhancement of I_{Ca} by taurine at low pCa is due to increased open time and decreased closed time of the Ca^{2+} channel. On the other hand, at high $[Ca]_o$, τ_{of} of the open time was almost unaffected. τ_{os} of the open time was increased, but τ_{cs} of the closed time was decreased. At high pCa, taurine inhibited I_{Ca}. The decrease in closed time and the increase in open time of Ca^{2+} channel could not explain the I_{Ca} inhibition. Therefore, these findings may suggest that the modulation by taurine of the Ca^{2+} channel is mainly responsible for the changes in the open probability.

Cardioprotective Actions

Our recent investigations show that taurine possesses potent cardioprotective actions in embryonic chick cardiomyocytes, rabbit SA nodal cells, and guinea pig ventricular myocytes[24,25,27-30,35,36,42,43]. Taurine enhances I_{Ca} at low pCa, and inhibits I_{Ca} at high pCa, suggesting that one of the actions of taurine is to normalize Ca^{2+} movement through the sarcolemma, dampening the effects of variation in perfusate Ca^{2+} concentration. The results indicate that taurine acts on I_{Ca} in a manner to keep $[Ca]_i$ levels relatively constant, and consequently taurine plays an important role in maintaining cell viability and cardiac functions.

Simultaneously, the findings that taurine inhibits I_{Na} (TTX-sensitive) current[35] may be one way to protect the heart against arrhythmias. Also, a decline in cellular Na^+ content may decrease $[Ca]_i$ level via Na^+-Ca^{2+} exchange. Thus, the resultant decline in $[Ca]_i$ induced by taurine should lead to a reduction in cellular Ca^{2+} overload during ischemia. Furthermore, taurine abolished arrhythmias due to a triggered activity[29,30]. Taurine promotes the sudden emergence of a dormant T-type Ca^{2+} current[36]. T-type Ca^{2+} current contributes to the generation of pacemaker activity during diastole[2,3], and automaticity is depressed under Ca^{2+} overload[34,37]. Thus, the stimulation of T-type current may also be one of the protective actions of taurine. But recent investigations reveal no contribution of T-type Ca^{2+} current to the generation of pacemaker depolarization[12,31]. High pCa itself shortened APD, whereas low pCa itself prolonged APD. In embryonic chick cardiomyocytes, taurine decreased the prolonged APD at low pCa[30]. Extreme prolongation of APD may elicit arrhythmias. Thus, the regulation by taurine of APD may also be one of the protective actions, since alteration of APD can regulate $[Ca]_i$ level[33]. These results indicate that the effects of high and low $[Ca]_i$ and $[Ca]_o$ on electrophysiological functions are antagonized by taurine application, indicative of a cardioprotective action of taurine.

SUMMARY AND CONCLUSION

Effects of taurine on the L-type Ca^{2+} channel in isolated guinea pig ventricular cardiomyocytes were examined at different Ca^{2+} concentrations, using a whole-cell and

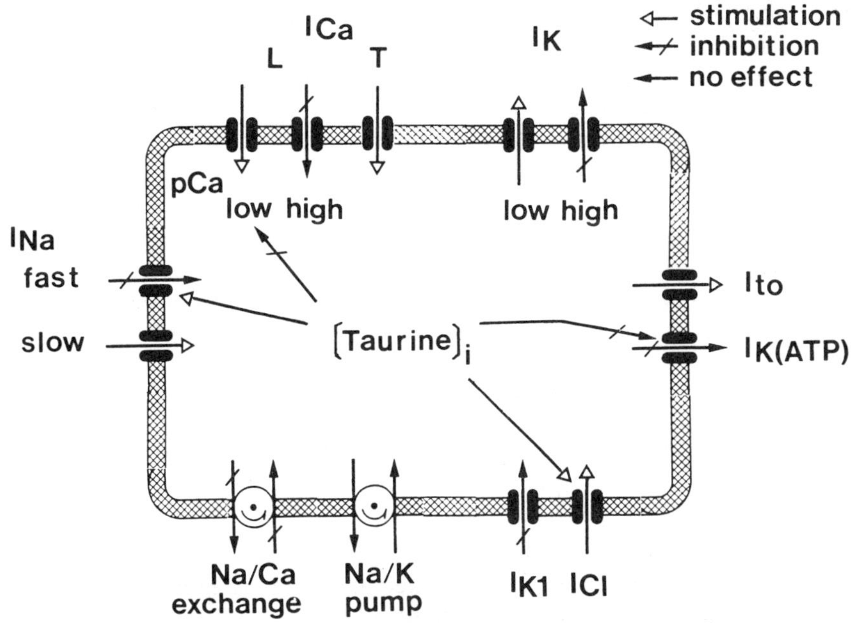

Figure 5. Summary for actions of taurine on the ionic channels in cardiac cells. The fast Na^+ current (I_{Na}) is depressed at 10 and 20 mM taurine, but is stimulated at low concentrations (1 to 5 mM). Intracellular application of taurine (10 and 20 mM) stimulates I_{Na}. In constast, the slow Na^+ (TTX-insensitive) current is stimulated by taurine application. L-type Ca^{2+} current (I_{Ca}) is increased at low pCa, but decreased at high pCa. Intracellular application of taurine inhibits I_{Ca} even at low pCa. Taurine stimulates T-type Ca^{2+} current. Delayed rectifier K^+ current is also stimulated at low pCa, but depressed at high pCa. Transient outward current (I_{to}) is enhanced during exposure to taurine. Taurine inhibits an ATP-sensitive K^+ current ($I_{K(ATP)}$), and also depresses the $I_{K(ATP)}$ channel by an inside-out patch-clamp mode. Intracellular and extracellular taurine enhances a Cl^- current (I_{Cl}), and inhibits an inwardly rectifying K^+ current (I_{K1}). Na^+/K^+ pump current is unaffected. Na^+/Ca^{2+} exchange current is inhibited or unaffected (still unclear).

cell-attached voltage-clamp mode. When $[Ca]_i$ was pCa 6, addition of 10 or 20 mM taurine to the bath solution reduced I_{Ca}. In contrast, when $[Ca]_i$ was pCa 8, I_{Ca} at +10 mV was enhanced. Taurine increased the time constants (τ_f and τ_s) of the inactivation phase of I_{Ca} current at both pCa 8 and 6. In cell-attached voltage-clamp experiments, taurine (20 mM) decreased the open probability of unitary Ba^{2+} current at 5.4 mM $[Ca]_o$, whereas taurine increased it at 0.9 mM $[Ca]_o$. Taurine did not affect the channel conductance. In addition, taurine (20 mM) increased the time constants (τ_{of} and τ_{os}) of the open time, and decreased τ_{cs} of the closed time at 0.9 mM $[Ca]_o$. At 5.4 mM $[Ca]_o$, the τ_{os} of open was increased and τ_{cs} decreased. τ_{of} and τ_{cf} were unaffected. These results indicate that taurine modulates the L-type Ca^{2+} channel opening dependent on $[Ca]_i$ and $[Ca]_o$, thereby resulting in regulation of the cell function to normalize $[Ca]_i$ level.

From our previous experiments, Figure 5 represents a summary for the actions of taurine on the ionic currents of cardiomyocytes.

Ca^{2+} and Na^+ Currents

Taurine inhibits L-type I_{Ca} at high $[Ca]_i$ and $[Ca]_o$, and enhances I_{Ca} at low $[Ca]_i$ and $[Ca]_o$ in embryonic chick and guinea pig cardiomyocytes. Intracellular application of taurine inhibited I_{Ca} even at low pCa. In contrast, T-type Ca^{2+} current was stimulated by taurine.

TTX-sensitive fast Na^+ current (I_{Na}) was enhanced by low concentrations of taurine (1-5 mM), but was inhibited by high concentrations of taurine (10-20 mM). Intracellular application of taurine at 10 and 20 mM enhanced I_{Na}. Young heart cells and tumor cells exhibit slow Na^+ current, TTX-insensitive and Mn^{2+}-sensitive. Taurine stimulates the slow Na^+ current in young (3-day-old) embryonic chick cells[43].

K^+ Currents

Effects of taurine on the delayed rectifier K^+ current (I_{Krec}) are also $[Ca]_i$-dependent. At high pCa, taurine inhibited I_{Krec}, and at low pCa, enhanced I_{Krec}[28]. Inwardly rectifying K^+ current (I_{K1}) and ATP-sensitive K^+ current ($I_{K(ATP)}$) in guinea pig ventricular myocytes were decreased by taurine[8,32]. Also, intracellular application of taurine inhibits I_{KATP}. In addition, taurine enhanced a transient outward current (I_{to}) in 10-day-old embryonic chick cardiomyocytes.

Pump and Exchange Currents

Taurine had no effect on Na^+/K^+-pump current[1,16,48], but Na^+/Ca^{2+} exchange current has been reported to decrease or not to be affected in guinea pig cardiomyocytes[8]. Finally, taurine enhances Cl^- current in hippocampus (but not in heart muscles)[46] and a background current in rabbit atrial myocytes[8].

Although many of the actions of taurine on cardiomyocytes still remain unclear, further studies should reveal possible therapeutic uses of taurine as a sulfur amino acid.

REFERENCES

1. Akera, T., Ku, D., and Brody, T.M. 1976, Alterations of ion movements as a mechanism of drug-induced arrhythmias and inotropic responses, in: "Taurine", Huxtable, R.J. and Barbeau, A. eds., Raven Press, New York, pp. 121-134.
2. Armstrong, C.M. and Matteson, D.R. 1985, Two distinct populations of calcium channels in a clonal line of pituitary cells, *Science*, 227:65-67.
3. Bean, B.P. 1985, Two kinds of calcium channels in canine atrial cells, *J.Gen.Physiol.* 86:1-30.
4. Benyajati, S. and Bay, S.M. 1992, Sodium-taurine cotransport in reptilian renal brush-border membrane vesicles, *Pflugers Arch.* 421:168-175.
5. Carbone, E. and Swandulla, D. 1989, Neuronal calcium channels: kinetics, blockade and modulation, *Prog.Biophys.Mol.Biol.* 54:31-58.
6. Chapman, R.A., Suleiman, M.-S., and Earm, Y.E. 1993, Taurine and the heart, *Cardiovascular Research*, 27:358-363.
7. Christensen, H.N. and Kilberg, M.S. 1987, Amino acid transport across the plasma membrane: role of regulation in interorgan flows, in: "Amino Acid Transport in Animal Cells," Yudilevich, D.L. and Boyd, C.A.R. eds., Manchester Univ. Press, Manchester, pp. 10-45.
8. Earm, Y.E., Ho, W.K., So, I.S., and Leem, C.H. 1992, Effect of taurine on the activation of background current in cardiac myocytes, in: "Cardiac Ion Channels and Effects of Taurine on the Heart," Earm, Y.E. and Noble, D. eds., Kluwer Academic Publishers, Boston,
9. Eckert, R. and Chad, J.E. 1984, Inactivation of Ca channels, *Prog.Biophys.Mol.Biol.* 44:215-267.
10. Fabiato, A. and Fabiato, F. 1979, Calculator programs for computing the composition of the solutions containing multiple metals and ligands used for experiments in skinned muscle cells, *J.Physiol.Paris*, 75:463-505.
11. Franconi, F., Martini, F., Stendardi, I., Matucci, R., Zilletti, L., and Giotti, A. 1982, Effect of taurine on calcium levels and contractility in guinea pig ventricular strips, *Biochem.Pharmacol.* 31:3181-3186.
12. Hagiwara, N., Irisawa, H., and Kameyama, M. 1988, Contribution of two types of calcium currents to the pacemaker potentials of rabbit sino-atrial node cells, *J.Physiol.(Lond)*, 395:233-253.

13. Hamil, O.P., Marty, A., Neher, E., Sakmann, B., and Sigworth, J. 1981, Improved patch-clamp techniques for high-resolution current recording from cells and cell-free membrane patches, *Pfluger Arch.* 391:85-100.
14. Huxtable, R.J. and Sebring, L.A. 1983, Cardiovascular actions of taurine, in: "Sulfur Amino Acids: Biochemical and Clinical Aspects," Kuriyama, K., Huxtable, R.J. and Iwata, H. eds., Alan R. Liss, New York, pp. 5-38.
15. Kass, R.S. and Sanguinetti, M.C. 1984, Inactivation of calcium channel current in the calf cardiac Purkinje fiber. Evidence for voltage- and calcium-mediated mechanisms, *J.Gen.Physiol.* 84:705-726.
16. Khatter, J.C., Soni, P.L., Hoeschen, R.J., Alto, L.E., and Dhalla, N.S. 1981, Subcellular effects of taurine on guinea pig heart, in: "The Effects of Taurine on Excitable Tissues," Schaffer, S.W., Baskin, S.I. and Kocsis, J.J. eds., Spectrum Publications, New York, pp. 281-294.
17. Kramer, J.H., Chovan, J.P., and Schaffer, S.W. 1981, Effect of taurine on calcium paradox and ischemic heart failure, *Am.J.Physiol.* 240:H238-H246.
18. Lee, K.S., Marban, E., and Tsien, R.W. 1985, Inactivation of calcium channels in mammalian heart cell: joint dependence on membrance potential and intracellular calcium, *J.Physiol.* 364:395-411.
19. Lombardini, J.B. 1980, Effects of ischemia on taurine levels, in: "Natural Sulfur Compounds," Cavallini, D., Gaull, G.T. and Zappia, V. eds., Plenum Press New York, pp. 295-306.
20. Lombardini, J.B. 1992, Effects of taurine on protein phosphorylation in mammalian tissues, *Adv.Exp.Med.Biol.* 315:309-318.
21. Lombardini, J.B. and Bricker, D.L. 1981, Effects of cardiovascular surgery on blood concentrations of taurine and amino acids, *Proc.Soc.Exp.Biol.Med.* 167:498-505.
22. Meldrum, M.J., Tu, R., Patterson, T.A., Dawson, R., and Petty, T. 1994, The effect of taurine on blood pressure, and urinary sodium, potassium and calcium excretion, in: "Taurine in Health and Disease," Huxtable, R. and Michalk, D.V. eds., Plenum Press, New York,
23. Satoh, H. 1992, Inhibition in L-type Ca^{2+} channel by stimulation of protein kinase C in isolated guinea-pig ventricular cardiomyocytes, *Gen.Pharmacol.* 23:1097-1102.
24. Satoh, H. 1994, Antagonistic actions of taurine on Ca^{2+} - induced responses in cardiac muscle cells, *Jpn.Heart J.* 35:457-458.
25. Satoh, H. 1994, Cardioprotective actions of taurine against intracellular and extracellular calcium-induced effects, in: "Taurine in Health and Disease," Huxtable, R.J. and Michalk, D. eds., Plenum Press, New York, pp. 181-196.
26. Satoh, H. 1994, Regulation of the action potential configuration by taurine in guinea-pig ventricular muscles, *Gen.Pharmacol.* 25:47-52.
27. Satoh, H. 1995, Taurine-induced hyperpolarizing shift of the reversal potential of the fast Na^+ current in embryonic chick cardiomyocytes, *Gen.Pharmacol.* 26:517-521.
28. Satoh, H. 1995, A dual action of taurine on the delayed rectifier K^+ current in embryonic chick cardiomyocytes, *Amino Acids*, 9:235-246.
29. Satoh, H. 1995, Electrophysiological actions of taurine on spontaneously beating rabbit sino-atrial nodal cells, *Jpn.J.Pharmacol.* 67:29-34.
30. Satoh, H. 1995, Regulation by taurine of the spontaneous activity in young embryonic chick cardiomyocytes, *J.Cardiovasc.Pharmacol.* 25:3-8.
31. Satoh, H. 1995, Role of T-type Ca^{2+} channel inhibitors in the pacemaker depolarization in rabbit sino-atrial nodal cells, *Gen.Pharmacol.* 26:581-587.
32. Satoh, H. 1996, Direct inhibition by taurine of the ATP-sensitive K^+ channel in guinea pig ventricular myocytes, *Gen.Pharmacol.* -in press.
33. Satoh, H. and Hashimoto, K. 1986, An electrophysiological study of amiloride on sino-atrial node cells and ventricular muscle of rabbit and dog, *Naunyn-Schmiedeberg's Arch.Pharmacol.* 333:83-90.
34. Satoh, H. and Hashimoto, K. 1988, On electrophysiological responses to phorbol esters which stimulate protein kinase C in rabbit sino-atrialnode cells, *Naunyn-Schmiedebergs Arch.Pharmacol.* 337:308-315.
35. Satoh, H. and Sperelakis, N. 1992, Taurine inhibition of fast Na^+ current in embryonic chick ventricular myocytes, *Eur.J.Pharmacol.* 218:83-89.
36. Satoh, H. and Sperelakis, N. 1993, Effects of taurine on Ca^{2+} currents in young embryonic chick cardiomyocytes, *Eur.J.Pharmacol.* 231:443-449.
37. Satoh, H., Tsuchida, K., and Hashimoto, K. 1989, Electrophysiological actions of A23187 and X-537A in spontaneously beating and in voltage-clamped rabbit sino-atrial node preparations, *Naunyn-Schmiedebergs Arch.Pharmacol.* 339:320-326.
38. Sawamura, A., Sada, H., Azuma, J., Kishimoto, S., and Sperelakis, N. 1990, Taurine modulates ion influx through cardiac Ca^{2+} channels, *Cell Calcium*, 11:251-259.

39. Schaffer, S.W., Allo, S., and Mozaffari, M. 1987, Potentiation of myocardial ischemic injury by drug-induced taurine depletion, in: "The Biology of Taurine: Methods and Mechanisms," Huxtable, R.J., Franconi, F. and Giotti, A. eds., Plenum Press, New York, pp. 151-158.
40. Schaffer, S.W., Kramer, J., and Chovan, J.P. 1980, Regulation of calcium homeostasis in the heart by taurine, *Federation Proceedings*, 39:2691-2694.
41. Segawa, T., Nomura, Y., and Shimazaki, I. 1985, Possible involvement of calmodulin in modulatory role of taurine in rat cerebral β-adrenergic neurons, in: "Taurine: Biological Actions and Clinical Perspectives," Oja, S.S., Ahtee, L., Kontro, P. and Paasonen, M.K. eds., Alan R. Liss, New York, pp. 321-341.
42. Sperelakis, N. and Satoh, H. 1993, Taurine effects on ion channels of cardiac muscle, in: "Ionic Channels and Effect of Taurine on the Heart," Noble, D. and Earn, Y. eds., Kluwer Academic Publishers, Boston, pp. 93-118.
43. Sperelakis, N., Satoh, H., and Bkaily, G. 1992, Taurine effects on ionic currents in myocardial cells, *Adv.Exp.Med.Biol.* 315:129-143.
44. Sperelakis, N., Yamamoto, T., Bkaily, G., Sada, H., Sawamura, A., and Azuma, J. 1989, Taurine effects on action potentials and ionic currents in chick myocardial cells, in: "Taurine and the Heart," Iwata, H., Lombardini, J.B. and Segawa, T. eds., Kluwer Academic Publishers, Boston, pp. 1-19.
45. Suleiman, M.-S., Rodrigo, G.C., and Chapman, R.A. 1992, Interdependence of intracellular taurine and sodium in guinea pig heart, *Cardiovasc.Res.* 26:897-905.
46. Taber, K.H., Lin, C.T., Liu, J.W., Thalmann, R.H., and Wu, J.Y. 1986, Taurine in hippocampus: localization and postsynaptic action, *Brain Res.* 386:113-121.
47. Tsien, R.Y. and Rink, T.J. 1980, Neutral carrier ion-selective microelectrodes for measurements of intracellular-free calcium, *Biochim.Biophys.Acta*, 559:623-638.
48. Welty, J.D. and Welty, M.C. 1981, Effects of taurine on subcellular calcium dynamics in the normal and cardiomyopathic hamster heart, in: "The Effects of Taurine on Excitable Tissues," Schaffer, S.W., Baskin, S.I. and Kocsis, J.J. eds., Spectrum Publications, New York, pp. 295-312.

EFFECT OF TAURINE ON ANGIOTENSIN II-INDUCED EXPRESSION OF IMMEDIATE EARLY RESPONSE GENES IN PRIMARY CULTURED NEONATAL RAT HEART CELLS

Kyoko Takahashi,[1] Hitoshi Hashimoto,[2] Akemichi Baba,[2]
Stephen W. Schaffer,[3] and Junichi Azuma[1]

[1] Department of Clinical Evaluation of Medicines and Therapeutics
[2] Department of Pharmacology
Osaka University, Faculty of Pharmaceutical Sciences, Osaka, Japan
[3] Department of Pharmacology
University of South Alabama, School of Medicine
Mobile, Alabama

INTRODUCTION

Taurine has been used successfully to treat patients with congestive heart failure[2]. Recently, it has been established that one of the important factors contributing to the development of cardiac hypertrophy and heart failure is angiotensin II[3,9,14]. Angiotensin II is a peptide hormone that plays a major role in the regulation of fluid balance, blood pressure and cardiac function. Its direct cardiac effects include modulation of Na^+, H^+ and Ca^{2+} transport, increasing heart rate and mechanical function, stimulation of protein synthesis and enhancing uptake and utilization of glucose[3,9].

The present study was designed to test the idea that the beneficial actions of taurine are linked to the modulation of angiotensin II activity. To test this, we examined the interaction between taurine and angiotensin II in two types of cultured heart cells (myocytes and cardiac fibroblasts) prepared from neonatal rats. The parameters examined were expression of immediate early gene (IEG) using Northern blot analysis, hyperplastic growth employing a hemocytometer and myocyte beating rate.

MATERIALS AND METHODS

Preparation of Cardiac Myocyte and Nonmyocyte Cultures

Preparation of primary cardiac myocyte and nonmyocyte cultures from 1-day old Wistar rats has been described by Sadoshima *et al.*[17]. For selective enrichment with cardiac

Taurine 2, edited by Huxtable et al.
Plenum Press, New York, 1996

myocytes, dissociated cells were preplated for 1 h, during which period the nonmyocytes attached readily to the bottom of the culture dish[13]. Non-adherent cells, mostly myocytes, were plated at a density of 2-5 x 10^6 cells/ml/dish. Bromodeoxyuridine (0.1 mM) was added during the first 2 days to prevent proliferation of nonmyocytes. This procedure yielded cultures with 90-95% myocytes, as assessed by microscopic observation of cell beating. Cells were kept in serum-containing culture medium for 48 h followed by serum-free medium. All experiments were performed 72 h after transferring the cells to serum-free medium.

Analysis of RNA

Total cellular RNA of cardiac cells was isolated using the acid guanidium thiocyanate-phenol-chloroform extraction method[4]. Northern blot analysis was performed according to the procedure described by Meinkoth and Carmichael[12]. The following probes were used for Northern blot analysis in this study: (i) c-*fos*: 2.1 kb EcoRI fragment of the human c-*fos* cDNA; (ii) c-*jun*: 2 kb EcoRI fragment of the rat c-*jun* cDNA.

Evaluation of Beating and Cell Number

The beating status of the cultured myocytes was monitored with an inverted phase-contrast microscope at 25°C. The beating rates of individual cells[20] were counted over periods of 20-60 s. Cells were detached with 1 ml of trypsin-EDTA and dissociated by trituration. Cell number was counted using a hemocytometer[17].

Statistics

Statistical significance was determined by Student's t-test. Values are expressed as mean ± SEM. Differences were considered significant when the calculated p value was less than 0.05.

RESULTS

The Direct Effect of Angiotensin II on Myocytes and Nonmyocytes

Monolayer sheets of myocytes beat synchronously at a mean beating rate of 20 beats/min. In the absence of exogenous stimuli, the rate remained constant for 2 h at 25°C in serum-free medium (Fig. 1). Exposure of the cells to angiotensin II (1 nM) significantly elevated the spontaneous beating rate within 15 min from 20 to 31 beats/min, a beating rate which was maintained for at least 2 h (Fig. 1).

In addition to modulating myocyte beating rate, angiotensin II induced IEG expression in both myocytes and nonmyocytes. Shown in Figure 2 is a temporal Northern blot analysis of c-*fos* and c-*jun* expression in cardiac cells following angiotensin II exposure. Maximal induction of both c-*fos* and c-*jun* occurred between 30 min and 1 h of angiotensin II exposure. As expected, the increase in myocyte and nonmyocyte IEG expression at 30 min was dependent on angiotensin II concentration over a range of 1 - 100 nM (data not shown).

Consistent with the IEG expression data, angiotensin II (1 and 10 nM) promoted hyperplasia of the nonmyocytes, increasing cell number by 75 - 109%. However, it caused no mitogenic effect on the myocytes (data not shown).

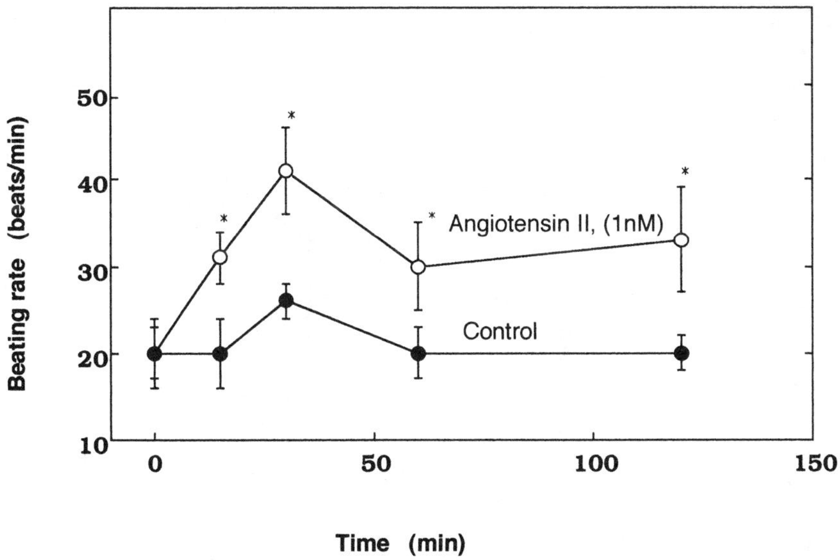

Figure 1. Time course of beating rate by cultured neonatal cardiac myocytes after addition of angiotensin II. Data represent mean ± SEM obtained from 11-20 experiments on 3 different cultures. Open circles: angiotensin II (1 nM); closed circle: control. *$p < 0.01$ compared to 0 min by paired t test.

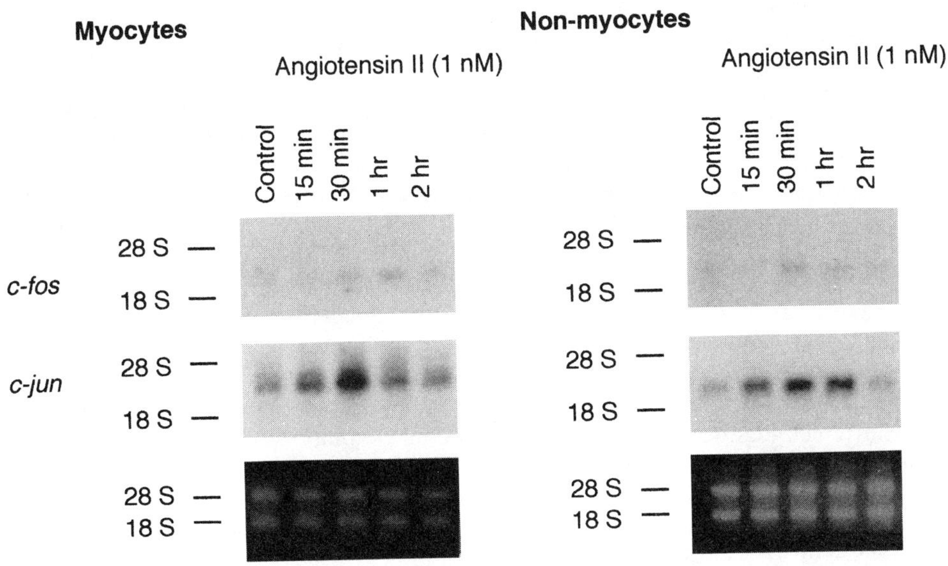

Figure 2. Angiotensin II-induced IEG expression of neonatal cardiac myocytes and nonmyocytes. Representative Northern blots of myocytes (left) and nonmyocytes (right) are shown. Cells were stimulated with angiotensin II (1 nM) for the indicated time. Ethidium bromide staining of 18S and 28S RNA served as a control (insert below the Northern blots) and showed that an equal amount of RNA was loaded in each lane. Similar results were obtained from two additional experiments.

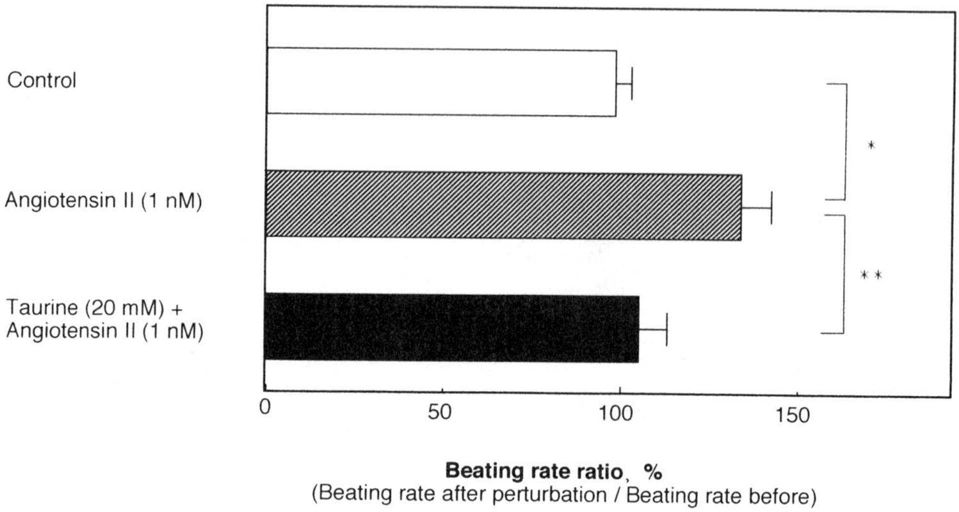

Figure 3. Effect of taurine on angiotensin II-induced changes in myocyte beating rate of neonatal rat heart. Cardiac myocytes were incubated with taurine (20 mM) and then stimulated with angiotensin II (1 nM) for 30 min in the presence of taurine. Data represent mean ± SEM obtained from 9 experiments on 3 cultures. Significant differences between indicated groups: $*p < 0.01$; $**p < 0.05$. Open bar: control; hatched bar: angiotensin II (1 nM); closed bar: taurine (20 mM) + angiotensin II (1 nM).

The Effect of Taurine on Angiotensin II-Mediated Cellular Responses

Figure 3 shows that taurine treatment altered angiotensin II-induced elevations in myocyte beating rate. After 30 min exposure, angiotensin II (1 nM) significantly increased the beating rate by about 35%. Taurine reversed this effect, decreasing the beating rate from 134% to 105%. Taurine alone had no effect on myocyte beating rate.

Cardiac cells incubated with taurine (20 mM) for 30 min and then stimulated with angiotensin II (1 nM) for 30 min in the presence of taurine exhibited reduced angiotensin II-induced IEG expression (Fig. 4). In the myocytes, taurine suppressed both c-*fos* and c-*jun* expression induced by angiotensin II. By comparison, taurine pretreatment for 30 min failed to alter angiotensin II-induced IEG expression in cultured nonmyocytes (Fig. 4). However, 24 h of taurine treatment attenuated angiotensin II-mediated stimulation of c-*jun* expression in the nonmyocytes (Fig. 5). Angiotensin II-induced c-*fos* expression in both nonmyocytes and myocytes was similarly suppressed by taurine treatment (data not shown). Taurine alone exhibited no effect on IEG expression in the two types of cardiac cells.

The effect of taurine on hyperplastic growth of nonmyocytes following angiotensin II treatment is shown in Figure 6. Angiotensin II (1 nM) caused a 60% increase in nonmyocyte cell number over 24 h. Taurine (20 mM) completely blocked this rise in cell number.

DISCUSSION

The development of heart failure is thought to proceed through several phases, one being cardiac hypertrophy. It has been established that most hypertrophic stimuli exhibit similar signal transduction pathways, in which $[Ca^{2+}]_i$ rises and protein kinase C is activated. The signal transduction pathway responsible for the hypertrophic effects of angiotensin II

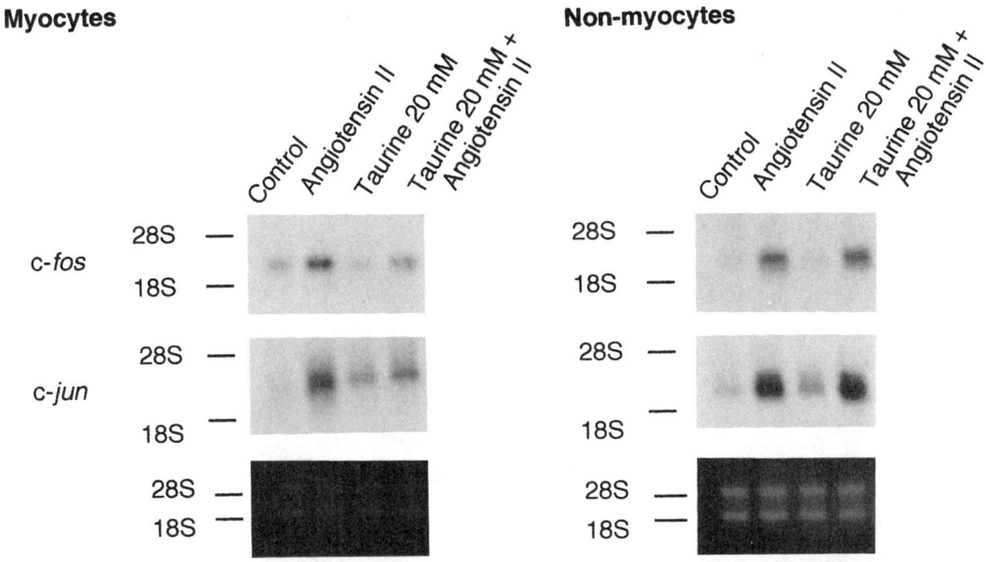

Figure 4. Effect of taurine (30 min-pretreatment) on angiotensin II-induced IEG expression of myocytes and nonmyocytes obtained from neonatal rat heart. Cardiac myocytes (left) and nonmyocytes (right) were incubated with taurine (20 mM) for 30 min and then stimulated with angiotensin II (1 nM) for 30 min in the presence of taurine. Ethidium bromide staining of 18S and 28S RNA (insert below the Northern blots) revealed that an equal amount of RNA was loaded in each lane. Similar results were obtained from two additional experiments.

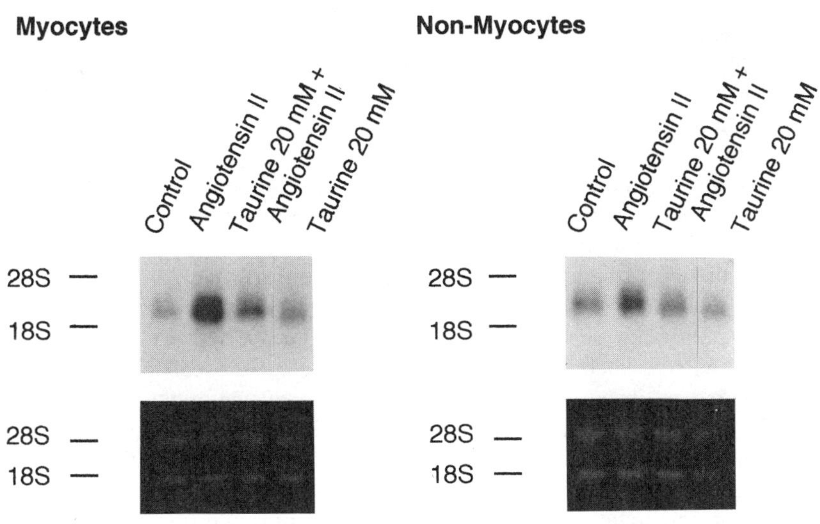

Figure 5. Effect of taurine (24 h-pretreatment) on angiotensin II-induced c-*jun* expression of myocytes and nonmyocytes prepared from neonatal rat heart. Cardiac myocytes (left) and nonmyocytes (right) were incubated with taurine (20 mM) for 24 h and then stimulated with angiotensin II (1 nM) for 30 min in the presence of taurine. Ethidium bromide staining 18S and 28S RNA (insert below the Northern blots) revealed that an equal amount of RNA was loaded in each lane. Similar results were obtained from two additional experiments.

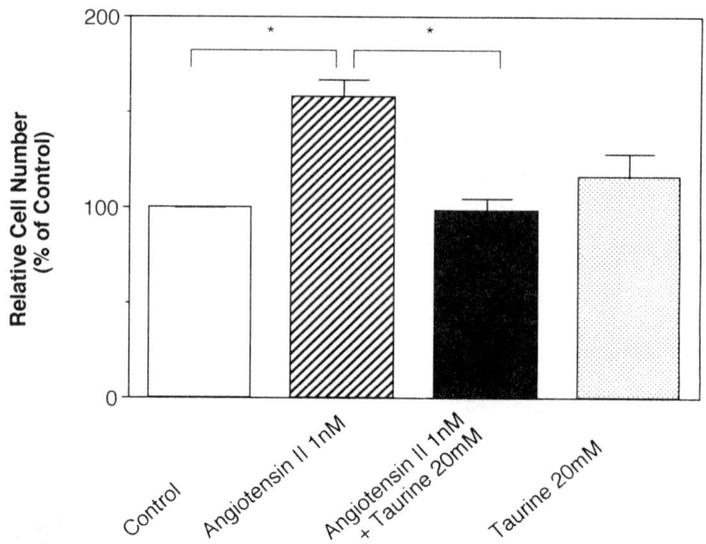

Figure 6. Effect of taurine on angiotensin II-induced hyperplastic growth of neonatal rat heart nonmyocytes. Cardiac nonmyocytes were incubated with taurine (20 mM) for 24 h and then stimulated with angiotensin II (1 nM) for 30 min in the presence of taurine. Data represent mean ± SEM obtained from 6-7 experiments on 3 cultures. Significant difference between indicated groups: $*p < 0.01$. Open bar: control; hatched bar: angiotensin II (1 nM): closed bar: taurine (20 mM) + angiotensin II (1 nM); dotted bar: taurine (20 mM).

begins with the interaction of angiotensin II with its receptor. The angiotensin II receptor is coupled through a G protein to a membrane bound phospholipase C isozyme[9], which is subsequently activated. The resulting products of phospholipase C action, IP_3 and diacylglycerol, are important second messengers in hypertrophic growth[1,9,15]. IP_3 promotes the release of Ca^{2+} from the sarcoplasmic reticulum, thereby increasing $[Ca^{2+}]_i$. The rise in $[Ca^{2+}]_i$ plays a potentiating role in the growth process by activating several signalling pathways[15,16]. Diacylglycerol serves as an activator of protein kinase C, which in turn promotes protein synthesis[1].

In this study, we have confirmed the effects of angiotensin II on: IEG expression; beating rate; and hyperplastic growth of cardiac cells. In our studies, angiotensin II induced c-*fos* and c-*jun* expression within 30 min in both cultured myocytes and nonmyocytes. Angiotensin II also increased the spontaneous beating rate of myocytes and promoted hyperplasia of the nonmyocytes, the latter being seen after 24 h.

We have previously reported that taurine exerts a beneficial effect in patients and animals with congestive heart failure[2,19]. However, the exact mechanism underlying this action of taurine has not been elucidated. Believing that taurine might function by altering angiotensin II activity, we examined the action of taurine on angiotensin II responsiveness in both cultured myocytes and nonmyocytes. The major findings of our study are as follows: First, exposure of the cells to taurine (20 mM) in the absence of angiotensin II had no effect on IEG expression, myocyte beating rate or hyperplastic growth of cultured cardiac cells. Secondly, cultured myocytes pretreated with 20 mM taurine had a significantly attenuated response to angiotensin II (1 nM). Following a 30 min pretreatment period with 20 mM taurine, the angiotensin II-induced positive chronotropic response was decreased by 30% while IEG expression was significantly suppressed. Thirdly, taurine pretreatment for 30 min failed to alter angiotensin II-induced IEG expression in cultured nonmyocytes, although the response was significantly attenuated after 24 h of taurine treatment. The longer period of taurine treatment also led to an inhibition in angiotensin II-mediated hyperplastic growth of nonmyocytes.

The mechanism underlying the attenuation of angiotensin II-induced actions by taurine remains an area of considerable interest. Several steps in the angiotensin II signal transduction pathway are logical sites of taurine action. First, taurine may interfere with the interaction of angiotensin II with its receptor. There is some precedence for this concept because taurine has been reported to interact with both the insulin receptor and the α-adrenergic receptor[6,11]. Secondly, the effects of taurine may be caused by alterations in $[Ca^{2+}]_i$, an important hypertrophic messenger. We have previously shown that taurine alters Ca^{2+} movement in cultured cardiac myocytes[21]. Thirdly, taurine may block the activation of protein kinase C by angiotensin II. It has been established that protein kinase C activation depends on membrane structure[5]. Since taurine modulates membrane structure by altering the phospholipid composition of the membrane[18], it could interfere with diacylglycerol-mediated activation of protein kinase C. Finally, taurine may directly inhibit phosphorylation of a key protein by protein kinase C. Lombardini and coworkers[8,10] have argued that many of the actions of taurine may involve inhibition of protein phosphorylation. Although the nature of the taurine-sensitive protein kinase has not been established, protein kinase C is a logical candidate[8].

The second question arising from the present study is: why does taurine pretreatment for 30 min fail to suppress angiotensin II-induced IEG expression in cultured nonmyocytes, but not in myocytes? We previously reported that at a dose of 20 mM, maximal taurine uptake by heart cells was reached by 15 min[20]. Thus, it is not surprising that most of the effects of taurine occur within 30 min. The basis for the delay in the response of nonmyocytes to taurine is therefore intriguing. One possibility is that the differences in temporal response between the myocyte and nonmyocyte relates to function and morphology. While myocytes in culture actively contract[7], nonmyocytes lack this characteristic. This may be important since $[Ca^{2+}]_i$ undergoes dramatic changes during the contraction-relaxation cycle. It is possible that the contracting cell may become more susceptible to the Ca^{2+}-modulating actions of taurine. Moreover, cell stretching of the contracting cell is known to initiate a series of reactions, which may alter the ability of taurine to regulate angiotensin II activity. An alternative hypothesis is that the temporal differences between the two cell types may reside with the protein kinase C step of the signal transduction pathway. Protein kinase C consists of a family of isozymes, whose distribution is cell-specific. Because each isozyme of the protein kinase C family has a unique substrate specificity, activator profile and intracellular localization, considerable flexibility is built into all protein kinase C-linked events. This flexibility may translate into different actions, or in the present study, differences in temporal response.

The present results indicate that taurine is an effective inhibitor of angiotensin II action. It is plausible that the beneficial effect of taurine in the treatment of heart failure could relate to its suppression of angiotensin II-mediated cellular responses.

REFERENCES

1. Allo S. N. Carl L.L. and Morgan H.E. 1992, Acceleration of growth of cultured cardiomyocytes and translocation of protein kinase C. *Am. J. Physiol.* 263:C319-C325.
2. Azuma J., Sawamura A., Awata N., Hasegawa H., Ogura K., Harada H., Ohta H., Yamauchi, K., Kishimoto S., Yamagami T., Ueda E. and Ishiyama T. 1983, Double-blind randomized crossover trial of taurine in congestive heart failure. *Curr. Therap. Res.,* 34:543-557.
3. Baker K.M., Booz G.W. and Dostal D.E. 1992, Cardiac actions of angiotensin II: Role of an intracardiac renin-angiotensin system. *Ann. Rev. Physiol.,* 54:227-241.
4. Chomczynski P. and Sacchi N. 1987, Single-step method of RNA isolation by acid guanidium thiocyanate-phenol-chloroform extraction, *Anal. Biochem.,* 162:156- 159.

5. Epand R.M. 1990, Relationship of phospholipid hexagonal phases to biological phenomena. *Biochem. Cell. Biol.*, 68:17-23.
6. Franconi F., Failli P., Bennardini F., Matucci R., Fazzini A., Stendardi I. and Giotti A. 1989, Taurine modulation of inotropism in guinea pig heart, in: "Taurine and the Heart", Iwata, H., Lombardini, J.B. and Segawa, T. eds., Kluwer Academic Publishers, Boston, Dordrecht & London, pp. 21-30.
7. Goshima K. and Tonomura Y. 1969, Synchronized beating of embryonic mouse myocardial cells mediated by FL cells in monolayer culture. *Exp.Cell.Res.* 56:387-392.
8. Li Y-P and Lombardini J.B. 1991, Taurine inhibits protein kinase C-catalyzed phosphorylation of specific protein in a rat cortical P_2 fraction. *J. Neurochem.* 56:1747-1753.
9. Lindpainter K. and Ganten D. 1991, The cardiac renin-angiotensin system. *Circ Res.*, 68:905-921.
10. Lombardini J.B. 1992, Effects of taurine on protein phosphorylation in mammalian tissues, in: "Taurine: Nutritional Value and Mechanism of Action", Lombardini, J.B., Schaffer, S.W. and Azuma, J. eds. Plenum Press, New York, pp. 309-318.
11. Maturo J. and Kulakowski E.C. 1987, Insulin-like activity of taurine, in: "The Biology of Taurine", Huxtable, R.J., Franconi F. and Giotti, A. eds., Plenum Press, New York, pp. 217-226.
12. Meinkoth J., and Carmichael G.G. 1984, Hybridization of nucleic acids immobilized on solid supports. *Anal. Chem.* 138:267-284.
13. Polinger I.S. 1970, Separation of cell types in embryonic heart cell cultures. *Exp. Cell. Res.*, 63:78-82.
14. Sadoshima J. and Izumo S. 1993, Molecular characterization of angiotensin II-induced hypertrophy of cardiac myocytes and hyperplasia of cardiac fibroblasts. *Circ. Res.*, 73:413-423.
15. Sadoshima J. and Izumo S. 1993, Signal transduction pathways of angiotensin II-induced c-*fos* gene expression in cardiac myocytes in vitro. *Circ. Res.*, 73:424-438.
16. Sadoshima J. and Izumo S. 1993, Mechanical stretch rapidly activates multiple signal transduction pathways in cardiac myocytes: potential involvement of an autocrine/paracrine mechanism. *The EMBO J.* 12:1681-1692.
17. Sadoshima J., Jahn L., Takahashi T., Kulik T.J. and Izumo S. 1992, Molecular characterization of the stretch-induced adaptation of cultured cardiac cells. *J. Biol. Chem.* 267:10551-10560.
18. Schaffer S.W., Azuma J. and Madura J.D. 1995, Mechanisms underlying taurine-mediated alterations in membrane function. *Amino Acids* 8:231-246.
19. Takihara K., Azuma J., Awata N., Ohta H., Hamagachi T., Sawamura A., Tanaka Y., Kishimoto S.and Sperelakis N. 1986, Beneficial effect of taurine in rabbits with chronic congestive heart failure. *Am. Heart J.* 112:1278-1284.
20. Takahashi K., Azuma J., Awata N., Sawamura A., Kishimoto S., Yamagami T., Kishi T., Harada H. and Schaffer S.W. 1988, Protective effect of taurine on the irregular beating pattern of cultured myocardial cells induced by high and low extracellular calcium ion. *J. Mol. Cell. Cardiol.* 20:397-403.
21. Takahashi K., Harada H., Schaffer S.W. and Azuma J. 1992, Effect of taurine on intracellular calcium dynamics of cultured myocardial cells during the calcium paradox, in: "Taurine; Nutritional Value and Mechanisms", Lombardini, J.B., Schaffer,S.W. and Azuma, J. eds., Plenum Press, New York, pp. 153-161.

32

INOTROPIC INTERVENTIONS IN THE ASSESSMENT OF MYOCARDIAL FAILURE ASSOCIATED WITH TAURINE DEFICIENCY IN DOMESTIC CATS

Mark J. Novotny and Patricia M. Hogan

Atlantic Veterinary College
University of Prince Edward Island
Charlottetown, Prince Edward Island, Canada, C1A 4P3

INTRODUCTION

Of the amino acids that exist freely in the cytoplasm of myocardial cells, taurine is in highest concentration. While the precise physiological function of taurine in the myocardium has not be defined, taurine may have antioxidant properties[8,9,29], modulating effects on myocardial calcium[23], or a positive inotropic effect[3,32]. In hearts isolated from taurine-depleted and control rats, Schaffer *et al.*[29] observed a significantly lower % recovery of mechanical performance in the depleted group than controls following ischemic-reperfusion injury. More recently, we[24,25] and others[17] have reported that depletion of myocardial taurine results in cardiac dysfunction characterized by decreased left ventricular (LV) contractility, suggesting that taurine is essential to the maintenance of normal myocardial contractile function. Because taurine is an essential dietary amino acid in the domestic cat, a taurine-deficient state can be achieved in this species through restriction of dietary taurine[14]. In the cat dietary taurine deficiency leads to a spectrum of changes in myocardial function associated with decreased contractility[24,25]. Dilated cardiomyopathy (DCM) appears to be one consequence of taurine deficiency in this species. However, not all cats maintained on taurine-deficient diets develop DCM[24,27,30].

In the present study the contractile response to changes in preload (end-diastolic pressure) was measured in hearts isolated from taurine-deficient cats that, under basal conditions, lacked significant changes in LV performance compared with hearts from taurine-supplemented cats. In addition, we studied the dose-response relationships between dobutamine and LV performance *in situ* in taurine-deficient cats that similarly, under basal conditions, lacked significant changes in LV pump function compared with taurine-supplemented cats. We now report that hearts from taurine-deficient cats: (i) have diminished ability to increase contractile function with increases in preload; and (ii) are less responsive to the increase in LV pump function that occurs in response to dobutamine challenge. Taurine

Taurine 2, edited by Huxtable et al.
Plenum Press, New York, 1996

replacement therapy restores both the basal functional deficit and the functional reserve in cats that develop taurine-deficient myocardial failure.

MATERIALS AND METHODS

Isolated Heart Studies

Twelve, colony-source, female domestic shorthair cats were used in isolated heart perfusion experiments following a feeding trial designed to create a taurine-deficient state. The cats ranged in age from 1-3 y. Prior to the onset of the feeding trial, cats had been maintained on dietary formulations that supported adequate plasma taurine concentrations. All cats were subjected to pre-study evaluation of cardiac function and structure as previously described[24]. All experiments were conducted according to guidelines of the "Guide to the Care and Use of Experimental Animals", Volumes 1 (2nd ed., 1993) and 2 (1984) of the Canadian Council on Animal Care.

At the onset of the feeding trial, the cats were divided into 2 groups. The first group (taurine-deficient; n=7) was placed on a commercially-prepared, taurine-deficient feline diet (<0.005% taurine; Results Brand Diet #F147, Bioserv, Frenchtown, NJ). The second group (taurine-supplemented; n=5) was placed on a commercially-prepared, taurine-supplemented feline diet containing 0.10% taurine (Results Brand Diet #2483, Bioserv, Frenchtown, NJ). Cats were maintained on respective dietary formulations for 15 months. Plasma taurine concentrations were measured as previously described[24] at the onset of the study, after 4 months, and 15 months.

The procedure for isolating and perfusing hearts *in vitro* has been previously described[25]. Perfusion of the coronary arteries was achieved with a modified Krebs-Henseleit bicarbonate-buffered solution by retrograde perfusion of the cannulated aortic stump at a constant perfusion pressure of 60 mm Hg. Perfusate effluent was not recirculated. Prior to its delivery to the aorta, the perfusate passed through a heating coil, maintained at 37°C, and an in-line bubble trap. Coronary perfusion pressure was monitored with a pressure transducer (Gould P23, Gould Inc., Oxnard, CA) connected to a side arm of the perfusion apparatus between the bubble trap and the heart.

Left ventricular function of perfused hearts was monitored using methods previously described[1,2,25,35]. Isovolumic-developed LV systolic pressure (LVP) was measured with a latex balloon inserted into the LV chamber through an incision in the left atrium. The balloon cannula was attached to a pressure transducer. All hearts were paced from the right atrial appendage at 150 beats/min with an electronic stimulator (Grass S88F, Grass Instruments Co., Quincy, MA) through miniature electrodes impaled in the epicardial surface. A square wave stimulus 2 volts in amplitude and 0.5 ms in duration was used[6].

Following a 30 min *in vitro* stabilization period at an end-diastolic pressure (EDP) of 8 mm Hg, balloon volumes were adjusted using a 0.5 ml Hamilton glass syringe (Hamilton Co., Reno, NV) to create an EDP of 0 mm Hg. EDP was maintained for 1 min at 0 mm Hg. LVP was measured at the end of the 1 min period. EDP was sequentially increased in 4 mm Hg steps over a range from 0 to 28 mm Hg through expansion of the LV balloon with water. One min following each step change in EDP, LVP was measured.

Dobutamine Challenge Studies

Domestic short hair cats (14 females and 8 males) were obtained from a research animal supplier (Harlan Sprague Dawley Inc., Madison, Wisconsin) and started on the study diets at 10 weeks of age. Cats (5 females and 3 males) were placed on the taurine-supplemented diet (0.10% taurine). The remaining 9 females and 5 males were placed on the taurine-deficient diet (< 0.005% taurine). Taurine-supplemented cats were maintained on the supplemented formulation for 30 months, whereas 7 taurine-deficient cats were changed to the taurine-supplemented diet after 20 months based on changes in myocardial function described below. Plasma taurine concentrations were measured at selected points during the feeding trials[24].

Left ventricular performance of taurine-supplemented and taurine-deficient cats was assessed using 2-dimensional guided M-mode echocardiography as previously described[24]. Echocardiography was performed with an ultrasound system (Genesis, Biosound Inc., Indianapolis, Indiana) using a 7.5 MHz mechanical sector transducer. The short-axis image of the LV chamber was captured for on-line measurement of LV end-systolic (LVDs) and end-diastolic (LVDd) internal dimensions using the leading edge method[36]. LV fractional shortening (FS) was calculated from dimension measurements. Each dimension was measured for 5 representative cardiac cycles and results were averaged.

To induce uniform restraint during echocardiographic procedures, cats were placed under general anesthesia. Prior to echocardiography, cats were fasted for 14 to 16 h, sedated with diazepam (Diazepam Injection USP, Sabex, Inc., Bouchevill, Quebec; 0.2 mg/kg of body weight, i.m.) and ketamine hydrochloride (Ketalean, MTC Pharmaceuticals, Cambridge, Ontario; 15 mg/kg of body weight, i.m.) and anesthetized with isoflurane (AErrane, Anaquest, Mississauga, Ontario). General anesthesia was induced with 5% isoflurane in O_2 and, after tracheal intubation, maintained at 1.5% isoflurane in O_2. A non-rebreathing system (Bain Breathing Circuit, Kendall Canada, Peterborough, Ontario) was used to deliver the isoflurane-oxygen mixture and cats were permitted to breath spontaneously during the procedure. Echocardiograms were recorded 30 min following induction of anesthesia. During dobutamine challenge studies conducted at 30 months, the left cephalic vein was cannulated for infusion of dobutamine hydrochloride (Dobutrex Solution, Eli Lilly Canada Inc., Toronto, Ontario).

Dobutamine hydrochloride was serially diluted in 0.9% NaCl solution so that for each drug infusion dose, saline + dobutamine was infused at a rate of 150 µl/kg/min. Each cat therefore received a total volume of 6.0 to 6.8 ml/kg of saline over the 40 to 45 min of the drug infusion period. In preliminary studies, this relatively small volume of saline without dobutamine had no effect on echocardiographic variables.

Starting at 30 min following induction of anesthesia, a stepwise infusion of dobutamine was initiated in increasing dosages of 0, 0.5, 1.5, 3.0, 5.0, 7.5, 10, and 15 µg/kg/min. The dose of dobutamine was increased every 5 min, with recording of M-mode echocardiographic variables during the last min of infusion of dobutamine at each concentration.

Statistical Analysis

Data are reported as the mean ± standard error of the mean. For isolated heart studies, between group comparisons were made using unpaired t-tests[11]. ANOVA and the Bonferroni t-test were applied to echocardiographic data sequentially collected during the 30 month feeding trails. For dobutamine experiments, within group statistical analysis was performed using repeated measures ANOVA and the Dunnett's multiple comparison test using sham drug (0 µg/kg/min dobutamine) as the control. P values less than 0.05 were considered significant. Where applicable, a statistical software computer program was used to assist in data analysis (Instat, Graphpad, San Diego, CA).

RESULTS

Plasma Taurine Concentrations

At 4 months and 15 months following the onset of the feeding trial, plasma taurine concentrations in all cats in the isolated heart studies and receiving the taurine-deficient diet were < 20 μM (range: < 5-19 μM). All taurine-supplemented cats maintained plasma taurine concentrations ≥ 82 μM (range: 82-158 μM) and were not considered at risk for developing taurine deficiency[14].

Cats enrolled in the dobutamine-challenge study and fed the taurine-deficient diet experienced similar reductions in plasma taurine concentrations (range at 20 months: < 5-13 μM). In most cases plasma taurine concentrations in the deficient group were below the detection limits of the assay (5 μM). Plasma taurine concentrations markedly increased within 1 month in taurine-deficient cats switched to the supplemented diet at 20 months (range: 48-195 μM). Within 3 months following the change to the taurine-supplemented diet, plasma taurine concentrations increased (range: 81-246 μM) to levels measured in cats maintained on the supplemented diet throughout the 30 months of the feeding trial (range: 103-200 μM).

Isolated Heart Study

Shown in Figure 1 is the response of isolated hearts to changes in end-diastolic pressure or preload. The LV function curve generated during changes in preload revealed significantly lower values for LVP in hearts from taurine-deficient cats at maximally effective preload levels. Also in Figure 1 are LV pressure-volume (chamber compliance) curves constructed from end-diastolic pressure-volume data. The compliance curve from taurine-deficient hearts is essentially identical to that of the taurine-supplemented group at higher LV volumes.

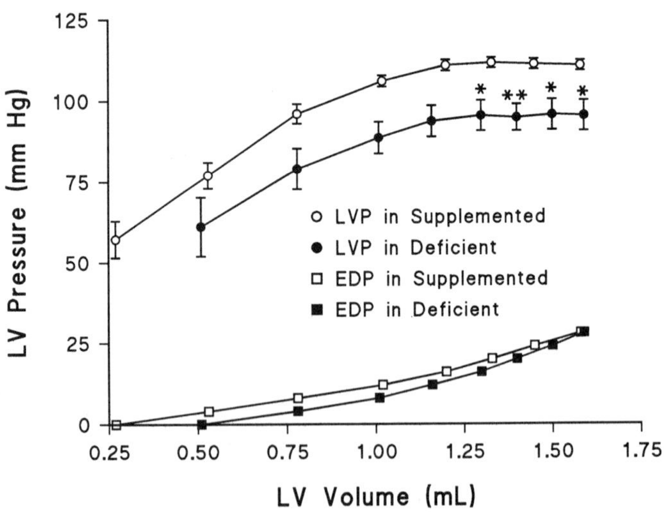

Figure 1. Composite developed LV systolic pressure (LVP) and end-diastolic pressure (EDP) relationships as a function of LV volume for perfused hearts. Each symbol represents the mean with SEM for LVP values. Between group statistical analysis was based on LVP responses to the independent variable EDP. * $p < 0.05$; ** $p < 0.01$.

Table 1. FS (%) and LVDs (mm) in taurine-supplemented cats (n = 8) and taurine-deficient cats (n = 7) following the change in the deficient group to the taurine-supplemented diet

	Supplemented		Deficient + Supplemented	
	FS	LVDs	FS	LVDs
20 months	43 ± 2	6.8 ± 0.4	26 ± 2***	11.6 ± 0.7***
21 months	43 ± 1	7.0 ± 0.6	26 ± 5***	10.6 ± 0.7***
23 months	42 ± 3	6.5 ± 0.6	26 ± 2***	10.1 ± 0.6**
25 months	38 ± 1	7.7 ± 0.7	30 ± 3*	9.9 ± 0.9
27 months	38 ± 2	7.0 ± 0.4	34 ± 1	8.9 ± 0.5
30 months	41 ± 2	7.7 ± 0.6	37 ± 3	8.7 ± 0.9

FS, fractional shortening; LVDs, LV end-systolic internal diameter. Each value represents the mean ± SEM. *p < 0.05; **p < 0.01; ***p < 0.001.

Echocardiographic Responses to the Dietary Change

The range of FS at 20 months for cats maintained on the taurine-supplemented diet was 36 to 50% (mean ± SEM: 43 ± 2%). Seven of the taurine-deficient cats had FS values < 35% (mean ± SEM: 26 ± 2%; $p < 0.001$) at 20 months and these cats were switched at 20 months to the taurine-supplemented diet. The remaining 7 taurine-deficient cats had FS values > 35% (41 ± 2%; $p > 0.05$) and were maintained on the taurine-deficient diet for 10 additional months. FS in the later group eventually decreased to 34 ± 2% at 30 months. However, as a group FS was not significantly different ($p > 0.05$) than that measured in the taurine-supplemented group at 30 months (41 ± 2%).

Summarized in Table 1 are the changes in echocardiographic variables, FS and LVDs, in taurine-deficient cats following the dietary change. LVDs and LVDd were significantly higher and FS was significantly lower in taurine-deficient cats at 20 months. While LVDd was no longer significantly different between the 2 groups of cats at 21 months (data not shown), it was not until 3 months and 5 months following the dietary change in previously deficient cats that LVDs and FS, respectively, were no longer significantly different between the 2 groups of cats.

Dobutamine Challenge Study

Dobutamine had no significant effect on LVDd in any of the 3 groups of cats. Illustrated in Figure 2 and Figure 3 are the effects of dobutamine on LVDs and FS, respectively. Dobutamine, at concentrations ≥ 1.5 μg/kg/min, caused significant decreases in LVDs and increases in FS in the taurine-supplemented group of cats and the group of cats that experienced the change to the supplemented diet at 20 months. Marked enhancement in FS was observed at dobutamine concentrations between 0.5 and 5.0 μg/kg/min. FS increased 40-42% in these 2 groups of cats at maximally-effective dobutamine concentrations. Cats in the taurine-deficient group experienced minor changes in LVDs and FS in response to dobutamine. FS increased only 21% in the taurine-deficient group at maximally-effective dobutamine concentrations.

DISCUSSION

Cats maintained on taurine-deficient diets showed significant diminution of plasma and myocardial taurine concentrations[24,26]. Taurine deficiency in the domestic cat can result

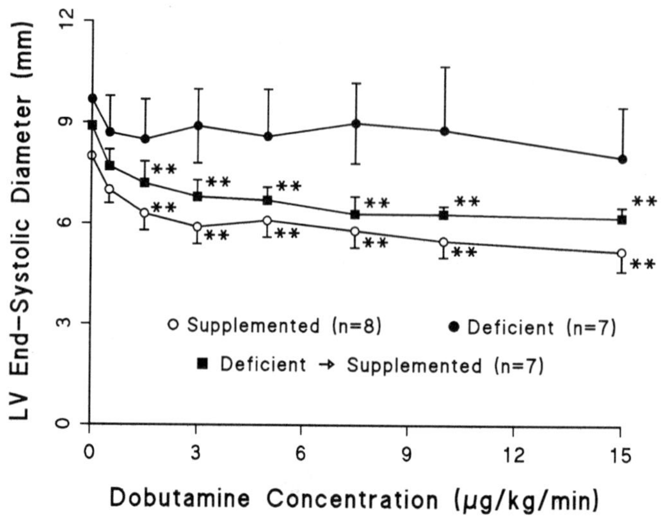

Figure 2. Effect of dobutamine on LV end-systolic diameter (LVDs). Each symbol represents the mean ± SEM. Within group statistical analysis was performed using repeated measures ANOVA and the Dunnett multiple comparison test with 0 μg/kg/min dobutamine as the control. $^{**}p < 0.01$.

in dilated cardiomyopathy[27,30] and depletion of taurine in rats can lead to diminished myocardial contractility[17]. We previously reported that taurine deficiency leads to a spectrum of echocardiographic changes in myocardial function in domestic cats[24]. It appears that dilated cardiomyopathy represents one type of cardiac response to taurine deficiency in the domestic cat. A significant relationship appears to exist between maintenance of adequate myocardial taurine concentrations and myocardial function.

The colony-source cats in the current study did not experience the degree of myocardial dysfunction and dilated cardiomyopathy we observed in the random-source cats previously studied[25]. Why certain cats or populations of cats seem more susceptible to taurine deficiency DCM is not known. Perhaps nutritional background prior to arrival in our animal care facility in someway influenced the response of random-source cats to the taurine-deficient diet. Other factors may contribute to differences in susceptibility to taurine deficiency between populations. For example, Lawler *et al.* recently reported on a genetic predisposition in the development of dilated cardiomyopathy in cats[18]. Perhaps some interaction between

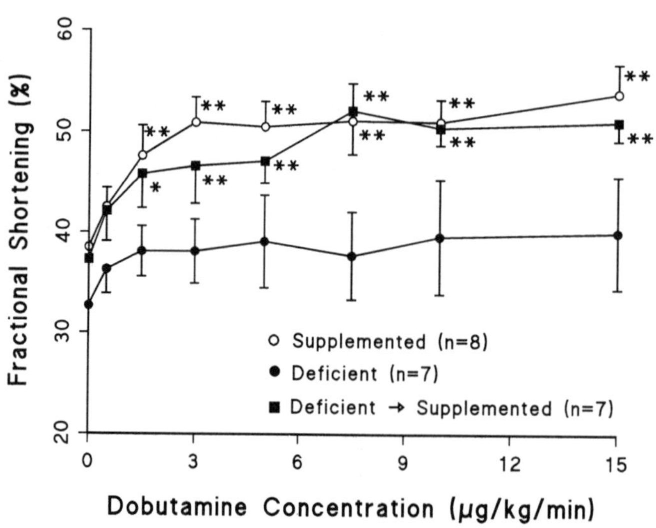

Figure 3. Effect of dobutamine on LV fractional shortening (FS). Each symbol represents the mean ± SEM. Within group statistical analysis was performed using repeated measures ANOVA and the Dunnett multiple compar-ison test with 0 μg/kg/min dobutamine as the control. $^{*}p < 0.05$; $^{**}p < 0.01$.

genetics, nutrition and other unknown factors influence susceptibility to myocardial failure and dilated cardiomyopathy in taurine-deficient cats.

We previously reported echocardiographic findings for the 12 cats subjected to *in vitro* isolated heart studies reported here. FS and LVDs in the 7 taurine-deficient cats (FS = 33 ± 2%; LVDs = 7.4 ± 0.4 mm) were not significantly different than corresponding values recorded from the 5 taurine-supplemented cats (FS = 35 ± 2%; LVDs = 6.6 ± 0.2 mm) after the 15 month feeding trial. As in our previous study[25], isolated ventricles from taurine-deficient cats in the current study retained the functional capacity to adjust systolic contractile performance in response to changes in diastolic muscle length analogous to intact hearts. However, only when EDP was increased to maximally effective preload was the contractile defect associated with taurine deficiency evident.

For *in vivo* echocardiographic studies we chose to place cats under general anesthesia with isoflurane at 1.5% to maintain uniform restraint. The dose of isoflurane is slightly below the minimum alveolar concentration (MAC) of 1.62% reported for cats[13]. All inhalant general anesthetic agents are capable of causing myocardial depression, hypotension and respiratory depression in a dose-dependent manner. Of the available inhalant agents, isoflurane has the least myocardial depressant properties and causes, on an equipotent dosage basis, the least effect on cardiac output, particularly at concentrations < 1-1.5 X MAC[13]. Since considerable individual variation can occur with any general anesthetic regimen, we chose to standardize the dose of isoflurane delivered and to limit statistical comparisons for dobutamine dose-response studies to within-group responses in echocardiographic variables such that each cat served as its own control (dobutamine dose = 0 μg/kg/min).

The response of the heart *in situ* to infusion of dobutamine has been described to assess functional reserve of the left ventricle[15,33] or to identify regional dysfunction following ischemic injury[5,28]. Echocardiography is frequently employed in such studies[28,31]. Dobutamine may be particularly useful in patients unable to perform conventional stress testing. Hurwitz *et al.*[15] used dobutamine-induced changes in LV ejection fraction to assess LV functional reserve in infants and children with different forms of myocardial dysfunction, including idiopathic, dilated cardiomyopathy. Steltzer *et al.* used dobutamine challenge along with transesophageal echocardiography in the assessment of left ventricular performance in septic patients[31]. In the current study we evaluated the myocardial response to dobutamine infusion using M-mode echocardiography. The most commonly reported echocardiographic indices of global LV function in the cat are fractional shortening along the short-axis and LV end-systolic internal diameter[24,27,30]. Ejection phase indices of LV performance are best considered indices of global LV pump function that are particularly valuable in assessing the response of individual animals to changes in global LV performance as in the current study[7]. However, FS and LVDs are determined not only by LV contractility, but also by LV loading conditions and heart rate. It was not the design of these experiments to distinguish between the effects of dobutamine on loading conditions and LV contractility in taurine-supplemented and deficient cats. Indeed, between group differences in FS and LVDs responses to dobutamine are probably the result of changes in all 3 influencing variables. However, data from our isolated heart studies suggest that LV contractile dysfunction contributes in a significant way to the cardiac response to taurine deficiency. Upon dobutamine challenge taurine-supplemented cats responded with significant decreases in LVDs and increases in FS while taurine-deficient cats failed to mount a significant response to dobutamine. Failure to respond significantly to dobutamine challenge suggests a lack of ventricular functional reserve in taurine-deficient cats.

The decreased ability of hearts from taurine-deficient cats to respond to inotropic measures employed in this study could be the result of any of several pathophysiological mechanisms. These include: (i) diminished Ca^{2+} availability to contractile proteins; (ii) impaired binding of Ca^{2+} to troponin; (iii) decreased ability to utilize energy reserves maximally or decreased availability of energy reserves for force generation; (iv) diminution of the basic sarcomere length-tension relationship in ventricular myocytes; (v) a loss of a significant fraction of contracting cells; or (vi) combinations of these mechanisms. The positive inotropic effects of dobutamine appear to be mediated through activation of postsynaptic β_1 and, to a lesser extent, α_1 adrenergic receptors[19]. The failure of taurine-deficient hearts to respond to dobutamine with increases in FS and decreases in LVDs may indicate a reduction in myocardial β_1 or α_1 adrenoceptors or attenuation of the inotropic response to adrenergic agonists. Malhotra *et al.*[20-22] demonstrated abnormalities in several components of the contractile protein apparatus in the genetic hamster model of dilated cardiomyopathy, including changes in myosin isoenzymes, diminished myosin ATPase, and reduced sensitivity of troponin-tropomyosin regulatory proteins to Ca^{2+}. Focal and diffuse losses of myocardium (and therefore contractile subunits) are features of various forms of cardiomyopathies[21]. More recently, Lake demonstrated a loss of cardiac myofibrils in hearts from rats treated with the taurine transport antagonist guanidinoethane sulfonate[16]. It is reasonable to hypothesize that similar changes may be associated with taurine-deficient myocardial failure in domestic cats.

While a taurine-deficient form of dilated cardiomyopathy has not been recognized in humans, taurine is a conditionally-essential nutrient in humans[4,10]. Taurine is supplemented in the formula diets of infants and in humans maintained on total parenteral nutrition[12,34]. Since taurine deficiency causes loss of contractile reserve without necessarily resulting in dilated cardiomyopathy, this raises the question of taurine deficiency causing a similar contractile deficit or loss of contractile functional reserve that is not clinically obvious in humans.

In summary, in the current study inotropic interventions were used to detect a loss of left ventricular contractile reserve in hearts from taurine-deficient cats that, under basal conditions, did not appear to have significant diminution of contractile function. While this finding does not explain mechanistically the relationship between taurine deficiency and myocardial failure, it is evident that taurine deficiency need not result in dilated cardiomyopathy in order to cause significant reduction in intrinsic LV contractile reserve.

ACKNOWLEDGMENTS

Supported by a grant from the Heart and Stroke Foundation of Canada. The authors thank Ms. Judy Furlong and Ms. Lisa Purcell for their technical assistance.

REFERENCES

1. Adams, H.R., Baxter, C.R., and Parker, J.L. 1985, Reduction of intrinsic contractile reserves of the left ventricle by Escherichia coli endotoxin shock in guinea pigs, *J.Mol.Cell.Cardiol.* 17:575-585.
2. Apstein, C.S., Mueller, M., and Hood, W.B. 1977, Ventricular contracture and compliance changes with global ischemia and reperfusion, and their effect on coronary resistance in the rat, *Circ.Res.* 4:206-217.
3. Azuma, J., Sawamura, A., Awata, N., Ohta, H., Hamaguchi, T., Harada, H., Takihara, K., Hasegawa, H., Yamagami, T., Ishiyama, T., Iwata, H., and Kishimoto, S. 1985, Therapeutic effect of taurine in congestive heart failure: a double-blind crossover trial, *Clin.Cardiol.* 8:276-282.
4. Chesney, R. 1988, Taurine: Is it required for infant nutrition? *J.Nutr.* 118:6-10.
5. Cigarroa, C.G., deFilippi, C.R., Brickner, E., Alvarez, L.G., Wait, M.A., and Grayburn, P.A. 1993, Dobutamine stress echocardiography identifies hibernating myocardium and predicts recovery of left ventricular function after coronary revascularization, *Circulation*, 88:430-436.
6. Doring, H.J. and Dehnert, H. 1987, Testing the Langendorff heart for perfect function, in: "The Isolated Perfused Heart According to Langendorff", Biomesstechnick-Verlag, March, FRG, pp. 60-74.

7. Feigenbaum, H. 1986, Echocardiographic evaluation of cardiac chambers, in: "Echocardiography", 4th edition, Lea & Febiger, Philadelphia and London, pp. 127-187.
8. Franconi, F., Stendardi, I., Failli, P., Matucci, R., Baccaro, C., Montorsi, L., Bandinelli, R., and Giotti, A. 1985, The protective effects of taurine on hypoxia (performed in the absence of glucose) and on reoxygenation (in the presence of glucose) in guinea-pig heart, *Biochem.Pharmacol.* 34:2611-2616.
9. Franconi, F., Stendardi, I., Matucci, R., Bennardini, F., Baccaro, C., and Giotti, A. 1985, A protective effect of taurine on "hypoxic" and "reoxygenated" guinea pig heart, in: "Taurine: biological actions and clinical perspectives", Oja, S.S., Ahtee, L., Kontro, P. and Paasonen, M.K. eds., Alan R. Liss, New York, pp. 177-182.
10. Gaull, G.E., Pasantes-Morales, H., and Wright, C.E. 1985, Taurine in human nutrition: overview, in: "Taurine: Biological Actions and Clinical Perspectives", Oja, S.S., Ahtee, L., Kontro, P. and Paasonen, M.K. eds., Alan R. Liss, New York, pp. 3-22.
11. Glantz, S.A. 1987, "Primer of Biostatistics", McGraw-Hill, New York.
12. Gray, G.E., Landel, A.M., and Meguid, M.M. 1994, Taurine-supplemented total parenteral nutrition and taurine status of malnourished cancer patients, *Nutrition*, 10:11-15.
13. Haskins, S.C. 1992, Inhalational anesthetics, *Vet.Clin.North Am.Small Anim.Pract.* 22:297-307.
14. Hayes, K.C. and Trautwein, E.A. 1989, Taurine deficiency syndrome in cats, *Vet. Clin. North Am. Small Animal Pract.*, 19:403-413.
15. Hurwitz, R.A., Siddiqui, A., Caldwell, R.L., Weetman, R.M., and Girod, D.A. 1990, Assessment of ventricular function in infants and children: response to dobutamine infusion, *Clin.Nuclear Med.* 8:556-559.
16. Lake, N. 1993, Loss of cardiac myofibrils: Mechanism of contractile deficits induced by taurine deficiency, *Am.J.Physiol.Heart Circ.Physiol.* 264:H1323-H1326.
17. Lake, N., Splawinski, J.B., Juneau, C. and Rouleau, J.L. 1990, Effects of taurine depletion on intrinsic contractility of rat ventricular papillary muscles, *Can. J. Physiol. Pharmacol.* 68:800-806.
18. Lawler, D.F., Templeton, A.J., and Monti, K.L. 1993, Evidence for genetic involvement in feline dilated cardiomyopathy, *J.Vet.Intern.Med.* 7:383-387.
19. Majerus, T.C., Dasta, J., F., Bauman, J.L., Danziger, L.H., and Ruffuolo, R.R. 1989, Dobutamine: ten years later, *Pharmacotherapy*, 9:245-259.
20. Malhotra, A. 1990, Regulatory proteins in hamster cardiomyopathy, *Circulation Research*, 66:1302-1309.
21. Malhotra, A., Karell, M., and Scheuer, J. 1985, Multiple cardiac contractile protein abnormalities in myopathic Syrian hamsters (BIO 53:58), *J.Mol.Cell.Cardiol.* 17:95-107.
22. Malhotra, A. and Scheuer, J. 1990, Troponin-tropomyosin abnormalities in hamster cardiomyopathy, *J.Clin.Invest.* 86:286-292.
23. McBroom, M.J. and Welty, J.D. 1985, Effects of taurine and verapamil on heart calcium in hamsters and rats, *Comp.Biochem.Physiol.* 82C(2):279-281.
24. Novotny, M.J., Hogan, P.M., and Flannigan, G. 1994, Echocardiographic evidence for myocardial failure induced by taurine deficiency in domestic cats, *Can.J.Vet.Res.* 58:6-12.
25. Novotny, M.J., Hogan, P.M., Paley, D.M., and Adams, H.R. 1991, Systolic and diastolic dysfunction of the left ventricle induced by dietary taurine deficiency in cats, *Am.J.Physiol.Heart Circ.Physiol.* 261:H121-H127.
26. Pion, P.D., Kittleson, M.D., Lewis, J., Greene, K., Morris, J.G., and Rogers, Q.R. 1991, Taurine deficient myocardial failure: incidence and relation to tissue taurine concentration, *J.Vet.Intern.Med.* 5:123.
27. Pion, P.D., Kittleson, M.D., Rogers, Q.R., and Morris, J.G. 1987, Myocardial failure in cats associated with low plasma taurine: a reversible cardiomyopathy, *Science*, 237:764-768.
28. Previtali, M., Poli, A., Lanzarini, L., Fetiveau, R., Mussini, A., and Ferrario, M. 1993, Dobutamine stress echocardiography for assessment of myocardial viability and ischemia in acute myocardial infarction treated with thrombolysis, *Am.J.Cardiol.* 72:124G-130G.
29. Schaffer, S.W., Allo, S., and Mozaffari, M. 1987, Potentiation of myocardial ischemic injury by drug-induced taurine depletion, in: "The Biology of Taurine: Methods and Mechanisms", Huxtable, R.J., Franconi, F. and Giotti, A. eds., Plenum Press, New York, pp. 151-158.
30. Sisson, D.D., Knight, D.H., Helinski, C., Fox, P.R., Bond, B.R., Harpster, N.K., Moise, N.S., Kaplan, P.M., Bonagura, J.D., Czarnecki, G., and Schaeffer, D.J. 1991, Plasma taurine concentrations and M-mode echocardiographic measures in healthy cats and in cats with dilated cardiomyopathy, *J.Vet.Intern.Med.* 5:232-238.
31. Steltzer, H., Simon, P., Owen, A.N., Thalmann, M., and Hammerle, A.F. 1994, The effects of dobutamine therapy in critically ill patients measured by transesophageal echocardiography and intracardiac monitoring, *Anaesthesia*, 49:432-437.

32. Takihara, K., Azuma, J., Awata, N., Ohta, H., Hamaguchi, T., Sawamura, A., Tanaka, Y., Kishimoto, S., and Sperelakis, N. 1986, Beneficial effect of taurine in rabbits with chronic congestive heart failure, *Am.Heart J.* 112:1278-1284.
33. Tan, L.B. and Littler, W.A. 1990, Measurement of cardiac reserve in cardiogenic shock: implications for prognosis and management, *Br.Heart J.* 64:121-128.
34. Vinton, N.E., Laidlaw, S.A., Ament, M.D., and Kopple, J.D. 1986, Taurine concentrations in plasma and blood cells of patients undergoing long-term parenteral nutrition, *Am. J. Clin. Nutr.*, 44:398-404.
35. Vogel, W.M., Lum, D., and Lucchesi, B.R. 1979, Methylprednisolone sodium succinate treatment in global ischemia of the cat isolated heart, *J.Cardiovasc.Pharmacol.* 1:53-68.
36. Wyatt, H.L., Haendchen, R.V., Meerbaum, S., and Corday, E. 1983, Assessment of quantitative methods for 2-dimensional echocardiography, *Am.J.Cardiol.* 52:396-401.

33

EFFECT OF TAURINE ON TOXICITY OF THE PYRROLIZIDINE ALKALOID MONOCROTALINE IN RATS*

Chong Chao Yan and Ryan J. Huxtable

Department of Pharmacology
College of Medicine
University of Arizona, Tucson, Arizona 85724

INTRODUCTION

The pyrrolizidine alkaloid (PA), monocrotaline, is found in a variety of *Crotalaria* species in the legume family[13]. It has been responsible for a large number of human intoxications, most dramatically in the Caribbean nations of Jamaica[3,11,31]. For many years, monocrotaline-induced veno-occlusive disease of the liver was endemic, due to the use of the leaves of *C. spectabilis* in the preparation of the so-called "bush teas" that were commonly consumed. It was not until the 1950s that the cause of the condition was established[2] and control measures instituted.

In rats, nonhuman primates, and other species, monocrotaline also produces pulmonary arterial hypertension and right ventricular hypertrophy of the heart[15]. Both hepatic and extrahepatic toxicities are due to hepatic metabolism of the parent alkaloid to the highly reactive alkylating compound, dehydromonocrotaline[25,34]. A number of pathways of further reaction are available to dehydromonocrotaline. It can alkylate macromolecular components in the liver, producing liver disease such as hepatomegaly or cancer. It can be released into the circulation, damaging the endothelium to cause veno-occlusive disease, or organs downstream, such as the lungs. It can be detoxified by conjugation with GSH to form 7-glutathionyl-6,7-dihydro-1-hydroxymethyl-5H-pyrrolizine (GSDHP)[5,23], which is excreted in high concentrations into bile[18,26]. Finally, it can undergo hydrolysis to the relatively innocuous compound 6,7-dihydro-7-hydroxy-1-hydroxymethyl-5H-pyrrolizine (DHP)[17].

Taurine has been shown to be protective against a number of insults to the heart, lungs and liver[16]. We have examined whether these protective effects can be extended to monocrotaline-induced toxicity.

---

* The data in this chapter are taken from Yan and Huxtable.[40]

Taurine 2, edited by Huxtable et al.
Plenum Press, New York, 1996

METHODS

Taurine, GSH, L-cysteine (Cys), glycine (Gly), L-glutamic acid (Glu), DL-dithiothreitol (DTT), γ-glutamylcysteine (γ-Glu-Cys), cysteinylglycine (Cys-Gly), glycylglycine, N-ethylmorpholine, NADPH, GSH reductase, oxidized GSH, 1-chloro-2,4-dinitrobenzene (CDNB), H_2O_2, and the kit for protein determination were purchased from Sigma (St. Louis, MO). Monobromobimane (Thiolyte Reagent; MB) was from Calbiochem (San Diego, CA). Monocrotaline was isolated from seeds of *Crotalaria spectabilis*[1] and was purified by counter-current distribution and identified by mass spectrometry.

Treatment Protocol

Male Sprague-Dawley rats (body weight 53-70 g) from the University of Arizona Division of Animal Resources were housed 2-3 to a cage under 12 h cycles of light and dark. After 3 days for adaptation, the animals were randomly divided into groups of 5 animals. Rats in the control and monocrotaline groups were maintained on fresh tap water throughout the study and rats in the monocrotaline plus taurine group were maintained on a solution of 1% taurine. On day 14, rats in the control group received an injection of 0.85% NaCl (i.p.). Rats in the other two groups each received an injection of monocrotaline (i.p., 65 mg/kg body weight) in the same volume as in the control group. Fourteen days after being injected, all rats were decapitated. To avoid differences due to diurnal variation in enzyme activities related to GSH metabolism, rats were killed between 9:00-10:00 a.m. Sera were collected for amino acid assay. Heart and lungs were removed for determination of tissue:body weight ratios. Wet weights were recorded immediately after washing (3x) in 0.85% NaCl. Dry weights were obtained after drying the organs at 50°C for three days. On the 11th day after monocrotaline injection, 2 rats in the monocrotaline-alone group died.

Determination of LD_{50}

Male rats (initial body weight 60-85 g) were maintained either on tap water (30 animals) or 1% taurine solution (30 animals). After 2 weeks, animals in both groups were injected i.p. with monocrotaline at 55, 82, 110, 165, or 220 mg/kg. Each dose was administered to a total of 6 animals in each group. Deaths of animals were recorded daily for the subsequent 2 weeks. LD_{50} values were calculated using the Litchfield and Wilcoxon II Pharmacologic Calculation System, version 4.2, computer program.

Collection of Serum

Whole blood was collected in untreated polypropylene tubes and stood at room temperature for 3 h. It was then centrifuged in a clinical centrifuge for 5 min, and the clear supernatant siphoned off as serum, and used for amino acid determination.

Tissue Cytosol and Microsome Preparation

Liver was removed and washed twice in saline at 4°C. Then the livers were minced and homogenized in 4 vols of 0.25 M sucrose and 20 mM Tris HCl containing 1 mM EDTA (pH 7.4). The homogenate was centrifuged at 3000 g for 10 min at 4°C and the resulting supernatant was centrifuged at 10,000 g for 20 min at 4°C. Finally, the supernatant was centrifuged at 105,000 g for 60 min at 4°C. The supernatant (cytosol) was used to determine

GSH and amino acid concentrations and the activities of γ-Glu-Cys synthetase, GSH synthetase, cytosolic GSH S-transferase, GSH reductase and GSH peroxidase. The microsomal pellet was used to determine the activities of dipeptidase, γ-glutamyl transpeptidase and microsomal GSH S-transferase.

Determination of GSH Concentration

Hepatic cytosolic fraction (100 μl) was mixed with 5% sulfosalicylic acid (SSA) and centrifuged at 10,000 g for 5 min to precipitate protein. The supernatant was derivatized with MB and analyzed by HPLC (see below).

Enzyme Assays

General Information. Incubations were performed at 37°C and terminated by addition of 100 μl incubate to 50 μl 5% SSA. The mixture was centrifuged for 3 min to precipitate protein. The supernatant was derivatized with MB (see below) and injected into the HPLC. Peaks were identified by comparison with the retention time of authentic peaks and quantified according to standard curves. Separation was achieved using a Beckman Ultrasphere™ ODS (250x4.6 mm I.D., 5 μm particle size) reversed-phase stainless-steel column (Beckman, San Ramon, CA, USA) maintained at room temperature. The analytical column was protected by a supelguard (5X4.6 mm, I.D.), pellicular reversed-phase cartridge precolumn (Supelco, Bellefonte, PA, USA). Elution was carried out using gradient program with 0.25% acetic acid as solvent A (pH 3.9) and methanol:0.25% acetic acid (30:70, v/v) as solvent B (pH 3.9). The flow rate was 1.5 ml/min and the starting condition was 35% B for 1 min followed by a linear gradient to 62% B within 6 min. This was maintained for another 2 min. The column was then washed with 100% B for 2 min and maintained under starting conditions for 5 min. With this program, the retention times for Cys, Cys-Gly, γ-Glu-Cys and GSH were 7.09, 8.34, 9.93 and 11.40 min, respectively.

γ-Glu-Cys Synthetase. Cytosolic γ-Glu-Cys synthetase activity was determined by following formation of γ-Glu-Cys. Incubations were performed in 1.5 ml microfuge tubes containing 10 mM Glu, 5 mM Cys, 10 mM ATP, 0.1 M Tris HCl (pH 8.2), 0.15 M KCl, 2 mM EDTA and 20 mM $MgCl_2$. After preincubation for 3 min at 37°C, cytosolic protein (about 1 mg) was added to initiate the reaction. The total volume was 1.0 ml and the incubation time was 20 min.

γ-Glutamyl Transpeptidase and Dipeptidase. The assay for microsomal γ-glutamyl transpeptidase was based on the transfer of glutamyl from GSH to Gly-Gly with concomitant release of Cys-Gly. Determination of the latter compound is used as an index of reaction rate. The standard assay solution (1.0 ml) contained 0.1 M Tris-HCl buffer (pH 8.0), 75 mM NaCl, 20 mM Gly-Gly, 5 mM GSH, 50 μM DTT and microsomal enzyme. The solution was incubated at 37°C for 20 min. To correct for hydrolytic release of Cys-Gly from GSH by γ-glutamyl transpeptidase, a blank tube lacking Gly-Gly was prepared at the same time. After equilibration at 37°C, the reaction was initiated by adding the enzyme. The assay for microsomal dipeptidase was based on the release of Cys from Cys-Gly. The incubation system contained 2 mM Cys-Gly, 75 mM NaCl, 0.1 M Tris-HCl (pH 8.0) and water (total volume 1 ml). The solution was brought to 37°C and incubated for 20 min. To correct for the nonenzymatic hydrolytic release of Cys from Cys-Gly, a blank tube lacking microsomal protein was run.

GSH Reductase and GSH Peroxidase. Cytosolic GSH reductase activity was assayed by the method of Massey and Williams[20]. The disappearance of NADPH absorption in the incubation was monitored at 340 nm. The activity is expressed as nmol/mg protein/min, based on an extinction coefficient for NADPH[10] of 6.22×10^6 mol/cm^{-2}. Cytosolic GSH peroxidase activity was measured spectrophotometrically by a modification of the procedure of Paglia and Valentine[27], as described by Lawrence *et al.*[19].

GSH S-Transferase. Cytosolic and microsomal GSH S-transferase activities were assayed with CDNB according to Habig *et al.*[9] Briefly, the incubation system contained 1 mM GSH, 0.13 M phosphate buffer (pH 7.0) and 50 μg of cytosolic or microsomal protein. Total incubation medium was 1 ml. The mixture was preincubated at 37°C for 3 min. The reaction was started by adding 1 mM CDNB (final concentration) and incubation continued for another 5 min. Under these conditions, the product formed was linear with respect to time and protein content. The reaction was stopped by adding 5% SSA, followed by centrifugation to precipitate protein. Appropriate controls were run for determination of spontaneous (nonenzymic) conjugation.

Amino Acid Analysis

Aliquots of hepatic cytosolic fraction were removed and deproteinized with 5% SSA. Deproteinized samples were diluted and mixed with *o*-phthaldehyde (OPA) solution and analyzed by HPLC as previously described[38].

Protein Assay

Protein assay was performed according to the method of Lowry (Sigma kit for protein determination) using bovine serum albumin as calibration standard.

Statistics

Statistical comparisons were obtained by MANOVA followed by a Student's t test (Bonferroni comparison) (Graphpad Software, version 1.14).

RESULTS

Following acute or subacute exposure of rats to monocrotaline, pulmonary hyperplasia and pulmonary arterial hypertension develop within 9 days, and right ventricular hypertrophy within 14 days. We have examined the effects of taurine on these processes using the protocol shown in Fig. 1. Rats were maintained on drinking water containing 1% taurine beginning two weeks prior to a single i.p. injection of monocrotaline (65 mg/kg). Animals were kept on taurine for a further two weeks, at which time surviving animals were killed and various biochemical determinations performed.

The effects of the three treatments on serum amino acid concentrations (Table 1), hepatic amino acid concentrations (Table 2), and hepatic enzymes involved in sulfur metabolism (Table 3) were determined. Compared to the monocrotaline-alone group, taurine increased serum levels of Glu, Ala and Met (Table 1). Liver Gln was increased and liver Ser slightly, but significantly, decreased by taurine treatment (Table 2). Monocrotaline lowered hepatic Cys and GSH levels 14 days post-administration (Fig. 2), in contradistinction to its acute effect[37]. Taurine restored GSH levels to control (Fig. 2), but was without significant effect on Cys levels.

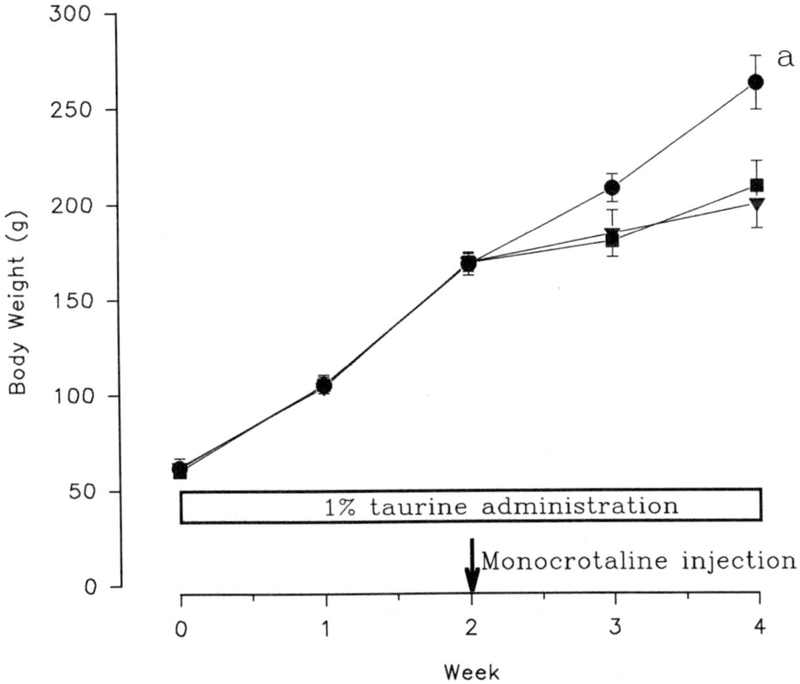

Figure 1. Protocol for effect of taurine on monocrotaline toxicity. Body weight gains of are shown for rats: maintained on tap water and given a saline injection(●); maintained on tap water and injected with monocrotaline (■); or maintained on 1% taurine solution and injected with monocrotaline (▼). Arrow indicates day of monocrotaline injection (65 mg/kg; i.p.). Data are means ± SD for 5 animals (3 for 4-week measurement on monocrotaline-alone group). [a]$p < 0.05$ compared to all other groups.

Table 1. Effect of taurine on serum amino acid concentrations (μmol/liter)

Amino Acid	Control*	Monocrotaline#	Monocrotaline + Taurine*
Asp	64.8 ± 8.1	88.4 ± 4.3[a]	79.6 ± 6.7
Glu	220.8 ± 17.3	185.5 ± 14.0	250.7 ± 32.4[b]
Asn	73.6 ± 2.3	67.1 ± 1.7	72.3 ± 5.6
Ser	303.6 ± 9.8	298.1 ± 6.3	293.2 ± 29.0
Gln	494.9 ± 35.6	429.1 ± 12.8	440.6 ± 46.5
Gly	646.6 ± 36.0	598.8 ± 28.6	568.6 ± 27.5[a]
Arg	305.3 ± 31.9	306.6 ± 25.8	268.6 ± 37.6
Tau	188.5 ± 27.0	241.4 ± 12.1[a]	859.6 ± 41.4[a,b]
Ala	439.8 ± 32.1	380.3 ± 13.8	444.3 ± 23.4[b]
Met	56.4 ± 5.3	58.5 ± 2.4	70.4 ± 4.0[a,b]
Val	176.1 ± 14.1	153.9 ± 10.4	161.0 ± 18.7
Phe	71.6 ± 3.6	71.7 ± 3.5	79.6 ± 8.4
Ile	119.8 ± 8.4	116.7 ± 6.4	125.8 ± 15.1
Leu	122.0 ± 7.6	111.5 ± 6.4	115.0 ± 13.9
Lys	118.1 ± 6.9	146.8 ± 8.3[a]	149.2 ± 15.9[a]

[a]$p < 0.05$ versus control; [b]$p < 0.05$ versus monocrotaline group. Data are means ± SD for 5* or 3# animals per group

Table 2. Effect of taurine on hepatic amino acid levels in rats treated with monocrotaline (μmol/g protein)

Amino Acid	Control*	Monocrotaline#	Monocrotaline + Taurine*
Asp	118.0 ± 8.7	133.6 ± 17.0	120.5 ± 9.8
Glu	108.6 ± 10.7	104.1 ± 12.1	114.2 ± 6.3
Asn	55.0 ± 3.1	54.2 ± 4.1	57.6 ± 5.3
Ser	30.7 ± 2.4	36.1 ± 0.4[a]	31.2 ± 1.6[b]
Gln	167.1 ± 11.4	148.2 ± 1.7[a]	180.8 ± 13.1[b]
Gly	107.6 ± 8.9	152.0 ± 1.7[a]	143.4 ± 4.8[a]
Arg	24.0 ± 2.8	26.6 ± 4.8	28.2 ± 5.9
Tau	26.7 ± 3.5	28.7 ± 1.1	230.7 ± 31.6[a,b]
Ala	123.6 ± 6.1	120.7 ± 14.3	123.9 ± 5.3
Met	11.8 ± 0.5	12.7 ± 1.0	12.0 ± 1.3
Val	16.0 ± 0.5	17.0 ± 1.1	17.0 ± 1.9
Phe	14.7 ± 0.8	15.3 ± 0.3	15.0 ± 1.1
Ile	20.5 ± 1.4	21.2 ± 1.7	19.7 ± 1.6
Leu	14.7 ± 2.3	14.3 ± 1.8	12.1 ± 1.4
Lys	19.4 ± 1.3	21.0 ± 1.1	21.6 ± 3.0

[a]$p < 0.05$ versus control; [b]$p < 0.05$ versus monocrotaline group. Data are means ± SD for 5* or 3# animals per group

Taurine treatment lessened, but did not prevent, monocrotaline-induced pulmonary hyperplasia (Fig. 2a). Right ventricular hypertrophy, however, was not seen in the taurine-treated group (Fig. 2b). The differences between the control and taurine-treated groups for right ventricular/body weight ratio or right ventricular/left ventricular weight ratios were not significantly different. The corresponding differences between the monocrotaline-alone and taurine-supplemented groups were significantly different.

In the lethality study, the highest dose of monocrotaline (220 mg/kg) killed all animals by the second day, regardless of whether or not they received taurine. For the lower doses, however, the taurine-treated groups had fewer deaths, and the deaths took longer to occur (Table 4). As PAs can produce death days or weeks after exposure, LD_{50} values decrease the longer the period over which the animals are observed. For all days for which LD_{50} values could be calculated and compared, values were lower in the monocrotaline-alone group (i.e. at a given dose, monocrotaline was more lethal when given to animals not receiving taurine) (Table 5). exceeded those in the taurine-supplemented group (Table 4).

Table 3. Effect of taurine on hepatic enzyme activities in monocrotaline-exposed rats

Enzyme	Control*	Monocrotaline#	Monocrotaline + Taurine*
γ-Glutamylcysteine Synthetase	8.55 ± 1.16	8.00 ± 0.66	10.54 ± 1.14[ab]
GSH Transferase (cystosolic)	1050 ± 80	1070 ± 100	1100 ± 30
GSH Transferase (microsomal)	494 ± 49	842 ± 153[a]	597 ± 95
GSh Reductase	6.12 ± 0.79	7.62 ± 0.38	5.67 ± 0.47[b]
GSH Peroxidase	9.57 ± 0.98	8.88 ± 0.78	10.06 ± 0.95
γ-Glutamyltranspeptidase	2.59 ± 0.24	5.02 ± 1.28[a]	4.83 ± 1.02[a]
Dipeptidase	3.46 ± 0.35	3.86 ± 0.17	3.82 ± 0.46

[a]$p < 0.05$ versus control; [b]$p < 0.05$ versus monocrotaline group. Data are means ± SD for 5* or 3# animals per group

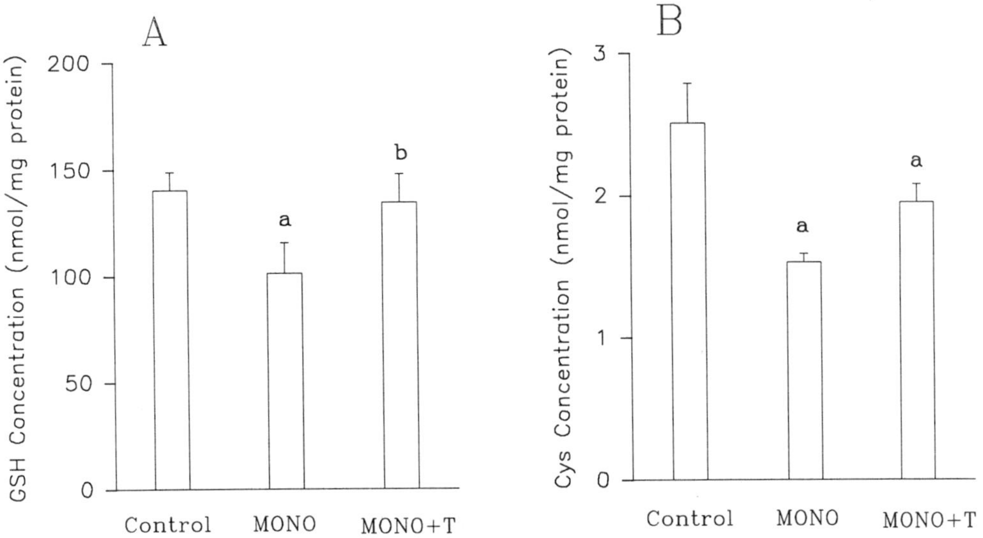

Figure 2. Effect of taurine on monocrotaline-induced depression in (A) hepatic GSH concentration; and (B) hepatic Cys concentration. $^{a}p < 0.05$ compared to control; $^{b}p < 0.05$ compared to monocrotaline-alone group.

DISCUSSION

Monocrotaline is a hepatotoxic PA, producing, in humans and other mammals, hepatomegalocytosis and veno-occlusive disease of the liver. In rat, dog, mouse, nonhuman primates, and other species, monocrotaline also produces pulmonary arterial hypertension and right ventricular hypertrophy[15]. Following initiation of subacute administration of

Table 4. Cumulative deaths in groups of rats given monocrotaline or monocrotaline plus taurine (1% in drinking water.)

	Days following monocrotaline injection													
Treatment	1	2	3	4	5	6	7	8	9	10	11	12	13	14
Monocrotaline (mg/kg)														
55	0	0	0	0	0	0	0	0	0	0	0	0	1	1
82	0	0	0	0	0	0	0	0	0	0	2	2	2	2
110	0	0	0	2	2	2	2	3	3	3	3	5	5	6
165	0	0	1	1	2	2	2	2	3	4	4	5	6	6
220	2	6	6	6	6	6	6	6	6	6	6	6	6	6
Monocrotaline (mg/kg) + Taurine														
55	0	0	0	0	0	0	0	0	0	0	0	0	0	0
82	0	0	0	0	0	0	0	0	0	0	1	1	1	1
110	0	0	0	0	0	0	0	1	1	1	1	2	3	3
165	0	0	0	0	0	0	0	0	1	2	2	3	3	4
220	0	6	6	6	6	6	6	6	6	6	6	6	6	6

Male rats (initial body weight 60-85 g) were maintained on either tap water or 1% taurine solution. After two weeks, the indicated doses of monocrotaline were given i.p. to groups of 6 animals in each treatment group, and deaths recorded daily for two weeks thereafter.

Table 5. LD_{50} values for various periods following administration of monocrotaline or monocrotaline plus taurine to rats

Days following monocrotaline injection	LD_{50} (mg/kg) with 95% confidence intervals	
	Monocrotaline Alone	Monocrotaline Plus Taurine
5	147 (101-214)	Insufficient data
9	130 (94-178)	175 (121-250)
10	122 (91-164)	161 (123-209)
11	113 (82-155)	153 (110-213)
12	95 (70-130)	135 (94-195)
13	82 (62-107)	129 (95-175)*
14	80 (62-104)	121 (93-156)*

The LD_{50} is the dose required to kill 50% of the animals within the period of days shown. *Significantly different from value for monocrotaline-alone group (Litchfield and Wilcoxon II test).

monocrotaline (20 mg/liter in drinking water), pulmonary hyperplasia occurs within 9 days, increased right ventricular to left ventricular weight is seen by 14 days, and pulmonary arterial blood pressure and right ventricular end diastolic pressure are significantly elevated by 22 days[17]. All these changes are presumed to be the consequence of hepatic conversion of monocrotaline to the reactive alkylating pyrrole, dehydromonocrotaline. This metabolic process was first suggested in 1968 by Mattocks[21]. Hepatic production of dehydromonocrotaline from monocrotaline was unequivocally demonstrated by Glowaz and Huxtable[8].

The biochemical concomitants of monocrotaline toxicity have not been so well explored as the pathophysiological effects. However, we have recently shown that monocrotaline and related PAs produce marked alterations in sulfur amino acid metabolism in both liver and extrahepatic organs[35-37,39]. Hepatic GSH levels are increased within 24 h of i.p. injection of 60% of an LD_{50} dose of monocrotaline[37], trichodesmine[37], or retrorsine[36]. Accompanying this are acute increases in the activity of enzymes that synthesize and remove GSH and increases in Glu concentrations in the liver (a substrate for GSH synthesis). There is increased excretion of GSH, Cys, and γ-Glu-Cys into bile, and marked decreases in hepatic taurine concentrations (an alternative product of Cys metabolism)[36]. The longer term effects of PAs on sulfur metabolism have not hitherto been examined.

Taurine has been shown in other circumstances to be hepato-, pneumo- and cardioprotective[16]. In view of these earlier findings, we have examined whether taurine affords protection against the subacute changes produced by monocrotaline in heart and liver. We find that 14 days after exposure to a dose of monocrotaline known to cause right ventricular hypertrophy, rats maintained on taurine show no such hypertrophy (Fig. 3). Although taurine-exposed rats have pulmonary hyperplasia, taurine may afford some protection, as dry lung weights were significantly below that of the monocrotaline-alone group (Fig. 4). In the monocrotaline-alone group, 2 of 5 animals died. None died in the other groups. The effects of taurine, therefore, are being compared to a survivors' group, in which the worst affected animals have been removed. Taurine also significantly raised the 14-day LD_{50} of monocrotaline (Table 5). The LD_{50} value for the monocrotaline-alone group at 14 days (80 mg/kg) compares with a value of 95 mg/kg reported for deaths within 7 days in male rats of initial body weight 90-200 g[22].

Taurine, given to monocrotaline-injected animals, normalizes a number of the biochemical changes caused by the PA (Fig. 2; Tables 2,3). It antagonizes the changes in hepatic Ser and Gln (Table 2), and the changes in serum Glu (Table 1). It shows a sparing effect on serum Met, and corrects the depression in hepatic GSH by stimulating γ-Glu-Cys synthetase and antagonizing the monocrotaline-induced changes in GSH transferase (Table 3).

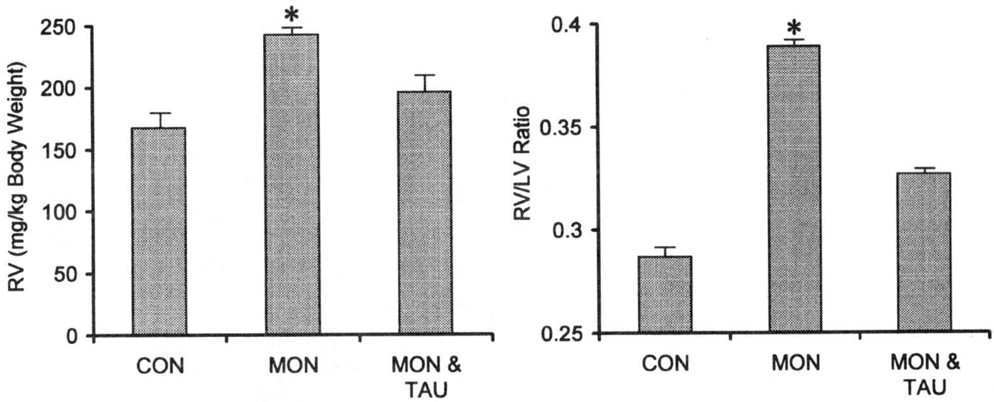

Figure 3. Effect of taurine on monocrotaline-induced right ventricular hypertrophy: Left: right ventricular weight (dry) to body weight ratio (mg/kg); Right: Right ventricular to left ventricular ratio. $^{*}p < 0.05$ compared to other two groups.

Taurine did not ameliorate the decreased body weight gain induced by monocrotaline. Decreased body weight gain is due to decreased eating[7]. Following administration of monocrotaline, pair-fed animals gain the same weight but still develop lung and right heart lesions[39]. The decreased weight gain, therefore, is unrelated to the cardiopulmonary toxicity of monocrotaline. Taurine itself, however, is known to depress eating behavior[33].

The "normalizing" effects of taurine are an example of the enantiostatic action that has been proposed as one of its major biological functions[14,16]. Our findings suggest that the

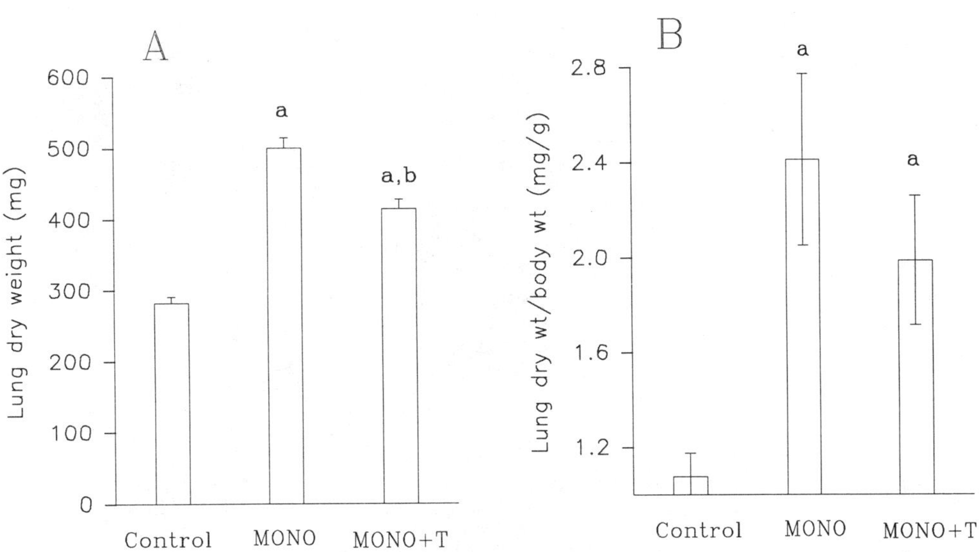

Figure 4. Effect of taurine on monocrotaline-induced pulmonary hyperplasia: (A) Lung dry weight; (B) Lung to body weight ratio. $^{a}p < 0.05$ compared to control; $^{b}p < 0.05$ compared to monocrotaline-alone group.

marked decrease in hepatic taurine concentrations immediately following retrorsine administration may be one of the mechanisms of toxicity of this intensely hepatotoxic PA.

The subacute actions of monocrotaline on sulfur metabolism reported herein differ somewhat from the acute effects, observed within 24 h of administration. Acutely, GSH concentrations are increased in liver[37], while subacutely they are decreased. Acutely, γ-Glu-Cys synthetase, dipeptidase and cytosolic GSH S-transferase activities are all elevated[37], while subacutely there are no significant changes. γ-Glutamyl transpeptidase and microsomal GSH S-transferase activities are, however, elevated both acutely and subacutely.

Humans have a limited capacity for synthesizing taurine, deriving it instead from meat and, to a lesser extent, dairy products. Our findings that taurine protects against both the morbidity and mortality of monocrotaline raise the possibility, therefore, that persons on a low dietary intake of taurine may be more susceptible to PA toxicity. Such a possibility has obvious relevance to the outbreaks of PA poisoning that occur in underdeveloped countries as a result of food contamination[4,13,32]. Populations in these countries typically have a low dietary intake of taurine-containing substances. Nutritional status may also be a factor in PA poisonings that occur in western nations as a result of herbal use[6,12,24,28-30].

ACKNOWLEDGMENTS

These studies were partially supported by USPHS HL 25258 and the Taisho Pharmaceutical Co. We thank P. Consroe for help with LD_{50} calculations.

REFERENCES

1. Adams, R. and Rogers, E.F. 1939, The structure of monocrotaline, the alkaloid in Crotalaria spectabilis and Crotalaria retusa, *J.Amer.Chem.Soc.* 61:2815-2819.
2. Bras, G., Berry, D.M., and Gyorgy, P. 1957, Plants as aetiological factor in veno-occlusive disease of the liver, *The Lancet*, 960-962.
3. Bras, G., Jelliffe, D.B., and Stuart, K.L. 1954, Veno-occlusive disease of liver with nonportal type of cirrhosis, occurring in Jamaica, *Archives of Pathology*, 57:285-300.
4. Chauvin, P., Dillon, J.-C., Moren, A., Talbak, S., and Barakaev, S. 1993, Heliotrope poisoning in Tadjikistan, *Lancet*, 341:1663.
5. Estep, J.E., Lamé, M.W., Jones, A.D., and Segall, H.J. 1990, N-acetylcysteine-conjugated pyrrole identified in rat urine following administration of two pyrrolizidine alkaloids, monocrotaline and senecionine, *Toxicol.Lett.* 54:61-69.
6. Fox, D.W., Hart, M.C., Bergeson, P.S., Jarrett, P.B., Stillman, A.E., and Huxtable, R.J. 1978, Pyrrolizidine (Senecio) intoxication mimicking Reye's syndrome, *J.Pediatr.* 93:980-982.
7. Gillis, C.N., Huxtable, R.J., and Roth, R.A. 1978, Effect of monocrotaline pretreatment of rats on removal of 5-hydroxytryptamine and norepinephrine by perfused lung, *Brit.J.Pharmacol.* 63:435-443.
8. Glowaz, S.L., Michnika, M., and Huxtable, R.J. 1992, Detection of a reactive pyrrole in the hepatic metabolism of the pyrrolizidine alkaloid, monocrotaline, *Toxicol.Appl.Pharmacol.* 115:168-173.
9. Habig, W.H., Pabst, M.J., and Jakoby, W.B. 1974, Glutathione S-transferases: The first enzymatic step in mercapturic acid formation, *J.Biol.Chem.* 249:7130-7139.
10. Horecker, B.L. and Kornberg, A. 1948, The extinction coefficients of the reduced band of pyridine nucleotides, *J.Biol.Chem.* 175:385-390.
11. Huxtable, R.J. 1979, Pyrrolizidine alkaloids and the lung endothelium: a paradigm of lung damage resulting from circulating toxins, in: "Pyrrolizidine (senecio) alkaloids: toxicity, metabolism and poisonous plant control measures", Cheeke, P.R. ed., Oregon State University, Corvallis, pp. 43-56.
12. Huxtable, R.J. 1980, Herbal teas and toxins: novel aspects of pyrrolizidine poisoning in the United States, *Perspect.Biol.Med.* 24:1-14.
13. Huxtable, R.J. 1989, Human health implications of pyrrolizidine alkaloids and herbs containing them, in: "Toxicants of Plant Origin, Vol I: Alkaloids", Cheeke, P.R. ed., CRC Press, Boca Raton, Florida, pp. 41-86.

14. Huxtable, R.J. 1992, Physiological actions of taurine, *Physiol.Rev.* 72:101-163.
15. Huxtable, R.J. 1993, Hepatic nonaltruism and pulmonary toxicity of pyrrolizidine alkaloids, in: "Metabolic Activation and Toxicity of Chemical Agents to Lung Tissue and Cells", Gram, T.E. ed., Pergamon Press, New York, pp. 215-239.
16. Huxtable, R.J. 1996, Taurine: Past, present and future, in: "Taurine: Basic and Clinical Aspects", Huxtable, R.J., Azuma, J., Nakagawa, M., Kuriyama, K. and Baba, A. eds., Plenum, New York.
17. Huxtable, R.J., Ciaramitaro, D., and Eisenstein, D. 1978, The effect of a pyrrolizidine alkaloid, monocrotaline, and a pyrrole, dehydroretronecine, on the biochemical functions of the pulmonary endothelium, *Mol.Pharmacol.* 14:1189-1203.
18. Lafranconi, W.M., Ohkuma, S., and Huxtable, R.J. 1985, Biliary excretion of novel pneumotoxic metabolites of the pyrrolizidine alkaloid, monocrotaline, *Toxicon*, 23:983-992.
19. Lawrence, R.A., Sunde, R.A., Schwartz, G.L., and Hoekstra, W.G. 1974, Glutathione peroxidase activity in rat lens and other tissues in relation to dietary selenium intake, *Exp.Eye Res.* 18:563-569.
20. Massey, V. and Williams, C.H. 1965, On the reaction mechanism of yeast glutathione reductase, *J.Biol.Chem.* 240:4470-4480.
21. Mattocks, A.R. 1968, Toxicity of pyrrolizidine alkaloids, *Nature (London)*, 217:723-728.
22. Mattocks, A.R. 1972, Acute hepatotoxicity and pyrrolic metabolites in rats dosed with pyrrolizidine alkaloids, *Chem.-Biol.Interact.* 5:227-242.
23. Mattocks, A.R., Croswell, S., Jukes, R., and Huxtable, R.J. 1991, Identity of a biliary metabolite formed from monocrotaline in isolated, perfused rat liver, *Toxicon*, 29:409-415.
24. McGee, J.O., Patrick, R.S., Wood, C.B., and Blumgart, L.H. 1976, A case of veno-occlusive disease of the liver in Britain associated with herbal tea consumption, *J.Clin.Path.* 29:788-794.
25. Miranda, C.L., Reed, R.L., Guengerich, F.P., and Buhler, D.R. 1991, Role of cytochrome P450IIIA4 in the metabolism of the pyrrolizidine alkaloid senecionine in human liver, *Carcinogenesis*, 12:515-519.
26. Nigra, L. and Huxtable, R.J. 1992, Hepatic glutathione concentrations and the release of pyrrolic metabolites of the pyrrolizidine alkaloid, monocrotaline, from the isolated perfused liver, *Toxicon*, 30:1195-1202.
27. Paglia, D.E. and Valentine, W.N. 1967, Studies on the quantitative and qualitative characterization of erythrocyte glutathione peroxidase, *J.Lab.Clin.Med.* 70:158-169.
28. Ridker, P.M., Ohkuma, S., McDermott, W.V., Trey, C., and Huxtable, R.J. 1985, Hepatic veno-occlusive disease associated with the consumption of pyrrolizidine-containing dietary supplements, *Gastroenterology*, 88:1050-1054.
29. Roulet, M., Laurini, R., Rivier, L., and Calame, A. 1988, Hepatic veno-occlusive disease in newborn infant of a woman drinking herbal tea, *J.Pediatr.* 112:433-436.
30. Stillman, A.E., Huxtable, R., Consroe, P., Kohnen, P., and Smith, S. 1977, Hepatic veno-occlusive disease due to pyrrolizidine poisoning in Arizona, *Gastroenterology*, 73:349-352.
31. Stuart, K.L. and Bras, G. 1957, Veno-occlusive disease of the liver, *Quart.J.Med.* 26:291-315.
32. Tandon, H.D., Tandon, B.N., and Mattocks, A.R. 1978, An epidemic of veno-occlusive disease of the liver in Afghanistan, *Am.J.Gastroenterology*, 70:607-613.
33. Thut, P.D., Hruska, R.E., Huxtable, R.J., and Bressler, R. 1976, Effect of taurine on eating and drinking behavior, in: "Taurine", Huxtable, R.J. and Barbeau, A. eds., Raven Press, New York, pp. 357-364.
34. Williams, D.E., Reed, R.L., Kedzierski, B., Dannan, G.A., Guengerich, F.P., and Buhler, D.R. 1989, Bioactivation and detoxication of the pyrrolizidine alkaloid senecionine by cytochrome P-450 enzymes in rat liver, *Drug Metab.Disp.* 17:387-392.
35. Yan, C.C. and Huxtable, R.J. 1995, Relationship between glutathione concentration and metabolism of the pyrrolizidine alkaloid, monocrotaline, in the isolated, perfused liver, *Toxicol.Appl.Pharmacol.* 130:132-139.
36. Yan, C.C. and Huxtable, R.J. 1995, Effects of the pyrrolizidine alkaloid, retrorsine, on sulfur metabolism in the liver, *Proc.Western Pharmacol.Soc.* 38:37-40.
37. Yan, C.C. and Huxtable, R.J. 1995, The effect of the pyrrolizidine alkaloids, monocrotaline and trichodesmine, on tissue pyrrole binding and glutathione metabolism in the rat, *Toxicon*, 33:627-634.
38. Yan, C.C. and Huxtable, R.J. 1995, Effect of the pyrrolizidine alkaloid, monocrotaline, on bile composition of the isolated, perfused rat liver, *Life Sci.* 57:617-626.
39. Yan, C.C. and Huxtable, R.J. 1996, Effect of pyrrolizidine alkaloids on taurine and sulfur metabolism in the rat liver, in: "Taurine: Basic and Clinical Aspects", Huxtable, R.J., Azuma, J., Nakagawa, M., Kuriyama, K. and Baba, A. eds., Plenum, New York.
40. Yan, C.C. and Huxtable, R.J. 1996, Effects of taurine and guanidinoethane sulfonate on right ventricular hypertrophy induced by the pyrrolizidine alkaloid, monocrotaline, *Biochem.Pharmacol.* 51:321-329.

Part 5

TAURINE AND THE AUDIOVISUAL SYSTEMS

34

ULTRASTRUCTURAL LOCALIZATION OF TAURINE IMMUNOREACTIVITY IN THE PINEAL ORGAN AND RETINA OF THE PIGEON

N. Lake,[1] M. Ueck,[2] A. Hach,[2] and C. Verdone-Smith[1]

[1] Departments of Physiology and Ophthalmology
McGill University
3655 Drummond Street
Montreal, Quebec, Canada H3G 1Y6
[2] Institut für Anatomie und Zytobiologie
Justus-Liebig-Universität Giessen
Aulweg 123, D35385 Giessen, Germany

INTRODUCTION

While taurine is ubiquitous in animal tissues, some of the highest levels occur in the hypothalamus, pituitary gland, retina and pineal organ[3,11,13]. This may indicate a role of taurine in processes that are more pronounced in neurosecretory and neurosensory cells than in other cell types. For example, taurine has been shown to interact with membrane phospholipids to alter their Ca^{2+} binding properties[7], and may thus influence Ca^{2+}-dependent processes such as ATPase ion pumps, ion channels flux, or Ca^{2+}-activated events[5,6], many of which contribute to neurosecretion and phototransduction. On the other hand, the large transient ionic fluxes associated with activity of pineal and retinal neurosensory/neurosecretory cells may also impose osmotic stresses that require use of taurine as an osmolyte; there is much evidence concerning a role for taurine in osmoregulation in tissues ranging from marine organisms, to the kidney and the brain[5,6]. The pineal organ contains many structures homologous to retinal photoreceptors. We hoped to gain information about the role of taurine in these tissues by examining and comparing the retinal and pineal distribution of endogenous taurine using immunocytochemistry.

MATERIALS AND METHODS

We have carried out these studies at the light (LM) and electron microscope (EM) level in tissues from the adult pigeon. Following fixation in 3.5% glutaraldehyde, retinas

Taurine 2, edited by Huxtable et al.
Plenum Press, New York, 1996

and pineal glands were postfixed in osmium tetroxide and embedded in Spurr resin. The method of generation and characterization of the taurine antibody (to taurine-BSA conjugates) has been described[9]. For LM immunocytochemistry, a biotinylated secondary antibody was detected using streptavidin conjugated to horseradish peroxidase, with diaminobenzidine or aminoethyl carbazol as chromogens. For ultrastructural studies the Protein A-gold method of Bendayan[2] was used as described[9]. Negative controls included omission of the primary antibody, or the use of antisera that had been preadsorbed with taurine-BSA conjugates.

RESULTS AND DISCUSSION

Taurine Immunoreactivity in Pigeon Retina

Our findings in the pigeon retina confirm our results in mammalian retinas[1,8,9], in that the highest levels of taurine-like immunoreactivity were found in the cone photoreceptors, in select cells of the inner nuclear layer and in sublaminae of the inner plexiform (synaptic) layer. In the EM studies a high density of immunogold particles was found within the presynaptic endings of cones, but not in the postsynaptic cells. The immunogold particles were not associated directly with the synaptic vesicles, suggesting that taurine is not present as a neurotransmitter. However, in view of its high cytoplasmic levels, it could be released by nonvesicular means. Alternatively it may be here in some role associated with the Ca^{2+} dependency of synaptic release mechanisms, or the operations of ATPase ion pumps, or in an osmoregulatory capacity[5,6].

The cone photoreceptor inner segment is divided into an ellipsoid region (characterized by its concentration of mitochondria and ergastoplasm) and a paraboloid region (containing smooth endoplasmic reticulum). At both LM and EM levels the ellipsoid region was much more intensely immunolabelled than the paraboloid. The step change in reactivity was clear-cut at the border of these two regions. This indicates that the conditions of our fixation (high glutaraldehyde and Mg^{2+} levels) limit diffusion of taurine, and that its function is probably related to mitochondrial and ergastoplasmic function, and but little associated with glycogen metabolism and smooth endoplasmic reticulum functions. The photopigment-containing photoreceptor outer segment membrane stacks were labelled with a moderate density of immunogold particles.

Taurine Immunoreactivity in Pigeon Pineal Organ

Observations using the light microscope indicated that there were rosettes of heavily immunostained pinealocytes surrounded by faintly stained pinealocytes. This taurine immunoreactivity was confined to the so-called dark pinealocytes identified at the EM level by their more dense cytoplasm. Regions of high taurine immunoreactivity common to both pigeon retina and pineal organ, seen at the EM level, were sites in the ergastoplasm, rich in polyribosomes and rough endoplasmic reticulum. The most striking examples were found in the pinealocytes: large clusters or chains of immunogold particles found adjacent to filamentous structures. These localizations support a direct or indirect functional relationship of taurine to ATP dependent processes.

Although pinealocytes share certain morphological similarities with retinal photoreceptors, structures that were highly reactive in retinal cones (synaptic terminal regions, outer segment membranous whorls) were unreactive in the pineal gland. It is possible that a circadian rhythm for taurine content, such as has been described for the rat[4,10], exists in the pigeon, and that our a.m. observation time was at a low point for the pigeon pineal cycle.

However, more recently another group reported the absence of cyclic alterations of rat pineal taurine content[12], so this point is controversial, and we need to make further studies to examine this matter in the pigeon pineal organ.

REFERENCES

1. Battista, G. and Lake, N. 1991, Taurine in developing rabbit retina, *Soc.Neurosci.Abst.* 17:559 (Abstract).
2. Bendayan, M. 1984, Protein A - gold electron microscopic immunochemistry: methods, application, and limitations, *J.Electron Microscope Techniques*, 1:243-270.
3. Crabai, F., Sitzia, A., and Pepeu, G. 1974, Taurine concentration in the neurohypophysis of different animal species, *J.Neurochem.* 23:1091-1092.
4. Grosso, D.A., Bressler, R., and Benson, B. 1978, Circadian rhythm and uptake of taurine by the rat pineal gland, *Life Sci.* 22:1789-1798.
5. Huxtable, R.J. 1989, Taurine in the central nervous system and the mammalian actions of taurine, *Prog.Neurobiol.* 32:471-533.
6. Huxtable, R.J. 1992, Physiological actions of taurine, *Physiol.Rev.* 72:101-163.
7. Huxtable, R.J. and Sebring, L.A. 1986, Towards a unifying theory for the action of taurine, *Trends Pharmacol.Sci.* 7:481-485.
8. Lake, N., De Marte, L., and Verdone-Smith, C. 1989, Immunocytochemical localization of taurine-like immunoreactivity (TLI) in the ferret retina, *J.Neurochem.* 52:S158D (Abstract).
9. Lake, N. and Verdone-Smith, C. 1989, Immunocytochemical localization of taurine in the mammalian retina, *Curr.Eye Res.* 8:163-173.
10. Leonard, B.E., Neuhoff, V., and Tongue, S.R. 1977, The effect of chronic administration of d-amphetamine upon circadian changes in amino acids in the pineal gland and pituitary glands of the rat, *J.Neurosci.Res.* 1:63-92.
11. Lombardini, J.B. 1976, Regional and subcellular studies on taurine in the rat central nervous system, in: "Taurine", Huxtable, R. and Barbeau, A. eds., Raven Press, New York, pp. 311-326.
12. McNulty, J.A., Kus, L., and Ottersen, O.P. 1992, Immunocytochemical and circadian biochemical analysis of neuroactive amino acids in the pineal gland of the rat: Effect of superior cervical ganglionectomy, *Cell Tissue Res.* 269:515-523.
13. Voaden, M.J., Lake, N., Marshall, J., and Morjaria, B. 1977, Studies on the distribution of taurine and other neuroactive amino acids in the retina, *Exp.Eye Res.* 25:249-257.

35

TAURET

A Taurine-Related Endogenous Substance in the Retina and Its Role in Vision

Andranik M. Petrosian,[1] Jasmine E. Haroutounian,[1] and Lida V. Zueva[2]

[1] Institute of Biochemistry of National Academy of Sciences of Armenia
Yerevan 44, Armenia
[2] Sechenov Institute of Evolutionary Physiology and Biochemistry of National Academy of Sciences of Russia
St.Peterbourg 223, Russia

INTRODUCTION

Since the first description of taurine in 1827[35] the range of biological phenomena involving this substance has progressively expanded. Efforts have been made to create a unifying theory for the actions of taurine[10,11,36]. It is believed that "A limited number of basic mechanisms of action may underlie this confusing plethora of biological effects"[10]. In this respect, the actions of taurine as an osmolyte, its conjugation with bile acids and lipids, and its interactions with Ca^{2+}, phospholipids, and chemoreceptors are of definite interest. Final understanding of the role of taurine in any organ or tissue is related to the elucidation of the relative contribution of these or a few other basic mechanisms in its function. It is necessary to clarify which is the primary action among the multiple actions of taurine in a given organ or tissue. For example, in some marine species the action of taurine as an osmolyte[2,6] seems primary. In liver, probably more central is taurine conjugation with bile acids and lipids, while in the heart taurine-Ca^{2+} interactions seem primary[3,5,29], although taurine may also be an osmoregulator in this organ[34]. To understand the role of taurine in the retina it seems best to consider each cell system separately. Large amounts of taurine (10-40 mM) are characteristic for the retina of all species tested, even for those which have low taurine content in other tissues[17,20,37]. Among the richer layers, photoreceptors are most abundant in taurine, containing 50-80% of the taurine pool of the retina[15,20,37]. High levels of taurine in photoreceptor cells are mainly supported by taurine transport, which is predominantly localized around the photoreceptor cell nuclei[14]. About a 50% drop of taurine levels in the retina can be obtained either by a taurine-deficient diet in animals with low liver capacity for taurine synthesis[9,12,31,33] or by treatment of rats (which can synthesize taurine) with the taurine transport antagonist guanidinoethane sulfonate (GES)[13,25]. Retinal depletion of taurine causes serious alterations in the structure. The most severe retinal damage appears in the regions of highest rhodopsin content[9,12,13,25,31,33]. Taurine deficiency causes a noticeable

Taurine 2, edited by Huxtable et al.
Plenum Press, New York, 1996

reduction of rhodopsin level in rat photoreceptors[13] and causes a drop of amplitude of the a- and b-waves of the electroretinogram (ERG)[13,31].

A parallel change of taurine content and the ERG in developing rat retina has been revealed[21]. A close relationship between taurine and retinal rod outer segments (ROS) have been also confirmed in light-mediated experiments. Surprisingly it appeared that light stimulation evokes taurine efflux from isolated frog ROS preloaded with [^{35}S]taurine[28]. Taurine and its natural occurring analog, homotaurine, exert powerful protective action against light-induced damage of isolated frog ROS[22,23].

Among the basic actions of taurine, three are known to act in the photoreceptors: It is one of the main organic osmolytes[24]; it has a dramatic action on Ca^{2+} fluxes[18,23]; and it exerts negative action on phosphorylation[18]. These actions do not explain the need for such high concentrations of taurine in the photoreceptors. It has not been clarified why taurine deficiency causes severe damage of the ROS structure or why light evokes release of taurine from ROS. The answer to these questions seems closely related to establishment of the primary role of taurine in photoreceptor cells. Structural investigations, whether there are connections between disk rims and plasma membrane in the retinal ROS, unexpectedly have revealed the possible central role of taurine. Supposition has been made regarding involvement of taurine in the transport of retinoids between photoreceptors and pigment epithelium[26,27]. In this paper we present new data in favor of this hypothesis.

MATERIAL AND METHODS

Animals

Frogs (*Rana ridibunda*) were used throughout this study. Animals were maintained at 10-20°C in transparent glass jars and subjected to standard light cycles. Frogs were dark-adapted overnight or at least 3 h before experiments. After dark-adaptation at 20°C all procedures of tissue preparation were performed in dim red light.

High-Pressure Liquid Chromatography

For analysis of tauret (retinylidentaurine) in the retina and pigment epithelium, thin-layer chromatography (TLC) and high-pressure liquid chromatography (HPLC) were applied. *Trans* and 11-*cis* tauret synthesized by a one-step reaction were used throughout this study. The retina and pigment epithelium were immediately detached in liquid nitrogen. Tissues were dried by lyophilization and then homogenized by a glass-glass homogenizer in absolute methanol. Homogenates of the retinas and pigment epithelium were extracted on ice for 30 min and then clarified by centrifugation. Supernatant was dried and kept O_2-free in the dark, below -10°C for further TLC and HPLC analysis. Dried extracts of the retina and pigment epithelium dissolved in absolute methanol and samples of synthetic tauret dissolved in methanol were separately spotted on Armsorb sheets of silica gel. The chromatograms were developed with chloroform:methanol:trifluoroacetic acid in ratios of 20:6:1 or 10:1:0.1. The blots were visualized by spraying with sulfuric acid and heating or by treatment with I_2 vapor. For HPLC, volumes of 10 μl of methanolic extracts of retina, pigment epithelium or methanolic solutions of synthetic tauret standards were injected on μ-Bondapack C_{18} columns for HPLC analysis. Elution was with methanol. Flow rate was 1 ml/min. The absorbance was monitored at 360 nm. The other HPLC system was based on Separon SGX columns (7 μm; 150 x 33 mm). The mobile phase was chloroform:methanol:trifluoroacetic acid in the ratio 87:13:0.1. Dried methanolic extracts of retina, pigment epithelium or synthetic *trans*-tauret were dissolved in eluent and volumes of 10 μl of samples were injected onto HPLC columns. Flow rate was 1.5 ml/min. The absorbance was monitored at 420 or 440 nm.

Absorbance Microspectrophotometry

Dark-adapted retina was detached in saline containing; NaCl 85 mM, $NaHCO_3$ 10 mM, KCl 2.5 mM, $CaCl_2$ l mM, $MgSO_4$ 0. 2 mM, glucose 10 mM, and Tris HCl 10 mM at pH 7.5. The saline was saturated with O_2/CO_2 (95:5). Microspectrophotometric measurements were carried out in dark-adapted and bleached retinae which were post-incubated in buffered saline containing 11-*cis*-tauret (≈1 mM). Bleaching of retinae was carried out for 2 min under 2000 lux illumination with a heat protective filter. Small pieces of retina, dark adapted, bleached or post-incubated with 1 mM 11-*cis*-tauret after bleaching were placed between two thin glasses in a saline drop. After slight pressure the glasses at their edges were stuck by vaseline to prevent the retina and floating ROS from drying. A microspectrophotometer was used, constructed on the basis of MCF-1 Lomo[8]. Absorbance of transmitted visible illumination, scanning a range from 400 to 650 nm was measured in single ROS.

RESULTS

A substantial shift about 80 nm in absorbance of the *trans*-tauret after protonation has been noticed (Fig. 1). Being easily water-soluble, tauret simultaneously is hydrolyzing with half time period about 20 min. TLC has been used to detect this transient substance in the retina and pigment epithelium. On TLC, R_f 0.65 corresponds to standard *trans*-tauret. On the basis of TLC analysis it can be concluded that endogenous tauret exists in the dark-adapted frog retina and pigment epithelium.

HPLC measurements have been done using methanol eluent to separate light molecular weight substances on a μ-Bondapack C_{18} column. The HPLC profile of the synthetic tauret is shown in Fig.2A. The major peak (arrow) corresponds to all *trans*-tauret. This peak disappears after hydrolysis, consistent with the identity of the major peak with all *trans*-tauret. Fig. 2B and C are HPLC profiles of methanol extracts of light molecular weight substances of the dark-adapted retina and pigment epithelium. Minor peaks (arrows) on the HPLC profiles of the retina and pigment epithelium indicate that there exists detectable endogenous *trans*-tauret in these tissues.

HPLC analysis based on Sepharon SGX column is shown in Figure 3. The only peak on the HPLC profile (Fig 3A) corresponds to synthetic *trans*-tauret. Figure 3B shows a profile of a methanol extract of substances in dark-adapted retina. This HPLC analysis also demonstrates the existence of endogenous tauret in the retina (arrow). Previous investigations done by electron microscopy have not marked revealed structural changes in retina incubated with low taurine concentrations.[26,27] In this study, about a 10% increase in ROS diameter has been noted by microspectrophotometry in isolated retina incubated 10-20 min in saline containing 1 mM 11-*cis*-tauret. This increase, probably related to an osmotic effect caused by active transport of taurine, will act on specific absorbance of ROS. Therefore it would be more expedient to take absorbance rather than specific absorbance as a measure of rhodopsin content to compare rhodopsin level in ROS at different conditions. Some typical results of microspectrophotometric measurements are shown in Figure 4. The absorbance spectrum of a single ROS detached from dark-adapted frog retina has a maxima at 500 nm, which corresponds to rhodopsin (Fig. 4A). After bleaching of the retina by 2000 lux illumination for 2 min, absorbance of a single ROS was decreased noticeably (Fig. 4B).

Following such bleaching, short or long lasting post-incubations of the isolated frog retina in darkness do not cause detectable regeneration of rhodopsin in the ROS. However,

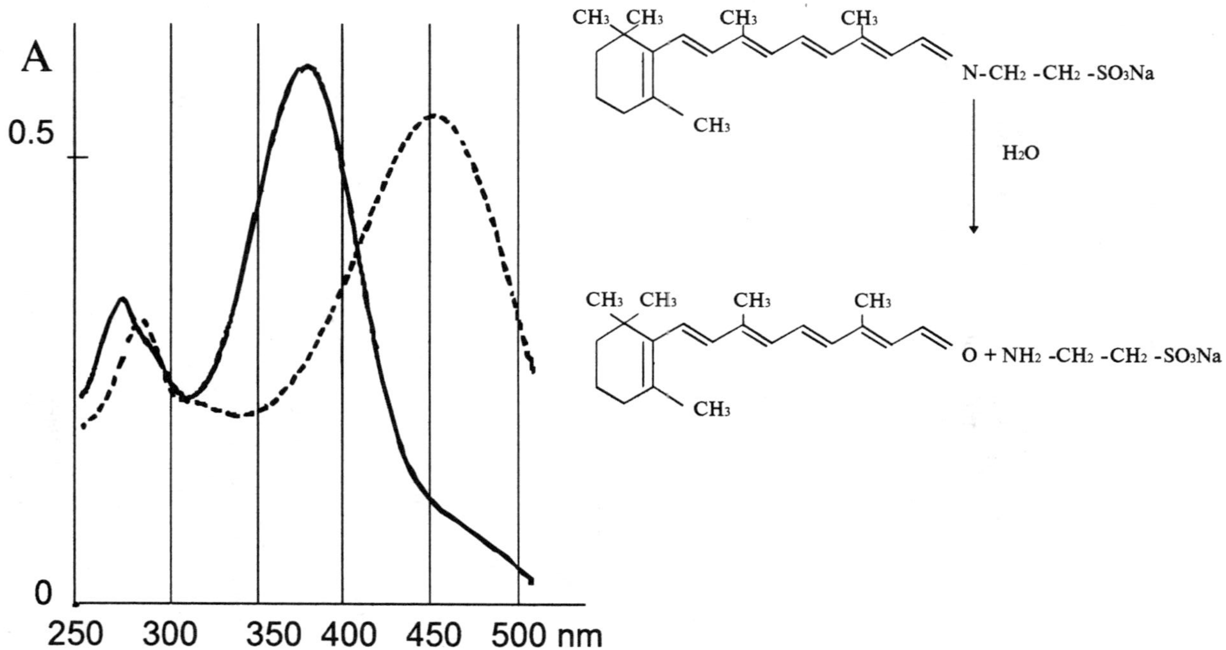

Figure 1. Structures of *trans*-tauret and its hydrolytic products. Left: absorption spectrum of tauret before (solid line) and after (dashed line) protonation.

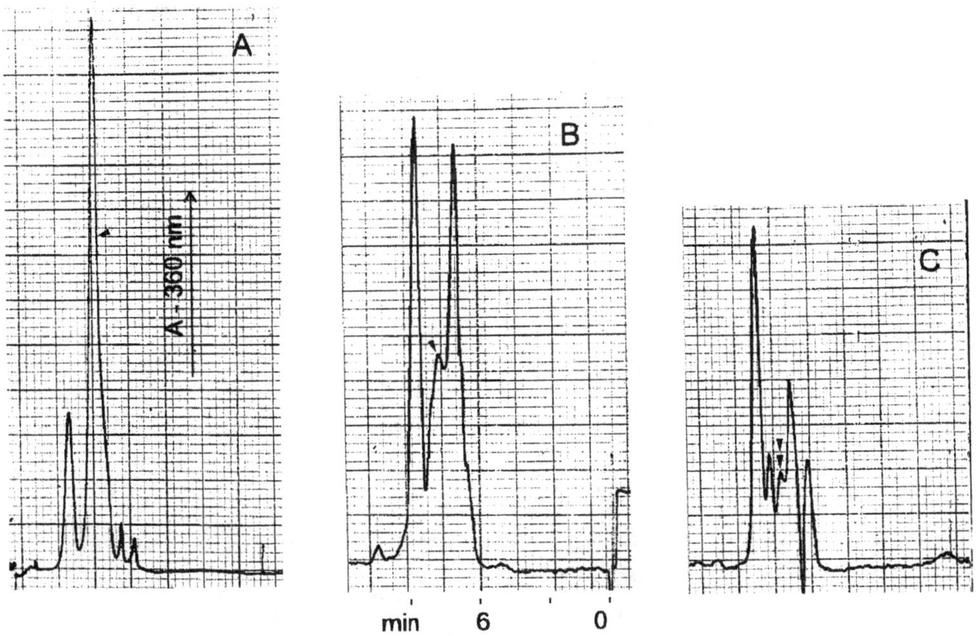

Figure 2. HPLC profiles on μ-Bondapack C_{18} columns. A: synthetic tauret; B: retinal extract; C: extract of pigment epithelium.

after such bleaching considerable recovery of rhodopsin has been obtained when retinas were post-incubated for 10 min or 20 min in darkness in the saline containing about 1 mM ll-*cis*-tauret. Typical absorbance spectra demonstrating the ability of 11-*cis*-tauret to regenerate rhodopsin in the ROS of isolated retinas incubated in aqueous solution is shown in Figure 4C. Figure 5 summarizes results of these experiments. Column heights reflect the mean absorbance of ROS at 500 nm wavelength.

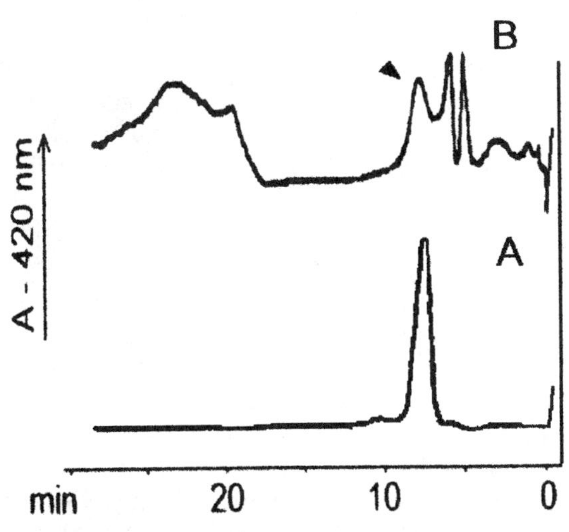

Figure 3. HPLC profiles on Separon SGX columns. A: synthetic tauret; B: retinal extract.

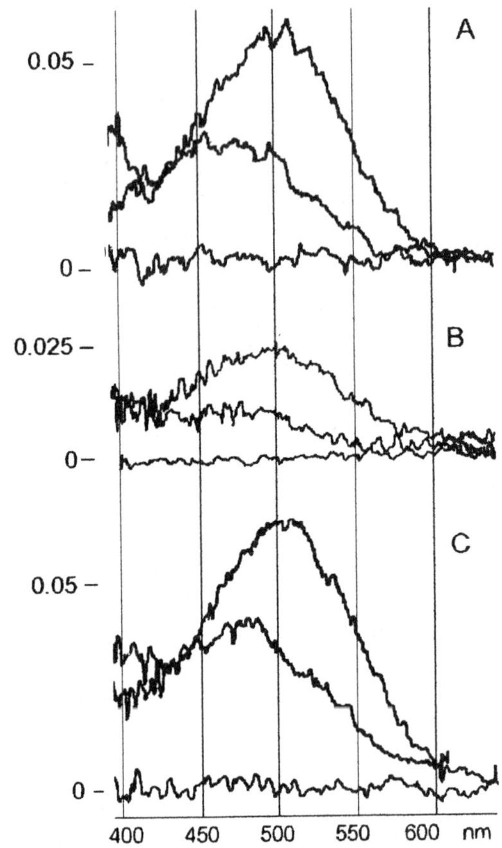

Figure 4. Absorbance spectra of single ROS. A: dark-adapted; B: bleached, C: 10 min exposure to 11-*cis*-tauret after bleaching.

DISCUSSION

We have tried to elucidate if tauret is an endogenous substance in the retina and pigment epithelium. For this propose TLC and two HPLC systems have been applied. There is no full separation of methanol extracts of retina and pigment epithelium in the systems used (Figs. 2 and 3). However, peaks characteristic for tauret were seen on HPLC and TLC. Their spectral characteristics and their disappearance after hydrolysis indicate that tauret is endogenous in the retina and pigment epithelium. It is necessary to note that special precautions have been taken during sample preparation procedures to protect tauret from hydrolysis. This last point probably was the reason why tauret has escaped previous observation.

It is of interest that previously a compound similar to tauret has been revealed among liver metabolites in the rats[32]. This compound, 2-[8-(6-hydroxymethyl)-2,6-dimethyl-cyclohexan-1-yl]-2,6-dimetoctadienamide ethane sulfonic acid, was shown to be a taurine conjugate of a retinoic acid metabolite and hence the trivial name retinotaurine is suggested for it. It was suggested that retinotaurine is biologically inactive and represents a metabolite in the inactivation pathway for retinoic acid[32].

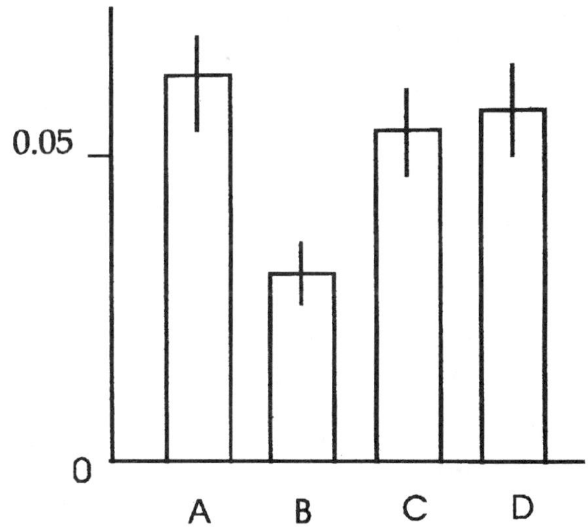

Figure 5. Mean absorbance of ROS at 500 nm. A and B as in Fig. 4. C and D after exposure to 11-*cis*-tauret for 10 and 20 min. Confidence interval 90%. 8 - 13 experiments.

Currently there are several other pieces of direct and indirect evidence indicating a close relationship between taurine and retinoids. Elevation of retinol levels in the liver of taurine-deficient kittens has been observed[16]. In cats with hypertrophic cardiopathy, hyperthyroidism or with cardiovascular disease, levels of taurine in plasma are noticeably decreased. In dilated cardiopathy, plasma taurine concentrations are only 38% of the normal level. In contrast, retinol levels in plasma are 15 to 40% higher than the levels in healthy cats[4]. Dietary supplementation with 13-*cis* retinoic acid of methionine-supplemented rats cause marked elevation in hepatic taurine levels[30]. These associations between taurine and retinoids in blood, liver, retina and pigment epithelium are in agreement with hypothesis that taurine may be involved in the transport of retinoids in vertebrate tissue[26,27]. By being conjugated with taurine, water-insoluble retinoids can more readily be delivered to target tissues and then freed by hydrolysis. This type of delivery may be an mechanisms in the transport of retinoids, which otherwise are known to be transported by retinoid binding proteins.

The nature of tauret formation in the retina and pigment epithelium is not clear. The significance of nonenzymatic chemical reactions of the amino group of taurine with 11-*cis* and *trans*-retinals in this process can not be excluded. Recently, it has been shown that taurine has a high reactivity with aldehydes, which suggests the possibility of an inhibiting effect of taurine against the modification of protein, as well as an antioxidative effect through the reactions of taurine with aldehydes *in vivo*[19].

Levels of tauret in the retina and pigment epithelium are probably around a few hundred μM. Free retinoid in ocular tissues are normally low. However, in situations, for example, such as strong illumination after prolonged dark adaptation, the level of all-*trans*-retinol rises rapidly. *trans*-Retinol can be quickly removed via formation of *trans*-tauret. In this way, damage to the ROS can be prevented. Following illumination, release of taurine from isolated ROS[28] is more likely a response to prompt formation of tauret and its release from ROS. The high concentrations of taurine found in ROS[20,37] is probably one determinant promoting to this process. Taurine-associated retinoid transport in the opposite direction, from pigment epithelium to photoreceptors, may be essential in processes related to regeneration of rhodopsin. To check this we have used 11-*cis*-tauret in experiments based on microspectrophotometric measurements. It has been shown that after bleaching no spontaneous recovery of rhodopsin can occur in ROS in isolated frog retina (Fig.4B). When retinas after bleaching were postincubated in darkness for 10 or 20 min in saline containing about 1 mM 11-*cis*-tauret, rhodopsin levels in ROS recovered to achieve 84 or 90% of the value in dark level, respectively (Fig. 4C, Fig. 5C, D). It is interesting that taurine depletion showed some features reminiscent of damage by constant light[13]. In a rat model of taurine depletion carried out under cyclic lighting, a marked enhancement in processes of functional disturbance occurs. Compared with dark maintenance, in cyclic lighting a sharp decrease of rhodopsin content in the retina and in ERG b wave amplitude occurred, which as proposed, implies that endogenous taurine may provide some protection against the damaging aspects of light and that at least part of the effects of taurine deficiency are dependent on, or mediated by, light[13]. It seems taurine is also involved in vitamin A-related processes in pigment epithelium and retina which are not linked with light. Recently it has been established that, *in vitro*, taurine stimulated retinal pigment epithelium proliferation[7], and induced regeneration of goldfish retina in culture[17]. Taurine-containing heat stable component (<l kDa; extracted from retina) has exerted a stimulatory action on rod photoreceptor development in culture[1].

On the base of these literature data plus our results, it can be concluded that probably taurine, via conjugation with all-*trans* and 11-*cis* retinals, is involved in processes of removal of *trans*-retinal from ROS in light as well as in the process of rhodopsin regeneration in darkness. In light of such interpretation concerning the role of taurine in the photoreceptors, it can be concluded that in part the severe alterations in photoreceptor structure caused either by a taurine free-diet[9,12,31,33] or by GES treatment[13,25] are initiated by the deterioration of the processes of removal of all-*trans* retinal from ROS and transport of 11-*cis* retinal from pigment epithelium to photoreceptors. In our previous investigations it has been established that tauret can penetrate intradisk spaces through Ca^{2+}-dependent connections between ROS plasma membrane and disk rims[26,27]. In other words, there exist vitamin A-selective molecular channels in the ROS and taurine seems to play a key role in their function. It is thought that the primary action of taurine in vision is associated with retinoids transportation. Further studies are needed for final elucidation.

REFERENCES

1. Altshuler, D., Lo Turco, J.J., Rush, J., and Cepko, C. 1993, Taurine promotes the differentiation of a vertebrate retinal cell type in vitro, *Development*, 119:1317-1328.
2. Amende, L.H. and Pirce, S.K. 1980, Cellular volume regulation in salinity stressed molluscs: the response of Noetia ponderosa (Arcidae) red blood cells to osmotic variations, *J.Comp.Physiol.* 138:283-289.
3. Azuma, J., Katsume, H., Kagoshima, T., Furukawa, K., Awata, N., Ishigama, T., Yamagami, T., Ishikawa, H., Iwata, H., Kishimoto, S., and Yamaura, Y. 1988, Clinical evaluation of taurine in congestive heart failure, in: "Taurine and the heart", Iwata, H., Lombardini, J.B. and Segawa, T. eds., Kluwer Acad. Publ. Boston, pp. 75-97.
4. Fox, P.R., Trautwein, E.A., Hayes, K.C., Bond, B.R., Sisson, D.D., and Moise, N.S. 1993, Comparison of taurine, α-tocopherol, retinol, selenium, and total triglycerides and cholesterol concentrations in cats with cardiac disease and in healthy cats, *Am.J.Vet.Res.* 54:563-569.
5. Franconi, F., Stendardi, I., Martini, F., Zilletti, L., and Giotti, A. 1982, Interaction between organic calcium-channel blockers and taurine in vitro and in vivo, *J.Pharm.Pharmacol.* 34:329-330.
6. Fugelli, K. and Thoroed, S.M. 1990, Taurine and volume regulation in fish cells, in: "Taurine: Functional Neurochemistry, Physiology and Cardiology", Pasantes-Morales, H., Martin, D.L., Shain, W. and del Rio, R.M. eds., Wiley-Liss, New York, pp. 481-488.
7. Gabrielian, K., Wang, H.M., Ogden, T.E. and Ryan, S.J. 1992, In vitro stimulation of retinal pigment epithelium proliferation by taurine, *Curr.Eye Res.* 11:481-487.
8. Govardovsky, V.I. and Zueva, L.V. 1988, A simple high-sensitive recording microspectrophotometer, *Cytology, Moscow*, 30:499-502.
9. Hayes, K.C., Carey, R.E., and Schmidt, S.Y. 1975, Retinal degeneration associated with taurine deficiency in the cat, *Science*, 188:949-951.

10. Huxtable, R.J. 1990, The interaction between taurine, calcium and phospholipids: Further investigations of a trinitarian hypothesis, in: "Taurine: Functional Neurochemistry, Physiology, and Cardiology", Pasantes-Morales, H., Martin, D.L., Shain, W. and del Río, R.M. eds., Wiley-Liss, New York, pp. 185-196.
11. Huxtable, R.J. and Sebring, L.A. 1986, Towards a unifying theory for the action of taurine, *Trends Pharmacol.Sci.* 7:481-485.
12. Imaki, H., Jacobson, S.G., Kemp, C.M., Knighton, R.W., Neuringer, M., and Sturman, J. 1993, Retinal morphology and visual pigment levels in 6- and 12-month-old rhesus monkeys fed a taurine-free human infant formula, *J.Neurosci.Res.* 36:290-304.
13. Lake, N. 1989, The role of taurine in retinal function, in: "Extracellular and intracellular messengers in the vertebrate retina", Redburn, D.H. and Pasantes-Morales, H. eds. Alan R. Liss, Inc. pp. 61-86.
14. Lake, N. and Orlowski, J. 1995, Cellular studies of the taurine transporter, *This volume.*
15. Lake, N. and Verdone-Smith, C. 1990, Immunocytochemical localization of taurine within glial cells in the optic nerve of adult albino rats, *Curr.Eye Res.* 9:1115-1120.
16. Lehmann, A., Knutsson, L., and Bosaeus, I. 1990, Elevation of retinol levels and suppression of alanine aminotransferase activity in the liver of taurine-deficient kittens, *J.Nutr.* 120:1163-1170.
17. Lima, L., Drujan, B., and Matus, P. 1990, Spatial distribution of taurine in the teleost retina and its role in retinal tissue regeneration, in: "Taurine: Functional Neurochemistry, Physiology, and Cardiology", Pasantes-Morales, H., Martin, D.L., Shain, W. and del Río, R.M. eds., Wiley-Liss, New York, pp. 103-112.
18. Lombardini, J.B. 1991, Taurine: Retinal function, *Brain Res.Rev.* 16:151-169.
19. Ogasawara, M., Nakamura, T., Koyama, I., Nemoto, M., and Yoshida, T. 1993, Reactivity of taurine with aldehydes and its physiological role, *Chem.Pharm.Bull.(Tokyo)*, 41:2172-2175.
20. Orr, H.T., Cohen, A.I., and Lowry, O.H. 1976, The distribution of taurine in the vertebrate retina, *J.Neurochem.* 26:609-611.
21. Parmer, R., Sheikh, K.H., Dawson, W.W., and Toskes, P.P. 1982, A parallel change in taurine and the ERG in the developing rat retina, *Comp.Biochem.Physiol.* 72:109-112.
22. Pasantes-Morales, H. and Cruz, C. 1985, Taurine and hypotaurine inhibit light-induced lipid peroxidation and protect rod outer segment structures, *Brain Res.* 330:154-157.
23. Pasantes-Morales, H., Lopez-Escalera, R., and Macedo, M.D. 1989, Taurine and ionic fluxes in photoreceptors, in: "Extracellular and Intracellular Messengers in the Vertebrate Retina", Redburn, D.H. and Pasantes-Morales, H. eds. Alan R. Liss, Inc. pp. 87-104.
24. Pasantes-Morales, H., Morán, J., and Schousboe, A. 1990, Taurine release associated to cell swelling in the nervous system, in: "Taurine: Functional Neurochemistry, Physiology, and Cardiology", Pasantes-Morales, H., Martin, D.L., Shain, W. and del Río, R.M. eds., Wiley-Liss, New York, pp. 369-376.
25. Pasantes-Morales, H., Quesada, O., Cárabez, A., and Huxtable, R.J. 1983, Effect of the taurine transport antagonists, guanidinoethane sulfonate and beta-alanine, on the morphology of the rat retina, *J.Neurosci.Res.* 9:135-143.
26. Petrosian, A., M. and Haroutounian, J.E. 1988, On the possibility of retinal transportation in rods by tauret through special channels, *13th All-Union conf.on Electron Microscopy*, (Abstract).
27. Petrosian, A.M. and Haroutounian, J.E. 1990, Tauret: Further studies of the role of taurine in retina, in: "Taurine: Functional Neurochemistry, Physiology, and Cardiology", Pasantes-Morales, H., Martin, D.L., Shain, W. and del Río, R.M. eds., Wiley-Liss, New York, pp. 471-475.
28. Salceda, R., Lopéz-Colomé, A.M., and Pasantes-Morales, H. 1977, Light-stimulated release of ^{35}S-taurine from frog retinal rod outer segments, *Brain Res.* 135:186-191.
29. Schaffer, S.W., Kramer, J., and Chovan, J.P. 1980, Regulation of calcium homeostasis in the heart by taurine, *Federation Proceedings*, 39:2691-2694.
30. Schalinske, K.L. and Steele, R.D. 1991, 13-cis-Retinoic acid alters methionine metabolism in rats, *J.Nutr.* 121:1714-1719.
31. Schmidt, S.Y., Berson, E.L., Watson, G., and Huang, C. 1977, Retinal degeneration in cats fed casein. III. taurine deficiency and ERG amplitudes, *Inves.Ophthal.Vis.Sci.* 16(7):673-678.
32. Skare, K.L., Schnoes, H.K., and De Luca, H.F. 1982, Biliary metabolites of *all-trans* retinoic acid in the rat: isolation and identification of a novel polar metabolite, *Biochemistry*, 21:3308-3317.
33. Sturman, J.A. 1990, Taurine deficiency, in: "Taurine: Functional Neurochemistry, Physiology, and Cardiology", Pasantes-Morales, H., Martin, D.L., Shain, W. and del Río, R.M. eds., Wiley-Liss, New York, pp. 385-395.
34. Thurston, J.H., Hauhart, R.E. and Naccarato, E.F. 1981, Taurine: Possible role in osmotic regulation of mammalian heart, *Science* 214:1373-1374.
35. Tiedemann, F. and Gmelin, L. 1827, Einige neue Bestandtheile der Galle des Ochsen, *Ann.Physik.Chem.* 9:326-337.

36. Van Gelder, N.M. 1990, Neuronal discharge hypersynchrony and the intracranial water balance in relation to glutamic acid and taurine redistribution: Migraine and epilepsy, in: "Taurine: Functional Neurochemistry, Physiology, and Cardiology", Pasantes-Morales, H., Martin, D.L., Shain, W. and del Río, R.M. eds., Wiley-Liss, New York, pp. 1-20.
37. Voaden, M.J., Lake, N., Marshall, J., and Morjaria, B. 1977, Studies on the distribution of taurine and other neuroactive amino acids in the retina, *Exp.Eye Res.* 25:249-257.

36

EFFECTS OF KINASE INHIBITORS AND TAURINE ANALOGUES ON THE PHOSPHORYLATION OF SPECIFIC PROTEINS IN MITOCHONDRIAL FRACTIONS OF RAT HEART AND RETINA

John B. Lombardini[1,2] and Cristi Props[1]

[1] Department of Pharmacology
[2] Department of Ophthalmology and Visual Sciences
Texas Tech University Health Sciences Center
Lubbock, Texas 79430

INTRODUCTION

We have demonstrated that taurine at physiological concentrations inhibits the phosphorylation of specific proteins in mitochondrial fractions of the rat heart and retina[2,5]. However, while the functions of these proteins, specifically an ~44 kDa phosphoprotein present in the heart and an ~20 kDa phosphoprotein present in the retina, are unknown we have considered it worthwhile to characterize these proteins with respect to a number of parameters. The concentration-response relationships of the effects of taurine and a series of taurine analogues on the phosphorylation of both the ~44 kDa[5] and ~20 kDa proteins have been studied. In addition, we have characterized both proteins with respect to the nature of the phosphate bond, the identification of the phosphorylated amino acid, the isoelectric point of the proteins, and the partition of the proteins into an aqueous phase rather than a lipid phase[3,4,5].

In the present studies we now report on the quantitative effects of various kinase inhibitors, specifically chelerythrine chloride, staurosporine, and W-7, and various taurine analogues (singly and in combination with taurine) on the phosphorylation of both the cardiac ~44 and retinal ~20 kDa proteins.

MATERIALS AND METHODS

Synthetic procedures for the taurine analogues have been described[5], as have preparation of the mitochondrial subcellular fraction of the rat heart and rat retina[3,5]. The phosphorylation assay and the polyacrylamide gel electrophoresis and autoradiography

Taurine 2, edited by Huxtable et al.
Plenum Press, New York, 1996

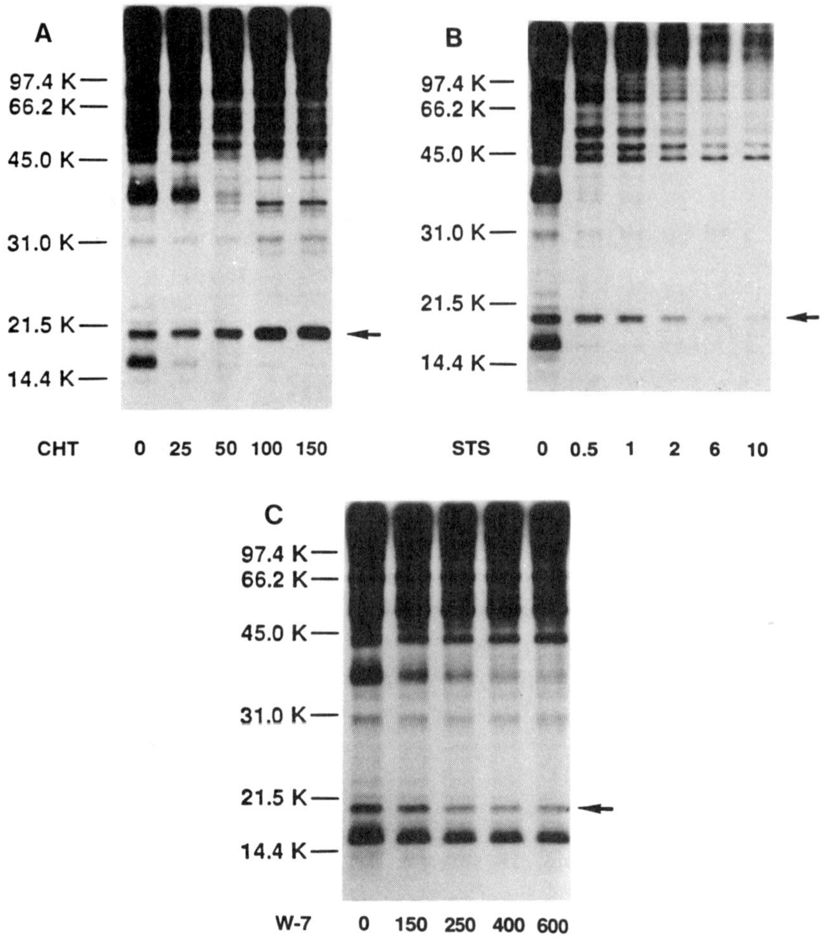

Figure 1. Effects of kinase inhibitors on the phosphorylation of a ~20 kDa protein present in the mitochondrial fraction of the rat retina. CHT = chelerythrine (0 to 150 μM), STS = staurosporine (0 to 10 μM), W-7 = calmodulin antagonist (0 to 600 μM). Molecular weight markers range from 14.4 to 97.4 kDa. Arrows designate the location of the ~20 kDa phosphoprotein.

techniques are described in Lombardini[5]. Quantitation of the [^{32}P]phosphate incorporated into either the ~44 or ~20 kDa proteins was determined by densitometry and the data are reported in arbitrary units.

RESULTS AND DISCUSSION

Effects of Kinase Inhibitors on the Phosphorylation of the Cardiac ~44 kDa and Retinal ~20 kDa Proteins Present in the Mitochondrial Fraction

The most interesting findings of these studies is that chelerythrine stimulates the phosphorylation of the retinal ~20 kDa protein (Fig. 1)[6] and that W-7, a calmodulin antagonist, stimulates the phosphorylation of the cardiac ~44 kDa protein (Fig. 2). Quantitative results of the effects of the kinase inhibitors, chelerythrine, staurosporine, and W-7,

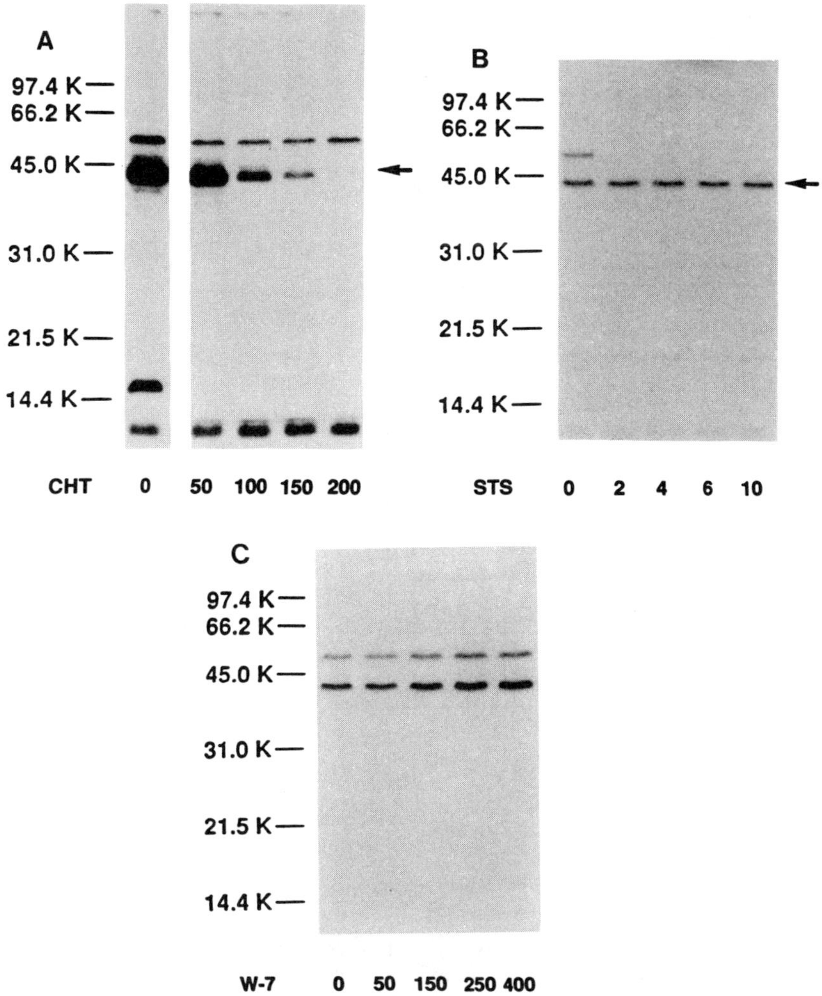

Figure 2. Effects of kinase inhibitors on the phosphorylation of a ~44 kDa protein present in the mitochondrial fraction of the rat heart. CHT = chelerythrine (0 to 200 μM), STS = staurosporine (0 to 10 μM), W-7 = calmodulin antagonist (0 to 400 μM). Molecular weight markers range from 14.4 to 97.4 kDa. Arrows designate the location of the ~44 kDa phosphoprotein.

are presented in Fig. 3. In the retinal preparation, the effect of chelerythrine was maximal at approximately 150 μM, attaining a 3-fold stimulation. This unique stimulatory action of chelerythrine could be eliminated by the addition of 10 mM dithiothreitol. Presumably, the iminium bond of chelerythrine reacts with the thiol group of dithiothreitol rather than the thiol group(s) of the kinase responsible for the phosphorylation reaction thereby protecting the kinase. It was also demonstrated that taurine, an inhibitor of the phosphorylation reaction, inhibited the stimulatory effect of chelerythrine[6]. A noncompetitive relationship (Fig. 4) was observed when chelerythrine was used as the variable activator and taurine as the fixed inhibitor (30 mM).

In the heart mitochondrial fraction the phosphorylation of the ~44 kDa protein was initially observed to be inhibited by chelerythrine (Fig. 2). However, it was subsequently determined that the time of preincubation of the mitochondrial preparation with chelerythrine

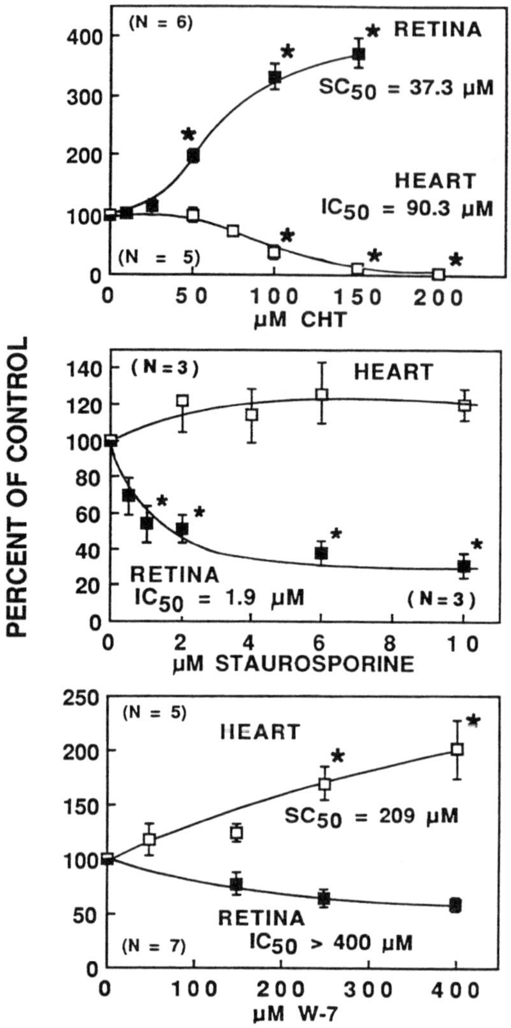

Figure 3. Quantitative analyses of the effects of known kinase inhibitors on the phosphorylation of a ~20 and ~44 kDa protein present in the mitochondrial fractions of the rat retina and heart. IC_{50} and SC_{50} = concentrations of kinase inhibitor necessary to inhibit or stimulate the phosphorylation of the designated protein by 50%. CHT = chelerythrine, W-7 = calmodulin antagonist. Number of experiments is denoted in the parentheses.

determined whether the effect of chelerythrine was going to be either inhibitory or stimulatory. At short preincubation periods (30 s to 1 min), chelerythrine stimulated the phosphorylation of the ~44 kDa protein by as much as 85% while longer preincubation times of 2-22 min produced a 40 to 95% inhibition (data not shown). Again as in the retinal preparation, dithiothreitol eliminated both the inhibitory and stimulatory effects of chelerythrine. When taurine was utilized to inhibit the phosphorylation of the ~44 kDa protein, dithiothreitol only partially protected the kinase reaction.

Effects of Taurine Analogues on the Phosphorylation of the Cardiac ~44 kDa and Retinal ~20 kDa Proteins Present in the Mitochondrial Fraction

Analogues of taurine were tested for their effects on the phosphorylation of the two specific proteins. Whereas most close analogues of taurine displayed inhibitory effects on the ~44 and ~20 kDa proteins, three analogues that contained unsaturated rings, pyridine-3-sulfonic acid (PyS), quinoline-8-sulfonic acid (QS), and 2-aminobenzenesulfonic acid

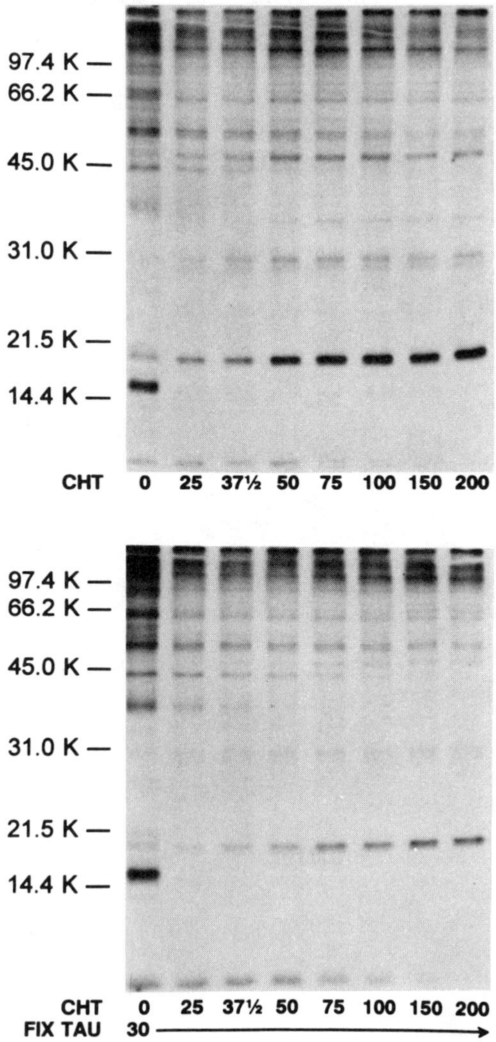

Figure 4. Inhibitory effect of a fixed concentration of taurine (30 mM) on the phosphorylation of an ~20 kDa protein in the presence of varying concentrations of chelerythrine. The mitochondrial fraction of the rat retina was used as the tissue source. CHT = chelerythrine (0 to 200 μM); TAU = taurine (30 mM). Molecular weight markers range from 14.4 to 97. 4 kDa. Arrows designate the location of the ~20 kDa phosphoprotein.

(ABS) were unique in that they stimulated the phosphorylation of the ~44 kDa protein (Fig. 5) while their saturated analogues [(±)piperidine 3-sulfonic acid (PiP) and 1,2,3,4-tetrahydroquinoline-8-sulfonic acid (THQS) (Fig. 5), and either (±)*trans*- or (±)*cis*-2-aminocyclohexanesulfonic acid (data not shown)] were inhibitory or had no effect[5].

When tested for activity in the retinal preparation the unsaturated analogues (PyS, QS) and their saturated counterparts (PiP, THQS) were inhibitors of the ~20 kDa protein. However, Pys and QS had considerably less activity than PiP and THQS (data not shown).

Effects of the Combination of Taurine and Taurine Analogues on the Phosphorylation of the Cardiac ~44 kDa and Retinal ~20 kDa Proteins Present in the Mitochondrial Fraction

Combinations of taurine plus the inhibitory taurine analogues in general exhibited antagonistic effects on the phosphorylation of the ~20 kDa protein in the retinal preparation.

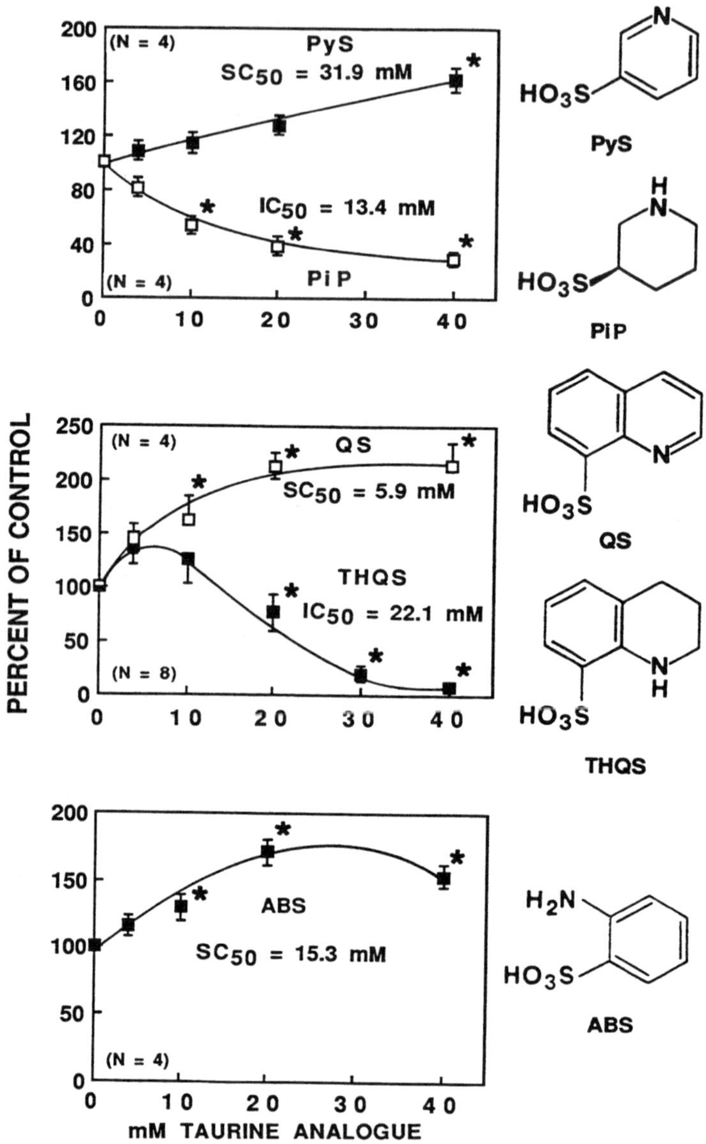

Figure 5. Quantitative analyses of the effects of taurine analogues on the phosphorylation of an ~44 kDa protein present in the mitochondrial fraction of the rat heart. IC_{50} or SC_{50} = concentrations of taurine analogues necessary to inhibit or stimulate the phosphorylation of the ~44 kDa protein by 50%. PyS = pyridine-3-sulfonic acid, PiP = piperidine-3-sulfonic acid, QS = quinoline-8-sulfonic acid, THQS = 1,2,3,4-tetrahydroquinoline-8-sulfonic acid, ABS = 2-aminobenzene-sulfonic acid. Number of experiments is denoted in the parentheses.

For example, the median-effect concentrations for taurine, quinoline-8-sulfonic acid (QS), and taurine plus QS (in a 2 to 1 ratio) as determined by the median-effect plots of Chou and Talalay[1] were 33.2, 13.3, and 67 mM, respectively. Figure 6 demonstrates a representative autoradiogram for this particular combination. The combination index values at 30, 50, and 70% saturation for the taurine plus QS combination were 2.43, 3.03 and 3.85, respectively. Combination index values greater than 1 indicate an antagonistic combination while those

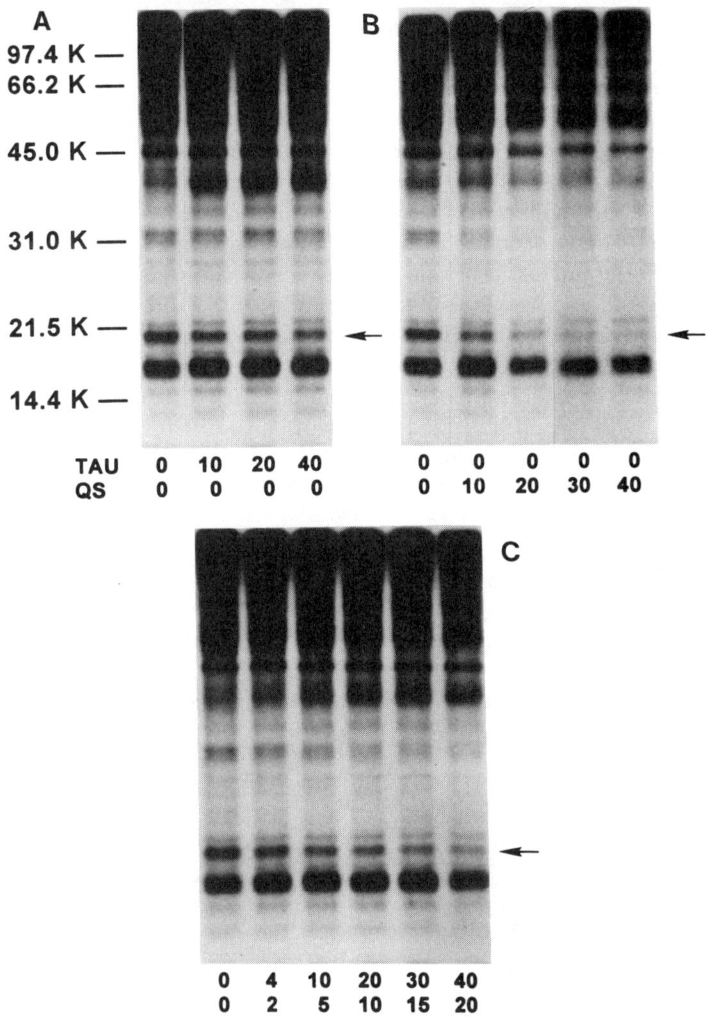

Figure 6. Effects of taurine (TAU), quinoline-8-sulfonic acid (QS), and TAU plus QS on the phosphorylation of the ~20 kDa protein present in the mitochondrial fraction of the rat retina. Concentrations of TAU and QS are listed in mmolarity. Molecular weight markers range from 14.4 to 97.4 kDa. Arrows designate the location of the 20 kDa phosphoprotein.

less than 1 signify synergism. Values equal to 1 indicate additive effects. Other combinations such as PyS plus taurine, PiP plus taurine, and ABS plus taurine were also highly antagonistic. However, the combination of taurine plus THQS (ratio 8:1) was additive with combination index values of approximately 1. No combination of taurine plus analogue that was tested was found to be synergistic.

When inhibitory analogues of taurine [such as PiP and (±)3-aminotetrahydrothiopyran-1,1-dioxide (APS)], were tested in combination with taurine, the combination index values suggested only antagonistic or additive effects on the phosphorylation of the cardiac ~44 kDa protein (data not shown). No combinations were synergistic. However, in the cardiac preparation when either PyS, QS, or ABS (stimulators) were varied in the presence of a fixed concentration of taurine, kinetic analyses indicated a competitive relationship in that the inhibitory effects of taurine could be negated by increasing the concentration of the stimulatory analogue (data not shown).

Our studies on the effects of analogues of taurine on the phosphorylation of both the ~44 kDa and ~20 kDa proteins suggest that the structural requirements for being either an inhibitor or stimulator are rather restricted. In the retina, it was concluded that: (i) the distance between the N and S atoms in the taurine structure was important for inhibitory activity; (ii) if the N atom is either within or attached to an unsaturated ring structure the inhibitory potency was significantly decreased; and (iii) if both the S and N atoms are present within the ring structure the analogue has no activity. In the heart, analogues of taurine which contained an unsaturated ring were found to be stimulators while their saturated counterparts were inhibitors[5]. Also, we have previously reported that sulfone derivatives of taurine have potent inhibitory activity against the phosphorylation of the ~44 kDa protein[5]. The inhibitory properties of these compounds are perplexing in that the sulfone moiety is neither isosteric nor isoelectronic with the sulfonic acid moiety of taurine. Thus, it has been suggested that the two classes of compounds (sulfones versus sulfonic acids) are acting at different sites on the kinase enzyme that phosphorylates the ~44 kDa protein.

ACKNOWLEDGMENTS

Supported in part by grants from the RGK Foundation of Austin, Texas, the South Plains Foundation of Lubbock, Texas, and the Taisho Pharmaceutical Co., Ltd. of Tokyo, Japan. We thank Ms. Josie L. Aleman for typing the manuscript.

REFERENCES

1. Chou, T.-C. and Talalay, P., 1986, Applications of the median-effect principle for the assessment of low-dose risk of carcinogens and for the quantitation of synergism and antagonism of chemotherapeutic agents, in: "New Advances in Developmental Cancer Therapy", Bristol-Meyers Cancer Symposia, Harrap, K.R. and Conners, T.A., eds., Academic Press, New York, pp. 37-64.
2. Lombardini, J.B., 1992, Effects of taurine on the phosphorylation of specific proteins in subcellular fractions of the rat retina, *Neurochem. Res.*, 17:821-824.
3. Lombardini, J.B., 1993, Partial characterization of an ~20 K M_r retinal protein whose phosphorylation is inhibited by taurine, *Biochem. Pharmacol.*, 46:1445-1451.
4. Lombardini, J.B., 1994, The inhibitory effects of taurine on protein phosphorylation: Comparisons of various characteristics of the taurine-affected phosphoproteins present in rat retina, brain and heart, in Adv. Exp. Med. Biol. *Taurine in Health and Disease*, Huxtable, R.J. and Michalk, D.V., eds., Plenum Press, New York, Vol. 359, pp 9-17.
5. Lombardini, J.B., 1994, Inhibitory effects of taurine and taurine analogues on the phosphorylation of an ~44 K M_r protein present in a mitochondrial subfraction of the rat heart: partial characterization of the ~44 K M_r phosphoprotein, *J. Mol. Cell. Card.*, 26:1675-1689.
6. Lombardini, J.B., 1995, Paradoxical stimulatory effect of the kinase inhibitor chelerythrine on the phosphorylation of an ~20 K M_r protein present in the mitoichondrial fraction of rat retina, *Brain Res.*, 46:1445-1451.

37

LONG-TERM EFFECTS ON RETINA OF RHESUS MONKEYS FED TAURINE-FREE HUMAN INFANT FORMULA

H. Imaki,[1] M. Neuringer,[2] and J. Sturman[1]

[1] Department of Developmental Biochemistry
New York State Institute for Basic Research in Developmental Disabilities
1050 Forest Hill Road, Staten Island, New York 10314
[2] Departments of Medicine and Ophthalmology
Oregon Health Sciences University, Portland
and Division of Neuroscience
Oregon Regional Primate Research Center, Beaverton, Oregon

INTRODUCTION

Research in the last 20 years has clearly established that taurine is essential for the normal development and maintenance of photoreceptor cell structure and function. This has been demonstrated best in cats as first described by Hayes *et al.*[5,6] These carnivores are dependent on a dietary supply of this amino acid, because of insufficient biosynthetic capacity due to the low activity of the enzyme, cysteine sulfinic acid decarboxylase. Cats become taurine-depleted when deprived of dietary taurine and develop central retinal degeneration. In rhesus monkeys and humans, dependency on dietary taurine appears to be age-dependent. In human infants, formulas low in taurine prompt a postnatal drop in plasma taurine concentration, unlike taurine-supplemented formulas or human milk, which is rich in taurine[2,9]. Human infants, children and adults receiving taurine-free parenteral nutrition for a prolonged time showed significant reductions of taurine in plasma, urine and blood cells[3,14,21,22,24], in some cases accompanied by ophthalmoscopic and electroretinographic abnormalities which could be reversed by taurine supplementation[1,3,4,20]. In rhesus monkeys, we have demonstrated that taurine-free human infant formula feeding from birth led to 50-60% reductions in plasma taurine concentration after 3 and 6 months in comparison to the same formula supplemented with taurine, and that these differences were no longer present at 12 months. Light and electron microscopic examinations revealed that taurine-deprived monkeys at 3 months of age show morphological changes in the retina and visual cortex compared with monkeys fed the taurine-supplemented formula[8,12]. The retinal changes regressed with increasing age, but were still present in most taurine-deprived monkeys at 12 months, while the densities of rod and cone pigments, measured by fundus reflectometry, were significantly reduced in the taurine-deprived monkeys at 6 months but

Taurine 2, edited by Huxtable et al.
Plenum Press, New York, 1996

not at 12 months[7]. The present study was undertaken to ascertain whether or not the morphological changes in the retina persists for a substantial period beyond the time at which tissue taurine concentrations appeared to approach normal levels in the taurine-deprived monkeys.

MATERIALS AND METHODS

Four rhesus monkeys were raised from birth on a soy protein-based human infant formula (Isomil, Ross Laboratories, Columbus, OH), which contained no taurine detectable by amino acid analysis. Three other monkeys were raised on the same formula supplemented with 70 µmol taurine per 100 ml, the approximate concentration found in rhesus milk[13]. At approximately 4 years of age, the monkeys were killed by an overdose of sodium pentobarbital (Nembutal, Abbott Laboratories, North Chicago, IL), followed by exsanguination. Right eyes were quickly excised and perfused with cold 3% glutaraldehyde in 0.1 M Sorensen's phosphate buffer, pH 7.4, using a syringe and needles. After storing the eyes in the same fixative for at least 24 h at 4°C, they were hemisected, and pieces of retinas sliced radially around the fovea were set in 2% agar and postfixed in 2% OsO_4 in 0.1 M Sorensen's buffer, dehydrated in acetone and embedded in EM-bed 812 (Electron Microscopy Sciences, Ft. Washington, PA). Semithin (0.75 µm) sections were stained with toluidine blue and light micrographs were taken on a Zeiss Photomicroscope III using Kodak TMX-100 film. Thin sections were stained with uranyl acetate and lead citrate and examined in a Philips EM 300 electron microscope operated at 80 kv.

The left retinas and other tissues, including various nervous system samples, muscles and all major internal organs, were rapidly dissected and frozen on dry ice. They were homogenized with 9 vol of 0.88 M trifluoroacetic acid, centrifuged at 10,000 x g for 30 min, and the clear supernatant was removed for determination of taurine concentrations by precolumn derivatization with phenylisothiocyanate and reverse-phase HPLC[10]. Taurine concentration data were analyzed by Student's t test, and differences were accepted as significant at $p \leq 0.05$.

RESULTS

There was no significant difference in taurine concentration between the two dietary groups in any of the tissues measured (Table 1). The concentrations in several brain regions in these monkeys were lower than those in 3-, 6- and 12-month-old monkeys reported previously[17,18], values generally resembling, or slightly higher than, those reported for adult monkeys consuming a stock diet[19]. The retinas showed the highest concentration of taurine in these monkeys, as in younger monkeys, but there was no difference between the groups, as in the 12-month-old monkeys. The composition of the steroid moiety of biliary bile acids and proportion of bile acids conjugated with taurine were similar to those reported for 12-month-old monkeys[17], and there were no differences between the two groups, unlike for 3- and 6-month-old monkeys, in which the proportions of bile acid conjugated with taurine were significantly greater in the supplemented group.

Light microscopic examination of retinas revealed that although they exhibit normal structure for the most part (Figs. 1a, b), certain signs of retinal degeneration were found in limited areas of retinas in all the taurine-deprived monkeys. These included clusters of photoreceptors with inner segments exhibiting vesicular appearance, massively swollen, pale axons and densely-stained nuclei (Fig. 1c), or those with markedly dense, or pyknotic nuclei, associated with densely-stained inner segments and disorganized outer segments (Fig. 1d).

Table 1. Concentration of taurine in tissues from 4-year-old taurine-deprived and taurine-supplemented rhesus monkeys (μmol/g wet weight)

Tissue	Formula Alone*	Formula + Taurine**
Occipital lobe	1.76 ± 0.24	2.36 ± 0.24
Cerebellum	2.55 ± 0.56	2.43 ± 0.20
Olfactory bulb	6.01 ± 1.72	7.40 ± 2.09
Optic nerve	2.36 ± 0.40	2.32 ± 0.30
Lateral geniculate nucleus	1.31 ± 0.53	1.20 ± 0.14
Superior colliculus	1.08 ± 0.16	1.10 ± 0.03
Retina	28.00 ± 5.60	28.70 ± 3.30
Liver	1.38 ± 0.37	1.34 ± 0.20
Kidney	9.03 ± 1.85	8.70 ± 0.43
Heart	13.00 ± 3.60	10.00 ± 2.70
Gastrocnemius	6.12 ± 1.26	4.67 ± 0.47
Plasma (μmol/liter)	164.00 ± 5.00	197.00 ± 25.00

*Each value represents the mean ± SD for four monkeys.
**Each value represents the mean ± SD for three monkeys.

The structure of other cell types in such areas were usually normal, but in some areas certain cells in the inner nuclear layer, most notably amacrine cells, and neuronal processes in the inner and outer plexiform layers and the nerve fiber layer appeared swollen (Fig. 1c).

Electron microscopy demonstrated that the outer segments in affected areas often showed generalized swelling or loosening of disc membranes, and occasional disorientation, in contrast to those with tightly-packed discs and straight alignment observed in the normal retina (Figs. 2a, b). In such areas most mitochondria in ellipsoids were swollen, often distorting the contour of inner segments in comparison to slender and straight forms seen in

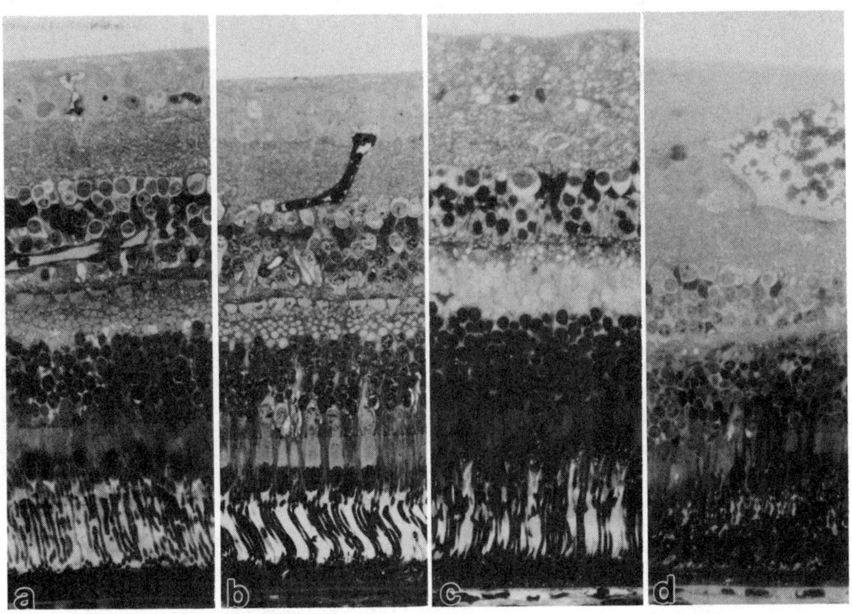

Figure 1. Light micrographs of 4-year-old monkey retinas in transverse section, stained with toluidine blue, X 400. a) Temporal retina of a taurine-supplemented (+T) monkey, b) temporal retina of a taurine-deprived (-T) monkey, c) and d) superior retinas of -T monkeys.

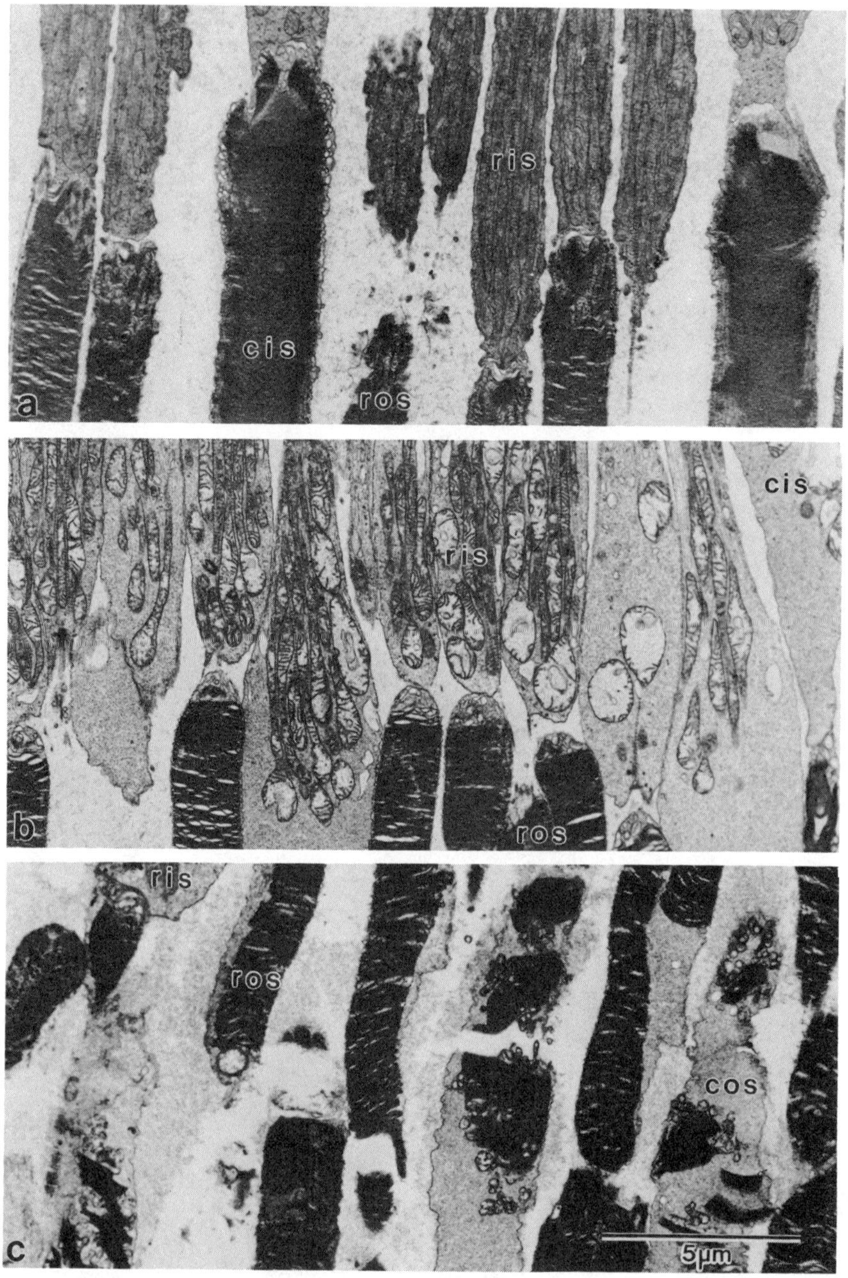

Figure 2. Electron micrographs (EMs) showing parts of inner and outer segments of cones (cis, cos) and rods (ris, ros) in a) temporal retina of a +T monkey, b) superior retina of a -T monkey (same area of retina as shown in Fig. 1c), and c) parafoveal retina of another -T monkey. All EMs are in the same magnification and the scale bar in Fig. 2c applies to all.

the normal areas. In other areas, increased number of vesicular profiles were found associated with the shorter, disjoined stacks of disc membranes in the cone outer segments (Fig. 2c). The myoids of inner segments in the affected photoreceptors contained vacuolar profiles representing distended cisternae of endoplasmic reticulum and Golgi complex in the matrix that often showed lower electron density, in comparison to the more uniform appearance noted in the healthy retina (Figs. 3a, b). The inner segments of affected photoreceptors in other areas, on the other hand, consisted of shrunken myoids with dense cytoplasm and equally dense ellipsoids expanded by broader but shorter mitochondria (Fig. 3c). Similarly, the nuclei of photoreceptors were either swollen or shrunken: in relatively severely affected areas, pyknotic nuclei surrounded by dense and extremely attenuated perikarya were interposed among markedly distended somata of neighboring photoreceptor cells, unlike the uniform population observed in the outer nuclear layer of the normal retina (Figs. 4a, b). In other areas, clusters of pyknotic nuclei were apposed by relatively normal looking nuclei (Fig. 4c). The most common abnormal structures observed in the axons and synaptic terminals of photoreceptor cells in the retinas of taurine-deprived monkeys were vacuoles of varying sizes and swollen mitochondria compared to the corresponding structures in the normal retinas (Fig. 5a, b). They were found both in pedicles and spherules which showed relatively normal synaptic structures despite deformation of their contours. In severely affected areas, however, synaptic structures appeared reduced, and the plasma membranes of some spherules, pedicles and axons partially dissolved (Fig. 5c).

DISCUSSION

The results in this report extend the earlier observations that taurine concentrations in various tissues of monkeys reached at 12 months of age remain relatively stable, except in some brain regions which showed slight reductions, even when they were deprived of dietary taurine throughout the experimental period. These results are consistent with the preliminary study finding that adult rhesus monkeys fed a taurine-free casein-based diet did not become taurine-depleted unless dietary protein also was restricted[11]. They differ from findings in 3- and 6-month-old monkeys raised on taurine-free formula which showed significant reductions in tissue taurine concentration. It appears then that the period of dependency on dietary taurine in rhesus monkeys lasts for at least 6 months but less than 12 months of age, and remains so at 48 months. Reductions of visual pigment levels in both cones and rods in taurine-deprived monkeys noted at 6 months of age but not at 12 months, on the other hand, indicates that the loss of the photoreceptor outer segment membranes, which contain these pigments, sustained earlier did not exist on a large enough scale to be detectable by fundus reflectometry at 12 months, when tissue taurine concentrations had become normalized.

Examination of the retinas by light and electron microscopy, however, revealed that the recovery of photoreceptors was not complete at 12 months, although the size and frequency of lesions observed were considerably reduced from earlier developmental stages. The photoreceptors of taurine-deprived monkeys at 3 and 6 months of age showed a variety of changes, including disorientation, fragmentation and vesiculation of photoreceptor outer segment membranes, mainly involving the foveal and parafoveal retina. The inner segments showed swollen mitochondria, dilated cisternae of endoplasmic reticulum and Golgi complex while the synaptic terminals contained large vacuoles and swollen mitochondria. Pyknotic photoreceptors were only rarely encountered. At 12 months the same types of changes were found in the retinas of taurine-deprived monkeys, but to lesser degrees. The pattern of photoreceptor degeneration observed in 48 month-old taurine-deprived monkeys was somewhat different. One commonly observed was the generalized swelling of photore-

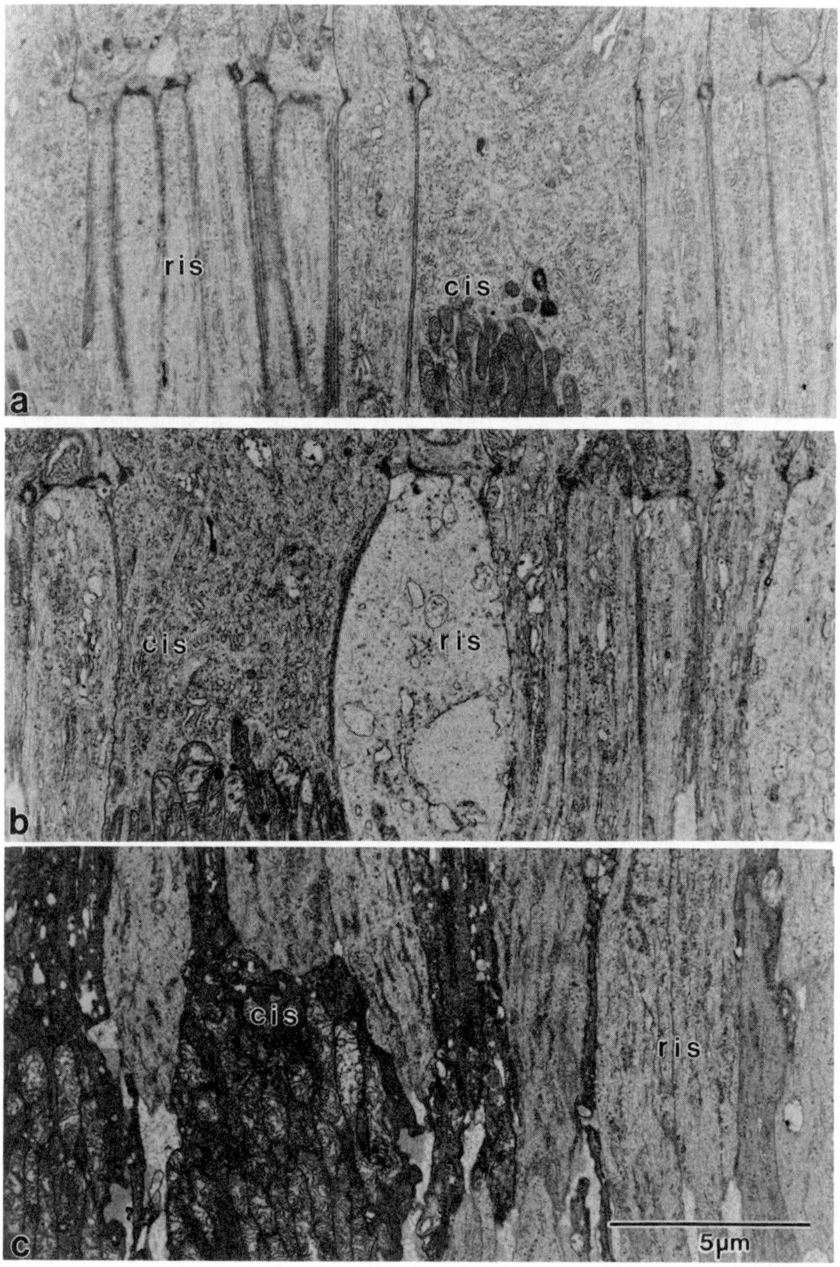

Figure 3. EMs of photoreceptors showing proximal parts of inner segments in a) nasal retina of a +T monkey, b) temporal retina of a -T monkey (same retina as shown in Fig. 2c), and c) superior retina of a -T monkey (same area as shown in Fig. 1d).

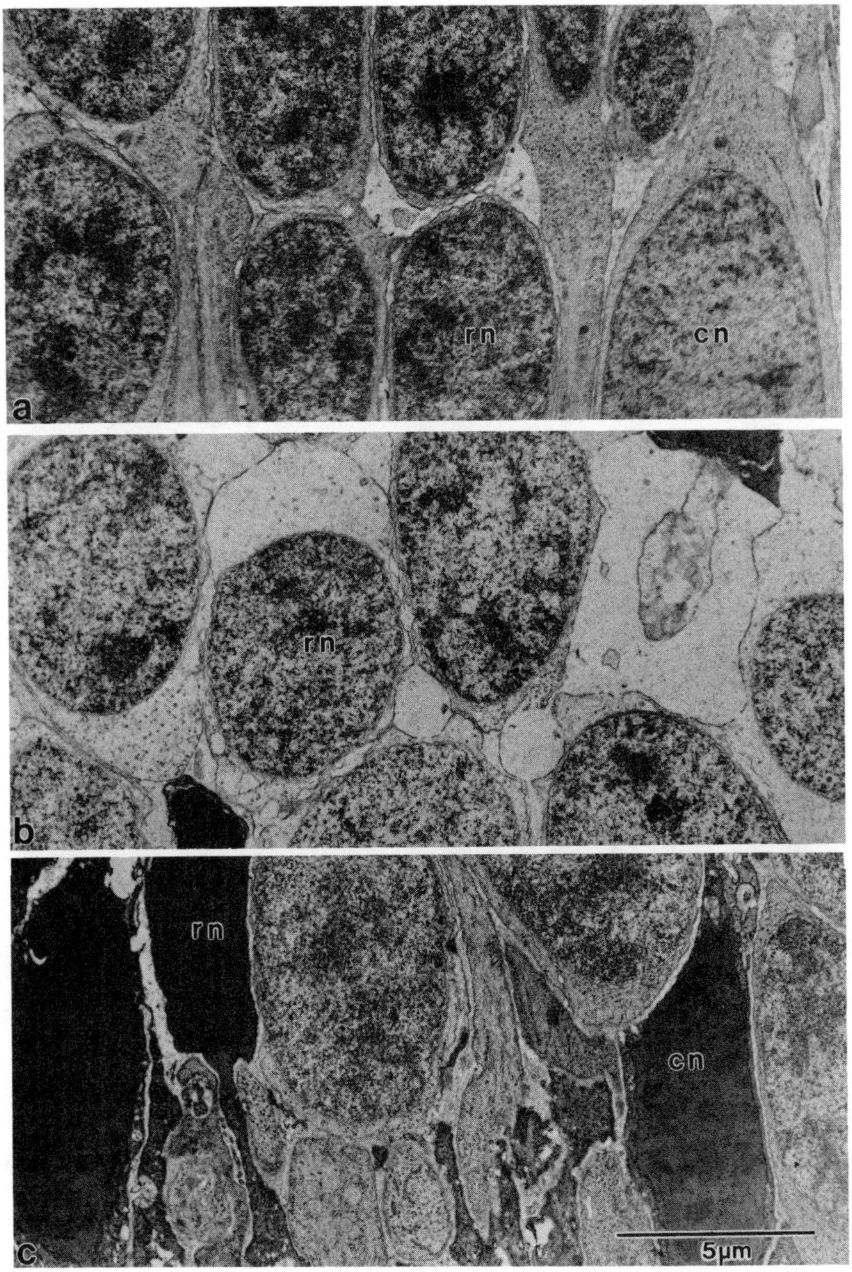

Figure 4. EMs of rod and cone nuclei (rn, cn) in a) inferior retina of a +T monkey (same area as in Fig. 3a), b) temporal retina of a -T monkey (same area as in Fig. 3b), and c) superior retina of a -T monkey (same area as shown in Figs. 1d and 3c).

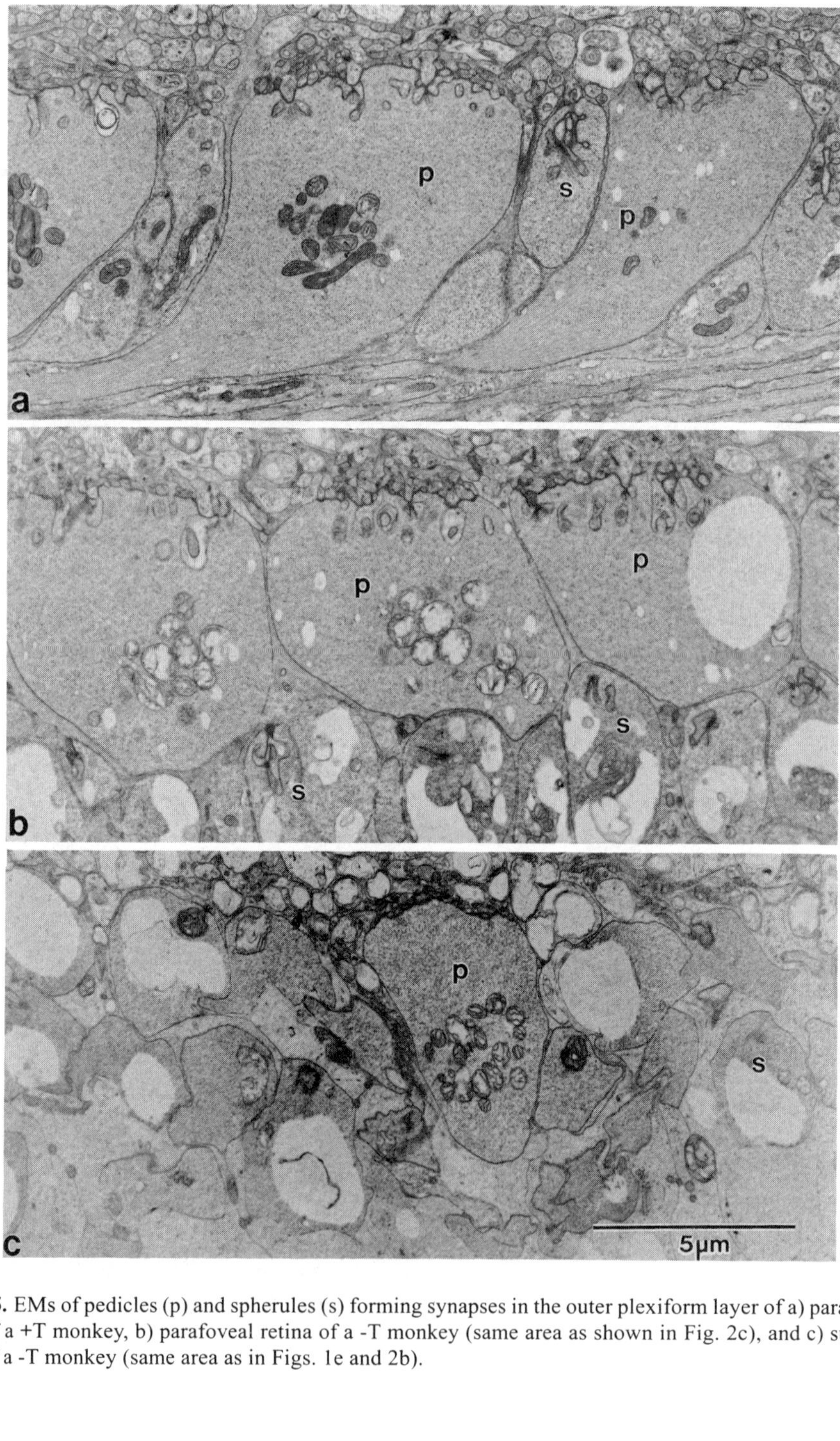

Figure 5. EMs of pedicles (p) and spherules (s) forming synapses in the outer plexiform layer of a) parafoveal retina of a +T monkey, b) parafoveal retina of a -T monkey (same area as shown in Fig. 2c), and c) superior retina of a -T monkey (same area as in Figs. 1e and 2b).

ceptors involving various parts i.e. outer and inner segments, somata and axons, but the organization of outer segments, the most susceptible units in younger retinas, were relatively undisturbed. Another prevalent pattern of change observed was markedly dense and shrunken or pyknotic photoreceptor cells occurring often in clusters. In such cells, mitochondria may be swollen if remaining, but the other organelles disappeared, or lost their fine structure. Isolated focal lesions involving these changes were almost always located in the peripheral region of the retina in the 48-month-old animals, in contrast to predominantly foveal-parafoveal localization of degenerative photoreceptors in younger, taurine-deprived monkeys. It is not clear whether such persistence of degenerative photoreceptors in 48-month-old monkeys represented a delay in normalization of the retinal morphology relative to tissue taurine concentrations, or a prolonged lack of reversibility. It is equally not clear whether present observation of relatively normal photoreceptors in the central retina and abnormal ones in the peripheral locations may indicate that photoreceptor degeneration proceeds from center to peripheral direction as the development of photoreceptor themselves is spread, while recovery or elimination of damaged photoreceptors in the central area takes place as the tissue taurine reaches normal levels, or the degenerative photoreceptors were present in the peripheral retina of younger taurine-deprived monkeys but escaped detection. One explanation for this pattern of results may be that some photoreceptor cells, particularly those in the foveal-parafoveal areas, sustain certain outer segment damage during the period of taurine depletion, but then, as the taurine levels normalize, are restored by the normal intrinsic process of outer segment membrane renewal. In other cells, located more peripherally, the renewal process may be slower or the damage may have gone too far to the point that repair is no longer possible. It would be this small population of permanently damaged cells which are present as abnormal photoreceptors at later ages. In taurine-deprived cats, which suffer more extensive retinal degeneration, taurine refeeding induced incomplete recovery[15,16].

In the previous study, taurine supplementation starting at 6 months of age resulted in virtually complete restoration of normal retinal morphology[7]. The monkeys in the present study remained on the taurine-free formula throughout 4 years of life. Thus it is possible that long-term dietary taurine deprivation had some continuing effect which is not related to blood or tissue taurine levels. It is more likely, however, that the increased population of more severely damaged photoreceptors generated between 6 and 12 months of reduced taurine levels is responsible for the persistence of abnormal photoreceptors at 4 years.

If the period of vulnerability to taurine deficiency is related to retinal maturity, this period would be considerably longer in human infants as their retinas are less mature than those of rhesus monkeys at birth, and develop more slowly. The photoreceptor cells in the fovea, the area of retina which develops most slowly, their structure is not yet adult-like in 4 year-old children[23]. Thus, the impact of taurine depletion on the development of retinal structure and function in human newborns, and especially premature infants, could be greater. In conclusion, this study reinforces the importance of supplementing taurine in human infant formulas, as has been recommended and available commercially in the U.S. since 1984.

ACKNOWLEDGMENTS

We are grateful to Jeff Messing for technical assistance and Anna Parese for secretarial assistance. This work was supported by the New York State Office of Mental Retardation and Developmental Disabilities, and by grants from the National Retinitis Pigmentosa Foundation and the Medical Research Foundation of Oregon and NIH grants HD-18678 and RR-00163.

REFERENCES

1. Ament, M.E. Geggel, H.S., Heckenlively, J.R., Martin, D.A. and Kopple, J. 1986, Taurine supplementation in infants receiving long-term parenteral nutrition, *J. Amer. Coll. Nutr.* 5:127-136.
2. Gaull, G.E., Rassin, D.K., Raiha, N.C.R. and Heinonen, K. 1977, Milk protein quantity and quality in low-birth-weight infants. III. Effects on sulfur amino acids in plasma and urine, *J. Pediat.* 90:348-355.
3. Geggel, H.S., Ament, M.E. Heckenlively, J.R., Martin, D.A. and Kopple, J.D. 1985, Nutritional requirement for taurine in patients receiving long-term parenteral nutrition, *New Engl. J. Med.* 312:142-146.
4. Geggel, H.S., Heckenlively, J.R., Martin, D.A., Ament, M.E. and Kopple, J.D. 1982, Human retinal dysfunction and taurine deficiency, *Docum. Ophthal. Proc. Series* 31:199-207.
5. Hayes, K.C., Carey, R.E. and Schmidt, S.Y. 1975a, Retinal degeneration associated with taurine deficiency in the cat, *Science* 188:949-951.
6. Hayes, K.C., Rabin, A.R. and Berson, E.L. 1975b, An ultrastructural study of nutritionally induced and reversed retinal degeneration in cats, *Am. J. Pathol.* 78:505-524.
7. Imaki, H., Jacobson, S.G., Kemp, C.M., Knighton, R.W., Neuringer, M. and Sturman, J. 1993, Retinal morphology and visual pigment levels in 6- and 12-month-old rhesus monkeys fed a taurine-free human infant formula, *J. Neurosci. Res.* 36:290-304.
8. Imaki, H., Moretz, R., Wisniewski, H., Neuringer, M. and Sturman, J. 1987, Retinal degeneration in 3-month-old rhesus monkey infants fed a taurine-free human infant formula, *J. Neurosci. Res.* 18:602-614.
9. Jarvenpaa, A.L., Rassin, D.K., Raiha, N.C.R. and Gaull, G.E. 1982, Milk protein quantity and quality in the term infant. II. Effects on acidic and neutral amino acids, *Pediatrics* 70:221-230.
10. Lippincott, S.S., Friedman, A.L., Siegel, F.L., Pityer, R.M. and Chesney, R.W. 1988, HPLC analysis of the phenylisothiocyanate (PITC) derivatives of taurine from physiological samples, *J. Am. Coll. Nutr.* 7:491-497.
11. Neuringer, M., Sturman, J.A., Wen, G.Y. and Wisniewski, H.M. 1985, Dietary taurine is necessary for normal retinal development in monkeys, In: "Taurine: Biological Actions and Clinical Perspectives", Oja, S.S., Ahtee, L., Kontro P. and Paasonen, M.K. eds., Alan R. Liss, NY, pp. 53-62.
12. Palackal, T., Kujawa, M., Moretz, R., Neuringer, M. and Sturman, J. 1991, Laminar analysis of the number of neurons, astrocytes, oligodendrocytes and microglia in the visual cortex (area 17) of 3-month-old monkeys fed a human infant soy-protein formula with or without taurine supplementation from birth, *Devel. Neurosci.* 13:20-33.
13. Rassin, D.K., Sturman, J.A. and Gaull, G.E. 1978, Taurine and other free amino acids in milk of man and other mammals, *Early Human Devel.* 2:1-3.
14. Rigo, J. and Senterre, J. 1977, Is taurine essential for the neonates?, *Biol. Neonate* 32:73-76.
15. Schmidt, S.Y., Berson, E.L. and Hayes, K.C. 1976a, Retinal degeneration in the taurine-deficient cat, *Trans. Amer. Acad. Ophthalmol. Otolaryn.* 81:687-693.
16. Schmidt, S.Y., Berson, E.L. and Hayes, K.C. 1976b, Retinal degeneration in cats fed casein. I. Taurine deficiency, *Invest. Ophthalmol.* 15:47-52.
17. Sturman, J.A., Messing, J.M., Rossi, S.S., Hoffmann, A.F. and Neuringer, M. 1991, Tissue taurine content, activity of taurine synthesis enzymes and conjugated bile acid composition of taurine-deprived and taurine-supplemented rhesus monkey infants at 6 and 12 mo of age, *J. Nutr.* 121:854-862.
18. Sturman, J.A., Messing, J.M., Rossi, S.S., Hofmann, A.F. and Neuringer, M.D. 1988, Tissue taurine content and conjugated bile acid composition of rhesus monkey infants fed a human infant soy-protein formula with or without taurine supplementation for 3 months, *Neurochem. Res.* 13:311-316.
19. Sturman, J.A., Rassin, D.K., Gaull, G.E. and Cote, L.J. 1980, Taurine in devleoping rhesus monkey brain, *J. Neurochem.* 35:304-310.
20. Vinton, M.E., Heckenlively, J.R., Laidlaw, S.A, Martin, D.A., Foxman, S.R., Ament, M.E. and Kopple, J.D. 1990, Visual function in patients undergoing long-term total parenteral nutrition, *Am. J. Clin. Nutr.* 52:895-902.
21. Vinton, N.E. Laidlaw, S.A., Ament, M.E. and Kopple, J.D. 1987, Taurine concentrations in plasma, blood cells, and urine of children undergoing long term total parenteral nutrition, *Pediat. Res.* 21:399-403.
22. Vinton, N.E., Laidlaw, S.A., Ament, M.E. and Kopple, J.D. 1986, Taurine concentrations in plasma and blood cells of patients undergoing long-term parenteral nutrition, *Am. J. Clin. Nutr.* 44:398-404.
23. Yuodelis, C. and Hendrickson, A. 1986, A qualitative and quantitative analysis of the human fovea during development, *Vis. Res.* 26:847-855.
24. Zelikovic, J., Chesney, R.W., Friedman, A.L. and Ahlfors, C.E. 1990, Taurine depletion in very low birth weight infants receiving prolonged total parenteral nutrition: role of renal immaturity, *J. Pediatr.* 116:301-306.

IMMUNOCYTOCHEMICAL LOCALIZATION OF TAURINE IN THE INNER EAR

Shin-ichi Usami[1] and Ole P. Ottersen[2]

[1] Department of Otorhinolaryngology
Hirosaki University School of Medicine, Japan
[2] Department of Anatomy
Institute of Basic Medical Sciences
University of Oslo, Norway

INTRODUCTION

Taurine is reported to be abundant in the inner ear[4]. However, data are scarce concerning the distribution and functional roles of taurine in this organ. In the CNS, taurine has been implicated in many functions including inhibitory neurotransmission, modification of Ca^{++} fluxes, and osmoregulation[9,10]. We recently demonstrated the cellular and subcellular localization of taurine in the organ of Corti by means of post embedding immunocytochemistry[27]. In the present paper, we describe the distribution of taurine in the inner ear of three mammalian species.

MATERIALS AND METHODS

Animals

Wistar rats, Hartley guinea pigs, and squirrel monkeys were used. The animals were deeply anesthetized with sodium pentobarbital (50 mg/kg) before perfusion. Animals were fixed and immunocytochemically processed as described[27].

Fixatives

For rat and guinea pig, 2.5% glutaraldehyde + 1% paraformaldehyde (in 0.1 M phosphate buffer pH 7.4) was used. In order to assure a good and rapid fixation for electron microscopic observation, the perilymphatic perfusion method was employed. The fixative was injected into the perilymphatic space as described elsewhere[27,28].

For guinea pig and squirrel monkey, 1% glutaraldehyde + 4% paraformaldehyde (in 0.1 M phosphate buffer pH 7.4) was used for whole mount preparation of cochlea. Local

Taurine 2, edited by Huxtable et al.
Plenum Press, New York, 1996

application of the fixative was achieved by injection through the tympanic membrane. The temporal bones were removed and post-fixed (4 h, room temperature) in the same fixatives.

Immunocytochemical Procedures

Post-embedding light and electron microscopic immunocytochemistry using Durcupan-embedded material was used for rat and guinea pig. Post embedding immunocytochemistry was performed as described elsewhere[27].

Post-embedding electron microscopic immunocytochemistry using the freeze substitution method was also used for guinea pig. Lowicryl HM 20 embedded material was processed as described in detail previously[7].

For whole mount preparation of guinea pig cochlea, the bony shell of the cochlea was opened and the ABC method[8] was performed. After the peroxidase reaction, cochleas were divided into individual turns and placed on the slides.

For squirrel monkeys, paraffin-embedded temporal bones were decalcified by 7% EDTA (in 0.1 M phosphate buffer) for 1 month. Serial paraffin sections (5 μm thick) were cut and placed on poly-L-lysine-coated slides. The ABC method was performed on the slides.

Antisera and Specificity Controls

The antisera used were raised against taurine-glutaraldehyde-bovine serum albumin complexes. Antisera No.20[16,20] and No. 486[1] used in this study have already been characterized. Adsorption control was performed by adding glutaraldehyde complexes of taurine (300 μM) to the primary antisera. For post-embedding immunocytochemistry, test sections containing taurine-glutaraldehyde-brain protein conjugates[19] were incorporated in each experiment for monitoring immunocytochemical selectivity.

RESULTS

No significant staining was found in the absorption controls. Test "sandwich" sections incubated together with the semithin and ultrathin tissue sections showed a selective pattern of immunolabelling. In spite of different fixatives, different species, and different immunocytochemical procedures, the immunostaining patterns of taurine in the inner ear were almost identical.

Cochlea

In the organ of Corti, supporting cells, including border cells, inner phalangeal cells, Deiters cells, pillar cells, and Boettcher cells were enriched in taurine-like immunoreactivity, contrasting sharply with inner and outer hair cells which did not show noteworthy immunolabelling (Fig. 1A). The cytoplasm of Hensen cells and inner sulcus cells exhibited low immunoreactivity except for their endolymphatic surfaces which were moderately labelled. In line with the results from semithin sections, ultrathin sections showed a high density of gold particles over supporting cells (Fig. 2A, B). Taurine-like immunoreactivity was evenly distributed in the cytoplasm including the mitochondria. In ultrathin sections, intense labelling occurred even in filament-rich areas and in the nuclei of the labelled supporting cells where only weak immunostaining was found at the light microscopic level (semithin sections).

The distribution of labelling in whole mount preparation of the guinea pig organ of Corti was compatible with that in the resin embedded material. In the whole mounts, the

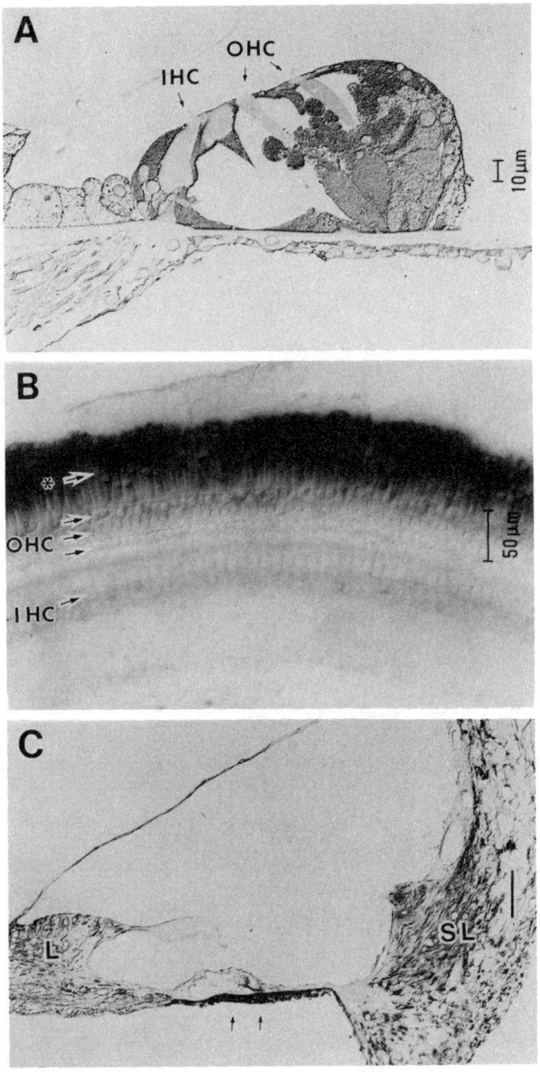

Figure 1. Light microscopic distribution of taurine in the cochlea. A: Light microscopic distribution of taurine in the guinea pig organ of Corti (Durcupan-embedded semithin section). Various types of supporting cells exhibit intense immunoreactivity whereas hair cells lack significant immunoreactivity. IHC: inner hair cell, OHC: outer hair cell. Bar=10μm. (From Usami and Ottersen[27], by permission.). B: Whole mount preparation of the guinea pig organ of Corti incubated with taurine antiserum. Immunoreactivity was found at the site of supporting cells (asterisk) lateral to the inner (IHC) and outer (OHC) hair cell region. Bar=50 μm. (From Usami and Ottersen[27], by permission.). C: Light microscopic distribution of taurine in the squirrel monkey cochlea (paraffin section). Intense immunoreactivity occurs in the limbus (L), sub-basilar tympanic cells (arrows), and the spiral ligament (SL). Bar=50 μm.

strongest immunoreactivity was found at the site of supporting cells. No significant staining was found in the inner and outer hair cell region (Fig. 1B). The immunostaining pattern in paraffin-embedded monkey cochlea resembles that in the rat and guinea pig cochlea. The lateral wall (stria vascularis and spiral ligament) of the cochlea was included in the monkey specimen. Fibrocytes in the spiral ligament showed pronounced labelling at both light (Fig. 1C) and electron microscopic levels (Fig. 2C).

Vestibular Endorgans

In the sensory epithelium of the crista ampullaris, immunoreactivity occurred in cup-like structures, corresponding to the nerve chalices and in dot-like immunoreactive structures suggestive of nerve terminals (Fig. 3A). In the stroma, fibrocytes and Schwann cells along the afferent nerve fibers were immunolabelled for taurine (Fig. 3A). The electron

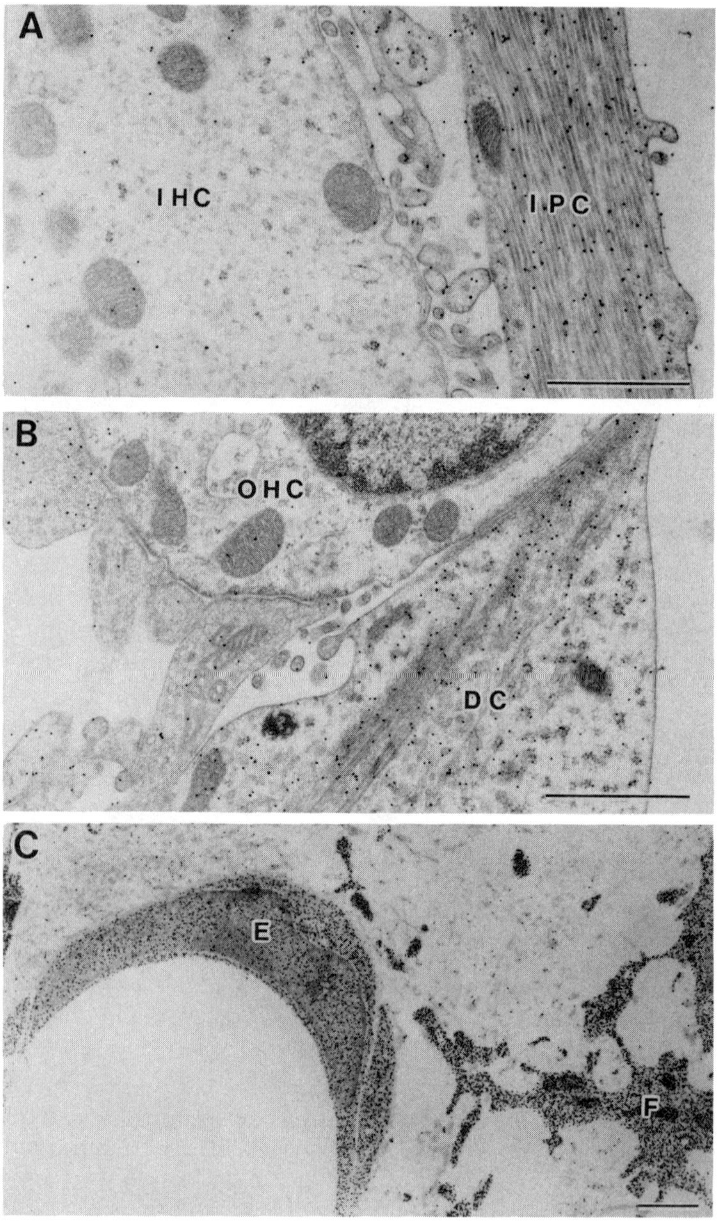

Figure 2. Electron microscopic distribution of taurine in the cochlea. A: Electronmicrograph of rat organ of Corti showing subcellular distribution of taurine (rat, Durcupan-embedding material). Supporting cells (IPC: inner pillar cell) show enrichment of taurine-like immunoreactivity. Note that the adjacent inner hair cell (IHC) was relatively weakly labelled. Bar=1 μm. B: Electron micrograph shows the distribution of taurine in the outer hair cell region (rat, Durcupan-embedding material). Deiters cell (DC) is intensely labelled, in contrast to the outer hair cell (OHC). Bar=1μm. (From Usami and Ottersen[27], by permission.). C: Electron micrograph showing fibrocytes in the guinea pig spiral ligament (freeze substituted, Lowicryl-embedded material). Immunolabelling is restricted to the fibrocytes (F) and endothelial cells of the capillary (E). Bar=1 μm.

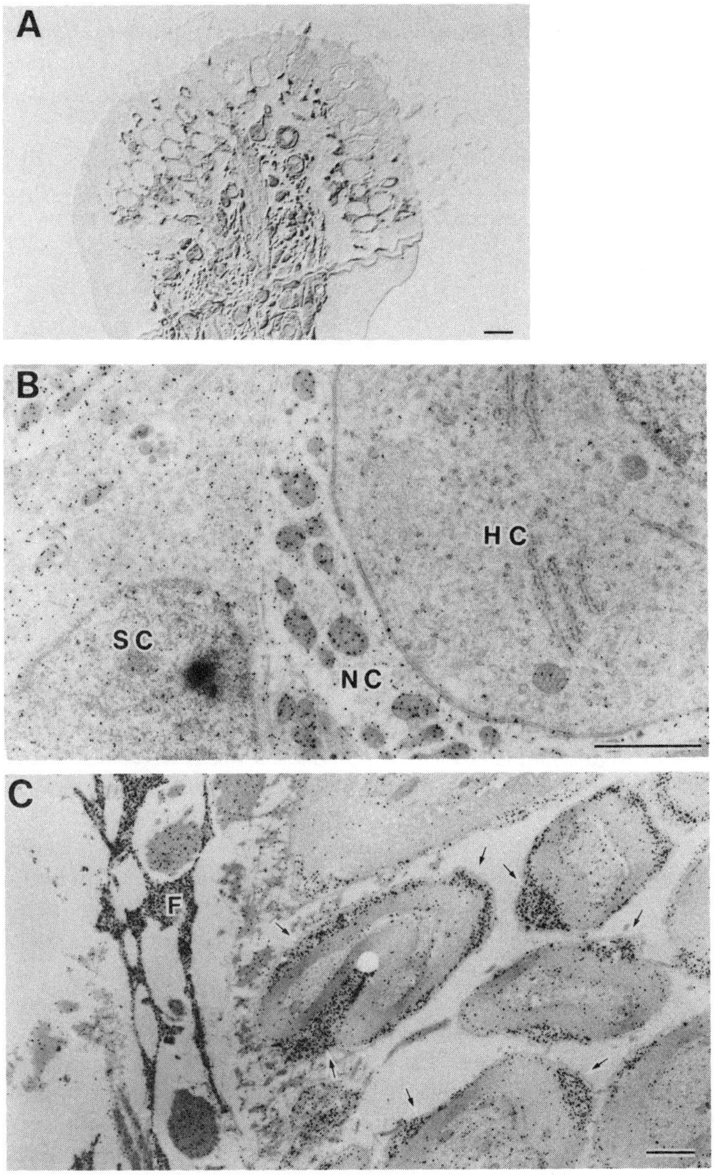

Figure 3. Distribution of taurine in the vestibular endorgan. A: Light microscopic distribution of taurine in the rat vestibular endorgan (Durcupan-embedded semithin section). In the crista ampullaris, taurine-like immunoreactivity is found around the hair cells. Immunoreactivity is also found in fibrocytes and myelin sheath (Schwann cell) in the stroma. Bar=10 μm. B: Electron micrograph showing the distribution of taurine in the guinea pig vestibular sensory epithelium. Immunolabelling in nerve chalices (NC) exceeds that in hair cells (HC). SC: supporting cell. Bar=1 μm. C: Electron micrograph of the guinea pig crista ampullaris showing afferent nerve fibers running in the stroma. Note that nerve fibers are surrounded by intensely immunoreactive Schwann cells (arrows) (freeze substituted/ Lowicryl embedded material). Fibrocytes (F) are also heavily labelled. Bar=1 μm.

microscopic results were compatible with the light microscopic findings. Notably, immunolabelling was more intense in the nerve chalice than in hair cells (Fig. 3B). Fibrocytes distributed in the stroma and Schwann cells distributed around the vestibular nerve were also enriched in taurine immunoreactivity (Fig. 3C).

Both in the cochlea and vestibular endorgans, capillary endothelial cells were heavily labelled (Fig. 2C).

To summarize, neuronal components did not show any significant taurine immunoreactivity except nerve chalices in the vestibular endorgans, while various types of supporting cells, Schwann cells, fibrocytes, and endothelial cells were strongly immunoreactive.

DISCUSSION

Taurine is reported to be abundant in the mouse cochlea[4], suggesting that taurine may possess a crucial role in this organ. HPLC analysis of perilymph demonstrated increased levels of taurine (as well as other amino acids) in response to sound exposure[5]. Potassium-induced release of taurine in perilymph of the guinea pig has also been demonstrated[2,13]. These observations pointed to a possible neurotransmitter role of taurine in the cochlea. However, this K^+-induced release of taurine continues after the destruction of the organ of Corti by the ototoxic agents, kanamycin and ethacrynic acid[2]. Electrophysiological studies also failed to show any effect of taurine on the compound action potential of cochlear nerve[3]. A high affinity uptake system for [^{3}H]taurine was not found in inner or outer hair cells, but rather in supporting cells such as Deiters cells and pillar cells[23,24]. The nonphysiological actions and the lack of an uptake system in the neuronal compartment speak against a neurotransmitter role of taurine in the cochlea.

One reason why it has proved difficult to resolve the functional significance of taurine in the inner ear is the lack of morphological data. The present immunocytochemical results indicated that taurine is enriched in various supporting cells, nerve chalices, Schwann cells, and fibrocytes and suggest that these compartments are the source of the sound/K^+-induced release of taurine in the cochlea[2,5,13]. This would be in agreement with the previous demonstration of taurine uptake in supporting cells and satellite (Schwann) cells[23,24].

What could be the functional significance of the release of taurine into the perilymph? Taurine is involved in osmoregulatory function[15,18,25,26,30], and volume regulatory fluxes of taurine have been demonstrated from astrocytes[14,21]. A recent immunocytochemical study showed that acute reductions in plasma osmolality cause a taurine flux from Purkinje cells to adjacent glial cells[17]. These observations suggested that taurine may be released from certain cell compartments to maintain their normal volume under hyposmotic conditions. In the organ of Corti, the volume of the supporting cells may be carefully regulated so as to maintain cochlear micromechanics in the face of perturbations of the extracellular fluid composition. The volume regulation of supporting cells could be achieved by release and uptake of taurine, which occurs in high concentrations in these cells. This assumption is compatible with the homogeneous intracellular distribution of taurine.

The contribution of taurine to Ca^{2+} modulation has been emphasized in the heart. Taurine appears to modulate Ca^{++} level through its effect on Ca^{++} pumps and channels[9,10]. It should be noted that fibrocytes are endowed with Ca^{++} ATPase as well as several Ca^{2+} binding proteins[11,12,22,29], and that they may be involved in Ca^{2+} regulation. Therefore it is conceivable that taurine may help maintain the cochlear microenvironment by controlling the Ca^{++} concentration in the perilymph.

Both in the cochlea and vestibular endorgans, capillary endothelial cells are heavily labelled. Previous immunocytochemical studies have demonstrated the presence of taurine in endothelial cells of the rat cerebellar cortex[6,20]. The functional role of taurine in capillary endothelia remains obscure.

In conclusion, the present study has revealed the cellular and subcellular localization of taurine in the inner ear. The differentiated distribution of this amino acid indicates that it could be involved in a variety of functions, the precise nature of which remains to be elucidated.

ACKNOWLEDGMENTS

We gratefully acknowledge the technical assistance of K.M.Gujord (immunocytochemistry), T. Nordby (electron microscopy), and N. Kuwano (photography). This work was supported by J.E.Isberg's Fund, Bergliot and Sigurd Skaugen's fund, the Sasakawa foundation, the Norwegian Council as part of a EU BIOMED concerted action, and a Grant-in-Aid for Scientific Research from the Ministry of Education, Science and Culture of Japan (05671413,05044145).

REFERENCES

1. Amiry-Moghaddam, H., Nagelhus, E. and Ottersen, O.P. 1994, Light- and electromicroscopic distribution of taurine, an organic osmolyte, in rat renal tubule cells, *Kidney International* 45:10-22.
2. Bobbin, R.P., Ceasar, G. and Fallon, M. 1990, Potassium induced release of GABA and other substances from the guinea pig cochlea, *Hearing Res.* 46:83-94.
3. Bobbin, R.P. and Thompson, M.H. 1978, Effects of putative transmitters on afferent cochlear transmission, *Ann. Otol. Rhinol. Laryngol.* 87:185-190.
4. Davies, W.E. and Owen, C. 1985, The nature of neurotransmitters in the mammalian lower auditory system, in: "Auditory Biochemistry", Drescher, D.G., ed. Charles C. Thomas, Springfield, IL. pp. 244-257.
5. Dresher, M., Dresher, D. and Medina, J. 1983, Effect of sound stimulation at several levels on concentrations of primary amines, including neurotransmitter candidates, in perilymph of the guinea pig inner ear, *J. Neurochem.* 41:309-320.
6. Gragera, R.R., Muniz, E. and Martinez-Rodgriguez, R. 1994, Neuromediators in the cerebellar blood-brain barrier and its microenvironment. Immunocytochemical demonstration of taurine, glycine, serotonin, thiamin and AATase, *J. Hirnforsch.* 35:31-38.
7. Hjelle, O., Farrukh, A. and Ottersen, O. 1994, Antisera to glutathione: characterization and immunocytochemical application to the rat cerebellum, *Eur. J. Neurosci.* 6:793-804.
8. Hsu, S.-M., Raine, L. and Franger, H. 1981, Use of avidin-biotin-peroxidase complex (ABC) in immunoperoxidase techniques: A comparison between ABC and unlabeled antibody (PAP) procedures, *J. Histochem. Cytochem.* 29:577-580.
9. Huxtable, R.J. 1989, Taurine in the cetral nervous system and the mammalian actions of taurine, *Prog. Neurobiol.* 32:471-533.
10. Huxtable, R.J. 1992, The physiological actions of taurine, *Physiol. Rev.* 72:101-163.
11. Ichimiya, I., Adamas, J. and Kimura, R. 1994, Changes in immunostaining of cochleas with experimentary induced endolymphatic hydrops, *Ann. Otol. Rhinol. Laryngol.* 103:457-468.
12. Ichimiya, I., Adams, J. and Kimura, R. 1994, Immunolocalization of Na+, K+-ATPase, calcium-binding proteins, and carbonic anhydrase in the guinea pig inner ear, *Acta Otolaryngol. (Stockh.)* 114:167-176.
13. Jenison, G.L., Bobbin, R.P. and Thalmann, R. 1985, Potassium-induced release of endogenous amino acids in the guinea pig cochlea, *J. Neurochem.* 44:1845-1853.
14. Kimelberg, H., Goderie, S., Higman, S., Pang, S. and Waniewski, R. 1990, Swelling-induced release of glutamate, aspartate and taurine from astrocyte cultures, *J. Neurosci.* 10:1583-1591.
15. Lehmann, A. 1989, Effects of microdialysis-perfusion with anisoosmotic media on extracellular amino acids in the rat hippocampus, *J. Neurochem.* 53:525-535.
16. Madsen, S., Ottersen, O.P. and Storm-Mathisen, J. 1985, Immunocytochemical visualization of taurine: neuronal localization in the rat cerebellum, *Neurosc. Lett.* 60:255-260.

17. Nagelhus, E.A., Lehmann, A. and Ottersen, O.P. 1993, Neuronal-glial exchange of taurine during hypo-osmotic stress: A combined immunocytochemical and biochemical analysis in rat cerebellar cortex, *Neuroscience* 54:615-631.
18. Nieminen, M.-L., Tuomisto, L., Solatunturi E., Eriksson, L. and Paasonen, M.K. 1988, Taurine in the osmoregulation of the Brattleboro rat, *Life Sci.* 42:2137-2143.
19. Ottersen, O.P. 1987, Postembedding light- and electron microscopic immunocytochemistry of amino acids: description of a new model system allowing identical conditions for specificity testing and tissue processing, *Exp. Brain Res.* 69:167-174.
20. Ottersen, O.P. 1988, Quantitaive assessment of taurine-like immunoreactivity in different cell types and processes in rat cerebellum: an electronmicroscopic study based on a postembedding immunogold labelling procedure, *Anat. Embryol.* 178:407-421.
21. Pasantes-Morales, H. and Schousboe, A. 1988, Volume regulation in astrocytes: A role for taurine as an osmoeffector, *J. Neurosci. Res.* 20:505-509.
22. Schulte, B.. 1993, Immunohistochemical localization of intracellular Ca-ATPase in outer hair cells, neurons and fibrocytes in the adult and developing inner ear, *Hearing Research* 65:262-273.
23. Schwartz, I.R. and Ryan, A.F. 1983, Differential labeling of sensory cell and neural populations in the organ of Corti following amino acid incubations, *Hearing Res.* 1983;9:185-200.
24. Schwartz, I.R. and Ryan, A.F. 1986, Uptake of amino acids in the gerbil cochlea, in: "Neurobiology of Hearing", Altschuler R.A., Hoffman, D.W. and Bobbin R,P., eds. Raven Press, New York, pp. 173-190.
25. Thurston, J.H., Hauhart, R.E. and Dirgo, J.A. 1980, Taurine: a role in osmotic regulation of mammalian brain and possible clinical significance, *Life Sci.* 26:1561-1568.
26. Trachtman, H., Barbour, R., Sturman, J.A. and Finberg, L. 1988, Taurine and osmoregulation: taurine is a cerebral osmoprotective molecule in chronic hypernatremic dehydration, *Pediat. Res.* 23:35-39.
27. Usami, S. and Ottersen, O. 1995, The localization of taurine-like immunoreactivity in the organ of Corti: A semiquantitative, post-embedding immuno-electron microscopic analysis in the rat with some observations in the guinea pig, *Brain Res.* 676:277-284.
28. Usami, S. and Ottersen, O. 1955, Differential cellular distribution of glutamate and glutamine in the rat vestibular endorgans: an immunocytochemical study, *Brain Res.* 676:285-292.
29. Usami, S., Shinkawa, H., Inoue, Y., Kanzaki, J. and Anniko, M. 1995, Calbindin-D28K localization in the primate inner ear, *J.Oto-Rhino-Laryng.* 57:94-95.
30. Wade, J.V., Olson, J.P., Samson, F.E., Nelson, S.R. and Pazdernik, T.L. 1988, A possible role for taurine in osmoregulation within the brain, *J. Neurochem.* 51:740-745.

Part 6

TAURINE AND THE CENTRAL NERVOUS SYSTEM

39

CELLULAR STUDIES OF THE TAURINE TRANSPORTER

Norma Lake[1,2] and John Orlowski[1]

[1] Departments of Physiology and
[2] Department of Ophthalmology
McGill University
3655 Drummond Street
Montreal, Quebec, Canada, H3G 1Y6

INTRODUCTION

It is generally recognized that the amino acid, taurine, is important for the proper functioning of a number of organs, such as the brain, retina, heart, liver and kidney. Individual tissues can maintain high intracellular taurine levels through local biosynthesis, and/or by the transmembrane transport of taurine synthesized either elsewhere in the body (such as in the liver) or obtained from external dietary sources. Taurine exists as a zwitterion at physiological pH and, therefore, diffuses poorly across lipid membrane bilayers[6,7]. Hence taurine requires a transporter for efficient entry into cells.

When one assays key enzymes in the biosynthesis of taurine, such as cysteine sulfinate decarboxylase (CSAD), there is sometimes a disparity between the activity of this enzyme and the taurine content of a tissue. For example, in retinal photoreceptors and heart, CSAD is undetectable yet taurine content is high[6]; in cerebellum, some classes of taurine-containing neurons appear to lack CSAD, whereas nearby glia are CSAD-positive[1]. While there may be alternative synthetic pathways, whose significance has not yet been demonstrated, a simpler possibility is that cells lacking the CSAD pathway may have a higher or absolute dependence on transmembrane transport to fulfill their taurine requirements. Other subclasses of taurine-containing cells appear to have both biosynthetic (CSAD) and transport capabilities (as deduced from autoradiographs of labelled taurine uptake).

The functional importance of the contribution of taurine transport across cell membranes is shown by the numerous pathological consequences that occur when the transport of taurine is curtailed, either by the use of transport antagonists, or by the lack of plasma taurine resulting from dietary deficiency. Such dysfunctions include blindness[5,12,13,16], cardiac dysfunction[4,14,15,25], growth and developmental defects[30], and immunological insufficiency[21,28].

In addition to the importance of taurine transport in (i) the maintenance of intracellular taurine content for its various cellular functions, the actual movement of taurine across

Taurine 2, edited by Huxtable et al.
Plenum Press, New York, 1996

membranes is also important in at least two other processes: (ii) adaptation of cells to changes in osmolarity where taurine is often the major organic osmolyte, and (iii) a specialized function in the nervous system: the salvaging of taurine released as a neurotransmitter or neuromodulator, and the termination of its synaptic action. Taurine has potent neuroinhibitory actions resembling those of GABA[6].

Within the last three years the application of molecular biological techniques has led to the isolation and characterization of cDNAs that encode taurine transporters from a number of tissue sources[10,22,26,27,29,32]. These molecules all show extremely high homology to each other, exhibiting >90% similarity at the amino acid level. They also show considerable homology to the multigene family of neurotransmitter transporters including GABA, glycine, dopamine and serotonin. These neurotransmitter transporters share a common requirement for Na^+ and Cl^-, and display considerable structural similarity, including 12 putative transmembrane domains. They are most different in their C-terminal cytoplasmic regions. There is some evidence for regulation of the taurine transporters by protein kinase C and calmodulin[3,26]. Northern blot analyses included in the cloning papers cited above have indicated the presence of taurine transporter messenger RNA (mRNA) in various tissue isolates and cell lines. However, information on their anatomical localization within tissue-specific cell types is lacking with the exception of one figure in a paper on mouse brain by Liu *et al.*[22].

MATERIALS AND METHODS

In order to perform localization studies using *in situ* hybridization, sense and anti-sense oligodeoxynucleotides (~50 nucleotides in length) to a unique portion of the C-terminal cytoplasmic domain of the taurine transporter that differs substantially amongst the glycine, GABA and taurine transporters cloned from rat brain[29] were synthesized.

In addition, an oligonucleotide to the initial 5' portion was made, and this, together with one of the 3' oligonucleotides, served as specific primers to allow amplification of the entire coding region (1.8 kilobases) by reverse transcriptase-polymerase chain reaction (RT-PCR).

A nonradioactive *in situ* hybridization procedure based on the steroid hapten, digoxigenin (DIG, Boehringer Mannheim), was developed from the manufacturer's application protocol, and used to localize this taurine transporter mRNA using the synthetic oligonucleotides. The oligonucleotide probes were labeled with DIG-dUTP in a tailing reaction using terminal transferase. Cryostat sections (10-12 μ) of rat tissues frozen either immediately, or following transcardial perfusion with 4% buffered paraformaldehyde, were hybridized with the labeled probes. The DIG-labeled hybrids were detected with an anti-DIG antibody conjugated to alkaline phosphatase. The immune complexes were visualized with chromogenic substrates (NBT/BCIP) that gave a blue-purple reaction product. DAPI, a fluorescent marker of DNA, was used to reveal the distribution of all cell nuclei for comparison with the *in situ* staining. Negative controls included omission of the anti-sense oligonucleotide probe, dilution of the DIG-oligonucleotide with excess unlabeled oligonucleotide prior to hybridization, and substitution by the sense oligonucleotide probe. All of these controls gave negligible reaction products. Precautions to eliminate RNAse activity (baking of glass at 260°C, DEPC-treated solutions) were used with all glassware, slides and solutions up to the post hybridization washes. A large number of protocol modifications were made either singly or in combination in order to evaluate and then optimize parameters such as the type and duration of fixation, compounds used to increase tissue adherence to the glass slides, prehybridization permeabilization to facilitate probe accessibility, and conditions required

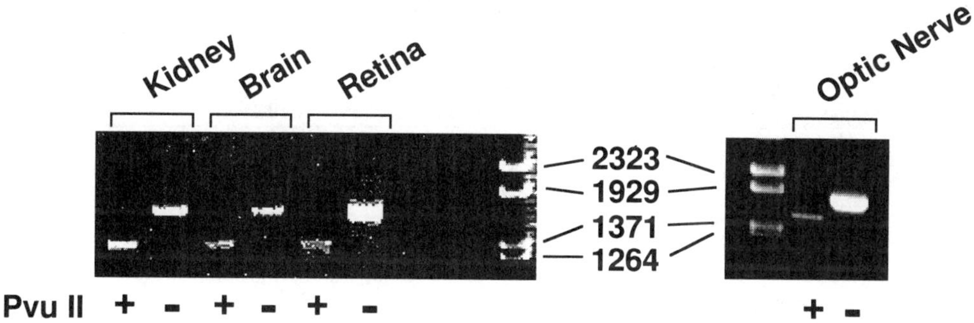

Figure 1. Numbers indicate DNA size markers. See text for additional explanation.

to minimize nonspecific hybridization, antibody binding or endogenous alkaline phosphatase activity.

RESULTS

Figure 1 shows the RT-PCR products derived from total RNA extracts of kidney, brain, retina and optic nerve, using the specific primers for the taurine transporter. Each of the tissues generated a product in the predicted molecular size range (1800 base pairs). This fragment of the taurine transporter cDNA also contains a unique site for the restriction enzyme *Pvu* II. Cleavage of this cDNA fragment with *Pvu* II generated the predicted 1500 and 300 base pair fragments and confirmed that the PCR product corresponds to the taurine transporter. The +*Pvu* II lanes show the unique 1500 base pair fragments in each of kidney, brain, retina, and optic nerve.

In Situ Hybridization Histochemistry

Observations were made on coronal and sagittal sections of brain, sections of the retina and pigment epithelium from an eyecup preparation, and transverse sections of the optic nerve. Sections adhered most successfully to glass slides coated with poly-L-lysine or 2% 3-aminopropyl-triethoxysilane in acetone. For the brain, there was little difference in signal intensity between sections from perfusion-fixed brain compared to those from fresh frozen material subsequently fixed on the slide. For the retina and optic nerve, however, perfusion fixation resulted in weaker signals. Permeabilization using proteinase K to digest away masking or interfering proteins required separate optimization for each tissue and each fixation variation, and could erode the signal. Incubation with 0.2M HCl was effective for most tissue/fixative combinations and had a wide margin of safety with respect to detectable damage or signal losses. To eliminate endogenous alkaline phosphatase activity, in some tissues treatment with 1 mM levamisole was found to be less effective compared to a brief submersion in 20% acetic acid.

Table 1 summarizes our data. Messenger RNA for the taurine transporter was found close to the nucleus in the cytoplasm of cells throughout the CNS, with particularly high levels in the retina, olfactory regions, the cerebral cortex, the hippocampus and the cerebellum. There was high correlation with cells previously identified by immunocytochemistry to contain taurine (i.e. publications of Ottersen, Magnusson, Lake, Sturman). In the usual procedure, colour development (arising from the alkaline phosphatase activity linked to the

Table 1. Localization of messenger RNA for the rat taurine transporter using in situ hybridization histochemistry

Region	mRNA Levels (Staining intensity)	Cell Type(s)
Neural retina	+++	Photoreceptors
	++	Inner retinal neurons
Pigment epithelium	+	Epithelial
Optic nerve	++	Glia, capillary endothelial cells
Cerebral cortex	++,+++	Neurons
	++	Glia, capillary endothelial cells
Olfactory cortex	+++	Neurons
Hippocampus	++,+++	Pyramidal and granule neurons
Cerebellum	+++	Neurons
	++	Glia
Brain stem	+++	Large neurons; e.g. vestibular nuclei

anti-DIG antibody) was allowed to proceed for 8 or 16 h. However it was found that extending this period to 72 or 96 h (with appropriate attention to avoid evaporation of the substrate solution) yielded an increased sensitivity that revealed sites of lower activity. This was true for the retinal pigment epithelium that appeared negative after 16 h, but had a signal at 96 h. This was of considerably lower intensity than that of the adjacent photoreceptors where signal was already present after 16 h.

DISCUSSION

Taurine is present at high mM levels in the retina and brain[8,34] where it has been proposed in various studies to function as an osmoregulator, neuromodulator, membrane stabilizer, regulator of Ca^{2+} binding and fluxes, trophic factor, antioxidant, etc.[6,7,13,30,33,35]. As discussed in the Introduction, the taurine transporter is clearly important, either directly or indirectly, in maintaining these functions, sometimes acting in conjunction with *de novo* biosynthesis. Our RT-PCR results confirm the global tissue distributions of the taurine transporter in rat reported by Smith *et al.*[29], and add the optic nerve to the list of tissues that contain this transcript. *In situ* hybridization histochemistry provides an additional experimental tool with inherent high specificity, sensitivity and cellular resolution to explore the role of taurine. However one should be judicious in interpreting this type of analysis, since it is designed to evaluate levels of mRNA, not expression of taurine transporter protein or activity. Nevertheless these data provide positive correlations with data obtained using independent techniques, such as immunocytochemistry of endogenous taurine, and autoradiographic descriptions of taurine uptake sites.

For example, in the retina our immunocytochemistry shows high levels of taurine in photoreceptors and bipolar cells[19], and autoradiography revealed high affinity uptake sites for exogenous taurine in the outer nuclear (photoreceptor cell) layer[18] (and N. Lake unpublished), presumably mediated in part by the taurine transporter that we have demonstrated here by RT-PCR and *in situ* hybridization. While taurine is unlikely to be a neurotransmitter in these cells[19], it may contribute to protecting cell viability by performing osmoregulatory and antioxidant functions[13] since photoreceptors are extremely active in terms of ionic flux and O_2-dependent metabolic processes that maintain those ion gradients.

In the optic nerve, we have localized taurine to glial cells, including astrocytes[20] where taurine movements may help to curtail the osmotic shifts induced by ion fluxes associated with axonal nerve impulse traffic. *In situ* hybridization located the taurine transporter mRNA in glia, including astrocytes, and in capillary endothelial cells in the optic nerve and in the brain, where these cell types form part of the blood-brain/nerve barrier and are involved in the transport of taurine from the blood.

The pigment epithelium is part of the blood-retinal barrier, has high affinity uptake sites for taurine and is involved in the transfer of taurine from the blood to the retina *in vivo*[17,35]. Here, we have shown that the transporter transcript is expressed at relatively low levels. Perhaps this is reflective of the iso-osmotic and taurine-replete state of the animals that we used (see below).

A final example (although many more are apparent) is in the cerebellum, where we find mRNA for the taurine transporter strongly expressed in neuronal and less intensely in glial cells. This is complementary to the finding that neurons likely import their taurine (visualized by immunocytochemistry[24]) since the CSAD (biosynthetic) activity appears confined to glia[1]. In addition, the transporter is probably involved in the neuronal-glial exchanges of taurine in response to hypo-osmotic stresses and their reversal, that favors neuronal rather than glial volume[23] (see Ottersen *et al.,* this volume).

Since there is indirect evidence that taurine transporter activity is modulated by substrate availability and osmolarity[2,9,11,23,31,33], additional studies are planned to examine these perturbations using *in situ* hybridization studies.

ACKNOWLEDGMENTS

These studies were supported by The Kidney Foundation of Canada. J.O. is supported by a Chercheur-Boursier from the Fonds de la Recherche en Santé du Québec.

REFERENCES

1. Almarghini, K., Barbagli, B., and Tappaz, M. 1994, Production and characterization of a new specific antiserum against the taurine putative biosynthetic enzyme cysteine sulfinate decarboxylase, *J.Neurochem.* 62:1604-1614.
2. Berkowitz, D., Hug, P., Sleight, R.G., and Bucuvalas, J.C. 1994, Fasting enhances taurine transport by rat liver plasma membrane vesicles, *Am.J.Physiol.* 267:G932-G937.
3. Brandsch, M., Miyamoto, Y., Ganapathy, V., and Leibach, F.H. 1993, Regulation of taurine transport in human colon carcinoma cell lines (HT-29 and Caco-2) by protein kinase C, *Am.J.Physiol.* 264:G939-G946.
4. Eley, D.W., Lake, N., and ter Keurs, H.E.D.J. 1994, Taurine depletion and excitation-contraction coupling in rat myocardium, *Circ.Res.* 74:1210-1219.
5. Hayes, K.C., Carey, R.E. and Schmidt, S.Y. 1975, Retinal degeneration associated with taurine deficiency in the cat. *Science* 188:949-951.
6. Huxtable, R.J. 1989, Taurine in the central nervous system and the mammalian actions of taurine, *Prog.Neurobiol.* 32:471-533.
7. Huxtable, R.J. 1992, Physiological actions of taurine, *Physiol.Rev.* 72:101-163.
8. Jacobsen, J.G. and Smith, L.H. 1968, Biochemistry and physiology of taurine and taurine derivatives, *Physiol.Rev.* 48:424-511.
9. Jayanthi, L.D., Ramamoorthy, S., Mahesh, V.B., Leibach, F.H., and Ganapathy, V. 1995, Substrate-specific regulation of the taurine transporter in human placental choriocarcinoma cells (JAR), *Biochim.Biophys.Acta Bio-Membr.* 1235:351-360.
10. Jhiang, S.M., Fithian, L., Smanik, P., McGill, J., Tong, Q., and Mazzaferri, E.L. 1993, Cloning of the human taurine transporter and characterization of taurine uptake in thyroid cells, *FEBS Lett.* 318:139-144.
11. Jones, D.P., Miller, L.A., and Chesney, R.W. 1995, The relative roles of external taurine concentration and medium osmolality in the regulation of taurine transport in LLC-PK1 and MDCK cells, *Pediatr.Res.* 37:227-232.

12. Lake, N. 1986, Electroretinographic deficits in rats treated with guanidinoethyl sulfonate, a depletor of taurine, *Exp.Eye Res.* 42:87-92.
13. Lake, N. 1989, The role of taurine in retinal function, in: "Extracellular and intracellular messengers in the vertebrate retina", Pasantes-Morales, H. and Redburn, D. eds. Alan R. Liss, Inc. pp. 61-86.
14. Lake, N. 1993, Loss of cardiac myofibrils: Mechanism of contractile deficits induced by taurine deficiency, *Am.J.Physiol.* 264:H1323-H1326.
15. Lake, N., De Roode, M., and Nattel, S. 1987, Effects of taurine depletion on rat cardiac electrophysiology: *in vivo* and *in vitro* studies, *Life Sci.* 40:997-1005.
16. Lake, N. and Malik, N. 1987, Retinal morphology in rats treated with a taurine transport antagonist, *Exp.Eye Res.* 44:331-346.
17. Lake, N., Marshall, J., and Voaden, M.J. 1977, The entry of taurine into the neural retinal and pigment epithelium of the frog, *Brain Res.* 128:497-503.
18. Lake, N., Marshall, J., and Voaden, M.J. 1978, High affinity uptake sites for taurine in retina, *Exp.Eye Res.* 27:713-718.
19. Lake, N. and Verdone-Smith, C. 1989, Immunocytochemical localization of taurine in the mammalian retina, *Curr.Eye Res.* 8:163-173.
20. Lake, N. and Verdone-Smith, C. 1990, Immunocytochemical localization of taurine within glial cells in the optic nerve of adult albino rats, *Curr.Eye Res.* 9:1115-1120.
21. Lake, N., Wright, E.D. and Lapp, W.S. 1992, Effects of taurine deficiency on immune function in mice, in: "Taurine: Nutritional Value and Mechanisms of Action", Lombardini, J.B., Schaffer, S.W. and Azuma, J. eds. Plenum Press, New York, pp. 241-243.
22. Liu, Q.-R., López-Corcuera, B., Nelson, H., Mandiyan, S., and Nelson, N. 1992, Cloning and expression of a cDNA encoding the transporter of taurine and β-alanine in mouse brain, *Proc.Natl.Acad.Sci.USA*, 89:12145-12149.
23. Nagelhus, E.A., Lehmann, A., and Ottersen, O.P. 1993, Neuronal-glial exchange of taurine during hypo-osmotic stress: A combined immunocytochemical and biochemical analysis in rat cerebellar cortex, *Neuroscience*, 54:615-631.
24. Ottersen, O.P. 1988, Quantitative assessment of taurine-like immunoreactivity in different cell types and processes in rat cerebellum: an electronmicroscopic study based on a postembedding immunogold labelling procedure, *Anat.Embryol.* 178:407-421.
25. Pion, P.D., Kittleson, M.D., Rogers, Q.R., and Morris, J.G. 1987, Myocardial failure in cats associated with low plasma taurine: a reversible cardiomyopathy, *Science*, 237:764-768.
26. Ramamoorthy, S., Del Monte, M.A., Leibach, F.H., and Ganapathy, V. 1994, Molecular identity and calmodulin-mediated regulation of the taurine transporter in a human retinal pigment epithelial cell line, *Curr.Eye Res.* 13:523-529.
27. Ramamoorthy, S., Leibach, F.H., Mahesh, V.B., Han, H., Yang-Feng, T., Blakely, R.D., and Ganapathy, V. 1994, Functional characterization and chromosomal localization of a cloned taurine transporter from human placenta, *Biochem.J.* 300:893-900.
28. Schuller-Levis, G., Mehta, P.D., Rudelli, R., and Sturman, J. 1990, Immunologic consequences of taurine deficiency in cats, *J.Leukocyte Biol.* 47:321-331.
29. Smith, K.E., Borden, L.A., Wang, C.-H.D., Hartig, P.R., Branchek, T.A., and Weinshank, R.L. 1992, Cloning and expression of a high affinity taurine transporter from rat brain, *Mol.Pharmacol.* 42:563-569.
30. Sturman, J.A. 1988, Taurine in development, *J.Nutr.* 118:1169-1176.
31. Sturman, J.A., Rassin, D.K., Hayes, K.C., and Gaull, G.E. 1978, Taurine deficiency in the kitten: exchange and turnover of 35S-taurine in brain, retina, and other tissues, *J.Nutr.* 108:1462-1476.
32. Uchida, S., Kwon, H.M., Yamauchi, A., Preston, A.S., Marumo, F., and Handler, J.S. 1992, Molecular cloning of the cDNA for an MDCK cell Na^+- and Cl^--dependent taurine transporter that is regulated by hypertonicity, *Proc.Natl.Acad.Sci.USA*, 89:8230-8234.
33. Uchida, S., Nakanishi, T., Kwon, H.M., Preston, A.S., and Handler, J.S. 1991, Taurine behaves as an osmolyte in Madin-Darby canine kidney cells. Protection by polarized, regulated transport of taurine, *J.Clin.Invest.* 88:656-662.
34. Voaden, M.J., Lake, N., Marshall, J., and Morjaria, B. 1977, Studies on the distribution of taurine and other neuroactive amino acids in the retina, *Exp.Eye Res.* 25:249-257.
35. Voaden, M.J., Oraedu, A.C.I., Marshall, J., and Lake, N. 1981, Taurine in the retina, in: "The effects of taurine on excitable tissues", Schaffer, S.W., Baskin, S.I. and Kocsis, J.J. eds., Spectrum Publications, New York, pp. 145-160.

40

HIGH-AFFINITY TAURINE UPTAKE AND ITS REGULATION BY PROTEIN KINASE C IN HUMAN GLIOMA CELLS

Hirohiko Nakamura,[1] Shen Hao Huang,[2] and Kintomo Takakura[3]

[1] Department of Neurosurgery
Tokyo Metropolitan Komagome Hospital
[2] Sankyo Pharmaceutical Co. Ltd.
[3] Department of Neurosurgery
Neurological Institute of Tokyo Women's Medical College

INTRODUCTION

In high-resolution ^{1}H NMR spectroscopy, malignant glioma tissues are shown to include higher concentration of taurine than normal brain tissues do. Although it is well known that mouse astrocyte[18], rat astrocyte[5] and glioma cells[12] have a high-affinity uptake system for taurine, little is known about such transport in human glioma cells.

In the present study, we investigated the properties of the taurine uptake system in human glioma cells, and the effect of protein kinase C activation on it. Protein kinase C activity is significantly increased in human malignant glioma cells[3], and the proliferation rates of gliomas may be modulated by the protein kinase C (PKC) signal transduction system[2].

MATERIALS AND METHODS

[2-^{3}H(N)]Taurine (specific radioactivity 740.0 GBq/mmol; 20.0 Ci/mmol) was purchased from DuPont-New England Nuclear, Boston, MA, USA. Unlabeled amino acids, phorbol esters and H7 were from Sigma Chemical Co., St. Louis, MO, USA. Eagle's minimum essential medium (MEM), fetal bovine serum, and other chemicals for cell cultures were obtained from Gibco Laboratories, Tokyo, Japan. All other chemicals were of analytical grade from Wako Chemical Co., Tokyo, Japan.

Culture of Human Glioma Cells

U251 cell line was established from the surgical material of malignant gliomas by Pontén[17]. U373 human glioma cells were purchased from the American Type Culture Collec-

Taurine 2, edited by Huxtable et al.
Plenum Press, New York, 1996

tion, Rockville, MD, USA. For the uptake studies, the cells were seeded in 35 mm^2 disposable Falcon Petri dishes at a density of approximately 2 x 10^6 cells/dish and allowed to grow as a monolayer. The cells received 2 ml of Minimum Essential Medium (MEM) supplemented with 10% fetal bovine serum and 50 units of penicillin and streptomycin per ml at 37° in a CO_2 incubator. The cells became confluent four days after the subculture, and the medium was replaced by 1.5 ml of the fresh culture medium for experiments 24 h after the medium change.

Measurement of Taurine Uptake

The medium was removed from confluent cultures and washed 3x with PBS. After the cells were incubated for 30 min with 1.5 ml of the uptake medium which contained 20 mM HEPES/Tris (PH 7.4), 130 mM NaCl, 5 mM KCl, 0.5 mM $MgSO_4$, 1 mM $CaCl_2$, and 5 mM D-glucose, measurement of taurine uptake was initiated by replacing 1.5 ml of the uptake medium with medium including 50.5 kBq (1.5 μCi) of [^{3}H]taurine. In most cases, the final concentration of [^{3}H]taurine was 50 nM. After incubation for 3 min, the buffer was removed and the cells were quickly washed 4x with the cold uptake buffer. After washing, 1 ml of 0.3 M NaOH was added to each dish for 60 min, after which time the contents from the dish were transferred to a scintillation vial and the radioactivity of the contents was measured by liquid scintillation spectrometry.

In the Na^+-free medium, NaCl was replaced by choline chloride. In the experiments determining the effects of Cl^- on taurine uptake, the uptake buffer contained 20 mM HEPES/Tris (PH 7.4), 5 mM potassium gluconate, 1 mM calcium gluconate, 0.5 mM magnesium gluconate, 5 mM D-glucose and 130 mM sodium gluconate.

Treatment of Cells with Phorbol Esters

Stock solutions of phorbol esters were prepared in dimethyl sulfoxide. These solutions were appropriately diluted with the culture medium and used for the treatment of the cells. The final concentration of dimethyl sulfoxide during treatment was 0.02%. The control cells were treated with an equal concentration of the solvent. After incubation for the desired time, the medium was removed from the dish, and the uptake buffer containing 50nM [^{3}H]taurine was added to the culture to initiate uptake measurement.

Protein Determination

Protein was assayed for each experiment by using duplicate dishes cultured concurrently and under conditions identical with those dishes used for uptake measurement. The protein concentration was determined by the method of Bradford[1], with bovine serum albumin as the standard.

Statistical Analysis

Data were analyzed using Student's t-test and a probability value of 5% or less was considered indicative of a significant effect.

RESULTS

Kinetics of Taurine Uptake

The kinetic parameters of taurine uptake in human glioma cells were analyzed by studying the relationship between the initial uptake rates (3 min incubation) and taurine concentration. The uptake rates were determined over a taurine concentration range of 10-150 μM. Non-carrier-mediated uptake was determined by measuring the uptake of [^{3}H]taurine in the presence of 1 mM unlabeled taurine. When compared with the values for radiolabel uptake in the presence of 10 μM and 150 μM unlabeled taurine, the radiolabel uptake in the presence of 1 mM unlabelled taurine was 7.5 ± 2.1% and 20.5 ± 3.4% for U251 glioma cells, and 6.9 ± 1.8% and 18.7 ± 2.9% for U373 cells. This value was subtracted from total uptake in each experiment to calculate carrier-mediated uptake. The relationship between the initial rates of carrier-mediated uptake and taurine concentration is illustrated in Fig. 1. When these data were transformed into a linear format by the Eadie-Hofstee method (initial uptake rate/taurine concentration versus initial uptake rate), linear plots (r=-0.97 for U251; r=-0.98 for U373) were obtained (Fig. 1a and 1b, inset). The K^i was 27.0 ± 3.2 μM and the V_{max} was 10.7 ± 2.5 pmol/3 min per mg of protein in U251 cells, and 30.3 ± 3.5 μM and 8.6 ± 1.8 pmol/3 min per mg of protein in U373 cells.

Na^+, Cl^--Dependence and Substrate Specificity of Taurine Uptake

Na^+-dependence and Cl^- requirement for the Na^+-dependent taurine uptake were investigated (Table 1). When NaCl in the medium was replaced by choline chloride, the uptake of taurine was drastically decreased. The initial uptake rates measured with a 3 min incubation were approximately 1 or 2% of those in the presence of NaCl. In the presence of sodium gluconate, the uptake rates were 4 or 5% of those in the presence of Cl^-. Therefore, there is a specific requirement for Cl^- as well as Na^+ for optimal transport of taurine in these glioma cells.

Table 1. Na^+-Dependence of [^{3}H]taurine uptake and inhibition by unlabelled amino acids

	[^{3}H]Taurine Uptake (%)	
Amino Acid (1 mM)	U251	U373
Control	100	100
Taurine	5 ± 2**	4 ± 3**
Hypotaurine	7 ± 4**	6 ± 3**
β-Alanine	9 ± 3**	8 ± 2**
GABA	75 ± 7*	74 ± 5*
Glutamate	81 ± 6*	77 ± 6*
Aspartate	82 ± 4*	79 ± 4*
Leucine	98 ± 5	96 ± 6
Alanine	94 ± 5	91 ± 5
α-Aminoisobutyrate	91 ± 6	92 ± 5
Na^+-free medium	1 ± 1	2 ± 1
Cl^--free medium	4 ± 2	5 ± 2

*p < 0.05; **p < 0.01. Values are means ± SD (n=4). Uptake of [^{3}H]taurine into U251 and U373 glioma cells was measured with a 3 min incubation. The concentration of [^{3}H]taurine was 50 nM and that of unlabelled amino acids was 1 mM. NaCl was replaced by choline chloride in the Na^+-free medium. In the experiments determining the effects of Cl^- on taurine uptake, the uptake buffer contained 20 mM HEPES/Tris (PH 7.4), 5 mM potassium gluconate, 1 mM calcium gluconate, 0.5 mM magnesium gluconate, 5 mM D-glucose and 130 mM sodium gluconate.

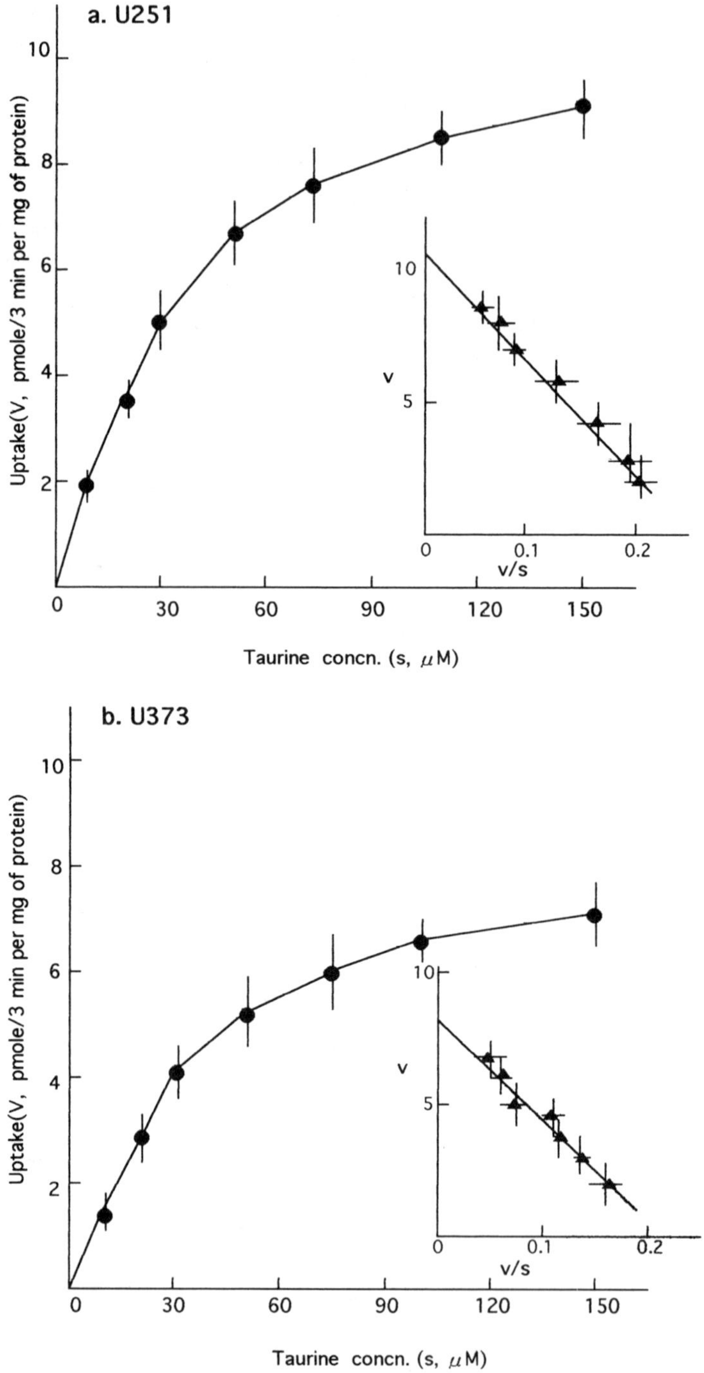

Figure 1. Kinetic analysis of Na^+-dependent taurine uptake in U251 (a) and U373 (b) glioma cells. Initial uptake rates of taurine were determined by using a 3 min incubation. The range of taurine concentrations employed was 10-150 μM. Inset: Eadie-Hofstee plot. The data represent means ± SD for 4 determinations.

The substrate specificity of the carrier system responsible for taurine uptake in human glioma cells was analyzed by studying the effects of a variety of unlabelled amino acids on the uptake of radiolabeled taurine. The following unlabelled amino acids were studied: taurine, hypotaurine, β-alanine, GABA, glutamic acid, aspartic acid, leucine, alanine, and α-aminoisobutyric acid.

At a concentration of 1 mM, leucine, alanine, and α-aminoisobutyric acid failed to inhibit the uptake of [^{3}H]taurine. Conversely, taurine, β-alanine, and hypotaurine almost completely inhibited [^{3}H]taurine uptake under similar conditions. GABA, glutamic acid, and aspartic acid, however, showed an appreciable competition with taurine for the uptake system.

Effects of Phorbol Esters on Taurine Uptake

We investigated the possible involvement of protein kinase C in the regulation of the taurine uptake by studying the influence of phorbol-12,13-dibutyrate (PDBu) on taurine uptake in human glioma cells. The cells were treated with PDBu (100 nM) for 10-120 min before initiation of uptake measurement. The data given in Fig. 2 show that the treatment of the cells with PDBu resulted in inhibition of taurine uptake, and the magnitude of inhibition was dependent on the duration of treatment of the cells with PDBu. A significant ($p < 0.05$) inhibition was observed within 20 min and maximal inhibition occurred at 60 min. H7, a potent inhibitor of protein kinase C, did not affect taurine uptake by glioma cells. 4α-Phorbol 12,13-didecanoate (4α-PDD), an analog of TPA that is unable to activate protein kinase C[9],failed to show any effect on taurine uptake under similar experimental conditions.

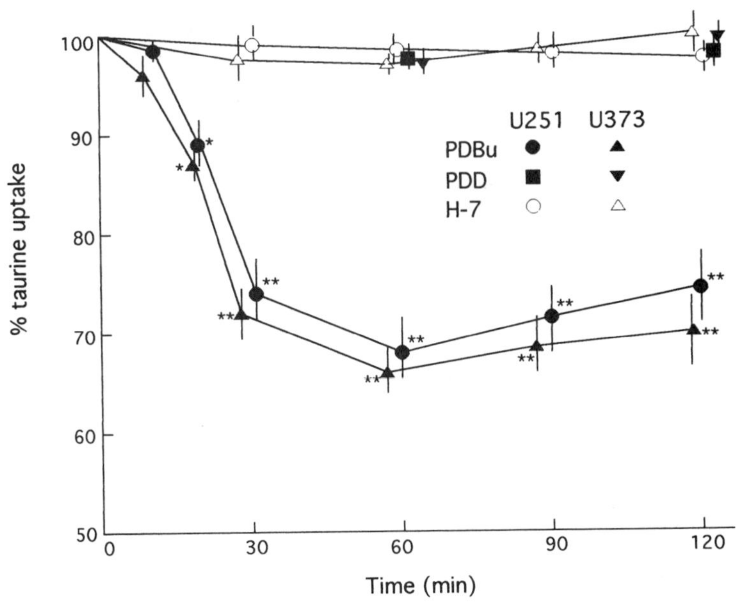

Figure 2. Effect of treatment time on the inhibition of taurine uptake induced by phorbol-12,13-dibutyrate (PDBu). The human glioma U251 and U373 cells were treated with dimethyl sulfoxide (DMSO; control), phorbol-12,13-dibutylate (PDBu), 4α-phorbol 12,13-didecanoate (4α-PDD) or H7 for different periods. The final concentrations of the reagents were: DMSO, 0.02%; PDBu, 200 nM; 4α-PDD, 1 μM; H7, 100 nM. [^{3}H]Taurine (50 nM) was added and incubated for 3 min. For each time period for the treatment with the phorbol esters and H7, the cells treated with DMSO for the corresponding time period served as the control. The values are means ± SD for 4 determinations. *$p < 0.05$; **$p < 0.01$.

The dose-response relationship of the inhibition by PDBu is described in Fig. 3. The inhibition was significant at a concentration as low as 10 nM. Pretreatment of H-7 (100 nM) for 30 min changed the EC_{50} of TPA from 41 nM to 83 nM in U251 cells and from 38 nM to 75 nM in U373 cells.

DISCUSSION

In this investigation, we have demonstrated that human glioma cells express a high-affinity taurine transporter with characteristics similar to those of the taurine transporter expressed in animal brain slices, membrane preparations, cell cultures and retina[6]. In addition, this study provides evidence for the regulation of the taurine transporter by protein kinase C in human glioma cells.

The presence of a high-affinity taurine transporter in rodent glia and glioma cells has been demonstrated[5,12,18], and cDNA clones encoding a taurine transporter have been isolated from mouse and rat brain[11,19]. However, the taurine transport system is still unknown in human glia or glioma cells. Since the characteristics of the taurine transport system exhibit

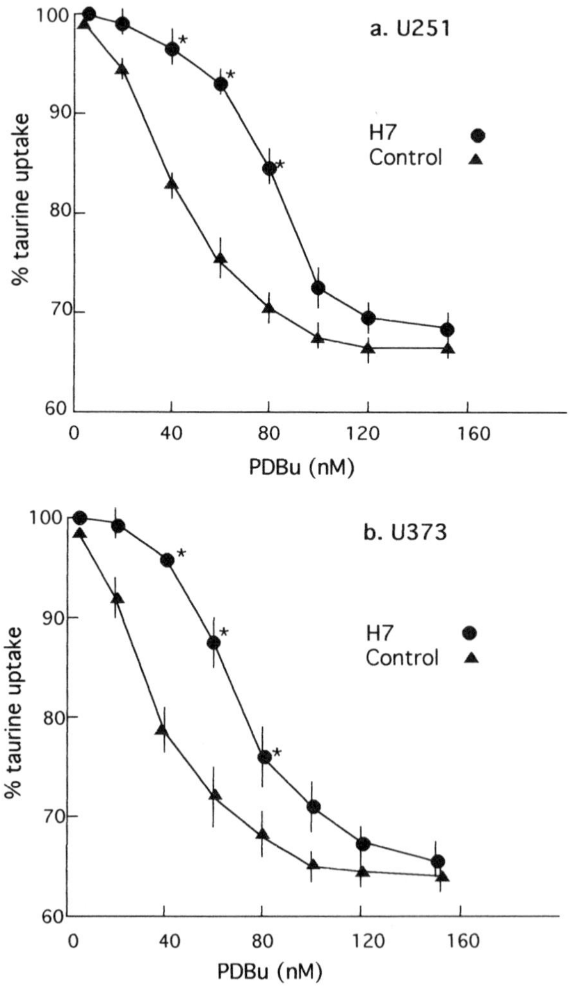

Figure 3. Concentration-dependent inhibition of taurine uptake by PDBu or PDBu plus H7 in U251 cells (a) and U373 cells (b). The cells were treated with 0.02% dimethyl sulfoxide (DMSO; control), the indicated concentrations of phorbol-12,13-dibutyrate (PDBu) with or without H7 (100 nM) for 60 min before the initiation of uptake measurement. Uptake of 50 nM [^{3}H]taurine was measured with a 3 min incubation. The values are means ± SD for 4 determinations. $^*p < 0.01$.

species differences[4,20], the present results that human glioma cells possess high-affinity taurine transporters can make these cell lines convenient models to investigate the biochemical and regulatory aspects of the taurine transport system in human glia.

The concentration of taurine in malignant glioma tissues is higher than that in normal brain tissues, as shown by high-resolution ^{1}H NMR spectroscopy[15]. Malignant astrocytomas (grade 3 and 4) are differentiated from benign astrocytomas (grade 2) by higher concentrations of taurine. Furthermore, malignant astrocytomas are considerably more heterogeneous than lower grade astrocytomas or meningiomas, corresponding to increased amounts of pathologically defined necrosis. It appears that the presence of a high-affinity taurine transporter in human glioma cells and its ability to accumulate taurine against a concentration gradient result in higher concentrations of taurine in malignant glioma tissues.

Taurine is more abundant in metabolically active tissues, such as retina, brain, heart, neutrophils and eosinophils[21]. Increased taurine levels have been found in other malignant tumors and taurine is regarded as a ^{1}H NMR spectroscopic malignancy marker for comparing normal and malignant colon[13]. Accordingly, high taurine levels in tumors could be the expression of a more active metabolism related to increased cellular proliferation and tumor aggressiveness. Considering the important roles that taurine plays in membrane stabilization, detoxification, antioxidation, and osmoregulation, higher taurine uptake may be essential to protect the proliferating cells from self-destruction, such as during processes that generate oxidants[7].

An association of taurine uptake with activation of PKC has not been studied in glia or glioma cells. There are only two reports on the association in JAR placental choriocarcinoma cell line[10] and LLC-PK1 renal cell line[8]. Pretreatment of both cultured cells with an active phorbol ester such as phorbol 12-myristate 13-acetate (TPA) results in the inhibition of the transport system of taurine. In this study, activation of PKC by the active phorbol ester, PDBu, reduced Na^{+}-dependent taurine uptake into U251 and U373 cells. The inhibitory effect of PDBu appears to be mediated through a PKC-dependent pathway because pretreatment of H7, a specific inhibitor of PKC, elevated the EC_{50} of PDBu. In addition to the reduction in taurine uptake observed after activation of PKC, efflux of taurine is increased in LLCPK1[8], rat cerebellar astrocytes[16] and C6 glioma cells[14]. Similarly, we also confirmed that active phorbol esters increase taurine efflux from U251 and U373 cells (unpublished data). PKC activity significantly increases in human malignant glioma cells[3]. Activation of PKC in human glia and glioma cells by various extracellular stimuli could, therefore, potentially play an important role in the regulation of taurine transport and intracellular taurine concentration.

REFERENCES

1. Bradford, M. 1976, A rapid and sensitive method for the quantitation of microgram quantities of protein utilizing the principle of protein-dye binding, *Anal.Biochem.* 72:248-254.
2. Couldwell, W.T., Antel, J.P., and Yong, V.W. 1992, Protein kinase C activity correlates with the growth rate of malignant gliomas: Part II. Effects of gioma mitogens and modulators of protein kinase C, *?* 31:717-724.
3. Couldwell, W.T., Uhm, J.H., Antel, J.P., and Yong, V.W. 1991, Enhanced protein kinase C activity correlates with the growth rate of malignant gliomas in vitro, *Neurosurgery*, 29:880-886.
4. Crabai, F., Sitzia, A., and Pepeu, G. 1974, Taurine concentration in the neurohypophysis of different animal species, *J.Neurochem.* 23:1091-1092.
5. Holopainen, L. and Kontro, P. 1986, High-affinity uptake of taurine and beta-alanine in primary cultures of rat astrocytes, *Neurochem.Res.* 11:207-215.
6. Huxtable, R.J. 1989, Taurine in the central nervous system and the mammalian actions of taurine, *Prog.Neurobiol.* 32:471-533.

7. Jhiang, S.M., Fithian, L., Smanik, P., McGill, J., Tong, Q., and Mazzaferri, E.L. 1993, Cloning of the human taurine transporter and characterization of taurine uptake in thyroid cells, *FEBS Lett.* 318:139-144.
8. Jones, D.P., Miller, L.A., Dowling, C., and Chesney, R.W. 1991, Regulation of taurine transporter activity in LLC-PK1 cells: role of protein synthesis and protein kinase C activation, *J.Am.Soc.Nephrol.* 2:1021-1029.
9. Kraft, A.S., Anderson, W.B., Cooper, H.L., and Sando, J.J. 1982, Decrease in cytosolic calcium/phospholipid-dependent protein kinase activity following phorbol ester treatment of EL4 thymoma cells, *J.Biol.Chem.* 257:13193-13196.
10. Kulanthaivel, P., Cool, D.R., Ramamoorthy, S., Mahesh, V.B., Leibach, F.H., and Ganapathy, V. 1991, Transport of taurine and its regulation by protein kinase C in the JAR human placental choriocarcinoma cell line, *Biochem.J.* 277:53-58.
11. Liu, Q.-R., López-Corcuera, B., Nelson, H., Mandiyan, S., and Nelson, N. 1992, Cloning and expression of a cDNA encoding the transporter of taurine and β-alanine in mouse brain, *Proc.Natl.Acad.Sci.USA*, 89:12145-12149.
12. Martin, D.L. and Shain, W. 1979, High affinity transport of taurine and β-alanine and low affinity transport of τ-aminobutyric acid by a single transport system in cultured glioma cells, *J.Biol.Chem.* 254:7076-7084.
13. Moreno, A., Rey, M., Montane, J.M., Alonso, J., and Arus, C. 1993, ^{1}H NMR spectroscopy of colon tumors and normal mucosal biopsies; elevated taurine levels and reduced polyethyleneglycol absorption in tumors may have diagnostic significance, *NMR Biomed*, 6:111-118.
14. Nakamura, H., Huang, S.H., and Takakura, K. 1993, Phorbol ester increases the efflux of endogenous taurine from C6 glioma cells, *Neurochem.Res.* 18:851
15. Peeling, J. and Sutherland, G. 1992, High-resolution 1H NMR spectroscopy studies of extracts of human cerebral neoplasms, *Magn.Reson.Med.* 24:123-136.
16. Philibert, R.A. and Dutton, G.R. 1988, Phorbol ester and dibutyryl cyclic AMP reduce content and efflux of taurine in primary cerebellar astrocytes in culture, *Neuroscience Letters*, 95:323-328.
17. Pontén, J. 1975, Neoplastic human glia cells in culture, in: "Human Tumor Cells *In vitro*", Fogh, J. ed., Plenum Publishing Corp. New York, pp. 175-206.
18. Schousboe, A., Fosmark, H., and Svenneby, G. 1976, Taurine uptake in astrocytes cultured from dissociated mouse brain hemispheres, *Brain Res.* 116:158-164.
19. Smith, K.E., Borden, L.A., Wang, C.-H.D., Hartig, P.R., Branchek, T.A., and Weinshank, R.L. 1992, Cloning and expression of a high affinity taurine transporter from rat brain, *Mol.Pharmacol.* 42:563-569.
20. Starr, M.S. 1978, Uptake of taurine by retina in different species, *Brain Res.* 151:604-608.
21. Wright, C.E., Tallan, H.H., Lin, Y.Y., and Gaull, G.E. 1986, Taurine: biological update, *Ann.Rev.Biochem.* 55:427-453.

41

SODIUM- AND CHLORIDE-DEPENDENT TRANSPORT OF TAURINE AT THE BLOOD-BRAIN BARRIER

Akira Tsuji and Ikumi Tamai

Faculty of Pharmaceutical Sciences
Kanazawa University
Kanazawa 920, Japan

INTRODUCTION

Sodium- and Cl^--dependent taurine transports have been clarified in kidney[12] and intestine[2]. However, there has been only one report, in 1989, for taurine transport at the blood-brain barrier (BBB). This study, using isolated brain capillaries, showed the existence of a Na^+-dependent high-affinity uptake system for taurine with a K_t value of about 30 μM at the abluminal membrane[4]. However, the polarity of taurine transport is difficult to establish with this classical isolated brain capillary technique, because overall uptake may include both luminal and abluminal components.

The purpose of the present study was to clarify the transport mechanism of taurine both at the luminal and abluminal membranes by using an *in vitro* experimental system of the primary cultured monolayers of brain capillary endothelial cells (BCECs), the *in situ* brain perfusion and brain capillary depletion techniques.

EXPERIMENTAL METHODS FOR BLOOD-BRAIN BARRIER TRANSPORT STUDY

In the *in vivo* methods, most determinations of cerebrovascular permeability of slowly penetrating compounds such as taurine have been made by using the brain vascular perfusion in the guinea pig or in the rat. The complexity of the whole animal as an experimental model results in certain limitation when specific events at the cellular level are to be examined. *In vitro* models such as isolated brain capillaries or cultured brain capillary endothelial cells offer an alternative for studying substrate transport at the BBB and clarifying the BBB function. However, the isolated brain capillary technique shows that the location of transport is difficult to assign and that the polarity and of any possible transport mechanism may be lost. A most disadvantageous point is that an isolated capillary preparation has a decreased cellular level of ATP to less than 10% of normal.

Taurine 2, edited by Huxtable et al.
Plenum Press, New York, 1996

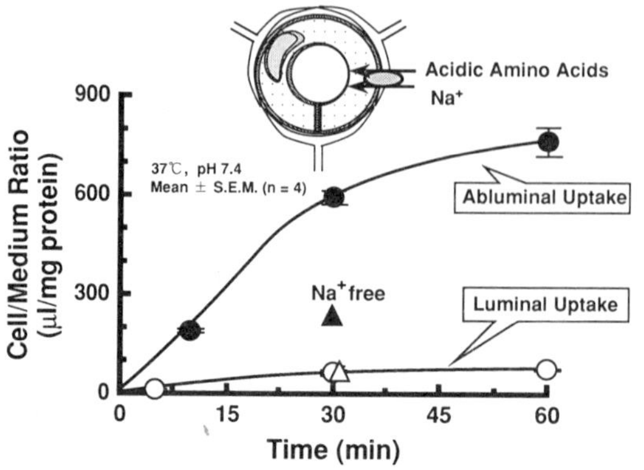

Figure 1. Time courses of luminal and abluminal uptake of L-glutamic acid by cultured BCEC. Luminal (open symbol) and abluminal (closed symbol) uptakes were measured in the presence (circle) or absence (triangle) of Na^+ ions. In the absence of Na^+, choline replacement was used. Concentration of [^{3}H]glutamic acid was 60 nM.

On the other hand, since 1987, *in vitro* studies on the BBB transport of compounds have been performed using primary cultured monolayers of BCECs, grown to confluence in culture[1]. Although the monolayer retains relatively a leaky tight junction compared to the physiological one, the use of primary cultured BCECs is a promising technique for studies on cellular- or molecular-characterization of carrier-mediated transport of compounds[1,5,6,9-11]. Additionally, monolayers of cultured cells retains functional polarity. Figure 1 illustrates a typical example for glutamate transport to evaluate the unidirectional transport ability of primary cultured BCECs (unpublished observation). It is well established using the primary cultured BCECs from the bovine that the Na^+-dependent transport system of glutamate is localized on the abluminal membrane. As can be seen in Fig. 1, for the time courses of luminal and abluminal uptakes of [^{3}H]glutamate into BCECs, in the presence of Na^+, the abluminal uptake of [^{3}H]glutamate was about 10-times greater than the luminal uptake. In the absence of Na^+, the abluminal uptake of [^{3}H]glutamate was remarkably reduced, whereas the luminal uptake was not changed. The present results indicate that BCECs were confirmed to retain the polarized transport function for nutrients. Therefore, we used the primary cultured BCECs for studying BBB transport mechanism of taurine[3].

TAURINE TRANSPORT AT THE BBB

Figure 2 shows the relationship between the initial uptake rate at the luminal and abluminal sides and the concentration of taurine. The luminal and abluminal uptakes of taurine were both saturable. When analyzed by means of the Eadie-Hofstee plot, as indicated in the insets, a single straight line was obtained in each case. We obtained K_t values of 12.1 μM for luminal uptake and 13.6 μM for abluminal uptake.

In order to explore the driving force for taurine uptake, we studied the effects of replacement of cations and anions on the luminal and abluminal uptakes. As shown in Fig. 3, substitution of Na^+ ion with choline, K^+, or N-methylglucamine with Cl^- as counter ion completely abolished both uptakes. The effects of anions on the taurine uptake were examined in the presence of Na^+ ion and the results are shown in Fig. 3. Substitution of Cl^- with nitrate and gluconate completely abolished both luminal and abluminal uptakes. Moreover, the substitution of NaCl with sucrose completely abolished both uptakes. These results suggest that Na^+ and Cl^- ions are both involved in the transport of taurine.

We tried to clarify whether the stimulation of taurine transport by Na^+ and Cl^- is caused by their binding to the carrier protein or by co-transport of the ions with taurine. Figure 4 shows the time-course of taurine uptake from the luminal side of ATP-depleted

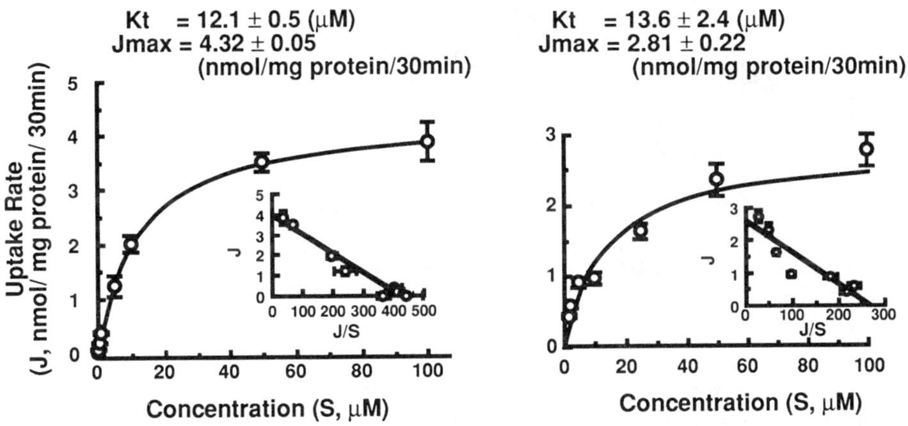

Figure 2. Concentration dependence of luminal (left) and abluminal (right) uptake of taurine by cultured BCEC. Uptakes at various concentrations of taurine were measured for 30 min. The solid line represents the saturable uptake rate estimated by the nonlinear least squares regression analysis. The inset represents Eadie-Hofstee plot of the saturable uptakes.

cultured BCECs. We can use ATP-depleted cells like membrane vesicles and can examine effects of ion gradient as the driving force. As shown in Fig. 4, a marked overshoot phenomenon occurred for the luminal uptake of taurine in the presence of inwardly directed gradients for both Na^+ and Cl^- gradients. A much lower effect was seen in the presence of only a Na^+ or only a Cl^- gradient. This indicates that taurine transport is caused by co-transport of both Na^+ and Cl^- with taurine.

As indicated by a Hill plot for Na^+ or Cl^- concentration dependencies on the luminal uptake of taurine (Fig. 5), a stoichiometric analysis indicates that two Na^+ and one Cl^- are associated with one taurine molecule.

To demonstrate that the taurine transport system observed in the cultured cells functions physiologically, we utilized an *in situ* rat brain perfusion technique. After perfusion of [^{3}H]taurine solution in the presence and absence of inhibitors, the brain was removed to determine the amount of taurine in the brain. Then, we calculated the cerebrovascular permeability. As shown in Fig. 6, the *in vivo* cerebrovascular permeability of [^{3}H]taurine

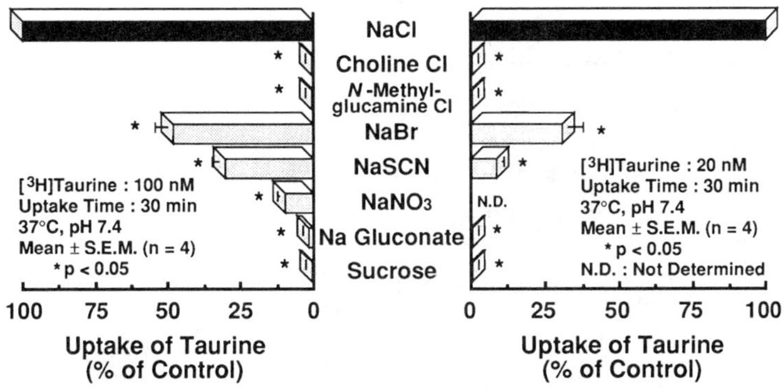

Figure 3. Effects of Na^+ and Cl^- ions on the luminal (left) and abluminal (right) uptake of taurine by cultured BCEC. Na^+ and Cl^- ions were replaced with respective cations and anions as shown.

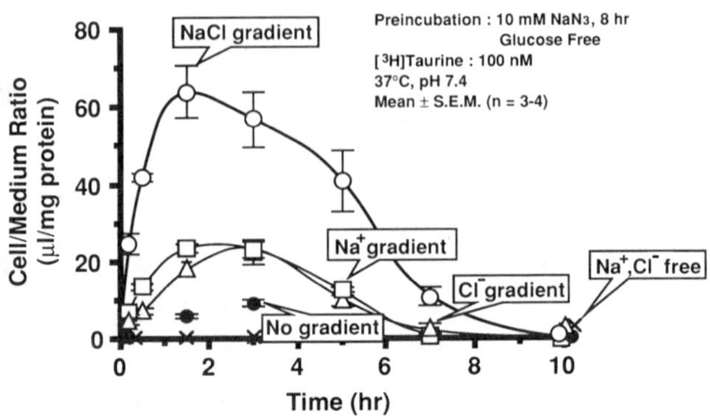

Figure 4. Effects of Na^+ and Cl^- ion gradients on the luminal uptake of [^{3}H]taurine by ATP-depleted cultured BCEC. ATP was depleted by preincubating the cells with NaN_3 (10 mM) in the absence of D-glucose. To study uptake in the presence of maximum driving forces, cells were preincubated with 141 mM potassium gluconate followed by uptake in the presence of NaCl (open circle). The effect of Na^+ gradient was studied in the presence of NaCl with cells preincubated with N-methylglucamine chloride (square). The effect of Cl^- gradient was studied in the presence of 141 mM NaCl with cells preincubated with 141 mM sodium gluconate (triangle). To produce a situation of no Na^+ or Cl^- gradient, cells were preincubated with 141 mM NaCl and the uptake was measured in the presence of NaCl (closed circle). For the study in the absence of NaCl, cells were preincubated with potassium gluconate and the uptake was studied in the presence of 141 mM potassium gluconate (cross).

was inhibited by unlabeled taurine and β-alanine at 1 mM concentration. The transport was not reduced by L-phenylalanine. Furthermore, the transport of [^{3}H]taurine was completely abolished by replacement of Na^+ or Cl^- with N-methylglucamine or gluconate, and by replacement of both Na^+ and Cl^- ions with sucrose. These observations suggest that taurine shares a Na^+- and Cl^--dependent carrier protein in the brain capillary endothelial cells with β-alanine, which is different from the L-system for neutral amino acids.

The polarized transport of taurine from blood to brain side was confirmed by the capillary depletion study. After brain perfusion, the brain was homogenized with dextran. Centrifugation of the homogenate gives the parenchymal fraction as supernatant and the capillary fraction as the pellet. When [^{3}H]taurine solution was perfused for 10-min in rat brain, as indicated in Fig. 7, the apparent parenchymal distribution volume of [^{3}H]taurine was twice that for [^{14}C]sucrose as the vascular space marker. This result indicates that taurine permeates into the brain interstitial fluid across the blood-brain barrier.

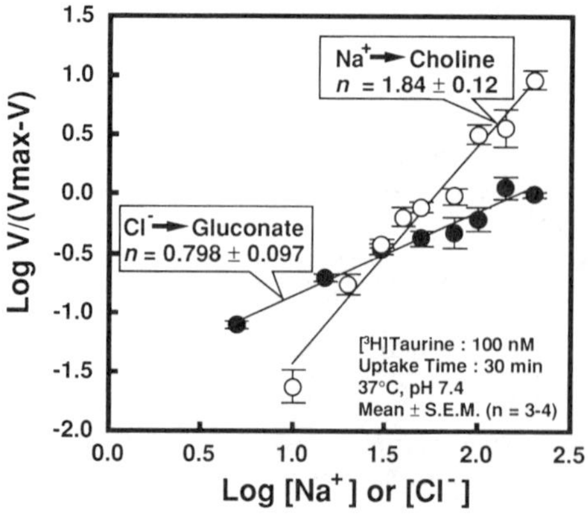

Figure 5. Na^+ and Cl^- dependencies of luminal uptake of taurine by cultured BCEC. Uptake of [^{3}H]taurine (100 nM) was measured in the presence of various concentrations of Na^+ (open circle) or Cl^- ions (closed circle) by replacing NaCl with choline chloride or sodium gluconate, respectively.

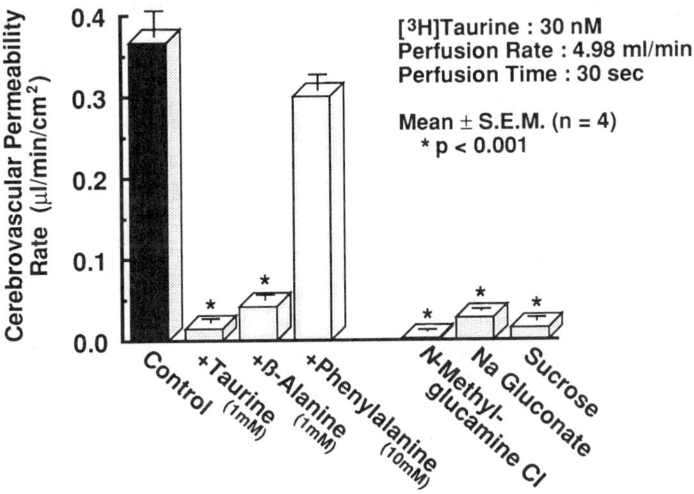

Figure 6. Effects of amino acids and ion replacement on the *in vivo* brain uptake of taurine in rats by brain perfusion method. *In vivo* brain uptake of [^{3}H]taurine (30 nM) was measured for 30 s by perfusing buffer containing each amino acid at a flow rate of 4.98 ml/min from the right external carotid artery. "N-methylglucamine Cl" means that Na^+ in the buffer was replaced with N-methylglucamine. "Na Gluconate" means that Cl^- ions in the buffer were replaced with gluconate ions. "Sucrose" means that NaCl and $NaHCO_3$ in the medium were replaced with sucrose and $KHCO_3$, respectively.

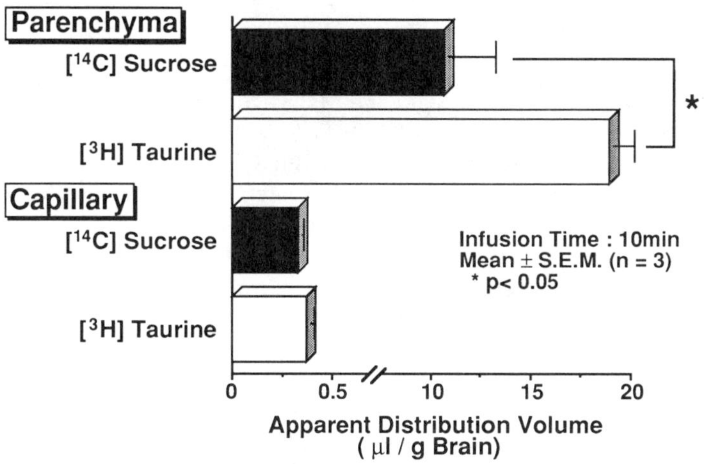

Figure 7. Apparent distribution volume of [^{3}H]taurine and [^{14}C]sucrose in the rat brain determined by the capillary depletion method. After 10-min infusion of [^{3}H]taurine or [^{14}C]sucrose into the carotid artery at a rate of 0.05 ml/min, the brain was isolated and treated by the capillary depletion technique. The result was expressed as the apparent distribution volume.

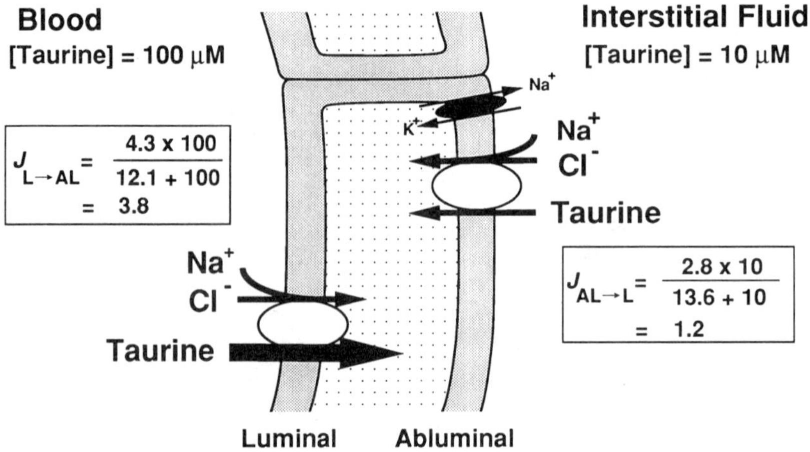

Figure 8. Schematic illustration of proposed blood-brain barrier transport of taurine.

CONCLUSION

As summarized in Fig. 8 for the taurine transport mechanism across the blood-brain barrier, carrier-mediated transport of taurine driven by inwardly directed Na^+ and Cl^- gradients occurs in both the luminal and abluminal membranes. This mechanism is in accordance with the characteristics of taurine transport observed in various tissues[2,12]. Considering the transport parameters of maximum transport rate, J_{max}, the Michaelis-Menten constant, K_t, obtained in the present *in vitro* study, and physiological conditions of plasma taurine concentration of 100 μM[7] and brain interstitial concentration of 10 μM[8], *in vivo* net transport of taurine is suggested to occur from blood to brain.

ACKNOWLEDGMENT

We thank to Dr. Tetsuya Terasaki and Ms. Mizuho Senmaru for their collaboration. This work was supported by a Grant-in-Aid for Scientific Research from the Ministry of Education, Science and Culture, Japan, Japan Health Sciences Foundation, Drug Innovation Science Project, the Grant-in-Aid for Cancer Research (5-14) from the Ministry of Health and Welfare and a grant from Taisho Pharmaceutical Co.

REFERENCES

1. Audus, K.L. and Borchardt, R.T. 1986, Characteristics of the large neutral amino acid transport system of bovine brain microvessel-endothelial cell monolayers, *J.Neurochem.* 47:484-488.
2. Miyamoto, Y., Tiruppathi, C., Ganapathy, V., and Leibach, F.H. 1989, Active transport of taurine in rabbit jejunal brush-border membrane vesicles, *Am.J.Physiol.* 257:G65-G72.
3. Tamai, I., Senmaru, M., Terasaki, T., and Tsuji, A. 1995, Na^+ and CL^--dependent transport of taurine at the blood-brain barrier, *Biochem.Pharmacol.* 50:1783-1793.
4. Tayarani, I., Cloëz, I., Lefauconnier, J.-M., and Bourre, J.-M. 1989, Sodium-dependent high-affinity uptake of taurine by isolated rat brain capillaries, *Biochim.Biophys.Acta*, 985:168-172.

5. Terasaki, T., Takakuwa, S., Moritani, S., and Tsuji, A. 1991, Transport of monocarboxylic acids at the blood-brain barrier: studies with monolayers of primary cultured bovine brain capillary endothelial cells, *J.Pharmacol.Exp.Ther.* 258:932-937.
6. Terasaki, T., Takakuwa, S., Saheki, A., Moritani, S., Shimura, T., Tabata, S., and Tsuji, A. 1992, Absorptive-mediated endocytosis of an adrenocorticotropic hormone (ATCH) analogue, ebiratide, into the blood-brain barrier: studies with monlaters of primary cultured bovine brain capillary endothelial cells, *Pharm.Res.* 9:529-534.
7. Törnquist, P. and Alm, A. 1986, Carrier-mediated tranport of amino acids through the blood-retinal and blood-brain barrier, *Graefe's Arch.Clin.Exp.Ophthalmol.* 224:21-25.
8. Tossman, U., Wieloch, T., and Ungerstedt, U. 1985, Gamma-aminobutyric acid and taurine release in the striatum of the rat during hypoglycemic coma, studied by microdialysis, *Neuroscience Letters*, 62:231-236.
9. Tsuji, A., Saheki, A., Tamai, I., and Terasaki, T. 1993, Transport mechanism of 3-hydroxy-3-methylglutaryl coenzyme A reductase inhibitors at the blood-brain barrier, *J.Pharmacol.Exp.Ther.* 267:1085-1090.
10. Tsuji, A., Tamai, I., Sakata, A., Tenda, Y., and Terasaki, T. 1993, Restricted transport of cyclosporin A across the blood-brain barrier by a multidrug transporter, P-glycoprotein, *Biochem.Pharmacol.* 46:1096-1099.
11. Tsuji, A., Terasaki, T., Takabatake, Y., Tenda, T., Tamai, I., Yamashima, T., Moritani, S., Tsuruo, T., and Yamashita, J. 1992, P-Glycoprotein as the durg efflux pump in primary cultured bovine brain capillary endothelial cells, *Life Sci.* 51:1427-1437.
12. Zelikovic, I., Stejskal-Lorenz, E., Lohstroh, P., Budreau, A., and Chesney, R.W. 1989, Anion dependence of taurine transport by rat renal brush-border membrane vesicles, *Am.J.Physiol.* 256:F646-F655.

42

CHARACTERIZATION OF THE VOLUME-ACTIVATED TAURINE PATHWAY IN CULTURED CEREBELLAR GRANULE NEURONS

H. Pasantes-Morales, C. Peña Segura, O. García, M. M. Morales Mulia, R. Sánchez Olea, and J. Morán

Institute of Cell Physiology
National University of Mexico
Mexico City, Mexico

INTRODUCTION

Early studies on the mechanism of cell taurine transport have consistently identified a component of taurine uptake which is nonsaturable, energy-independent and clearly corresponding to a diffusional mechanism. For long time this component was normally discarded in all studies of taurine uptake in order to obtain the saturable curves of the energy-dependent, high affinity component, which was associated with a presumed neurotransmitter function for taurine. The recent studies on a role for taurine as an osmolyte have praised the diffusional component of the transport system as it has been identified as the mechanism allowing a rapid extrusion of the amino acid to regulate, together with other osmolytes, the cell water content. The occurrence and features of the two components of the taurine transport system fit very adequately this osmolyte role: a diffusional, rapid release of intracellular pools corrects almost immediately an excess of cell water and after this regulatory function, the intracellular taurine pools are replenished by the energy-dependent, high affinity component, which is able to accumulate taurine against large concentration gradients.

Most animal cells, when exposed to hyposmotic solutions, first swell as perfect osmometers but immediately after, an active process of volume regulation occurs, by which cells tend to recover their normal value, despite the persistence of the hyposmotic condition[6]. This regulatory process occurs by extrusion of intracellular osmolytes together with osmotically obligated water. Osmolytes involved in this regulatory process include the inorganic ions K^+ and Cl^- and organic molecules such as amino acids, notably taurine, polyalcohols and polyamines[7]. Taurine efflux occurs immediately in response to cell swelling and the time course of this release closely follows that of the volume regulatory process[7]. Taurine release in response to swelling has been observed in most animal cells, suggesting a widespread taurine involvement in this cellular response.

Taurine 2, edited by Huxtable et al.
Plenum Press, New York, 1996

The activation of taurine efflux associated to swelling is now documented in a large variety of animal cells but the mechanism of taurine release is only starting to be characterized. Cells in culture represent a simple and convenient preparation for the study of this mechanism. In the present work we report the properties of taurine efflux from neurons in culture. Results are compared with those we previously reported in astrocytes and those found in other cell types.

TAURINE AS AN OSMOLYTE IN BRAIN

The involvement of taurine in osmoregulatory mechanisms in invertebrates has been known for a long time but it was only recently that the possibility of an action of taurine as osmolyte in vertebrate tissues, including nervous tissue, has been seriously considered[15]. Yet the high concentration of taurine in animal tissues and its metabolic inertness, make it the perfect osmolyte. The suggestion of an osmoregulatory role for taurine in brain was first supported by the work of Thurston *et al.* showing that hyponatremia results in decrease of taurine brain content[23]. It was later shown that changes of taurine content in brain and brain cells increase or decrease according to external osmolarity[24,25]. The enhanced taurine levels during hypernatremia seems to be due to an increase in the V_{max} of the Na^+-dependent transporter[19,24], whereas the decrease in taurine content during hyponatremia is presumably a consequence of the activation of the diffusion pathways that will be described later. Subsequent studies by Wade *et al.*[27] and Martín del Río *et al.*[22] reported an increase of taurine efflux from discrete brain regions, monitored by microdialysis, upon superfusion with hyposmotic solutions. Taurine efflux associated to cell swelling in brain occurring in pathological states such as hyperexcitability or ischemia has been consistently observed[12].

SWELLING-INDUCED TAURINE RELEASE OCCURS BY DIFFUSION

The diffusional nature of the swelling-activated taurine efflux in neurons and astrocytes in culture was conclusively demonstrated by experiments showing that efflux or influx of taurine occurs upon swelling depending on the concentration gradient[17,21]. Thus, in the physiological situation when the intracellular concentration of taurine is higher than the extracellular concentration, swelling results in taurine release whereas if experimentally the gradient is reversed and extracellular taurine is higher than intracellular taurine, swelling causes taurine influx. In keeping with its diffusional character, the swelling-activated release of taurine is energy- and Na^+-independent. We examined the effect of volume increase on the kinetic constants of taurine uptake in cultured neurons. As shown in Table 1, hyposmolarity and the subsequent cell swelling did not affect any of the kinetic parameters of the Na^+-dependent component whereas the K_{diff} of the Na^+- and energy-independent component was markedly increased.

TAURINE AND CHLORIDE FLUXES ASSOCIATED TO CELL VOLUME ARE CLOSELY INTERCONNECTED

As mentioned above, the inorganic ions Cl^- and K^+ together with a number of organic molecules are involved in volume regulation in most cells including neurons. The K^+ and Cl^- extrusion associated with the regulatory cell response to swelling occurs either via electroneutral cotransporters or through separate conductive channels for each ion. In order

Table 1. Kinetic parameters for taurine uptake in cerebellar granule neurons under isosmotic and hyposmotic conditions

	K_m (μM)	V_{max} nmol.min^{-1}.mg^{-1}	K_{diff} ml.min^{-1}.mg^{-1}
Isosmotic	111	0.64	1.5×10^{-4}
Hyposmotic	87	0.59	4.6×10^{-4}

Cells were incubated for 5 min in isosmotic or hyposmotic (50%) solutions prepared as in described in Fig. 1, containing different concentrations of taurine and subsequently for 2 min in analogous solutions containing [^{3}H]taurine. Uptake was terminated by rapid wash of the cultures with taurine-free medium. Radioactivity and protein in the cultures were determined after solubilization of cells in 1 M KOH. Radioactivity present in cultures incubated at 0°C was substracted from the experimental values obtained at 37°C[21].

to maintain electroneutrality, the ion movement through these channels is interdependent. A simple experiment shows that K^+ and Cl^- transport during regulatory volume decrease in neurons occurs through separate channels. The membrane permeability for K^+ can be increased by addition of gramicidin, a cation selective ionophore, which in the absence of Na^+, will allow K^+ to move between the intra and extracellular compartments simply by following its concentration gradient. However, these K^+ fluxes must be accompanied by Cl^- to maintain electroneutrality. We observed that addition of gramicidin to neurons in isosmotic conditions does not result in any noticeable K^+ efflux nor volume change. Since intracellular K^+ concentration is high, the explanation for this result is that K^+ transport cannot occur because the permeability to Cl^- is restricted. Upon cell swelling, in contrast, addition of gramicidin greatly accelerates the volume regulatory process, an indication that the permeability to Cl^- has increased to such an extent that K^+ permeability is now rate-limiting for Cl^- release and consequently for cell volume regulation (Fig. 1A). These observations point to a key role for Cl^- permeability during the volume adaptive response of neurons and also demonstrate that the K^+ and Cl^- transport activated by swelling of neurons occurs through separate conductive pathways. The interest of these results to the subject of the role of taurine in the volume regulatory process is the finding that taurine efflux in response to cell swelling is markedly sensitive to compounds with known effects as Cl^- channel blockers[16]. A number of these drugs have potent effects on volume regulation. NPPB and DFSK, at concentrations of 75-100 μM, almost block the regulatory volume decrease (Fig.1A). Others like DIDS and niflumic acid also exert a strong inhibition on volume regulation[14] (Fig.1A). If these drugs were acting solely on Cl^- fluxes, this strong inhibitory effect on volume regulation would mean that the contribution of other osmolytes besides Cl^- (and K^+) to the volume regulatory process is negligible. An alternate explanation is that those inhibitors were also blocking the transport of organic osmolytes, which was found to be the case, as these compounds effectively inhibit the volume-associated release of amino acids in astrocytes[9,20] and in neurons as well as that of ^{125}I (Fig. 2).

The polyunsaturated fatty acids, arachidonic, linoleic and linolenic acids, are among the most potent inhibitors of RVD (Fig. 1B) and of both taurine and Cl^- fluxes in neurons (Table 2). The monounsaturated oleic acid and the saturated fatty acids did not affect taurine or Cl^- release (Table 2). This effect appears to result from a direct action of the fatty acids on the transport pathway[13] as no change in the inhibitory effect is observed after blockade of the metabolic cascade of arachidonic acid[18].

The similar pharmacological sensitivity of taurine and Cl^- fluxes may result from (i) a common pathway for Cl^- and taurine, presumably a chloride channel, or (ii) different

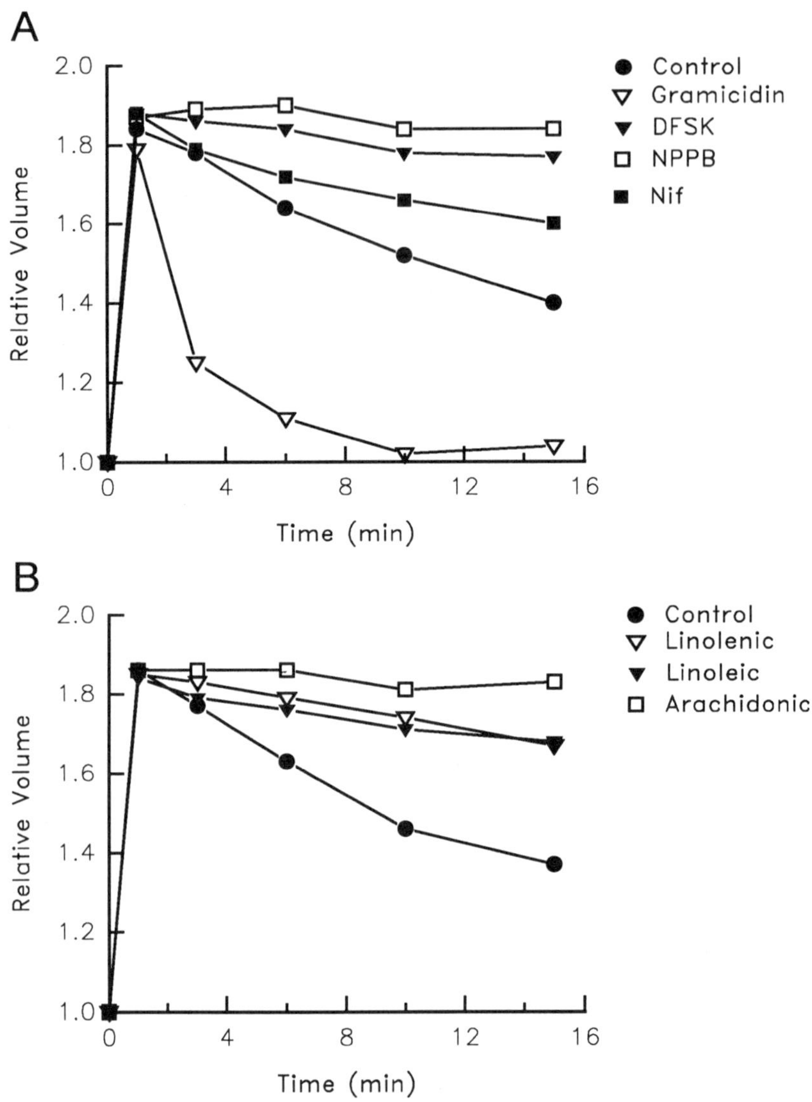

Figure 1. Effect of gramicidin and blockade of regulatory volume decrease by inhibitors of Cl^- channels (A) or by polyunsaturated fatty acids (B) in cerebellar granule neurons. Cells were preincubated during 5 min in isosmotic medium with the inhibitors: niflumic acid (300 μM), DIDS (200 μM), NPPB (100 μM) and DDFKS (75 μM) and the polyunsaturated fatty acids (5 μM). Gramicidin (5 μM) was added at 1 min. Volume measurements were carried out as described in Pasantes-Morales *et al.*[14]. Cultured neurons (1-2 days in culture) were detached by treatment with a Ca^{2+}-free saline containing 1 mM EDTA and 0.01% trypsin. The cell suspension was centrifuged and resuspended in isosmotic medium. At zero time, a sample of the cell suspension was diluted with isosmotic or hyposmotic media with or without the indicated drug. Cell volume was measured at the indicated times by electronic sizing. Isosmotic medium contains (in mM): 135 NaCl, 5 KCl, 0.6 $MgSO_4$, 1 $CaCl_2$, 10 glucose, and 10 HEPES (pH 7.4). Hyposmotic media is prepared by reducing NaCl concentration. Results are expressed as % of increase over the isosmotic volume at the indicated times after exposing cells to a 50% hyposmotic medium at time zero. Results are means of 4-6 experiments.

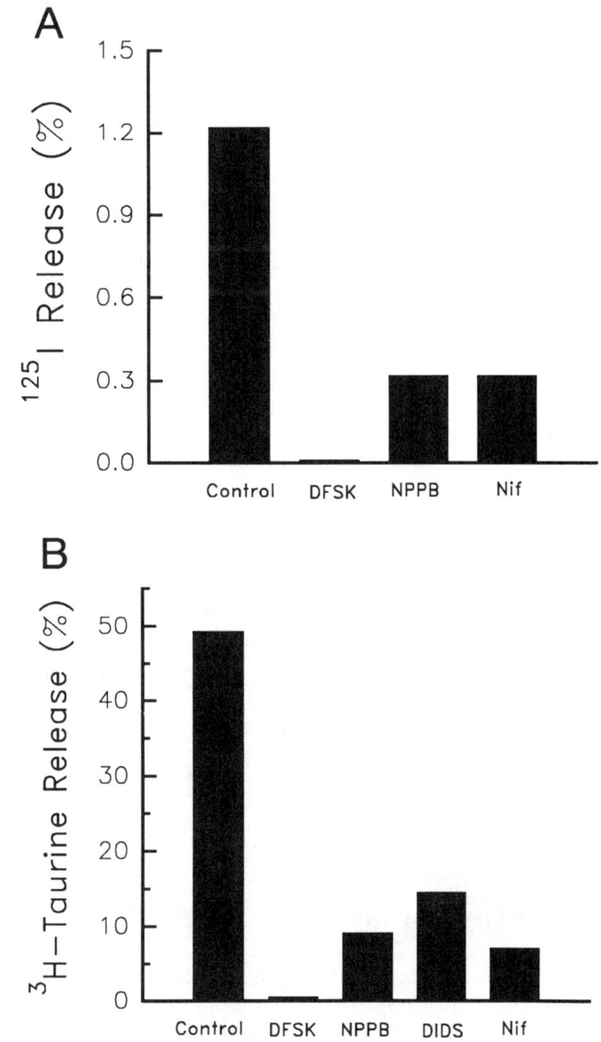

Figure 2. A. Effect of the Cl^- channel blockers on the volume-associated release of [^{3}H]taurine (A) and ^{125}I (B) from cerebellar granule neurons. Drugs were used at the following concentrations: 100 μM NPPB, 200 μM DIDS, 50 μM DDFSK, 300 μM niflumic acid. [^{3}H]Taurine efflux was sequentially measured in isosmotic and 50% hyposmotic media as in Fig. 1 during a 5 min period. ^{125}I efflux rate constants were calculated from radioactivity collected every 20 s as described in Sánchez-Olea *et al.*[18]. Results are expressed as percentage inhibition from the release in hyposmotic solutions in the absence of drugs (control) and are the means of 6-8 experiments.

pathways for the two osmolytes, but so closely interconnected that inhibition of one of them leads to blockade of the other. There is some evidence in support of the first possibility; i.e. that taurine released by swelling occurs through a Cl^- channel-like structure. The different possibilities so far considered are next discussed on the basis of available evidence.

A NUMBER OF ANION CHANNEL-LIKE STRUCTURES COULD CORRESPOND TO THE TAURINE PATHWAY ACTIVATED BY CELL VOLUME

Chloride Channels

The sensitivity of taurine efflux activated by swelling to blockers of Cl^- channels raised the question of whether taurine could permeate an anion channel. Recent electrophysiological evidence provides some support to this possibility. In MDCK cells[1] and in cultured

Table 2. Inhibitory effect of saturated and unsaturated fatty acids on the swelling-stimulated efflux of ^{125}I and [^{3}H]taurine from cultured cerebellar granule neurons

Fatty acids (20 μM)	Efflux (% control)	
	[^{3}H]Taurine	^{125}I
Control	100	100
Ricinoleic	98	95
Oleic	96	98
Arachidonic	9	18
Linoleic	2	12
Linolenic	2	10

^{125}I and [^{3}H]taurine efflux were measured, calculated and expressed as described in Fig. 2. Results are means of 6-8 experiments.

astrocytes[9], it has been shown that taurine and other amino acids can be transported through an anion channel which, similarly to taurine and Cl^- transport evoked by swelling, is blocked by NPPB, DFSK and the polyunsaturated fatty acids. The *P*taurine/*P*Cl^- is 0.75 in MDCK cells. It remains to be demonstrated whether this taurine-transporting anion channel is also present in neurons. Research on Cl^- channels has rapidly developed in recent years. Several Cl^- channels has been cloned, that are similar in molecular structure, ionic selectivity and electrophysiological features[10,16]. One of them, the ClC2, is ubiquitously distributed, being present in most cell types and tissues, and is relevant to mechanisms of volume regulation, and hence to the mechanism of taurine transport, as it is activated by hyposmolarity[16]. It is still unclear whether this channel corresponds to the anion channel transporting taurine in astrocytes and MDCK cells.

The P-Glycoprotein

Another candidate for the taurine transport pathway during volume regulation is the P-glycoprotein (PG). This belongs to the ATP-binding cassette family of transport proteins and its expression is responsible for multidrug resistance in cancer cells. Swelling of cells that express GP causes the activation of a Cl^- conductance suggesting that cell swelling may switch GP from its transport mode to a mode in which it activates Cl^- channels[26]. The suggested function of GP as the transport pathway for taurine and Cl^- during volume regulation comes from the finding that a number of blockers of the Cl^- conductance through GP (quinidine, dihydropyridines, diltiazem, trifluoperazine)[8] are also potent inhibitors of volume regulation and taurine efflux associated with swelling in astrocytes and neurons and possibly in other cells as well. There is no evidence so far indicating that PG is expressed in brain cells.

The Anion Exchanger

The anion exchanger of the band 3 type of red cells has also been implicated as transport mechanism for the volume-associated taurine efflux[5]. The anion exchanger molecules normally contain a structure which may function as an unidirectional Cl^- channel[4] and since it is assumed that Cl^- and taurine share the transport pathway, this Cl^- channel structure of the anion exchanger may also function as a taurine extrusion pore. In support of this possibility is the fact that most inhibitors of taurine efflux elicited by swelling are blockers of the anion exchanger. Band 3 type molecules have been found in a number of nonerythroid tissues including brain[11]. Anion transport in brain vesicles containing a band 3-type molecule is also inhibited by the Cl^- channel blockers which inhibit taurine transport[11].

Porins

Another molecule type suggested for taurine release associated with cell swelling is a porin. Porins are channel-forming proteins present in the bacterial cell membrane and in the outer mitochondrial membrane of eukaryotic cells. In mitochondria, porin channels provide the pathway for transport of nucleotides and a variety of metabolites[2]. Mitochondrial porins exhibit the properties of the Cl^- channel activated by hyposmolarity in astrocytes, such as large unitary conductance (>400 pS) and voltage-dependent gating for Cl^-. Moreover, a recent report describes the cloning of a brain-derived porin that functions as a large conductance anion channel, localized in the astrocytic plasma membrane[3]. Therefore, this type of channel may be involved in the transport of Cl^- associated to volume regulation in nervous cells.

In conclusion, the present findings showing remarkable similarities in the pharmacological sensitivity of taurine and Cl^- fluxes and the suggestion that taurine and Cl^- are transported through the same pathway, extend our view about the mechanisms involved in taurine transport in neurons and in many other cell types as well. The different possible mechanisms suggested to act as Cl^--taurine pathway may not be mutually exclusive. They can all function in different cells or two or more be present and functional in one type of cell. Clarifying these points must await the demonstration that taurine effectively permeates through these different transport systems.

ACKNOWLEDGMENTS

The technical assistance of Ms. Claudia Rodriguez is highly appreciated. This work was supported in part by grant No. IN202093-DGAPA, UNAM.

REFERENCES

1. Banderali, U. and Roy, G. 1992, Anion channels for amino acids in MDCK cells, *Am.J.Physiol.* 263:C1200-C1207.
2. Benz, R. 1994, Permeation of hydrophilic solutes through mitochondrial outer membranes: review on mitochondrial porins, *Biochim.Biophys.Acta*, 1197:167-196.
3. Dermietzel, R., Hwang, T., Buettner, R., Hofer, A., Dotzler, E., Kremer, M., Deutzmann, R., Thinnes, F.P., Fishman, G.I., Spray, D.C., and Siemen, D. 1994, Cloning and in situ localization of a brain-derived porin that constitutes a large conductance anion channel in astrocytic plasma membranes, *Proc.Natl.Acad.Sci.USA*, 91:499-503.
4. Falke, J.J. and Chan, S.I. 1986, Molecular mechanisms of band 3 inhibitors: I transport site inhibitors, *Biochemistry*, 25:7888-7894.
5. Goldstein, L. and Brill, S.R. 1991, Volume-activated taurine efflux from skate erythrocytes: Possible band 3 involvement, *Am.J.Physiol.* 260:R1014-R1020.
6. Hallows, K.R. and Knauff, P.A. 1994, Principles of cell volume regulations, in: "Cellular and Molecular Physiology of Cell Volume Regulation", Strange, K. ed., CRC Press, Boca Raton, pp. 3-30.
7. Hoffmann, E.K. and Simonsen, L.O. 1989, Membrane mechanisms in volume and pH regulation in vertebrate cells, *Physiol.Rev.* 69:315-382.
8. Hollt, V., Kouba, M., Dietel, M., and Vogt, G. 1992, Steroisomers of calcium antagonists which differ markedly in their potencies as calcium blockers are equally effective in modulating drug transport by p-glycoprotein, *Biochem.Pharmacol.* 3:2601-2608.

9. Jackson, P.S. and Strange, K. 1993, Volume-sensitive anion channels mediate swelling-activated inositol and taurine efflux, *Am.J.Physiol.* 265:C1489-C1500.
10. Jentsch, T. 1994, Molecular physiology of anion channels, *Curr.Opin.Cell Biol.* 6:600-606.
11. Kay, M.M.B., Hughes, J., Zagón, I., and Lin, F. 1991, Brain membrane protein 3 performs the same functions as erythrocyte band 3, *Proc.Natl.Acad.Sci.USA*, 88:2778-2782.
12. Lehmann, A. 1990, Derangements in cerebral osmohomeostasis: A common denominator for stimulation of taurine and phosphoethanolamine release, in: "Taurine: Functional Neurochemistry, Physiology, and Cardiology", Pasantes-Morales, H., Martin, D.L., Shain, W. and del Río, R.M. eds., Wiley-Liss, New York, pp. 337-347.
13. Ordway, R.W., Singer, J.J., and Walsh, J.V. 1991, Direct regulation of ion channels by fatty acids, *Trends Neurosci.* 14:96-100.
14. Pasantes-Morales, H., Chacón, E., Murray, R.A., and Morán, J. 1994, Properties of osmolyte fluxes activated during regulatory volume decrease in cultured cerebellar granule neurons, *J.Neurosci.Res.* 37:720-727.
15. Pasantes-Morales, H. and Del Río, R.M. 1990, Taurine and mechanisms of cell volume regulation, in: "Taurine: Functional Neurochemistry, Physiology, and Cardiology", Pasantes-Morales, H., Martin, D.L., Shain, W. and del Río, R.M. eds., Wiley-Liss, New York, pp. 317-328.
16. Paulmich, M., Gschwentner, M., Woll, E., Schmarda, A., Ritter, M., Kanin, G., Ellemunter, H., Waitz, W., and Deetjen, P. 1993, Insight into the structure-function relation of chloride channels, *Cell Physiol.Biochem.* 3:374-387.
17. Sanchez-Olea, R., Morán, J., Schousboe, A., and Pasantes-Morales, H. 1991, Hyposmolarity-activated fluxes of taurine in astrocytes are mediated by diffusion, *Neurosci.Lett.* 130:233-236.
18. Sánchez-Olea, R., Morales-Mulia, M., Morán, J., and Pasantes-Morales, H. 1995, Inhibition by polyunsaturated fatty acids of cell volume regulation and osmolyte fluxes in astrocytes, *Am.J.Physiol.* 269:C96-C102.
19. Sánchez-Olea, R., Morán, J., and Pasantes-Morales, H. 1992, Changes in taurine transport evoked by hyperosmolarity in cultured astrocytes, *J.Neurosci.Res.* 32:86-92.
20. Sánchez-Olea, R., Peña, C., Morán, J., and Pasantes-Morales, H. 1993, Inhibition of volume regulation and efflux of osmoregulatory amino acids by blockers of Cl^- transport in cultured astocytes, *Neurosci.Lett.* 156:141-144.
21. Schousboe, A., Sánchez Olea, R., Morán, J., and Pasantes-Morales, H. 1991, Hyposmolarity-induced taurine release in cerebellar granule cells is associated with diffusion and not with high-affinity transport, *J.Neurosci.Res.* 30:661-665.
22. Solís, J.M., Herranz, A.S., Herreras, O., Lerma, J., and Del Río, R.M. 1988, Does taurine act as an osmoregulatory substance in the rat brain, *Neurosci.Lett.* 91:53-58.
23. Thurston, J.H., Hauhart, R.E., and Dirgo, J.A. 1980, Taurine: a role in osmotic regulation of mammalian brain and possible clinical significance, *Life Sci.* 26:1561-1568.
24. Trachtman, H., Futterweit, S., and Del Pizzo, R. 1992, Taurine and osmoregulation. IV. Cerebral taurine transport is increased in rats with hypernatremic dehydration, *Pediatr.Res.* 32:118-124.
25. Tratchman, H., Barbour, R., and Sturman, J.A. 1988, Taurine and osmoregulation: taurine is a cerebral osmoprotective molecule in chronic hypernatremic dehydration, *Pediatr.Res.* 23:35-41.
26. Valverde, M.A., Díaz, M., Gill, F.V., Hyde, S.C., and Higgins, C.F. 1992, Volume-regulated chloride channels associated with the human multidrug resistance p-glycoprotein, *Nature*, 355:830-833.
27. Wade, J.V., Olson, J.P., Samson, F.E., Nelson, S.R., and Pazdernik, T.L. 1988, A possible role for taurine in osmoregulation within the brain, *J.Neurochem.* 51:740-745.

43

TAURINE IS A SUBSTRATE OF THE ANION EXCHANGER TRANSPORT SYSTEMS

Rafael Martín del Río, Mario Galarreta, Nieves Menéndez,
Concepción Conejero, and José M. Solís

Servicio de Neurobiología
Dpto. Investigación
Hospital Ramón y Cajal
28034 Madrid, Spain

INTRODUCTION

It is now generally accepted that a number of physiological actions of taurine in the central nervous system (CNS) depend on its intracellular level. The very high resting intracellular taurine concentration in brain mainly results from the functional equilibrium between the rate of its Na^+-dependent uptake system and that of its release pathways, since it is known that there is no taurine degradation and the *de novo* synthesis of the amino acid in this tissue is very poor[3]. Uptake transport systems of β-amino acids including taurine have been extensively studied and characterized[12], and also several genes coding for the responsible proteins have been clonned[16]. The nature of the pathway allowing the diffusion efflux of taurine from nerve cells, however, is still controversial.

Chloride channels[9] as well as the anion-exchanger (band-3) carrier[2], both involved with Cl^- transport, have been proposed as membrane pathways allowing the evoked release of taurine in different types of cells from several tissues. As a first approach, we have performed *in vitro* experiments in rat erythrocytes intended to demonstrate a direct role of the anion-exchanger on taurine transport across the plasma membrane, taking advantage of the elevated number of band-3 copies in these cells (10^6 per cell[13]). Since the presence of at least two isoforms of band-3 in brain tissue with functional properties similar to the erythrocyte anion-exchanger has been demonstrated[5], we wanted to further extend the erythrocyte studies on the above hypothesis to other experimental CNS preparations such as *in vitro* primary cultures of neurons and astrocytes, and also to *in situ* rat hippocampal tissue.

Taurine 2, edited by Huxtable et al.
Plenum Press, New York, 1996

METHODS

Whole blood samples from albino rats were obtained in heparinized test tubes and the cells were recovered by centrifugation. Erythrocyte suspensions, at a 40% hematocrit in KRB, were used to measure taurine transport.

Taurine uptake by cell suspensions incubated at 37°C in Krebs Ringer bicarbonate (KRB), containing amino acid concentrations as indicated, was allowed to proceed for different time periods and eventually arrested by diluting the samples 1:1 with ice-chilled KRB containing 1 mM flufenamic acid, a potent anion-exchanger inhibitor[1]. After thorough washing (4x, 2 vol. each) with the inhibitory solution, cells were lysed with 10 vol of distilled water and both the protein and the free amino acid content[10] measured, after centrifugation, in the supernatant.

Since the endogenous taurine content of rat erythrocyte is low (≈100 μM), amino acid release was estimated from preloaded red blood cell suspensions. After a loading period (10 mM taurine solution in KRB, 37°C, 1 h), the taurine remaining in the medium was rapidly discarded by centrifugation, and the cells, resuspended in fresh medium, were extensively washed, allowing the spontaneous release of the preloaded amino acid to proceed in the presence or in the absence (control) of anion-exchanger inhibitors. Release was evaluated by measuring the taurine content in both the incubation medium, as well as in the cells after different efflux periods.

Taurine release from primary cultures of hippocampal neurons and astrocytes was evoked by changing the incubation medium from a standard KRB to a hypotonic fluid (KRB with 25% less NaCl), both in the presence and in the absence of a Cl^- transport inhibitor. After 10 min of hypotonic stimulation, the taurine content in the incubation medium, as well as that remaining in the cells were measured as above.

Primary cultures of neurons and astrocytes were prepared as reported[11] with minor modifications.

In situ hippocampal microdialysis experiments were conducted as published elsewhere[15]. Taurine-evoked release was induced by decreasing the osmotic strength of the perfusion fluid in the presence and in the absence of Cl^- transport inhibitors. The content of several amino acids, including taurine, was evaluated on 10 min perfusate samples (20 μl) obtained during a period control of KRB perfusion (60 min), followed by a stimulation period in which the perfusion fluid was a modified KRB (hypotonic) containing 50 mM less NaCl. The inhibitors were applied during the last 30 min of the control period and were present throughout the hypotonic condition.

RESULTS

Uptake of both GABA and taurine by rat red blood cells occurred by low affinity processes which were linear in the concentration range used (0.5-10 mM, 30 min incubation) and independent of Na^+ and Cl^- gradients. At the higher concentration assayed (10 mM), taurine uptake, but not that of GABA, was also linear with incubation time for at least 2 h (data not shown).

Since it is claimed that taurine transport through the plasma membrane is related with some Cl^- transport systems[6], we studied taurine uptake by erythrocytes in the presence of several Cl^- transport inhibitors. For comparison with a chemically related amino acid, we conducted similar experiments with GABA. Figure 1 shows the effect of two Cl^- transport inhibitors, furosemide and the stilbene compound, DIDS (4,4'-diisothiocyanatestilbene-2-2'-disulfonic acid), on taurine and GABA uptake by erythrocyte suspensions. Furosemide,

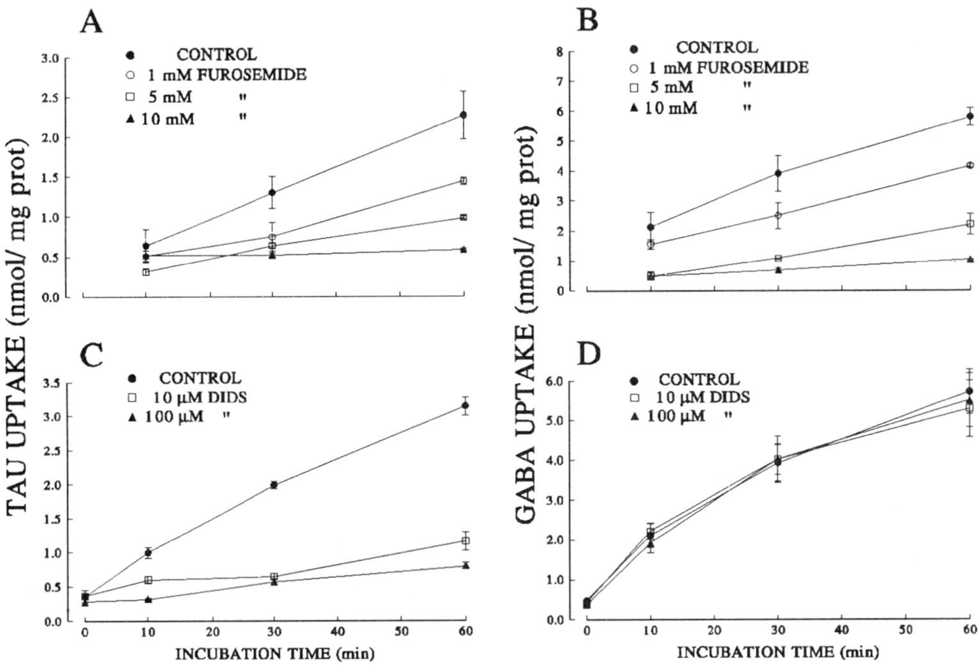

Figure 1. Taurine and GABA uptake inhibition by furosemide and DIDS. Erythrocytes suspensions were incubated in KRB with 10 mM taurine (A, C) or 10 mM GABA (B, D) in KRB at 37°C for 10, 30 or 60 min in the absence (control) or in the presence of furosemide (A, B) or DIDS (C, D), after which intracellular content of the amino acids were evaluated. Symbols represent means ± SEM of data from 3 experiments, each in triplicate.

which is an inhibitor of several Cl^- transport systems, including the anion-exchanger[1], blocked both taurine and GABA uptake to a similar extent and in a dose-dependent manner (Fig.1A and 1B). By contrast, DIDS, which when used at low μM concentration is considered a rather specific band-3 inhibitor[1], almost totally supressed the taurine uptake when applied at 100 μM or even at 10 μM concentrations (Fig.1C), while GABA transport was unaltered by this inhibitor (Fig.1D). Another drug, bumetanide, which inhibits a different choride transport, the $Cl^-/Na^+/K^+$ cotransport, did not affect either GABA or taurine uptake even when present at 100 μM, a concentration high enough to block this Cl^- transport totally (data not shown). Further experiments were designed to show if the erythrocyte anion-exchanger could also constitute an efflux pathway for taurine. Figure 2 shows the time course of taurine release from preloaded cells, expressed as the ratio of extra- to intracellular amounts of taurine, in the presence of different furosemide concentrations (Fig. 2A). This compound was able to block the spontaneous taurine release in a dose-dependent manner. A 0.5 mM furosemide solution, a concentration of the drug that is active when acting on the anion exchanger, clearly inhibited taurine release, whereas that of 2 mM totally abolished the amino acid efflux. Similar inhibitory results were obtained when using DNDS (4-4'-dinitro stilbene-2-2'disulfonic acid), a stilbene compound similar to DIDS, which although as potent, is by contrast a reversible and noncompetitive inhibitor of anion-exchanger[1]. A concentration of 0.1 mM of this compound was able to completely supress taurine release (Fig. 2B). These uptake and release results obtained from inhibition experiments strongly suggest that the anion-exchanger is not only an uptake route for taurine but also constitutes an efflux pathway for this amino acid through the rat erythrocyte membrane.

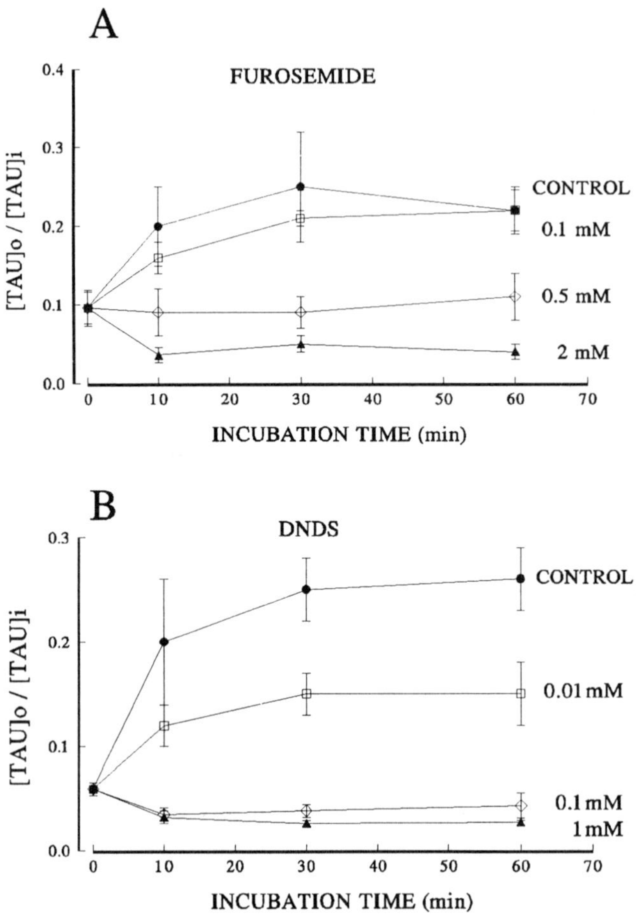

Figure 2. Effect of furosemide and DNDS on taurine release from rat erythrocytes. Red blood cells were preloaded with taurine (10 mM, 37°C, 1 h) and the spontaneous release of the amino acid allowed to proceed at 37°C for 1 h in the presence of furosemide (A) or DNDS (B). Symbols represent the means ± SEM of data from 3 experiments expressed as the ratio of the amount of taurine present in the incubation medium and that remaining inside the cells at the indicated times. Data from 0.1 mM furosemide experiments and that from time 10 min from 0.01 mM DNDS were not statistically significant when compared to control. Student "t" test, $p \leq 0.01$.

Once the capability of taurine as a substrate by the anion-exchanger in the rat erythrocyte model was known, our interest was to assess if this amino acid could also be a substrate for the anion-exchanger isoform (AE3) expressed in brain tissue[7]. To test this hypothesis we studied whether anion-exchanger inhibitors blocked the taurine release evoked by hypotonic stimulus on *in vitro* preparations of brain tissue such as primary cultures of astrocytes, as well as hippocampal neurons. At 1 mM, both DIDS (Fig. 3A) and niflumic acid (Fig. 3C), a noncompetitive band-3 inhibitor, greatly diminished the taurine release elicited by a 10 min hypotonic stimulus applied to cultured astrocytes (Fig. 3). Accordingly, both agents avoided the intracellular taurine concentration decrease induced by the hypotonic conditions (Fig. 3B, 3D). An identical procedure was used in the case of cultured hippocampal neurons which also responded to a hypotonic shock with taurine release. This response was similarly inhibited by 1 mM DIDS or niflumic acid (data not shown).

In addition, we wanted to know if taurine release induced by osmotic stimuli could occur through the anion-exchanger in a CNS preparation, but in a more physiological situation than cell culture. For this purpose we have used an *in situ* preparation, the *in vivo* hippocampal structure, in which taurine release was induced by a local hypotonic stimulus applied through a microdialysis probe. Figure 4 shows that induced taurine release from hippocampal tissue *in vivo* was partially blocked by the presence of 1mM niflumic acid in

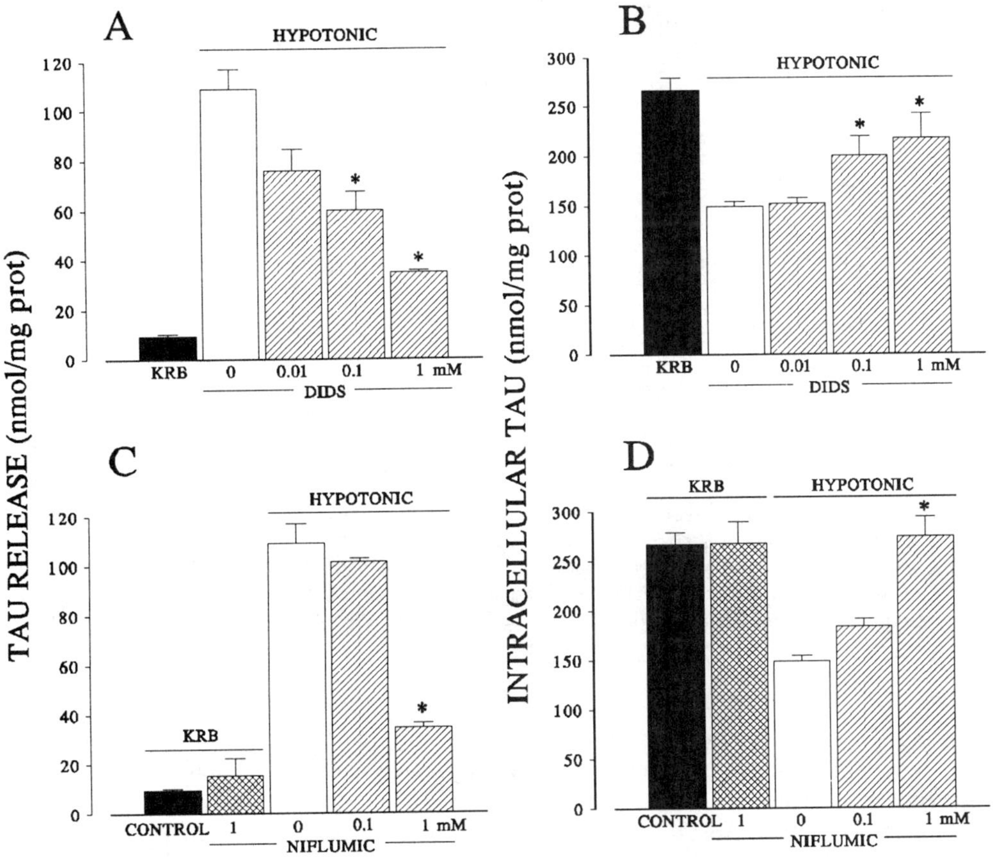

Figure 3. Effect of DIDS and niflumic acid on the release of taurine evoked by hypotonic conditions from cultured astrocytes. DIDS was applied at different concentrations in KRB at 37°C for 30 min, then it was thoroughly washed out and the hyposmotic condition established for 10 min thereafter. Taurine release (A) was evaluated as the amino acid content in the incubation medium at the end of this period, or as the amount remaining inside the cells (B). Niflumic acid (C, D) was present 10 min before and also during the hyposmotic stimulation. Bars represent the means ±SEM of 4 experiments in triplicate. Comparisons are made (Student "t" test) between data obtained at 0 and the other inhibitor concentration in the hypotonic condition. $*p \leq 0.01$.

the perfusion fluid. A similar result was obtained when the hyposmotic shock was delivered after a 30 min washing of a 1 mM DIDS solution applied for 30 min (data not shown).

DISCUSSION

Results obtained from red blood cell suspensions show that taurine in these cells is taken up through a low affinity process which is independent of Na^+ and Cl^- gradients. Thus, this taurine uptake is different from other classical β-amino acid transport systems, which have high affinity for the substrate and are Na^+/Cl^- dependent. Moreover, the inhibition studies strongly indicate that this taurine uptake process is carried out by the erythrocyte anion-exchanger, since the amino acid influx is totally inhibited by furosemide and more important by very low concentrations of DIDS, a specific band-3 inhibitor in this condition. The transport system is rather specific for taurine, although it has been previously reported

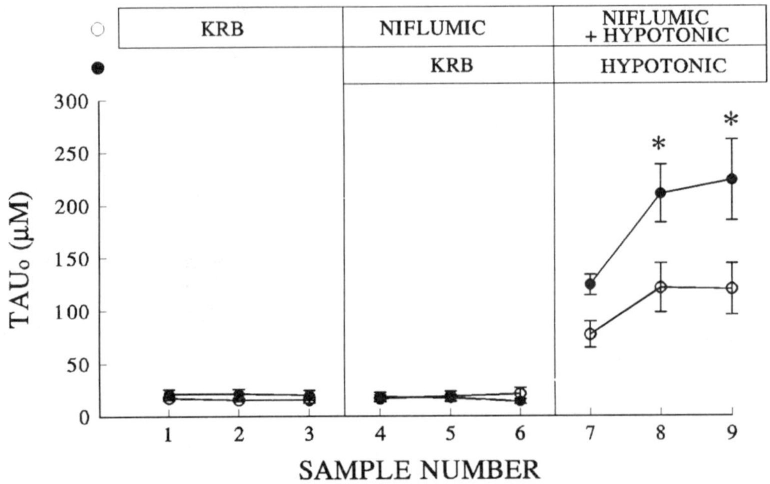

Figure 4. Effect of niflumic acid on taurine release induced by hypotonic stimulation on *in vivo* rat hippocampus. The inhibitor (1 mM in KRB) was delivered through the microdialysis probe and was present 30 min before and also during the hyposmotic shock administration. Symbols represent the means ± SEM of 6 experiments. Statistical differences between mean pairs in the presence and absence of niflumic acid were estimated by the Newman-Keuls-Hartley test after performing the ANOVA test (* $p < 0.01$).

that glycine is also transported by band-3[4]. It should be noted that GABA, an analog of taurine in other transport systems[12], was not a substrate of anion-exchanger, as shown in the present experiments.

Taurine release from different types of cells occurs in many cases through an ill-defined diffusional pathway that is Ca^{2+}-independent[3]. This is important in the case of taurine since this amino acid could use this transport system to decrease its intracellular concentration; e.g. during the volume decrease regulatory process. The diffusional transport could also contribute to regulate the actual intracellular concentration of the amino acid and consequently some of its putative physiological actions. Using red blood cell suspensions from rats as a model for taurine release, we show that the spontaneous efflux of this amino acid from preloaded erythrocytes occurs through band-3, as judged by the data obtained from inhibition experiments. Similar results were obtained by other authors using hypotonic stimuli on erythrocytes of several fish species[2]. However, since taurine efflux from rat erythrocytes occurs spontaneously and is not induced by hypotonic conditions, we could discard the possibility that taurine diffusional release is via a volume-activated transport system, blocked by DIDS and independent of the anion-exchanger. It is known that taurine release from nerve cells under hyposmotic conditions proceeds through a diffusional pathway[14]. Recently, several isoforms of band-3 called AEs (anion-exchanger isoforms) have been shown to be present in different tissues, including several brain structures[8], with amino acid sequence and kinetic properties similar to those of band-3, although with a lower sensitivity for some typical inhibitors such as niflumic acid[5,7]. As shown in Figures 3 and 4 and as cited in the text, astrocytes, neurones, as well as intact brain tissue all released taurine when challenged by a hyposmotic shock. The amino acid efflux was also inhibited in all cases by band-3 blockers albeit with an apparently lower potency than when used in the case of the erythrocyte transporter. We can then speculate that taurine release from cultured brain cells induced by hypotonic conditions probably occurs through the isoform of anion-exchanger expressed in these cells and so that this pathway could be relevant for the physiologycal role of taurine in the CNS.

ACKNOWLEDGMENTS

We thanks A. Latorre for technical assistance, Dr. C.L. Paino for his help with cell cultures, and S. McGrath for editorial help. This paper is based on the thesis work of Dr. Conejero.

REFERENCES

1. Cabantchik, Z.I. and Greger, R. 1992, Chemical probes for anion transporters of mammalian cell membranes, *Am.J.Physiol.* 262:C803-C827.
2. Goldstein, L. and Brill, S.R. 1991, Volume-activated taurine efflux from skate erythrocytes: Possible band 3 involvement, *Am.J.Physiol.Regul.Integr.Comp.Physiol.* 260:R1014-R1020.
3. Huxtable, R.J. 1989, Taurine in the central nervous system and the mammalian actions of taurine, *Prog.Neurobiol.* 32:471-533.
4. Joung, J.D., Jones, S.E.M., and Ellory, J.C. 1981, Amino acid transport via the red cell anion transport system, *Biochem.Biophys.Acta*, 645:157-160.
5. Kay, M.M.B., Hughes, J., Zagón, I., and Lin, F. 1991, Brain membrane protein band-3 performs the same functions as erythrocyte band-3, *Proc.Natl.Acad.Sci.* 88:2778-2782.
6. Kirk, K. and Kirk, J. 1993, Volume-regulatory taurine release from a human lung cancer cell line: Evidence for amino acid transport via a volume-activated chloride channel, *FEBS Lett.* 336:153-158.
7. Kopito, R.R., Lee, B.S., Simmons, D.M., Lindesey, A.E., Morgans, C.W., and Schneider, L. 1989, Regulation of intracellular pH by neuronal homolog of the erythrocyte anion exchanger, *Cell*, 59:927-937.
8. Kudrycki, K.E., Newman, P.R., and Shull, G.E. 1990, cDNA cloning and tissue distribution of mRNAs for two proteins that are related to the band-3 Cl^-/HCO_3^-exchanger, *J.Biol.Chem.* 265:462-471.
9. Lambert, I.H. and Hoffmann, E.K. 1994, Cell swelling activates separate taurine and chloride channels in Ehrlich mouse ascites tumor cells, *J.Membr.Biol.* 142:289-298.
10. Lerma, J., Herranz, A.S., Herreras, O., Abraira, V., and Del Rio, M. 1986, In vivo determination of extracellular concentration of amino acids in the rat hippocampus. A method based on brain dialysis and computerized analysis, *Brain Res.* 384:145-155.
11. MacCarthy, K.D. and DeVellis, J. 1980, Preparation of separate astroglial and oligodendroglial cultures from rat cerebral tissue, *J.Cell.Biol.* 85:890-902.
12. Oja, S.S. and Kontro, P. 1983, Taurine, in: "Handbook of Neurochemistry, Vol. 3.", 2nd edition, Lajtha, A. ed., Plenum Press, New York, pp. 501-533.
13. Salhany, J.M. 1990, Erythrocyte Band-3 Protein, CRC Press, Boca Raton, Florida.
14. Sanchez-Olea, R., Morán, J., Schousboe, A., and Pasantes-Morales, H. 1991, Hyposmolarity-activated fluxes of taurine in astrocytes are mediated by diffusion, *Neuroscience Letters*, 130:233-236.
15. Solís, J.M., Herranz, A.S., Herreras, O., Lerma, J., and Martín del Río, R. 1988, Low chloride-dependent release of taurine by a furosemide-sensitive process in the *in vivo* rat hippocampus, *Neuroscience*, 24:885-891.
16. Uchida, S., Kwon, H.M., Yamauchi, A., Preston, A.S., Marumo, F., and Handler, J.S. 1992, Molecular cloning of the cDNA for an MDCK cell Na^+- and Cl^--dependent taurine transporter that is regulated by hypertonicity, *Proc.Natl.Acad.Sci.USA, 89:8230-8234.*

44

THE ROLE OF SULPHYDRYL GROUPS IN EFFLUX OF TAURINE AND GABA FROM CEREBRAL CORTICAL CELLS

Emma L. Upton and R. O. Law

Department of Cell Physiology and Pharmacology
University of Leicester
Leicester, United Kingdom

INTRODUCTION

A reduction of plasma osmolality due to a fall in Na^+ concentration (hyponatraemia) is the disorder of electrolyte homeostasis that is most frequently encountered in clinical medicine[32]. Hyponatraemia commonly presents as disordered CNS function[24]. However, ample experimental evidence indicates that brain cells, both *in vitro* (e.g. cultured preparations) or *in vivo*, are able to preserve near-normal volumes (fluid contents) when exposed to hyposmotic media. This is accomplished through a reduction of their osmotic potential mediated by loss of inorganic and organic intracellular solutes. The latter comprise mainly straight-chain, aliphatic, non-essential (in most species) amino acids — taurine in particular — and myo-inositol[13,14,16] and their contribution to the limitation of cell swelling has been quantified in severely hyponatraemic rats[31]. The role of inorganic ions has been reviewed elsewhere[4].

Recently, evidence has begun to accumulate which suggests that the hyposmotically stimulated efflux of amino acids from brain cells is in some manner dependent upon the presence of sulphydryl (-SH) groups, presumably in or adjacent to the efflux pathway(s). The latter, both in brain cells and in a wide variety of other cell types, are believed to be anionic diffusive channels that are also permeable to small neutral molecules and are assumed to be activated by cell swelling (for relevant references, as well as channel characterization, see ref. 8).

Evidence for the involvement of -SH groups derives from studies using a variety of SH modifying reagents including N-ethylmaleimide (NEM) and *p*-chloromercuribenzenesulphonic acid (pCMBS) or its phenyl analogue pCMPS. Both are well recognized non-specific inhibitors of amino acid transport, but they also interfere with a number of other membrane transport processes, some of which will be referred to later in this paper. NEM has been shown strongly to inhibit the volume-sensitive efflux of taurine from cultured astrocytes[19], in which it appears to enter into a simple bimolecular interaction with -SH groups. A similar stoichiometry has previously been reported in renal brush border membrane

Taurine 2, edited by Huxtable et al.
Plenum Press, New York, 1996

vesicles[30]. Thiol reagents also block taurine release from hyposmotically swollen skate hepatocytes[2]. In astrocytes it seems that NEM (which is hydrophobic) penetrates the cell and exerts its inhibitory effect at the cytoplasmic face of the membrane, since this effect cannot be elicited with impermeant 5-eosin maleimide[19]. pCMBS, which penetrates cells relatively slowly, had an inhibitory effect comparable with that of NEM in hyposmotic media but—seemingly paradoxically—caused an enhancement of efflux under isosmotic conditions, a response which could not be elicited with NEM[19]. A similar effect of pCMBS in isosmotic media has been reported in shark rectal gland cells[10]. In other recent studies the sulphydryl reagent mercuric chloride accelerated the release of D-aspartate from cultured astrocytes under isosmotic conditions[21], whereas neither NEM nor the hydrophilic reagent iodoacetamide elicited this response, although both attenuated hyposmotic D-aspartate release[1].

One possibility to account for the effects of -SH reagents during hyposmotic incubation is that cell swelling may induce conformational alterations in membrane -SH groups which are otherwise inaccessible to blocking effects[1], although clearly alternative hypotheses must be sought to account for the acceleratory effects of mercurial (but not alkylating) reagents in isosmotic media. The present study was designed primarily to examine the relationship between the efflux of taurine and GABA from pre-loaded cells in superficial slices of rat cerebral cortex, the susceptibility of efflux to inhibition by NEM and pCMPS, and the accompanying steady-state cell volumes, assessed as slice non-inulin spaces. Rates of efflux, and cell volumes, were also studied under isosmotic conditions and in hyposmotic media in which cell swelling was prevented by the addition of 25 mM sucrose to the incubation medium. Also examined were the effects of pre-exposure of cells to the -SH reducing agent dithiothreitol (DTT, Cleland's reagent[3]). DTT has been reported to block the inhibitor effects of pCMBS, but not of NEM, on hyposmotic release of taurine from cultured astrocytes[18,19].

METHODS

The methods employed in this study, including details of the iterative curve-fitting package used to determine amino acid efflux rate constants, as well as statistical methods, have been published in detail[15]. In brief, minislices of rat superficial cerebral cortex (thickness approx. 200-300 μm) were loaded with amino acids for 120 min in Krebs phosphate-bicarbonate Ringer containing 1 mM taurine or GABA labelled with 200 kBq/ml [^{3}H]taurine or [^{14}C]GABA. Subsequent efflux of amino acids into amino acid-free media at 315 (isosmotic) or 265 (hyposmotic) mosmol/kg was followed for 60 min by serial sampling. Efflux of both acids was resolved into 3 mono-exponential components, presumed to represent 1 extracellular and 2 intracellular pools[15]. The reason for the intracellular duality is not clear. It probably reflects compartmentalization of the amino acids, possibly into diffuse cytosolic and concentrated vesicular pools. For taurine, at least, the rates of efflux do not vary greatly between neurons and glia[29], and since under basal conditions in the present study the rate constants for the 2 intracellular pools vary by 50 to 100-fold it is unlikely that the duality is due to efflux from these two cell types. Moreover, triple exponential washout of taurine has been observed in shark rectal gland, which is known to comprise a homogeneous cell population[32]. The 3rd (slowest) phase of efflux is the only one for which it is practicable to perform meaningful kinetic analysis, and this is the phase considered in this report. Steady-state cell volumes were calculated from the equilibrium distribution of [^{14}C]inulin within slices over the period (30-50 min) of 3rd phase efflux[15].

The -SH reagent NEM or pCMPS, and the -SH reducing agent DTT, were added to the medium, as appropriate (all 100 μM), DTT also being present in the loading medium.

Table 1. The rate constants (k) for the 3rd (presumed cellular) phase of efflux of taurine and GABA into isosmotic and hyposmotic media, and hyposmotic media incorporating 25 mM sucrose, and the accompanying steady-state cell volumes

	$k \times 10^{-2} min^{-1}$	Cell volume (µl/mg dry wt)
Taurine		
Isosmotic	0.64 ± 0.04 (6)	2.33 ± 0.11 (12)
Hyposmotic	0.99 ± 0.05 (6)	2.55 ± 0.09 (10)
Hyposmotic + sucrose	0.89 ± 0.04 (6)	2.26 ± 0.11 (10)
GABA		
Isosmotic	0.55 ± 0.05 (6)	2.31 ± 0.12 (10)
Hyposmotic	0.76 ± 0.04 (6)	2.65 ± 0.10 (13)
Hyposmotic + sucrose	0.69 ± 0.03 (6)	2.30 ± 0.14 (12)

Values are means ± SEM (n).

As has been found using cultured astrocytes, pre-exposure of cells to DTT is necessary to demonstrate its reversal of the inhibitory effects of -SH blockade[19]. In hyposmotic media containing 25 mM sucrose, osmolality was maintained by equimolar reduction of medium NaCl. All values are quoted as means ± SEM (n).

RESULTS

The principal findings of the present study are summarized in Tables 1 - 3.

Control Values

Control values for efflux rate constants, and cell volumes, are shown in Table 1. Reduction of medium osmolality from 315 to 265 mosmol/kg led to acceleration of taurine ($p < 0.001$ vs. isosmotic control) and GABA ($p < 0.01$) effluxes, accompanied by limited (nonosmometric) cell swelling (for taurine $0.1 > p > 0.05$ and for GABA $p < 0.05$ vs. isosmotic control). Addition of sucrose restored cell volumes to isosmotic levels without significantly affecting the acceleration of amino acid effluxes.

Table 2. Effects of N-ethylmaleimide (100 µM) on rate constants (k) for the 3rd (presumed cellular) phase of efflux of taurine and GABA into isosmotic and hyposmotic media, and hyposmotic media incorporating 25 mM sucrose, and the accompanying steady-state cell volumes

	$k \times 10^{-2} min^{-1}$	Cell volumes (µl/mg dry wt)
Taurine		
Isosmotic	0.71 ± 0.06 (6)	2.42 ± 0.09 (13)
Hyposmotic	0.79 ± 0.04 (6)	2.94 ± 0.12 (12)
Hyposmotic + sucrose	0.94 ± 0.04 (6)	2.38 ± 0.12 (10)
GABA		
Isosmotic	0.49 ± 0.05 (6)	2.41 ± 0.11 (11)
Hyposmotic	0.76 ± 0.04 (6)	2.92 ± 0.09 (13)
Hyposmotic + sucrose	0.70 ± 0.05 (6)	2.27 ± 0.14 (20)

Values are means ± SEM (n).

N-Ethylmameimide

Table 2 shows the effects of NEM on effluxes and cell volumes. In isosmotic media NEM caused very small (not significant) increases in taurine efflux and cell volume. The acceleration of GABA efflux by NEM in isosmotic media approached statistical significant ($0.1 > p > 0.05$) although cell swelling remained negligible. For both amino acids, NEM significantly impaired the acceleration of efflux in hyposmotic media (for taurine $p < 0.01$ and for GABA $p < 0.001$ vs. hyposmotic media without NEM) with significant increases in cell volume (taurine $p < 0.025$ and GABA $p < 0.01$). NEM had no effect on efflux of either acid when swelling was abolished by the presence of sucrose.

p-Chloromercuriphenylsulphonic Acid

Results using pCMPS are summarized in Table 3. The pattern of responses is essentially similar to that with NEM and, as with NEM, effects on taurine and GABA effluxes, and cell volumes, are qualitatively comparable. The effects of pCMPS on these variables under isosmotic conditions, however, are more pronounced than those of NEM. Effluxes of both acids was accelerated (for taurine $0.1 > p > 0.05$ and for GABA $p < 0.005$) with increased cell swelling which, however, only attained significance ($p < 0.025$) in media containing GABA. In hyposmotic media pCMPS depressed efflux (taurine $p < 0.001$ AND GABA $p < 0.025$) with cell swelling (taurine $p < 0.005$ and GABA $p < 0.001$). As with NEM, hyposmotic efflux was restored when cell volumes were reduced to control levels by sucrose.

Dithiothreitol

Pre-exposure of cells to DTT reversed the effects of NEM and pCMPS on GABA efflux in hyposmotic media (with slight but not significant acceleration) and restored volumes to normal, but was without effect on taurine efflux or associated cell swelling in the presence of these -SH reagents. The effects of DTT were not examined in isosmotic media or in the presence of sucrose.

DISCUSSION

A perennial problem associated with studies that depend on the effects of inhibitory reagents is that of specificity, both of the agents themselves and of their target sites in relation

Table 3. Effects of p-chloromercuriphenylsulphonic acid (100 μM) on rate constants (k) for the 3rd (presumed cellular) phase of efflux of taurine and GABA into isosmotic and hyposmotic media, and hyposmotic media incorporating 25 mM sucrose, and the accompanying steady-state cell volumes

	k x 10^{-2}min^{-1}	Cell volume (μl/mg dry wt)
Taurine		
Isosmotic	0.73 ± 0.02 (6)	2.50 ± 0.10 (10)
Hyposmotic	0.64 ± 0.04 (6)	3.15 ± 0.15 (9)
Hyposmotic + sucrose	0.92 ± 0.06 (6)	2.28 ± 0.20 (10)
GABA		
Isosmotic	0.84 ± 0.04 (6)	2.67 ± 0.14 (10)
Hyposmotic	0.56 ± 0.06 (6)	3.21 ± 0.12 (11)
Hyposmotic + sucrose	0.76 ± 0.05 (6)	2.40 ± 0.10 (10)

Values are means ± SEM (n).

to the overt effects of inhibition. Thiol groups are widespread and reactive ligands within both the membrane and cytosol of animal cells. One cannot always be certain that effects produced by exogenous reagents necessarily reflect their reaction with the target molecules directly concerned in the biological event under investigation. The observed responses might be due to effects on reactive groups elsewhere, leading, for example, to a general depression of cellular metabolism, a consideration which Rothstein has termed "geographical specificity"[27]. Thus while it may be tentatively predicated that what is being monitored in the present study is the activity of -SH groups located within some part of the amino acid efflux pathway, this remains presumptive rather than proven. Moreover, membrane -SH groups are known to be involved in a number of organic molecule transport processes (uptake and release) including tryptophan-induced glutamine release from cerebral capillaries[7], hexose transport in fat cells[5], organic cation transport in renal brush border membrane vesicles[30] and glucose transport in thymocytes[25]. Reference has already been made to -SH involvement in astrocytic taurine and D-aspartate release[1,21], and these groups are also known to be implicated in the uptake of 2-aminoisobutyric acid and GABA by membrane vesicles prepared from rat brain synaptosomes[9,26].

Interpretation of results obtained using -SH reagents is further complicated by their differing rates of cellular penetration (e.g. NEM penetrates far more readily than pCMBS) and the reversible (e.g. pCMBS and other mercurials) or irreversible (e.g. NEM) nature of their binding. It has been suggested that in certain -SH dependent transport processes (KCl co-transport in low K^+ sheep red cells) both activating and inhibitory -SH groups could be involved, whose participation may be revealed by differing concentrations of -SH reagent[12].

This brief overview gives some idea of the difficulties inherent in the interpretation of studies using -SH reagents. With that in mind, and also having regard to the fact that in the present study only single concentrations of such reagents were used, a number of tentative conclusions can be drawn. Firstly, both reagents tend to enhance amino acid efflux and cause cell swelling in isosmotic media. This effect was more noticeable with pCMPS that NEM, and with GABA- rather than taurine-media. A probable explanation of this effect is that these reagents react with -SH groups in the membrane Na^+/K^+-ATPase (the Na^+ pump)[6]. The resultant increase in intracellular Na^+ should lead to cell swelling and associated enhancement of amino acid efflux, such as has been seen in shark rectal gland cells[10,33]. Paradoxically, cell shrinkage was observed in pCMBS-treated astrocytes[19], which was ascribed to the cytoskeletal disruption typical of mercurial -SH reagents but not NEM[11]. Increased taurine efflux has been reported in astrocytes treated with cytochalasin B[23]. A decrease in the cytoskeletal component F-actin adjacent to the inner face of the shark rectal gland cell membrane has been detected using fluorescence techniques[10]. The discrepancy between swelling and shrinkage responses to pCMBS (or pCMPS) may reflect nothing more than differing experimental models. NEM, while reacting with the Na^+ pump, does not lead to cytoskeletal disruption, thus minimizing the effects of Na^+ pump inhibition upon cell volume.

Perhaps the most significant findings in the present study are that hyposmotically stimulated taurine and GABA efflux persist when cell swelling is prevented by the presence of sucrose, and that following the prevention of swelling cells become refractory to the inhibitory effects of NEM and pCMPS. This suggest two things. Firstly, that cellular geometry is a factor in the responsiveness of -SH groups to inhibition of NEM or pCMPS. This interpretation is consistent with the idea that volume-related conformational changes in -SH group availability play a role in hyposmotically-activated amino acid efflux[1]. Secondly, they suggest that, while hyposmotic stress is normally associated with cell swelling, it is the hyposmolality *per se* and not swelling (or accompanying membrane stretch) that determines the rate of efflux. This interpretation of volume responses to hyposmolality has been advances in trout red cells[20]. While it does not necessarily preclude a role for stretch-activated pathways in hyposmotic cell volume regulation[28], it offers an alternative (or complementary) explanation of solute loss which may be relevant to cells in whole animals, which are not subject to sudden hyposmotic shock such as is commonly applied *in vitro*, do not swell significantly or abruptly (the isovolumetric response[17]), and thus do not need to activate mechanosensitive channels in order to shed internal solutes in response to gradually advancing hyposmotic perturbation[13,14].

The present findings on the effects of DTT are at variance with those reported in cultured astrocytes, in which inhibition of taurine efflux by pCMBS, but not NEM, was prevented by pre-treatment with this reagent[18,19]. DTT restored GABA efflux in the presence of both inhibitors, while taurine efflux was unaffected. As with the effects of these reagents on cell volume (see above), the discrepancy may be due to differences in the cell preparations studied. One potentially important distinction is that amino acids and other solutes effluxing from the mixed cell population in slices (but not from cells in culture) may be subject to re-uptake from the tissue interstitium. This may itself be -SH dependent, and the possibility of re-accumulation is one which needs to be born in mind when considering all aspects of effluxes measured in tissue slices or other whole tissue preparations. Of particular relevance in the present context is the recent demonstration of neuronal-glial taurine exchange during hyposmotic stress in rat cerebellar cortex[22].

SUMMARY

1. Effluxes of taurine and GABA from pre-loaded cells in slices of rat cerebral cortex in isosmotic media is enhanced by the -SH reagent *p*-chloromercuriphenylsulphonic acid (pCMPS) (100 μM) accompanied by moderate swelling. Those effects were more pronounced with GABA than with taurine. N-ethylmaleimide (NEM) (100 μM) had only slight affects on these variables.
2. The acceleration of effluxes that occurs when medium osmolality is reduced from 315 to 265 mosmol/kg is blocked by NEM and by pCMPS, with pronounced cell swelling.
3. The inhibitory effects of these reagents on efflux is abolished when cell swelling is prevented by the addition of 25 mM sucrose to hyposmotic incubation media (with equimolar reduction in NaCl concentration).
4. Pre-exposure of slices to dithiothreitol (DTT) (100 μM) blocks the effects of NEM and pCMPS on GABA efflux in hyposmotic media, but has no effect on taurine efflux.
5. The cell membrane and cytoskeletal responses which may underlie these effects are briefly discussed.

ACKNOWLEDGMENT

Supported in part by Wellcome Trust Grant No. 037688/Z/93/Z to ROL.

REFERENCES

1. Aschner, M., Mullaney, K.J., Fehm, M.N., Vitarella, D., Wagoner, D.E. and Kimelberg, H.K., 1994, The role of sulfhydryl groups in D-aspartate and rubidium release from neonatal rat primary astrocyte cultures. *Brain Research* 648: 16-23.

2. Ballatori, N. and Boyer, J.L., 1992, Taurine transport in skate hepatocytes. II. Volume activation, energy, and sulfhydryl dependence. *Am. J. Physiol.* 262: G451-G460.
3. Cleland, W.W., 1964, Dithiothreitol, a new protective reagent for SH groups. *Biochemistry* 3: 480-482.
4. Cserr, H.F. and Patlak, C.S., 1991, Regulation of brain volume under isosmotic and anisosmotic conditions, *in*: "Volume and Osmolality Control in Animal Cells", Gilles, R., Hoffmann, E.K. and Bolis, L. eds., Springer-Verlag, Berlin and Heidelberg, pp. 61-80.
5. Czech, M.P., 1976, Differential effects of sulfhydryl reagents on activation and deactivation of the fat cell hexose transport systems. *J. Biol. Chem.* 251: 1164-1170.
6. Glick, N.B., 1972, *in*: "Metabolic Inhibitors, Vol. III", Hochster, R.M., Kates, M. and Quastel, J.H., eds, Academic Press, New York, pp. 1-45.
7. Hilgier, W., Puka, M. and Albrecht, J., 1992, Characteristics of large neutral amino acid-induced release of preloaded L-glutamine from rat cerebral capillaries in vitro: effects of ammonia, hepatic encephalopathy, and gamma-glutamyl transpeptidase inhibitors. *J. Neurosci. Res.* 32: 221-226.
8. Jackson, P.S. and Strange, K., 1995, Single-channel properties of a volume-sensitive anion conductance: current activation occurs by abrupt switching of closed channels to an open state. *J. Gen. Physiol.* 105: 643-660.
9. Kanner, B.I., 1978, Active transport of γ-aminobutyric acid by membrane vesicles isolated from rat brain. *Biochemistry* 17: 1207-1211.
10. Kleinzeller, A., Booz, G.W., Mills, J.W. and Ziyadeh, F.N., 1990, pCMBS-induced swelling of dogfish (*Squalus acanthias*) rectal gland cells: role of the Na^+, K^+-ATPase and the cytoskeleton. *Biochim. Biophys. Acta* 1025: 21-31.
11. Kunimoto, M., Shibata, K. and Miura, T., 1987, p-Chloromercuribenzoate-induced dissociation of cytoskeletal proteins in red blood cells. *Biochim. Biophys. Acta* 905: 257-267.
12. Lauf, P.K., 1988, Kinetic comparison of ouabain-resistant K:Cl fluxes (K:Cl [Co]transport) stimulated in sheep erythrocytes by membrane thiol oxidation and alkylation. *Mol. Cell. Biochem.* 80: 97-106.
13. Law, R.O., 1991, Amino acids as volume-regulatory osmolytes in mammalian cells. *Comp. Biochem. Physiol.* 99A: 263-277.
14. Law, R.O., 1994, Regulation of mammalian brain cell volume. *J. Exp. Zool.* 268: 90-96.
15. Law, R.O., 1994, Taurine efflux and the regulation of cell volume in incubated slices of rat cerebral cortex. *Biochim. Biophys. Acta* 1221: 21-28.
16. Law, R.O. and Burg, M.B., 1991, The role of organic osmolytes in the regulation of mammalian cell volume, *in*: "Volume and Osmolality Control in Animal Cells", Gilles, R., Hoffmann, E.K. and Bolis, L., eds., Springer-Verlag, Berlin and Heidelberg, pp. 189-225.
17. Lohr, J.W. and Grantham, J.J., 1986, Isovolumetric regulation of isolated S_2 proximal tubules in anisotonic media. *J. Clin. Invest.* 78: 1165-1172.
18. Martinez, A., Guerra, G. and Pasantes-Morales, H., 1993, Inhibition by sulfhydryl blocking reagents of the volume-sensitive taurine release in cultured astrocytes. *J. Neurochem.* 61 Suppl.: S47.
19. Martinez, A., Munoz-Clares, R.A. Guerra, G., Moran, J. and Pasantes-Morales, H., 1994, Sulfhydryl groups essential for the volume-sensitive release of taurine from astrocytes. *Neurosci. Lett.* 176: 239-242.
20. Motais, R., Guizouarn, H. and Garcia-Romeu, F., 1991, Red cell regulation: the pivotal role of ionic strength in controlling swelling-dependent transport systems. *Biochim. Biophys. Acta* 1975: 169-180.
21. Mullaney, K.J. Vitarella, D., Albrecht, J., Kimelberg, H.K. and Aschner, M., 1993, Stimulation of aspartate efflux by mercuric chloride from rat primary astrocyte cultures. *Dev. Brain Res.* 75: 261-268,
22. Nagelhus, E.A., Lehmann, A. and Ottersen, O.P., 1993, Neuronal-glial exchange of taurine during hypo-osmotic stress: a combined immunocytochemical and biochemical analysis in rat cerebellar cortex. *Neuroscience* 54: 615-631.
23. Pasantes-Morales, H., Moran, J. and Schousboe, A., 1990, Volume-sensitive release of taurine from cultured astrocytes: properties and mechanisms. *Glia* 3: 427-432.
24. Pollock, A.S. and Arieff, A.I., 1980, Abnormalities of cell volume regulation and their functional consequences. *Am. J. Physiol.* 239: F195-F205.
25. Regen, D.M. Schraw, W.P., Tarpley, H.L. and Juliao, S.F., 1981, Effects of thiol reagents on glucose transport in thymocytes. *Biochem. Biophys. Acta* 644: 62-68.
26. Robillard, G.T., Schaaf, J.M. and Teelken, A.W., 1987, Dithiols and monothiols are linked with GABA transport in membrane vesicles of rat brain synaptosomes. *FEBS Lett.* 224: 391-395.
27. Rothstein, A., 1970, Sulfhydryl groups in membrane structure and function. *Curr. Topics Membr. Trans.* 1: 135-176.
28. Sackin, H., 1995, Review of mechanosensitive channels. *Ann. Rev. Physiol.* 57: 333-353.
29. Schousboe, A., Moran, H. and Pasantes-Morales, J., 1990, Potassium-stimulated release of taurine from cultured cerebellar granule neurons is associated with cell swelling. *J. Neurosci. Res.* 27: 71-77.

30. Sokol, P.P., Holohan, D. and Ross, C.R., 1986, Essential disulfide and sulfl,ydryl groups for organic cation transport in renal brush-border membrane membranes. *J. Biol. Chem.* 261: 3282-3287.
31. Sterns, R.H., Baer, J., Ebersol, S., Thomas, D., Lohr, J.W. and Kamm, D.E., 1993, Organic osmolytes in acute hyponatraemia. *Am. J. Physiol.* 264: F833-F836.
32. Verbalis, J.G., 1993, Hyponatraemia: epidemiology, pathophysiology, and therapy. *Curr. Opinion Nephrol. Hypertens.* 2: 636-652.
33. Ziyadeh, F.N., Feldman, G.M., Booz, G.W. and Kleinzeller, A., 1988, Taurine and cell volume maintenance in the shark rectal gland: cellular fluxes and kinetics. *Biochim. Biophys. Acta* 943: 43-52.

45

NITRIC OXIDE-EVOKED [^{3}H]TAURINE RELEASE IS MEDIATED BY REVERSAL OF THE Na^+-DEPENDENT CARRIER-MEDIATED TAURINE TRANSPORT SYSTEM

Seitaro Ohkuma, Masahi Katsura, Da-Zhi Chen, and Kinya Kuriyama

Department of Pharmacology
Kyoto Prefectural University of Medicine
Kamikyo-Ku, Kyoto 602, Japan

SUMMARY

The pharmacological characteristics of [^{3}H]taurine release evoked by nitric oxide (NO) were investigated using mouse cerebral cortical neurons in primary culture. *N*-Methyl-D-aspartate (NMDA) and *S*-nitroso-N-acetylpenicillamine (SNAP) dose-dependently increased [^{3}H]taurine release from neurons. The NMDA-evoked release of [^{3}H]taurine was reduced to the basal level by N^{ω}-nitro-L-arginine, a NO synthase inhibitor, and MK-801, a noncompetitive antagonist for NMDA receptors. The NMDA- and SNAP-evoked releases of [^{3}H]taurine were completely abolished by hemoglobin, indicating that these [^{3}H]taurine releases were evoked by NO produced by NMDA receptor activation and liberated from SNAP. Withdrawal of Na^+ from incubation buffer significantly inhibited the NMDA- and SNAP-induced [^{3}H]taurine releases, whereas removal of Ca^{2+} produced no alteration in the SNAP-evoked [^{3}H]taurine release. In addition, β-alanine and guanidinoethane sulfonate, antitransporters of the carrier-mediated taurine transport system, reduced the NMDA- and SNAP-evoked releases of [^{3}H]taurine in a dose-dependent manner. These results indicate that the NO-evoked [^{3}H]taurine release from cerebral cortical neurons is mediated by a reversal of the Na^+-dependent carrier-mediated taurine transport system.

INTRODUCTION

Taurine is abundantly distributed in the mammalian central nervous system (CNS) and has been assumed to serve as a neurotransmitter or a neuromodulator in the CNS[10]. Over the last two decades, the pharmacological characteristics of taurine release evoked by the application of electrical and high K^+ stimulations have been reported to be both Ca^{2+}-dependent and Ca^{2+}-independent[30]. These characteristics of taurine release in response to

Taurine 2, edited by Huxtable et al.
Plenum Press, New York, 1996

several types of stimulation are assumed to support a neurotransmitter function, although the exact role of this sulfur-containing amino acid in the CNS remains to be elucidated.

Among factors inducing release of neurotransmitters, the accumulated data support that NO, an unstable radical formed from L-arginine via oxidative deamination catalyzed by Ca^{2+} -dependent NO synthase[4], can evoke release of several neurotransmitters such as acetylcholine[15,23,28], catecholamines[8,17,21,31] and GABA[24]. However, there is scant data on taurine release induced by NO. In the present study, therefore, we have attempted to clarify the pharmacological characteristics of taurine release evoked by NO which is liberated from SNAP, a NO generator, and produced by NMDA receptor activation. For this purpose, we have used primary cultured cerebral cortical neurons capable of biosynthesizing[25] and transporting taurine[11].

MATERIALS AND METHODS

Isolation and Primary Culture of Cerebral Cortical Neurons

Isolation and primary culture of cerebral cortical neurons were performed as reported previously[25]. In brief, the neopallium free of meninges was dissected from 15-day-old fetuses of ddY strain mice, minced, trypsinized and centrifuged. The resultant pellet was suspended in Dulbecco's modified Eagle medium (DMEM) containing 15% fetal calf serum and incubated at 37°C in humidified 95% air-5% CO_2. After incubation for 3 days, the cells were exposed to 10 μM cytosine arabinoside with DMEM supplemented with 10% horse serum for 24 h. Thereafter, the neurons were cultured in DMEM with 10% horse serum under the same conditions described above. The culture medium was exchanged every 4 days and the neurons of 13-day-old in culture were used for the following experiments.

Measurement of [^{3}H]Taurine Released from Neurons

The measurement of [^{3}H]taurine released from the cerebral cortical neurons was carried out as reported previously[23] with a minor modification. After the incubation of the neurons in Krebs-Ringer bicarbonate buffer (KRB; pH 7.4) with 5 μM [^{3}H]taurine (1 μCi [^{3}H]taurine/dish) at 37°C for 30 min, the neurons were rinsed 5x with ice-cold KRB followed by the incubation with KRB at 37°C for 10 min. Thereafter, the neurons were incubated with fresh KRB at intervals of 5 min for a total of 25 min at 37°C. NMDA and SNAP were added into a dish at the beginning of the 4th interval of the incubation. In order to examine the effect of NMDA on [^{3}H]taurine release, Mg^{2+}-free KRB was used as the incubation buffer. An aliquot of the incubation buffer collected during each interval was used to measure the radioactivity released from the neurons by liquid scintillation spectrometry. After the experiments, the neurons were digested with 0.5 M NaOH. The release of [^{3}H]taurine in the presence of NMDA or SNAP was defined as the evoked release and expressed as a percentage of the basal release determined during the 3rd interval. Sodium-free buffer was prepared by the substitution of Na^+ with equimolar choline.

Protein Measurement

The protein content in alkali-digested neurons was measured according to the method of Lowry *et al.*[16] using bovine serum albumin as standard.

Statistical Analysis

Data were expressed as the mean ± SEM and statistical significance was analyzed as described in figure and table legends following the application of the one-way ANOVA.

Materials

[^{3}H]Taurine (810.3 GBq/mmol) was purchased from New England Nuclear (Boston, USA). Hemoglobin, N^G-methyl-L-arginine (M-Arg), N^ω-nitro-L-arginine (N-Arg), and β-alanine were obtained from Sigma Chemicals (St. Louis, US). Research Biochemicals Inc. (Natick, US) was the source of NMDA, MK-801, and SNAP. Fetal calf and horse sera were purchased from Hazelton Research Products Inc. (Lenexa, US). Guanidinoethane sulfonate was a kind gift of Prof. Ryan J. Huxtable (Dept. of Pharmacology, University of Arizona Health Sciences Center, Tucson, Arizona, US). Other chemicals used were available locally and of analytical grade.

RESULTS

Characteristics of [^{3}H]Taurine Release from Neurons by NMDA

As shown in Fig. 1, NMDA increased [^{3}H]taurine release in a dose-dependent manner, reaching a plateau at 30 μM NMDA. The NMDA (10 μM)-induced increase in [^{3}H]taurine release was significantly suppressed by MK-801, an antagonist specific for NMDA receptors, and hemoglobin (Table 1). In addition, NO synthase inhibitors, N-Arg and M-Arg, significantly reduced the stimulatory action of NMDA on neuronal [^{3}H]taurine

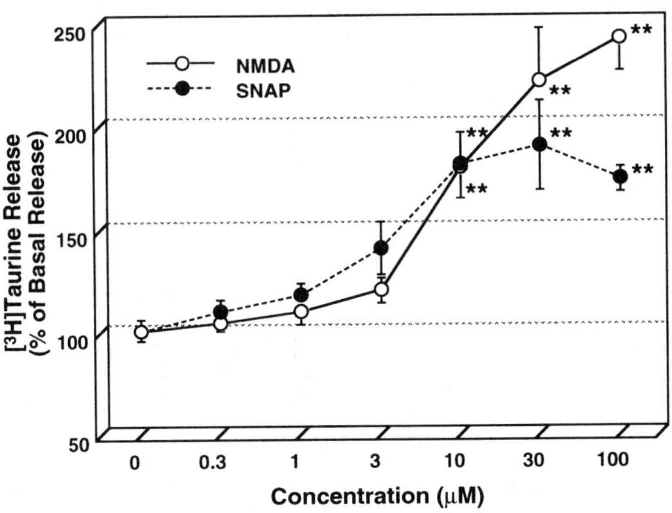

Figure 1. Effects of N-methyl-D-aspartate (NMDA) and S-nitroso-N-acetylpenicillamine (SNAP) on [^{3}H]taurine release from mouse cerebral cortical neurons. The basal release of [^{3}H]taurine was 5,585 ± 89 dpm/mg protein/interval. Each value represents the mean ± SEM obtained from 4 separate experiments. **p < 0.01, compared with the release of [^{3}H]taurine determined in the absence of NMDA and SNAP (Dunnett's test).

Table 1. Effects of hemoglobin and NO synthase inhibitors on NMDA-induced [^{3}H]taurine release from mouse cerebral cortical neurons

	[^{3}H]Taurine Release (% of Basal Release)	
	No NMDA	10 μM NMDA
Control	100.0 ± 5.0	186.2 ± 4.9$^{\#\#}$
MK-801 (10 μM)	103.2 ± 6.4	96.5 ± 4.0**
Hemoglobin (40 μM)	100.1 ± 4.1	107.3 ± 6.1**
N-Arg (100 μM)	99.2 ± 3.1	103.2 ± 4.1**
M-Arg (100 μM)	101.5 ± 1.9	127.9 ± 8.1$^{\#}$,**

The basal release of [^{3}H]taurine was 5,892 ± 279 dpm/mg protein/fraction. N-Arg: N$^{\omega}$-nitro-L-arginine, M-Arg; N^{G}-methyl-L-arginine. $^{\#}p < 0.05$, $^{\#\#}p < 0.01$, compared with the basal release (Bonferroni's test. N=5). **$p < 0.01$, compared with each release determined in the presence of NMDA alone (Bonferroni's test, N=5).

release (Table 1). These data indicate that the stimulatory action of NMDA on [^{3}H]taurine release is mediated by the formation of NO subsequent to NMDA receptor activation.

Characteristics of [^{3}H]Taurine Release from Neurons by SNAP

SNAP dose-dependently elevated [^{3}H]taurine release from cerebral cortical neurons (Fig. 1). Fig. 2 shows that hemoglobin significantly reduces the increase in [^{3}H]taurine release evoked by 5 μM SNAP, and that hemoglobin does not alter the basal release (Fig. 2). These results indicate that the increase in [^{3}H]taurine release by SNAP is induced by NO liberated from SNAP.

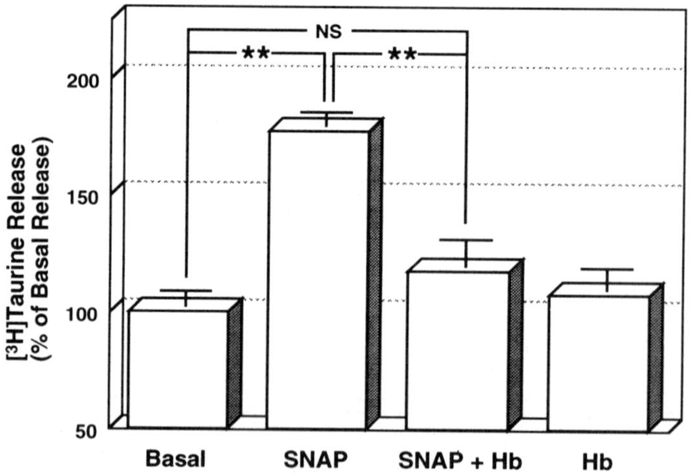

Figure 2. Effect of hemoglobin on SNAP-induced [^{3}H]taurine release from mouse cerebral cortical neurons. Hemoglobin was added into a culture dish immediately before the addition of SNAP. The concentrations of SNAP and hemoglobin were 10 and 40 μM, respectively. The basal release of [^{3}H]taurine was 5,989 ± 371 dpm/mg protein/interval. Each value represents the mean ± SEM obtained from 4 separate experiments and expressed as a percentage of the basal release. **$p < 0.01$, NS: not significant (Bonferroni's test).

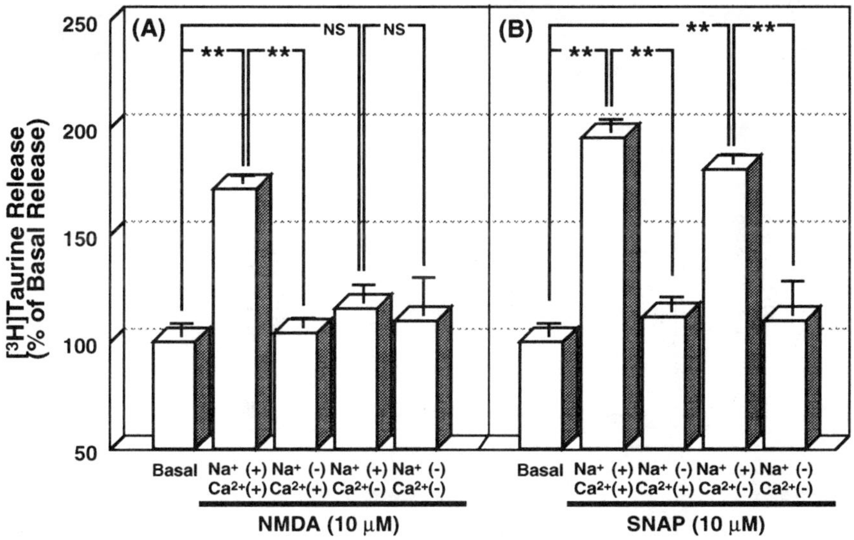

Figure 3. Effects of (A) withdrawal of Na^+ on NMDA (10 μM)-induced [^{3}H]taurine release and (B) withdrawal of Na^+ and Ca^{2+} on SNAP (10 μM)-induced [^{3}H]taurine release from mouse cerebral cortical neurons. The basal releases of [^{3}H]taurine for (A) and (B) were 6,439 ± 393 and 6,254 ± 357 dpm/mg protein/interval, respectively. Each value represents the mean ± SEM obtained from 4 separate experiments and expressed as a percentage of each basal release. $^{**}p < 0.01$, NS; not significant (Bonferroni's test).

Effect of Withdrawal of Ca^{2+} and Na^+ on [^{3}H]Taurine Releases Evoked by NMDA and SNAP

Fig. 3A shows the effect of Na^+ withdrawal from the incubation buffer on the NMDA- and SNAP-induced increases in [^{3}H]taurine release from cerebral cortical neurons. The Na^+ withdrawal significantly reduced the NMDA-evoked [^{3}H]taurine release to the level of the basal release (Fig. 3A). The NMDA-induced release of [^{3}H]taurine was not observed in the absence of Ca^{2+} in the incubation buffer (Fig. 3A). Similarly, SNAP-induced [^{3}H]taurine release was reduced by the removal of Na^+ from the incubation buffer to a level similar to the basal release. Release was unaffected by removal of Ca^{2+} (Fig. 4B). These results suggest that the NO-induced [^{3}H]taurine release is mediated by a Ca^{2+}-independent and Na^+-dependent release mechanism.

Effects of Taurine Transport Inhibitors on [^{3}H]Taurine Releases Evoked by NMDA and SNAP

β-Alanine dose-dependently decreased the NMDA-induced [^{3}H]taurine release from the neurons (Fig. 4). The maximal inhibition of the NMDA-induced [^{3}H]taurine release was observed in the presence of 10 μM β-alanine (Fig. 4). GES also showed a dose-dependent inhibition of the NMDA-induced [^{3}H]taurine release (Fig. 4). In the case of GES, the inhibition of the release attained to a plateau at 1 μM. The IC_{50} values of β-alanine and GES for the NMDA-evoked [^{3}H]taurine release were 0.72 and 0.04 μM (the mean of 5 separate experiments), respectively. Similarly, these inhibitors dose-dependently reduced the SNAP-induced [^{3}H]taurine release (Fig. 5). The IC_{50} values of β-alanine and GES for the SNAP-evoked [^{3}H]taurine release were 1.06 and 0.45 μM (the mean of 5 separate experiments),

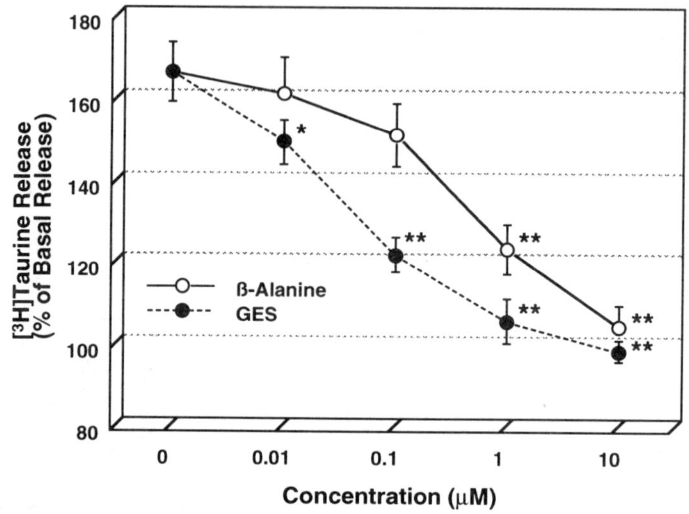

Figure 4. Effects of β-alanine and guanidinoethane sulfonate (GES) on NMDA (10 μM)-induced [^{3}H]taurine release from mouse cerebral cortical neurons. β-Alanine and GES were added into a culture dish immediately before the addition of NMDA. The basal release of [^{3}H]taurine was 7,124 ± 362 dpm/mg protein/interval. Each value represents the mean ± SEM obtained from 4 separate experiments and expressed as a percentage of the basal release. **$p < 0.01$, compared with the release of [^{3}H]taurine determined in the presence of NMDA alone (Dunnett's test).

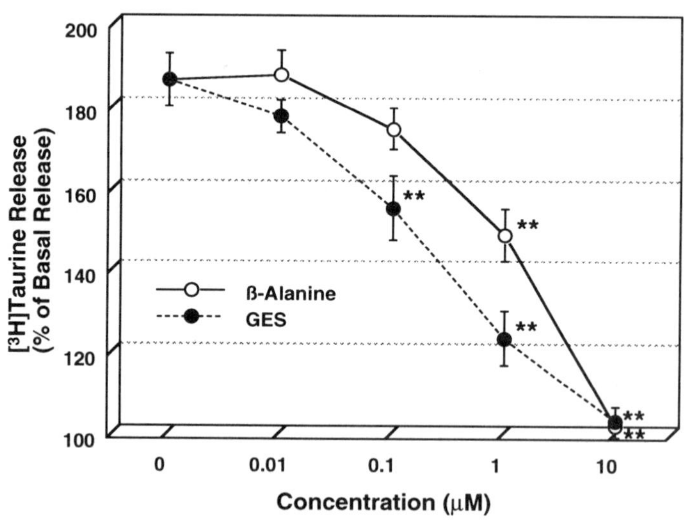

Figure 5. Effects of β-alanine and GES on SNAP (10 μM)-induced [^{3}H]taurine release from mouse cerebral cortical neurons. β-Alanine and GES were added into a culture dish immediately before the addition of SNAP. The basal release of [^{3}H]taurine was 6,777 ± 561 dpm/mg protein/interval. Each value represents the mean ± SEM obtained from 4 separate experiments and expressed as a percentage of the basal release. The concentration of SNAP was 10 μM. *$p < 0.05$, **$p < 0.01$, compared with the release of [^{3}H]taurine determined in the presence of SNAP alone (Dunnett's test).

respectively. The maximal inhibitory actions of β-alanine and GES on the SNAP-induced [^{3}H]taurine release were found at concentrations of 10 and 3 μM, respectively. However, both taurine transport inhibitors at 10 μM showed no effect on the basal release of [^{3}H]taurine from the neurons (data not shown).

DISCUSSION

In the present study, we have attempted to examine whether NO releases taurine and to clarify the mechanism of the NO-evoked taurine release, using mouse cerebral cortical neurons with NMDA receptors4[22]. To produce NO, the direct application of a NO generator, SNAP, and the stimulation of NMDA receptors by its specific agonist, NMDA, were employed. The reason for activating NMDA receptors for generating NO is that the activation of NMDA receptors has been reported to induce NO formation via increased Ca^{2+} influx subsequent to the stimulation of Ca^{2+}-dependent NO synthase[5,6].

The stimulation of NMDA receptors induces release of neurotransmitters[3,7,12,19,26,27]. Such release of neurotransmitters evoked by NMDA receptor activation is mediated by the formation of NO[9,21,23,24]. Similarly, NMDA has been reported to release taurine from the hippocampus[20,29], although these investigations do not refer to the role of NO on the release of taurine evoked by NMDA. In this study, we have demonstrated that the release of [^{3}H]taurine evoked by NMDA is mediated by NO production via the stimulation of NMDA receptors, which is supported by the findings that NO synthase inhibitors and hemoglobin abolish NMDA-evoked [^{3}H]taurine release.

The present study demonstrates that NO liberated from a NO generator, SNAP, evokes [^{3}H]taurine release. Release of neurotransmitter induced by direct application of NO generators, such as nitroglycerin[17], hydroxylamine[15], sodium nitroprusside[23,31] or SNAP[23], has been also reported.

Several lines of evidence have revealed that the release of endogenous taurine induced by electrical stimulation and high K^+ application is predominantly Ca^{2+}-dependent[30]. In addition, taurine release evoked by NMDA is dependent on extracellular Ca^{2+} [13,18] and/or Ca^{2+} mobilized from intracellular Ca^{2+} stores[20]. In the present study, NMDA showed no effects on [^{3}H]taurine release when extracellular Ca^{2+} was absent. However, such Ca^{2+}-dependency of the NMDA-evoked release of taurine appears to be involved in the activation of Ca^{2+} -dependent NO synthase rather than in the stimulatory action of NMDA for releasing neurotransmitter, because hemoglobin abolished completely the NMDA-evoked [^{3}H]taurine even in the presence of extracellular Ca^{2+}.

We have shown in this study that the release of [^{3}H]taurine induced by NO produced by the NMDA receptor activation is Na^+-dependent, because the withdrawal of Na^+ from the incubation buffer completely abolished NMDA-evoked [^{3}H]taurine release from the neurons. SNAP-induced [^{3}H]taurine release is also revealed to be driven by a Na^+-dependent and Ca^{2+}-independent process. Similarly, Belhage *et al.*[1] reported that glutamate induced [^{3}H]GABA release from mouse cerebral cortical neurons in only a Na^+-dependent manner and that this type of release was Ca^{2+}-independent.

Recent investigations have emphasized that Ca^{2+}-independent release of neurotransmitters is dominantly mediated through reversal of the Na^+-dependent carrier-mediated transport system for neurotransmitters[2,14]. Therefore, in the present study we examined whether the NMDA- and SNAP-evoked [^{3}H]taurine releases were mediated by reversal of Na^+-dependent carrier-mediated transport by use of cerebral cortical neurons which have been confirmed to possess the ability to transport taurine by a Na^+-dependent carrier-mediated transport system[11]. The enhancement of [^{3}H]taurine release by both NMDA and SNAP was dose-dependently reduced by two taurine transport inhibitors, β-alanine and GES. These data indicate that NO-evoked [^{3}H]taurine release operates by reversal of Na^+-dependent carrier-mediated transport.

In summary, both NMDA and SNAP dose-dependently increased [^{3}H]taurine release from the cerebral cortical neurons via the production of NO. This NO-induced release of [^{3}H]taurine was abolished by the withdrawal of Na^+ from the incubation buffer, suggesting that the NO-induced [^{3}H]taurine release is Na^+-dependent. The removal of Ca^{2+} did not affect the NO-evoked [^{3}H]taurine release. In addition, β-alanine and GES, taurine transport inhibitors, dose-dependently reduced the NO-evoked [^{3}H]taurine release. These results indicate that NO-induced [^{3}H]taurine release operates by reversal of Na^+-dependent carrier-mediated taurine transport.

ACKNOWLEDGMENT

The authors wish to express their thanks to Prof. R. J. Huxtable (Dept. of Pharmacology, University of Arizona Health Sciences Center, US) for his kind gift of guanidinoethane sulfonate used in this study.

REFERENCES

1. Belhage, B., Hansen, G.H., and Schousboe, A. 1993, Depolarization by K^+ and glutamate activates different neurotransmitter release mechanisms in GABAergic neurons: vesicular versus non-vesicular release of GABA, *Neuroscience*, 54:1019-1034.
2. Bernath, S. 1992, Calcium-independent release of amino acid neurotransmitters: fact or artifact? *Prog.Neurobiol.* 38:57-91.
3. Blandina, P., Johnson, D., Walcott, J., and Goldfarb, J. 1992, Release of endogenous norepinephrine from rat hypothalamus by stimulation of N-methyl-D-aspartic acid receptors, *J.Pharmacol.Exp.Ther.* 263:61-68.
4. Bredt, D.S. and Snyder, S.H. 1990, Isolation of nitric oxide synthetase, a calmodulin-requiring enzyme, *Proc.Natl.Acad.Sci.* 87:682-685.
5. Collier, J. and Vallance, P. 1989, Second messenger role for NO widens to nervous and immune systems, *Trends Pharmacol.Sci.* 10:427-431.
6. Dawson, T.M., Dawson, V.L., and Snyder, S.H. 1992, A novel neuronal messenger molecule in brain: the free radical, nitric oxide, *Ann.Neurol.* 32:297-311.
7. Drejer, J., Honoré, T., and Schousboe, A. 1987, Excitatory amino acid-induced release of ^{3}H-GABA from cultured mouse cerebral cortex interneurons, *J.Neurosci.* 7:2910-2916.
8. Guevara-Guzman, R., Emson, P.C., and Kendrick, K.M. 1994, Modulation of in vivo striatal transmitter release by nitric oxide and cGMP, *J.Neurochem.* 62:807-810.
9. Hanbauer, I., Wink, D., Osawa, Y., Edelman, G.M., and Gally, J.A. 1992, Role of nitric oxide in NMDA-evoked release of [^{3}H]-dopamine from striatal slices, *Neuroreport*, 3:409-412.
10. Huxtable, R.J. 1989, Taurine in the central nervous system and the mammalian actions of taurine, *Prog.Neurobiol.* 32:471-533.
11. Kishi, M., Ohkuma, S., Kimori, M., and Kuriyama, K. 1988, Characteristics of taurine transport system and its developmental pattern in mouse cerebral cortical neurons in primary culture, *Biochem.Biophys.Acta.* 939:615-623.
12. Krebs, M.O., Desce, J.M., Kemel, M.L., Gauchy, C., Godeheu, G., Cheramy, A., and Glowinski, J. 1991, Glutamatergic control of dopamine release in the rat striatum: evidence of presynaptic *N*-methyl-D-aspartate receptors on dopaminergic nerve terminals, *J.Neurochem.* 56:81-85.
13. Lehmann, A., Lazarewicz, J.W., and Zeise, M. 1985, N-Methylaspartate-evoked liberation of taurine and phosphoethanolamine in vivo: Site of release, *J.Neurochem.* 45:1172-1177.
14. Levi, G. and Raiteri, M. 1993, Carrier-mediated release of neurotransmitters, *Trends Pharmacol.Sci.* 16:415-419.
15. Lonart, G., Wang, J., and Johnson, K.M. 1992, Nitric oxide induces neurotransmitter release from hippocampal slices, *Eur.J.Pharmacol.* 220:271-272.

16. Lowry, O.J., Rosebrough, N.J., Farr, A.L., and Randall, R.J. 1951, Protein measurement with Folin phenol reagent, *J.Biol.Chem.* 193:265-275.
17. Ma, S. and Long, J.P. 1991, Effects of nitroglycerin on release, synthesis and metabolism of norepinephrine and activation of tyrosine hydroxylase in guinea-pigs, *Eur.J.Pharmacol.* 199:27-33.
18. Magnusson, K.R., Koerner, J.F., Larson, A.A., Smullin, D.H., Skilling, S.R., and Beitz, A.J. 1991, NMDA-, kainate- and quisqualate-stimulated release of taurine from electrophysiologically monitored rat hippocampal slices, *Brain Res.* 549:1-8.
19. Martínez-Fong, D., Rosales, M.G., Góngora-Alfaro, J.L., Hernández, S., and Aceves, J. 1992, NMDA receptor mediates dopamine release in the striatum of unanesthetized rats as measured by brain microdialysis, *Brain Res.* 595:309-315.
20. Menéndez, N., Solís, J.M., Herreras, O., Galarreta, M., Conejero, C., and Martín del Río, R. 1993, Taurine release evoked by NMDA receptor activation is largely dependent on calcium mobilization from intracellular stores, *Eur.J.Neurosci.* 5:1273-1279.
21. Montague, P.R., Gancayco, C.D., Winn, M.J., Marchase, R.B., and Friedlander, M.J. 1994, Role of NO production in NMDA receptor-mediated neurotransmitter release in cerebral cortex, *Science*, 263:973-977.
22. Ohkuma, S., Chen, D.-Z., Katsura, M., Chen, S.-H., and Kuriyama, K. 1994, $GABA_A$ receptor stimulation enhances NMDA-induced Ca^{2+} influx in mouse cerebral cortical neurons in primary culture, *Mol.Brain Res.* 27:145-151.
23. Ohkuma, S., Katsura, M., Guo, J.-L., Hasegawa, T., and Kuriyama, K. 1995, Participation of peroxynitrite in acetylcholine release induced by nitric oxide generators, *Neurosci.Lett.* 183:151-154.
24. Ohkuma, S., Narihara, H., Katsura, M., Hasegawa, T., and Kuriyama, K. 1995, Nitric oxide (NO)-induced [^{3}H]GABA release from cerebral cortical neurons is mediated by peroxynitrite, *J.Neurochem.* 65:1109-1114.
25. Ohkuma, S., Tomono, S., Tanaka, Y., Kuriyama, K., and Mukainaka, T. 1986, Development of taurine biosynthesizing system in cerebral cortical neurons in primary culture, *Int.J.Devel.Neurosci.* 4:383-395.
26. Ping, J.-P., Van-Vliet, B.J., and Bockaert, J. 1988, NMDA- and kainate-evoked GABA release from striatal neurones differentiated in primary culture: differential blocking by phencyclidine, *Neurosci.Lett.* 87:87-92.
27. Pittaluga, A. and Raiteri, M. 1992, N-methyl-D-aspartic acid (NMDA) and non-NMDA receptors regulating hippocampal norepinephrine release. III. Changes in the NMDA receptor complex induced by their functional cooperation, *J.Pharmacol.Exp.Ther.* 263:327-333.
28. Prast, H. and Philippu, A. 1992, Nitric oxide releases acetylcholine in the basal forebrain, *Eur.J.Pharmacol.* 216:139-140.
29. Saransaari, P. and Oja, S.S. 1991, Excitatory amino acids evoke taurine release from cerebral cortex slices from adult and developing mice, *Neuroscience*, 45:451-459.
30. Saransaari, P. and Oja, S.S. 1992, Release of GABA and taurine from brain slices, *Prog.Neurobiol.* 38:455-482.
31. Zhu, X.-Z. and Luo, L.-G. 1992, Effect of nitroprusside (nitric oxide) on endogenous dopamine release from rat striatal slices, *J.Neurochem.* 59:932-935.

46

IN VIVO RELEASE OF TAURINE FROM RAT NEOSTRIATUM AND SUBSTANTIA NIGRA

Loria Bianchi,[1] John P. Bolam,[2] Francesca Galeffi,[1] Maria Frosini,[3] Mitri Palmi,[3] Gianpietro Sgaragli,[3] and Laura Della Corte[1]

[1] Dipartimento di Farmacologia Preclinica e Clinica "M. Aiazzi Mancini"
Università degli Studi di Firenze
Viale G.B. Morgagni 65, 50134 Florence, Italy
[2] MRC Anatomical Neuropharmacology Unit
Mansfield Road, Oxford, OX1 3TH, United Kingdom
[3] Istituto di Scienze Farmacologiche
Università degli Studi di Siena
Via E.S. Piccolomini 170, 53100 Siena, Italy

INTRODUCTION

Taurine has been shown to fulfil many of the criteria of a neurotransmitter in the basal ganglia. The neostriatum (STR) and substantia nigra (SN) contain high levels of taurine and its synthetic enzyme sulfinoalanine decarboxylase (EC 4.1.1.29, commonly referred to as cysteine sulfinic acid decarboxylase, CSDI)[19,29,34,41,43,45]. The presence of a high affinity uptake system for taurine has been detected in both the STR[11,29] and SN[14]. Furthermore, uptake and release studies of exogenous radiolabelled taurine suggest that neurones identified as medium-size densely spiny striatonigral neurones contain taurine and release it at their terminals in the SN[15] (see Table 1).

The medium-size spiny neurones represent the major projection neurone of the STR, sending their axons to the globus pallidus or the entopeduncular nucleus/SN. They also possess extensive local axonal collaterals that ramify within the neostriatum[38]. These neurones, together with a small population of neostriatal interneurones[4,5,12], use GABA as the main neurotransmitter. Anatomical data have provided the evidence that in the neostriatum, both the medium-size spiny GABAergic neurones[38] and the GABA interneurones receive direct synaptic excitatory input from the cortex[27].

Recent studies using dual probe microdialysis have demonstrated that it is possible to selectively stimulate the release of endogenous GABA from the local axon collaterals of striatonigral neurones by intrastriatal application of the excitatory amino acid non-NMDA receptor agonist, kainic acid (KA), and detect at the same time the release of GABA from their axon terminals in the SN[3]. If taurine is also present in striatonigral neurones then one would expect that following stimulation of striatal neurones there would be a simultaneous

Taurine 2, edited by Huxtable et al.
Plenum Press, New York, 1996

Table 1. Taurine and the striatonigral system

Presence in STR and SN of:	
	Taurine and it synthetic enzyme cysteine sulfinic acid decarboxylase I[18,19,29,34,41,43,45]
	High affinity uptake system[11,14,29]
	Ca^{2+}-dependent and TTX-sensitive release of taurine induced by veratridine from slices[2,15]
Presence in striatonigral neurons:	
	Autoradiography of local [^{3}H]taurine uptake[10,15]
	KA lesion of STR leads to a reduction in SN of taurine levels[26], taurine-immunolabelled terminals[42], cysteine sulfinic acid decarboxylase I[28], veratridine-induced [^{3}H]taurine release in slices[15]
Taurine injection into SN induces	
	Glycine-like inhibition of dopaminergic neurons[21]
	GABA-like contralateral rotation[23,24,31]
Taurine interacts in vitro with a $GABA_A$ receptor subclass which is enriched in SN[6]	

release of taurine both in the STR and SN. The aim of the present investigation was to test this hypothesis.

METHODS

Experiments were performed on male Sprague-Dawley rats. Single cannula microdialysis probes were implanted vertically into the right STR (3 mm exposed tip) and the ipsilateral SN pars reticulata (2 mm exposed tip), as previously described[3] (Figure 1, right panel). Twenty four hours later, the probes, of the now freely moving rats, were perfused with artificial CSF at a rate of 1.2 μl/min. After a perfusion period of 2 h the STR was exposed to either 100 mM K^+ or to 100 μM KA in osmotically balanced CSF, alone or in the presence of 10 μM DNQX (6,7-dinitroquinoxaline-2,3-dione) or 3 μM TTX (tetrodotoxin) throughout the experiment. Fractions of the perfused fluid (20 min) were collected for the 2 h period before the KA perfusion and for up to 3 h afterwards. Taurine in the perfusate fractions was derivatized, separated by HPLC and the orthophthalaldehyde derivative was detected fluorimetrically[13] and expressed either as pmol of taurine/μl of perfusate or as % of mean basal outflow prior to stimulation. Data were evaluated by analysis of variance of the original values as described previously[3].

RESULTS

Basal Efflux of Taurine from the Neostriatum and the Substantia Nigra

The average level of taurine (± SEM, n=14) in basal perfusates of STR and SN was 0.995 ± 0.128 pmol/μl and 1.085 ± 0.130 pmol/μl, respectively. In both regions the output was stable for several hours.

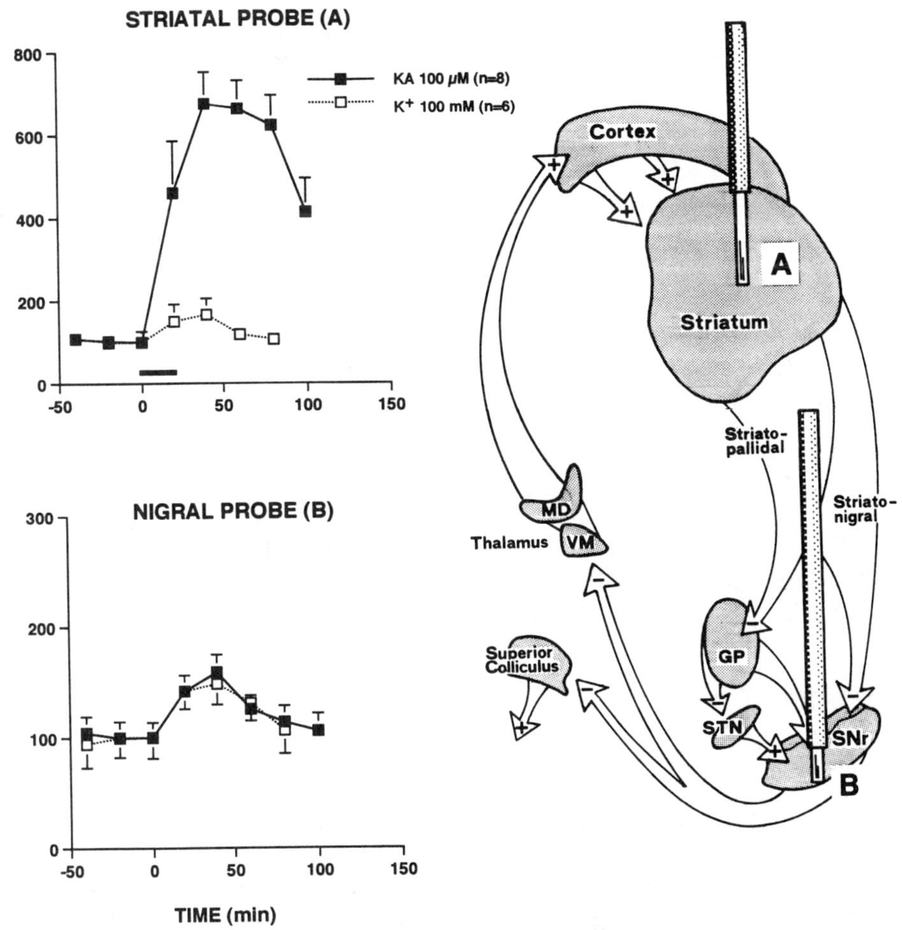

Figure 1. The time course of the release of endogenous taurine from the STR (A) and SN(B) (left panel) following stimulation of the STR with 100 mM K^+ (open squares) and 100 μM KA (closed squares) Data (% of basal outflow) are presented as mean ± SEM of the number of samples indicated by n. In the right panel a simplified diagram (not to scale) shows the position of the probes in relation to basal ganglia circuitry.

Effect of Intrastriatal Application of Potassium on the Release of Taurine from the Neostriatum and the Ipsilateral Substantia Nigra

Perfusion of the neostriatum with 100 mM K^+ caused a statistically significant increase in the levels of taurine both in the striatal and nigral perfusates (Figure 1, left panel). The release detected in the SN had a similar time course and magnitude to that observed in the STR; i.e. peak levels (n=6) were 148 ± 18% and 166 ± 19% of basal output in SN and STR, respectively (Fig. 1).

Effect of Intrastriatal Application of Kainic Acid on the Release of Taurine from the Neostriatum and the Ipsilateral Substantia Nigra

Intrastriatal application of the selective non-NMDA agonist, KA (100 μM, 20 min), also induced an increase of taurine output from both the STR and ipsilateral SN (Fig. 1). In

Table 2. Effect of DNQX and TTX on taurine output, induced by striatal KA application, simultaneously detected in striatum and substantia nigra

		Taurine Output (% of Basal Output)	
Intrastriatal Treatment	n	Striatal Probe	Nigral Probe
KA 100 μM	8	675 ± 77	159 ± 16
KA 100 μM + DNQX 10 μM	7	564 ± 156	108 ± 20
KA 100 μM + TTX 3 μM	5	895 ± 110	102 ± 19

Data are means ± SEM

contrast to the K^+-stimulated release, the KA-stimulated release of taurine induced locally in the STR differed significantly from that detected distally in the SN. The magnitude was far greater in the STR, showing a peak increase of 675 ± 77% as opposed to 159 ± 16% in SN and showed a longer time course (Fig. 1).

In order to verify the specificity of the increase in taurine output induced by KA, the stimulation was also performed in the presence of a selective non-NMDA receptor antagonist, DNQX (10 μM) or the specific Na+ channel blocker, TTX (3 μM) (Table 2). The release of taurine at the site of stimulation, i.e. the STR, again showed a marked difference to that in the SN. Neither DNQX nor TTX affected the release of taurine within the STR but both abolished the increased release in the SN.

DISCUSSION

Our aim was to ascertain whether taurine, which has been suggested to be present in striatonigral neurones, behaves like GABA; i.e. whether it is simultaneously released from both STR and ipsilateral SN following striatal stimulation. Our results demonstrate that local application of a depolarizing concentration of K^+ induces a simultaneous increase in the output of endogenous taurine from STR and from the ipsilateral SN. The release of taurine at the nigral site had a time course and magnitude similar to that detected in STR. Thus, these data are consistent with the hypothesis that stimulation of the STR by K^+ induces the release of taurine from local axon collaterals of the medium spiny projection neurones and, simultaneously, from their axon terminals in SN. These observations are thus similar to those observed in earlier experiments in which an increase in the release of GABA induced by KA application to the STR was observed in both the STR and SN[3].

Stimulation of the STR with KA elicited a statistically significant, DNQX- and TTX-sensitive increase of taurine release at the distal site in the SN. The release, although smaller in magnitude, followed a pattern similar to that of GABA, in that it was abolished by the addition of DNQX to the striatal perfusate which implies that the release of taurine in the SN is dependent upon the activity of striatonigral neurones triggered by stimulation of striatal DNQX-sensitive receptors. This is consistent with the observed immunocytochemical localization of subunits of the KA/AMPA receptor on striatal neurones[32]. The block induced by DNQX, which is presumed to act via postsynaptic non-NMDA receptors[22], suggests a postsynaptic action of KA directly on the striatal projection neurones. The anatomical evidence that these neurones are the major target of corticostriatal fibers[38] and the observation that they respond electrophysiologically to excitatory amino acids[25] suggests that the effect on taurine release observed in the SN following striatal KA stimulation is largely due to direct stimulation of the striatonigral output neurones. The fact that the application of TTX to the STR blocks the release in the SN demonstrates that the distal release was dependent on the propagation of action potentials along the striatonigral pathway.

In contrast to the release detected in the SN, stimulation of the STR with KA produced an increase in taurine release that was not affected by DNQX or TTX. The magnitude of the release was far greater than in the SN and time course was different. The lack of effect of both DNQX and TTX suggests the involvement of mechanism(s) different from that responsible for either the release of taurine evoked by K^+ or of GABA induced by KA. Such mechanisms appear to be independent of DNQX-sensitive receptors and do not involve the propagation of action potentials. One possibility that may underlie the large increase observed in the STR may be excitotoxicity exerted by the locally applied KA. Excitatory agonists have been reported to cause toxic actions in different regions of the mammalian brain through activation of specific receptors[9,20]. Cellular swelling is an early component of the toxicity produced by these agonists, due to Na^+ entry through receptor-operated channels and subsequent entry of Cl^- plus water[8,36]. Since taurine release has been observed following several stimuli inducing osmotic stress in the rat brain[35,40,44], taurine released by excitatory amino acids may correct osmotic imbalances, thus behaving as an osmolyte. In the present experiments, the KA-induced release of taurine observed in STR could occur partly as the result of normal synaptic activity and partly as a result of excitotoxicity. It is interesting in this context that a part of the KA-induced release of taurine from rat hippocampus, either in *in vitro*[30] or in *in vivo*[33] conditions, is related to excitotoxicity/osmotic stress.

The approximately 50% increase in taurine release seen in the SN, which is presumed to relate to neuronal activity, raises the possibility that in the STR also there was also a DNQX- and TTX-dependent release but that this was so small that it was masked by the massive release (approximately 500%) that was independent of DNQX-sensitive receptors and axon potential propagation.

The observation that taurine, like GABA, can be released from the terminals of striatonigral neurones by stimulation of non-NMDA receptors in the striatum provides further support to the hypothesis that taurine may play a neurotransmitter role in this neuronal system. The physiological significance of the release and the nature of its targets remains to be established. In the SN, however, taurine has been shown: (i) to behave like GABA, when locally applied, causing contralateral turning in the rat[23,24,31], an effect that was blocked by the putative[46] taurine antagonist, 6-aminomethyl-3-methyl-4H-1,2,4-benzothiadiazine-1,1-dioxide (AMBD)[31]; (ii) to behave like glycine, another established inhibitory neurotransmitter, in inhibiting the firing rate of nigral dopaminergic neurones by increasing a specific membrane Cl^- conductance[21]; and, (iii) to bind to a $GABA_A$ receptor subclass, which is particularly enriched in the SN[6].

The striatonigral/striatoentopeduncular inhibitory neurones represent the major striatal projection which conveys the information received from the cortex, via the "direct pathway", along the nigro-thalamo-cortical portion of the cortical-basal ganglia-thalamo-cortical "motor" circuit, which is responsible for the control of movement[1,16] (Figure 1, right panel). Activation of this "direct pathway" is believed to lead to a decrease in the firing rate of the inhibitory output neurones and thus to disinhibition of basal ganglia targets[7], whereas activation of the "indirect pathway", along the pallido/subthalamic-nigro-thalamo-cortical portion of this loop[1,16], would have the opposite effect[1,38,39]. Imbalance between the two pathways is thought to underlie disorders of movement. Indeed overactivity of the "indirect pathway" is associated with the development of parkinsonian motor signs characterized by slowness of movement with increased muscular tone, as in parkinsonian monkeys a lesion of the "indirect pathway" leads to marked reduction of muscular rigidity[17]. It is interesting that the i.c.v. injection of taurine induces a decrease of muscular tone in the rabbit[37]

Although the final targets of taurine released from the striatonigral neurones could be GABA- or glycine-sensitive ion channels, structure-activity correlation (Sgaragli *et al.*, this volume) suggests the existence of specific taurine recognition sites in the brain.

ACKNOWLEDGMENTS

This work was supported by the European Community (BMH1 CT94-1402), CNR-Bilateral Project (95.00833 CT04) and MPI-60%, Rome, Italy.

REFERENCES

1. Albin, R.L., Young, A.B. and Penney, J.B. 1989, The functional anatomy of basal ganglia disorders. *TINS*, 12:366-375.
2. Bianchi, L., Della Corte, L., Federico, S. and Bolam P. 1991, Release of taurine from rat striatum in vitro. *Eur. J. Neurosci.*, Suppl. 4:151.
3. Bianchi, L., Sharp, T., Bolam, J.P. and Della Corte, L. 1994, The effect of kainic acid on the release of GABA in rat neostriatum and substantia nigra. *NeuroReport*, 5:1233-1236.
4. Bolam, J.P., Clarke, D.J., Smith, A.D. and Somogyi, P. 1983, A type of aspiny neurone in the rat neostriatum accumulates [^{3}H]γ-aminobutyric acid: combination of Golgi-staining, autoradiography, and electron microscopy. *J. Comp. Neurol.* 213:121-134.
5. Bolam, J.P., Powell, J.F., Wu J.-Y. and Smith, A.D. 1985, Glutamate decarboxylase-immunoreactive structures in the rat neostriatum: a correlated light and electron microscopic study including a combination of Golgi impregnation with immunocytochemistry. *J. Comp. Neurol.* 237:1-20.
6. Bureau, M.H. and Olsen, R.W. 1991, Taurine acts on a subclass of $GABA_A$ receptors in mammalian brain in vitro. *Eur. J. Pharmacol.* 207:9-16.
7. Chevalier, G. and Deniau, J.M. 1990, Disinhibition as a basic process in the expression of striatal functions. *TINS*, 13:277-280.
8. Choi, D.W. 1987, Ionic dependence of glutamate neurotoxicity. *J. Neurosci.* 7:369-379.
9. Choi, D.W., Koh, J. and Peters, S. 1988, Pharmacology of glutamate neurotoxicity in cortical cell culture: attenuation by NMDA antagonists. *J. Neurosci.* 8:185-196.
10. Clarke, D.J., Smith, A.D. and Bolam, J.P. 1983, Uptake of [^{3}H]taurine into medium-size neurones and into identified stratonigral neurones in the rat neostriatum. *Brain Res.* 289:342-348.
11. Collins, G.G.S. 1974, The rates of synthesis, uptake and disappearance of [^{14}C]taurine in eight areas of the rat central nervous system. *Brain Res.* 76:447-459.
12. Cowan, R.L., Wilson, C.J., Emson, P.C. and Heizmann, C.W. 1990, Parvalbumin-containing GABAergic interneurones in the rat neostriatum. *J. Comp. Neurol.* 302:197-205.
13. Della Corte L., Bianchi L., Federico S. and Michelassi S. 1993, In vivo HPLC estimation of extracellular aspartate, glutamate, taurine and GABA in rat striatum: improved methodology. *Eur. J. Neurosci.* Suppl. 6:261.
14. Della Corte, L., Bolam, J.P. and Smith, A.D. 1987, Uptake, localisation and release of taurine in the rat basal ganglia. In: "The Biology of Taurine", Huxtable, R.J., Franconi, F. and Giotti, A. (eds), pp. 285-294. Plenum Publ. Corp., New York.
15. Della Corte, L., Bolam, J.P., Clarke, D.J., Parry, D.M. and Smith, A.D. 1990, Sites of [^{3}H]taurine uptake in the rat substantia nigra in relation to the release of taurine from striatonigral pathway. *Eur. J. Neurosci.* 2:50-61.
16. De Long, M.R. 1990, Primate models of movement disorders of basal ganglia origin. *TINS*, 13:281-285.
17. De Long, M.R. and Wichmann, T. 1993, Basal ganglia-thalamocortical circuits in parkinsonian signs. *Neuroscience*, 1:18-26.
18. Dray, A. and Straughan, D.W. 1976, Synaptic mechanisms in the substantia nigra. *J. Pharm. Pharmacol.* 28:400-405.
19. Farrant, M. and Webster, R.A. 1989, Compartmental distribution of endogenous amino acids in the substantia nigra of the rat. *Brain Res.* 480:344-348.
20. Frandsen, A., Drejer, J. and Schousboe, A. 1989, Direct evidence that excitotoxicity in cultured neurones is mediated via N-methyl-D-aspartate (NMDA) as well as non-NMDA receptors. *J. Neurochem.* 53:297-299.
21. Häusser, M.A., Yung, W.H. and Lacey, M.G. 1992, Taurine and glycine activate the same Cl^- conductance in substantia nigra dopamine neurones. *Brain Res.* 571:103-108.
22. Honoré, T., Davies, S.N., Drejer, J., Fletcher, E.J., Jacobsen, P., Lodge, D. and Nielsen, D. 1988, Quinaxolinediones: potent competitive non-NMDA glutamate receptor antagonists. *Science*, 201:701-703.

23. Kaakkola, S. and Kääriäinen, T. 1980, Contralateral circling behaviour induced by intranigral injection of taurine in rats. *Acta Pharmacol. Toxicol.* 46:293-298.
24. Kamata, K., Kameyana, T., Okuyama, S., Hashimoto, S. and Aihara, H. 1985, Contralateral circling behaviour induced by intranigral micro injections of taurine and GABA in rats. *Brain Res.* 343:275-282.
25. Kitai, S.T., Kocsis, J.D., Preston, R.J. and Sugimori, M. 1976, Monosynaptic inputs to caudate neurones identified by intracellular injection of horseradish peroxidase. *Brain Res.* 109:601-606.
26. Korf, J. and Venema, K. 1983, Amino acids in the substantia nigra of rats with striatal lesions produced by kainic acid. *J. Neurochem.* 40:1171-1173.
27. Lapper, S.R., Smith, Y., Sadikot, A.F., Parent, A. and Bolam, J.P. 1992, Cortical input to parvalbumin-immunoreactive neurones in the putamen of squirrel monkey. *Brain Res.* 580:215-224.
28. Legay, F., Lecestre, D. and Tappaz, M. 1987, Taurine biosynthesis in rat brain in vivo: lack of relationship with cysteinesulfinate decarboxylase glutamate decarboxylase-associated activity (GAD/CSDII). *J. Neurochem.* 48:340-344.
29. Lombardini, J.B. 1976, Regional and subcellular studies on taurine in the rat central nervous system. In: "Taurine", Huxtable, R.J. and Barbeau, A., eds., pp. 311-326, Raven Press, New York.
30. Magnusson, K.R., Koerner, J.F., Larson, A.A., Smullin, D.H., Skilling, S.R. and Beitz, A.J. 1991, NMDA-, kainate- and quisqualate-stimulated release of taurine from electrophysiologically monitored rat hippocampal slices. *Brain Res.* 549:1-8.
31. Martin, G.E., Bendesky, R.J. and Williams, M. 1981, Further evidence for selective antagonism of taurine by 6-aminomethyl-3-methyl-4H-1,2,4-benzothiadiazine-1,1-dioxide. *Brain Res.* 299:530-535.
32. Martin, L.J., Blackstone, C.D., Levey, A.J., Huganir, R.L. and Price, D.L. 1993, AMPA glutamate receptor subunits are differentially distributed in rat brain. *Neuroscience*, 53:327-358.
33. Menendez, N., Solìs, J.M., Herreras, O., Herranz, A.S. and Martin del Rìo, R. 1990, Role of endogenous taurine on the glutamate analogue-induced neurotoxicity in the rat hippocampus in vivo. *J. Neurochem.* 55:714-717.
34. Palkovits, M., Elekes, I., Lang, T. and Patthi, A. 1986, Taurine levels in discrete brain nuclei of rats. *J. Neurochem.* 47:1333-1335.
35. Pasantes-Morales, H. and Schousboe, A. 1988, Volume regulation in astrocytes: a role for taurine as an osmoeffector. *J. Neurosci.* 20:505-509.
36. Rothman, S.M. 1985, The neurotoxicity of excitatory amino acids is produced by passive chloride influx. *J. Neurosci.* 5:1483-1489.
37. Sgaragli, G.P., Carlà, V., Magnani, M. and Giotti, A. 1978, Homotaurine and muscimol mimic taurine and GABA effects on muscle tone and temperature regulation. *N.-S. Arch. Pharmacol.* 305:155-158.
38. Smith, A.D. and Bolam, J.P. 1990, The neural network of the basal ganglia as revealed by the study of synaptic connections of identified neurones. 1990. *TINS*, 13:259-285.
39. Smith, A.D. and Bolam, J.P. 1991, Convergence of synaptic inputs from the striatum and the globus pallidus onto identified nigrocollicular cells in the rat: a double anterograde labelling study. *Neuroscience*, 44:45-73.
40. Solis, J.M., Herranz, A.S., Herreras, O., Lerma J. and Martin del Rio, R. 1988, Does taurine act as an osmoregulatory substance in the rat brain? *Neurosci. Lett.* 91:53-58.
41. Spears, R.M. and Martin, J.P. 1982, Resolution and brain regional distribution of cysteine sulfinate decarboxylase isoenzyme from hog brain. *J. Neurochem.* 38:985-991.
42. Strømhaugh, J., Skumlien, S., Storm-Mathisen, J. and Ottersen, O.P. 1987, Immunocytochemical demonstration of putative amino acid neurotransmitters in the striatonigral pathway. *Neuroscience*, Suppl. 22:S36.
43. Tossman, U., Jonsson, G. and Ungerstedt, U. 1986, Regional distribution and extracellular levels of amino acids in rat central nervous system. *Acta Physiol. Scand.* 127:533-545.
44. Wade, J.V., Olson, J.P., Samson, F.E., Nelson, S.R. and Pazdernik, T.L. 1988, A possible role for taurine in osmoregulation within the brain. *J. Neurochem.* 51:740-745.
45. Wu, J.Y. 1982, Purification and characterisation of cysteic acid and cysteine sulfinic acid decarboxylase and L-glutamate decarboxylase from bovine brain. *Proc. Natl. Acad. Sci. USA*, 79:4270-4274.
46. Yarbrough, G.G., Singh, D.K. and Taylor, D.A. 1981, Neuropharmacological characterisation of a taurine antagonist. *J. Pharmacol. Exp. Ther.* 219:604-613.

47

DISTRIBUTIONS OF TAURINE, GLUTAMATE, AND GLUTAMATE RECEPTORS DURING POST-NATAL DEVELOPMENT AND PLASTICITY IN THE RAT BRAIN

Kathy R. Magnusson

Department of Anatomy and Neurobiology
Colorado State University
Fort Collins, Colorado 80523

INTRODUCTION

Taurine and glutamate both play roles in the development of the mammalian nervous system[10,17,46]. Taurine is involved in granule cell migration in the developing cerebellum[37,47,48,50] and is in high concentrations within axons of the optic tract before and during the establishment and refinement of synapses[12,34,45]. Glutamate, and one subtype of glutamate receptor, the N-methyl-D-aspartate (NMDA) receptor, have also been shown to be involved in the process of granule cell migration in the cerebellum[10]. NMDA receptors also appear to influence neurite outgrowth and survival of these granule cells[2,32,35]. In the hippocampus, glutamate promotes branching of neurites and synapse formation[3,26,28]. These studies indicate that there is an optimal range of glutamate concentration during development that promotes complexity, but higher concentrations are lethal to the developing neuron[26]. An inhibitory transmitter, γ-aminobutyric acid (GABA), is able to reduce the glutamate-induced regression of dendrites in culture[25], suggesting that a modulatory influence is necessary during development to encourage optimal growth, while preventing excitotoxicity.

Taurine is also a candidate for this modulatory role. The ratio of the concentrations of taurine and glutamate appears to be important in development in many species[36,37]. Both glutamate and taurine are reduced in mutant weaver mice during the period when migration of cerebellar granule cells occurs in normal mice[50,51]. The relationship of their roles is not entirely clear yet, but may be related to the fact that glutamate can be excitotoxic to developing neurons, if not carefully controlled[17,28]. In the adult brain, glutamate toxicity is associated with both Na^+ and Ca^{2+} influxes[4]. Taurine is known to be protective against NMDA-induced Ca^{2+} influxes[14] and in disease states, such as epilepsy[31,49], hypoxia/ischemia[40], and excitotoxicity[8]; all of which seem to involve excitatory amino acids[5,8]. Taurine

Taurine 2, edited by Huxtable et al.
Plenum Press, New York, 1996

is released in adult hippocampus by infusions of glutamate agonists[15,16,22] and our study showed that it is released primarily in response to toxic concentrations of these agonists[22]. Both NMDA and non-NMDA receptor agonists induce neurotoxicity to a greater degree in development than adulthood[29,41] and glutamate-induced release of taurine from a developing brain also is greater than from the adult brain[11,38]. Glutamate appears to be important in many aspects of development and it is possible that at least a part of the role of taurine in development is to protect growing neurons from excess excitation induced by glutamate. The present study was designed to examine changes in glutamate localization and glutamate receptor localization and density and determine whether they correlate with changes in taurine localization during development of the cerebellum and hippocampus. Emphasis was placed on the molecular layers of the cerebellum and CA1 hippocampal region because of the changes in taurine that are distinguishable at the light level in these regions.

One would also expect that because taurine and glutamate are important in developmental processes in the nervous system, they might also be a necessary component of regenerative processes within the adult nervous system. If this is true, taurine and glutamate should increase in concentration in the regenerative region and reestablish a developmental localization. A model of regeneration in the adult brain has been developed and characterized and involves lesioning the entorhinal cortex[6,19,24]. This removal of synaptic inputs induces axonal sprouting in the other, nonlesioned pathways that normally synapse in the molecular layer of the dentate gyrus. This model was used in the present study to determine whether taurine and/or glutamate localizations change in the presence of regenerative processes. The information gained by these studies should give us insights into the potential relationships between taurine and glutamate during periods of intensive neuronal growth.

METHODS

Development Study

Immunohistochemistry. Male and female Sprague-Dawley rats at 7, 14, 28 or 56-60 days of age were anesthetized with halothane and sacrificed by transcardial perfusion of Ca^{2+}-free Tyrode's buffer solution, followed by 4% paraformaldehyde - 2% glutaraldehyde. Brains were sectioned (50 μm) in the coronal plane through the hippocampus and cerebellum with the use of a vibratome. Free-floating sections were processed with the Vectastain avidin-biotin-peroxidase method (Vector Labs, Burlingame, CA) as follows, with three rinses in PBS (pH 7.2) between each incubation step. Tissue sections were incubated overnight with either Tau1 culture supernate (1:100) or Glu2 ascites (1:50,000)[23,27] and normal horse serum (NHS; 1:66), diluted in PBS + 0.3% Triton. Sections were incubated for 30 min in secondary biotinylated antibody (1:1600 in PBS) and 30 min in avidin-biotin-peroxidase complex (1:400 in PBS), followed by reaction with diaminobenzidine (DAB) and hydrogen peroxide to produce a homogeneous brown reaction product. Absorption controls were performed by incubating the Tau1 antibody with taurine conjugated to BSA with glutaraldehyde.

Autoradiography. Littermates of the pups used for immunohistochemistry were anesthetized with Halothane and decapitated. The brains were removed, frozen in dry ice, stored in a -70°C freezer and were sectioned (6 μm) on a Zeiss cryostat. NMDA-displaceable [^{3}H]glutamate, [^{3}H]AMPA, and [^{3}H]kainate binding assays were performed as previously described[21]. [^{3}H]Glutamate binding to Met1 and Met2 sites was performed as described by

Dure and coworkers[7]. Autoradiographic analysis was performed with a Macintosh computer and NIH Image software.

Regeneration Study

Male Sprague-Dawley rats (2-3 months old) received electrolytic lesions in the right entorhinal cortex or were sham-operated and were allowed to survive for 7 or 56 days post-lesioning. Animals were anesthetized with injections of ketamine (85 mg/kg; i.m.), Xylazine (12.5 mg/kg) and atropine (0.2 ml). Entorhinal lesions were performed as described by Scheff and coworkers[39]. Briefly, a small hole was drilled in the calvaria and an electrode lowered into the entorhinal cortex (EC) with the use of a stereotaxic apparatus. The right EC was lesioned at nine separate sites (1 mA; 45 s @ site). Sham-operated animals received identical pre-operative and postoperative treatment except for passage of the electrode and current into the EC. Following the appropriate survival periods, each animal was anesthetized with ketamine and Xylazine (see above) and transcardially perfused with Ca^{2+}-free Tyrode's solution and 4% paraformaldehyde - 0.5% glutaraldehyde. Immunohistochemical staining of the hippocampus and dentate gyrus was performed as described above in the development study. Acetylcholinesterase staining[30] was performed to verify the effectiveness of the lesion in promoting axonal sprouting[39].

RESULTS AND DISCUSSION

Changes in Taurine, Glutamate and Glutamate Receptors in the Developing Cerebellum

Taurine-like immunoreactivity (Tau-LI) was present in Purkinje cell bodies (arrowheads; Fig. 1B) by post-natal day 7 (PN7) in the rat. There was a homogeneous band of Tau-LI that was present above the Purkinje cell bodies and covered the inner half of the molecular layer (ML) on PN7. The external granule cell layer (EGL) was relatively unstained on PN7 (Fig. 1B). By PN14 (Fig. 1C), the Purkinje cell bodies (arrowheads; Fig. 1C) still contained Tau-LI, and presumed Purkinje cell dendrites (arrows; Fig. 1C) were stained more intensely than the rest of the neuropil of the ML. A larger proportion of the ML was labeled with Tau-LI on PN14 than on PN7. The EGL was still relatively unstained on PN14. The entire ML, extending from the Purkinje cell layer to the surface of the cerebellum (dashed line), was immunostained with Tau-LI by PN28 (Fig. 1D) and PN60 (Fig 1D inset), with presumed Purkinje cell dendrites (arrows; Fig. 1D and inset) labeled more intensely than the surrounding neuropil. Purkinje cell bodies (arrowhead; Fig. 1D) were still labeled with Tau-LI. The more intensely labeled profiles in the ML were revealed to be Purkinje cell dendrites in the adult (arrow; Fig. 1D inset). All of the Tau-LI was absorbed out by incubation of the primary antibody with a taurine-carrier protein antigen in all age groups (Fig. 1A; PN7 shown).

Glutamate-like immunoreactivity (Glu-LI) was present in the inner ML and was relatively absent from the EGL on PN7 (Fig. 1F) and PN14 (Fig. 1G), similar to Tau-LI. By PN28 (Fig. 1H) and continuing to adulthood, Glu-LI was homogeneously present in the ML, and extended from the Purkinje cell layer (arrowhead) to the external surface (Fig. 1H, dashed line). There was little evidence of darker staining with Glu-LI in any particular structures in the ML from PN7 to adult (Fig. 1F-H). The Purkinje cells (arrowheads) did not appear to be labeled relative to the ML throughout the post-natal period examined. Glu-LI was completely absorbed out by incubation of sections with glutamate conjugated to carrier

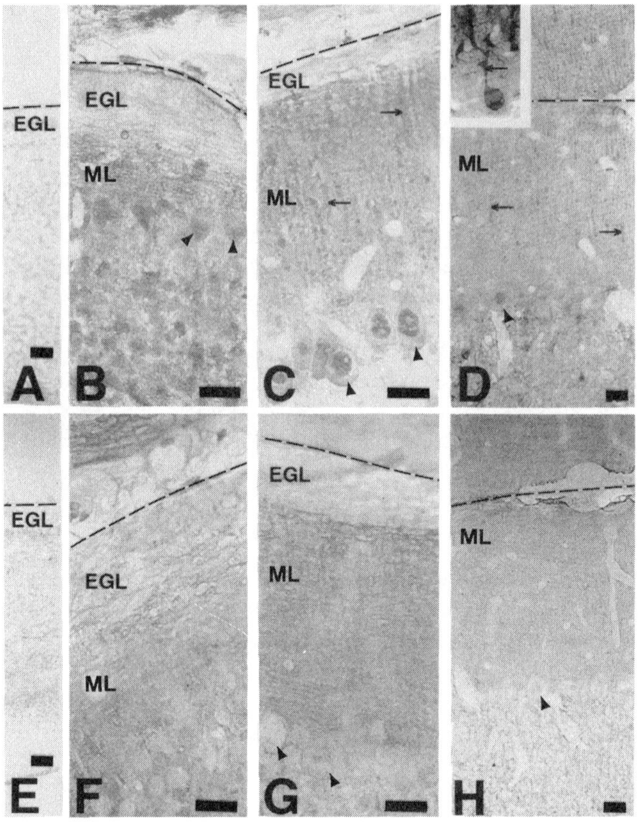

Figure 1. Localization of Tau-LI and Glu-LI in the cerebellum of the rat during development. A) Absorption control for taurine-like immunostaining in the seven day old cerebellum. B-D) Tau-LI in the cerebellum of rats on PN7 (B), PN14 (C), PN28 (D), and PN60 (adult; D inset). E) Absorption control for glutamate-like immunostaining in the seven day old cerebellum. F-G) Glu-LI in the cerebellum of rats on PN7 (F), PN14 (G), and PN28 (H) EGL, external granule cell layer; ML, molecular layer; dashed line, edge of cerebellar folia small arrows, presumed Purkinje cell dendrites, arrowheads, Purkinje cell bodies. Bars = 25 μm.

proteins (Fig. 1E) and was eliminated by omission of any step in the ABC protocol (not shown).

[^{3}H]AMPA binding was heaviest in the molecular layer of the cerebellum (Fig. 2A-D; darkest regions) and was present from PN7 to adulthood. The pattern of binding appeared to widen progressively from PN7 to adulthood (Fig. 2A-D). The EGL (arrow), which was located between the ML (darkest regions) and the edge of the cerebellar folia, generally lacked binding on PN7 and PN14 (Fig. 2A,B). AMPA binding sites appeared to be denser on PN14 and PN28-29 than in the adult, but the trend was not significant (Fig. 2E). Kainate binding was also present in the molecular layer and appeared to be missing from the EGL on PN7 and PN14 (not shown). The Met2 binding site appeared to decrease in density with increasing age in the cerebellar ML. but the change was not significant (Fig. 2E).

The band of Tau-LI widened in the molecular layer over the same time course as the Purkinje cell dendrites are expanding out from the Purkinje cell layer to the surface of the molecular layer[1]. Thus it appeared that taurine may be involved in the development of the dendritic tree of the Purkinje cells. A similar band of Tau-LI is present in kittens during a comparable period of Purkinje cell development[18]. Taurine appeared to be present within the Purkinje cell dendrites by PN14 and was in the Purkinje cell bodies from PN7 to PN60, suggesting that fairly early on in post-natal development, taurine obtained a localization resembling the adult, at least with respect to the Purkinje cells. Immunohistochemical localization of glutamate also followed the outgrowth of the Purkinje cells, but was absent from the cells themselves. The glutamate was likely to be localized within excitatory parallel

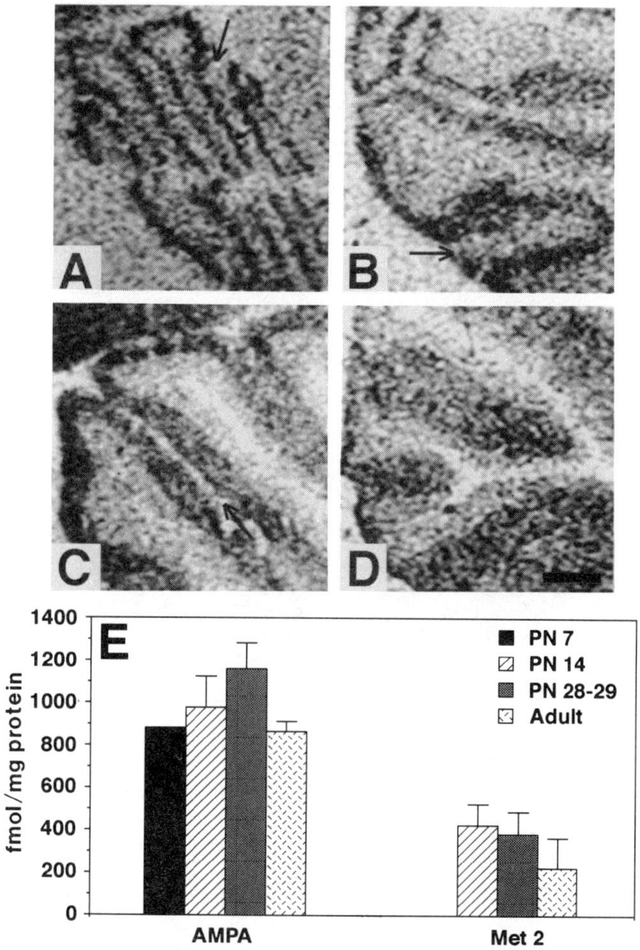

Figure 2. Development of AMPA and Met2 glutamate receptors in the cerebellum. A-D) AMPA receptor autoradiograms in the cerebellum on PN7 (A), PN14 (B), PN28 (C), and PN60 (D). E) Graph depicting changes in density of specific binding (fmol/mg protein) for AMPA and Met2 glutamate receptor subtypes during development of the molecular layer of the cerebellum. arrow, external granule cell layer. Bar = 500 μm.

fibers and synapses, which are formed on these Purkinje cell dendrites as early as PN7[1]. AMPA receptors showed some evidence of remodeling during development, although these changes did not reach significance. The presence of these receptors and glutamate in the ML by PN7 suggests that taurine may be present early on in the Purkinje cells to serve some of the same functions as it does in the adult, such as protection from excitotoxicity[8,13,15,16,22,50]. Release of taurine stimulated by excitatory amino acids is greater in the developing mouse than the adult[38], providing further support for a protective role of taurine during development. The Met2 receptors appeared to decrease in density over the same time course that taurine decreases its concentration[46] and, therefore, may be of interest to study for interrelationships.

Changes in Taurine, Glutamate and Glutamate Receptors in the Developing Hippocampus

The pyramidal cell bodies within the pyramidal cell layer (PCL) and proximal dendrites (large arrows) were labeled with Tau-LI throughout the ages examined (Fig. 3A,B). Tau-LI was fairly homogeneous in intensity of staining within the molecular layer of CA1

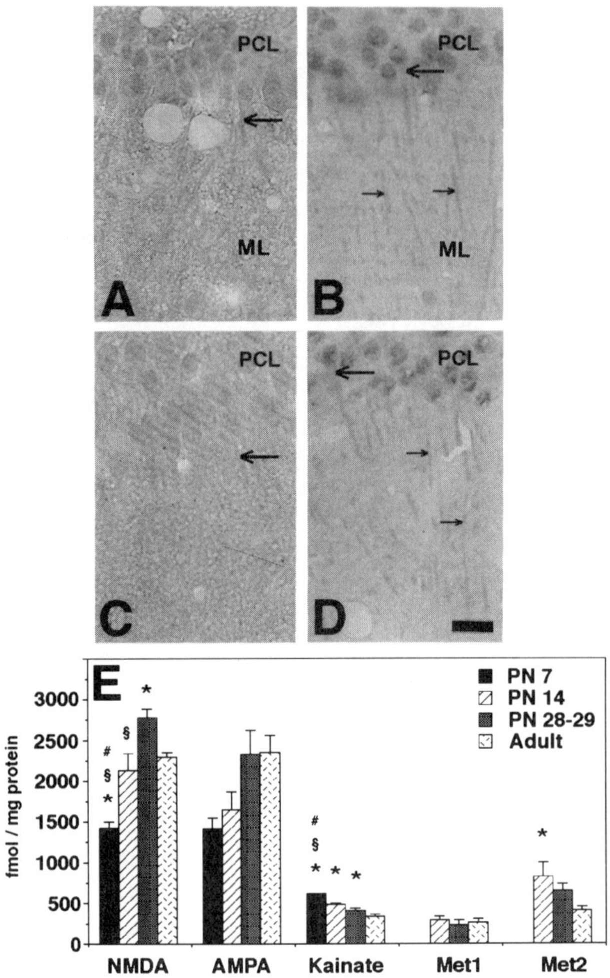

Figure 3. Changes in Tau-LI, Glu-LI and glutamate receptors during development in the CA1 region of the hippocampus. A-D) Changes in localization of Tau-LI (A,B) and Glu-LI (C,D) in CA1 during development at PN7 (A,C) and PN28 (B,D). E) Graph depicting changes in specific binding density (fmol/mg protein) for glutamate receptor subtypes during development in the molecular layer of CA1. PCL, CA1 pyramidal cell layer; ML, CA1 molecular layer; large arrow, proximal dendrite; small arrow, presumed distal dendrite of pyramidal cell. Bars = 25 μm. * = $p < 0.05$ for difference from adult; § = $p < 0.05$ for difference from PN28; # = $p < 0.05$ for difference from PN14.

on PN7 (Fig. 3A) and PN14 (not shown). By PN28, labeled pyramidal cell dendrites (Fig. 3B; small arrows) became apparent above the neuropil staining throughout the ML, which is the pattern seen in adults[20]. All of the Tau-LI was absorbed out by incubation of the primary antibody with a taurine-carrier protein antigen in all age groups (not shown). Glutamate-like immunoreactivity (Glu-LI) exhibited the same spatio-temporal pattern as Tau-LI in the PCL and ML of hippocampal CA1 during the developmental periods studied (Fig. 3C,D).

Taurine and glutamate appeared to be present in pyramidal cell bodies and at least the proximal dendrites by PN7, suggesting that they establish an adult localization within these cells early in development and may already serve some of the same functions as in the adult. The period of homogenous staining for Tau-LI and Glu-LI coincides with a period of outgrowth of branches along the primary dendrite, which begins before PN7, but the majority of branches are formed by PN24[33]. Taurine and glutamate may be localized to other structures besides the dendrites during this period and, if so, the timing suggests that taurine and/or glutamate may be involved in the processes of dendritic branching and/or synapse formation in this region.

NMDA and AMPA receptors were increasing in density in the ML of CA1 between PN7 and PN28-29 (Fig. 3E), presumably setting up the adult synapses, but they also may be

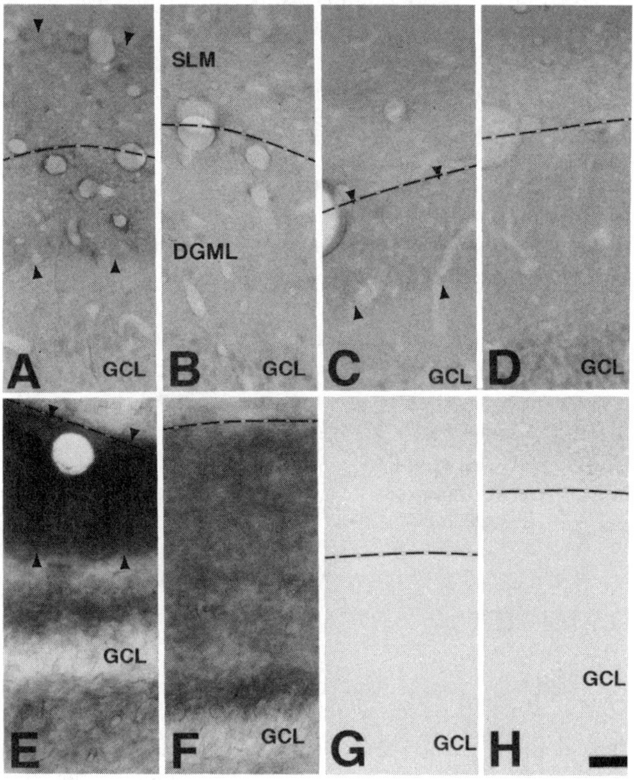

Figure 4. Changes in localization of Tau-LI and Glu-LI during regeneration in the dentate gyrus following unilateral lesioning of the entorhinal cortex. A-D) Tau-LI (A,B) and Glu-LI (C,D) in the dentate gyrus on the lesioned side (A,C) and the control side (B,D) of rats that survived for 7 days post-lesioning. E,F) Acetylcholinesterase staining in the dentate gyrus on the lesioned side (E) and on the control side (F) of a rat following lesioning. G,H) Absorption control for Tau-LI (G) and Glu-LI (H) in the dentate gyrus on the lesioned side. SLM, CA 1 stratum lacunosum/moleculare; DGML, dentate gyrus molecular layer; GCL, granule cell layer; dashed line, hippocampal fissure; arrowheads indicate edge of increased staining. Bar = 50 μm.

involved in developmental processes[3,26,28] at the same time. Kainate and Met2 receptors appeared to also be candidates for involvement in developmental processes in this region and should be investigated with taurine (Fig. 3E). Met1 receptors appeared to show little change during development (Fig. 3G).

Localization of Taurine and Glutamate in the Regenerating Dentate Gyrus

Following unilateral entorhinal lesions, the pattern of acetylcholinesterase (AChE) staining in the molecular layer of the dentate gyrus changed on the lesioned side (Fig. 4E), as compared to the nonlesioned side (Fig. 4F), which indicated that sprouting of cholinergic axons had occurred and verified the success of the lesions[6,42]. The pattern of Tau-LI (Fig. 4A,B) and Glu-LI (Fig. 4C,D) in the molecular layer of the dentate gyrus (DGML) also changed on the lesion side (Fig. 4A,C), as compared to the nonlesioned side (Fig. 4B,D). The intensity of staining in the outer DGML was increased over the staining in the inner DGML on the lesioned side for both taurine and glutamate. Tau-LI also exhibited an increase in the intensity of staining in the stratum lacunosum/moleculare of CA1 (SLM) on the lesioned side (Fig. 4A), that was not seen with Glu-LI (Fig. 4C). These changes in both Tau-LI and Glu-LI were present by 7 days post-lesioning and persisted for at least 56 days after surgery. All the Tau-LI and Glu-LI could be absorbed out by incubation with a taurine-carrier protein antigen (Fig. 4G,H). The pattern of Tau-LI and Glu-LI on the "lesion side" of sham-operated controls (not shown) resembled the staining on the nonlesioned side of lesioned animals (Fig. 4B,D).

Both Tau-LI and Glu-LI appeared to increase in intensity of staining in the regenerative region of the dentate gyrus relative to the rest of the molecular layer, suggesting that they were involved in the axonal sprouting and synaptogenesis processes in this region. The pattern of staining appeared homogeneous with light microscopy, which was similar to the pattern of early post-natal development in the CA1 region. It is interesting that only taurine appeared to increase in the stratum lacunosum/moleculare, which also receives inputs from the entorhinal cortex and undergoes some changes following removal of the perforant pathway inputs[43]. It is possible that, because the pyramidal cell dendrites in CA1 are not as depleted of synapses as the granule cells, the region undergoes different changes that only require an increase in taurine. It is also possible that these increases in staining had more to do with degeneration than regeneration, but degenerating fibers are primarily cleared by 10 days post-lesioning[24], while new synapses are seen up to 7 months after surgery[44].

The Glu-LI changes were opposite to the findings of Grandes and Streit[9], who describe decreased Glu-LI in the regenerative region at 10 days post-lesioning. Our methods differed in many aspects including; method of lesioning, pre- versus post-embedding staining, ABC versus PAP visualization, and thickness of section. We are currently working on confirming our results with post-embedding staining on semi-thin sections to clear up this controversy.

SUMMARY

In summary, taurine and glutamate distributions seemed to be related spatio-temporally during development in the hippocampus and cerebellum and during plasticity of the adult dentate gyrus. In some cases, the amino acids appeared to be setting up adult localizations, while others involved a change in distribution from early development to adulthood that may indicate a related role for taurine and glutamate in dendritic outgrowth and synapse formation. Further elucidation of the subcellular localizations should provide some insight into the functions of taurine and glutamate during these critical periods in development. In addition, there appeared to be developmental patterns of decreased density of kainate and Met2 glutamate receptors that may be worth exploring in terms of interrelationships with taurine.

REFERENCES

1. Altman, J. 1972, Postnatal development of the cerebellar cortex in the rat. II. Phases in the maturation of Purkinje cells and of the molecular layer, *J. Comp. Neurol.* 145:399-464.
2. Balazs, R., Jorgensen, O.S. and Hack, N. 1988, Stimulation of the N-methyl-D-aspartate receptor has a trophic effect on differentiating cerebellar granule cells, *Neuroscience*, 27:437-451.
3. Brewer, G.J. and Cotman, C.W. 1989, NMDA receptor regulation of neuronal morphology in cultured hippocampal neurons, *Neurosci. Lett.*, 99:268-273.
4. Choi, D.W. 1987, Ionic dependence of glutamate neurotoxicity, *J. Neurosci.* 7:369-379.
5. Collins, R.C. 1986, Selective vulnerability of brain: New insights from the excitatory synapse, *Metab. Brain Dis.* 1:231-240.
6. Cotman, C.W., Matthews, D.A., Taylor, D. and Lynch, G. 1973, Synaptic rearrangement in the dentate gyrus: Histochemical evidence of adjustments after lesions in immature and adult rats, *Proc. Natl. Acad. Sci. USA*, 70:3473-3477.
7. Dure, L.S., Young, A.B. and Penney, J.B. 1991, Excitatory amino acid binding sites in the caudate nucleus and frontal cortex of Huntington's disease, *Ann. Neurol.*, 30:785-793.
8. French, E.D., Vezzani, A., Whetsell Jr., W.O. and Schwarcz. R. 1986, Antiexcitotoxic actions of taurine in the rat hippocampus studied in vivo and in vitro, *Adv. Exp. Med. Biol.* 203:349-362.

9. Grandes, P. and Streit, P. 1991, Effect of perforant path lesion on pattern of glutamate-like immunoreactivity in rat dentate gyrus, *Neuroscience*, 41:390-400.
10. Komuro, H. and Rakic, P. 1993, Modulation of neuronal migration by NMDA receptors, *Science*, 260:95-97.
11. Kontro, P. and Oja, S.S. 1989, Release of taurine and GABA from cerebellar slices from developing and adult mice, *Neuroscience,* 29:413-423.
12. Lake, N. 1992, Taurine, GABA and GFAP immunoreactivity in the developing and adult rat optic nerve, *Brain Res.,* 596:124-132.
13. Lehmann, A., Hagberg, H. and Hamberger, A. 1984, A role for taurine in the maintenance of homeostasis in the central nervous system during hyperexcitation, *Neurosci.Lett,* 52:341-346.
14. Lehmann, A., Hagberg, H., Lazarewicz, J.W., Jacobson, I. and Hamberger, A. 1986, Alterations in extracellular amino acids and Ca2+ following excitotoxin administration and during status epilepticus, *Adv. Exp. Med. Biol.,* 203:363-373.
15. Lehmann, A., Isacsson, H. and Hamberger, A. 1983, Effects of in vivo administration of kainic acid on the extracellular amino acid pool in the rabbit hippocampus, *J. Neurochem.* 40:1314-1320.
16. Lehmann, A., Lazarewicz, J.W. and Zeise, M. 1985, N-Methylaspartate-evoked liberation of taurine and phosphonoethanolamine in vivo: Site of release, *J. Neurochem.,* 45:1172-1177.
17. Lipton, S.A. and Kater, S.B.. 1989, Neurotransmitter regulation of neuronal outgrowth, plasticity and survival, *Trends Neurosci.* 12:265-270.
18. Lu, P., Schuller-Levis, G. and Sturman, J.A. 1991, Distribution of taurine-like immunoreactivity in cerebellum of kittens from taurine-supplemented and taurine-deficient mothers, *Int. J. Dev. Neurosci.* 9:621-629.
19. Lynch, G., Deadwyler, S. and Cotman, C. 1973, Postlesion axonal growth produces permanent functional connections, *Science* 180:1364-1366.
20. Magnusson, K.R., Clements, J.R., Wu, J.-Y. and Beitz, A.J. 1989, Colocalization of taurine- and cysteine sulfinic acid decarboxylase-like immunoreactivity in the hippocampus of the rat, *Synapse,* 4:55-69.
21. Magnusson, K.R. and Cotman, C.W. 1993, Age-related changes in excitatory amino acid receptors in two mouse strains, *Neurobiol. Aging,* 14:197-206.
22. Magnusson, K.R., Koerner, J.F., Larson, A.A., Smullin, D.H., Skilling, S.R. and Beitz, A.J. 1991, NMDA-, kainate- and quisqualate-stimulated release of taurine from electrophysiologically-monitored rat hippocampal slices, *Brain Res.,* 549:1-8.
23. Magnusson, K.R., Madl, J.E., Clements, J.R., Wu, J., Larson, A.A. and Beitz, A.J. 1988, Colocalization of taurine- and cysteine sulfinic acid decarboxylase-like immunoreactivity in the cerebellum of the rat with monoclonal antibodies against taurine, *J. Neurosci.,* 8:4551-4564.
24. Matthews, D.A., Cotman, C. and Lynch, G. 1976, An electron microscopic study of lesion-induced synaptogenesis in the dentate gyrus of the adult rat. I. Magnitude and time course of degeneration, *Brain Res.* 115:1-21.
25. Mattson, M.P. and Kater, S.B. 1989, Excitatory and inhibitory neurotransmitters in the generation and degeneration of hippocampal neuroarchitecture, *Brain Res.,* 478:337-348.
26. Mattson, M.P., Lee, R.E., Adams, M.E., Guthrie, P.B. and Kater, S.B. 1988, Interactions between entorhinal axons and target hippocampal neurons: A role for glutamate in the development of hippocampal circuitry, *Neuron,* 1:865-876.
27. McDonald, A.J., Beitz, A.J., Larson, A.A., Kuriyama, R., Sellitto, C. and Madl, J.E. 1989, Colocalization of glutamate and tubulin in putative excitatory neurons of the hippocampus and amygdala: An immunohistochemical study using monoclonal antibodies, *Neuroscience,* 30:405-421.
28. McDonald, J.W. and Johnston, M.V., 199O, Physiological and pathophysiological roles of excitatory amino acids during central nervous system development, *Brain Res. Rev.* 15: 41-70.
29. McDonald, J.W., Silverstein, F.S. and Johnston, M.V. 1988, Neurotoxicity of N-methyl-D-aspartate is markedly enhanced in developing rat central nervous system, *Brain Res.,* 459:200-203.
30. Naik, N.T. 1963, Technical variations in Koelle's histochemical method for demonstrating cholinesterase activity, *Quart. J. Microsc. Sci.* 104:89-100.
31. Oja, S.S. and Kontro, P. 1983, Free amino acids in epilepsy: Possible role of taurine, *Acta Neurol. Scand.,* 67(Suppl. 93):5-20.
32. Pearce, I.A., Cambray-Deakin, M.A. and Burgoyne, R.D. 1987, Glutamate acting on NMDA receptors stimulates neurite outgrowth from cerebellar granule cells, *FEBS Lett.,* 223:143-147.
33. Pokorny, J. and Yamamoto, T. 1981, Postnatal ontogenesis of hippocampal CA1 area in rats. I. Development of dendritic arborisation in pyramidal neurons, *Brain Res. Bull.,* 7:113-120.
34. Politis, M.J. and Ingoglia, N.A. 1979, Axonal transport of taurine along neonatal and young adult rat optic axons, *Brain Res.* 166:221-231.

35. Rashid, N.A. and Cambray-Deakin, M.A. 1992, N-methyl-D-aspartate effects on the growth, morphology, and cytoskeleton of individual neurons in vitro, *Dev. Brain Res.*, 67:301-308.
36. Rassin, D.K. 1982, Taurine, cysteinesulfinic acid decarboxylase and glutamic acid in brain, *Adv. Exp. Med. Biol.* 139:257-268.
37. Roffler-Tarlov, S. and Turey, M. 1982, The content of amino acids in the developing cerebellar cortex and deep cerebellar nuclei of granule cell deficient mutant mice, *Brain Res.*, 247:65-73.
38. Saransaari, P. and Oja, S.S. 1991, Excitatory amino acids evoke taurine release from cerebral cortex slices from adult and developing mice, *Neuroscience*, 45:451-459.
39. Scheff, S.W., Benardo, L.S. and Cotman, C.W. 1980, Decline in reactive fiber growth in the dentate gyrus of aged rats compared to young adult rats following entorhinal cortex removal, *Brain Res.* 199:21-38.
40. Schurr, A., Tseng, M.T., West, C.A. and Rigor, B.M. 1987, Taurine improves the recovery of neuronal function following cerebral hypoxia: An in vitro study, *Life Sci.*, 40:2059-2066.
41. Silverstein, F.S., Chen, R.C. and Johnston, M.V. 1986, The glutamate agonist quisqualic acid is neurotoxic in striatum and hippocampus, *Neurosci. Lett.*, 71:13-18.
42. Stanfield, B.B. and Cowan, W.M. 1982, The sprouting of septal afferents to the dentate gyrus after lesions of the entorhinal cortex in adult rats, *Brain Res.*, 232:162-170.
43. Steward, O. 1977, Topographic organization of the projections from the entorhinal area to the hippocampal formation of the rat, *J. Comp. Neurol.* 167:285-314.
44. Steward, O. and Vinsant, S.L. 1983, The process of reinnervation in the dentate gyrus of the adult rat: A quantitative electron microscopic analysis of terminal proliferation and reactive synaptogenesis, *J. Comp. Neurol.*, 214:370-386.
45. Sturman, J.A. 1979, Taurine in the developing rabbit visual system: Changes in concentration and axonal transport including a comparison with axonally transported proteins, *J. Neurobiol.* 10:221-237.
46. Sturman, J.A. 1993, Taurine in development, *Physiol. Rev.*, 73:119-147.
47. Sturman, J.A., Moretz, R.C., French, J.H. and Wisniewski, H.M. 1985, Postnatal taurine deficiency in the kitten results in a persistence of the cerebellar external granule cell layer: Correction by taurine feeding, *J. Neurosci. Res.* 13:521-528.
48. Sturman, J.A., Moretz, R.C., French, J.H. and Wisniewski, H.M. 1985, Taurine deficiency in the developing cat: Persistence of the cerebellar external granule cell layer, *J. Neurosci. Res.* 13:405-416.
49. Toth, E., Lathja, A., Sarhan, S. and Seiler, N. 1983, Anticonvulsant effects of some inhibitory neurotransmitter amino acids, *Neurochem. Res.*, 8:291-302.
50. Trenkner, E. 1990, The role of taurine and glutamate during early postnatal cerebellar development of normal and weaver mutant mice, *Adv. Exp. Med. Biol.*, 268:239-244.
51. Trenkner, E., Gargano, A., Scala, P. and Sturman, J. 1992, Taurine synthesis in cat and mouse in vivo and in vitro, *Adv. Exp. Med. Biol.*, 315:7-14.

48

RELEASE OF TAURINE AND ITS EFFECTS ON RELEASE OF NEUROTRANSMITTER AMINO ACIDS IN RAT CEREBRAL CORTEX

Yoshinori Kamisaki, Kouichirou Wada, Kentaro Nakamoto, and Tadao Itoh

Department of Clinical Pharmacology
Faculty of Medicine
Tottori University,Yonago 683, Japan

SUMMARY

Taurine has been postulated to function as a neurotransmitter or neuromodulator. The possibility of depolarization-evoked release of taurine from nerve terminals, and the effects of taurine on release of endogenous glutamate (Glu), aspartate (Asp), and γ-aminobutyrate (GABA) were examined using a superfusion of crude synaptosomes prepared from rat cerebral cortex. Taurine contents in cerebral cortex and its synaptosomes were 31.7 and 25.2 nmol/mg protein, respectively. Although the basal rate of taurine release was 35.3 pmol/min/mg protein of synaptosomes (second highest releasing rate), the 2-min stimulation with KCl (30 mM) evoked only a 1.3-fold increase in release of taurine (47.3 pmol/min/mg). The increase was largely Ca^{2+}-dependent. The addition of taurine to the perfusion medium significantly reduced the depolarization-evoked increases in Glu, Asp, and GABA release. The taurine-induced reduction in GABA release was attenuated by phaclofen, a $GABA_B$ antagonist, but not by bicuculline, a $GABA_A$ antagonist. However, these antagonists did not block the effects on Glu and Asp release. These data suggest that taurine may be only partly released from nerve terminals by depolarization in the cerebral cortex, but that taurine may act upon nerve terminals to regulate the release of excitatory and inhibitory amino acid transmitters.

INTRODUCTION

As taurine is the second most abundant amino acid after glutamate (Glu) in the CNS[19], it has been postulated to function as a neurotransmitter. Application of taurine onto brain slices produces GABA-like inhibitory currents in neurons[8,26]. Taurine has been reported to interact with both $GABA_A$ and $GABA_B$ receptor binding sites[2,18,22,24]. Therefore, the membrane-stabilizing effects of taurine, such as anticonvulsant activity, may be mediated through GABA receptors. However, several findings demonstrate that taurine possesses many

Taurine 2, edited by Huxtable et al.
Plenum Press, New York, 1996

distinct actions from those of GABA, such as development and survival of neuronal cells[7], modulation of Ca^{2+} fluxes and transmitter release[9,25], and regulation of osmolarity[31] in the mammalian brain.

Recently, perfusion experiments with synaptosomes indicated the existence of presynaptic heteroreceptors which regulate release of distinct neurotransmitters, in addition to autoreceptors. For example, α_2-adrenoceptors are present on nerve terminals of excitatory amino acid neurons in various regions where they inhibit depolarization-evoked release of aspartate (Asp) and Glu[10,11,13]. Therefore, superfusion of synaptosomes is a suitable procedure for investigating the possibility of stimulus-induced release of taurine from nerve terminals. For this purpose, we developed an assay system to detect pmol levels of taurine in superfusate of synaptosomes prepared from rat cerebral cortex. In addition, in order to clarify the roles of taurine in rat brain, we also investigated the effects of taurine on synaptosomes of amino acidergic nerve terminals in regulating release of Asp, Glu and GABA.

MATERIALS AND METHODS

Determination of Amino Acids

Endogenous Asp, Glu, taurine and GABA content was measured by high performance liquid chromatography using a gradient elusion system (Fig. 1). The analytical method involved precolumn derivatization of amino acids with *o*-phthaldialdehyde and 2-mercaptoethanol in borate buffer (pH 10.5)[15,16]. Fluorescent derivatives of amino acids were separated on a reversed-phase column (Wakosil 5C18) with two mobile phases of A: 40 mM NaH_2PO_4 (pH 7.00), acetonitrile and methanol (90:9:1, v/v) and B: 100 mM Na^+ citrate (pH 3.50), acetonitrile and methanol (59:32:9, v/v) using a programmable pump (L6200, Hitachi) at a flow rate of 1.5 ml/min. Amino acid concentrations were determined from peak

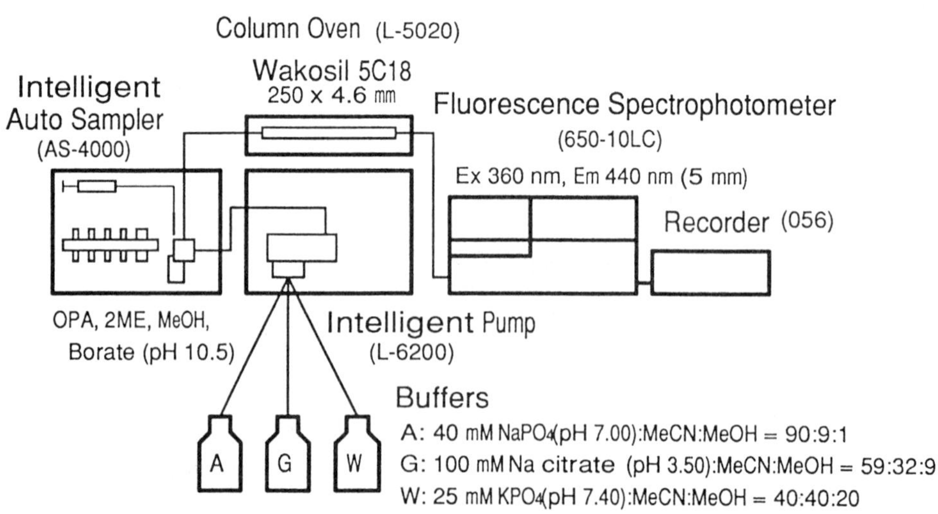

Figure 1. Schematic diagram of assay system for amino acids. Samples were automatically derivatized with *o*-phthaldialdehyde (OPA) and 2-mercaptoethanol (2ME) into fluorescent substances, which were subsequently separated on a reversed phase column by a gradient elusion system with 2 buffers (A, G, and W for wash).

fluorescent intensities and were corrected using 5-aminopentanoic acid (AVA) as internal standard. The protein content was measured with the method of Lowry *et al.*[21] using bovine serum albumin as a standard.

Preparation of Synaptosomes

Crude synaptosomes (P_2 fraction) were prepared according to Gray and Whittaker[5] with minor modifications[12,14,15]. Male Wistar rats, 170-200 g (SLC Inc., Shizuoka), were killed by decapitation. The cerebral cortex was separated on an ice-cold stage as described by Glowinski and Iversen[4] and homogenized in 20 vol of 0.32 M sucrose buffered with 20 mM Tris-HCl (pH 7.5). After centrifugation at 1,000 g for 5 min, synaptosomes were prepared from the supernatant by centrifugation at 12,000 g for 20 min. The pellet was resuspended in a Krebs Ringer buffer (KRB), pH 7.4 (composition in mM: NaCl 135; KCl 4.8; $CaCl_2$ 2; $MgSO_4$ 1.2; KH_2PO_4 1.2; $NaHCO_3$ 25; HEPES, 12.5; glucose, 10), previously equilibrated with O_2/CO_2 (95:5).

Release of Amino Acids

P_2 synaptosomes were superfused as described previously[14,15,17]. The synaptosomes (approximately 1.0 mg protein) were mixed with a gel (0.2 ml) of Sephadex G-10 (Pharmacia) and layered on a Whatman GF/C filter in a perfusion chamber. Superfusion was conducted at a flow rate of 0.5 ml/min with a standard medium of KRB aerated continuously with O_2/CO_2 (95:5) at 37°C. After a 25-min equilibration period, fractions were subsequently collected at 5 min intervals into a tube containing 1.5 nmol of AVA as an internal standard and kept frozen at -80°C until assayed. Forty min after the start of superfusion the synaptosomes were stimulated by 30 mM KCl for 2 min, replacing an equimolar concentration of NaCl. Taurine and/or various agents were added to KRB 10 min before the high KCl stimulation. In Ca^{2+}-free experiments, KRB containing 1.0 mM EGTA instead of $CaCl_2$ was used.

RESULTS

Analysis of Amino Acids

Fig. 2 shows representative chromatograms for the analysis of authentic standards, a perfusate sample of synaptosomes, and a mixture of 30 biological amino acids, which were analyzed by the gradient elusion system (Fig. 1). In addition to Glu, Asp, GABA, and AVA, taurine peaks were completely separated from other contaminants. When various amounts of taurine were analyzed by the system, the peak heights were linearly correlated with the concentration of taurine. Sensitivity was at the subpmol level.

Content of Amino Acids in Whole Tissue and Synaptosomes of Cerebral Cortex

Taurine content in rat cerebral cortex was 31.7 nmol/mg protein, a concentration exceeded only by Glu (87.1 nmol/mg protein), and much higher than that of Asp or GABA (ca. 8 nmol/mg protein)[17]. Although the preparation of the nerve terminal fractions from the cerebral cortex showed lower concentrations than the whole tissue, taurine and Glu were still present in the high concentrations (25.2 and 66.2 nmol/mg protein, respectively)[17].

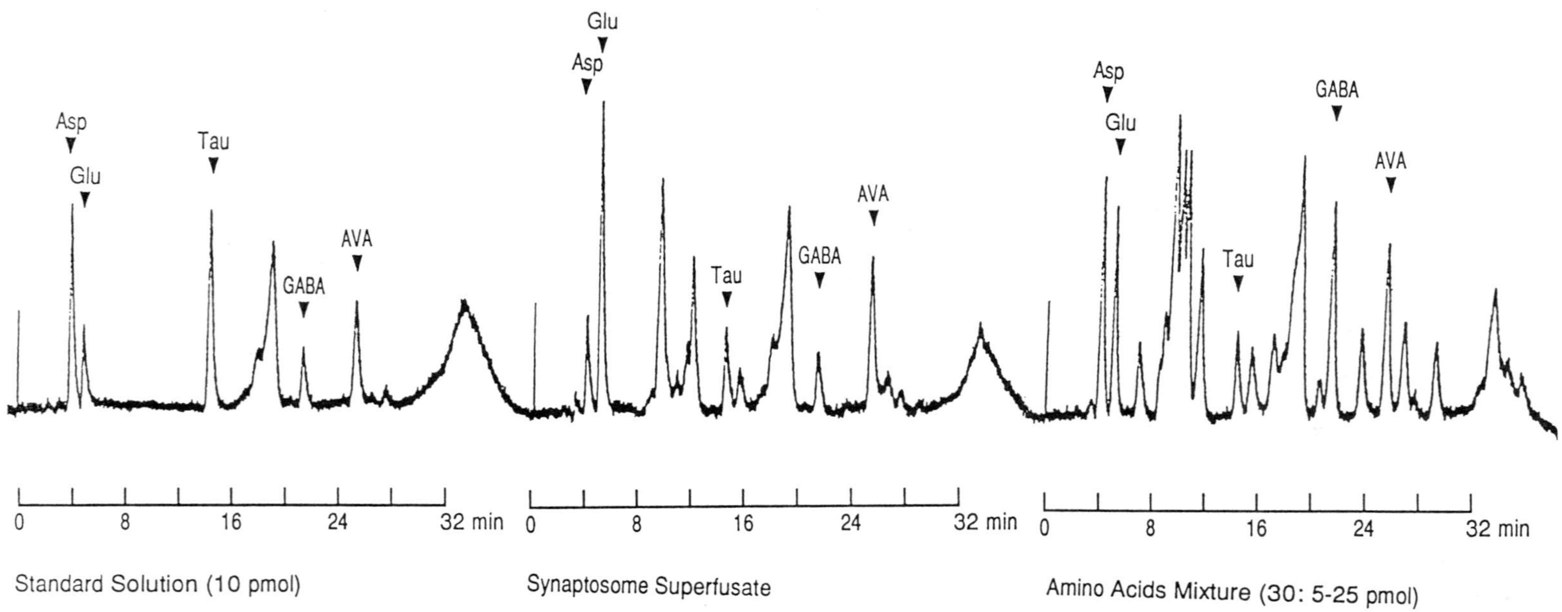

Figure 2. Typical chromatograms of a mixture of 5 authentic amino acids [aspartate (Asp), glutamate (Glu), taurine (Tau), GABA, and 5-aminopentanoic acid (AVA, an internal standard); 10 pmol each], a representative superfusate, and a mixture of 30 biological amino acids (5-25 pmol each). Note the separation of 5 amino acids from other biological contaminants.

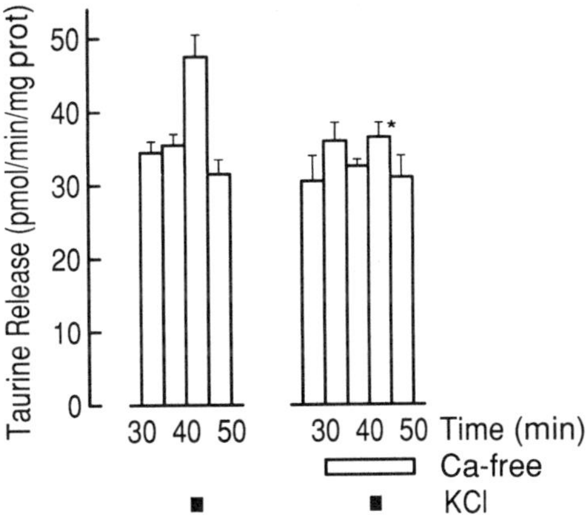

Figure 3. Taurine release from rat cortical synaptosomes. Crude synaptosomes were prepared from rat cerebral cortex and superfused with the standard KRB medium (left columns). Depolarization was evoked by 2-min exposure to 30 mM KCl (solid bars) at 40 min of the perfusion. In Ca^{2+}-free experiments (right columns), 1.0 mM EGTA was added to the perfusion medium containing no $CaCl_2$ 10 min before the depolarization. Sets of columns represent amino acid contents in fractions of before, during and after the depolarization. Release rates of taurine are expressed as means ± SEM from 6 separate experiments. *$p < 0.05$ versus the evoked taurine release in medium containing Ca^{2+}.

Release of Taurine from Synaptosomes

The left columns in Fig. 3 demonstrate release rates of endogenous taurine from superfused rat cortical synaptosomes before, during and after depolarization evoked by 2-min superfusion with medium containing 30 mM KCl. The superfusion with KRB revealed constant basal release of taurine (35.3 pmol/min/mg protein) and depolarization evoked an approximately 30% increase in taurine release (47.3 pmol/min/mg protein). When Ca^{2+} was omitted from perfusion medium, the evoked increase was completely diminished, but the basal release rates were unchanged (right columns in Fig. 3), indicating that the evoked increase in taurine release is mostly Ca^{2+}-dependent.

Effects of Taurine on Release of Other Amino Acids

Fig. 4A-C illustrates release of endogenous Glu, Asp and GABA from superfused rat cortical synaptosomes before, during and after depolarization[14]. The superfusion with KRB revealed constant basal release of Glu, Asp, and GABA from synaptosomes (69.6, 18.4 and 14.6 pmol/min/mg protein, respectively), and their release was raised by the KCl stimulation to 243.7, 49.0, and 28.6 pmol/min/mg protein, respectively. The addition of taurine (10 μM) to KRB did not affect the basal Glu, Asp and GABA release, but significantly reduced the evoked increases of these amino acids by 40, 30 and 60%, respectively (second sets of columns in Fig. 4A-C). These effects of taurine were concentration-dependent and it was more effective on GABA release than on Glu and Asp release[14].

Effects of GABA Receptor Antagonists on the Taurine-Induced Inhibition

Since taurine is reported to interact with GABA receptors[2], the effect of GABA antagonists on taurine-induced reduction was also examined (3rd and 4th sets of columns in Fig. 4A-C). The co-perfusion of synaptosomes with taurine (10 μM) and bicuculline ($GABA_A$ receptor antagonist, 10 μM) or phaclofen ($GABA_B$ receptor antagonist, 10 μM) failed to block the effects of taurine on the evoked release of Glu and Asp. On the other hand, phaclofen partly, but significantly, reversed the inhibition by taurine of the evoked GABA release, although bicuculline did not, suggesting that taurine may regulate GABA release from nerve terminals in cerebral cortex though $GABA_B$ autoreceptors.

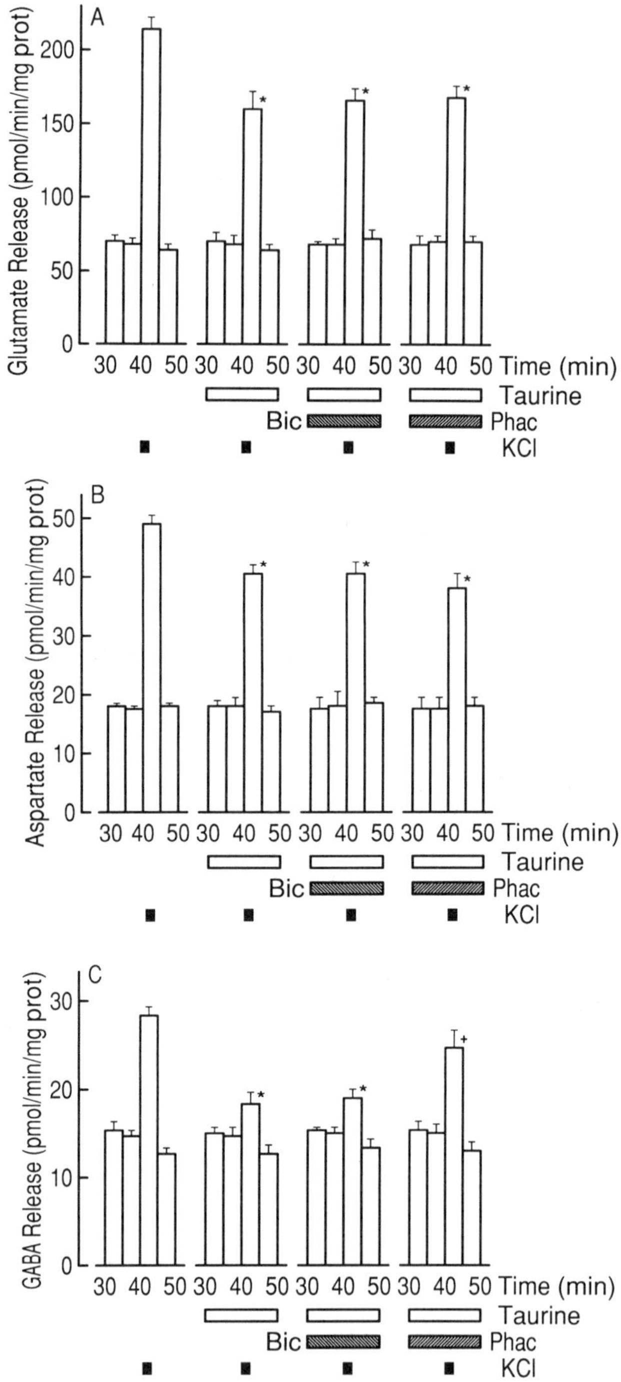

Figure 4. Effects of taurine on glutamate (A), aspartate (B), and GABA (C) release from rat cortical synaptosomes[14]. Crude synaptosomes were perfused with standard KRB and depolarized with 2-min exposure to 30 mM KCl (solid bars). Taurine (10 μM) was added to the superfusion medium (open bars) 30 min after the start of perfusion as described under Materials and Methods. GABA receptor antagonists (10 μM, shaded bars), bicuculline (Bic) or phaclofen (Phac) were also added 10 min before the depolarization. Release rates are expressed as means ± SEM from 6-8 separate experiments. $^{*}p < 0.01$ vs. the evoked release without taurine; $^{\dagger}p < 0.01$ vs. that with taurine alone.

Effects of Taurine on Amino Acid Release in Ca^{2+}-Free Medium

When Ca^{2+} ions were omitted from the superfusion medium, the KCl-evoked increases in Glu, Asp and GABA release were reduced by 70, 60 and 40%, respectively (second sets of columns in Fig. 5A-C)[14]. Their basal release was not affected. Moreover, the addition of taurine (10 μM) to the Ca^{2+}-free medium did not significantly alter the evoked release of Glu, Asp and GABA. Their basal release was not altered either. These data suggest that taurine may inhibit the Ca^{2+}-dependent portion of amino acid release in cerebral cortex.

DISCUSSION

In the present experiments, crude synaptosomes (P_2 fractions) were prepared from rat cerebral cortex. Since taurine has been reported to be released from glial cells by receptor-mediated stimulation[23], it is rather difficult to discuss the subcellular localization of taurine without further fractionation of the crude synaptosomes. However, the concentration of taurine was second highest after Glu in the synaptosome fraction, suggesting that taurine may be stored not only in glial and neuronal cell bodies, but also at nerve terminals in the rat cerebral cortex. In addition, there is a possibility that taurine may play a role similar to Glu as a neurotransmitter.

Similarly to Glu, Asp and GABA, taurine was released from synaptosomes into the superfusate (Figs. 1-2). Release rate under basal conditions was exceeded only by the release of Glu (Fig. 3). However, high K^+ stimulation evoked a significant, but only a 30%, increase in taurine release from synaptosomes in a Ca^{2+}-dependent manner. To date, it has not been established whether depolarization may elicit taurine release from brain tissue. An *in vivo* dialysis experiment revealed that depolarization with either high K^+ or veratridine evoked a 2.5-3.5 fold increase in taurine release in the hypothalamus and that taurine release was reduced by treatment with tetrodotoxin[29]. On the other hand, stimulation with KCl caused no effect on taurine efflux from cerebral cortex slices[20]. Together with these data, the present results suggest that taurine may be only partly released from nerve terminals by a depolarizing nerve impulse.

On the other hand, HEPES in KRB has been reported to interfere with uptake of taurine into tissues[28]. However, our continuous perfusion system prevented reuptake of released substances from synaptosomes[14,15,17]. In addition, endogenous taurine released into perfusate was determined in our present studies, differing from previous procedures in which radioactive taurine was taken up and released. Therefore, the possibility that taurine exists as a neurotransmitter in nerve terminals may be low in rat cerebral cortex, although taurine is abundantly stored in and released from nerve terminals. However, taurine may play an important role as a neuromodulator whose release is probably regulated through the activation of presynaptic receptors located in taurine-containing nerve terminals.

Because of the high levels of taurine in cerebral cortex, we investigated whether taurine may regulate release of other amino acid transmitters from nerve terminals. The synaptosomes were superfused with KRB in the absence or presence of taurine. Taurine failed to affect basal release of Asp, Glu and GABA. However, the depolarization-evoked increases in Glu, Asp and GABA release were significantly inhibited by taurine (Fig. 4). In addition, taurine did not cause further inhibition of the evoked release of Glu, Asp and GABA under Ca^{2+}-free conditions (Fig. 5), suggesting that taurine may modify the Ca^{2+}-related mechanism of amino acid release through taurine-specific sites. The potency of inhibition by taurine was also greater for GABA release (IC_{50}, 5 nM) than for Asp and Glu release (IC_{50} 0.8 μM and 1 μM, respectively)[14]. The inhibitory effects of taurine on GABA release is in accordance with previous reports that taurine inhibits [^{3}H]GABA release from brain slices[9,25]. Moreover, the present experiments using

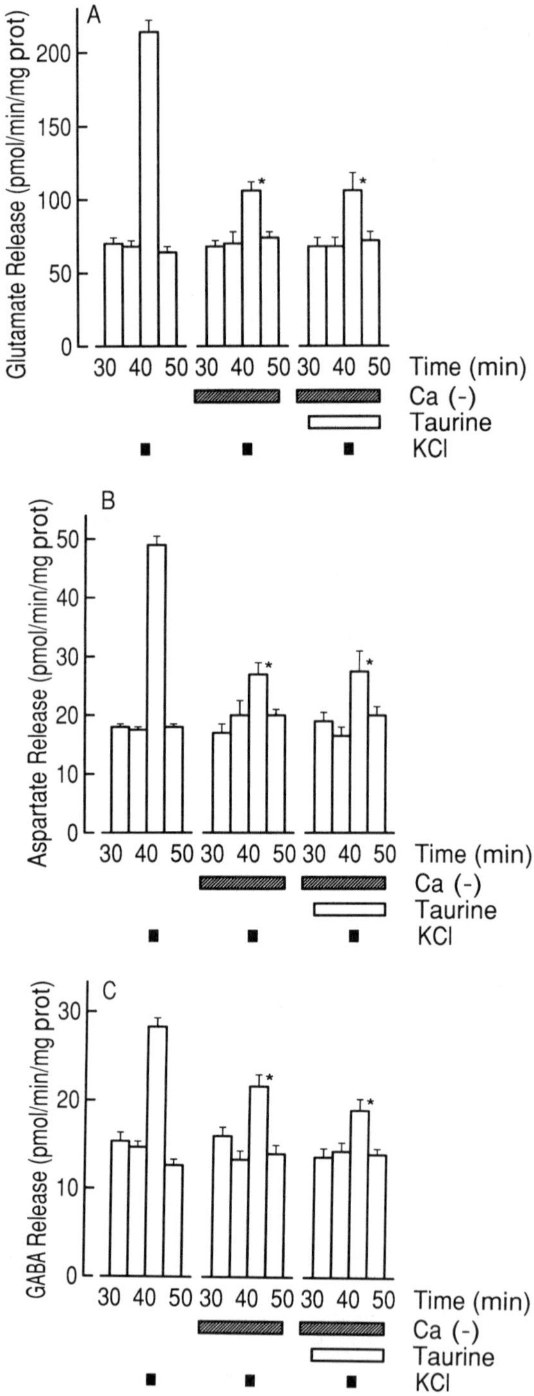

Figure 5. Effects of taurine on Ca^{2+}-dependent release of glutamate (A), aspartate (B) and GABA (C)[14]. Synaptosomes were superfused by standard KRB medium or KRB without Ca^{2+} (shaded bars) and stimulated with 30 mM KCl (solid bars). Taurine (10 μM) was added to the medium 10 min before the depolarization. Values are means ± SEM of 4-8 experiments. *$p < 0.01$ vs. the evoked release in KRB containing Ca^{2+}.

synaptosomes of nerve terminals indicate that the taurine-induced inhibition is mediated through the presynaptic sites of GABAergic nerves. Since taurine has been reported to interact with GABA receptors which are classified as $GABA_A$ and $GABA_B$ subtypes[2,18,22,24], we tried to distinguish these inhibitory actions of taurine from those of GABA by the addition of GABA receptor antagonists to the superfusion medium containing taurine. The effect of taurine on GABA release was counteracted by phaclofen (a $GABA_B$ receptor antagonist) but not by bicuculline (a $GABA_A$ receptor antagonist) (Fig. 4C). However, there have been several reports indicating that GABA release is regulated through presynaptic $GABA_B$ autoreceptors. For example, electrophysiological studies revealed that baclofen (a $GABA_B$ receptor agonist) reduced the inhibitory synaptic conductance of postsynaptic membranes and that this effect is not antagonized by phaclofen[3,6]. However, it was also reported that phaclofen-sensitive receptors regulate [^{3}H]GABA release from synaptosomes preincubated with radioactive substances[1], although endogenous GABA release was not measured. In the present studies, taurine potently inhibited depolarization-evoked GABA release and was antagonized by phaclofen. Therefore, together with previous reports, these data suggest that the inhibitory effect of taurine on GABA release may be mediated through $GABA_B$ autoreceptors, which can be bound with and activated by taurine.

On the other hand, μM concentrations of taurine were necessary to inhibit the depolarization-evoked release of Glu and Asp[14]. The inhibition was not antagonized by either bicuculline or phaclofen (Fig. 4A-B). These results suggest that the inhibition may be mediated through taurine-specific sites but not through GABA receptors. However, there are phaclofen-insensitive subtypes among $GABA_B$ receptors[3,6]. In addition, it was reported that GABA heteroreceptors regulating release of neurotransmitters are present on nerve terminals of excitatory amino acid and peptide neurons[27,30,32]. Therefore, it may be difficult to confirm which sites mediate the regulation of excitatory amino acid release for lack of specific antagonists against presynaptic receptors.

In conclusion, the present observations provide evidence that taurine may be only partly released from nerve terminals by depolarization in the cerebral cortex, but that taurine may play an important roles as a neuromodulator in regulating release of excitatory and inhibitory amino acid transmitters through different mechanisms.

ACKNOWLEDGMENT

This study was supported in part by a grant from Taisho Pharmaceutical Co. (Tokyo, Japan).

REFERENCES

1. Bonanno, G., Pellegrini, G., Asaro, D., Fontana, G., and Raiteri, M. 1989, $GABA_B$ autoreceptors in rat cortex synaptosomes: response under different depolarizing and ionic conditions, *Eur.J.Pharmacol.* 172:41-49.
2. Bureau, M.H. and Olsen, R.W. 1991, Taurine acts on a subclass of $GABA_A$ receptors in mammalian brain in vitro, *Eur.J.Pharmacol.Mol.Pharmacol.* 207:9-16.
3. Davies, C.H., Davies, S.N., and Collingridge, G.L. 1990, Paired-pulse depression of monosynaptic GABA mediated inhibitory postsynaptic responses in rat hippocampus, *J.Physiol.* 242:513-531.
4. Glowinski, J. and Iversen, L.L. 1966, Regional studies of catecholamines in rat brain. I. The disposition of [^{3}H]norepinephrine, [^{3}H]dopamine and [^{3}H]dopa in various regions of the brain, *J.Neurochem.* 13:655-669.
5. Gray, E.G. and Whittaker, V.P. 1962, The isolation of nerve endings from brain: an electron microscopic study of cell fragments derived by homogenization and centrifugation, *J.Anat.* 96:79-88.

6. Harrison, N.L. 1990, On the presynaptic action of baclofen at inhibitory synapses between cultures rat hippocampal neurons, *J.Physiol.* 422:433-446.
7. Hayes, K.C., Carey, R.E., and Schmidt, S.Y. 1975, Retinal degeneration associated with taurine deficiency in the cat, *Science*, 188:949-951.
8. Häusser, M.A., Yung, W.H., and Lacey, M.G. 1992, Taurine and glycine activate the same Cl^- conductance in substantia nigra dopamine neurones, *Brain Res.* 571:103-108.
9. Huxtable, R.J. 1989, Taurine in the central nervous system and the mammalian actions of taurine, *Prog.Neurobiol.* 32:471-533.
10. Kamisaki, Y., Hamada, T., Maeda, K., Ishimura, M., and Itoh, T. 1993, Presynaptic α_2 adrenoceptors inhibit glutamate release from rat spinal cord synaptosomes, *J.Neurochem.* 60:522-526.
11. Kamisaki, Y., Hamahashi, T., Hamada, T., Maeda, K., and Itoh, T. 1992, Presynaptic inhibition by clonidine of neurotransmitter amino acid release in various brain regions, *Eur.J.Pharmacol.* 217:57-63.
12. Kamisaki, Y., Hamahashi, T., Mita, C., and Itoh, T. 1991, D-2 dopamine receptors inhibit release of aspartate and glutamate in rat retina, *J.Pharmacol.Exp.Ther.* 256:634-638.
13. Kamisaki, Y., Hamahashi, T., Okada, C.M., and Itoh, T. 1991, Clonidine inhibition of potassium-evoked release of glutamate and aspartate from rat cortical synaptosomes, *Brain Res.* 568:193-198.
14. Kamisaki, Y., Maeda, K., Ishimura, M., Omura, H., and Itoh, T. 1993, Effects of taurine on depolarization-evoked release of amino acids from rat cortical synaptosomes, *Brain Res.* 627:181-185.
15. Kamisaki, Y., Maeda, K., Ishimura, M., Omura, H., Moriwaki, Y., and Itoh, T. 1994, No enhancement by nitric oxide of glutamate release from P2 and P3 synaptosomes of rat hippocampus, *Brain Res.* 644:128-134.
16. Kamisaki, Y., Takao, Y., Itoh, T., Shimomura, T., Takahashi, K., Uehara, N., and Yoshino, Y. 1990, Determination of gamma-aminobutyric acid in human cerebrospinal fluid by isocratic high-performance liquid chromatography, *J.Chromatogr.* 529:417-423.
17. Kamisaki, Y., Wada, K., Nakamoto, K., and Itoh, T. 1996, Effects of taurine on GABA release from synaptosomes of rat olfactory bulb, *Amino Acids*, in press.0
18. Kontro, P. and Oja, S.S. 1990, Interactions of taurine with $GABA_B$ binding sites in mouse brain, *Neuropharmacology*, 29:243-247.
19. Lombardini, J.B. 1976, Regional and subcellular studies on taurine in the rat central nervous system, in: "Taurine", Huxtable, R. and Barbeau, A. eds., Raven Press, New York, pp. 311-326.
20. Low, R.O. 1994, Effects of extracellular bicarbonate ions and pH on volume-regulatory taurine efflux from rat cerebral cortical slices in vitro: evidence for separate neutral and anionic transport mechanisms, *Biochim.Biophys.Acta*, 1224:377-383.
21. Lowry, O.J., Rosebrough, N.J., Farr, A.L., and Randall, R.J. 1951, Protein measurement with folin phenol reagent, *J.Biol.Chem.* 193:265-275.
22. Malminen, O. and Kontro, P. 1986, Modulation of the GABA-benzodiazepine receptor complex by taurine in rat brain membranes, *Neurochem.Res.* 11:85-94.
23. Martin, D.L. and Shain, W. 1993, β-Adrenergic-agonist stimulated taurine release from astroglial cells is modulated by extracellular $[K^+]$ and osmolarity, *Neurochem.Res.* 18:437-444.
24. Medina, J.H. and De Robertis, E. 1984, Taurine modulation of the benzodiazepine-γ-aminobutyric acid receptor complex in brain membranes, *J.Neurochem.* 42:1212-1217.
25. Namima, M., Okamoto, K., and Sakai, Y. 1983, Modulatory action of taurine on the release of GABA in cerebellar slices of the guinea pig, *J.Neurochem.* 40:1-9.
26. Okamoto, K., Kimura, H., and Sakai, Y. 1983, Taurine-induced increase of the Cl^- conductance of cerebellar Purkinje cell dendrites in vitro, *Brain Res.* 259:319-323.
27. Pende, M., Lanza, M., Bonanno, G., and Raiteri, M. 1993, Release of endogenous glutamic and aspartic acids from cerebrocortex synaptosomes and its modulation through activation of a γ-aminobutyric $acid_B$ ($GABA_B$) receptor subtype, *Brain Res.* 604:325-330.
28. Petegnief, V., Lleu, P.-L., Gupta, R.C., Bourguignon, J.-J., and Rebel, G. 1995, Taurine analog modulation of taurine uptake by two different mechanisms in cultured glial cells, *Biochem.Pharmacol.* 49:399-410.
29. Singewald, N., Guo, L., and Philippu, A. 1993, Taurine release in the hypothalamus is altered by blood pressure changes and neuroactive drugs, *Eur.J.Pharmacol.* 240:21-27.
30. Travagli, R.A., Ulivi, M., and Wojcik, S.J. 1991, γ-Aminobutyric acid-B receptors inhibit glutamate release from cerebellar granule cells: consequences of inhibiting cyclic AMP formation and calcium influx, *J.Pharmacol.Exp.Ther.* 258:903-909.
31. Wade, J.V., Olson, J.P., Samson, F.E., Nelson, S.R., and Pazdernik, T.L. 1988, A possible role for taurine in osmoregulation within the brain, *J.Neurochem.* 51:740-745.
32. Zhang, S.J. and Jackson, M.B. 1993, GABA-activated chloride channels in secretory nerve endings, *Science*, 259:531-534.

TAURINE-INDUCED SINGLE-CHANNEL CURRENTS IN CULTURED RAT CEREBELLAR GRANULE CELLS

M.-L. Linne[1], T. O. Jalonen[2], P. Saransaari,[1] and S. S. Oja[1,3]

[1] Tampere Brain Research Center
Medical School, University of Tampere
Box 607, FIN-33101 Tampere, Finland
[2] Division of Neurosurgery
Albany Medical College
A-60, Albany, New York 12208
[3] Department of Clinical Physiology
Tampere University Hospital
Box 2000, FIN-33521 Tampere, Finland

INTRODUCTION

Taurine increases Cl^- conductance across neuronal membranes, leading to changes in neuronal excitability[12,25]. However, the exact mechanisms of these taurine effects are not yet known. For instance, taurine may act via metabotropic receptors coupled to a second-messenger system or via ionotropic receptors gating ion channels. Indeed, taurine has been shown to interfere with the binding of GABA to both ionotropic $GABA_A$[22] and metabotropic $GABA_B$[19] receptors and to displace strychnine from ionotropic glycine receptors[17]. Moreover, the existence of a specific taurine receptor is not excluded[18,20,27].

Our objective was to investigate the properties of taurine-induced single-channel currents in cultured rat cerebellar granule cells. The existence of voltage-gated K^+ and Cl^- channels and ligand-gated (GABA and glycine) Cl^- channels has been demonstrated in cultured rat cerebellar granule cells[2,6,14,16] and in inhibitory interneurons in a mouse cerebellar granule cell culture[11]. So far, the only published studies on the action of taurine on the single-channel currents in neuronal membranes have been made on mouse spinal cord neurons[24]. To our knowledge, the present study is the first on taurine-sensitive single-channel activity in brain neurons. Preliminary results of this work have been presented in abstract form[15,21].

Taurine 2, edited by Huxtable et al.
Plenum Press, New York, 1996

MATERIALS AND METHODS

Cell Preparation

Primary cultures of granule cells were prepared from the cerebella of 7-day-old Wistar rats[9]. The cerebella were cut into pieces and dissociated by mild trypsinization (0.025%, w/v), followed by trituration in DNAase solution (0.004%, w/v) containing a trypsin inhibitor (0.03%, w/v). The cells were then suspended in an enriched medium (Dulbecco) and cultured on cover slips coated with poly-L-lysine in 35-mm tissue culture dishes. After 48 h in culture 10 μM cytosine-arabinoside was added. The cells were used for normal control recordings after 5-8 days and for taurine experiments after 6-7 days in culture. Cultured granule cells start frequent firing of action potentials after 7 days in culture. The cell resting potentials become increasingly negative during the first days, being likewise stabilized to around -67 mV after 7 days in culture[6].

Solutions

The extracellular solution (EC) bathing the cells and in the pipette contained (in mM): NaCl 137, KCl 5, $MgCl_2$ 1.2, $CaCl_2$ 1, KH_2PO_4 0.44, *N*-tri(hydroxymethyl)methyl-2-aminoethanesulfonic acid (TES) 20, pH 7.4. Taurine and strychnine were dissolved in EC and pH adjusted with 1 mM NaOH.

Electrophysiology

The cell-attached (CA) and inside-out (IOP) patch configurations of the single-channel patch-clamp technique were used to record ionic currents[7]. The signals were recorded using a List EPC-7 patch-clamp amplifier (List-Medical, Darmstadt-Eberstadt, Germany) (bandwidth 100 kHz) and filtered with an EPC-7 built-in 3-pole Bessel filter at 10 kHz. The signal was stored on a Racal Store 4DS FM tape recorder (Racal Recorders Ltd, Southampton, UK) (with bandwidth 1.25 kHz, Tchebychev). Glass microelectrodes were pulled with the Mecanex BB-CH microelectrode puller (Nyon, Switzerland) from borosilicate glass capillaries (GC120F-10; Clark Electromedical Instruments, Reading, UK). The electrodes had resistances of 5-12 MΩ when filled with EC solution and immersed in the bath. The liquid junction potentials were corrected by the offset compensation of the amplifier. All experiments were performed at room temperature (22-24°C).

Analysis

For analysis the recorded signals were transferred to a computer hard drive using a TL-1 DMA interface (sampling rate 10 kHz; Axon Instruments, Inc., Foster City, CA, USA). The analysis was made using the pCLAMP software (version 5.1.1; Axon Instruments, Inc.). In the more detailed analysis, the burst-type activity was defined as a group of channel openings where the durations of channel closed times within a burst are shorter than could be resolved (<0.8 ms).

RESULTS

Taurine Application to the Bath

Each recording was first followed for 5 min to test whether any spontaneous activation of ion channels occurred (pre-exposure recording). The potential was kept at 0 mV

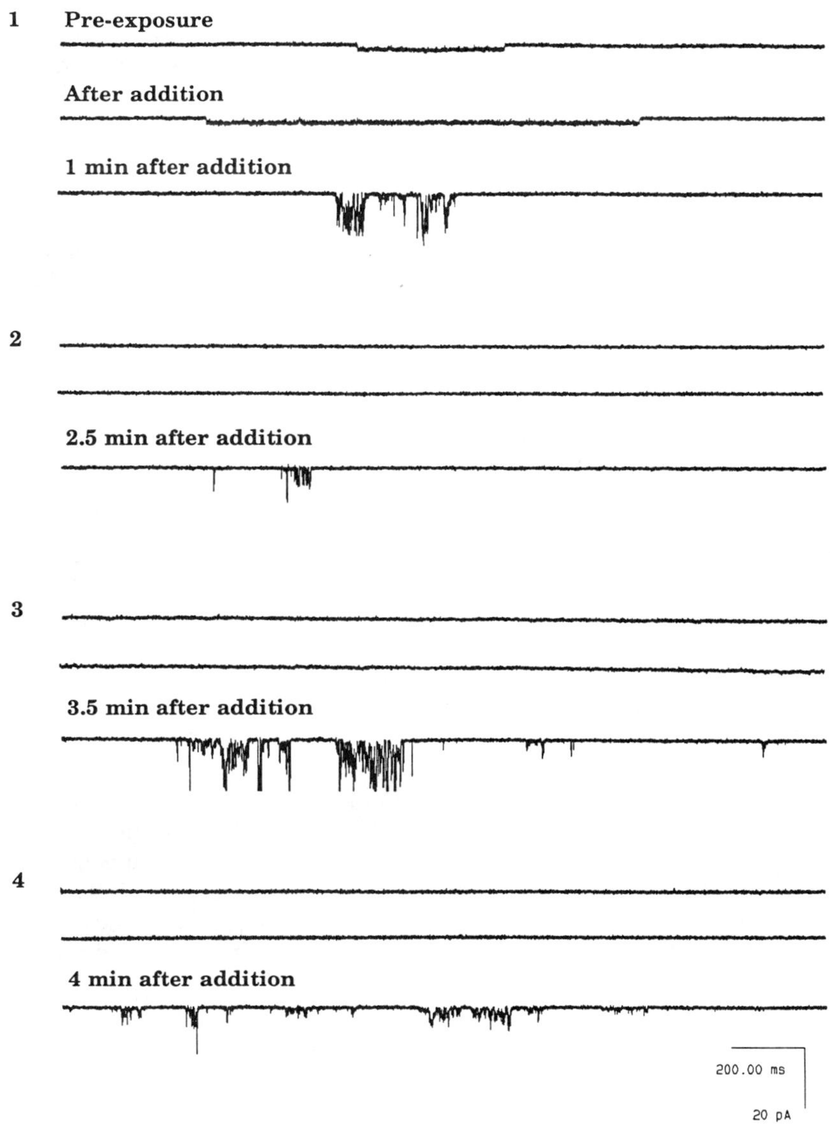

Figure 1. Delayed activation evoked by taurine in the bath. After a 5-min pre-exposure recording, 10 μM taurine (pH 7.4) added to the chamber induced a large inward current (CA recording at $V_{pipette}$ = 0 mV, EC in the bath and pipette). Traces of 2 s from 4 recordings are shown. The channels open downwards. The first trace in each case represents the pre-exposure recording, the second is taken immediately after the taurine addition and the third shows the first activation. The burst-like activity normally lasted 0.5-2.5 min followed by complete inactivation. In the first recording, channel activity different from that induced by taurine is also seen (both in the pre-exposure trace and in the trace after the application of taurine).

pipette potential ($V_{pipette}$), which reflects the cell resting potential in the CA recording configuration (typically between -50 and -60 mV in the granule cells)[6,14]. In some recordings single-channel activity different from the taurine-induced activity was already seen during the 5-min pre-exposure recording (Fig. 1). Spontaneous impulse current waveforms due to action potential generation were occasionally seen in 7-day-old cultures in the CA voltage clamp mode at $V_{pipette}$ = 0 mV. Their maximum amplitudes were 56-58 pA. In control experiments in the CA configuration the application of plain EC (the same volume as in subsequent taurine experiments) to the recording chamber after 1-5 min pre-exposure recording did not induce current activation within 5 min (not shown).

In the CA configuration the application of 10 μM to 10 mM taurine at pH 7.4 to the recording chamber induced a delayed burst-type activation of an inward current in 18 cells out of a total 25 tested. The maximum burst-like current activation was obtained with taurine concentrations ≤100 μM. With 10 μM taurine (pH 7.4), currents were activated in 7 out of 8 recordings. Samples of them are shown in Fig. 1. The delay before the onset of activity was typically 1-4 min. Single-channel events occurred in bursts with rapid transitions between different open channel levels. Bursts (channel openings) also appeared in clusters. The burst-like activity lasted 0.5-2.5 min, normally followed by complete inactivation. The maximum burst current amplitudes were from 15 to 40 pA. One burst seemed to consist of many individual channels opening simultaneously.

In experiments in which taurine was added to the cytoplasmic face of an inside-out patch (IOP, i.e., an intact cell was not present), no burst-type current activation was detected with symmetrical solutions (EC in the bath and pipette) at $V_{pipette}$ = 0 mV (n=9) or in a separate experiment at $V_{pipette}$ = +60 mV. In a separate set of experiments the cells were first incubated with 10 μM strychnine for 5-10 min. The CA configuration was then formed and 10 μM taurine (pH 7.4) applied to the recording chamber. No burst-type activity was detected within 5 min (n=3, not shown).

Taurine in the Pipette

In the CA recordings at $V_{pipette}$ = 0 mV, from 10 μM to 10 mM taurine (pH 7.4) in the recording pipette induced a burst-type transient activation of an inward current in 8 cells out of a total 10 tested. Taurine at pH 7.4 seemed to be most effective in activating the current flow (Figs. 2a and 3). Alkaline taurine solution (pH 8.0) in the pipette induced one burst of openings in 1/5 cells tested (Fig. 2b). Acidic taurine solution (pH 6.8) in the pipette did not induce activity (n=2). There was no taurine-induced activation in other experiments in which the patch was excised immediately after seal formation (n=2).

In a separate set of experiments the cells were first incubated for 5 min with 10 μM strychnine in EC solution (pH 7.4). After seal formation (CA) no channel activation was detected with 10 μM taurine (pH 7.4) in the recording pipette (n=4) (Fig. 2c).

DISCUSSION

In spinal cord neurons 2 μM GABA and 5-40 μM taurine activate a Cl^--dependent single-channel current which is blocked by 10 μM strychnine but not by 50 μM bicuculline[24]. This current was inwardly directed at $V_{pipette}$ = -80 mV in the outside-out patch-clamp configuration. Both GABA and taurine were found to activate Cl^- channels with multiple conductance states. In these same neurons 6-aminomethyl-3-methyl-4H-1,2,4-benzothiadiazine-1,1-dioxide (TAG) antagonized taurine-activated Cl^--permeable ion channels[23]. Single-channel currents activated by 0.5 μM GABA in outside-out patches of cultured rat cerebellar granule cells appear in bursts and show two conductance states (19 and 31 pS)

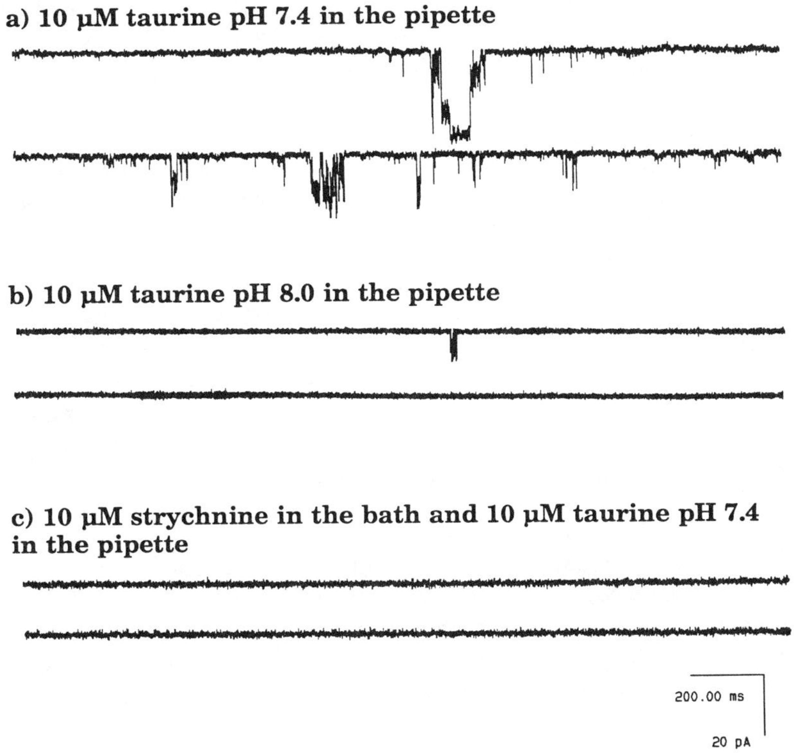

Figure 2. (a) Transient activation evoked by taurine in the pipette. A CA recording with EC plus taurine (pH 7.4) in the pipette and EC solution in the bath (pH 7.4). After seal formation a burst-type transient inward current was detected. (b) With alkaline taurine solution (pH 8.0) in the pipette only one burst was seen in one recording. (c) After the 5-min incubation with 10 µM strychnine 10 µM taurine induced no current activation.

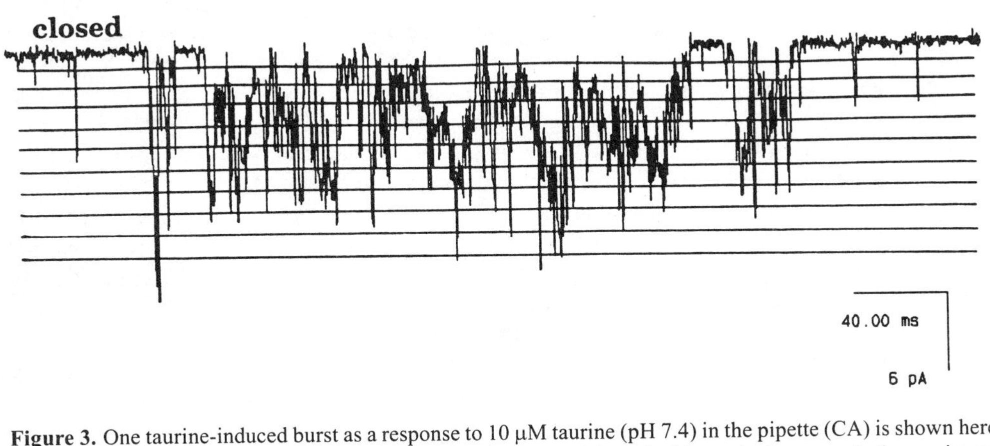

Figure 3. One taurine-induced burst as a response to 10 µM taurine (pH 7.4) in the pipette (CA) is shown here on an expanded time scale to illustrate the rapid open-close transitions. In this particular example the maximum burst amplitude was 16 pA, which consisted of approximately 10 individual channels opening at the same time (individual channel amplitude 1.6 pA). The burst lengths vary from 10 to 50 ms; the shortest were 1-5 ms and the longest 500-700 ms. The burst-like activity lasted 15 s in this example.

and an occasional subconductance state[16]. Huck and Lux[11] have reported that in inhibitory interneurons from the mouse cerebellum GABA and glycine (≥ 5 μM) applied to the outside-out patches induce the opening of at least 5-15 channels simultaneously, after which the effect is desensitized. Our findings share some common aspects with these latter two studies, since the taurine-induced channel activity was burst-like and many channels were opening simultaneously. Current amplitudes varied from 15 to 40 pA and consisted of at least 10 channels opening in bursts. Such taurine-induced channel kinetics are thus consonant with the known kinetics of Cl^- channels.

In spontaneously firing substantia nigra neurons, 300 μM-20 mM taurine reduces or completely inhibits action potentials, accompanied by alterations in membrane potential; biphasic changes in the membrane potential have also been observed[8]. In extracellular microelectrode recordings microiontophoretic application of taurine has increased the spike frequency near the soma and decreased it near the dendritic region of the cerebellar Purkinje cells, whereas GABA has had a depressive effect in both regions[4]. Our preliminary results have shown an increase in the frequency of spontaneous impulse current waveforms after taurine application in the soma. A long-lasting enhancement of field excitatory postsynaptic potentials with bath application of 10 mM taurine also occurs in the CA1 region of rat hippocampal slices (5). In mudpuppy olfactory neurons, Cl^-, K^+ and nonselective cation conductances have been modulated by 10 nM to 100 μM taurine (inhibitory and excitatory effects) when studied by a resistive whole-cell method[3].

Taurine is a zwitterion and thus bears no net charge in physiological solutions at pH 7.4. As taurine is more easily permeable through ion channels in alkaline solutions[1], we sought to test whether the change in pH would modulate the effects. Our results indicated, however, that neutral taurine solution (pH 7.4) in the pipette is the most potent in inducing burst-type channel activity.

In intracellular recordings from substantia nigra neurons, taurine and glycine have increased membrane Cl^- conductance, which effects are blocked by strychnine[8]. In a further characterization, intracellular application of cyclic AMP inhibited the responses to high concentrations of taurine in nystatin-perforated and conventional patch-clamp whole-cell current recordings from acutely isolated nigral neurons[13]. Experiments on poly(A)$^+$ mRNA, isolated from cultured mouse cerebellar granule cells and injected into *Xenopus* oocytes, have shown taurine to be a weak agonist which activates glycine receptors when measured by the two-electrode voltage-clamp method[26]. In a separate study where *Xenopus* oocytes were injected with mouse poly(A)$^+$ mRNA from three brain regions (cerebrum, cerebellum and brain stem) the taurine responses were found to be mediated mainly through $GABA_A$ receptors in the cerebrum and cerebellum and through both $GABA_A$ and glycine receptors in the brain stem[10]. These earlier investigations thus suggest that taurine acts through either $GABA_A$ or glycine receptors. In our study the taurine-induced current activation was not seen after incubation with strychnine (a glycine receptor antagonist).

As in our present studies activation was induced by both bath and pipette application of taurine, the effect cannot be explained solely by the known ionotropic receptor-channel mechanisms. Three types of taurine receptors may exist in spinal cord tissue. One of these types, which pharmacologically resembles the glycine receptor, is present in the ventral roots of the spinal cord and is suggested by Mathers *et al.*[24] to be also expressed in spinal cord neurons in culture. The other two taurine receptor subtypes may mediate the depolarizing and hyperpolarizing responses in the primary afferent terminals of the frog spinal cord[20]. They are pharmacologically distinguishable from both GABA and glycine receptors. It thus remains unsettled whether or not the present electrophysiological responses are explicable by the actions of taurine on the known $GABA_A$ and glycine receptors or whether a specific taurine receptor is responsible. Only the discovery of a truly specific taurine receptor antagonist could give the definite answer to this question.

CONCLUSIONS

The basic findings of the present study are: (i) inward currents are induced by taurine in cultured rat cerebellar granule cells; (ii) low concentrations of taurine induce the above effect, most potently at neutral pH; (iii) an ionotropic receptor-channel mechanism does not alone explain the effect; a metabotropic receptor-coupled second-messenger system needs to be involved; and (iv) it still remains open whether or not a specific taurine receptor is in charge.

ACKNOWLEDGMENTS

We thank Oili Pääkkönen for the skillful preparation of cell cultures and Paula Kosonen for technical advice. We are grateful to Harold K. Kimelberg for critical comments. This work was financially supported by the Alfred Kordelin and Pirkanmaa Cultural Foundations (M.-L.L.), by a project grant from the University of Tampere (P.S.) and by the Medical Research Fund of Tampere University Hospital and the Academy of Finland (S.S.O.).

REFERENCES

1. Banderali, U., and Roy, G., 1992, Anion channels for amino acids in MDCK cells, *Am. J. Physiol.,* 263: C1200-C1207.
2. Cull-Candy, S.G., and Ogden, D.C., 1985, Ion channels activated by L-glutamate and GABA in cultured cerebellar neurons of the rat, *Proc. R. Soc. Lond. B,* 224: 367-373.
3. Dubin, A.E., and Dionne, V.E., 1993, Modulation of Cl^-, K^+, and nonselective cation conductances by taurine in olfactory receptor neurons of the mudpuppy *Necturus maculosus, J. Gen. Physiol.,* 101: 469-485.
4. Frederickson, R.C.A., Neuss, M., Morzorati, S.L., and McBride, W.J., 1978, A comparison of the inhibitory effects of taurine and GABA on identified Purkinje cells and other neurons in the cerebellar cortex of the rat, *Brain Res.,* 145: 117-126.
5. Galarreta, M., Bustamante, J., Martin del Rio, R., and Solis, J.M., 1994, Taurine induces a long lasting increase of synaptic transmission in rat hippocampal slices, *Soc. Neurosci. Abstr.,* 20: 1342.
6. Galdzicki, Z., Lin, F., Moran, O., Novelli, A., Puia, G., and Sciancalepore, M., 1991, Development of voltage-dependent ionic currents in rat cerebellar granule cells grown in primary culture, *Int. J. Neurosci.,* 56: 193-200.
7. Hamill, O.P., Marty, A., Neher, E., Sakmann, B., and Sigworth, F.J., 1981, Improved patch-clamp techniques for high-resolution current recording from cells and cell-free membrane patches, *Pflügers Arch.,* 391: 85-100.
8. Häusser, M.A., Yung, W.H., and Lacey, M.G., 1992, Taurine and glycine activate the same Cl^- conductance in substantia nigra dopamine neurones, *Brain Res.,* 571: 103-108.
9. Holopainen, I., Malminen, O., and Kontro, P., 1987, Sodium-dependent high-affinity uptake of taurine in cultured cerebellar granule cells and astrocytes, *J. Neurosci. Res.,* 18: 479-483.
10. Horikoshi, T., Asanuma, A., Yanagisawa, K., Anzai, K., and Goto, S., 1988, Taurine and ß-alanine act on both GABA and glycine receptors in *Xenopus* oocyte injected with mouse brain messenger RNA, *Mol. Brain Res.,* 4: 97-105.
11. Huck, S., and Lux, H.D., 1987, Patch-clamp study of ion channels activated by GABA and glycine in cultured cerebellar neurons of the mouse, *Neurosci. Lett.,* 79: 103-107.
12. Huxtable, R.J., 1989, Taurine in the central nervous system and the mammalian actions of taurine, *Prog. Neurobiol.,* 32: 471-533.
13. Inomata, H., Nabekura, J., and Akaike, N., 1993, Suppression of taurine response in acutely dissociated substantia nigra neurons by intracellular cyclic AMP, *Brain Res.,* 615: 347-350.
14. Jalonen, T., Johansson, S., Holopainen, I., Oja S.S., and Århem P., 1990, Single-channel and whole-cell currents in rat cerebellar granule cells, *Brain Res.,* 535: 33-38.

15. Jalonen, T., Kivelä, A., Johansson, S., Holopainen, I., Oja, S.S., and Århem, P., 1990, Electrophysiological properties of cerebellar granule cells in culture, *Abstracts of the 5th Finnish Neurochemistry Meeting, Espoo,* p. 20.
16. Kilic, G., Moran, O., and Cherubini, E., 1993, Currents activated by GABA and their modulation by Zn^{2+} in cerebellar granule cells in culture, *Eur. J. Neurosci.,* 5: 65-72.
17. Kontro, P., and Oja, S.S., 1987, Glycinergic systems in the brain stem of developing and adult mice: effects of taurine, *Int. J. Dev. Neurosci.,* 5: 461-470.
18. Kontro, P., and Oja, S.S., 1987, Co-operativity in sodium-independent taurine binding to brain membranes in the mouse, *Neuroscience,* 23: 567-570.
19. Kontro, P., and Oja, S.S., 1990, Interactions of taurine with $GABA_B$ binding sites in mouse brain, *Neuropharmacology,* 29: 243-247.
20. Kudo, Y., Akiyoshi, E., and Akagi, H., 1988, Identification of two taurine receptor subtypes on the primary afferent terminal of frog spinal cord, *Br. J. Pharmacol.,* 94: 1051-1056.
21. Linne, M.-L., Oja, S.S., and Jalonen, T.O., 1995, The effects of taurine on single ion channel activity in rat cerebellar granule cells, *Biophys. J.,* 68: A379.
22. Malminen, O., and Kontro, P., 1987, Actions of taurine on the GABA-benzodiazepine receptor complex solubilized from rat brain, *Neurochem. Int.,* 11: 113-117.
23. Mathers, D.A., 1993, Effects of benzothiadiazine TAG on channel activation at mammalian glycine receptors, *Neurosci. Lett.,* 149: 23-26.
24. Mathers, D.A., Grewal, A., and Wang, Y., 1989, Membrane channels activated by taurine in cultured mouse spinal cord neurons, *Neurosci. Lett.,* 98: 229-233.
25. Oja, S.S., Korpi, E.R., and Saransaari, P., 1990, Modification of chloride flux across brain membranes by inhibitory amino acids in developing and adult mice, *Neurochem. Res.,* 15: 797-804.
26. Wahl, P., Elster, L., and Schousboe, A., 1994, Identification and function of glycine receptors in cultured cerebellar granule cells, *J. Neurochem.,* 62: 2457-2463.
27. Wu, J.-Y., Tang, X.W., and Tsai, W.H., 1992, Taurine receptor: kinetic analysis and pharmacological studies, *Adv. Exp. Med. Biol.,* 315: 263-268.

50

A NEW NEUROMODULATORY ACTION OF TAURINE: LONG-LASTING INCREASE OF SYNAPTIC POTENTIALS

Mario Galarreta,[1] Julián Bustamante,[2] Rafael Martín del Río,[1] and José M. Solís[1]

[1] Servicio de Neurobiología
Depto. Investigación
Hospital Ramón y Cajal
28034 Madrid, Spain
[2] Depto. Fisiología
Facultad de Medicina
Universidad Complutense
28040 Madrid, Spain

INTRODUCTION

The physiological role of taurine, one of the most abundant free amino acids in the mammalian brain, is still poorly understood. A solid body of electrophysiological studies has demonstrated that taurine application causes neuroinhibitory actions in different regions of the CNS. In this sense, taurine has been shown to reduce the spontaneous neuronal firing, to hyperpolarize the resting membrane potential, to diminish the membrane input resistance and to increase the membrane Cl^- conductance in different CNS neurons[2,4,7,19,23,24]. On the basis of these observations, taurine has been proposed as a putative neuroinhibitory transmitter in the CNS. However, because of the lack of selective taurine antagonists, this possibility has not been unambiguously demonstrated so far. In fact, many of the taurine-induced neuroinhibitory effects have been shown to be blocked by GABA and/or glycine receptor antagonists[4,7,20].

Alternative to a receptor-mediated role as a neurotransmitter, taurine could also have a modulatory action on the neurotransmission process performed by other substances[8]. This concept would be consistent with a number of observations indicating that taurine regulates Ca^{2+} fluxes in nerve terminals[21], appears to stabilize excitable membranes[6], inhibits protein phosphorylation in synaptosomes[15] and reduces neurotransmitter release[10,18].

In this context, the purpose of our work has been to investigate the possible modulatory actions of taurine application on synaptic transmission in the hippocampus. To accomplish this we have used the well characterized glutamatergic pathway of the Schaffer collateral-commisural (SCC) fibers, whose terminals make monosynaptic excitatory con-

Taurine 2, edited by Huxtable et al.
Plenum Press, New York, 1996

tacts with the dendrites of the pyramidal cells of the CA1 region. We have found that taurine application, in addition to an inhibitory action mediated via $GABA_A$ receptors, induces a long-lasting increase of synaptic potentials. This long-lasting potentiation of synaptic responses is caused by at least two factors: (i) an increase in axon excitability and (ii) an enhancement of synaptic efficacy.

METHODS

The experiments were carried out in transverse hippocampal slices (400 μm thick) obtained from adult female Sprague-Dawley rats following standard procedures. Briefly, after decapitation the brain was rapidly removed and placed in ice-cold standard solution (composition in mM: NaCl 119, $NaHCO_3$ 26.2, KCl 2.5, KH_2PO_4 1, $MgSO_4$ 1.3, $CaCl_2$ 2.5 and glucose 11) bubbled with 95% O_2 and 5% CO_2. The hippocampi were dissected out, and sliced with a manual chopper. The slices were placed in an interface holding chamber for at least 1 h (20-25°C) and then transferred to a submersion-type chamber for recording. The standard solution (composition as above) was continuously perfused (~2 ml/min), pre-gassed with 95% O_2 and 5% CO_2. Experiments were performed at 30-32°C unless otherwise indicated.

Synaptic responses were evoked by stimulating the SCC fibers with electrical pulses (0.05-0.3 mA, 20-40 μs, 0.05-0.066 Hz) through a bipolar microelectrode (Fig. 1A). Field excitatory postsynaptic potentials (fEPSPs) were recorded by a micropipette (filled with 2 M NaCl) placed at the stratum radiatum of the CA1 subfield (Fig. 1A). These potentials were usually preceded by a small deflection, the presynaptic fiber volley (FV), that corresponds to the compound action potential of the SCC fibers (Fig. 1A). Intracellular excitatory postsynaptic potentials (EPSP) were obtained at the somatic level using conventional recordings with micropipettes filled with 3 M potassium acetate (resistence 70-100 MΩ). The apparent input resistance (Rin) of the neurons was calculated from the voltage deflection produced by hyperpolarizing current pulses (0.1-0.3 nA, 50 ms). The stimulus strength was adjusted to give ~30% maximal response. Recordings were filtered between 1-3000 Hz, amplified, digitized at 25 or 100 kHz and collected on a 386 IBM compatible computer for off-line analysis.

All the compounds tested were applied by addition to the perfusion solution. To prevent epileptiform discharges when $GABA_A$ antagonists (BMI or picrotoxin) were applied, a cut was made between CA3 and CA1 regions, and the concentration of $CaCl_2$ and $MgSO_4$ were increased to 4 mM. These conditions did not modify taurine effects on synaptic potentials. For the experiments in low extracellular Na^+ concentration (Fig. 5A), all the NaCl in the standard solution was equimolarly replaced by N-methyl-D-glucamine chloride during taurine application and a prior period of 25 min. This substitution abolished all the evoked potentials that were reverted when the standard solution was perfused during the taurine washout period. Field EPSP increase due to taurine perfusion was calculated substrating the fEPSP enhancement due to low Na^+ "per se" (122 ± 10%, n=4 control experiments).

Statistical differences were assessed by one or two analyses of variance and two-tailed Student's *t*-test. Data are presented as mean ± SEM.

RESULTS

Synaptic transmission between Schaffer collateral-commisural fibers and CA1 pyramidal cells was studied monitoring the evoked fEPSP (Fig.1A). After establishing a base line with a control period, bath application of taurine (10 mM, for 30 min) produced a

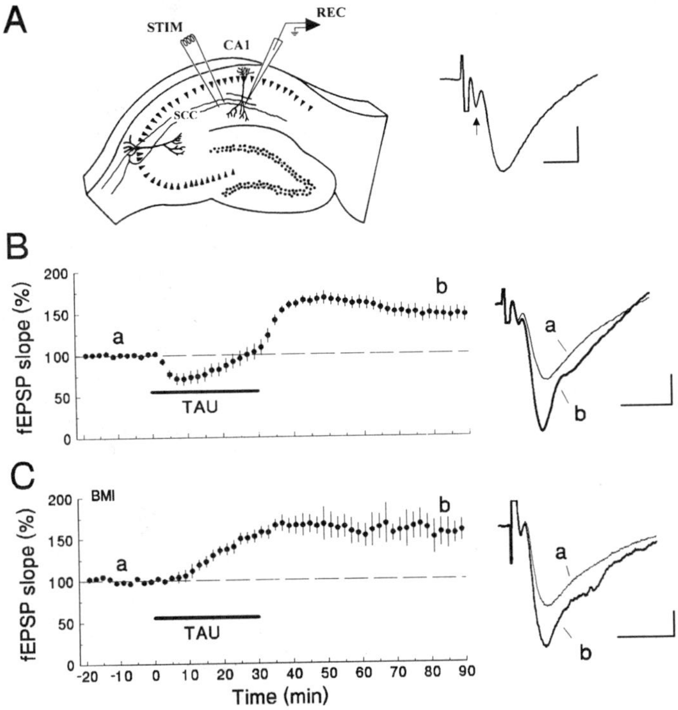

Figure 1. Diagram of hippocampus and effects of taurine on extracellular synaptic potentials. (A) Left: diagram showing a hippocampal slice with the location of the stimulating (STIM) and extracellular recording (REC) electrodes. Right: a typical fEPSP preceded by the presynaptic FV (arrow). (B) and (C) These and subsequent similar graphs plot the slope of the fEPSP versus time. Taurine (10 mM, TAU) was bath applied during the period indicated by the black horizontal bar in standard medium (n = 15) (A), or in the presence of BMI (50 μM, n = 6) (B). Data from each experiment were normalized with respect to the mean fEPSP slope of the control period. Each point represents the mean ± SEM. For the sake of clarity, points have been plotted every two min. Sample traces on the right depict the average of 4 consecutive fEPSPs taken at the indicated times from a representative experiment, and show the increase in the synaptic response after taurine application. Calibration bars: 0.5 mV, 5 ms.

biphasic effect on the fEPSP slope (Fig. 1B). Thus, whereas the fEPSP was transitorily reduced during taurine perfusion (70 ± 8% of control value, n = 15), it was remarkably increased after taurine withdrawal (167 ± 7% of control, 15 min of washout). Interestingly, this potentiation of the synaptic responses induced by taurine persisted throughout the washout period (146 ± 9% of control, 60 min of washout).

Taurine has been previously shown to interact with $GABA_A$ receptors[1,17], and many of the previously described neuroinhibitory effects of taurine are blocked by $GABA_A$ receptor antagonists[4,20]. On the basis of these facts we next considered if taurine inhibition of fEPSP could also be mediated by $GABA_A$ receptor activation, and we performed a series of experiments in the presence of $GABA_A$ antagonists, such as bicuculline (BMI) or picrotoxin. Fig. 1C shows that BMI (50 μM) completely blocked the inhibitory effect of taurine on the fEPSP (n = 6), while a progressive increase of the fEPSP slope could be observed from the beginning of taurine perfusion. This taurine induced potentiation was, after taurine withdrawal, similar to that obtained in standard medium (see Fig 1B). Identical results were obtained in the presence of picrotoxin (n = 10).

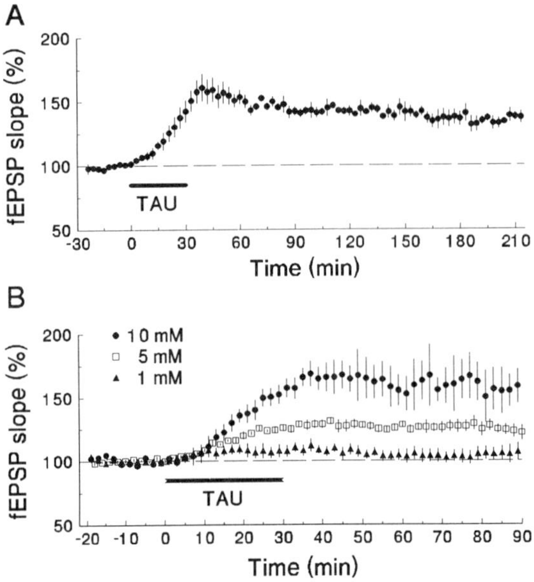

Figure 2. Taurine induced fEPSP potentiation is long-lasting and dose-dependent. (A) Taurine (10 mM) induced fEPSP potentiation persisted at least 3 h after taurine withdrawal. The experiments were performed in picrotoxin (100 μM). Data are the average from four experiments to min 150, and from three experiments to min 210. (B) Time course of fEPSP slope regarding taurine perfusion at 10 mM (n = 6), 5 mM (n = 5), and 1 mM (n = 5). The experiments were carried out in the presence of BMI (50 μM) or picrotoxin (100 μM). Points have been plotted every 3 (A) and 2 min (B).

These results indicate that taurine application produces two opposite actions on the fEPSP in the CA1 region: (i) a reversible inhibition due to $GABA_A$ receptor activation, and (ii) a long-lasting potentiation independent of the activation of these receptors.

To characterize further this taurine induced fEPSP potentiation, we carried out two other groups of experiments in the presence of $GABA_A$ antagonists. First, we prolonged the washout period to truly confirm that the taurine induced potentiation was a long-lasting phenomenon (Fig. 2A). The fEPSP slope remained potentiated 3 h after taurine withdrawal (137 ± 4% of control, n = 3). Secondly, we applied different concentrations of taurine for a common 30 min period, to study the dose-dependence of taurine induced fEPSP potentiation. Taurine perfusion at a 5 mM concentration induced a long-lasting potentiation of lower magnitude than that produced by 10 mM taurine perfusion. On the other hand, taurine application at 1 mM did not produce any significant long-lasting change on fEPSP slope (Fig. 2B).

We next examined the effect of taurine application on intracellularly recorded EPSPs. Our purpose with these experiments was to determine whether individual neurons show the same taurine induced long-lasting changes in synaptic responses as the whole pyramidal cell population. With this approach, we could also study if the taurine induced potentiation could be associated with changes in the basic electrical properties of the neuronal membrane such as membrane potential (Vm) and input resistance (Rin). The EPSP was isolated performing the experiments in the presence of BMI, and the Rin was monitored throughout the experiment. The actual value of the Rin was estimated by measuring the amplitude of the recorded voltage deflection induced by the application of a hyperpolarizing current pulse (see Fig. 3). In this condition, taurine perfusion (10 mM, for 20 min) induced a progressive increase in the EPSP amplitude, that persisted potentiated at least 50 min after taurine withdrawal (126 ± 14%, n = 6 cells). This EPSP potentiation, however, was not associated to any significant change in Rin (see Fig. 3) or Vm. Intracellular recordings revealed, therefore, that taurine application increases the synaptic responses without concomitant changes in the basic electrical properties of neuronal membrane.

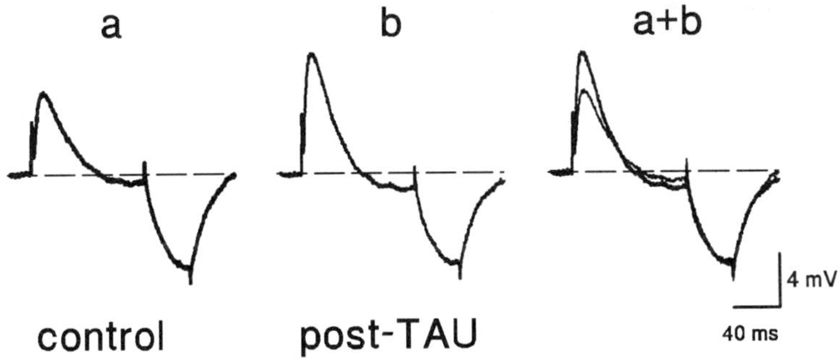

Figure 3. Taurine induces a long-lasting increase of intracellular EPSP without changing basic electrical properties of the neuronal membrane. Average of 4 consecutive traces obtained at control period (a) and 15 min after taurine (10 mM, for 20 min) perfusion (b) from a sample cell (Vm = -70 mV) (n = 6). The experiments were carried out in BMI (50 μM). The stimulus artifact is followed by the EPSP. The downward deflection after the EPSP corresponds to the hyperpolarizing current pulse used to measure the Rin. Note in (a+b) that this deflection did not change when the EPSP was clearly increased.

To investigate the possible factors underlying the taurine-induced potentiation of synaptic responses, we first examined the possibility that taurine perfusion could modify axon excitability. In order to accomplish this issue we took advantage of the presynaptic FV, that reflects the compound action potential of the activated SCC fibers (Fig. 1A, right), and measured its amplitude before and 1 h after taurine perfusion (10 mM, for 30 min) (n = 4). This study (input-output curves, Fig. 4A) revealed that after taurine application a larger presynaptic FV was produced by a given stimulus, and therefore, that a long-lasting increase of the SCC axon excitability had occurred. Reasonably, the taurine induced potentiation of synaptic responses could be at least partially explained if a larger number of axons are activated by the same stimulus.

With the purpose of discovering if in addition to its effect on axon excitability taurine could also enhance the efficacy of the synapses, we next compared the relation between the fEPSP slope and the FV amplitude before and 60 min after taurine application (10 mM, 30 min) (Fig. 4B). The shift to the left in this relation indicated that for a given presynaptic FV a larger postsynaptic response was evoked, and consequently, that the synaptic efficacy had also been increased after taurine perfusion. These results, in summary, are consistent with the idea that both an increase in axon excitability and in synaptic strength contribute to the long-lasting fEPSP produced by taurine.

Considering the slow development of the taurine-induced potentiation of the fEPSPs and that the taurine concentrations used in this study (mM range) would have a physiological significance regarding the intracellular compartment, we considered if taurine needs to be transported into the cellular interior as a previous step to develop this action. To investigate this possibility we administered taurine under two conditions: we decreased the temperature of the recording chamber (from 30-32°C to 20-21°C), and also we reduced the extracellular Na^+ concentration. Both of these circumstances are known to interfere with the taurine uptake systems[8], and in both of them, the taurine-induced potentiation of the fEPSP was significantly less than in standard conditions (Fig. 5A). Finally, we also studied the effects of a group of structural taurine analogs on the fEPSP, to compare their capacity to potentiate the fEPSP and their ability as subtrates of taurine transport systems. Thus, while β-alanine and GABA induced during their washout a similar potentiation as taurine, GES (2-guanidinoethane sulfonic acid) produced a significantly smaller one, and N-methyltaurine did not modify the

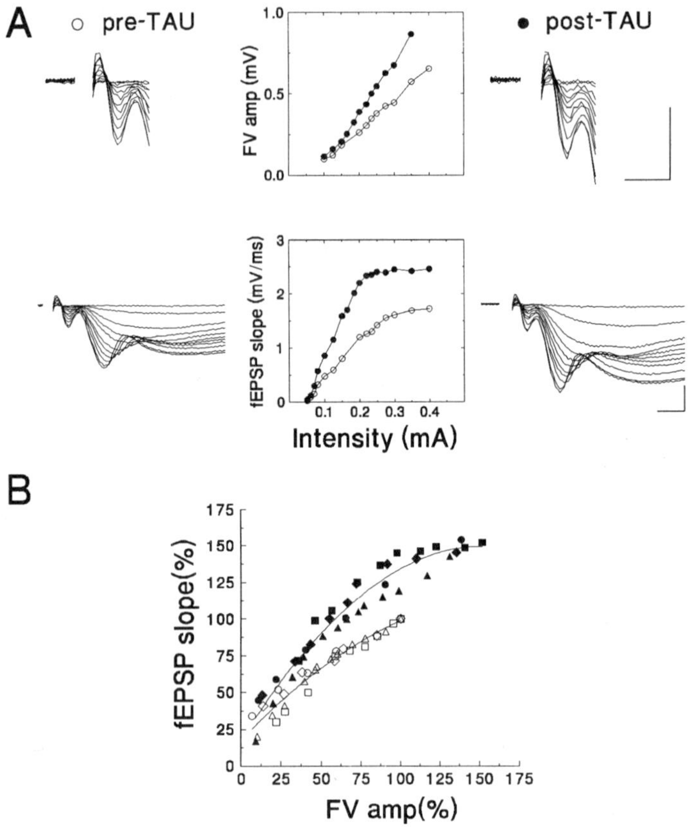

Figure 4. Input-output curves show that both an increase of axon excitability and synaptic efficacy account for the taurine induced fEPSP potentiation. (A) Data in this figure correspond to a sample experiment out of four with similar results. Left and right columns present the superposition of extracellular potentials evoked by increasing stimulus intensity (FV in upper row and fEPSP in bottom row). Central column shows the input-output curves for FV and fEPSP. Data recorded during control period (○) are depicted on the left column, and data obtained one hour after taurine application (10 mM, 30 min) (●) on the right column. Calibration bars: 1 mV, 2 ms. (B) fEPSP/FV relation. Open and closed symbols (each symbol represents an individual experiment) correspond to data obtained before and one hour after taurine perfusion (10 mM, 30 min), respectively. Normalized data from four experiments.

synaptic potentials (Fig. 5B). This order of potency is similar to their ability for competing with taurine for the taurine transport systems[12,13].

DISCUSSION

We present in this study a new neuromodulatory action of taurine on synaptic transmission, that fundamentally consists of a long-lasting potentiation of both synaptic efficacy and fiber excitability. These actions of taurine are not mediated by $GABA_A$ receptor activation, and thus differ from other electrophysiological actions of this amino acid.

Although it is logical to assume that the taurine-induced potentiation of fiber excitability should be due to a still undefined effect of this amino acid on the axon itself, the subcellular localization of the mechanism that induces the increase in the synaptic efficacy

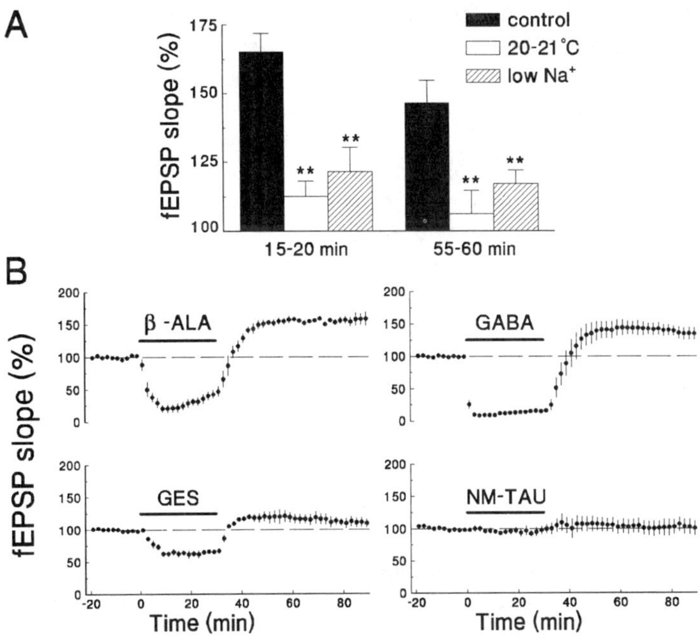

Figure 5. Taurine induced fEPSP potentiation could involve a transport system. (A) Taurine induced fEPSP potentiation is temperature and Na^+-dependent. Histogram bars represent the fEPSP potentiation at 15-20 and 55-60 min of washout, after taurine perfusion (10 mM, 30 min) under three conditions: control (30-32°C, normal Na^+) (n = 15, same data shown in Fig. 1B), low temperature (20-21°C) (n = 6), and low Na^+ (see methods) (n = 5). Asterisks indicate significant differences compared to control values (**$p < 0.01$). (B) Effects on the fEPSP slope time course of the perfusion of some structural taurine analogs: β-alanine (β-ALA), GABA, guanidinoethane sulfonate (GES) and N-methyltaurine (NM-TAU). All these substances were applied at 10 mM, for 30 min (n = 4 for each one).

is not clear. The fact that taurine potentiates EPSPs without changing the membrane potential or the input resistance of the cell could be interpreted as an indication of taurine acting on the presynaptic terminals. Since taurine does not change the paired-pulse facilitation ratio[5], a phenomenon that reflects a presynaptic increase in transmitter release[25], its potentiation is not easily compatible with a presynaptic action. We must remember, however, that the paired-pulse facilitation ratio might not be able to detect all the alterations related with the presynaptic release mechanisms[11], and consequently a presynaptic mechanism cannot be totally discarded.

Taurine-induced synaptic potentiation seems to be dependent on its previous uptake into the intracellular compartment, as judged by the dependence of the phenomenon on temperature and extracellular Na^+ concentration. These results could explain the slow time course of the taurine-induced potentiation, as well as the requirement of a high concentration of this amino acid (well above its reported extracellular concentration around 20 μM[14]) to evoke the increase in synaptic potentials. The idea of taurine acting on an intracellular locus has also been proposed for different preparations[3,22].

Although the synthesis of taurine in brain tissue is poor, this amino acid is highly concentrated in the intracellular compartment of almost all brain structures. This fact is congruent with the existence of specific taurine transport systems that have been recently identified[16]. At this point it is worth asking, why cells have developed a highly selective machinery to accumulate an amino acid that seemingly does not have extracellular receptors?

In view of the data presented in this paper it could be proposed that taurine accumulation in the intracellular compartment modulates the synaptic transmission strength. Intracellular taurine levels could modulate a number of cellular functions by its action on the phosphorylation of specific proteins by protein kinases[15] or by its interaction with phospholipids of the plasma membrane[9]. Thus, we hypothetize that changes in the intracellular content of taurine, induced by cell swelling or by any other situation stimulating the release of taurine, could be very important in modulating synaptic transmission. This possibility would be meaningful for the physiological role of taurine in the nervous system.

CONCLUSION

The main finding presented in this paper is that taurine can induce a long-lasting potentiation of synaptic transmission acting on axon excitability and synaptic efficacy.

ACKNOWLEDGMENTS

We thank Dr. N. Menéndez for their helpful comments and Dr. R.J. Huxtable for his generous gift of GES. We also thank A. Latorre for technical assistance and S. McGrath for editorial help. This work was supported by a grant from the Fondo de Investigaciones Sanitarias de la Seguridad Social (93/0565). M.G. is the recipient of predoctoral fellowship from the Dirección General de Investigación Científica y Técnica.

REFERENCES

1. Bureau, M.H. and Olsen, R.W. 1991, Taurine acts on a subclass of $GABA_A$ receptors in mammalian brain in vitro, *Eur. J. Pharmacol.,* 207:9-16.
2. Curtis, D.R., Phillis, J.W. and Watkins, J.C. 1961, Actions of amino acids on the isolated hemisected spinal cord of the toad, *Br. J. Pharmacol.,* 16:262-283.
3. Franconi, F., Martini, F., Stendardi, I., Matucci, R., Zilleti, L. and Giotti, A. 1982, Effect of taurine on calcium levels and contractility in guinea-pig ventricular strips, *Biochem. Pharmacol.,* 31:3181-3185.
4. Frederickson, R.C.A., Neuss, M., Morzorati, S.L. and McBride, W.J. 1978, A comparison of the inhibitory effects of taurine and GABA on identified Purkinje cell and other neurons in the cerebellar cortex of the rat, *Brain Res.,* 145:117-126.
5. Galarreta, M., Bustamente, J., Martín del Río, R. and Solís, J.M. 1996, Taurine induces a long-lasting increase of synaptic efficacy and axon excitability in the hippocampus, *J.Neurosci.* 16:92-102.
6. Gruener, R., Bryant, H., Markovitz, D., Huxtable, R.J. and Bressler, R. 1976, Ionic actions of taurine on nerve and muscle membranes: electrophyisologic studies, in: "Taurine", Huxtable, R.J. and Barbeau A. (Eds.), Raven Press, New York, pp. 225-242.
7. Häusser, M.A., Yung, W.H. and Lacey, M.G. 1992, Taurine and glycine activate the same Cl^- conductance in the substantia nigra dopamine neurons, *Brain Res.,* 571:103-108.
8. Huxtable, R.J. 1989, Taurine in the central nervous system and the mammalian actions of taurine, *Prog. Neurobiol.,* 32:471-533.
9. Huxtable, R.J. and Sebring, L.A. 1986, Towards a unifying theory for the action of taurine, *Trends Pharmac. Sci.,* 7:481-485.
10. Kamisaki, Y., Maeda, K., Ishimura, M., Omura, H. and Itoh, T. 1993, Effects of taurine on depolarization-evoked release of amino acids from rat cortical synaptosomes, *Brain Res.,* 627:181-185.
11. Kamiya, H. and Zucker, R.S. 1994, Residual Ca^{2+} and short-term synaptic plasticity, *Nature,* 371:603-606.
12. Kontro, P. and Oja, S.S. 1981, Hypotaurine transport in brain slices: comparison with taurine and GABA, *Neurochem. Res.,* 6:1179-1191.
13. Kontro, P. and Oja, S.S. 1983, Binding of taurine to brain synaptic membranes. In: "CNS Receptors: From Molecular Pharmacology to Behavior", Mandel, P. and DeFeudis, F.V. (Eds.), Raven Press, New York, pp. 23-34.

14. Lerma, J., Herranz, A.S., Herreras, O., Abraira, V. and Martín del Río, R. 1986, In vivo determination of extracellular concentrations of amino acids in the rat hippocampus. A method based on brain dialysis and computerized analysis, *Brain Res.*, 384:145-155.
15. Li, Y.-P. and Lombardini, J.B. 1991, Taurine inhibits protein kinase C-catalyzed phosphorylation of specific proteins in a rat cortical P2 fraction, *J. Neurochem.*, 56:1747-1753.
16. Liu, Q.-R., López-Corcuera, B., Nelson, H., Mandiyan, S. and Nelson, N. 1992, Cloning and expression of cDNA encoding the transporter of taurine and β-alanine in mouse brain, *Proc. Natl. Acad. Sci. USA*, 89:12145-12149.
17. Malminen, O. and Kontro, P. 1986, Modulation of the GABA-benzodiazepine receptor complex by taurine in rat brain membranes, *Neurochem. Res.*, 11:85-94.
18. Namima, M., Okamoto, K. and Sakai, Y. 1983, Modulatory action of taurine on the release of GABA in cerebellar slices of guinea pig, *J. Neurochem.*, 40:1-9.
19. Okamoto, K., Kimura, H. and Sakai, Y. 1983, Effects of taurine and GABA on Ca spikes and Na spikes in cerebellar Purkinje cells in vitro: intrasomatic study, *Brain Res.*, 260:249-259.
20. Okamoto, K. and Sakai, Y. 1981, Inhibitory actions of taurocyamine, hypotaurine, homotaurine, taurine and GABA on spike discharges of Purkinje cells, and localization of sensitive sites, in guinea pig cerebellar slices, *Brain Res.*, 206:371-386.
21. Pasantes-Morales, H. and Gamboa, A. 1980, Effect of taurine on $^{45}Ca^{2+}$ accumulation in rat brain synaptosomes, *J. Neurochem.*, 34:244-246.
22. Sebring, L.A. and Huxtable, R.J. 1985, Taurine modulation of calcium binding to cardiac sarcolemma, *J. Pharmacol. Exp. Therap.*, 232:445-451.
23. Taber, K.H., Lin, C.-T., Liu, J.-W., Thalmann, R.H. and Wu, J.-Y. 1986, Taurine in hippocampus: localization and postsynaptic action, *Brain Res.*, 386:113-121.
24. Zeise, M. 1985, Taurine on hippocampal slices: comparison to GABA and glycine, and antagonism by 4-aminopyridine, in: "Taurine: Biological Actions and Clinical Perspectives", Oja, S.S., Ahtee, L., Kontro, P. and Paasonen, M.K., (Eds.), Alan R. Liss, Inc., pp. 281-287.
25. Zucker, R.S. 1989, Short-term synaptic plasticity, *Ann. Rev. Neurosci.*, 12:13-31.

51

INTERACTION OF γ-L-GLUTAMYLTAURINE WITH KAINATE-INDUCED CYCLIC AMP FORMATION IN THE RAT HIPPOCAMPUS

V. Varga,[1,4] Réka Janáky,[1] A. Takáts,[2] S. S. Oja,[1,3] R. Dohovics,[4] and Leena Rechardt[1]

[1] Tampere Brain Research Center
University of Tampere Medical School
Box 607, FIN-33101 Tampere, Finland
[2] Frederic Joliot-Curie National Research Institute for Radiobiology and Radiohygiene
Box 101, H-1775 Budapest, Hungary
[3] Department of Clinical Physiology
Tampere University Hospital
Box 2000, FIN-33521 Tampere, Finland
[4] Department of Animal Physiology
Kossuth Lajos University of Sciences
Box 18, H-4010 Debrecen, Hungary

INTRODUCTION

γ-L-Glutamyltaurine (LGT) is the most abundant taurine-containing peptide in the CNS[19,31], being located mainly in neurons[30]. LGT is synthesized by γ-glutamyltransferase (EC 2.3.2.2) and can be degraded to glutamate and taurine by γ-glutamyltransferase or to 5-oxoproline and taurine by γ-glutamylcyclotransferase (EC 2.3.2.4)[17,31]. Some intriguing central effects (e.g. anticonflict and antiepileptic actions) of LGT have been reported[10,16]. They may be attributed to the interaction of LGT with different steps of glutamatergic neurotransmission[32-34].

The ionotropic glutamate receptors are involved in epilepsy and long-term potentiation[5,9]. The rigid glutamate analogue kainate (KA) is epileptogenic and selectively destroys neurons[25]. In the limbic system the CA3 pyramidal cells in the stratum lucidum (the mossy fiber terminal zone) are the most sensitive[22]. This selective vulnerability matches the distribution of KA binding sites[8,29] and the expression pattern of KA receptors[12,36]. The excitotoxic and epileptogenic effects of KA may thus be a result of direct stimulation of KA receptors and the consequent depolarization and Ca^{2+} influx[6].

In cultured hippocampal neurons, forskolin, an effective adenylate cyclase (AC) activator that acts at the catalytic site, enhances the whole-cell current responses to glutamate

Taurine 2, edited by Huxtable et al.
Plenum Press, New York, 1996

and KA. A single channel analysis has revealed that forskolin, through the cyclic AMP-dependent protein kinase A, increases the opening frequency and mean opening time of the KA receptor-gated ion channels[13]. This indicates that these channels are subject to neuromodulatory regulation through the adenylate cyclase cascade. It is also known that excitatory amino acids stimulate the formation of cyclic AMP[28], particularly in the hippocampus[2,3]. An increase in the level of cyclic AMP may likewise contribute to the neurotoxicity and epileptogenic effect of KA. To our knowledge the involvement of cyclic AMP formation in the action of KA has not yet been tested and ultrastructurally located in the hippocampus.

Both D- (DGT) and L-forms of γ-glutamyltaurine have been found to be antagonists of KA receptors[14,32]. We therefore now studied by biochemical and histochemical electron microscopic methods the possible interactions of LGT and DGT with the effects of KA on the formation of cyclic AMP and the activity of AC in the hippocampus.

MATERIAL AND METHODS

Slices (0.4 mm thick) were prepared from the hippocampi of Sprague-Dawley rats with a McIlwain tissue chopper. For both cyclic AMP and histochemical assays, the slices were preincubated at 37°C for 60 min (recovery period) in Krebs-Ringer-glucose (KRG) medium (with changes to new medium at 30 and 60 min and continuous bubbling with 95% O_2/5% CO_2). At 60 min the slices were rinsed and further incubated in KRG or Tricine (5 mM)-supplemented KRG (Tricine-KRG). In the histochemical reactions, NaH_2PO_4 was omitted from the media to avoid nonspecific precipitation of inorganic phosphate (see later).

In order to activate glutamate receptors, glutamate (1.0 mM) or KA (0.01 to 1.0 mM) was added and the incubation continued for an additional 5 or 10 min. DGT (Tocris Cookson, Bristol, UK), LGT (synthesized by us) (both 0.01-1.0 mM) and guanosine 5'-(b,g-imido-triphosphate [Gpp(NH)p], Sigma, St. Louis, MO) (0.01-0.1 mM) were added to medium 2 min before the agonists. When the cyclic AMP content was to be measured, the incubation was terminated by trichloracetic acid (final concentration 5% w/v). After homogenization and centrifugation, trichloracetic acid was extracted from the supernatants with water-saturated ethylether, after which aliquots were taken from the water phase for cyclic AMP determination with radioimmune assay kits (Amersham International, Bristol, UK). Protein content was measured from the precipitated pellets by the method of Lowry *et al.*[18].

Histochemical demonstration of AC activity was carried out by the method of Rechardt and Hervonen[23], based on the formation of cerium precipitate of inorganic pyrophosphate released by AC. After incubation with agonists and peptides (AC activation period) the slices were immersed for 5 min in ice-cold 1% paraformaldehyde fixative freshly prepared in 0.1 M Tricine-KRG solution (pH 7.3). The slices were then thoroughly washed several times with this buffer to remove free paraformaldehyde. The histochemical reaction for AC was carried out at 37°C for 60 min in medium containing (mM) Tricine buffer 50, adjusted with NaOH to pH 8.4, $MgCl_2$ 3, adenosine 5'-(b,g-imido)triphosphate lithium salt 0.5 (Sigma), 3-isobutyl-1-methylxanthine 10 (Sigma) and cerium chloride 2. The reaction was terminated by several rinses in physiological saline. The specimens were fixed in 1% OsO_4 for 1 h, dehydrated through a series of ethanol rinses and embedded in Epon-Araldite mixture. Ultrathin sections without poststaining were viewed and photographed in a Jeol-100S electron microscope operated at 60 kV.

The above method allows treatment of the slices with agonists and antagonists of glutamate receptors *in vitro*. Our control experiments indicated that the activity and the agonist-induced active conformation of the enzyme were preserved after paraformaldehyde fixation. This was confirmed by measuring the cyclic AMP content in paraformaldehyde-fixed slices after incubation for 1 h.

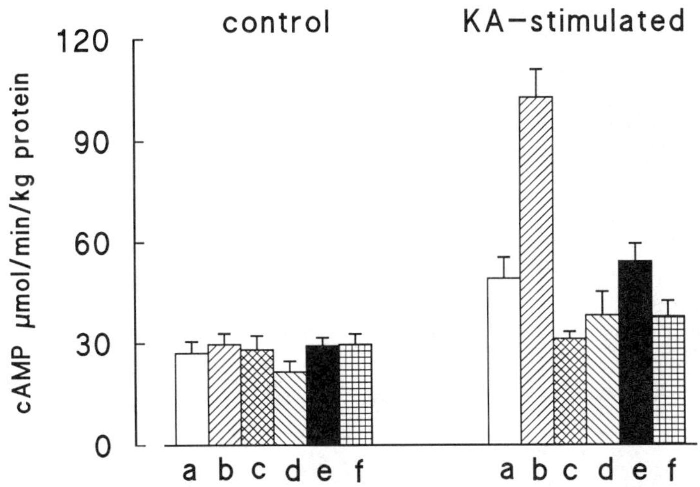

Figure 1. Effects of kainate, 5'-guanylylimidodiphosphate [Gpp(NH)p], γ-L- and γ-D-glutamyltaurine on the formation of cyclic AMP. Hippocampal slices were incubated in Krebs-Ringer-glucose buffer for 60 min. Thereafter the compounds to be tested, Gpp(NH)p and/or peptides, were added for 2 min followed by an additional incubation for 5 min with or without kainate. The reaction was stopped with 5% trichloracetic acid and the cyclic AMP measured as described in the text. Number of experiments was three. (a) control, (b) 0.1 mM Gpp(NH)p, (c) 1.0 mM LGT, (d) 1.0 mM DGT, (e) Gpp(NH)p+LGT, (f) Gpp(NH)p+DGT. Kainate (0.1 mM) significantly ($p < 0.01$) increased the level of cyclic AMP. This effect was significantly ($p < 0.01$) enhanced by Gpp(NH)p. The enhancement was inhibited ($p < 0.01$) by both LGT and DGT.

RESULTS

KA dose-dependently enhanced the cyclic AMP content in hippocampal slices. At the highest concentration tested (1 mM) the cyclic AMP content was increased by 81 ± 10% (mean ± SEM, n=3) (Fig. 1). To test whether or not G-proteins are involved in the formation of cyclic AMP as a response to KA, Gpp(NH)p, a GTP analogue and an activator of G-protein-dependent AC, was added to some incubation media. The effect of KA on the cyclic AMP content was enhanced by Gpp(NH)p by 245 ± 19%, (mean ± SEM, n=3), while this GTP analogue alone did not have any effect.

In control histochemical experiments in the absence of adenosine 5'-(b,g-imido)triphosphate (an ATP analogue resistant to hydrolysis by ATPases), no electron-dense cerium pyrophosphate precipitates were detectable (data not shown). In nonstimulated control specimens in the presence of the ATP analogue only faint precipitations, indicating AC reaction, were seen at synaptic densities and in a few myelin sheaths but not in other non-neuronal elements (Fig. 2A). Glutamate (1.0 mM) had no significant effect on AC activity (not shown). KA increased the amount of electron-dense precipitates in a dose-dependent manner. The KA-stimulated AC reaction in the hippocampus was confined to axodendritic and axospinous synapses. The precipitates were predominantly postsynaptically located (Figs 2B and C). Some cell processes and the cytoplasm around synaptic vesicles and tubuli also exhibited the reaction, but less intensely than synaptic densities (Fig. 2C, arrows). The perikarya of neurons, in turn,

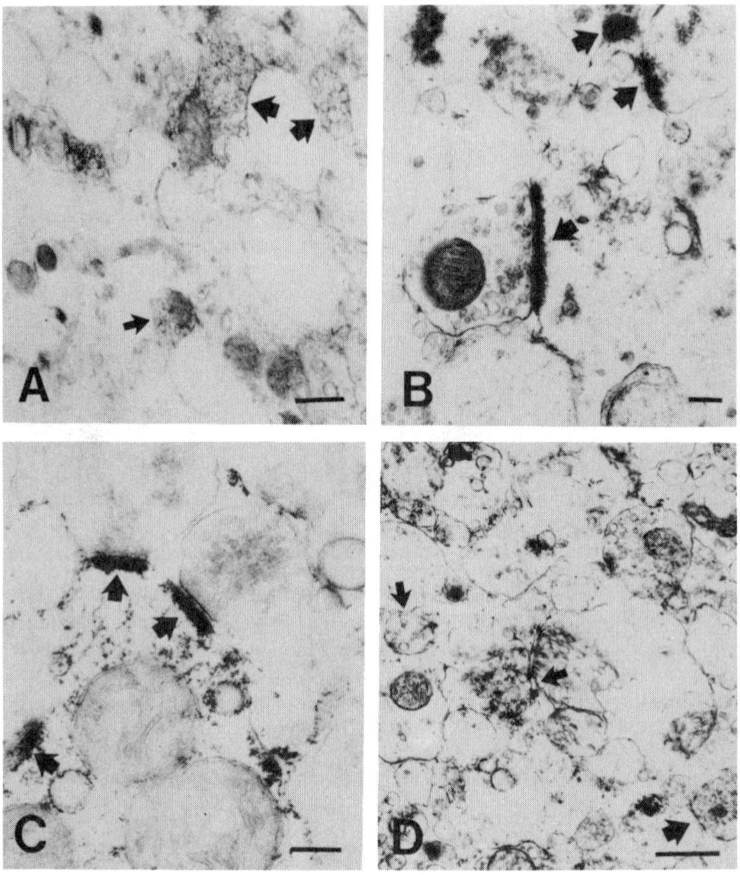

Figure 2. Histochemical demonstration of adenylate cyclase activity in rat hippocampal slices. Slices were preincubated in Krebs-Ringer-glucose medium for 60 min and thereafter the substances to be tested were added 2 min before kainate in Tricine-Krebs-Ringer-glucose buffer, pH 7.3. After the final 5-min incubation the slices were fixed with paraformaldehyde. The adenylate cyclase reaction was carried out in plain 50 mM Tricine buffer at pH 8.4. (A) Control slices. Only fine precipitates around presynaptic vesicles and synaptic junctions were observed (arrows, bar 500 nm). (B) and (C) Kainate stimulation of adenylate cyclase in the presence of 5'-guanylylimidodiphosphate [Gpp(NH)p]. The reaction product is seen predominantly postsynaptically in synaptic thickenings, around vesicles and in the cytoplasm (arrows, bar 200 nm). (D) Inhibition of the kainate + Gpp(NH)p stimulation of adenylate cyclase by γ-L-glutamyltaurine. A randomly scattered reaction product with a faintly synaptic localization is visible (bar 500 nm).

were devoid of enzyme activity. The large number of nonreactive postsynaptic thickenings in the vicinity of reactive ones bespeaks the specificity of the cytochemical reaction and a specific function of AC in these synapses. The effect of KA on the activation of AC was significantly enhanced by 0.1 mM Gpp(NH)p. Gpp(NH)p alone failed to activate AC (data not shown).

The KA-induced formation of cyclic AMP enhanced by Gpp(NH)p was attenuated by LGT. In this respect DGT was even more potent than LGT (Fig. 1). The peptides alone had no significant effects on AC, but both of them prevented the activation of AC by KA (Fig. 2D).

DISCUSSION

Many hormones, neurotransmitters and neuromodulators stimulate the rate of cyclic AMP synthesis[15]. They act at three distinct types of plasma membrane-associated proteins: receptors, guanine nucleotide-binding proteins (G-proteins) and adenylate cyclase (AC) itself. Acidic amino acids (glutamate, aspartate and cysteine sulfinate) also markedly increase the formation of cyclic AMP in brain slices[11,27], but their effects have been assumed to be indirect, occurring via release of adenosine and/or biogenic amines[21]. A direct excitatory amino acid-evoked cyclic AMP accumulation has also been suggested by Shimizu *et al.*[27,28]; however, the receptor involved has not been characterized. More recently, it has been shown that the stimulation of cyclic AMP synthesis may occur via direct activation of G-protein-coupled metabotropic mGluR1 glutamate receptors[1,20,24]. The present results demonstrate a KA-enhancement of AC activity in the synaptic structures and a KA-induced formation of cyclic AMP in the hippocampus. A subpopulation of KA receptors is coupled to a pertussis toxin-sensitive G protein in the goldfish brain[4,38]. The potentiation of the KA effect by Gpp(NH)p in the present experiments indicates that at least part of the observed effect of KA is coupled to G-proteins also in the mammalian brain. These G-protein-regulated receptors mediate the formation of cyclic AMP via activation of AC. However, it remains open whether they are ionotropic KA sites or identical to the metabotropic receptors of mGluR1 type[1]. They may even represent a new subpopulation of metabotropic glutamate receptors. Moreover, indirect mechanisms of AC activation via release of biogenic amines or adenosine cannot be excluded.

The glutamyl and taurine components possibly invest LGT with a neuromodulatory role. The present results yield further support to such a hypothesis, showing prevention of the KA-evoked G-protein-dependent AC activation and cyclic AMP accumulation both by DGT and LGT. The KA-induced depolarization of neurons has also been seen to be antagonized by DGT in electrophysiological studies[14]. Both LGT and DGT displace glutamate agonists in binding assays. The D-form shows some preference for the KA and 2-amino-3-hydroxy-5-methyl-4-isoxazole propionate (AMPA) receptors, whereas the L-form is less selective and also affects the release and uptake of glutamate[32,33,34].

The KA receptor subunit GluR6 is phosphorylated by a cyclic AMP-dependent protein kinase A[26], which increases the opening frequency and the mean opening time of the receptor-gated ionophore and enhances the whole-cell current responses to glutamate and KA in cultured hippocampal neurons[13,35]. Since KA itself activates AC, a potent positive feedback in the response to KA may obtain in the hippocampus, the most vulnerable part of the brain[22], where both GluR6 subunits and different types of AC are present[7,36,37]. Such a positive feedback without inhibition of cyclic AMP formation via stimulation of metabotropic glutamate receptors that are negatively coupled to cyclic AMP formation[20] may contribute to the KA-induced long-term potentiation[5] and may be responsible for the epileptic and toxic effects of KA. On the other hand, the antiepileptic and anxiolytic properties of LGT and DGT[10,16] are also explicable by an interference with the above effect of KA.

CONCLUSIONS

The results here indicate that in certain hippocampal neuronal circuits glutamate receptors of the KA type are functionally coupled to adenylate cyclase in a G-protein-regulated manner. The KA antagonist γ-D-glutamyltaurine and the endogenous L enanthiomorph γ-L-glutamyltaurine prevent the activation of adenylate cyclase and attenuate the formation of cyclic AMP caused by KA in the presence of a GTP analogue. Our findings corroborate the neuromodulatory role of LGT and may explain its anticonvulsant and anxiolytic actions.

ACKNOWLEDGMENTS

The excellent technical assistance of Ms. Oili Pääkkönen and Ms. Ulla Jukarainen and the financial support of the Academy of Finland and the Medical Research Fund of Tampere University Hospital are gratefully acknowledged.

REFERENCES

1. Aramori, I., and Nakanishi, S. 1992, Signal transduction and pharmacological characteristics of a metabotropic glutamate receptor, mGluR1, in transfected CHO cells, *Neuron,* 8: 757-765.
2. Baba, A., Lee, E., Tatsuno, T., and Iwata, H. 1982, Cysteine sulfinic acid in the central nervous system: antagonistic effect of taurine on cysteine sulfinic acid-stimulated formation of cyclic AMP in guinea pig hippocampal slices, *J. Neurochem.,* 38, 1280-1285.
3. Baba, A., Nishiuchi, Y, Uemura, A., and Iwata, H., 1988, Mechanism of excitatory amino acid-induced accumulation of cyclic AMP in hippocampal slices: role of extracellular chloride, *J. Pharmacol. Exp. Ther.,* 245: 299-304.
4. Barnes, J.M., Murphy, P.A., Kirkham, D., and Henley, M. 1993, Interaction of guanine nucleotides with [^{3}H]kainate and 6-[^{3}H]cyano-7-nitroquinoxaline-2,3-dione binding in goldfish brain, *J. Neurochem.,* 61: 1685-1691.
5. Ben-Ari, Y., and Represa, A. 1990, Brief seizure episodes induce long-term potentiatiom and mossy fibre sprouting in the hippocampus, *Trends Neurosci.,* 13: 312-318.
6. Bird, S.J., and Gulley, R.L. 1979, Evidence against a presynaptic mechanism for kainate neurotoxicity in the cochlear nucleus, *Neurosci. Lett.,* 15: 55-60.
7. Cali, J.J., Zwaagstra, J.C., Mons, N., Cooper, D.M.F., and Krupinski, J. 1994, Type VIII adenylyl cyclase. A Ca^{2+}/calmodulin-stimulated enzyme expressed in discrete regions of rat brain, *J. Biol. Chem.,* 16: 12190-12195.
8. Choi, D.W. 1988, Glutamate neurotoxicity and diseases of the nervous system, *Neuron,* 1: 623-634.
9. Collingridge, G.L., and Lester, A.J. 1989, Excitatory amino acid receptors in the vertebrate central nervous system, *Pharmacol. Rev.,* 40: 143-210.
10. Eder, T., Rieder, P., and Weiser, M. 1982, Über die Hemmwirkung von Gamma-L-Glutamyl-Taurin und einiger anderer Antikonvulsiva auf durch Umweltreize auslösbare epileptiforme Anfälle mongolischer Wustenrennmäuse (Meriones unguiculatus), *Wiener Tierärztl. Monatschr.,* 69: 16-18.
11. Ferrendelli, J.A., Chang, M.M., and Kinscherf, D.A., 1974, Elevation of cyclic GMP levels in central nervous system by excitatory and inhibitory amino acids, *J. Neurochem.,* 22: 535-540.
12. Foster, A.C., Mena, E.E., Monaghan, D.T., and Cotman, C.W. 1981, Synaptic localization of kainic acid binding sites, *Nature,* 289: 73-75.
13. Greengard, P., Jen, J., Nairn, A.C., and Stevens, C.F. 1991, Enhancement of the glutamate response by cAMP-dependent protein kinase in hippocampal neurons, *Science,* 253: 1135-1138.
14. Jones, A.W., Smith, D.A.S., and Watkins, J.C. 1984, Structure-activity relations of dipeptide antagonists of excitatory amino acids, *Neuroscience,* 13: 573-581.
15. Kendall, D.A., Mills, P.J., and Firth, J.L. 1992, Direct and indirect stimulations of cyclic AMP formation in human brain, *Br. J. Pharmacol.* 105: 899-902.
16. Kuribara, H., and Tadokoro, S. 1982, An anticonflict effect of gamma-L-glutamyltaurine (Litoralon) in rats, *Japan J. Pharmacol.,* 32: 1067-1074.
17. Lähdesmäki, P. 1987, Biosynthesis of taurine peptides in brain cytoplasmic fraction in vitro, *Int. J. Neurosci.,* 37: 79-84.
18. Lowry, O.H., Rosebrough, N.J., Farr, A.L., and Randall, R.J. 1951, Protein measurement with the Folin phenol reagent, *J. Biol. Chem.* 193: 265-275.
19. Marnela, K.M., Morris, M.R., Panico, M., Timonen, M., and Lähdesmäki, P. 1985, Glutamyl-taurine is the predominant synaptic taurine peptide, *J. Neurochem.,* 44: 752-754.
20. Nakanishi, S. 1992, Molecular diversity of glutamate receptors and implications for brain function, *Science,* 258: 597-603.

21. Newman, M., and McIlwain, H. 1977, Adenosine as a constituent of brain and of isolated cerebral tissues, and its relationship to the generation of adenosine 3':5'-cyclic monophosphate, *Biochem. J.* 164: 131-146.
22. Olney, J.W., Feller, T., and DeGubareff, T. 1979, Acute dendrotoxic changes in the hippocampus of kainate treated rats, *Brain Res.* 176: 91-100.
23. Rechardt, L., and Hervonen, H. 1985, Cytochemical demonstration of adenylate cyclase activity with cerium, *Histochemistry,* 82: 501-505.
24. Schoepp, D.D., and Conn, P.J. 1993, Metabotropic glutamate receptors in brain function and pathology, *Trends Pharmacol. Sci.* 14: 13-20.
25. Schwarcz, R., and Coyle, J.T. 1977, Neurochemical sequelae of kainate injections in corpus striatum and substantia nigra of the rat, *Life Sci.,* 20: 431-436.
26. Seeburg, P.H. 1993, The TINS/TIPS lecture. The molecular biology of mammalian glutamate receptor channels, *Trends Neurosci.* 16: 359-365.
27. Shimizu, H., Ichishita, H., and Odagiri, H. 1974, Stimulated formation of cyclic adenosine 3',5'-monophosphate by aspartate and glutamate in cerebral cortical slices of guinea pig, *J. Biol. Chem.,* 249: 5955-5962.
28. Shimizu, H., Ichishita, H., and Umeda, I. 1975, Inhibition of glutamate-elicited accumulation of cyclic adenosine 3', 5'-monophosphate in brain slices by α,ω-diaminocarboxylic acids, *Mol. Pharmacol.* 11: 866-873.
29. Tanaka, T., Tanaka, S., Fujita, T., Takano, K., Fukuda, H., Sako, K., and Yonemasu, Y. 1991, Experimental complex partial seizures induced by a microinjection of kainic acid into limbic structures, *Prog. Neurobiol.,* 38: 317-334.
30. Tomida, Y., and Kimura, H. 1987, Immunohistochemical and biochemical studies of substances with taurine-like immunoreactivity in the brain, *Acta Histochem. Cytochem.,* 20: 31-40.
31. Varga, V., Török, K., Feuer, L., Gulyás, J., and Somogyi, J. 1985, γ-Glutamyltransferase in the brain and its role in formation of γ-L-glutamyl-taurine, *in:* "Taurine: Biological Actions and Clinical Perspectives", S.S. Oja, L. Ahtee, P. Kontro, and M.K. Paasonen, eds., Alan R. Liss, New York, pp. 115-125.
32. Varga, V., Janáky, R., Marnela, K.-M., Gulyás, J., Kontro, P., and Oja, S.S. 1989, Displacement of excitatory amino acid receptor ligands by acidic oligopeptides, *Neurochem. Res.* 14: 1223-1227.
33. Varga, V., Janáky, R., Marnela, K.-M., Saransaari, P., and Oja, S.S. 1994, Interactions of γ-L-glutamyltaurine with excitatory aminoacidergic neurotransmission, *Neurochem. Res.* 19: 243-248.
34. Varga, V., Janáky, R., Saransaari, P., and Oja, S.S. 1994, Endogenous γ-L-glutamyl and β-L-aspartyl peptides and excitatory aminoacidergic neurotransmission in the brain, *Neuropeptides,* 27: 19-26.
35. Wang, L.-Y., Salter, M.W., and MacDonald, J.F. 1991, Regulation of kainate receptors by cAMP-dependent protein kinase and phosphatases, *Science,* 253: 1132-1135.
36. Werner, P., Voigt, M., Keinänen, K., Wisden, W., and Seeburg, P.H. 1991, Cloning of a putative high-affinity kainate receptor expressed predominantly in hippocampal CA3 cells, *Nature,* 351: 742-744.
37. Worley, P.F., Baraban, J.M., De Souza, E.B., and Snyder, S.H. 1986, Mapping second messenger systems in the brain: differential localization of adenylate cyclase and protein kinase C, *Proc. Natl. Acad. Sci. USA,* 83: 4053-4057.
38. Ziegra, C.J., Willard, J.M., and Oswald, R.E. 1992, Coupling of a purified goldfish brain kainate receptor with a pertussis toxin-sensitive G protein, *Proc. Natl. Acad. Sci. USA,* 89: 4134-4138.

52

TAURINE AND NEURAL CELL DAMAGE

Transport of Taurine in Adult and Developing Mice

Pirjo Saransaari[1] and Simo S. Oja[1,2]

[1] Tampere Brain Research Center
University of Tampere Medical School
Box 607, FIN-33101 Tampere
[2] Department of Clinical Physiology
Tampere University Hospital
Box 2000, FIN-33521 Tampere, Finland

INTRODUCTION

There is evidence that taurine protects neural cells from excitotoxicity induced by excitatory amino acids[8], forestalls harmful metabolic cascades evoked by ischemia or hypoxia[40], and attenuates Ca^{2+} influx during ischemia[20]. Taurine also ameliorates symptoms in epilepsy[29]. The mechanism of this neuroprotection is not known, but it may be related, in addition to neuromodulation[38], to osmoregulatory, antioxidant and Ca^{2+} regulatory effects[16].

Of all neurons, hippocampal cells are the most vulnerable to transient ischemia causing subsequent cell death[36]. On the one hand, the hippocampus in the newborn is particularly sensitive to excitotoxic agents[7], but on the other more resistant to hypoxic insults than in the adult[4]. We have recently demonstrated that excitatory amino acid agonists evoke taurine release from both adult and developing mouse hippocampus[39] and cerebral cortex[37], the effects being more pronounced in the latter. Moreover, hypoxic conditions also induce marked taurine release in the hippocampus[39]. We have now further investigated how the uptake and release of taurine in developing and adult mice respond to such experimental conditions which are known to be damaging to the brain, e.g., to hypoxia, hypoglycemia, ischemia and exposure to free radicals and oxidative stress.

MATERIAL AND METHODS

Material

NMRI mice of both sexes aged 6-8 days and 3 months (adults) were used throughout. [1,2-^{3}H]Taurine (specific activity 1.07 PBq/mol) was obtained from Amersham Interna-

Taurine 2, edited by Huxtable et al.
Plenum Press, New York, 1996

tional, Bristol, UK. Dizocilpine (MK-801) was a gift from Merck, Sharp & Dohme (Rahway, NJ). Other reagents were from common commercial sources.

Uptake Experiments

Synaptosomal preparations were obtained from the cerebral cortex of developing and adult mice and incubated in Krebs-Ringer-Hepes-glucose medium (pH 7.4) as described in detail by Kontro[17]. The synaptosomes were first preincubated under O_2 for 10 min and then for 10 min with 30 MBq/l of [^{3}H]taurine. The kinetics of the uptake were assessed in the concentration range from 1 µM to 10 mM and the parameters estimated by nonlinear regression analysis using the simplex algorithm. The protein content of the synaptosomes was measured according to Lowry et al.[22].

Efflux Experiments

Slices 0.4 mm thick weighing 15-20 mg were prepared from the hippocampi with a Stadie-Riggs tissue slicer and used immediately in efflux experiments. The slices were preloaded for 30 min with 10 µM (50 MBq/l) [^{3}H]taurine in preoxygenated Krebs-Ringer-Hepes-glucose medium under O_2 and superfused with the same medium as described in detail in Kontro and Oja[18]. The medium was pooled during the first 20 min of superfusion and thereafter 2-min fractions (0.5 ml) were collected. At 30 min the medium was in many experiments changed to another modified medium. After superfusion the slices were weighed, homogenized in ice-cold 5 % (w/v) trichloracetic acid solution and centrifuged, and the clear supernatants used for scintillation counting. The effluent samples were likewise counted for radioactivity.

Estimation of Efflux Rate Constants

Desaturation curves of labeled taurine from the slices were plotted as a function of time on the basis of the radioactivities remaining in the slices after superfusion and recovered in the collected superfusate fractions[18]. The efflux rate constants of taurine for the time intervals of 20 to 30 min (k_1) and 34 to 50 min (k_2) were computed as negative slopes for the regression lines of the logarithm of radioactivity remaining in the slices vs. superfusion time.

Experimental Conditions

Neural cell damage was induced by modified experimental conditions in both uptake and release experiments. The medium for metabolic blockade contained 5 mM NaCN, 5 mM iodoacetate or 1 mM 2,4-dinitrophenol. In hypoglycemia, glucose was omitted from the medium and in hypoxia the medium was bubbled with N_2 gas for 1 h before the experiments and then during experiments as detailed in the results section. In ischemia the glucose-free medium was bubbled with N_2 gas. Lipid peroxidation[43] (oxidative stress) was induced by $FeSO_4$ (7.5 µM) together with ascorbate (1.5 mM)[1]. Free radical production was achieved by exposure to hydrogen peroxide (0.01 %) or hydrogen peroxide together with $FeSO_4$[10].

Statistical Calculations

The presence of statistically significant differences between the sample means was detected by variance analysis. Comparisons of individual means were made by Hartley's sequential method of testing.

Table 1. [^{3}H]Taurine uptake by synaptosomal preparations from the cerebral cortex from adult and developing mice in various experimental conditions

	Uptake (% of control) ± SEM	
Experimental conditions	Adult	7-day-old
Control	100.0 ± 3.5 (8)	100.0 ± 4.1 (8)
NaCN	75.8 ± 3.8** (4)	84.4 ± 3.2* (4)
Iodoacetate	79.1 ± 4.2** (4)	83.1 ± 2.9* (4)
Hypoglycemia	98.7 ± 2.8 (4)	56.9 ± 4.2** (4)
Hypoxia	106.2 ± 3.6 (4)	99.2 ± 2.4 (4)
Ischemia	99.8 ± 1.5 (4)	94.9 ± 3.8 (4)
Ischemia, 10 min	63.0 ± 5.1** (4)	81.7 ± 3.8* (4)
Ischemia, 30 min	53.1 ± 3.0** (4)	51.6 ± 2.9** (4)
Ischemia, 60 min	11.0 ± 0.8** (4)	32.3 ± 1.0** (4)
Oxidative stress	102.0 ± 3.2 (4)	108.1 ± 5.2 (4)
Free radicals	51.6 ± 2.7** (4)	87.9 ± 2.4 (4)
Oxidative stress + free radicals	51.2 ± 3.4** (4)	86.8 ± 3.2 (4)

The synaptosomes were incubated for 10 min with 10 μM [^{3}H]taurine in Krebs-Ringer-Hepes medium (pH 7.4) under the various experimental conditions, explained in detail in Material and Methods. Number of independent experiments in parenthesis. Significance of differences compared to normal conditions (control): $^{*}p<0.05$, $^{**}p<0.01$.

RESULTS

The uptake of [^{3}H]taurine by synaptosomal preparations isolated from the cerebral cortices of adult and developing mice was measured in various experimental conditions at a concentration of 10 μM. The metabolic poisons cyanide and iodoacetate (both 5 mM) moderately inhibited the uptake, the inhibition being more marked in the adults (Table 1). When glucose was omitted from the incubation medium (hypoglycemia) the uptake was reduced only in developing mice, but the 10-min incubation under N_2 (hypoxia) had no effect at either age. When glucose was omitted under N_2 (ischemic conditions) during the final 10-min incubation the uptake was unaffected, being however gradually inhibited more and more as the ischemia was prolonged. The reduction was more pronounced in the adults (Table 1).

In order to assess how the different kinetic parameters of transport were altered, the uptake of taurine was analyzed in detail in normal and ischemic conditions. In normal incubation conditions the uptake was saturable at both ages, though considerable nonsaturable penetration was also discernible (Fig. 1). The saturable uptake consisted of two components, high- and low-affinity (Table 2). The maximal velocities (V) of both high- and low-affinity uptake components were greater in developing than in adult mice, whereas the transport constants (K_m) were smaller (Table 2). In ischemic conditions no high-affinity component could be extracted from the total influx in the adults, whereas the maximal velocity of low-affinity uptake and nonsaturable diffusion were increased. In the immature cerebral cortex only the diffusion component was markedly increased in ischemic conditions.

Oxidative stress induced by $FeSO_4$ and ascorbate had no effect on taurine uptake, and production of free radicals by H_2O_2 reduced the uptake only in the adults (Table 1). Furthermore, taurine uptake in the immature cerebral cortex was not affected by oxidative stress together with free radicals (Table 1) or by hypoxic conditions together with oxidative stress and/or free radicals (data not shown), in which conditions the uptake was strongly reduced in the adults, to 45 ± 3.2% (n = 4), 15 ± 1% (n = 4) and 28 ± 1% (n = 4) of the control, respectively.

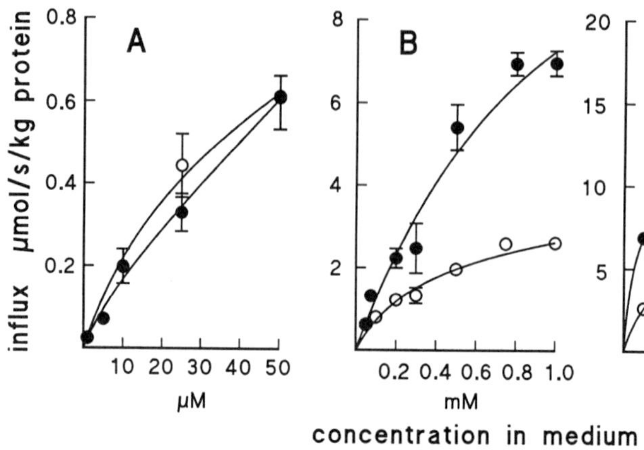

Figure 1. Taurine uptake by cerebral cortical synaptosomal preparations from adult mice as a function of the taurine concentration in control (-O-) and ischemic (-●-) conditions. (A) Low, (B) medium and (C) high concentration range. The results are mean values ± SEM of 4-8 experiments. The calculated kinetic parameters are shown in Table 1.

The basal release of [^{3}H]taurine from hippocampal slices from adult and developing mice was generally greatly increased by the experimental conditions studied (initiated from the outset of superfusion); i.e., in the presence of 2,4-dinitrophenol (1.0 mM) and cyanide (5.0 mM), in hypoglycemic, ischemic and hypoxic conditions and in medium inducing free radical production (Figs 2-3, Table 3). Oxidative stress did not affect taurine release. Potassium stimulation (50 mM) potentiated taurine release in all conditions except for hypoglycemia and ischemia in the adults (Table 3). The unstimulated release of taurine was enhanced relatively even more substantially in various experimental conditions. The only exception was the oxidative stress in developing mice, in which the unstimulated release was unaltered and the magnitude of potassium stimulation slightly diminished.

Table 2. Kinetic parameters of taurine uptake in synaptosomal preparations from the mouse cerebral cortex

Parameter	Control	Ischemic conditions
Adult		
Km_1, µM	18 ± 2	—
V_1, nmol/s/kg	28 ± 3	—
Km_2, mM	1.1 ± 0.2	0.80 ± 0.12
V_2, µmol/s/kg	4.5 ± 0.6	11 ± 1*
K_d, $s^{-1} \times 10^{-3}$	0.39 ± 0.8	0.74 ± 0.19*
Developing		
Km_1, µM	20 ± 5	23 ± 8
V_1, nmol/s/kg	44 ± 13	69 ± 18
Km_2, mM	0.21 ± 0.01	0.21 ± 0.03
V_2, µmol/s/kg	6.5 ± 3.0	6.7 ± 1.0
K_d, $s^{-1} \times 10^{-3}$	0.89 ± 0.04	2.78 ± 0.10*

Mean values ± SEM. Significance of difference between control and ischemic conditions: *$p<0.01$.

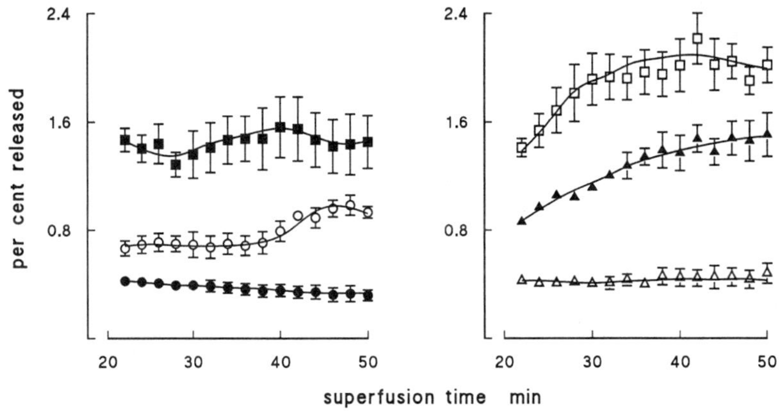

Figure 2. Time-course of taurine release from hippocampal slices from adult mice in control (-●-) and in cell-damaging conditions: hypoxia (-○-), hypoglycemia (-■-), ischemia (-□-), oxidative stress (-▲-), and free radicals (-△-). The slices were superfused in the above conditions as described in Material and Methods. The results are mean values ± SEM of 4 independent experiments.

Table 3. [^{3}H]Taurine release from mouse hippocampal slices in various experimental conditions

Experimental conditions	Efflux rate constants ± SEM k_2 (34-50 min) x 10^{-3} min^{-1}	
	Adult	7-day-old
Control	1.47 ± 0.03 (71)	0.30 ± 0.01 (73)
+ 50 mM K^+	2.06 ± 0.07 (6)	2.13 ± 0.21 (8)
2,4-DNP	6.95 ± 0.24 (6)	7.68 ± 0.07 (4)
NaCN	5.87 ± 0.38 (4)	0.51 ± 0.08 (4)
Hypoglycemia	4.55 ± 0.50 (10)	2.21 ± 0.21 (16)
+ 50 mM K^+	4.99 ± 0.64 (3)	3.49 ± 0.14 (8)
Ischemia	8.81 ± 0.68 (8)	2.56 ± 0.10 (4)
+ 50 mM K^+	6.77 ± 0.15 (4)	3.79 ± 0.49 (4)
Hypoxia	3.80 ± 0.30 (5)	0.77 ± 0.07 (9)
+ 50 mM K^+	4.74 ± 0.63 (3)	2.32 ± 0.09 (7)
Free radicals	6.18 ± 0.51 (4)	2.37 ± 0.13 (4)
+ 50 mM K^+	7.01 ± 0.51 (4)	3.31 ± 0.35 (4)
Oxidative stress	1.67 ± 0.16 (4)	0.27 ± 0.03 (4)
+ 50 mM K^+	2.85 ± 0.08 (4)	1.53 ± 0.19 (4)
+ hypoxia	5.21 ± 0.35 (4)	2.13 ± 0.54 (4)

The slices were preloaded for 30 min with 10 μM [^{3}H]taurine in Krebs-Ringer-Hepes-glucose medium (pH 7.4) and superfused for 50 min in the experimental conditions described in detail in Material and Methods. Results are the rate constants ± SEM for the period of 34-50 min. Number of independent experiments in parenthesis. The release was significantly ($p<0.01$) increased by all experimental conditions except for oxidative stress. The potassium stimulation further significantly ($p<0.01$) enhanced the release, except for hypoglycemia and ischemia in the adults.

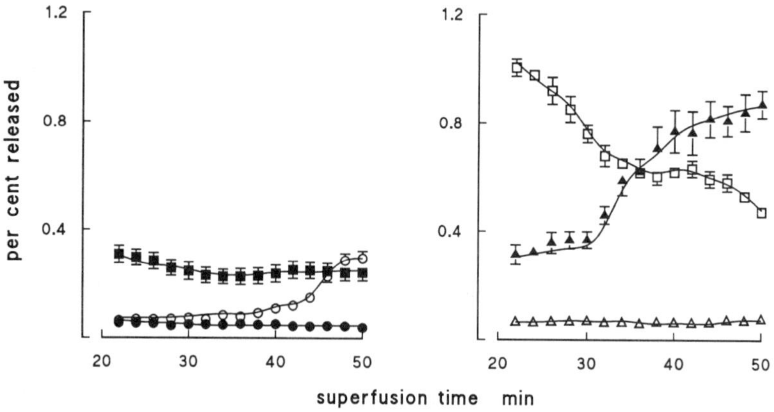

Figure 3. Time-course of taurine release from hippocampal slices from 7-day-old mice in control (-●-) and in cell damaging conditions: hypoxia (-○-), hypoglycemia (-■-), ischemia (-□-), oxidative stress (-▲-), and free radicals (-△-). The slices were superfused in the above conditions as described in Material and Methods. The results are mean values ± SEM of 4 independent experiments.

The effects of glutamate receptor agonists on taurine release were studied in hypoxic conditions (Fig. 4). In the adult hippocampus, kainate and 2-amino-3-hydroxy-5-methylisoxazole-4-propionate (AMPA) (both 0.1 mM) potentiated taurine release. These effects were not influenced by their respective antagonists 6-cyano-7-nitroquinoxaline-2,3-dione (CNQX) and 6-nitro-7-sulphamoylbenzo(f)quinoxaline-2,3-dione (NBQX) (both 0.1 mM). In the hippocampus of developing mice N-methyl-D-aspartate (NMDA), kainate and AMPA (all 0.1 mM) greatly stimulated the release, NMDA and AMPA being the most potent. These actions were greatly attenuated by their respective antagonists, dizocilpine (MK-801), CNQX and NBQX.

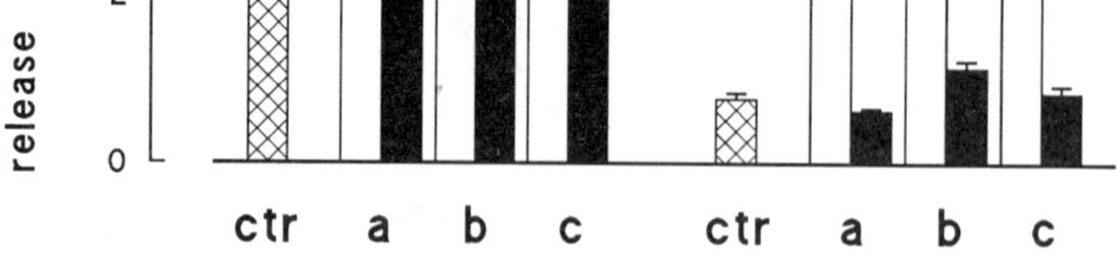

Figure 4. Effect of glutamatergic substances (all 0.1 mM) on taurine release from hippocampal slices from adult (A) and developing (B) mice in hypoxic conditions. The results show efflux rate constants (34-50 min) k_2 (x 10^{-3} min^{-1}) in hypoxic medium (ctr) and in the presence of NMDA (a), kainate (b) and AMPA (c). The open columns represent the effects of NMDA, kainate and AMPA alone and the black columns the effects of the same substances in the presence of their respective antagonists MK-801, CNQX and NBQX. The results are mean values ± SEM of 4-8 experiments. There were no significant differences in adult mice, but the antagonists significantly ($p < 0.01$) reduced the effects of the agonists in developing mice.

DISCUSSION

The conditions which are known to cause neural cell damage[12,14] now affected the uptake and release of taurine in brain tissue. The release processes were, however, the more profoundly affected. Metabolic poisons, hypoglycemia, hypoxia and ischemia all distort the oxidative metabolism apparently needed for active taurine transport across cell membranes. Taurine possesses a saturable, Na^+-dependent transport system operating at neuronal and glial cell membranes, comprising both high- and low-affinity components[16,28]. Short-term exposure to hypoxia and ischemia had seemingly no effect on uptake of 10 μM taurine uptake. Kinetic analyses of the uptake revealed, however, that nonsaturable diffusion was greatly increased under ischemic conditions in both age groups. Apparently membrane structures were disrupted, thus increasing the leakage of taurine molecules. The carrier proteins were still operative, however. Long-lasting exposure to ischemia gradually reduced taurine uptake, more profoundly in the adults. The uptake systems seem to be more resistant to disturbances in oxidative metabolism in developing than in adult mice. The immature brain is also generally very resistant to hypoxia[27].

Biological systems produce free radicals as intermediate by-products of oxidative metabolism. Under certain conditions, including hypoxia, ischemia and ageing, free radical levels and membrane lipid peroxidation may become exaggerated, damaging nervous system functions by perturbing the membrane structure[6,34,42]. The free radicals are also thought to be involved in a number of neurological disorders[3,13]. Experimental conditions inducing production of free radicals and oxidative stress now affected taurine uptake only in the adults. The uptake of GABA by sheep brain synaptosomes is also decreased after membrane lipid peroxidation[32].

Both basal and K^+-stimulated release of taurine were modified by cell-damaging conditions in the adult and developing hippocampus. The considerable increase in the basal release in hypoxic and ischemic conditions could be due partly to inhibition of reuptake, but other mechanisms must also be involved. Hypoxic and ischemic conditions greatly enhance the release of excitatory amino acids[2,5,11,21,31,35], the release probably being caused by reversed operation of uptake carriers[41]. It has been proposed that the released glutamate activates glutamate receptors, particularly of the NMDA type, increases the intracellular concentration of Ca^{2+} and triggers a long-lasting potentiation of NMDA receptor-gated currents[41]. The massive amount of taurine released simultaneously could counteract by several mechanisms these noxious effects which lead to neuronal death. Taurine could attenuate the depolarizing effects of excitatory amino acids by increasing membrane Cl^- conductance[30], acutely inhibit Ca^{2+} influx and antagonize depolarization-evoked Ca^{2+} efflux[19]. Furthermore, taurine still continues to inhibit Ca^{2+} influx during the posthypoxic period[20]. The activated NMDA receptors also trigger a substantial release of taurine in the hippocampus[20,23,25,26,39], in the immature hippocampus in particular[39]. Now we have shown for the first time that in developing mice the release evoked by NMDA in hypoxia is even greater than in normoxia. Moreover, the release evoked by the kainate and AMPA types of glutamate receptors was also greater in hypoxic than in normoxic conditions in developing mice. In the adult hippocampus the NMDA-evoked release[39] seems, however, to be lost in hypoxic conditions. AMPA and kainate still stimulated taurine release, but the release was not receptor-mediated. Taurine seems thus to play an important role in the maintenance of homeostasis in the developing hippocampus upon impending hyperexcitation.

The properties of enhanced release of taurine were also fairly similar in free radical-containing medium and hypoxia. The effects of ischemia and exposure to free radicals share many common features[10]. The release of glutamate and D-aspartate has also been enhanced by exposure to peroxide-generated free radicals in rat hippocampal slices[33] and cerebral cortical synaptosomes[9,10], leading to accumulation of extracellular excitatory amino acids and contributing to excitotoxicity. The increase in the release could be due to disruption of Na^+-Ca^{2+} exchange by peroxide-induced free radicals[15], resulting in an elevated concentration of intracellular Ca^{2+} and consequently in enhanced amino acid release.

Taurine has been shown to improve the recovery of neuronal functions after cerebral hypoxia[40] and taurine-containing neurons in the hippocampus are more resistant to damage induced by ischemia than other neurons[24,44]. On the one hand cell-damaging conditions inhibit the uptake of taurine and on the other greatly increase the release of taurine. The increase in the extracellular levels of taurine in cell-damaging conditions may thus be an important endogenous protective mechanism particularly efficient in the immature brain.

ACKNOWLEDGMENTS

The skillful technical assistance of Mrs. Irma Rantamaa and Ms. Paula Kosonen and the financial support of the Medical Research Fund of Tampere University Hospital are gratefully acknowledged.

REFERENCES

1. Agostinho, P., Duarte, C.B., Carvalho, A.P., and Oliveira, C.R., 1994, Effect of oxidative stress on the release of [^{3}H]GABA in cultured chick retina cells, *Brain Res.,* 655: 213-221.
2. Benveniste, H., Drejer, J., Schousboe, A., and Diemer, N.H, 1984, Elevation of the extracellular concentrations of glutamate and aspartate in rat hippocampus during transient cerebral ischemia monitored by intracerebral microdialysis, *J. Neurochem.,* 43: 1369-1374.
3. Bondy, S.C., 1995, The relation of oxidative stress and hyperexcitation to neurological disease, *Proc. Soc. Exp. Biol. Med.,* 208: 337-345.
4. Cherubini, E., Ben-Ari, Y., and Krnjevic, K., 1989, Anoxia produces smaller changes in synaptic transmission, membrane potential and input resistance in immature rat hippocampus, *J. Neurophysiol.,* 62: 882-895.
5. Collard, K.J., and Menon-Johansson, A.S., 1993, Effects of short-term hypoxia on [^{3}H]glutamate release from preloaded hippocampal and cortical synaptosomes, *Neurochem. Res.,* 8: 165-170.
6. Colton, C.A., and Gilbert, D., 1985, Effect of H_2O_2 on excitatory transmission in the lobster neuromuscular junction, *Fed. Proc.,* 44: 1273.
7. Cook, T.M., and Crutcher, K.A., 1986, Intrahippocampal injection of kainic acid produces significant pyramidal cell loss in neonatal rats, *Neuroscience,* 18: 79-92.
8. French, E.D., Vezzani, A., Whetsell, W.O., Jr., and Schwarcz, R., 1986, Antiexcitotoxic actions of taurine in the rat hippocampus studied in vivo and in vitro, *Adv. Exp. Med. Biol.,* 203: 349-362.
9. Gilman, S.C., Bonner, M.J., and Pellmar, T.C., 1992, Peroxide effects on [^{3}H]L-glutamate release by synaptosomes isolated from the cerebral cortex, *Neurosci. Lett.,* 140: 157-160.
10. Gilman, S.C., Bonner, M.J., and Pellmar, T.C., 1994, Free radicals enhance basal release of D-[^{3}H]aspartate from cerebral cortical synaptosomes, *J. Neurochem.,* 62: 1757-1763.
11. Globus, M.Y.-T., Busto, R., Dietrich, W.D., Martinez, E., Valdes, I., and Ginsberg, M.D., 1988, Effect of ischemia on the in vivo release of striatal dopamine, glutamate and γ-aminobutyric acid studied by intracerebral microdialysis, *J. Neurochem.,* 51: 1455-1464.
12. Haddad, G.G., and Jiang, C., 1993, O_2 deprivation in the central nervous system: on mechanisms of neuronal response, differential sensitivity and injury, *Prog. Neurobiol.,* 40: 277-318.
13. Halliwell, B., 1992, Reactive oxygen species and the central nervous system, *J. Neurochem.,* 59: 1609-1623.
14. Hara, H., Sukamoto, T., and Kogure, K., 1993, Mechanism and pathogenesis of ischemia-induced neuronal damage, *Prog. Neurobiol.,* 40: 645-670.
15. Hayashi, H., Miyata, H., Watanabe, H., Kobayashi, A., and Yamazaki, N., 1989, Effects of hydrogen peroxide on action potentials and intracellular Ca^{2+} concentration of guinea pig heart, *Cardiovasc. Res.,* 23: 767-773.
16. Huxtable, R.J., 1992, The physiological actions of taurine, *Physiol. Rev.,* 72: 101-163.

17. Kontro, P., 1984, Comparison of taurine, hypotaurine, and ß-alanine uptake in brain synaptosomal preparations from developing and adult mouse, *Int. J. Dev. Neurosci.*, 2: 465-470.
18. Kontro, P., and Oja, S.S., 1987, Taurine and GABA release from mouse cerebral cortex slices: potassium stimulation releases more taurine than GABA from developing brain, *Dev. Brain Res.*, 37: 277-291.
19. Kontro, P., and Oja, S.S., 1988, Effects of taurine on the influx and efflux of calcium in brain slices of adult and developing mice, *Int. J. Neurosci.*, 38: 103-109.
20. Lehmann, A., Hagberg, H., Nyström, B., Sandberg, M., and Hamberger, A., 1985, In vivo regulation of extracellular taurine and other neuroactive amino acids in rabbit hippocampus. *Prog. Clin. Biol. Res.*, 179: 289-311.
21. Lekieffre, D., Callebert, J., Plotkine, M., and Boulu, R.G., 1992, Concomitant increases in the extracellular concentrations of excitatory and inhibitory amino acids in the rat hippocampus during forebrain ischemia, *Neurosci. Lett.*, 137: 78-82.
22. Lowry, O.H., Rosebrough, N.J., Farr, A.L., and Randall, R.J., 1951, Protein measurement with the Folin phenol reagent, *J. Biol. Chem.*, 193: 265-275.
23. Magnusson, K. R., Koerner, J.F., Larson, A.A., Smullin, D.H., Skilling, S.R., and Beitz, A.J., 1991, NMDA-, kainate- and quisqualate-stimulated release of taurine from electrophysiologically monitored rat hippocampal slices, *Brain Res.*, 549: 1-8.
24. Matsumoto, K., Ueda, S., Hashimoto, T., and Kuriyama, K., 1991, Ischemic neuronal injury in the rat hippocampus following transient forebrain ischemia: evaluation using in vivo microdialysis, *Brain Res.*, 543: 236-242.
25. Menéndez, N., Herreras, O., Solis, J.M., Herranz, A.S., and Martin del Rio, R., 1989, Extracellular taurine increase in rat hippocampus evoked by specific glutamate receptor activation is related to the excitatory potency of glutamate agonists, *Neurosci. Lett.*, 102: 64-69.
26. Menéndez, N., Solis, J.M., Herreras, O., Herranz, A.S., and Martin del Rio, R., 1990, Role of endogenous taurine on the glutamate analogue-induced neurotoxicity in the rat hippocampus in vivo, *J. Neurochem.*, 55: 714-717.
27. Nabetani, M., and Okada, Y., 1994, Developmental and regional differences in the vulnerability of rat hippocampal slices to brief and prolonged periods of hypoxia, *Dev. Neurosci.*, 16: 301-306.
28. Oja, S.S., and Kontro, P., 1983, Taurine, *in:* "Handbook of Neurochemistry", Vol. 3, 2nd edn, A. Lajtha, ed., Plenum Press, New York, pp. 501-533.
29. Oja, S.S., and Kontro, P., 1983, Free amino acids in epilepsy: possible role of taurine, *Acta Neurol. Scand.*, 67, Suppl., 93: 5-20.
30. Oja, S.S., Korpi, E.R., and Saransaari, P., 1990, Modification of chloride flux across brain membranes by inhibitory amino acids in developing and adult mice, *Neurochem. Res.*, 15: 797-804.
31. O'Regan, M.H., Smith-Barbour, M., Perkins, L.M., and Phillis, J.W., 1995, A possible role for phospholipases in the release of neurotransmitter amino acids from ischemic rat cerebral cortex, *Neurosci. Lett.*, 185: 191-194.
32. Palmeira, C.M., Santos, M.S., Carvalho, A.P., and Oliveira, C.R., 1993, Membrane lipid peroxidation induces changes in γ-[^{3}H]aminobutyric acid transport and calcium uptake by synaptosomes, *Brain Res.*, 609: 117-123.
33. Pellegrini-Giampietro, D.E., Cherici, G., Alesiani, M., Carla, V., and Moroni, F., 1990, Excitatory amino acid release and free radical formation may cooperate in the genesis of ischemia-induced neuronal damage, *J. Neurosci.*, 10: 1035-1041.
34. Pellmar, T.C., Neel, K.L., and Lee, K.H., 1989, Free radicals mediate peroxidative damage in the guinea pig hippocampus in vitro, *J. Neurosci. Res.*, 24: 437-444.
35. Phillis, J.W., Walter, G.A., and Simpson, R.E., 1991, Brain adenosine and transmitter amino acid release from the ischemic rat cerebral cortex: effect of the adenosine deaminase inhibitor deoxycoformycin, *J. Neurochem.*, 56: 644-650.
36. Pulsinelli, W.A., Brierley, J.B., and Plum, F., 1982, Temporal profile of neuronal damage in a model of transient forebrain ischemia, *Ann. Neurol.*, 11: 491-498.
37. Saransaari, P., and Oja, S.S., 1991, Excitatory amino acids evoke taurine release from cerebral cortex slices from adult and developing mice, *Neuroscience*, 45: 451-459.
38. Saransaari, P., and Oja, S.S., 1992, Release of GABA and taurine from brain slices, *Prog. Neurobiol.*, 38: 455-482.
39. Saransaari, P., and Oja, S.S., 1994, Taurine release from mouse hippocampal slices: effects of glutamatergic substances and hypoxia, *Adv. Med. Exp. Biol.*, 359: 279-287.
40. Schurr, A., Tseng, M.T., West, C.A., and Rigor, B.M., 1987, Taurine improves the recovery of neuronal function following cerebral hypoxia: an in vitro study, *Life Sci.*, 40: 2059-2066.

41. Szatkowski, M., and Attwell, D., 1994, Triggering and execution of neuronal death in brain ischemia: two phases of glutamate release by different mechanisms, *Trends Neurosci.*, 17: 359-365.
42. Watson, B.D., Busto, R., Goldberg, W.J., Santiso, M., Yoshida, S., and Ginsberg, M.D., 1984, Lipid peroxidation in vivo induced by reversible global ischemia in rat brain, *J. Neurochem.*, 42: 268-274.
43. Wills, E.D., 1969, Lipid peroxide formation in microsomes. General considerations, *Biochem. J.*, 113: 315-324.
44. Wu, J.-Y., Johansen, F.F., Lin, C.-T., and Liu, J.-W., 1987, Taurine system in the normal and ischemic rat hippocampus, *Adv. Exp. Med. Biol.*, 217: 265-274.

PROTECTIVE EFFECT OF TAURINE AGAINST REPERFUSION INJURY IN CULTURED RAT ASTROCYTES

Toshio Matsuda,[1] Kazuhiro Takuma,[1] Yoko Kishida,[1] Junichi Azuma,[2] and Akemichi Baba[1]

[1] Department of Pharmacology
[2] Department of Clinical Evaluation of Medicine and Therapeutics
Faculty of Pharmaceutical Sciences
Osaka University, Suita, Osaka 565, Japan

INTRODUCTION

Astrocytes, the most abundant glial cells in the brain, are considered to have physiological and pathological roles in neuronal activities. Goldman *et al.*[3] and we[11] independently reported that the Na^+/Ca^{2+} exchanger is present in cultured astrocytes. The Na^+/Ca^{2+} exchanger plays an important role in regulation of $[Ca^{2+}]_i$, the antiporter being bi-directional: the forward mode causes Ca^{2+} efflux, and the reverse mode causes Ca^{2+} influx.

We[9] have recently found that a Ca^{2+} paradox-like phenomenon occurred in cultured rat astrocytes. We demonstrated, using the exchange inhibitor ascorbic acid[10], the exchange stimulator nitric oxide[1] and the antisense oligomers, that the Na^+/Ca^{2+} exchanger might be involved in the phenomenon. In this study, we examined the effect of taurine on reperfusion injury in cultured rat astrocytes, in view of the previous finding that this sulfur amino acid had a protective effect on Ca^{2+} paradox-induced heart failure[6]. This paper demonstrates that taurine reduces reperfusion injury and the effect is possibly mediated by an inhibition of the Na^+/Ca^{2+} exchanger in the reverse mode.

METHODS

Drugs were obtained from the following sources: $^{45}CaCl_2$, Amersham (Tokyo, Japan); fetal calf serum (FCS), JRH Biosciences, Lenexa, KS); fura-2 acetoxymethylester, Dojindo Lab. (Kumamoto, Japan); amiloride, 3-(4,5-dimethylthiazol-2-yl)-2,5-diphenyl tetrazolium bromide (MTT), Sigma Chemical Co. (St. Louis, MO); and Eagle's minimum essential medium (MEM), Nissui Pharmaceutical 'Co., Ltd. (Tokyo, Japan). All other chemicals used were of the highest purity commercially available.

Taurine 2, edited by Huxtable et al.
Plenum Press, New York, 1996

Astrocytes were isolated from cerebral cortices of 1-day-old Sprague-Dawley rats as reported previously[11]. The secondary cultures on plastic plates and glass coverslips were grown for 14-20 days and for 1-3 days, respectively. The cells on the plastic plates consisted of more than 95% flat polygonal astrocytes (type-1 astrocytes), as confirmed by phase contrast microscopy and positive immunostaining with anti-glial fibrillary acidic protein antibody.

Hanks balanced saline solution (HBSS) containing (mM): NaCl 137; KCl 5.4; $CaCl_2$ 1.3; $MgCl_2$ 0.49; $MgSO_4$ 0.41; Na_2HPO_4 0.34; KH_2PO_4 0.44; $NaHCO_3$ 4.17 and D-glucose 5.55 (pH 7.3 equilibrated with air) and Earle's solution containing (mM): NaCl 116; KCl 5.4; $CaCl_2$ 1.8; $MgSO_4$ 0.81; NaH_2PO_4 1.01; $NaHCO_3$ 26.2 and D-glucose 5.55 (pH 7.3 under CO_2/air) were used for $[Ca^{2+}]_i$ measurement in an air atmosphere and viability measurement of cells in a humidified CO_2/air atmosphere, respectively. Ca^{2+}-free solution was made by replacing $CaCl_2$ with 10 mM ethylene glycol bis(β-aminoethyl ether)*N,N,N',N'*-tetraacetic acid. In some experiments, Na^+ was replaced with choline to make a low Na^+ solution. The cells were washed and exposed to Ca^{2+}-containing solution (control) and Ca^{2+}-free solution for the indicated time (usually 30 min). After 2 rapid washes, the cells were reperfused with Ca^{2+}-containing solution for the indicated time.

The method for measurement of $[Ca^{2+}]_i$ has been published[11]. Briefly, fura-2 fluorescence (510 nm emission excited by 340 and 380 nm illumination) from astrocytes, as well as background fluorescence, were imaged using a Nikon microscope system. The fluorescence ratios of fura-2-loaded astrocytes were calculated and calibrated to express $[Ca^{2+}]_i$. Measurements (usually 4-5 cells per coverslip) were limited to non-nuclear areas to simplify interpretation of the data.

Cell survival was measured by a colorimetric assay using MTT, as reported previously[8]. The survival of cells was expressed as % of control. Taurine content in the cells was measured using HPLC as described previously[5]. Na^+/Ca^{2+} exchange activity in astrocytes was determined using ouabain and monensin[11].

RESULTS

Fig. 1 shows that 30 min exposure of astrocytes to Ca^{2+}-free medium causes taurine release. Taurine content in the cells decreased after Ca^{2+} depletion and did not return to control level on reperfusion (Fig. 1 A). Consistent with this observation, taurine content in the medium was markedly increased by Ca^{2+} depletion, and it was not changed by reperfusion (Fig. 1 B).

Reperfusion after Ca^{2+}-depletion increased $[Ca^{2+}]_i$, but this increase was significantly reduced by reperfusion with taurine-containing medium (Fig. 2). Fig. 3 shows the effect of taurine on reperfusion injury in cultured astrocytes. Taurine at 3-30 mM reduced significantly the injury. In contrast to taurine, glycine, GABA and β-alanine were ineffective in protecting against the injury (Fig. 4). When astrocytes were exposed to ouabain and monensin for 30 min or 4 h, cell death was observed after reperfusion. Cell toxicity was reduced by reperfusion with a taurine-containing medium (Fig. 5). Fig. 6 shows the effect of taurine added at different times on reperfusion injury. When taurine was present during Ca^{2+}-depletion or Ca^{2+}-repletion, the protective effect was observed. However, taurine added 3 h after reperfusion was ineffective in protection.

The effect of taurine on Na^+/Ca^{2+} exchange activity in astrocytes was examined using ouabain and monensin as described previously[11]. Taurine inhibited the exchange activity in a dose-dependent manner: a significant effect was observed at concentrations above 3 mM (Fig. 7).

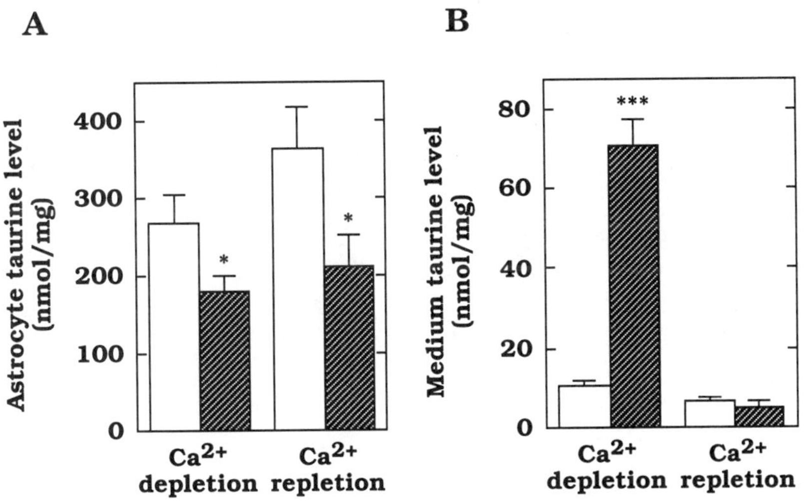

Figure 1. Effects of Ca^{2+} depletion and Ca^{2+} repletion on taurine release in cultured rat astrocytes. Astrocytes were exposed to normal medium (open, control) and Ca^{2+}-free medium (hatched) for 30 min and then reperfused with normal medium for 6 h. Taurine contents in the cells (A) and medium (B) were determined. Results are means ± SE of 6 wells obtained from 2 separate preparations. $*p < 0.05$: $***p < 0.001$, compared with the corresponding value without Ca^{2+}-depletion (Student's t-test).

DISCUSSION

The present study shows that taurine protects astrocytes from reperfusion injury (Fig. 3). The reperfusion injury was similar to the Ca^{2+} paradox phenomenon which was originally found in the heart. In both the heart and astrocytes, Ca^{2+}-repletion after Ca^{2+}-depletion causes an increase in $[Ca^{2+}]_i$ and cell injury. It should be noted that there was a difference in the period for expression of cell injury between the heart and astrocytes. Cell toxicity of astrocytes was observed 3-5 days after reperfusion, indicating delayed cell death. Although the previous study[9] shows that Ca^{2+} entry via Na^+/Ca^{2+} exchanger in the reverse mode is involved in reperfusion injury, the exact mechanism for cell death is not known.

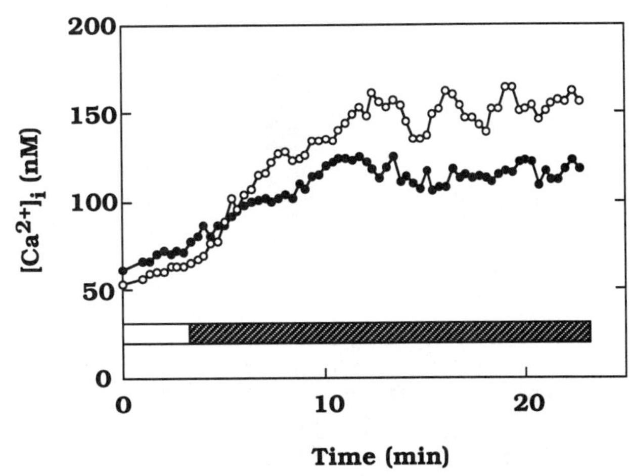

Figure 2. Effect of taurine on reperfusion-induced increase in $[Ca^{2+}]_i$ in single astrocytes. The cells were exposed to Ca^{2+}-free medium and reperfused in the presence (closed) and absence (open) of 10 mM taurine. Results are means ± SE of 56 cells obtained from 12 separate coverelips. Two-way ANOVA shows that the effect of taurine is significant at $p < 0.01$.

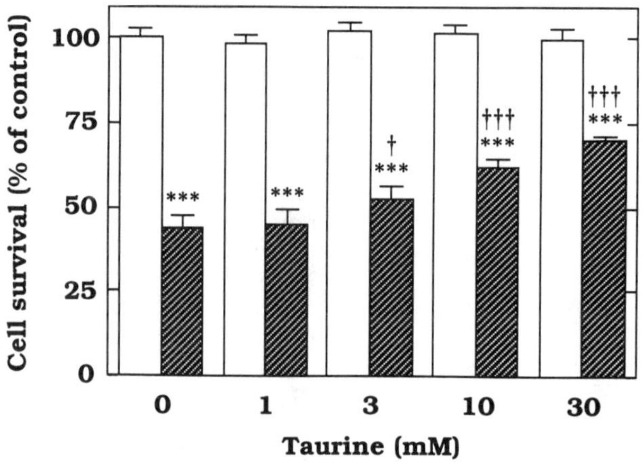

Figure 3. Effect of taurine on reperfusion-induced cell toxicity in cultured astrocytes. The cells were exposed to normal medium (open) and Ca^{2+}-free medium (hatched) for 30 min, and reperfused for 3 days. Cell survival was determined by the MTT assay. Results are means ± SE of 16-18 wells obtained from 4-6 separate preparations. $***p < 0.001$, compared with the corresponding value without Ca^{2+}-depletion (Student's t-test): $^{+}p < 0.05$, $^{+++}p < 0.001$, compared with the value of Ca^{2+}-depletion without taurine (ANOVA followed by Duncan test).

Kramer *et al.*[6] reported that taurine had a protective effect on Ca^{2+} paradox-induced heart failure, but the mechanism was not elucidated. The present study provides the first evidence that taurine reduces reperfusion-induced increase in $[Ca^{2+}]_i$ (Fig. 2). Moreover, we demonstrated that taurine inhibited Na^+/Ca^{2+} exchange activity in the reverse mode in cultured astrocytes (Fig. 7), in agreement with the previous finding in cardiac membrane preparations[7]. The effective doses for the effect of taurine on reperfusion injury (Fig. 3) were in agreement with those for the effect of the amino acid on Na^+/Ca^{2+} exchange activity (Fig. 7). These findings suggest that taurine may protect astrocytes against reperfusion injury by inhibiting Ca^{2+} entry via the reverse mode of the Na^+/Ca^{2+} exchanger.

Fig. 8 summarizes our idea on the mechanisms for reperfusion injury and its protection by taurine. Under Ca^{2+}-free conditions, $[Na^+]_i$ probably increases, and taurine release occurs. The change in $[Na^+]_i$ appears to be related to taurine release, but the exact mechanism is not known. Reperfusion causes an increase in $[Ca^{2+}]_i$ via the reverse mode of

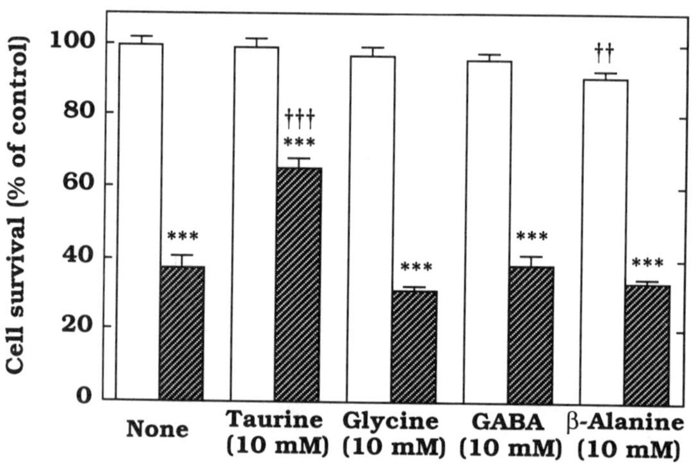

Figure 4. Effects of some amino acids on reperfusion-induced cell toxicity. The conditions for cell viability measurement are the same as those in Fig. 3. Results are means ± SE of 14 wells obtained from 4 separate preparations. $***p < 0.001$, compared with the corresponding value without Ca^{2+}-depletion (Student's t-test). $^{++}p < 0.01$; $^{+++}p < 0.001$, compared with the corresponding value without amino acids (ANOVA followed by LSD test).

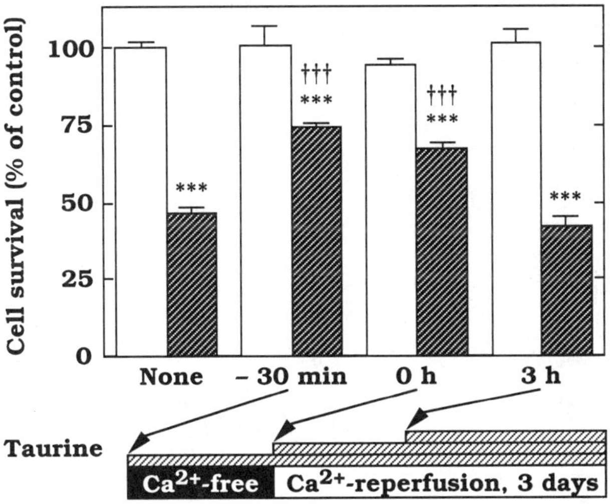

Figure 5. Effect of taurine added at the different time on reperfusion-induced cell toxicity. The conditions for cell viability measurement are the same as those in Fig. 3, and taurine was added at the indicated time. Results are means ± SE of 9-34 wells obtained from 3-7 separate preparations. $^{***}p < 0.001$, compared with the corresponding value without Ca^{2+}-depletion (Student's t-test);. $^{+++}p < 0.001$, compared with Ca^{2+}-depletion (ANOVA followed by LSD test).

Na^+/Ca^{2+} exchanger, and ouabain plus monensin mimics Ca^{2+} paradox-induced cell toxicity. It is likely that taurine inhibits Na^+/Ca^{2+} exchange activity in the reverse mode and then attenuates Ca^{2+} influx and cell toxicity. It is not known how taurine inhibits Na^+/Ca^{2+} exchange activity.

Reperfusion-induced cell toxicity reported here is considered to be an *in vitro* ischemic injury model, because trauma or ischemia of central nervous tissues produces a rapid and profound depression of $[Ca^{2+}]_o$[2,4,12]. Then, it is considered that not only delayed neuronal death, a well known phenomenon, but also delayed glial cell death reported here

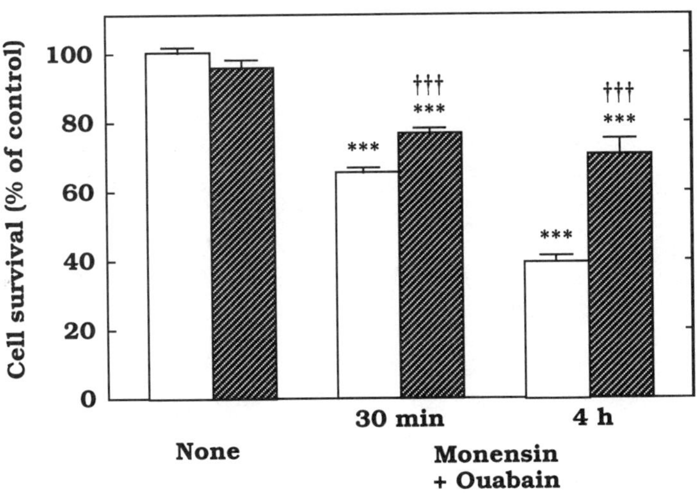

Figure 6. Effect of taurine on cell toxicity induced by ouabain and monensin in astrocytes. The cells were exposed to 20 μM monensin and 1 mM ouabain for 30 min and 4 h, and reperfused with normal medium in the presence (hatched) and absence (open) of 30 mM taurine for 24 h. Results are means ± SE of 9 wells obtained from 3 separate preparations. $^{***}p < 0.01$, compared with the corresponding value without monensin plus ouabain (ANOVA followed by Duncan test); $^{+++}p < 0.001$, compared with the corresponding value without taurine (ANOVA followed by Bonferroni test).

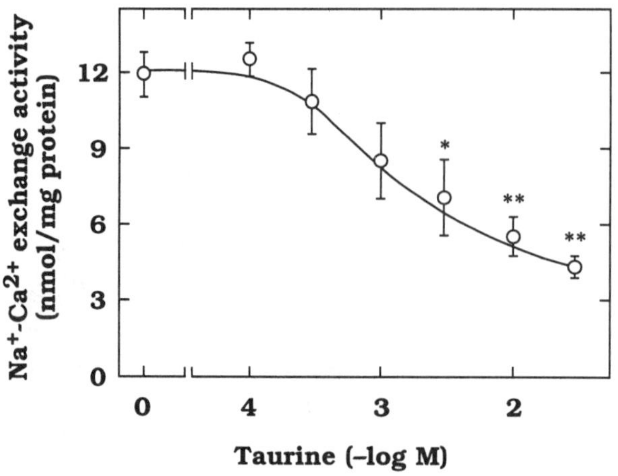

Figure 7. Effect of taurine on Na^+/Ca^{2+} exchange activity in astrocytes. Na^+/Ca^{2+} exchange activity was expressed as the difference in $^{45}Ca^{2+}$ uptake between the presence and absence of 1 mM ouabain and 20 μM monensin. Results are means ± SE of 6 wells obtained from 2 separate preparations. $^*p < 0.05$; $^{**}p < 0.01$, compared with control (one-way ANOVA followed by Duncan test).

contribute to the complex pathology of brain ischemic damage. The present finding implies that taurine may be beneficial for treatment of brain ischemic damage.

SUMMARY

Reperfusion of cultured rat astrocytes with Ca^{2+}-containing medium after exposure to Ca^{2+}-free medium for a short time caused an increase in intracellular Ca^{2+} ($[Ca^{2+}]_i$), and delayed cell death (Ca^{2+} paradox-like injury). Exposure of astrocytes to Ca^{2+}-free medium caused a marked release of taurine. Taurine (3-30 mM) reduced the reperfusion-induced increase in $[Ca^{2+}]_i$ and attenuated the delayed glial cell death. Glycine, GABA and β-alanine did not affect reperfusion-induced cell toxicity. The protective effect of taurine required addition at an early time during reperfusion. Ouabain and monensin mimicked reperfusion injury and their toxicity was also reduced by taurine. Taurine (3-30 mM) inhibited dose-dependently $^{45}Ca^{2+}$ uptake stimulated by ouabain and monensin in astrocytes. These findings

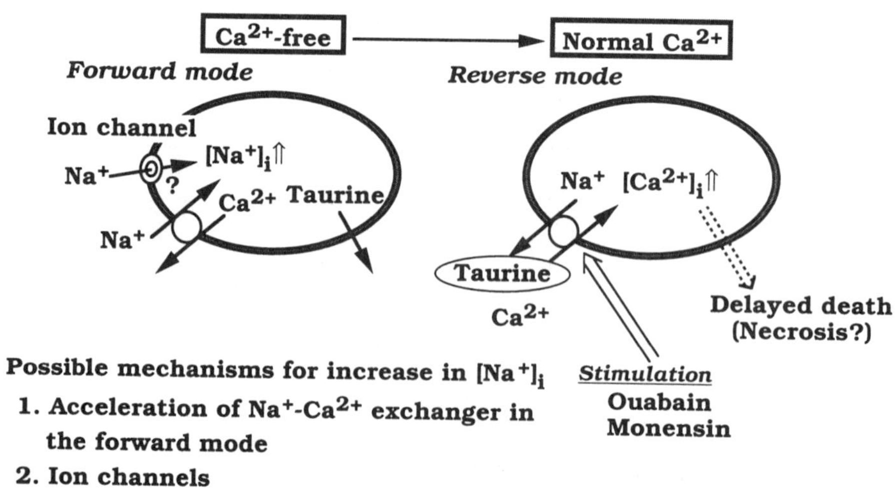

Figure 8. Reperfusion injury and protection by taurine in astrocytes.

suggest that taurine has a protective effect against reperfusion injury via an inhibition of Na^+/Ca^{2+} exchange activity in the reverse mode.

ACKNOWLEDGMENT

This study was supported by a grant from Taisho Pharmaceutical Co., Ltd.

REFERENCES

1. Asano, S., Matsuda, T., Takuma, K., Kim, H.S., Sato, T., Nishikawa, T., and Baba, A. 1995, Nitroprusside and cyclic GMP stimulate Na^+-Ca^{2+} exchange activity in neuronal preparations and cultured rat astrocytes, *J.Neurochem.* 64:2437-2441.
2. Bredt, D.S. and Snyder, S.H. 1990, Isolation of nitric oxide synthetase, a calmodulin-requiring enzyme, *Proc.Natl.Acad.Sci.* 87:682-685.
3. Goldman, W.F., Yarowsky, P.J., Juhaszova, M., Krueger, B.K., and Blaustein, M.P. 1994, Sodium/calcium exchange in rat cortical astrocytes, *J.Neurosci.* 14:5834-5843.
4. Harris, R.J., Symon, L., Branston, N.M., and Bayhan, M. 1981, Changes in extracellular calcium activity in cerebral ischaemia, *J.Cereb.Blood.Flow.Metabol.* 1:203-209.
5. Koyama, Y., Ishibashi, T., Tanaka, K., and Baba, A. 1994, L-Glutamate-stimulated taurine release from rat cerebral cultured astrocytes, *J.Neurosci.Res.* 38:75-80.
6. Kramer, J.H., Chovan, J.P., and Schaffer, S.W. 1981, Effect of taurine on calcium paradox and ischemic heart failure, *Am.J.Physiol.* 240:H238-H246.
7. Matsuda, T., Gemba, T., Baba, A., and Iwata, H. 1989, Inhibition by taurine of Na^+-Ca^{2+} exchange in sarcolemmal membrane vesicles from bovine and guinea pig hearts, *Comp.Biochem.Physiol.* 94C:335-339.
8. Matsuda, T., Murata, Y., Kawamura, N., Hayashi, M., Tamada, K., Takuma, K., Maeda, S., and Baba, A. 1993, Selective induction of $\alpha 1$ isozyme of $(Na^+ + K^+)$ - ATPase by insulin/insuin-like growth factor-I in cultured rat astrocytes, *Arch.Biochem.Biophys.* 307:175-182.
9. Matsuda, T., Takuma, K., Nishiguchi, E., Asano, S., Hashimoto, H., Azuma, J. and Baba, A. 1996, Involvement of Na^+-Ca^{2+} exchanger in reperfusion-induced delayed cell death of cultured rat astrocytes, *Eur.J.Neurosci.* in press.
10. Takuma, K., Matsuda, T., Asano, S., and Baba, A. 1995, Intracellular ascorbic acid inhibits the Na^+-Ca^{2+} exchanger in cultured rat astrocytes, *J.Neurochem.* 64:1536-1540.
11. Takuma, K., Matsuda, T., Hashimoto, H., Asano, S., and Baba, A. 1994, Cultured rat astrocytes possess Na^+-Ca^{2+} exchanger, *Glia*, 12:336-342.
12. Young, W., Yen, V., and Blight, A. 1982, Extracellular calcium activity in experimental spinal cord contusion, *Brain Res.* 253:115-130.

54

BIPHASIC EFFECT OF TAURINE ON EXCITATORY AMINO ACID-INDUCED NEUROTOXICITY

X. W. Tang, D. L. Deupree, Y. Sun, and J.-Y. Wu

Department of Physiology and Cell Biology
University of Kansas
Lawrence, Kansas 66045-2106

INTRODUCTION

In the past, the biological role of taurine as an end product of methionine metabolism and conjugation with bile acids in the liver was considered relatively trivial. However, the physiological role of taurine has received considerable attention since the report by Hayes *et al.*[7] and Pion *et al.*[21] that cats fed a taurine-deficient diet developed a central retinal degeneration and a cardiomyopathy. Now, taurine has been proposed to be involved in many important physiological functions such as serving as a trophic factor in the development of the CNS (at least in some species)[19,23,24], maintaining the structural integrity of the membrane[17,20], regulating Ca^{2+} binding and transport[12,16], and serving as an osmoregulator[22,27], a neuromodulator[11] and neurotransmitter[14,15,18,25]. It is well accepted now that taurine plays an important role in maintaining the integrity of retina and the viability of photoreceptor cells, since absence of dietary taurine in the cat leads to a decrease in taurine concentration in the retina, followed by defects in the electroretinogram, degeneration of the photoreceptors and eventual blindness[7,23].

It is generally believed that neuronal degeneration is triggered in part by excessive stimulation by L-glutamate (L-Glu) or related excitatory amino acids (EAA), presumably through the opening of Ca^{2+} channels[1-3]. Since taurine has been shown to be involved in the regulation of Ca^{2+} homeostasis[12,16] and maintaining the integrity of cell membrane[17,20], it will be of interest to determine whether taurine has protective function against EAA-induced neurotoxicity. In this communication, evidence is presented to show that taurine has biphasic effects on EAA-induced neurotoxicity.

MATERIALS AND METHODS

Preparation of Neuronal Culture

Whole brain neuronal cultures were prepared as described[13] with some modifications. Brains dissected from fetal rats (16-17 days gestation) were mechanically dissociated in

Taurine 2, edited by Huxtable et al.
Plenum Press, New York, 1996

Basal Medium Eagle media (BME supplemented with 13.1 mM sodium bicarbonate, 2.0 mM D-glucose, and 20% heat-inactivated fetal calf serum (4 ml/brain). The cell suspensions were then plated on tissue culture dish (2 ml/10 mm dish) precoated with 5 μg/ml poly-L-lysine, and incubated at 37°C for 30-45 min in a humidified incubator equilibrated with 5% CO_2. The incubation medium was then replaced with serum-free BME and the cultures were maintained in the incubator until use. Cultures prepared under these conditions usually contain about 80-85% neurons[13].

Treatment with Taurine and EAA

Cultures of 14-day *in vitro* (DIV) were used in all the neurotoxicity experiments. Prior to the experiment, the original culture media were replaced with fresh serum-free BME, and the cultures were incubated for 45 min. In the experiments which taurine or GABA was tested, the cultures were first preincubated with these compounds for 10 min. The cultures were then stimulated with Glu or other EAAs for 5 min. After stimulation, the cultures were then washed twice with BME, and further incubated for 20-24 h in BME. In all experiments, compounds to be tested, such as taurine or GABA, were present throughout the incubation steps unless otherwise specified.

Analysis of Neurodegeneration by LDH Assay

Neuronal degeneration was assayed 20-24 h following incubation with EAAs by measuring the amount of lactate dehydrogenase (LDH) released into the culture media as described[9]. The amount of LDH was quantified spectrophotometrically using an LDH diagnostic kit from Sigma according to the manufacture's instruction. In order to reduce variations among dishes due to differences in cell numbers, LDH released into the media was normalized by expression as the % of total LDH, which is the sum of LDH activity in the media and the activity remaining in the cells. [LDH released = LDH media / (LDH media + LDH cells)].

RESULTS

Effect of Excitatory Amino Acids on Cultured Neurons

Among EAA agonists tested, trans-(±)-1-amino 1,3-cyclopentanedicarboxylic acid (ACPD) and α-amino-hydroxy-5-methylisoxazole-4-propionic acid (AMPA), two specific agonists for metabotropic EAA receptors, produced no toxic effects even at 1 mM. Other EAAs including kainic acid (KA), Glu and N-methyl-D-aspartate (NMDA), did produce toxic effects at concentrations as low as 50 μM. L-Glu and NMDA were the most toxic agonists tested producing maximum neurotoxicity at 0.5 mM. NMDA and L-Glu produced 80% and 50% of maximum toxicity, respectively, at 0.1 mM. Thus, it appears that NMDA was the most effective neurotoxin in these experiments. The results of EAA-induced neurotoxicity are summarized in Fig. 1. The LDH release ratios for L-Glu and NMDA at 0.5 mM were both about 300% above control and did not increase above these values when the concentration was increased to 5.0 mM, indicating the presence of surviving neurons within the experimental cultures. This was expected, since whole brain tissue cultures that contain neurons that are not sensitive to EAAs were used in these experiments.

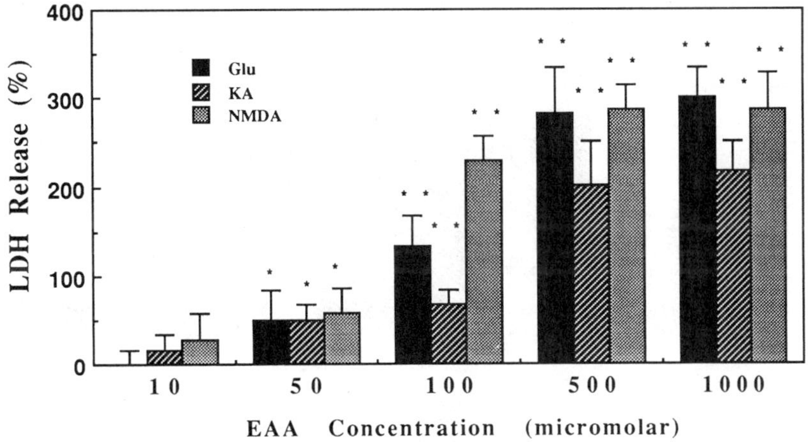

Figure 1. Effect of EAA agonists on cultured neurons. Experiments were conducted as described in MATERIALS AND METHODS. Each set of data is the summary of 5 separate experiments, each with its own control group (n = 5 for all groups). The results are expressed as percentage of LDH release above the control. Bars represent means (± sd) of percentage increases of experimental groups above control levels. All the agonists tested at 50 μM or higher produced a significant increase in LDH release over control levels. $^{**}p < 0.01$; $^{*}p < 0.05$.

Effect of Taurine on L-Glu-Induced Neurotoxicity

Taurine was found to potentiate neurotoxicity induced by 0.25 mM L-Glu in a dose-dependent manner. A maximum potentiation of Glu-induced neurotoxicity was obtained 0.5-1 mM with taurine at 0.5-1 mM, which represents about a 100% increase in neurotoxicity. At high concentrations, 5-25 mM, taurine was found to protect against L-Glu-induced neurotoxicity. Again, the protective function of taurine is also dose-dependent and reaches a maximum protection (almost complete protection) at 25 mM, as shown in Fig. 2.

Effect of Taurine on KA-Induced Neurotoxicity

Taurine was also found to potentiate KA-induced neurotoxicity. At 0.5 mM, taurine increased neurotoxicity induced by 0.5 mM KA by about 75%. No further increase of KA-induced neurotoxicity was observed when taurine concentration was increased from 0.5 to 1.0 mM. Similarly, taurine protected from KA-induced neurotoxicity in a dose-dependent manner at 5 mM or higher, reaching a maximum protection (>90%) at 25 mM (Fig. 3).

The protective function of taurine against EAA-induced neurotoxicity appears to be only slightly affected by an increase of EAA concentration, at least within the concentrations used (Fig. 4). Furthermore, taurine is most effective in protection against KA-induced neurotoxicity.

Effect of GABA on EAA-Induced Neurotoxicity

GABA was also found to potentiate EAA-induced neurotoxicity. For instance, at 1 mM, GABA increased neurotoxicity induced by 0.25 mM L-Glu and 0.5 mM KA by about 30% and 25%, respectively (Fig. 5A). However, unlike taurine, GABA did protect against KA-induced neurotoxicity, although the degree of GABA-induced potentiation of neuro-

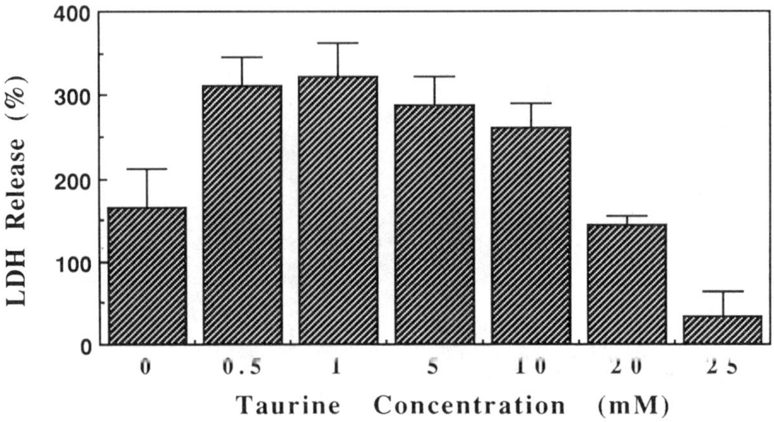

Figure 2. Effect of taurine on Glu-induced neurotoxicity in cultured neurons. Treatment of cultured neurons with L-Glu in the presence or absence of taurine was conducted as described in MATERIALS AND METHODS. In all experiments, stimulation was carried out with 0.25 mM L-Glu while taurine concentrations varied from 0-25 mM. Each set of data is the summary of 5 separate experiments, each with its own control group (n = 5 for all groups). Bars represent means ± SD of % increases of experimental groups above control levels.

toxicity increased with GABA concentration from 0 to 0.1 mM and then decreased from 0.1 to 25 mM, as shown in Fig. 5B.

DISCUSSION

EAA-induced neurotoxicity is thought to be produced via a breakdown in ionic homeostasis[1-3,6] mediated via the EAA ionotropic receptor subtypes, e.g., KA, NMDA, and quisqualic acid (QA). The initial "excitotoxic" phase, lasting about one hour, is characterized by neuronal swelling. This phase is thought to be mediated via excessive activity of non-NMDA receptors, as neuronal swelling is eliminated when incubations are performed

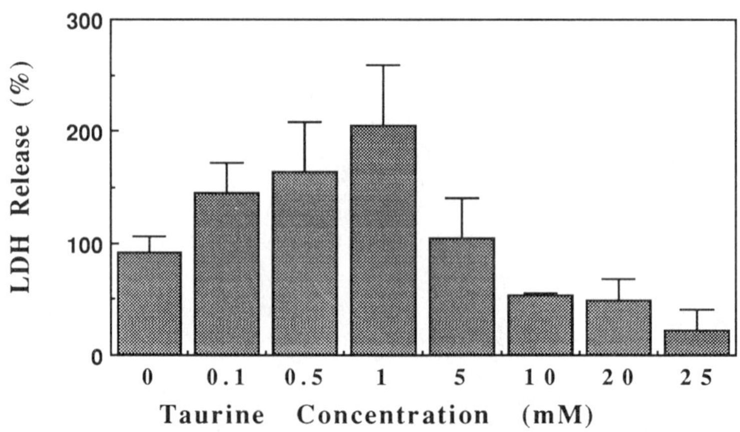

Figure 3. Effect of taurine on KA-induced neurotoxicity in cultured neurons. Experiments were performed as described in Fig. 2 except that cultured neuron were stimulated with 0.5 mM KA instead of 0.25 mM L-Glu.

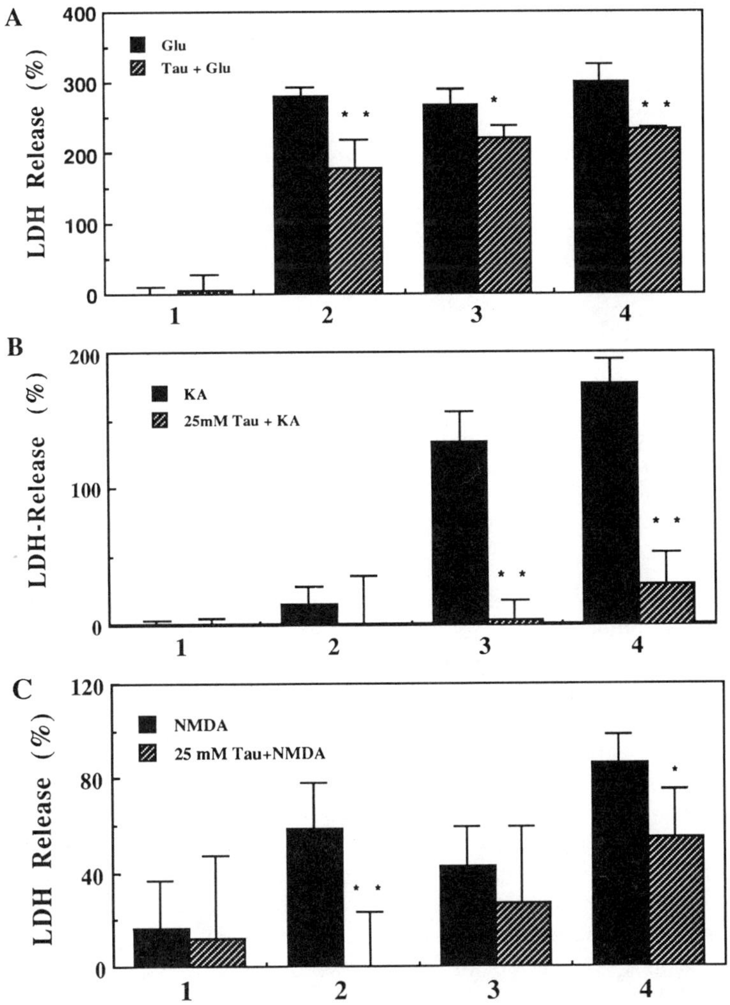

Figure 4. Protection of taurine against neurotoxicity induced by various concentrations of EAAs. Experiments were carried out as described in Fig. 2 except that the concentration of taurine used was kept at 25 mM and the concentration of EAA varied as indicated (A) Cultured neurons were stimulated with L-Glu at 1) 0, 2) 0.1, 3) 0.35 and 4) 0.5 mM; (B) The same as (A) except KA used was 1) 0, 2) 0.05, 3) 0.2 and 4) 0.5 mM; (C) The same as (B) except NMDA used was 1) 0.005, 2) 0.05, 3) 0.2 and 4) 0.5 mM. $^{**}p < 0.01$; $^{*}p < 0.05$.

in Na^+-free culture media (1). There is also a late "delayed" phase of neurodegeneration seen in cultured neurons following brief EAA agonist exposure, which covers several hours and is characterized by darkening and increased granularity of the soma, with eventual degeneration of the membrane (lysis) and cell death. Because this delayed phase of degeneration is Ca^{2+}-dependent and can be blocked by NMDA antagonists, it is hypothesized that this is to be mediated via excessive activation of the NMDA receptor[1,5].

The potentiation of EAA-induced neurotoxicity by either taurine or GABA reported here could be due to its effect on the "initial excitotoxic" phase. Both GABA and taurine are known inhibitory neurotransmitters inhibiting neuronal firing by opening Cl^--channel on the cell membrane[10,18,25,26]. It is conceivable that GABA and taurine may potentiate neurotoxic-

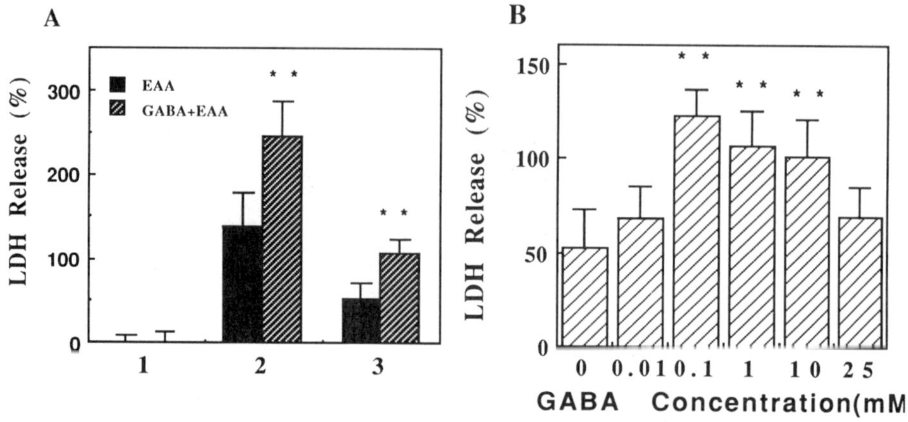

Figure 5. Effect of GABA on EAA-induced neurotoxicity. The conditions were the same as described in Fig. 2 except that cultured neurons were stimulated either with 0.25 mM Glu or 0.5 mM KA in the presence or absence of 1 mM GABA. (A) Effect of 1 mM GABA on L-Glu and KA-induced neurotoxicity. (B) Dose-dependence of GABA effect on KA-induced neurotoxicity.

ity through their action on Cl^--channel resulting in increase of Cl^- influx and swelling of the cell. However, these results are somewhat unexpected since one would expect that the extent of neuronal excitation and the resulting neuronal degeneration would be reduced in the presence of inhibitory inputs such as GABA or taurine. Perhaps the influx of Cl^- and associated cell swelling outweighs the inhibitory inputs and hence GABA and taurine at low concentrations (1 mM or less) actually potentiate neurotoxicity. It is of interest that GABA or its agonists has been reported to accelerate L-Glu-induced neurotoxicity in rat brain cultures[4]. However, this differs from the present results in that the previous report of GABA action is to accelerate the process but not the extent of neurotoxicity as observed in the present studies. The discrepancy may be due to differences of the experimental conditions; e.g., type and age of cultures, incubation conditions, etc.

Although both GABA and taurine at low concentrations potentiate neurotoxicity, presumably through the same mechanism, namely facilitation of opening of Cl^--channel, the mechanisms for their action at high concentrations appear to be different. Taurine has strong protective function against EAA-induced neurotoxicity whereas GABA has no effect. The mechanism underlying the protective function of taurine is not clear. One possible mechanism is that taurine may be taken up inside the cell and act as an osmolyte to regulate cell volume and prevent cell swelling. It is also possible that taurine may regulate intracellular Ca^{2+} concentration through mechanisms yet to be determined and hence can protect cells from lysis. The known properties of taurine as an osmolyte[22,27] and as a Ca^{2+}-regulator[12,16] are compatible with the above interpretations.

ACKNOWLEDGMENT

This study was supported in part by grants from the Office of Naval Research (N00014-94-1-0457), the National Science Foundation (BNS-8820581), NIH (PHS grant #NS 20978), the Marion Merrell Dow Foundation, and the general research fund (GRF) from the University of Kansas. The word processing of the manuscript done by Sharon Hopkins and Judy Wiglesworth is much appreciated.

REFERENCES

1. Choi, D. W. 1987, Ionic dependence of glutamate neurotoxicity, *J. Neurosci.*, 7:369-379.
2. Choi, D. W. 1994, Glutamate receptors and the induction of excitotoxic neuronal death, *Prog. Br. Res.*, 100:47-51.
3. Choi, D. W., Maulucci-Gedde, M., and Kreigstein, A. R. 1987, Glutamate neurotoxicity in cortical cell culture, *J. Neurosci.*, 7:357-368.
4. Erdö, S. L., Angelika, M. and Wolff, J. R. 1991, GABA accelerates excitotoxic cell death in cortical cultures: protection by blockers of GABA-gated chloride channels, *Brain Res.*, 542:254-258.
5. Foster, A. C., Gill, R. and Woodruff, G. N. 1988, Neuroprotective effecs of MK-801 in vivo: selectivity and evidence for delayed degeneration mediated by NMDA receptor activation, J. Neurosci., 8:4745-4754.
6. Garthwaite, J. and Garthwaite, G. 1990, Mechanisms of excitatory amino acid neurotoxicity in rat brain slices, in "Excitatory Amino Acids and Neuronal Plasticity", Ben-Ari, Y. ed., Plenum Press, New York, pp. 505-518.
7. Hayes, K. C., Carey, R. E. and Schmidt, S. Y. 1975, Retinal degeneration associated with taurine deficiency in the cat, *Science*, 188:949-951.
8. Jacobsen, J. G. and Smith, L. H. 1968, Biochemistry and physiology of taurine and taurine derivatives, *Physiol. Rev.*, 48:424-511.
9. Koh, J.-Y. and Choi, D. W. 1987, Quantitative determination of glutamate mediated cortical neuronal injury in cell culture by lactate dehydrogenase efflux assay, *J. Neurosci. Meth.*, 20:83-90.
10. Krnjevic, K. 1974, Chemical nature of synaptic transmission in vertebrates, *Physiol. Rev.*, 54:418-540.
11. Kuriyama, K. 1980, Taurine as a neuromodulator, *Fed. Proc.*, 39:2680-2684.
12. Lazarewicz, J. W., Noremberg, K., Lehmann, A. and Hamberger, A. 1985, Effects of taurine on calcium binding and accumulation in rabbit hippocampal and cortical synaptosomes, *Neurochem. Int.*, 7:421-428.
13. Lee, Y.-H., Deupree, D. L., Chen, S.-C., Kao, L.-S. and Wu, J.-Y. 1994, Role of Ca^{2+} in AMPA-mediated poly-PI turnover in primary neuronal cultures, *J. Neurochem.*, 62:2325-2332.
14. Lin, C.-T., Su, Y. Y. T., Song, G.-X. and Wu, J.-Y. 1985, Is taurine a neurotrans-mitter in rabbit retina? *Brain Res.*, 337:293-298.
15. Lin, C.-T., Song, G.-X. and Wu, J.-Y. 1985, Ultrastructural demonstration of L-glutamate decarboxylase and cysteinesulfinic acid decarboxylase in rat retina by immunocytochemistry, *Brain Res.*, 331:71-80.
16. Lombardini, J. B. 1985, Effects of taurine on calcium ion uptake and protein phosphorylation in rat retinal membrane preparations, *J. Neurochem.*, 45:268-275.
17. Moran, J., Salazar, P. and Pasantes-Morales, H. 1988, Effect of tocopherol and taurine on membrane fluidity of retinal rod outer segments, *Experimental Eye Research*, 45:769-776.
18. Okamoto, K., Kimura, H. and Sakai, Y. 1983, Evidence for taurine as an inhibitory neurotransmitter in cerebellar stellate interneurons: Selective antagonism by TAG (6-aminomethyl-3-methyl-4H,1,2,4-benzothiadiazine-1,1-dioxide), *Brain Res.*, 265:613-168.
19. Palackal, T., Kujawa, M., Moretz, R. C., Neuringer, M. and Sturman, J. 1991, Laminar analysis of the number of neurons, astrocytes, oligodendrocytes and microglia in the visual cortex (area 17) of 3-month-old rhesus monkeys fed a human infant soy-protein formula with or without taurine supplementation from birth, *Dev. Neurosci.*, 13:20-33.
20. Pasantes-Morales, H. and Cruz, C. 1985, Taurine and hypotaurine inhibit light-induced lipid peroxidation and protect rod outer segment structures, *Brain Res.*, 330:154-157.
21. Pion, P. D., Kittleson, M. D., Rogers, Q. R. and Morris, J. G. 1987, Myocardial failure in cats associated with low plasma taurine: A reversible cardiomyopathy, *Science*, 237:764-768.
22. Solis, J. M., Herranz, A. S., Herreras, O., Lerma, J. and Del Rio, R. M. 1988, Does taurine act as an osmoregulatory substance in the rat brain, *Neurosci. Lett.*, 91:53-58.
23. Sturman, J. A., Gargano, A. D., Messing, J. M. and Imaki, H. 1986, Feline maternal taurine deficiency:effect on mother and offspring, *J. Nutr.*, 116:655-667.
24. Sturman, J. A., Moretz, R. C., French, J. H. and Wisniewski, H. M. 1985, Postnatal taurine deficiency in the kitten results in a persistence of the cerebellar external granule cell layer: correction by taurine feeding, *J. Neuro. Res.*, 13:521-528.
25. Taber, T. C., Lin, C.-T., Song, G.-X., Thalman, R. H. and Wu, J.-Y. 1986, Taurine in the rat-hippocampus-localization and postsynaptic action, *Brain Res.*, 386:113-121.
26. Takeuchi, A. 1976, Studies of inhibitory effects of GABA in invertebrate nervous systems, in: "GABA in Nervous System Function," Roberts, E., Chase, T. N. and Tower, D. B., eds., Raven Press, New York, pp. 255-281.
27. Wade, J. V., Olson, J. P., Samson, F. E., Nelson, S. R. and Pazdernik, T. L. 1988, A possible role for taurine in osmoregulation within the brain, *J. Neurochem.*, 51:740-745.

55

BALANCED INTERACTION OF GROWTH FACTORS AND TAURINE REGULATE ENERGY METABOLISM, NEURONAL SURVIVAL, AND FUNCTION OF CULTURED MOUSE CEREBELLAR CELLS UNDER DEPOLARIZING CONDITIONS

E. Trenkner, A. El Idrissi, and C. Harris

New York State Institute for Basic Research in Developmental Disabilities
1050 Forest Hill Road
Staten Island, New York 10314

ABSTRACT

The development of neuronal cells in a given cellular environment requires mechanisms that dynamically regulate the balanced interactions of multiple factors which are known to control maintenance and plasticity in function of neurons throughout constantly changing extracellular conditions. Periodic release of excitatory amino acids from both developing glial and neuronal cells into the extracellular environment and their uptake has been shown to stimulate neuronal function in concert with growth factors that control the degree of depolarization and, therefore, neuronal function. This study attempts to characterize the critical concentrations of these factors either alone or together in relation to energy metabolism, cell survival and function. We demonstrate a close correlation between energy metabolism of neuronal cells, controlled by the combination of growth-factors (βFGF, BDNF), and glutamate-taurine as well as K^+ in depolarizing concentrations (10-25 mM), during the balancing act of neuronal survival or death, and neuronal function. These functions depend on medium conditions (energy sources, ion composition), the ratio of glial cells versus neurons and cell density. Granule cell migration as a measure of developmental neuronal function was analyzed in the presence of various combinations of growth factors and taurine under various depolarizing conditions (glutamate, K^+). We found that K^+ concentrations > 7 mM in BME and 10% horse serum blocked migration in less than 30 min. Taurine did not prevent this effect. However, in the presence of HEPES as well as in F12-medium with HEPES, taurine restored granule cell migration. On the other hand, glutamate- or NMDA-mediated depolarization stopped migrating granule cells while NMDA antagonists extended the period of migration. Taurine amplified the stop-signal in the

Taurine 2, edited by Huxtable et al.
Plenum Press, New York, 1996

presence of glutamate agonists but increased the number of migrating cells in the absence of glutamate. Thus, the mechanisms of glutamate receptor-mediated depolarization and K^+-depolarization appear to be different. Since taurine prevents glutamate-mediated excitotoxicity, possibly by reducing Ca^{2+} influx under depolarizing conditions, but amplifies the stop-signal, Ca^{2+} levels may not control granule cell migration.

INTRODUCTION

A balanced interaction of many factors including growth factors and neuroactive amino acids control survival and function of neurons. However, little is known about which of the many factors cooperate and how their optimal range of concentrations, when added together, affect the vulnerability of neurons towards extracellular excitatory amino acids. Most of the detailed evidence describing the biological effects of various growth factors, excitatory amino acids and media conditions with or without sera from heterologous and homologous sources have been described in tissue culture systems in a one-dimensional manner, by adding one factor at a time to more or less defined media. Cell survival and possibly neuronal function were used to assess the role of these factors in an artificial cellular environment. This study attempts to demonstrate that conclusions reached under one set of conditions cannot necessarily be compared with results of another set of conditions. Therefore, we are questioning the generalization of results and conclusions obtained under different conditions. Based on our studies we conclude that little information from *in vitro* studies can be used to describe the *in vivo* conditions at a given time. However, the results described by others together with our comparative results clearly suggest a remarkable ability of neurons to adapt and cope with changing environments in order to survive and function.

Cerebellar granule cell migration is one of the fundamental processes that define cerebellar morphogenesis. Postmitotic granule cells migrate along Bergman glial fibers through the molecular layer around Purkinje cells into the internal granule cell layer where, with time, they assume their final position in glomeruli[18,19,21]. There, they form synaptic connections with incoming mossy fibers. The receptor systems and substrates involved in the process of migration begin to unravel, although contradictory results have been reported *in vitro* as compared to *in vivo* conditions[4,7,8,12,13]. Little is known of how the spatial distribution of granule cells is determined, specifically or randomly, how granule cells stop migrating, whether the stop-signal for migrating granule cells acts extracellularly or intracellularly, and what the molecular nature of a stop-signal may be.

We have demonstrated that granule cell migration is controlled by an internal clock *in vitro*[24]. Based on this study, at least two events have to take place in order to stop granule cell migration: (i) Granule cells have to express NMDA-specific glutamate receptors or combinations therefore, at the end of their differentiation period. The expression may be regulated by an internal clock; (ii) Incoming glutamatergic mossy fibers must form synaptic contacts (transient and/or permanent) with differentiated granule cells in the developing glomerulus, and depolarize migrating granule cells. The depolarization of migrating granule cell neurons will lead to a loss of mitochondrial energy, increase of intracellular Ca^{2+}, and inactivation of the migratory mechanism(s), probably through second messenger-systems.

Taurine appears to be directly involved in this process. Taurine plays an essential role in the development and survival of neuronal cells. Multiple functions have been linked to taurine, from membrane stabilization to the synthesis of bile salts. It is synthesized in mouse brain but only minimally in cat. Taurine occurs in concentrations similar to those of glutamate in rodents[11,22]. We have demonstrated its protective role in the process of glutamate excitotoxicity in cultured mouse cerebellar cells, and have suggested that taurine is the balancing factor to glutamate in the excitatory amino acid transmitter system[25]. This study shows that the role of taurine as a balancing factor may be based on its ability to modulate Ca^{2+} influx in the presence of glutamate. We also compared the effects of some of the conventional neuronal tissue culture conditions on neuronal cell migration (function) and survival in relation to the mitochondrial energy metabolism under depolarizing and non-depolarizing conditions. Secondly, this study analyses how various concentrations of a growth factor (bFGF), described as factor that can prevent cell death *in vitro*, interact with taurine in preventing excitotoxicity[14,15,16]. Thirdly, we report the effects these factors have on $^{45}Ca^{2+}$ uptake under various conditions. Fourthly, we discuss results that identify depolarization of cerebellar granule cells as stop-signal for cell migration.

MATERIAL AND METHODS

Glutamate, taurine and carbonylcyanide p-tri-gluoro-methoxy-phenyl-hydrazone (FCCP) were purchased from Sigma. AP5, Mk801 were obtained from Tocris Cookson. BME and F12, with and without HEPES, came from GIBCO.

Tissue Culture

Early postnatal cerebellar granule cells were cultured as described elsewhere[23,26]. Briefly, cerebella from 7 day old mice (P7) were dissected, trypsinized for 12 min (1 mg Trypsin/ml; 0.05 mg/DNase ml), triturated in 0.05 mg/ml DNase, pelleted and resuspended in BME, supplemented with 0.1% glucose, 0.2 mg glutamine, 10% horse serum (HS) or foetal calf serum (FCS) and 25 mg/ml of penicillin/streptomycin. Cells were resuspended in F12 medium with and without HEPES. Other additions were as indicated. Cells were cultured in microwell plates for the analysis of migration and in 24 well plates for $^{45}Ca^{2+}$ uptake studies.

Serum-Free Medium Conditions

P7 cerebellar cells were first cultured in DMEM + 10% horse serum (HS) for 24 h, medium was replaced with DMEM and N-2 supplement instead of serum[3]. Under these culture conditions neuronal cells survive for more than 7 days after medium exchange. Fibroblasts and glial cells do not proliferate under these conditions but appear healthy and functional.

$^{45}Ca^{2+}$ Uptake

$^{45}Ca^{2+}$ uptake was measured according to the method of Eimerl and Schramm[6]. The culture medium was removed and cells were washed twice with 0.5 ml of Locke's solution. To the final volume of 0.25 ml ~2 x 10^5 cpm of $^{45}CaCl_2$, were added ~10 s before addition of the agonist. Taurine was added to the cultures 24 h before $^{45}Ca^{2+}$, as this pre-treatment with taurine has been shown to be essential for the Ca^{2+}-modulatory role of taurine. After 20-30 min at room temperature, the cells were rapidly washed 3x times with 0.5 ml of Locke's solution containing 2 μM MK-801. A relatively high concentration of MK-801 was used to ensure fast channel block. The cells were lysed in 0.1 M NaOH (0.5 ml) and the lysate counted in a Beckman scintillation counter.

$^{45}Ca^{2+}$ uptake was determined in triplicate cultures. Experiments were repeated at least twice. The amount of $^{45}Ca^{2+}$ uptake was related to the number of cells/well (~8.5 x 10^5 cells/well). This number was determined in at least 10 cultures before the addition of $^{45}Ca^{2+}$. The variations from culture to culture were less than 12%.

RESULTS

Taurine Restores Mitochondrial Energy Metabolism

The mitochondrial respiratory chain provides energy as ATP that is required for neuronal survival and function. On the other hand, the reduction of the mitochondrial energy metabolism or the inhibition of particular complexities of the electro-chemical gradient has been shown to increase intracellular Ca^{2+}, followed by an accumulation of free radicals, depletion of ATP and neuronal cell death[2,17]. We have analyzed the role of taurine in maintaining or restoring energy levels in cerebellar granule cells in culture. FCCP was used to uncouple the electro-chemical gradient and subsequently to deplete the cellular energy levels as measured with Rhodamine 123 transport through the outer membrane. The survival of early postnatal cerebellar granule cells, pre-treated with 4 μM FCCP alone, was affected only marginally in serum-containing medium. However, when glutamate was added in physiological concentrations (10^{-4} M, 10^{-5} M) to FCCP-pretreated cultures, neuronal cell survival was significantly reduced, suggesting that low mitochondrial energy levels increase the vulnerability of neurons to the extent that physiological concentrations of glutamate became excitotoxic. On the other hand, physiological concentrations of taurine (10^{-2} M) restored the energy levels and promoted neuronal survival (Fig. 1).

Taurine Modulates Ca^{2+} Uptake during Glutamate Depolarization

It has been established in recent years that intracellular Ca^{+2} concentrations determine in part, neuronal cell survival and functions[5,10,17]. Therefore, the mechanism by which taurine prevents excitotoxic cell death may function through regulating the amount of Ca^{2+} influx into neuronal cells during receptor-mediated depolarization by glutamate. We have reported earlier that Ca^{2+}-dependent PKC activity was significantly increased in the presence of 1 μM

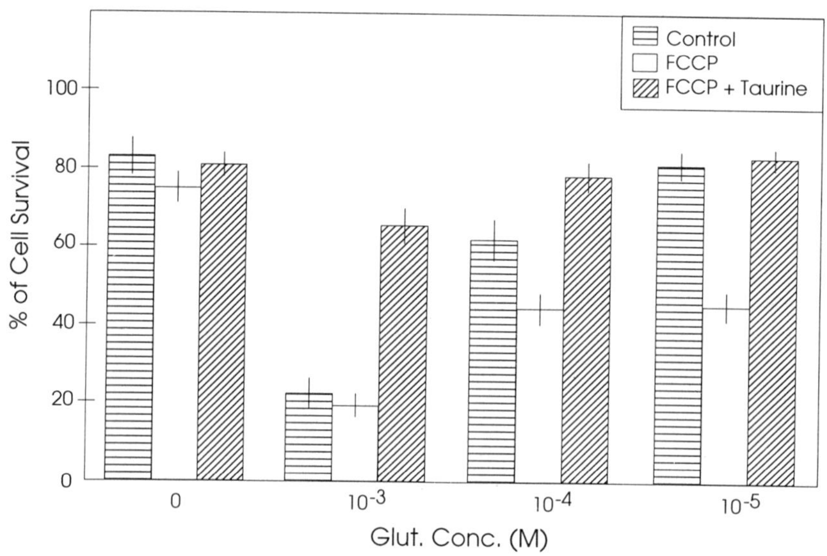

Figure 1. Mitochondrial energy levels affect neuronal cell survival. 7-day-old mouse cerebella cells were cultured in BME + HS for 3 days. Cultures were pre-incubated with 4 μM FCCP for 30 min Before glutamate (10^{-3} - 10^{-5} M) and taurine (10^{-2} M) were added. The number of living cells was counted after 6-10 h.

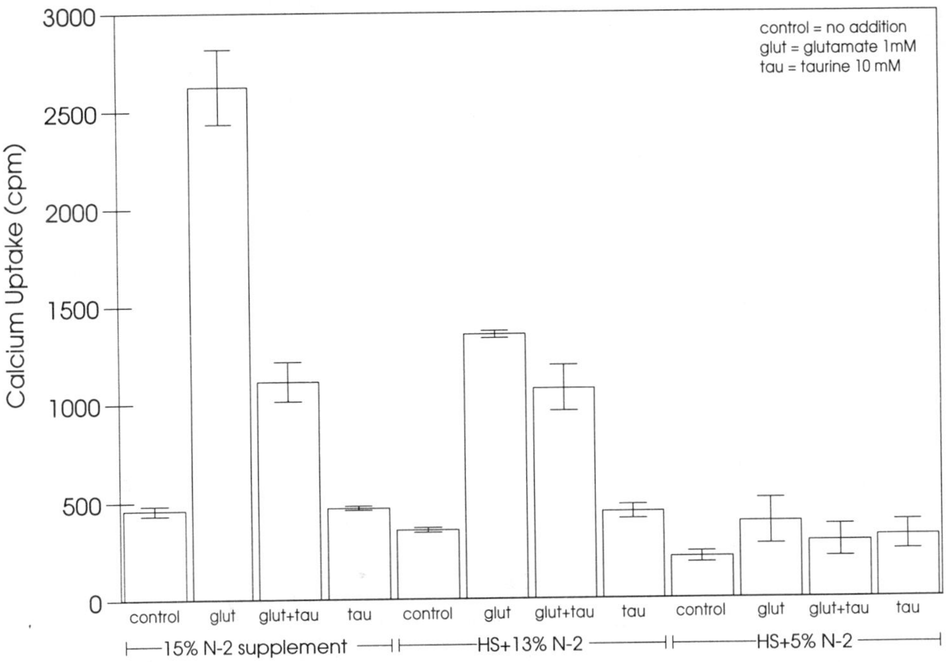

Figure 2. Taurine regulates intracellular Ca^{2+} concentrations after glutamate depolarization in different media. The role of taurine (10^{-2} M) was analysed after glutamate depolarization in serum free medium supplemented with 15% N-2, 13% N-2 + 2% HS and 5% N-2 + 10% HS respectively. $^{45}Ca^{2+}$ was added for 30 min, cultures were washed and intracellular $^{45}Ca^{2+}$ measured as described.

glutamate or its agonist NMDA but was down-regulated to control levels when 10 μM taurine was added. In addition, the mitochondrial energy metabolism was restored[24]. These results strongly suggest that taurine is involved in the control of intracellular Ca^{2+} levels. We confirm this by demonstrating directly $^{45}Ca^{2+}$ uptake into mouse cerebellar granule cells which have been treated for 1 h with 10^{-3} M glutamate alone or together with 10^{-2} M taurine.

We found that the extent of the glutamate depolarization expressed as $^{45}Ca^{+2}$ uptake depended on the concentration of horse serum in the medium (Fig. 2). Similarly, the function of taurine was affected by horse serum in a concentration-dependent manner: Taurine reduced glutamate-induced $^{45}Ca^{2+}$ uptake by more than half in serum-free medium, whereas in serum-containing media $^{45}Ca^{2+}$ uptake was significantly reduced, although the reduction was significantly less than in serum-free medium.

Glutamate and K^+ Depolarization Affect Granule Cell Migration Depending on Media Compositions

Glutamate Depolarization. Early postnatal mouse cerebella granule cells migrate *in vitro* under 'non-depolarizing' conditions[24]. Here we have studied granule cell migration after two days under various culture conditions in the presence of the depolarizing reagents NMDA and L-glutamate together with their antagonists AP5 and MK 801 and taurine. We found that NMDA (>10^{-4} M) or glutamate (5 x 10^{-3} M) arrested approximately 85% of migrating granule cells in less than 30 min. However, after pre-treatment of cultures with AP5 (10^{-4} M), and MK 801 (10^{-4} M) with and without NMDA, granule cells continued to

Table 1. Depolarization of cerebellar granule cells can act as stop signal during migration. Taurine modulates ratio of migrating to nonmigrating cells

Condition	Ratio
Control	0.95
10^{-4}M NMDA	0.54
10^{-5}M NMDA	0.83
10^{-5}M NMDA + 10^{-4}M AP5	1.15
10^{-4}M AP5	1.42
10^{-4}M NMDA + 10^{-4}M MK-801	1.10
10^{-4}M MK-801	2.06
10^{-4}M NMDA + 10^{-2}M taurine	0.48
10^{-2}M taurine	1.23

4 x 10^4 P7 cerebellar cells were plated into microwells to form reaggregates that subsequently were connected with fiber bundles. Different generations of granule cells migrated onto these bundles as described in detail[23,24]. The number of granule cells per fiber bundle, therefore, reflect the migratory activity. Migrating cells were distinguished from non-migrating cells by measuring the distance of granule cells from both ends of these fiber bundles over a period of 2 h, starting after addition of agonists and antagonists as indicated.

migrate for the duration of the experiments (up to 16 h) (Table 1). With taurine, under depolarizing conditions (5 x 10^{-3} M glutamate (not shown) or 10^{-4} M NMDA), the number of non-migrating cells was increased. On the other hand, taurine alone increased the number of migrating cells, underscoring our hypothesis that taurine mediates glutamate excitation in the CNS.

K^+ Depolarization. K^+ concentrations higher than 10 mM have been used to increase neuronal survival in primary cultures of mouse and rat cerebellum from days to weeks[1,9,20]. These artificially high concentrations of K^+ require particular medium compositions that balance the K^+ concentrations osmotically, and provide intracellular conditions that will regulate signal transduction cascades, mitochondrial energy metabolism, Ca^{2+} homeostasis, etc., in response to K^+ depolarization. We have begun to compare some of these media, which promote neuronal survival, with our media which supports basic developmental functions such as synapse formation and neuronal migration[23,24].

Table 2 summarizes these results. In BME + 10% HS, significant numbers of granule cells migrated, as described earlier[23], whereas in F12 medium without HEPES with 10% FCS, granule cell migration was reduced although the survival rate in this medium was similar or better than in BME + 10% HS or BME + 10% FCS (not shown). The addition of HEPES blocked migratory activity depending on the media composition. As expected, excitotoxic concentrations of glutamate eliminated migratory activity in all media. However, unexpectedly taurine had no effect on restoring migratory activity in either medium, although cell survival, intracellular Ca^{+2} concentrations and PKC activity reached control levels (Fig. 2; Trenkner et al. 1994). Granule cells disintegrated in K^+-containing (10 mM, 25 mM), osmotically adjusted BME + HS or the BME + FCS (not shown). F12 with or without HEPES, on the other hand, promoted granule cell migration in the presence of FCS but not in the presence of HS. HEPES had no effect. In contrast to glutamate depolarization, taurine increased the number of migrating cells significantly in HEPES containing BME or F12. It remains to be determined, however, why the number of migrating granule cells increased

Table 2. K^+ depolarization and granule cell migration in different media

	BME + HS	BME + HEPES + HS	F12+ FCS	F12 w/o HEPES + FCS	F12 w/o HEPES + HS
Control	35	9	0	10	0
Glutamate	3	3	0	0	0
Glutamate + Taurine	7	0	0	1	0
10 mM K^+	0	0	15	18	0
10 mM K^+ + Taurine	2	38	9	3	4
25 mM K^+	0	0	0	0	6
25 mM K^+ + Taurine	0	27	12	41	48

Glutamate concentrations were 1 mM; taurine concentrations were 10 mM. The numbers of presumably migrating granule cells found on 5 randomly selected fiber bundles per experiment were counted after 3 days in culture in various media, as indicated. Taurine and K^+ were added after two days *in vitro*. Three experiments were performed.

without HEPES but with 25 μM K^+ and taurine present. These results, although preliminary, clearly demonstrate that the medium conditions determine the effectiveness of media components and consequently cellular survival and function.

Interaction of bFGF and Taurine to Prevent Excitotoxicity

The survival of early postnatal cerebellar cells in cultures with mixed cell populations[21], particularly with astroglial cells present, was compared to the survival of purified granule cells in serum-containing and serum-free medium supplemented with various concentrations of bFGF, and combinations of bFGF + 10^{-2} M taurine, bFGF + glutamate, and bFGF + glutamate + taurine, in order to characterize the cooperation between these factors in the control of neuronal survival.

In general, we found that serum-supplemented cultures containing neurons as well as glial cells (mixed populations) exhibited a higher neuronal survival rate (Fig. 3A) than those without astroglial cells (Fig. 3B&C). Significant differences were observed in the presence and absence of serum with and without factors. In cultures of mixed cell populations (Fig. 3 A), bFGF, or bFGF + taurine had little effect on the survival of neurons, reaching control levels. As expected, 10^{-3} M glutamate reduced the number of surviving granule cell neurons to approximately half in both mixed cultures and isolated neurons, when cells were cultured with horse serum. While no significant improvements occurred in mixed cerebellar cultures in the presence of glutamate and bFGF and/or taurine, purified granule cells survived with taurine and/or b-FGF in numbers similar to control levels. In serum-free medium, on the other hand, purified neurons responded to bFGF in a concentration-dependent matter, as reported earlier by others: higher concentrations of bFGF were less effective in promoting cell survival - in some cases even proving toxic - than lower concentrations. However, all concentrations improved cell survival. Without serum, taurine appeared to act independently from bFGF by significantly increasing the number of surviving granule cells with or without b-FGF to the levels of b-FGF alone, suggesting separate mechanisms.

Collaboration of bFGF and taurine was observed, however, when glutamate was added. Glutamate (10^{-3} M) added to these cultures significantly reduced neuronal survival in the absence of bFGF. bFGF (100 nM) increased the number of surviving neurons to control levels. Under these conditions taurine had little effect on its own but in collaboration with lower concentrations of bFGF (10 nM) neuronal survival was significantly increased over levels obtained with either bFGF or taurine alone, suggesting that taurine under certain conditions may support bFGF function.

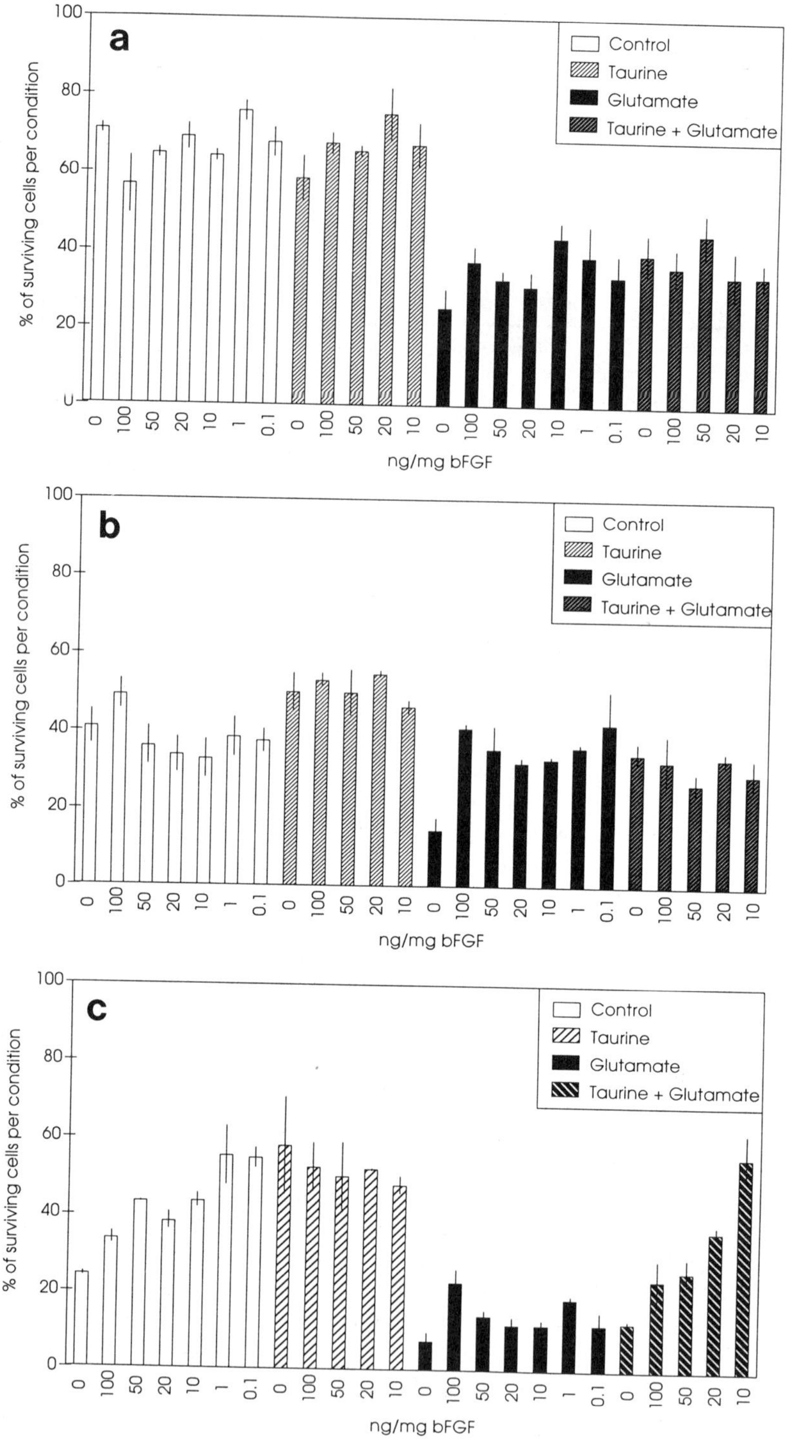

Figure 3. The number of surviving 7-day-old mouse cerebellar cells, expressed as % of the total cell number, was determined *in vitro* as a function of bFGF (100-0.1 ng), taurine (10 μM), and glutamate (10^{-3} M) and combinations thereof, after 3 days in culture. Three conditions were compared: A: mixed cerebellar cell populations including astrocytes in 10% HS containing DMEM. B: Enriched cerebellar granule cells maintained in DMEM with serum. C: Cerebellar granule cells maintained in serum free medium with N2-supplement.

We conclude from these experiments that bFGF and taurine do not act synergistically or additively but that both factors are capable of stimulating optimal cell survival, by, however, different mechanisms. Such alternative survival strategies may indicate the variety of mechanisms that control neuronal plasticity.

DISCUSSION

Culturing CNS neurons under different artificial conditions supplemented with heterologous sera, growth factors and neuroactive amino acids underscores the ability of neurons to respond to changing cellular environments throughout development, after injury, and during the process of aging. This capacity for plasticity is regulated through mitochondrial energy metabolism and through the co-operation of a variety of neurotrophic factors at a given time throughout life.

In this study we have begun to analyze whether and how a minimal reduction of mitochondrial energy metabolism can reduce neuronal survival under physiological conditions, how taurine relates to mitochondrial energy metabolism, and to growth factors, characterized as neurotrophic factors under tissue culture conditions, and whether these factors interact with excitatory amino acids in the control of survival and function of cerebellar cell populations.

Energy metabolism determines the vulnerability of neurons to excitotoxicity induced by extracellular glutamate during development and aging[2]. Here, we have shown that neurons treated with sub-optimal concentrations of FCCP, an uncoupler of the mitochondrial electrochemical gradient, are significantly more vulnerable to physiological concentrations of glutamate than untreated cells. Taurine restores the mitochondrial electrochemical gradient, and neurons survive (Fig. 1). Since intracellular Ca^{2+} concentrations determine, in part, the viability of neurons[5,10,17], partly through controlling mitochondrial functions, our results suggest that the mechanism underlying the preventive role of taurine is the modulation of Ca^{2+} influx and possibly the distribution and activation of different intracellular Ca^{2+} pools (Fig. 2). By measuring $^{45}Ca^{2+}$ uptake into cerebellar cells directly, we found that both Ca^{2+} uptake and the modulatory response of taurine was serum concentration-dependent: the less serum the more Ca^{2+} uptake was observed and the more effective taurine down-regulated Ca^{2+} uptake. The intracellular Ca^{2+} concentrations were the same after taurine treatment in serum-free medium, supplemented with N-2, and medium supplemented with 2% HS and 13% N-2, while in the presence of 10% HS + 5% N-2 both Ca^{2+} uptake and the reduction through taurine were significantly lower, suggesting that taurine can maintain a threshold of intracellular Ca^{2+} concentration that may be more physiological, for significantly more neurons survive under these conditions. The mechanism is unknown.

Similarly, serum affected the activity of bFGF, when added alone or together with taurine to serum-containing medium. With serum the viability of neuronal cells increased to a maximal level. This was independent of bFGF concentrations, suggesting that bFGF added to already existing factors in the serum may have reached saturation (Fig. 3A,B). On the other hand, in serum-free medium bFGF alone or with taurine improved neuronal survival significantly in a concentration-dependent manner when compared to untreated cells (Fig. 3C). This might be explained by the fact that serum contains various growth factors in unknown and varying quantities that may change the balanced relationship of those tested, whereas in serum-free medium only IGF could interact.

Astroglial cells together with neurons in mixed cultures responded to bFGF alone or together with taurine by significantly increasing the survival rate of neurons (Fig. 3A) when compared to cultures without astroglia cells (Fig. 3B). This difference was not obvious, however, in the presence of depolarizing concentrations of glutamate, suggesting that glial cells produce and release, via depolarization, combinations of factors (including bFGF and taurine) that could modulate the survival of neurons in addition to those factors added to the cultures and thus changing the balance of these factors, as illustrated in Fig. 3C. Under serum-free conditions without glial cells, on the other hand, bFGF and taurine co-operated concentration-dependently. With decreasing concentrations of growth factor and 10^{-2} M taurine, increasing survival of granule cells was obtained. Since the co-operation between bFGF and taurine did not result in an increase of surviving cells above the observed maximum, we are suggesting that both bFGF and taurine might use the same second messenger system to control cell-survival, possibly through balancing Ca^{2+} influx to physiological concentrations. This is supported by the indirect finding that NCAM binding to the bFGF receptor causes Ca^{2+} influx through Ca^{2+} channels (Filbin, pers. communication), suggesting that bFGF as well as taurine modulates the increase of intracellular Ca^{2+} concentrations. Furthermore, ongoing experiments suggest that taurine can assume different configurations in different media, as suggested by Huxtable (personal communication), that could affect influx and re-distribution of Ca^{2+} to various intracellular pools.

Depending on the medium composition, taurine can control the activity of granule cell migration. In BME, under depolarizing conditions, after treatment with NMDA (Table 1) granule cell migration was inhibited while in the presence of NMDA antagonists glutamate-induced depolarization was prevented and granule cells continued to migrate (Table 1). This suggests that NMDA-type receptors are involved in delivering the stop-signal during granule cell migration. Taurine appeared to intensify this signal since fewer cells migrated, although more cells survived. As demonstrated above, taurine reduced Ca^{2+} uptake during glutamate depolarization to physiological intracellular concentrations, (Fig. 2). Therefore, taurine could adjust intracellular Ca^{2+} concentrations available for Ca^{2+}-dependent processes like PKC-activity[27]. This result may also suggest that granule cell migration is Ca^{2+}-independent. Therefore, taurine might affect the distribution of Ca^{2+} into different intracellular pools which might explain why taurine enforces the stop signal delivered by NMDA receptors.

In BME and HS supplemented with HEPES buffer, only 30% of the granule cells migrated. In HEPES-containing F12 medium, cell migration was abolished while cells in F12 without HEPES showed some activity (Table 2). The survival rate of granule cells in these media was approximately the same. These differences in migratory activity based on medium composition is not yet understood.

K^+ depolarization presented here is another example that media conditions determine whether or not particular cellular functions are activated. In BME, granule cells died when more than 7 mM K^+ was added to osmotically adjusted BME. In other media[1,9,20], however, granule cells survive *in vitro* for a considerable length of time but do not support granule cell migration (data not shown). The presence of taurine restored granule cell migration in K^+-treated cultures which contained HEPES. On the other hand, cells survived and migrated in F12 medium without HEPES. Although we cannot explain these contradicting results at the moment, these examples clearly demonstrate that different medium conditions lead to different results and conclusions. However, the described differences may help to identify the molecular mechanisms that control neuronal survival and function in a changing tissue environment *in vivo*.

ACKNOWLEDGMENTS

We are grateful to Diane Cocozza for preparing this manuscript and we thank Dr. Huxtable who made this manuscript comprehensible. This study was supported by a grant from NYS Office of Mental Retardation and Developmental Disabilities.

REFERENCES

1. Balazs, R., Gallo, V. and Kingsbury, A. 1988, Effect of depolarization on the maturation of cerebellar granule cells in culture. *Dev. Brain Res.* 40:269-279.
2. Beal, M.F. 1994, Energy, oxidative damage,, and Alzheimer's disease: Clues to the underlying puzzle. *Neurobiol.Aging* 15:171-174.
3. Bottenstein, J.E, Skaper, S.D., Varon, S.S. and Sato, G.H. 1980, Selective survival of neurons from chick embryo sensory ganglionic dissociates utilizing serum-free supplemented medium. Exp. Cell. Res. 125:183-290.
4. Cameron, R.S. and Rakic, P. 1994, Identification of membrane proteins that comprise plasmalemmal junction between migrating neurons and radial glial cells. *J. Neurosci.* 14:3139-3155.
5. Choi, D.W. 1988, Glutamate neurotoxicity and disease of the nervous system. *Neuron* 1:623.
6. Eimerl, S. and Schramm M. 1993, Resuscitation of brain neurons in the presence of Ca^{2+} after toxic NMDA receptor activity. *J.Neurochem.* 61:518-525.
7. Feng, L. Hatten, M.E. and Heintz, N. 1994, Brain lipid-binding protein (BLBP): A novel signalling system in the developing mammalian CNS. *Neuron* 12:895-908.
8. Fishman, R.B. and Hatten, M.E. 1993, Multiple receptor systems promote CNS neural migration. *J. Neurosci.* 13:3485-3495.
9. Gallo, V., Kingsbury, A., Balazs, R. and Jorgensen, O.S. 1987, The role of depolarization in the survival and differentiation of cerebellar granule cells in culture. *J.Neurosci.* 7:2203-2213.
10. Ghosh, A. and Grennberg, M.E. 1995, Calcium signalling in neurons: molecular mechanisms and cellular consequences. *Science* 268:239-247.
11. Huxtable, R.J. 1992, The physiological actions of taurine. *Physiol. Rev.* 72:101-163.
12. Komuro, H. and Rakic, P. 1995, Dynamics of granule cell migration: A confocal microscopic study in acute cerebellar slice preparations. *J. Neurosci.* 15:1110-1120.
13. Liesi, P., Seppala, I. and Trenkner, E. 1992, Neuronal migration in cerebellar microcultures is inhibited by antibodies against a neurite outgrowth domain of laminin. *J. Neurosci. Res.* 33:170-176.
14. Lindholm, D., Dechant, G., Heisenberg, C.-P. and Thoemen, H. 1993, Brain-derived neurotrophic factor is a survival factor for cultured rat cerebellar granule neurons and protects them against glutamate-induced neurotoxicity. *Eur.J.Neurosci.* 5:1455-1464.
15. Mattson, M.P., Lovell, M.A., Furakawa, K. and Markesberry, W. 1995, Neurotrophic factors attenuate glutamate-induced accumulation of peroxides, elevation of intracellular Ca2+ concentration, and neurotoxicity and increased antioxidant enzyme activities in hippocampal neurons. J.Neurochem. 65:1740-1751.
16. Mattson, M.P., Zhang, Y. and Bose, S. 1993, Growth factors prevent mitochondrial dysfunction, loss of calcium homeostasis, and cell injury, but not ATP depletion in hippocampal neurons deprived of glucose. *Exp. Neurol.* 121:1-18.
17. Mattson, M.P., Guthrie, P.B., Murrain, M. and Kates, S.B. 1989, Fibroblast growth factor and glutamate: Opposing actions in the generation and degeneration of hippocampal neuroarchitecture. *FABEB J.* 3:2519-2526.
18. Miale, I. and Sidman, R.L. 1961, An autoradiographic analysis of histogenesis in mouse cerebellum. *Exp. Neurol.* 4:277-296.
19. Ramon y Cajal, S. 1960, Studies on Vertebrate Neurogenesis Transl. L. Guth, Springfield, IL, p. 423.
20. Schousboe, A., Meier, E., Dreier, J. and Hertz, L. 1989, Preparation of primary cultures of mouse (rat) cerebellar granule cells, in: "A Dissection and Tissue Culture Manual of the Nervous System", Shahar, A., DeVellis, J., Vernadakis, A. and Haber, B. eds. Alan R. Liss, New York, pp. 203-206.
21. Sidman, R.L. and Rakic, P. 1973, Neuronal migration, with special reference to developing human brain. *Brain Res.* 62:1-35.
22. Sturman, J.A. 1993, Taurine in development. *Physiol. Rev.* 73:119-147.
23. Trenkner, E. and Sidman, R.L. 1977, Histogenesis of mouse cerebellum in microwell cultures: Cell reaggregation and migration, fiber and synapse formation. *J. Cell Biol.* 75:915-940.
24. Trenkner, E., Smith D. and Segil, N. 1984, Is cerebellar granule cell migration regulated by an internal clock? *J. Neurosci.* 4:2850-2855.
25. Trenkner, E. 1990, The role of taurine and glutamate during early postnatal cerebellar development of neonatal and weaver mutant mice, in: "Excitatory Amino Acids and Neuronal Plasticity", Ben-Ari, Y. ed. Plenum Press, New York, pp. 239-244.
26. Trenkner, E. 1991, Cerebellar cells in culture, in: "Culturing Nerve Cells", Banker, G. and Goslin, K. eds. MIT Press, pp. 283-307.
27. Trenkner, E., Liu, DJ, Harris, C. and Sturman, J.A. 1994, Regulation of protein kinase C activity by taurine and β-alanine during excitotoxicity in cat and mouse cerebellar cultures, in: "Taurine in Health and Disease", Huxtable, R.J. and Michalk, D.V. eds. Plenum Press, New York, pp. 309-316.

56

TAURINE DEFICIENCY AND NEURONAL MIGRATION

Julio Morán,[1] Thomas Maar,[1-3] Georgi Gegelashvili,[3] Elisabeth Bock,[3] Arne Schousboe,[2] and Herminia Pasantes-Morales[1]

[1] Department of Neurosciences
Institute of Cellular Physiology
UNAM, Apartado Postal 70-253
04510 Mexico City, Mexico D.F.
[2] PharmaBiotec Research Center
Department of Biological Sciences
Royal Danish School of Pharmacy
DK-2100 Copenhagen, Denmark
[3] Protein Laboratory
Panum Institute
DK-2200 Copenhagen, Denmark

INTRODUCTION

Cerebellar granule neurons migrate from the external layer to the internal granule layer along the processes of Bergmann fibers[14,15]. It is known from a number of studies that this is a finely modulated process of cell-cell interaction controlled by a variety of external conditions. This event involves cell surface molecules on both neurons and glia, including extracellular matrix molecules, neural cell adhesion molecules (NCAM), and some members of the integrin and cadherin families[1,3,6,13,19,20]. Numerous observations *in vitro* and *in vivo* document the involvement of taurine in the process of neuronal migration. This process may be disturbed or even prevented by reducing the cerebellar taurine content which normally is high during early development[17,18]. This relationship between taurine and granule cell migration has been particularly well studied in the cat[17,18], a species especially vulnerable to taurine deficiency by removal of taurine from the diet[4]. Taurine depletion, however, is a slow process, requiring several months to be completed. We have previously reported a simple and rapid model of taurine deficiency in an *in vitro* system by preventing cellular taurine uptake by guanidinoethane sulfonate (GES)[5,10]. The present study has utilized an *in vitro* system to investigate the role of taurine for cell migration, aggregate formation, and the expression of NCAM in cerebellar cells employing GES to render the cultures taurine deficient.

Taurine 2, edited by Huxtable et al.
Plenum Press, New York, 1996

CELL MIGRATION

The basic properties of the migration process can be studied in an *in vitro* system consisting of microwell cultures of dissociated cerebellum. This allows quantification of the migration process to be performed[3,21]. Morphological inspection of the cultures during a period of 3 h after 72 h in culture revealed that a migrational process of the neuronal cell bodies on the glial processes was essentially linear with time over the 3 h period with an average total migration of about 100 μm (Fig. 2). The migration rate observed in the present study of about 30 μm x h^{-1} is close to that reported by previous investigations[2,3]. As pointed out by Hatten[3], the migrational behavior varies greatly among individual neurons and in some cases neurons may remain in the same position on the glial fiber for long periods before finally migrating, often with high rates.

The morphological appearance of the cells in the microwell cultures of dissociated cerebella of 6-day-old mice showed neuronal-like cell bodies attached to fibrous structures. These fibrous structures represent relatively thick process that stained positively for GFAP in most cases and always stained positive for γ-enolase indicating that they are of both glial and neuronal or of pure neuronal origin. In addition to the cell bodies attached to the fibers, aggregating cell bodies were frequently observed. These structures contained some cells that were positive for GFAP labeling, particularly around the periphery, indicating the participation of glial elements in the organization of these aggregates. All cell bodies were found to

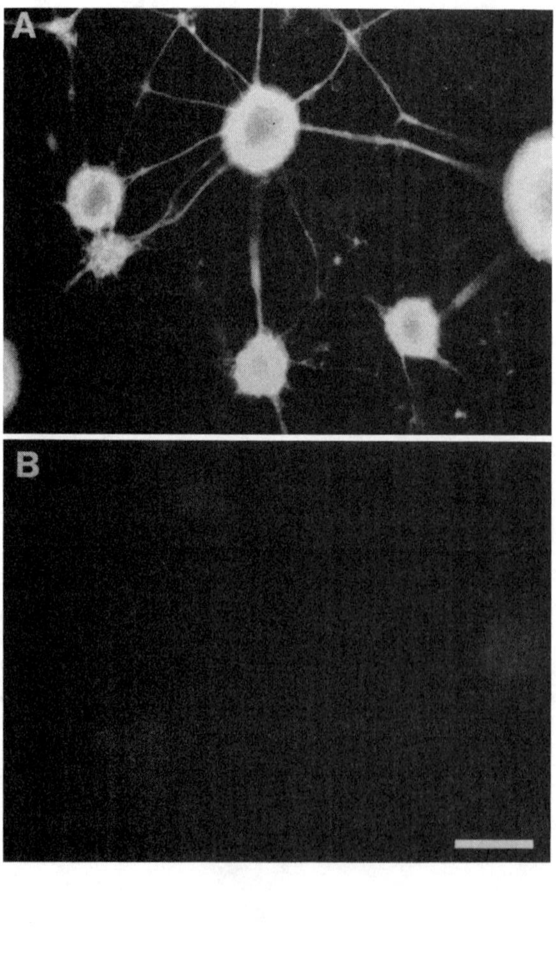

Figure 1. Immunofluorescence of NCAM in a microwell culture of dissociated cerebellar cells of mice. Primary microwell cultures of dissociated cerebellum were obtained as previously described[21]. Cells from cerebella of 6-day-old mice were plated in 60-well microtiter plates at a density of 5 x 10^6 cells/ml in a volume of 10 μl/well. Cells were grown for 72 h in a culture medium supplemented with 10% (v/v) horse serum with or without 2 mM GES. The cultures were washed 3x in PBS, fixed in 2% paraformaldehyde, blocked with 10% swine serum and subsequently immunostained with a polyclonal rabbit anti-NCAM antibody and visualized using an FITC-conjugated secondary anti-rabbit IgG antibody (A). In B, the micrograph shows the corresponding control blocked with 10% swine serum, with the primary antibody omitted. The bar indicates 100 μM.

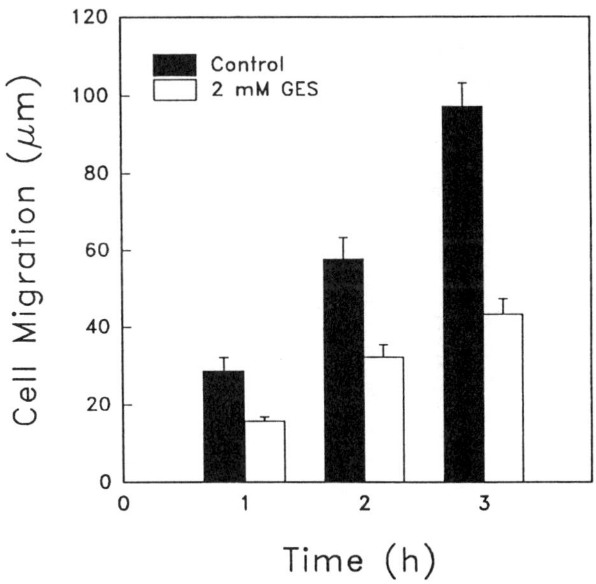

Figure 2. Effect of GES (2mM) on migration of neuronal cell bodies on glial fibers in microwell cultures of dissociated cerebella. Primary microwell cultures of dissociated cerebellum were obtained as detailed in Fig. 1. Determination of migration was performed by measuring neuronal movement and other parameters in a series of slides taken every 30 min during a 3 h period. The migration rate was calculated as the distance (in µm) of cell body movement. Results are averages of 8-10 separated experiments ± SEM. GES results are statistically significant different compared to control values using Student's t-test ($p < 0.001$). Modified from Maar *et al.*[7]

be labeled with the γ-enolase antibody identifying these structures as neuronal cell bodies. Moreover, Figure 1 shows that cell bodies and processes stained positively for the cell adhesion molecule NCAM, which is expressed in high amounts by neurons.

TAURINE AND CELL MIGRATION

Treatment of microcultures with 2 mM GES during 72 h caused a reduction of 61.5 ± 7.3% in intracellular taurine levels. This treatment was specific for taurine, without affecting any other amino acid tested, as previously shown[5,10]. A reduction of the cellular taurine content clearly led to a highly significant ($p < 0.001$; Student's t-test) reduction in cell migration (Fig. 2). This effect of taurine depletion could be fully prevented by increasing taurine concentration in the culture medium from the basal level (9.06 ± 1.1 µM) to 2 mM taurine in the presence of GES (Fig. 3). It should be emphasized that the ameliorating action of taurine depended upon the extent of the restoration of the cellular taurine level. It should also be noted that the effect of GES was dependent upon the time period of exposure to GES, emphasizing the mechanism of action of GES: i.e. a depletion of taurine levels. Figure 3 clearly shows a close correlation between the intracellular taurine levels and the migration rate, regardless the presence of GES in the medium. These results are in line with the observation that migration *in vivo* in cerebellum during early development is retarded by taurine deficiency[17,18].

Cell viability in the cultures, assessed by the ability of cells to transform the stain nitroblue tetrazolium into formazan blue[11], was unaffected by treatment with GES during both a 3- and a 5-day culture period (not shown). In addition, it was found in experiments monitoring Ca^{++}-dependent, K^{+}-stimulated transmitter release measured by the aid of ^{3}[H]-D-aspartate that this parameter was also not affected by the presence of 2 mM GES in the medium during a culture period of 7 days (not shown). These results, together with the fact that cell migration was found to be identical in cultures exposed to GES plus taurine or taurine alone, indicate that GES had no adverse effect on the cells.

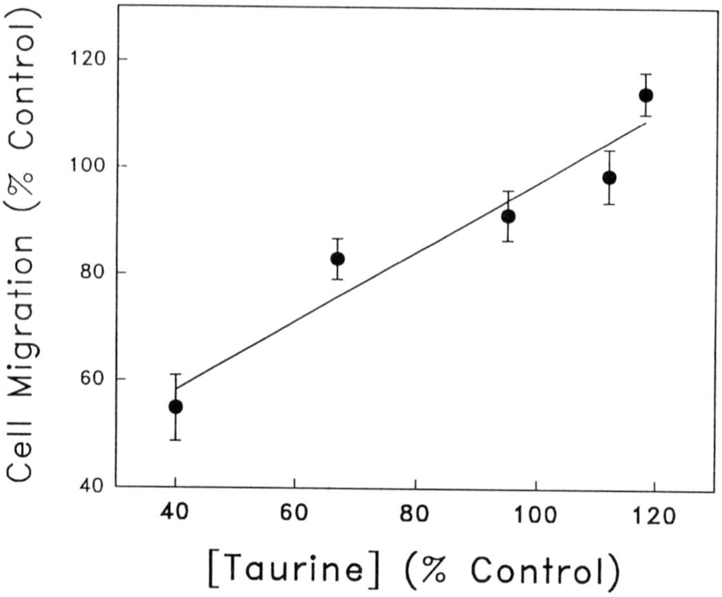

Figure 3. Correlation between intracellular taurine levels and neuronal migration in cerebellar microwell cultures. Preparation of cultures and determination of neuronal migration were performed as detailed in Fig.1. The taurine content was measured by HPLC in the cells grown under the different conditions and is expressed as % of the control cultures (56 ± 4.4 nmol/mg protein). The migration in control cultures was 92.4 ± 8.0 μm during 3 h. Both cell migration and taurine content were measured in cell cultures treated with 2 mM GES ± 2 mM taurine during 24-72 h. Results are averages ± SEM of 10-30 different experiments.

MORPHOLOGY OF CULTURES

A quantification, expressed as the number of a particular structure per visual field, of the different morphological characteristics of the cultures is shown in Figure 4. It was calculated that on the average one of the thick fibers has one neuronal cell body attached. With regard to aggregating neuronal cell bodies, the number of these aggregates per visual field was similar to the number of either one of the two types of fibers. The number per visual field of the different morphological parameters measured except aggregates was found to be decreased by GES treatment (Fig. 4). It should be noted that the decrease in the density of neurons attached to fibers was much larger than the decrease in the density of the number of thick fibers. This means that the number of neurons per fiber was dramatically reduced by the taurine deficiency induced by GES. In case of the density of aggregates, addition of GES during the entire culture period led to an increase whereas addition of GES during the last 48 h of the culture period had no effect. Addition of taurine in the presence of GES prevented the decrease caused by GES in the density of both types of fibers and that of neurons attached to fibers. However, the values observed in the presence of GES and taurine were marginally but consistently lower than those in the corresponding controls (Fig. 4). In case of the aggregates, the density seen after treatment with GES plus taurine was similar to the control, albeit slightly higher, regardless of the exposure period. Addition of taurine alone in both experimental paradigms had essentially no effect on the density of any of the different morphological structures investigated (Fig. 4).

It should also be noted that the glial fibers not associated with neurons were reduced in number by GES and that taurine could only partly prevent this effect. This may indicate

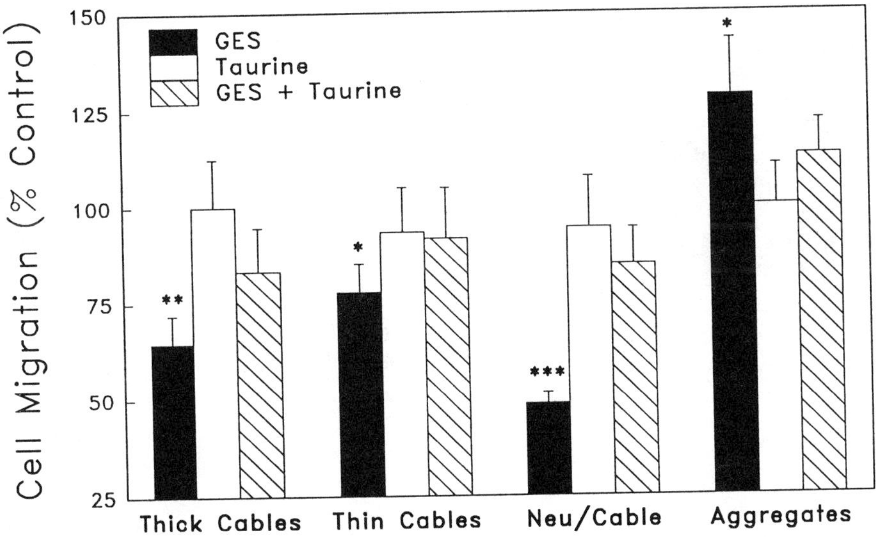

Figure 4. Effect of GES and taurine on the morphological characteristics of cultured cerebellar cells. Preparation of cultures was performed as detailed in Fig. 1. Cultures were exposed to GES (2 mM), taurine (2 mM) or the two compounds together for 72 h. Subsequently the different morphological structures , i.e., thin and thick fibers, neuronal cell bodies attached to thick fibers (neu/cable) and aggregates of cell bodies, were identified by phase contrast microscopy and their density as numbers per visual field determined. Results are averages ± SEM values of 10-30 visual fields in at least 3 different culture preparations. Asterisks indicate statistically significant difference from controls using Student's t-test for paired samples (*$p < 0.05$; **$p < 0.01$; ***$p < 0.001$) (modified from Maar *et al.*[7])

that glial proliferation may be influenced by taurine depletion in a manner which may not be fully reversible. Whether this may affect formation of the thick fibers remains to be established. At the present time it is unknown whether or not the surface properties of the glial processes might be different under these conditions. Any such differences could be of importance for the interaction with neurons[19].

The effect of GES on aggregate formation may be interesting as the ability of the GES-induced taurine depletion to increase the number of aggregates was time-dependent. It appeared that taurine depletion during the first 24 h of the culture period might be critical, as exposure to GES during the last 48 h had no effect. A full explanation for this cannot be provided at the present but it is conceivable that expression of neuronal adhesion molecules responsible for neuron-neuron recognition might be influenced by taurine during the initial phases of the recognition process (cf. below). On the other hand, once the cell contacts have been established, taurine may no longer be able to prevent the aggregation which subsequently interferes with attachment of neuronal cell bodies to the fibers as reflected in the decrease in the number of neurons per fiber (Fig. 4).

CELL ADHESION AND MIGRATION

The decrease in cell migration caused by the GES-induced taurine deficiency could be fully prevented by addition of taurine. It may be important to note that, in spite of this, taurine did not fully prevent the decrease in the density of neurons attached to the processes caused by GES treatment. This may indicate that different factors may regulate attachment

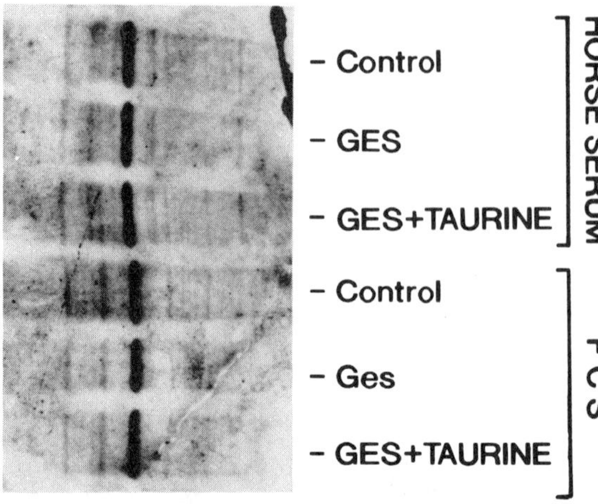

Figure 5. SDS-PAGE of homogenatesfrom microwell cultures treated with GES and taurine. Cultures were prepared as detailed in Fig. 1. Cells were grown in a culture medium supplemented with either 10% fetal calf serum or horse serum. Taurine (2 mM) or taurine plus GES (2 mM) was added to the media. After 3 days *in vitro*, cells were lysed and prepared for SDS-GAGE, with subsequent staining with Coomasie blue.

of the neurons and the subsequent migration process and that taurine may play different roles in these events. Studies of the ability of taurine to alter expression of proteins including NCAM were carried out in cultured cerebellar granule cells using SDS-PAGE combined with either Coomasie blue staining or Western blot analysis utilizing an NCAM-specific antibody. Figure 5 shows the Coomasie blue staining and reveals major differences in protein expression when using horse and fetal calf serum. However, regardless of the type of serum, the presence of 2 mM GES or GES plus 2 mM taurine had no effect on the protein pattern (Fig. 5). Other experiments in which the expression of NCAM was investigated by Western blot analysis indicated that this expression was not seriously affected by taurine depletion (results not shown). Nevertheless, preliminary results from these experiments hint at possible effects of taurine depletion on the expression of certain slice variants of the NCAM molecule (results not shown). This possibility as well as a possible influence of taurine on other cell adhesion molecules is currently being investigated.

CONCLUDING REMARKS

The mechanism by which taurine influences neuronal migration on fibers is not presently understood. It should be emphasized that the neuronal migration is likely to reflect the cell surface properties of both cell types[3]. Thus, antibodies to the neuronal antigen astrotactin prevents binding of neurons to glial cells[3] and astrotactin is missing from granule cells in the weaver mutant mouse where granule cell migration is impaired[1]. It may be important in this context that addition of taurine to cultures of weaver mouse granule cells promotes cell migration[22]. The neuronal cell adhesion molecules NCAM and L1 do not, on the other hand, appear to be directly involved in the attachment of neurons to glia and the subsequent migration process[3,13] but they do play an important role in the shaping of neuronal networks *in vivo*[6]. Treatment with GES will lead to a reduction of the cellular taurine content in both neurons and glial cells. Therefore if indeed taurine plays a role in expression of surface molecules it is likely that both neurons and glial cells would be affected. The observed effect of GES on formation of thick fibers may indicate that glial cells are affected by taurine deficiency. It may also be hypothesized that taurine deficiency could alter the osmoregulation in the cells as taurine is likely to function as an important osmoregulator[11,23,24]. Whether a deficiency in osmoregulation can have an effect on cell surface properties remains to be established. Finally it should be pointed out that taurine to some extent can mimic actions of GABA which in addition to its function as an inhibitory neurotransmitter acts as a stimulator of neuronal growth and differentiation[16]. In this context it may be particularly important that GABA has been shown to affect expression of the cell adhesion molecule NCAM in neurons *in vitro* as well as *in vivo*[8,9].

ACKNOWLEDGMENTS

The excellent technical assistance of Ms. Claudia Rodríguez and Guadalupe Domínguez is gratefully acknowledged. Financial support from the following granting agencies is acknowledged: DGAPA-UNAM, Mexico (Project IN202093); the Lundbech Foundation; the Danish State Biotechnology Program (1991-95); the CEC BIOTEC Programs (BIO2CT 930260 and BIOTCT 030224).

REFERENCES

1. Edmondson J.C., Liem R.K.H., Kuster J.E. and Hatten M.E. 1988, Astrotactin: a novel neuronal cell surface antigen that mediates neuron-astroglial interactions in cerebellar microcultures. *J. Cell Biol.* 106:505-517.
2. Hatten M.E., Lien M.K.H. and Mason C.A. 1986, Weaver granule neurons fail to migrate on wild-type cerebellar astroglial processes in vitro. *J. Neurosci.* 6:2676-2683.
3. Hatten M.E. and Mason C.A. 1990, Mechanisms of glial-guided neuronal migration *in vitro* and *in vivo*. *Experientia*. 46:907-916.
4. Hayes K.C., Carey R.E. and Schmidt S.Y. 1975, Retinal degeneration associated with taurine deficiency in the cat. *Science*. 188:949-951.
5. Huxtable R.J., Laird H.E. and Lippincott S.E. 1979, The transport of taurine in the heart and the rapid depletion of tissue taurine content by guanidinoethyl sulfonate. *J. Pharmacol. Exp. Ther.* 211:465-471.
6. Lindner J., Rathjen F.G. and Schachner M. 1983, L1 mono- and polyclonal antibodies modify cell migration in early postnatal mouse cerebellum. *Nature*. 305:427-430.
7. Maar T., Morán J., Schousboe A. and Pasantes-Morales H. 1995, Taurine deficiency in dissociated mouse cerebellar cultures affects neuron migration. *Int. J. Devel. Neurosci.* 13:491-502.
8. Meier E. and Jorgensen O.S. 1986, Gamma-aminobutyric acid affects the developmental expression of neuron-associated proteins in cerebellar granule cell cultures. *J. Neurochem*. 46:1256-1262.
9. Meier E., Jorgensen O.S. and Schousboe A. 1987, Effect of repeated treatment with a gamma-aminobutyric acid receptor agonist on postnatal neural development in rats. *J. Neurochem.* 49:1462-14:70.
10. Morán J. and Pasantes-Morales H. 1991, Taurine-deficient cultured cerebellar astrocytes and granule neurons obtained by treatment with guanidinoethane sulfonate. *J. Neurosci. Res*. 29:533-537.
11. Mossman T. 1983, Rapid colorimetric assay for cellular growth and survival: application to proliferation and cytotoxic assays. *J. Immunol. Meth*. 65:55-63.
12. Pasantes-Morales H. and Schousboe A. 1988, Volume regulation in astrocytes: A role for taurine as an osmoeffector. *J. Neurosci. Res*. 20:505-509.
13. Persohn E. and Schachner M. 1987, Immunoelectron microscopic localization of the neural cell adhesion molecules L1 and N-CAM during postnatal development of the mouse cerebellum. *J. Cell Biol.* 105:569-576.
14. Rakic P., Cameron R.S. and Komuro H. 1994, Recognition, adhesion, transmembrane signaling and cell motility in guided neuronal migration. *Curr. Opin. Neurobiol.* 4:63-69.
15. Rakic P. 1971, Neuron-glia relationship during granule cell migration in developing cerebellar cortex. A Golgi and electron microscopy study in *Macacus rhesus*. *J. Comp. Neurol.* 141:283-312.
16. Redburn D.A.R. and Schousboe A. Eds., 1987, Neurotrophic Activity of GABA during Development. pp 1-277, New York: Alan R. Liss.
17. Sturman J.A., Moretz R.M., French J.H. and Wisniewski H.M. 1985a, Taurine deficiency in the developing cat: persistence of the cerebellar external granule cell layer. *J. Neurosci. Res.* 13:405-416.
18. Sturman J.A., Moretz R.M., French J.H. and Wisniewski H.M. 1985b, Postnatal taurine deficiency in the kitten results in a persistence of the cerebellar external granule cell layer: correction by taurine feeding. *J. Neurosci. Res.* 13:521-528.

19. Tomaselli K.J., Reichardt L.F. and Bixby J.L. 1986, Distinct molecular interactions mediate neuronal process outgrowth on non-neuronal cell surfaces and extracellular matrices. *J. Cell Biol.* 103:2659-2672.
20. Tomaselli K.J., Neugebauer K.M., Bixby J.L., Lilien J. and Reichardt L.F. 1988, N-Cadherin and integrins: Two receptor systems that mediate neuronal process outgrowth on astrocyte surfaces. *Neuron.* 1:33-43.
21. Trenkner E. and Sidman R.L. 1977, Histogenesis of mouse cerebellum in microwell cultures, cell reaggregation and migration, fiber amd synapse formation. *J. Cell Biol.* 75:915-940.
22. Trenkner E. 1990, The role of taurine and glutamate during early posnatal cerebellar development. In: Excitatory Amino Acids and Neuronal Plasticity. Y. Ben-Ari, ed) *Plenum Press*, New York.
23. Van Gelder N. 1983, A central mechanism of action of taurine: osmoregulation, bivalent cations and excitation threshold. *Neurochem. Res.* 8:687-699.
24. Van Gelder N. 1989, Brain taurine content as a function of cerebral metabolic rate: osmotic regulation of glucose derive water production. *Neurochem. Res.* 14:495-497.

57

ROLE OF TAURINE IN THERMOREGULATION AND MOTOR CONTROL

Behavioural and Cellular Studies

Gianpietro Sgaragli,[1] Maria Frosini,[1] Mitri Palmi,[1] Henry B. F. Dixon,[2] Nicholas Desmond-Smith,[2] Loria Bianchi,[3] and Laura Della Corte[3]

[1] Istituto di Scienze Farmacologiche
Università degli Studi di Siena
Via E.S. Piccolomini 170, 53100 Siena, Italy
[2] Department of Biochemistry
University of Cambridge
Cambridge, CB2 1QW United Kingdom
[3] Dipartimento di Farmacologia Preclinica e Clinica "M. Aiazzi-Mancini"
Università degli Studi di Firenze
Viale G.B. Morgagni 65, 50134 Florence, Italy

INTRODUCTION

Despite extensive studies on the synthesis and possible neuronal or extraneuronal uptake and release of taurine, there is still no consensus about its functions[23,24,35]. It has variously been designated as a neurotransmitter, neuromodulator, a modulator of Ca^{2+} fluxes, an osmoregulator or membrane stabilizer and an antioxidant (Figure 1). These effects may, at least in some cases, be interlinked. Thus, for example, the neuroprotective effects of taurine may be related to its osmoregulatory, antioxidant and Ca^{2+}-modulating functions.

Experiments have been performed in this laboratory to clarify the possible function of taurine in the brain as a neurotransmitter or a neuromodulator. These studies have been mainly focused on the role of taurine in such integrated functions as thermoregulation and motor control. Regulation of body temperature involves a delicate balance between the production and loss of heat. The hypothalamus regulates the set point at which body temperature is maintained. One of the most striking and consistent effects of central (i.c.v.) administration of taurine in mammals is hypothermia. A major part of this effect is the heat dissipation caused by a reduction in peripheral vasomotor tone and by the inhibition of shivering[49]. Furthermore, a decrease in metabolic heat production and an increase in evaporative heat loss, both under serotoninergic control, may also play a role in taurine-induced hypothermia[48].

Taurine 2, edited by Huxtable et al.
Plenum Press, New York, 1996

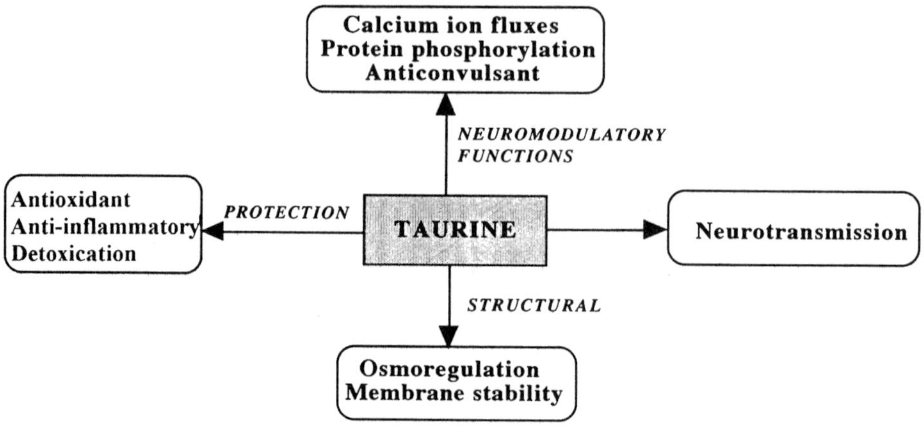

Figure 1. Possible functions of taurine.

The basal ganglia are involved in the control of movement. The role of basal ganglia is mediated by the cortico-basal ganglia-thalamocortical circuit. Striatal projections receive central information and transmit this via "direct" and "indirect" pathways, along the pallido/nigro-thalamo-cortical pathways[1,13]. The "direct" transmission of corticostriatal information is made by the monosynaptic inhibitory projection from the striatum to the output of the substantia nigra plus entopeduncular neurons, whereas the "indirect" pathway is mediated by at least two polysynaptic pathways; i.e. striatum to globus pallidus to subthalamic nucleus to output neurons[1] and striatum to globus pallidus to output nuclei[3,50]. Activation of the "direct" pathway is believed to lead to a decrease in the firing rate of output neurons and thus disinhibition of basal ganglia targets[8], whereas activation of the "indirect" pathways would have the opposite effect[1,51]. The imbalance between the two is thought to underlie disorders of movement. The "overactivity" of the "indirect" pathway is indeed important for the development of parkinsonian motor signs characterized by slowness of movement with increased muscular tone. In parkinsonian monkeys a subthalamic lesion leads to a marked reduction of muscular rigidity. It is interesting in this context that the i.c.v. injection of taurine, which has been associated with striatonigral neurons[9,17], induces a decrease of muscular tone in the rabbit[47]. Furthermore, taurine behaves like GABA when locally applied to the substantia nigra of the rat, causing the animals to rotate in a contralateral direction[26,27,40].

Bianchi *et al.* (this volume) report additional evidence that taurine may have a role in the striatonigral system of the rat. Considering that, whether a neurotransmitter or a neuromodulator, taurine must be recognized before initiating its action, in the present contribution a structure-activity relationship (SAR) study is presented which gives some information on the nature of a possible recognition site in the rabbit brain, helping to explain the role of endogenous taurine in the central mechanisms of thermoregulation.

METHODS AND MATERIALS

Adult, male New Zealand albino rabbits weighing 2.0-2.5 kg were supplied with food and water ad libitum and individually caged at an ambient temperature of 15-20°C, with natural day and night cycles. Under sodium pentobarbital anaesthesia, a cannula guide was chronically implanted, using a stereotaxic apparatus, into the cisterna magna for the withdrawal of cerebrospinal fluid[44]. For administration of drugs, an inflow cannula guide was inserted stereotaxically into a lateral cerebral ventricle. Samples of cerebrospinal fluid, drawn at a constant flow rate of 5 µl/min from the cisterna magna through a polyethylene tubing connected to a peristaltic pump, were collected into plastic tubes at 20 min interval (fractions of 100 µl). Fractions were refrigerated at 4°C and analysed on the same day. Compounds dissolved in pyrogen-free water were administered i.c.v. in a 10 µl volume by a micrometer syringe. Conscious rabbits were individually housed in a thermostated chamber and then restrained in stainless steel cages. The ambient temperature was set at 20°C. Core and ear skin temperatures were recorded using a thermocouple thermometer (Columbus Instruments, Columbus, OH USA). Temperatures were monitored for at least 1 h before any drug was injected. Cerebrospinal fluid samples were analysed for taurine content using reversed-phase HPLC and o-phthalaldehyde pre-column derivatisation[15]. Compounds were obtained as follows: piperidine-3-sulfonic acid was prepared according to Freifelder *et al.*[19], with slight modifications; 2-aminoethylarsonic acid was prepared as described by Geoghegan & Dixon[20]; N-methyltaurine was obtained from the aldehyde, 2-oxoethanesulfonic acid, mixing it with methylamine in ethanol to form the imine, $Me\text{-}NH^+{=}CH\text{-}CH_2\text{-}SO_3^-$, which was reduced by addition of borohydride, to the desired compound; N,N-dimethyltaurine was synthesised by complete methylation of taurine, heating it in a solution of formaldehyde and formic acid, to the stage of tertiary amine[10]. The protonated imine between formaldehyde and taurine, $CH_2{=}NH^+\text{-}CH_2\text{-}CH_2\text{-}SO_3^-$, is reduced by formate, which becomes CO_2, to form $CH_3\text{-}NH_2^+\text{-}CH_2\text{-}CH_2\text{-}SO_3^-$; this in turn forms the imine $CH_2{=}N^+(\text{-}CH_3)\text{-}CH_2\text{-}CH_2\text{-}SO_3^-$ with more formaldehyde, and this is further reduced to $(CH_3)_2NH^+\text{-}CH_2\text{-}CH_2\text{-}SO_3^-$.

RESULTS AND DISCUSSION

Taurine and eight analogues (Figure 2) have been tested in an SAR study for their effects on thermoregulatory behaviour of the rabbit. Taurine administered into the lateral ventricle in µM amounts caused a transient dose-dependent fall in body temperature, which was accompanied by vasodilatation at the ear vascular bed and reduction in muscle tone. This was concomitant with an elevation of taurine concentrations in the cisterna magna cerebrospinal fluid from 5-7 µM (basal values) up to 1-2 mM, which peaked 40-60 min after injection.

The effects elicited by the analogues are shown in Table 1. Guanidinoethane sulphonate and piperidine-3-sulphonic acid, like taurine, caused hypothermia. However, they were less potent than taurine. N-Methyltaurine exhibited a bimodal effect according to the dose. At low doses it caused a moderate hyperthermia, but at the highest dose it induced a mild hypothermia. N,N-dimethyltaurine caused hyperthermia at all doses tested. The same effect was induced by 2-aminoethylphosphonic acid and 2-aminoethylarsonic acid. The latter compound, however, was markedly toxic. Also pyridine-3-sulphonic acid caused hyperthermia, whereas β-alanine did not induce any significant effects. To summarise, some of the tested compounds elicited taurine-like effects, others had the opposite effect, with N-methyltaurine sharing the two properties.

Taking into account the conformation of these compounds, their effects can be explained by postulating the existence of a receptor possessing two recognition sites, one negatively and the other positively charged, interacting with the amino and the sulphonic group, respectively, of taurine (Figure 3, panel A). The interaction of taurine with the receptor at both sites is essential, but the receptor binding of the sulphonate group, is specific for its tetrahedral shape, since β-alanine, with its planar carboxylate group, had no effect. When the S atom is replaced with As or P, the resulting arsono and phosphono groups are similar in tetrahedral shape and size to the sulphonic group of taurine. A receptor that has a site for

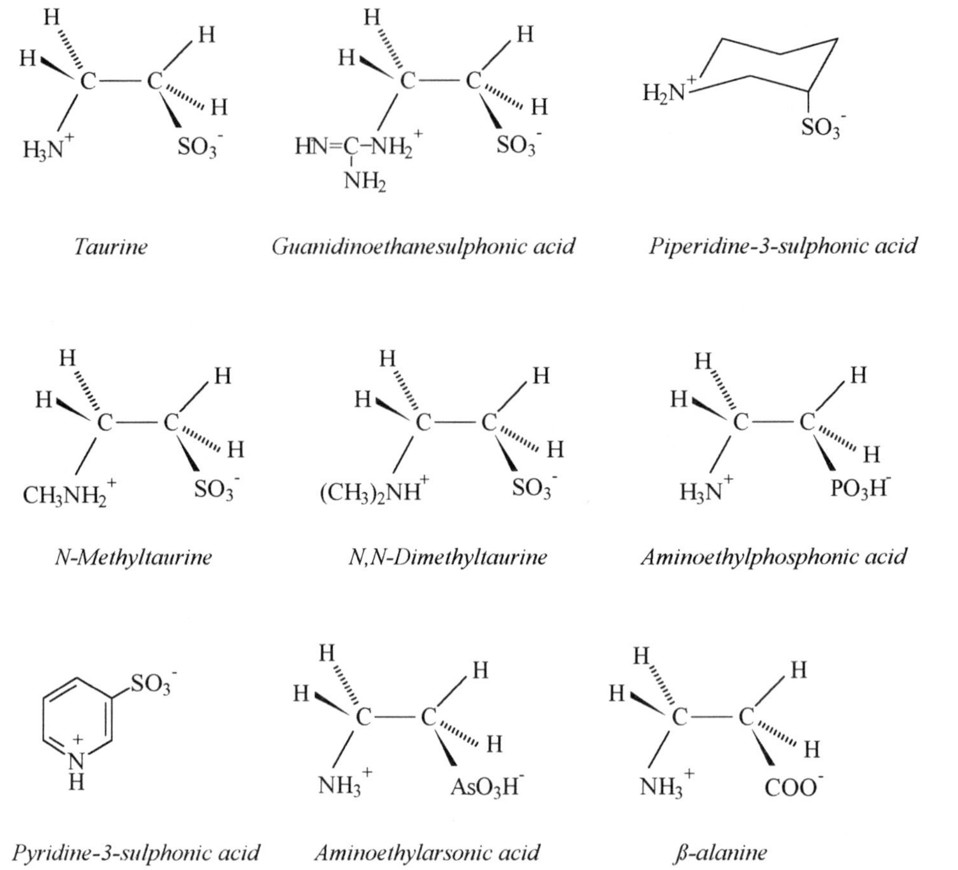

Figure 2. Conformation of taurine and some of its analogues.

a singly charged anion could force the arsono and the phosphono groups into the singly charged forms $-AsO_3H^-$ and $-PO_3H^-$, so that they can bind. The addition of H^+ removes their rotational symmetry, altering the position of the negative charge with respect to the rest of the molecule; this may mean that the rest of the molecule is slightly displaced with respect to the negative receptor site (Figure 3, panel B). This has been noted by Kluger *et al.*[28] who

Table 1. Effects of taurine and some of its analogues ICV injected on rabbit thermoregulation

Compound	Dose (μmol)	Effect
Taurine	12-48	Hypothermia
GES	24	Hypothermia
Piperidine-3-sulfonic acid	48	Hypothermia
N-Methyltaurine	24-48	Hyperthermia; Hypothermia
N,N-Dimethyltaurine	0.01-12	Hyperthermia
Aminoethane phosphonic acid	1.2-48	Hyperthermia
Pyridine-3-sulfonic acid	48	Hyperthermia
Aminoethane arsonic acid	0.002-24	Hyperthermia; Toxicity
β-Alanine	48	No effect

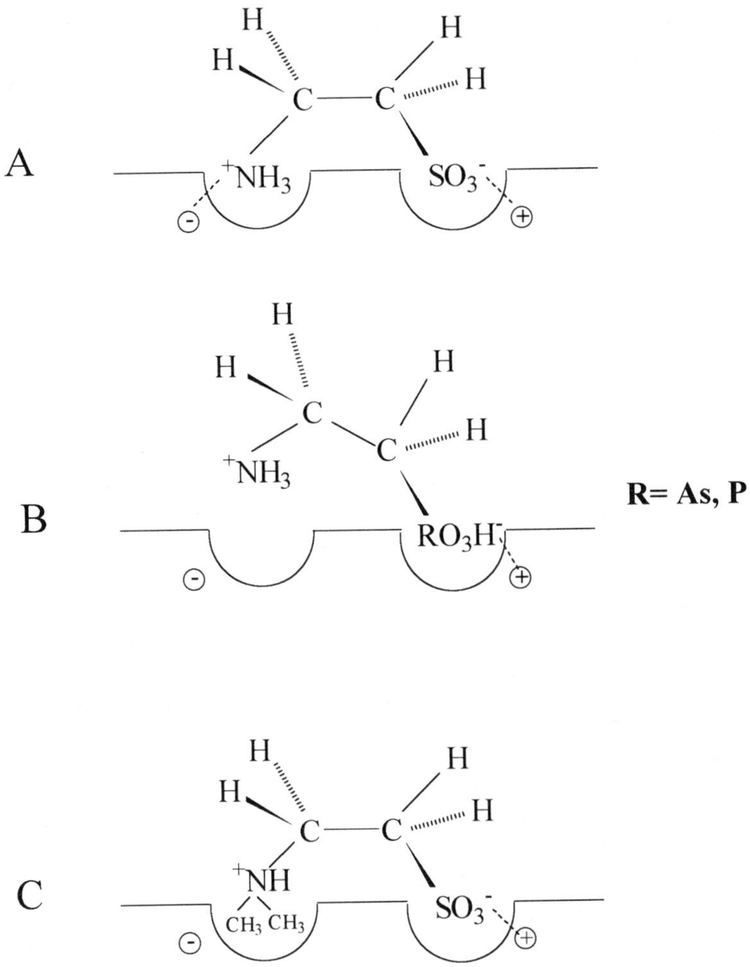

Figure 3. Interaction of taurine and some of its analogues with an hypothetical receptor in rabbit brain.

found that CH_3-CO-CH_2-SO_3H, but not CH_3-CO-CH_2-PO_3H_2, was a substrate for reduction by 3-hydroxybutyrate dehydrogenase, which normally acts on CH_3-CO-CH_2-COOH. Similarly, with N,N-dimethyltaurine the steric hindrance from the two methyl groups does not allow the positively charged nitrogen to interact with the negatively charged receptor site (Figure 3, panel C).

The hypothesis of a taurine receptor might help to explain why some taurine analogues elicit hyperthermia. If we admit that they are able to block the receptor without activating it, then they will not allow endogenous taurine to exert a tonic influence on thermoregulatory centres directed towards dissipation of body heat. Without this influence the system will be unbalanced and hyperthermia will develop.

As outlined previously[6], the effect of taurine on thermoregulatory behaviour may depend on its interaction with hypothalamic cells situated in the vicinity of the third ventricle. Neurochemistry of taurine in the hypothalamus is essentially based on studies performed on the rat. Taurine and sulfinoalanine decarboxylase (EC 4.1.1.29, commonly referred to as cysteine sulfinic acid decarboxylase, CSDI) activity have been detected by immunocytochemistry in various regions of the brain using different antibodies,[5,7,25,36,38,42,45,46,57] revealing distribution differences in the patterns of labeling obtained. In contrast to cerebellum, cortex and caudate-putamen, where labeling was mainly neuronal, the hypothalamus displayed a predominant glial taurine-like immunoreactivity[37,53].

Unilateral kainic acid (KA) injections in the hypothalamus induce a strong increase in taurine concentration in the lesioned side, indicating a predominant localization of this amino acid in the glial cells that proliferate after such neurotoxic lesion[21]. Furthermore, it has been demostrated[14] that taurine can be released from supraoptic nucleus glia, by a small decrease in osmolarity or by a receptor-mediated mechanism, during conditions of low hormonal (vasopressin and/or oxytocin) needs. The authors suggested that such released taurine could then act on pre- or post-synaptic sites, or both, to exert its neuromodulatory actions.

The present SAR study suggesting the existence of a taurine receptor provides support to the hypothesis that taurine may act as a neurotransmitter in the CNS. In the basal ganglia taurine seems to fulfil many of the criteria of a neurotransmitter. Thus, (i) high concentrations of the enzyme that synthesises taurine, sulfinoalanine decarboxylase, and of taurine itself have been found in the neostriatum and substantia nigra[18,30,32,43,52,54,55] and the decarboxylase shown to be enriched in brain synaptosomal fractions[31]; (ii) a high affinity uptake system has been described in the striatum[11,33,39] and in the substantia nigra[16], using brain slices or synaptosomes[34,41]; (iii) locally applied exogenous taurine was shown to accumulate selectively in the striatum and substantia nigra, at the level of perikarya and of axons and terminals, respectively, of neurons identified as medium size densely spiny striatonigral neurons[9,17]; (iv) preloaded [^{3}H]taurine was released from superfused slices of substantia nigra by veratridine in a Ca^{2+}-dependent and TTX-sensitive manner[17]; and (v) access by exogenous taurine to its releasable pool is dependent upon the presence of striatonigral neurons, since taurine levels[29], taurine-immunolabeled terminals[54] and the levels of the decarboxylase activity[31] are reduced in the substantia nigra by lesions of the striatum.

Consistent with its possible neurotransmitter role in the striatonigral system, locally applied taurine inhibits the firing rate of nigral dopaminergic neurons[22]. Furthermore, taurine has been found to bind to a $GABA_A$ receptor subclass of which the substantia nigra is particularly enriched[4].

In a recent study[2], where dual microdialysis was carried out in the neostriatum and substantia nigra of freely moving rats, it was demonstrated that KA stimulation of the neostriatum causes a DNQX-sensitive release of GABA locally in the neostriatum and, simultaneously, in the ipsilateral substantia nigra. This indicated that GABA was released from the local axon collaterals of the medium spiny GABAergic projection neurons and that their stimulation also induced a simultaneous release from their axon terminals in the substantia nigra. If, in addition to GABA, taurine is also present in striatonigral neurons, then one would expect that following their stimulation there would also be a simultaneous release of taurine from the striatum and the ipsilateral substantia nigra. Bianchi *et al.* (this volume) have shown that the release of taurine evoked by K^+ or KA stimulation in the neostriatum is likely to be derived from the local axon collaterals of the spiny projection neurons, adding support to the possible neurotransmitter role of this amino acid in the striatonigral system.

CONCLUSION

It is conceivable that taurine may act in the brain as a neurotransmitter. For a substance to be identified as a synaptic transmitter the following criteria must be fulfilled[12]:

(i) Localization: A suspected synaptic transmitter must be proven to reside in the presynaptic terminal of a selected nerve pathway associated to synaptic vesicles, along with the enzymes for its synthesis and catabolism. Taurine and its biochemical machinery have been localized in the striatonigral system, although the existence of taurine-containing synaptic vescicles has not yet been proven; (ii) Releasibility: A suspected synaptic transmitter must be demonstrated to be released from prejunctional nerve endings when the nerve fibers are selectively stimulated. Bianchi *et al.* (this volume) present evidence that taurine is released into the substantia nigra upon pharmacological stimulation of striatum; (iii) Neuroeffectiveness, synaptic mimicry: An exogenous substance suspected of being a neurotransmitter must mimic the action of the transmitter released by nerve stimulation. The application of taurine to the substantia nigra causes controlateral turning, which is inhibited by the concurrent application of 6-aminomethyl-3-methyl-4H-1,2,4-benzothiadiazine-1,1-dioxide (AMBD)[40], a putative taurine antagonist[56]; (iv) Pharmacology of synaptic effects: Identical pharmacological effects are elicited by drugs potentiating or blocking postsynaptic responses to both the neurally released or administered suspected neurotransmitter. Although no taurine receptor has hitherto been demonstrated in brain, this may have been because taurine seems to interact with only a subpopulation of $GABA_A$ receptors[4] or because selective taurine agonists/antagonists have been lacking. To this end the present SAR study is promising and we hope that the synthesis of new analogues of taurine will allow the identification and characterization of the brain receptors responsible for its activity.

ACKNOWLEDGMENTS

This work was supported by the European Community (BHMT1 CT-94-1402) and CNR-bilateral project (95.00833 CT04). The authors thank B. Gorelli and A. Benocci for technical assistance.

REFERENCES

1. Albin, R.L., Young, A.B. and Penney, J.B., 1989. The functional anatomy of basal ganglia disorders. *Trend Neurosci.*, 12:366-375.
2. Bianchi, L., Sharp, T., Bolam, J.P. and Della Corte, L., 1994, The effect of kainic acid on the release of GABA in rat neostriatum and substantia nigra. *Neuroreport*, 5:1233-1236.
3. Bolam, J.P. and Smith, A.D., 1992. The striatum and the globus pallidus send convergent synaptic inputs onto single cells in the entopeduncular nucleus of the rat: a double anterograde labeling study combined with post-embedding immunocytochemistry for GABA. *J. Comp. Neurol.*, 321:456-476.
4. Bureau, M.H. and Olsen, R.W., 1991. Taurine acts on a subclass of GABAA receptors in mammalian brain in vitro. *Eur. J. Pharmacol.*, 207:9-16.
5. Campistron, G., Geffard, M. and Buijs, R.M., 1986. Immunological approach to the detection of taurine and immunocytochemical results. *J. Neurochem.*, 46:862-868.
6. Carlà, V., Dacke, C.G., Davidson, N., Giotti, A., Magnani, M., and Sgaragli, G.P., 1982. Taurine and thermoregulation: behavioral and cellular studies, in: "Taurine in Nutrition and Neurology". R.J. Huxtable, H. Pasantes-Morales, eds, Plenum Publ. Corp., New York, pp 361-372.
7. Chan-Palay, V., Palay, S.L. and Wu, J.-Y., 1982. Sagittal cerebellar microbands of taurine neurons: immunocytochemical demonstration by using antibodies against the taurine synthesizing enzyme cysteine sulphinic acid decarboxylase. *Proc. Natl. Acad. Sci. USA*, 79:4221-4225.
8. Chevalier, G. and Deniau, J.M., 1990. Disinhibition as a basic process in the expression of striatal function. *Trend Neurosci.*, 13:277-281.
9. Clarke, D.J., Smith, A.D., and Bolam, J.P., 1983. Uptake of [^{3}H]taurine into medium-size neurons and into identified stratonigral neurons in the rat neostriatum. *Brain Res.*, 289:342-348.
10. Clarke, H.T., Gillespe, H.B., and Weisshaus, S.Z., 1933. The action of formaldehyde on amines and amino acids. *J. Am.Chem.Soc.*, 55:4571-4587 .

11. Collins, G.G.S., 1974. The rates of synthesis, uptake and disappearance of [^{14}C]taurine in eight areas of the rat central nervous system. Brain Res., 76:447-459.
12. Cooper, J.R., Bloom, F.E., and Roth, R.H., 1991. Identification of synaptic transmitters, in, "The Biochemical Basis of Neuropharmacology", pp. 39-43, Oxford University Press, New York, Oxford.
13. De Long, M.R., Wichmann, T., 1993. Basal ganglia-thalamocortical circuits in parkinsonian signs. *Clin.Neurosci.*, 1:18-26.
14. Decavel, C., and Hatton, G.I., 1995. Taurine immunoreactivity in the rat supraoptic nucleus: prominent localization in glial cells. *J. Comp. Neurology,* 354:13-26.
15. Della Corte, L., Bianchi, L., Federico, S., and Michelassi, S., 1991. In vivo HPLC estimation of extracellular aspartate, glutamate, taurine and GABA in rat striatum. *Eur.J.Neurosci.*, suppl.4, p 151.
16. Della Corte, L., Bolam, J.P., and Smith, A.D., 1987. Uptake, localization and release of taurine in the rat basal ganglia, In: "The Biology of Taurine", R.J. Huxtable, F. Franconi and A. Giotti, eds, pp. 285-294, Plenum Publ. Corp., New York.
17. Della Corte, L., Bolam, J.P., Clarke, D.J., Parry, D.H., and Smith, A.D., 1990. Sites of [^{3}H]taurine uptake in the rat substantia nigra in relation to the release of taurine from the striatonigral pathway. *Eur. J. Neurosci.,* 2:50-61.
18. Dray, A., and Straughan, D.W., 1976. Synaptic mechanisms in the substantia nigra. *J. Pharm. Pharmacol.,* 28:400-405.
19. Freifelder, M., Howard, B., and Wright, H.B., 1964. Hypocholesteremic agents. II. The hydrogenation of some pyridine sulfonic acids. *J. Med. Chem.,* 7:664-665.
20. Geoghegan, K.F., and Dixon, H.B.F., 1989. Synthesis of 2-aminoethylarsonic acid: A new synthesis of amines from haloalkanes. *Biochem. J.,* 260:295-296.
21. Hanretta, A.T., and Lombardini, J.B., 1985. Effects of intrahypothalamic kainic acid injection on taurine levels, binding and uptake. *Brain Res.,* 338:351-354.
22. Häusser, M.A., Yung, W.H., and Lacey, M.G., 1992. Taurine and glycine activate the same Cl^- conductance in substantia nigra dopamine neurons. *Brain Res.,* 571:103-108.
23. Huxtable, R.J., 1989. Taurine in the central nervous system and the mammalian actions of taurine. *Prog. Neurobiol.,* 32:471-533.
24. Huxtable, R.J., 1992. Physiological actions of taurine. *Physiol. Rev.,* 72:101-159.
25. Ida, S., Kuriyama, K., Tomida, Y., and Kimura, H., 1987. Antisera against taurine: quantitative characterization of the antibody specificity and its application to immunohistochemical study in the rat brain. *J. Neurosci. Res.,* 18:626-631.
26. Kaakkola, S., and Kaariainen, T., 1980. Contralateral circling behaviour induced by intranigral injection of taurine in rats. *Acta Pharmacol. Toxicol.,* 46:293-298.
27. Kamata, K., Kameyana, T., Okuyama, S., Hashimoto, S., and Aihara, H., 1985. Contralateral circling behaviour induced by intranigral microinjections of taurine and GABA in rats. *Brain Res.,* 343:275-282.
28. Kluger, R., Nakaoka, K., and Tsui, W.C., 1978. Substrate analogue studies of the specificity and catalytic mechanism of D-3-hydroxybutyrate dehydrogenase. *J. Am. Chem. Soc.,* 100:7388-7392.
29. Korf, J., and Venema, K., 1983. Aminoacids in the substantia nigra of rats with striatal lesions produced by kainic acid. *J. Neurochem.,* 40:1171-1173.
30. Legay, F., Lecestre, D., and Tappaz, M., 1987. Taurine biosynthesis in rat brain in vivo: lack of relationship with cystein sulfinate decarboxylase glutamate decarboxylase-associated activity (GAD/CSDII). *J. Neurochem.,* 48:340-344.
31. Legay, F., Waise, V.K., Oertel, W.H. and Tappaz, M.L., 1987. Taurine biosynthesis in rat brain: a new specific and sensitive microassay of cysteine sulfinate decaboxylase (CSDI) activity through selective immunotrapping and its use for distribution studies. *J. Neurochem.,* 48:345-351.
32. Lombardini, J.B., 1976. Regional and subcellular studies on taurine in the rat central nervous system. *Acta Physiol. Scand.,* 412:1-48.
33. Lombardini, J.B., 1977. High affinity uptake systems for taurine on tissue slices and synaptosomal fractions prepared from various brain regions of the rat central nervous system. *J. Neurochem.,* 29:305-312.
34. Lombardini, J.B., 1976. Regional and subcellular studies on taurine in the rat nervous system, in "Taurine", R. Huxtable and A. Barbeau, eds, pp. 311-326, Raven Press, New York.
35. Lombardini, J.B., 1991. Taurine and retinal function. *Brain Res. Rev.,* 16:151-169.
36. Madsen, S., Ottersen, O.P., and Storm-Mathisen, J., 1985. Immunocytochemical visualization of taurine: neuronal localization in the rat cerebellum. *Neurosci. Lett.,* 60:255-260.
37. Madsen, S., Ottersen, O.P., and Storm-Mathisen, J., 1987. Immunocytochemical demonstration of taurine, in "The biology of taurine: methods and mechanisms", R.J. Huxtable, F. Franconi and A. Giotti, eds, pp. 275-284, Plenum Press, New York and London.

38. Magnusson, K.R., Madl, J.E., Clements, J.R., Wu, J.-Y., Larson, A., and Beitz, A.J., 1988. Colocalization of taurine- and cysteine sulfinic acid decarboxylase-like immunoreactivity in the cerebellum of the rat with the use of monoclonal antibodies against taurine. *J. Neurosci.*, 8:4551-4564.
39. Mandel, P., and Pasantes-Morales, H., 1978. Taurine in the central nervous system. *Rev. Neurosci.*, 3:157-193.
40. Martin, G.E., Bendesky, R.J., and Williams, M., 1981. Further evidence for selective antagonism of taurine by 6-aminomethyl-3-methyl-4H-1,2,4-benzothiadiazine-1,1-dioxide. *Brain Res.*, 299:30-535.
41. Oja, S.S., and Kontro, P., 1979. Neurotransmitter actions of taurine in the central nervous system, in, "Taurine and neurological disorders", A. Barbeau and R. J. Huxtable, eds, pp 181-200, Raven Press, New York.
42. Ottersen, O.P., Madsen, S., Meldrum, B.S., and Storm-Mathisen, J., 1985. Taurine localization in the hippocampal formation of the senegalese baboon Papio papio: an immunocytochemical study with an antiserum against conjugated taurine. *Exp. Brain Res.*, 59:457-462.
43. Palkovits, M., Elekes, I., Lang, T., and Patthy, A., 1986. Taurine levels in discrete brain nuclei of rats. *J. Neurochem.*, 47:1333-1335.
44. Palmi, M., Frosini, M., and Sgaragli , G.P., 1992. Calcium changes in rabbit cerebrospinal fluid during endotoxin, IL-1, and prostaglandin fever. *Pharmacol.Biochem.Behav.*, 43:1253-1262.
45. Pirvola, U., and Panula, P., 1992. Distribution of taurine in the rat cerebellum and insect brain: application of a new antiserum against carbodiimide-conjugated taurine. *Histochem. J.*, 24:266-274.
46. Remy, A., Henry, S., and Tappaz, M., 1990. Specific antiserum and monoclonal antibodies against taurine biosynthesis enzyme cysteine sulfinate decarboxylase: identity of brain and liver enzyme. *J. Neurochem.*, 54:870-879.
47. Sgaragli, G.P., Carlà, V., Magnani, M. and Giotti, A., 1978. Homotaurine and muscimol mimic taurine and GABA effects on muscle tone and temperature regulation. *Naunyn Schmiedeberg's Arch. Pharmacol.*, 305:155-158.
48. Sgaragli, G.P, Carlà, V., Magnani, M., and Galli, A., 1981. Hypothermia induced in rabbits by intracerebroventricular taurine: specificity and relationship with central serotonin (5HT) systems. *J. Pharmacol. Exp. Ther.*, 219:778-785.
49. Sgaragli, G.P., and Palmi, M., 1985. The role and mechanism of action of taurine in mammalian thermoregulation, in, "Taurine: biological action and clinical perspectives", S.S. Oja, L. Athee, P. Kontro. and M.K. Paasonen, eds., pp. 343-357, Alan R. Liss, Inc., New York.
50. Smith, A.D., and Bolam, J.P., 1989. Neurons of the substantia nigra reticulata receive a dense GABA-containing input from the globus pallidus in the rat. *Brain Res.*, 493:160-167.
51. Smith, A.D., and Bolam, J.P., 1991. Convergence of synaptic inputs from the striatum and the globus pallidus onto identified nigrocollicular cells in the rat: a double anterograde labelling study. *Neuroscience*, 44:45-73.
52. Spears, R.M., and Martin, J.P., 1982. Resolution and brain regional distribution of cysteine sulfinate decarboxylase isoenzyme from hog brain. *J. Neurochem.*, 38:985-981.
53. Storm-Mathisen, J., and Ottersen, O.P., 1986. Antibodies against aminoacid neurotransmitters, in: "Neurohistochemistry: modern methods and applications", P. Panula, H. Paivarinta and S. Soinila, eds, pp. 107-136, Alan R. Liss, New York.
54. Stromhaugh, J., Skumlien, S., Storm-Mathisen, J. and Ottersen, O.P., 1987. Immunocytochemical demonstration of putative aminoacid neurotransmitters in the striatonigral pathway. *Neuroscience*, Suppl. 22:S36.
55. Wu, J-Y., Bird, E.D., Chen, M.S., and Huang, W.M., 1979. Abnormalities of neurotransmitter enzymes in Huntington's chorea. *Neurochem. Res.*, 4:575-586.
56. Yarbrough, G.G., Singh, D.K., and Taylor, D.A., 1981. Neuropharmacological characterization of a taurine antagonist. *J. Pharmacol. Exp. Ther.*, 219:604-613.
57. Yoshida, M., Karasawa, N., Ito, M., Sakai, M., and Nagatsu, I., 1986. Demonstration of taurine immunoreactivity in the rat brain. *Neurosci. Res.*, 3:356-363.

Part 7

TAURINE AND DIABETES

CARDIOPROTECTIVE EFFECT OF TAURINE ON CALCIUM PARADOX IN STREPTOZOTOCIN-INDUCED DIABETIC RAT HEARTS

Tetsuya Tatsumi, Satoaki Matoba, Akira Kawahara, Miyuki Kobara, Kouki Tsuruyama, Tetsuya Tanaka, Chiaki Nakagawa, Bon Ohta, Jun Asayama, and Masao Nakagawa

Second Department of Medicine
Kyoto Prefectural University of Medicine
Kawaramachi-Hirokoji, Kamigyo-ku, Kyoto 602, Japan

INTRODUCTION

Through the modification of Ca^{2+} metabolism, taurine is known to have several beneficial physiological actions, including antiarrhythmic[18], positive inotropic[2-4], and membrane stabilizing effects[7]. In addition, it has been reported that taurine protects the heart against Ca^{2+} paradox-induced myocardial injury[9]. However, it is unclear whether taurine has a similar cardioprotective effect in diabetic hearts, which exhibit a number of alterations in metabolism. Diabetes mellitus is a disorder of carbohydrate, lipid, and protein metabolism affecting many organs. In addition to contractile abnormalities, previous studies revealed disturbances in function of subcellular organelles in diabetic hearts, including impaired glucose utilization, mitochondrial dysfunction, and depressed Na^+-Ca^{2+} Na^+-H^+ exchange activities[12,17,22].

The objective of the present study was to examine the effect of taurine on myocardial energy metabolism, mitochondrial function, and Ca^{2+} metabolism in Ca^{2+} paradox-induced myocardial injury in streptozotocin-induced diabetic rat hearts.

METHODS

Animals

All animals were handled in accordance with the Guide for the Care and Use of Laboratory Animals (NIH publication #85-23, revised 1985). The protocol was approved by the Bioethics Committee of the Kyoto Prefectural University of Medicine. Male Sprague-

Taurine 2, edited by Huxtable et al.
Plenum Press, New York, 1996

Dawley rats weighing 150-160 g were made diabetic with an i.p. injection of streptozotocin (70 mg/kg). These animals were maintained for 4 weeks and exhibited blood glucose levels higher than 400 mg/dl. As a normal control, age-matched male Sprague-Dawley rats were used. Heparin (1000 IU/kg body weight) was injected i.v., and the hearts were rapidly excised. Then the hearts were retrogradely perfused, at a constant perfusion pressure of 80 mm Hg, with Krebs-Henseleit buffer, consisting of (mM) 118 NaCl, 4.7 KCl, 1.2 $MgSO_4$, 1.2 KH_2PO_4, 1.8 $CaCl_2$, 25 $NaHCO_3$, and 11 glucose, pH 7.4, oxygenated with 95% O_2 and 5% CO_2 at 37°C. The atria were removed and the hearts were paced at 270 beats/min via two pacing wires attached to the right ventricular myocardium. A water-filled latex balloon was inserted into the left ventricle for the measurement of left ventricular developed pressure (LVDP). The left ventricular end-diastolic pressure (LVEDP) was adjusted to 5-10 mm Hg during the initial equilibration. LVDP was recorded continuously on a chart recorder (Sekonic SS-250-F, Kishimoto Co., Ltd., Kyoto, Japan).

Langendorff Perfusion Study

Four groups were studied: (i) control normal rat group; (ii) taurine-treated normal rat group; (iii) control diabetic rat group; and (iv) taurine-treated diabetic rat group. Langendorff-perfused rat hearts were subjected to 8 min arrest with Ca^{2+}-free conditions followed by 15 min perfusion with Krebs-Henseleit buffer containing 1.8 mM Ca^{2+}. In taurine-treated groups, 10 mM taurine was added to the perfusate during all phases of the experiment. Coronary effluent after Ca^{2+} paradox was collected and creatine kinase (CK) activity was measured as an index of myocardial damage. High energy phosphates and mitochondrial enzyme activities were measured at the end of the protocol.

Assay of Creatine Kinase (CK) Release

Coronary effluent after Ca^{2+} paradox-induced myocardial injury was pooled and CK activity was measured in the four groups. CK activity was assayed spectrophotometrically at 37°C according to Rosalki's procedure. The activity of CK was expressed as IU/g dry weight[26].

Measurement of Adenine Nucleotides

Hearts were freeze-clamped with Wollenberger tongs cooled in liquid N_2 at the end of the protocol. The hearts were kept in liquid N_2 and were lyophilized overnight. For the measurements of creatine, creatine phosphate, and adenine nucleotides, 40-60 mg of lyophilized tissue was homogenized in 0.6 N ice-cold perchloric acid and centrifuged at 2000 rpm for 10 min at 4°C. The supernatants were neutralized with KOH to pH 5.0-7.0. After 10 min the extracts were centrifuged to remove the $KClO_4$, and the supernatant was used for the assays. Creatine, creatine phosphate, ATP, ADP, and AMP, were measured by HPLC (LC-9A liquid chromatograph, Shimadzu, Kyoto, Japan) with a column of STR ODS-M (Shimadzu)[1].

Mitochondrial Isolation

Heart mitochondria were isolated at the end of the protocol by the method of Sordahl *et al.*[20], with some modification. Following perfusion, the heart was rapidly immersed in ice-cold phosphate buffered saline (PBS) containing 15 mM Na_2HPO_4, 5 mM NaH_2PO_4, and 145 mM NaCl (pH 7.3). The ventricular myocardium was isolated, rinsed in ice-cold potassium phosphate buffer, consisting of 25 mM K_2HPO_4, 25 mM KH_2PO_4, and 1 mM EDTA (pH 7.2), and minced into small pieces with scissors in 1.6 ml of potassium phosphate buffer containing 0.5 mM phenylmethanesulfonyl fluoride (PMSF). The minced heart tissue was homogenized with a Polytron PT 10-35 tissue processor (Kinematica Instrument) 3x at a setting of 7 for 10 s, and the homogenate was centrifuged at 27,000 g for 10 min. The supernatant was discarded and the pellet was suspended in isolation medium (5% w/v), consisting of 0.18 mM KCl, 5 mM Tris-HCl, and 10 mM EGTA (pH 7.2), using a Teflon-glass homogenizer. The suspension was centrifuged at 1000 g for 10 min, and the supernatant was centrifuged at 12,000 g for 10 min. The mitochondrial pellet was washed once and resuspended in 550 μl low salt medium, consisting of 20 mM HEPES, 1.0 mM $MgCl_2$, and 2.0 mM EGTA (pH 7.0). The protein concentration in the mitochondrial suspension was determined by the Lowry method.

Mitochondrial Enzyme Activities

Mitochondrial NADH-cytochrome C oxidoreductase (complex I-III) and succinate-cytochrome C oxidoreductase (complex II-III) activities were assayed spectrophotometrically at 30°C by measuring the initial rate of cytochrome C reduction at 550 nm. Oligomycin-sensitive mitochondrial ATPase (complex V) activity was assayed spectrophotometrically at 30°C by coupling the reaction to pyruvate kinase and lactate dehydrogenase systems and measuring NADH oxidation at 340 nm.

Isolation of Cardiomyocytes

Cardiomyocytes were isolated as described previously[24]. Briefly, hearts were perfused in Tyrode solution, consisting of 145 mM NaCl, 5 mM KCl, 1.2 mM $MgSO_4$, 5 mM HEPES, 1.37 mM Na_2HPO_4, 0.43 mM NaH_2PO_4, 1 mM $CaCl_2$, and 10 mM glucose (pH 7.4), oxygenated with 100% O_2 at 37°C. The perfusate was changed to Ca^{2+}-free Tyrode solution and recirculated after addition of collagenase, 100 units/ml (Sigma Type I) for 20 min. The ventricles were cut from the softened heart, minced, then transferred to fresh Ca^{2+}-free Tyrode solution containing collagenase (250 U/ml), and shaken vigorously for 20 min at 37°C.

Measurement of $^{45}Ca^{2+}$ Uptake in Isolated Myocytes

The rapidly exchangeable $^{45}Ca^{2+}$ in isolated myocytes with intact sarcolemmal membranes was measured to examine the impairments in Ca^{2+} homeostasis in diabetic heart cells. The myocytes were incubated in the presence of 0.2 mM Ca^{2+} for 0-90 min. The cells were separated by millipore filters, and their radioactivities were determined in a liquid scintillation counter.

$^{45}Ca^{2+}$ uptake in isolated myocytes in Ca^{2+} paradox was measured to examine the cardioprotective effect of taurine on Ca^{2+} homeostasis. The myocytes were incubated in Ca^{2+}-free buffer and Ca^{2+} paradox was induced with 2 mM or 4 mM Ca^{2+} for 5 min. The cells were separated by millipore filters, and their radioactivities were determined in a liquid scintillation counter.

Statistics

Differences in time course of LVDP (Fig. 1) between control and taurine-treated groups and Ca^{2+} uptake in between normal and diabetic myocytes (Fig. 8) are analyzed using one-way analysis of variance for repeated measurement. Unpaired Student's t-test was applied to determine the significance of difference in CK release (Fig. 2), the effect of taurine on CK release (Fig. 3), adenylates (Figs. 4-6), and mitochondrial ATPase activity (Fig. 7) between control and taurine-treated groups. Differences in $^{45}Ca^{2+}$ uptake in myocytes (Fig. 9) among taurine-treated groups

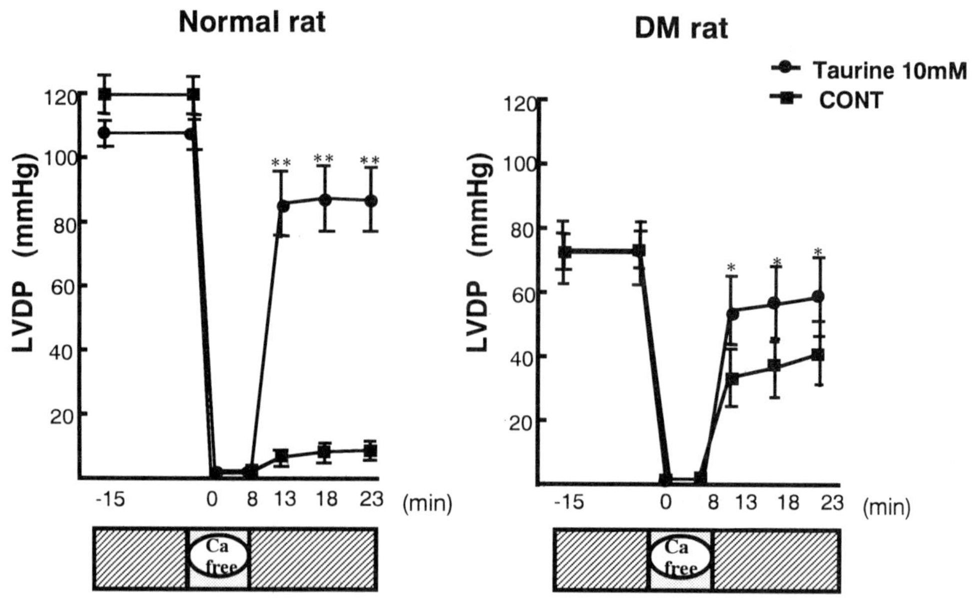

Figure 1. Time course change in left ventricular developed pressure (LVDP). Langendorff-perfused rat hearts were subjected to 8 min arrest with Ca^{2+}free condition followed by 15 min perfusion with Krebs-Henseleit buffer containing 1.8 mM Ca^{2+}. ■: control group; ●: taurine-treated group. *$p < 0.05$; **$p < 0.01$.

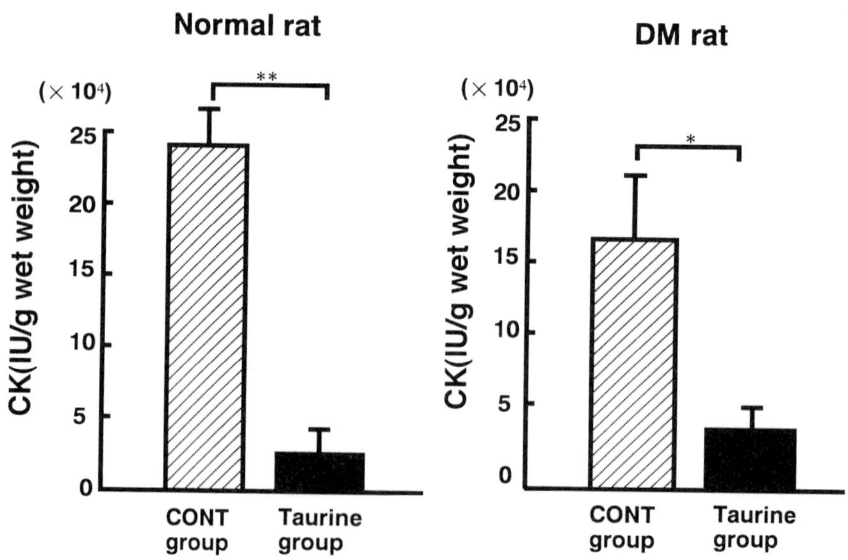

Figure 2. Creatine kinase release after Ca^{2+} paradox. The coronary effluent from reperfused control and taurine-treated hearts was collected, and CK activity was measured in the pooled samples as described in Methods. Shaded column indicates the control group and filled column indicates the taurine-treated group. *$p < 0.05$; **$p < 0.01$.

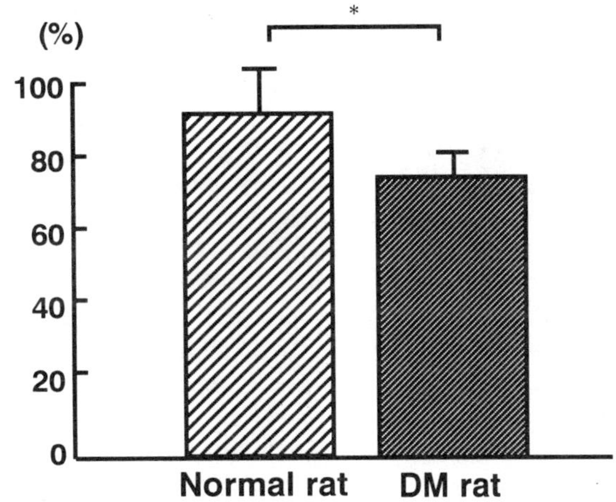

Figure 3. Suppression of creatine kinase release by taurine. The cardioprotective effect of taurine was evaluated between the control and diabetic hearts. $^{*}p < 0.05$.

were analyzed using two-way analysis of variance. Each value is expressed as mean ± SEM, and a p-value of less than 0.05 was considered to be statistically significant.

RESULTS

The time course changes in LVDP of normal and diabetic rat groups are shown in Fig. 1. At baseline, LVDP in the diabetic groups was smaller than in the normal groups. Calcium-free conditions induced rapid cardiac arrest in both groups. Although at baseline

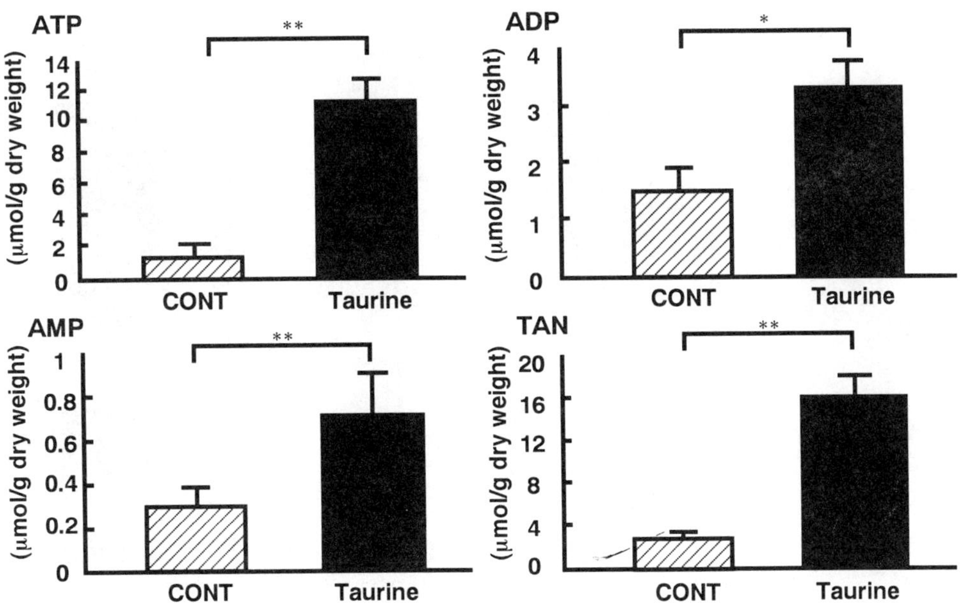

Figure 4. Adenine nucleotides after Ca^{2+} paradox in normal rat hearts subjected to the perfusion protocol. Shaded column indicates the control group and filled column indicates the taurine-treated group. $^{*}p < 0.05$; $^{**}p < 0.01$.

there were no differences in LVDP between the control and taurine-treated groups, recovery of LVDP was significantly greater in both the normal and diabetic groups treated with taurine (normal control 8.0 ± 2.9 vs normal taurine-treated: 84.9 ± 9.7 [$p < 0.01$]; diabetic control 38.0 ± 9.4 vs diabetic taurine-treated 55.1 ± 11.6 mm Hg [$p < 0.05$]). Especially marked recovery of LVDP was seen in normal taurine-treated group.

Fig. 2 shows CK activity in the coronary effluent released after Ca^{2+} paradox. In both the normal and diabetic groups, the activity of CK released after Ca^{2+} paradox was significantly lower in the taurine-treated groups (normal control 24.6 ± 2.6 vs normal taurine-treated 2.4 ± 1.8 [$p < 0.01$]; diabetic control 17.5 ± 24.5 vs diabetic taurine-treated 4.2 ± 1.5 $\times 10^4$ IU/g wet weight [$p < 0.05$]). The suppression of CK release by taurine between the normal and diabetic rat groups is shown in Fig. 3. A significantly greater suppression of CK release was seen in the normal rat group. In addition, in the histological examination, contraction band necrosis and vacuole degeneration necrosis were seen in myocardial tissue after Ca^{2+} paradox in the control groups, though such changes were not detected in the taurine-treated groups (data not shown).

Adenylates in the heart tissue after Ca^{2+} paradox-induced myocardial injury in the normal groups and the diabetic rat group are shown in Fig. 4 and Fig. 5, respectively. In both normal and diabetic rat groups, ATP, ADP, and TAN were all markedly and significantly preserved in the taurine-treated group.

The CP and creatine in the heart tissue after Ca^{2+} paradox in the normal and the diabetic rat groups are shown in Fig. 6. In both groups, CP and creatine were markedly preserved in the taurine-treated group.

Fig. 7 shows the mitochondrial oligomycin-sensitive ATPase activity after Ca^{2+} paradox-induced myocardial injury. In both normal and diabetic rat groups, ATPase activity was significantly preserved in taurine-treated groups (normal control 1.21 ± 0.06 vs normal taurine-treated 1.57 ± 0.12 [$p < 0.01$]; diabetic control 1.12 ± 0.06 vs diabetic taurine-treated 1.37 ± 0.09 μmol/min/mg protein [$p < 0.01$]). On the other hand, there were no significant

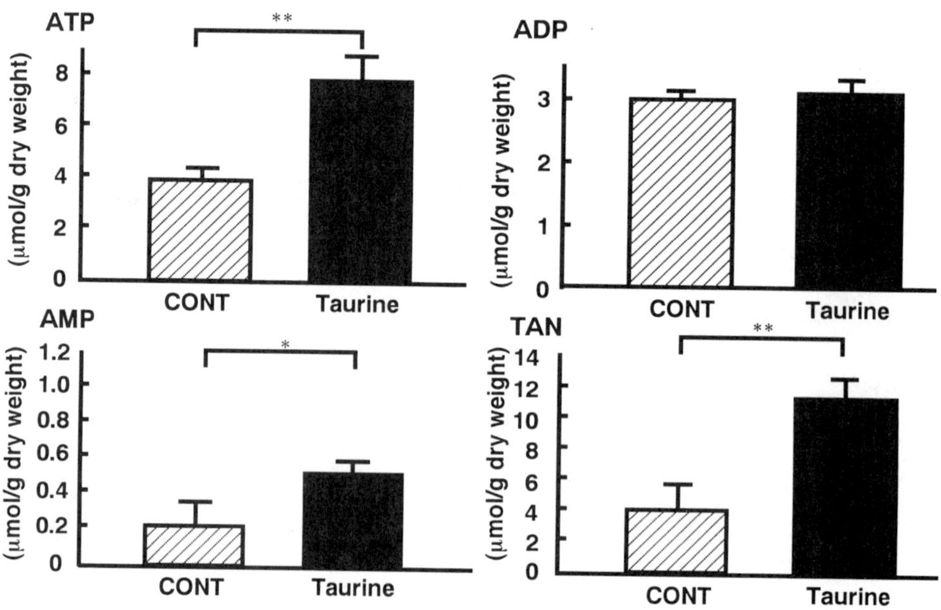

Figure 5. Adenine nucleotides after Ca^{2+} paradox in diabetic rat hearts subjected to the perfusion protocol. Shaded column indicates the control group and filled column indicates the taurine-treated group. *$p < 0.05$; **$p < 0.01$.

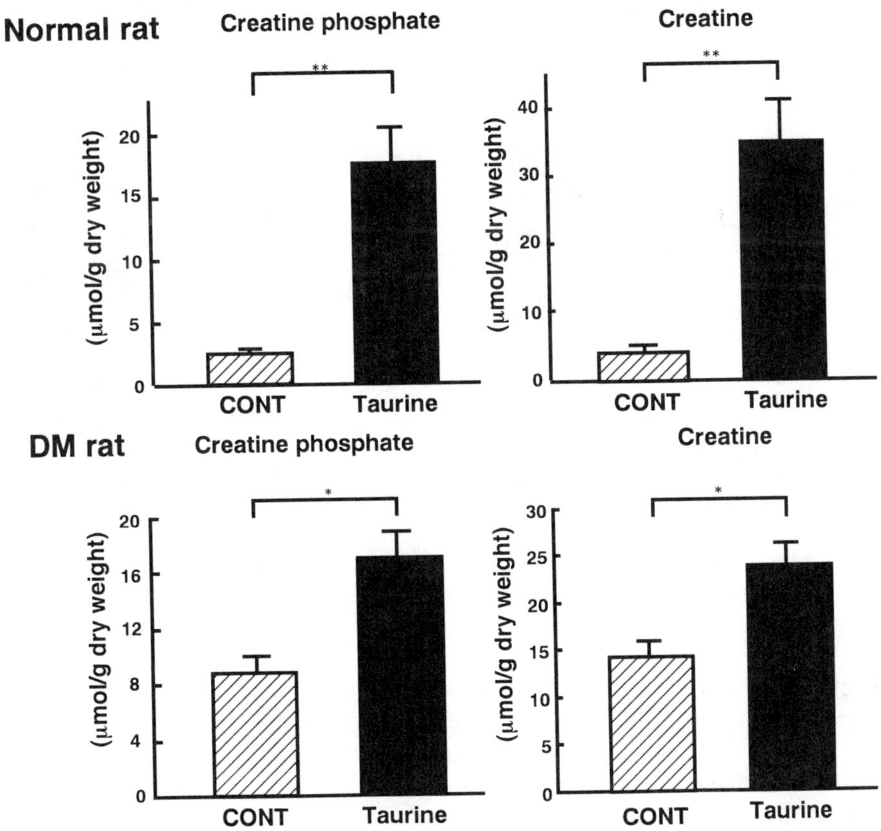

Figure 6. Creatine phosphate and creatine after Ca^{2+} paradox. Isolated hearts of either normal or diabetic rats were subjected to the perfusion protocol. The concentrations of CP and creatine were determined. Shaded column indicates the control group and filled column indicates the taurine-treated group. $^{*}p < 0.05$; $^{**}p < 0.01$.

differences in the mitochondrial NADH-cytochrome C oxidoreductase or succinate-cytochrome C oxidoreductase activities after Ca^{2+} paradox between the control and taurine-treated groups (data not shown).

The rapidly exchangeable $^{45}Ca^{2+}$ uptake curves in the normal and diabetic myocytes are shown in Fig. 8. The rapidly exchangeable $^{45}Ca^{2+}$ content in the myocytes with intact sarcolemmal membrane was significantly diminished in the diabetic myocytes (group differences $p < 0.01$). The $^{45}Ca^{2+}$ uptake in isolated myocytes after Ca^{2+} paradox-induced myocardial injury in normal and diabetic rat myocytes is shown in Fig. 9. Calcium paradox was induced with 2 mM or 4 mM Ca^{2+}. Although $^{45}Ca^{2+}$ uptake was lower in diabetic rat myocytes compared with the normal rat myocytes, taurine inhibited $^{45}Ca^{2+}$ uptake in a dose-dependent manner in both groups. Especially, 10 mM taurine significantly inhibited $^{45}Ca^{2+}$ uptake in Ca^{2+} paradox with 4 mM Ca^{2+} in both normal and diabetic rat myocytes.

DISCUSSION

In the present study, we have confirmed that: (i) in Ca^{2+} paradox-induced myocardial injury, taurine preserves myocardial function, high energy phosphates such as ATP and CP,

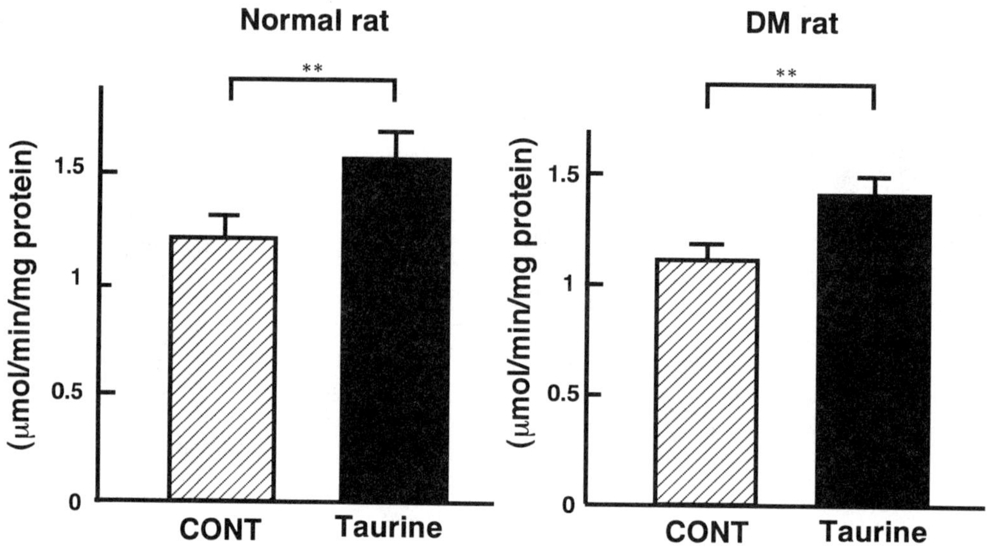

Figure 7. Oligomycin-sensitive ATPase activity after Ca^{2+} paradox. Isolated hearts of either normal or diabetic rats were subjected to the perfusion protocol and mitochondrial ATPase activity was assayed. Shaded column indicates the control group and filled column indicates the taurine-treated group. $^{**}p < 0.01$.

and mitochondrial ATPase activity in both normal and diabetic rats; (ii) in taurine-treated groups, greater recovery of LVDP and suppression of CK release were seen in the normal rat group; and (iii) taurine significantly inhibited $^{45}Ca^{2+}$ uptake in Ca^{2+} paradox in diabetic rat myocytes as well as normal rat myocytes in a dose-dependent manner.

Taurine is known to have important physiological actions on many organs. It has been reported that the myocardial taurine levels are changeable in many clinically pathological

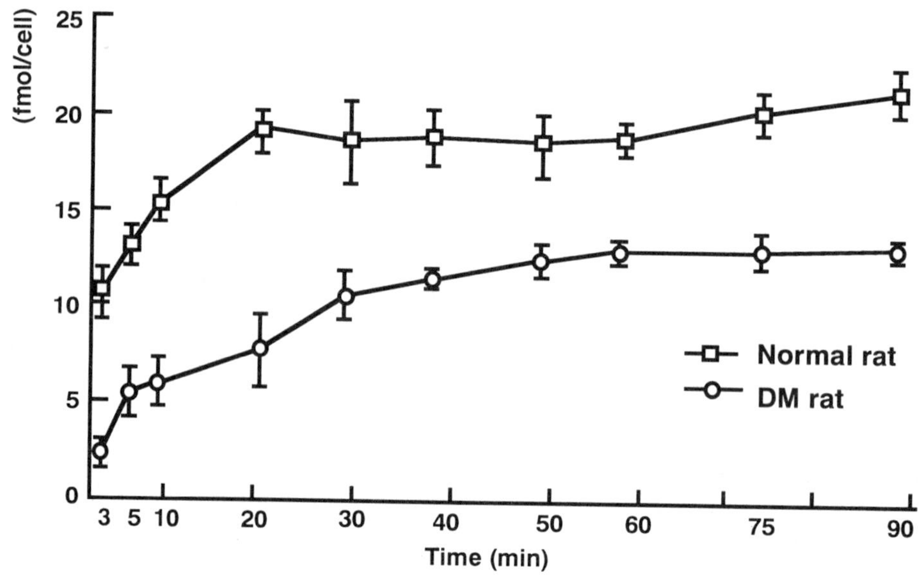

Figure 8. Time course change in rapidly exchangeable $^{45}Ca^{2+}$ uptake. Isolated myocytes of either normal (□) or diabetic (○) rats were incubated in the presence of 0.2 mM Ca^{2+} for 0-90 min, and $^{45}Ca^{2+}$ uptake was determined.

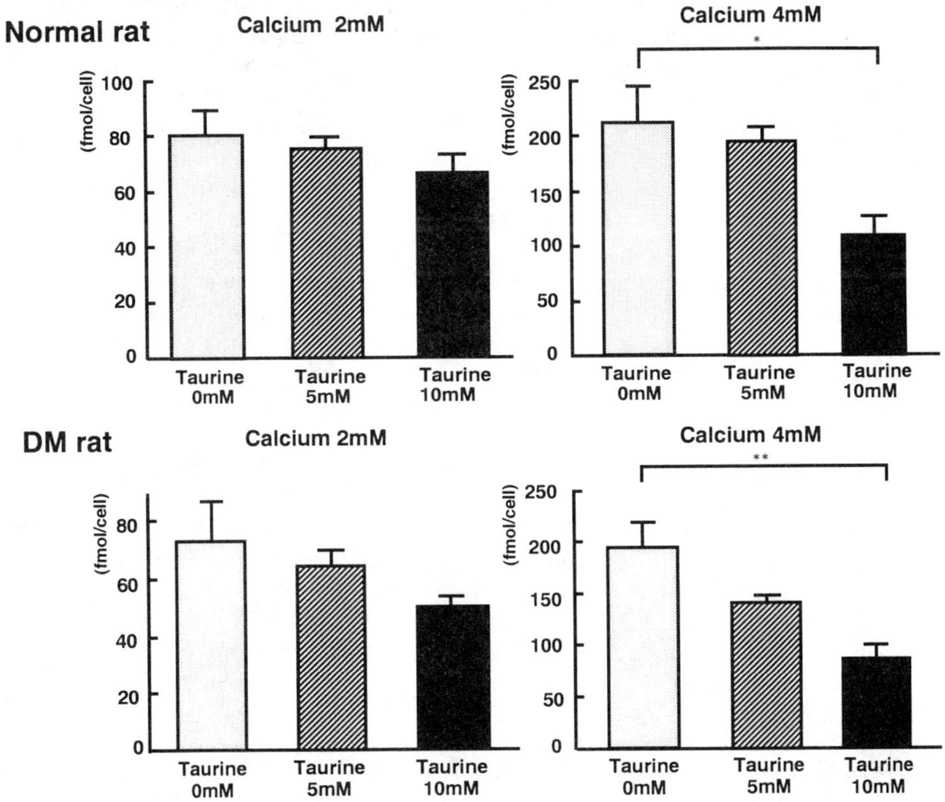

Figure 9. $^{45}Ca^{2+}$ uptake in myocytes of either normal or diabetic rats after Ca^{2+} paradox. Calcium paradox was induced with 2 mM or 4 mM Ca^{2+}, and $^{45}Ca^{2+}$ uptake was determined. Dotted column indicates the effect of 0 mM taurine on $^{45}Ca^{2+}$ uptake, shaded column indicates the effect of 5 mM taurine, and filled column indicates the effect of 10 mM taurine. *$p < 0.05$; **$p < 0.01$.

states, such as cardiac hypertrophy[14,16], cardiac myopathy[6], and congestive heart failure[8], and that depletion of taurine correlates with myocardial mechanical dysfunction[6,9]. Read and Welty[18] observed that taurine administration suppresses arrhythmias resulting from toxic doses of either epinephrine or the digitalis glycosides. It is thought the antiarrhythmic and positive inotropic action of taurine is based on the alteration of sarcolemmal ion fluxes[2-4,18]. In addition, it has been recently reported that treatment with taurine decreases the lipoperoxidation and the mitochondrial damage in coronary artery bypass grafting, through a free-radical scavenging action[13]. These reports suggest that taurine plays an important role in cardiac function and metabolism.

It is well-known that Ca^{2+} paradox induces ultrastructural myocardial damages through intracellular Ca^{2+} overload. It has been reported that electron-dense granule of Ca^{2+} phosphates accumulate in the mitochondria in Ca^{2+} paradox. The depletion of high energy phosphates is thought to be a major factor in cardiac necrosis[19,25]. In the present study, high energy phosphates such as ATP and CP were significantly and severely depleted in the control groups of both normal and diabetic rats, and it is reasonable to speculate that this depletion of high energy compounds is mainly derived from the Ca^{2+} overload.

It has been already reported that taurine improves myocardial function in Ca^{2+} paradox-induced injury in working normal rat heart model[9]. However, little is known about whether taurine has a similar cardioprotective effect in diabetic hearts. Our results demonstrate that taurine preserves myocardial function, estimated by LVDP, not only in normal rats but also in diabetic rats. There are some reports describing impaired Ca^{2+} transport of sarcolemma, sarcoplasmic reticulum, and mitochondria from diabetic hearts[5,15,22]. The present results show that the rapidly exchangeable $^{45}Ca^{2+}$ content in the myocytes with intact sarcolemmal membranes was significantly diminished in diabetic rat hearts, and that taurine significantly inhibited $^{45}Ca^{2+}$ uptake in Ca^{2+} paradox in diabetic rat myocytes as well as normal myocytes in a dose-dependent manner, suggesting that the major cardioprotective effect of taurine is derived from the inhibition of Ca^{2+} overload. We have measured Ca^{2+} uptake by using an isotope technique. Since Ca^{2+} efflux from subcellular organelles was not determined, the observed $^{45}Ca^{2+}$ uptake is thought to represent a balance of influx and efflux in the intact sarcolemmal myocytes and to reflect the effect of taurine on Ca^{2+} homeostasis through the sarcolemmal ion flux. Recently, Kramer *et al.*[9] have suggested that taurine alters the sarcolemmal low affinity Ca^{2+} binding sites, which is associated with the slow inward current, resulting in the prevention of Ca^{2+} overload. Thus it has been demonstrated that taurine protects the normal heart against Ca^{2+} overload through the modification of sarcolemmal Ca^{2+} flux. Therefore, it is reasonable to expect that similar interaction of taurine to sarcolemmal protein can be involved in the cardioprotective effect on the diabetic hearts.

In the present study, mitochondrial ATPase activity was significantly preserved in taurine-treated groups. Previous reports have revealed that the impairment of sarcolemmal function, including the degradation of Na^+/K^+-, Mg^{2+}-, and Ca^{2+}-ATPase activities, is observed in Ca^{2+} paradox-induced myocardial injury, and that taurine improves these enzyme degradation[9]. Though the precise mechanism for the preservation of these enzyme activities is unknown, it may be a secondary effect of taurine via the inhibition of Ca^{2+} overload. In addition, it is possible to speculate that taurine interacts with specific sarcolemmal membrane proteins and exhibits membrane stabilizing activity, including on mitochondria, through the prevention of phospholipase-mediated membrane damage[7,10].

In taurine-treated groups, greater recovery of LVDP and suppression of CK release were seen in the normal rat group, compared with the diabetic rat group, suggesting that the cardioprotective effect of taurine is more efficient in former group. We have no good explanation for this phenomenon. However, there is a possibility that Ca^{2+} paradox-induced damage is relatively less in the diabetic rats. Indeed, there are several reports suggesting that diabetic heart is tolerant to ischemia-reperfusion injury, which may be derived from the lower Ca^{2+} handling capacity[11,23]. Another possibility is that it has been reported that myocardial taurine levels correlate the myocardial dysfunction, and that taurine levels decrease in peripheral nerve and taurine uptake increases in synaptosomes in diabetes mellitus[21]. Though in this study we did not measure myocardial taurine contents, taurine levels in diabetic rat hearts may modify the cardioprotective effect in Ca^{2+} paradox-induced myocardial injury.

In conclusion, the present results suggest that in Ca^{2+} paradox taurine preserves myocardial function, high energy phosphates, and mitochondrial ATPase activity not only in normal rats but also in diabetic rats in which Ca^{2+} handling is impaired, and that the cardioprotective effect of taurine is mainly derived from the inhibition of Ca^{2+} overload.

REFERENCES

1. Asayama, J., Yamahara, Y., Ohta, B., Miyazaki, H., Tatsumi, T., Matsumoto, T., Inoue, D., and Nakagawa, M. 1992, Release kinetics cardiac troponin T in coronary effluent from isolated rat hearts during hypoxia and reoxygenation, *Basic.Res.Cardiol.* 87:428-436.
2. Chovan, J.P., Kulakowski, E.C., Benson, B.W., and Schaffer, S.W. 1979, Taurine enhancement of calcium binding to rat heart sarcolemma, *Biochim.Biophys.Acta*, 551:129-136.

3. Chovan, J.P., Kulakowski, E.C., Sheakowski, S., and Schaffer, S.W. 1980, Calcium regulation by the low affinity taurine binding sites of cardiac sarcolemma, *Mol.Pharmacol.* 17:295-301.
4. Dolara, P., Agresti, A., Giotti, A., and Pasquini, G. 1973, Effect of taurine on calcium kinetics of guinea pig heart, *Eur.J.Pharmacol.* 24:352-358.
5. Heyliger, C.E., Prakash, A., and McNeill, J.H. 1987, Alterations in cardiac sarcolemmal Ca^{2+} pump activity during diabetes mellitus, *Am.J.Physiol.* 252:540-544.
6. Huxtable, R.J. 1976, Metabolism and function of taurine in the heart, in: "Taurine", Huxtable, R.J. and Barbeau, A. eds., Raven Press, New York, pp. 99-119.
7. Huxtable, R.J. and Bressler, R. 1973, Effect of taurine on a muscle intracellular membrane, *Biochim.Biophys.Acta*, 323:573-583.
8. Huxtable, R.J. and Bressler, R. 1974, Elevation of taurine in human congestive heart failure, *Life Sci.* 14:1353-1359.
9. Kramer, J.H., Chovan, J.P., and Schaffer, S.W. 1981, Effect of taurine on calcium paradox and ischemic heart failure, *Am.J.Physiol.* 240:H238-H246.
10. Kulakowski, E.C., Maturo, J., and Schaffer, S.W. 1978, The identification of taurine receptors from rat heart sarcolemma, *Biochem.Biophys.Res.Commun.*, 80:936-941.
11. Kusama, Y., Hearse, D.J., and Avkiran, M. 1992, Diabetes and susceptibility to reperfusion-induced ventricular arrhythmias, *J.Mol.Cell.Cardiol.* 24:411-421.
12. Makino, N., Dhalla, K.S., Elimban, V., and Dhalla, N.S. 1987, Sacrolemmal Ca^{2+} transport in streptozotocin-induced diabetic cardiomyopathy in rats, *Am.J.Physiol.* 253:202-207.
13. Milei, J., Ferreira, R., Llesuy, S., Forcada, P., Covarrubias, J., and Boveris, A. 1992, Reduction of reperfusion injury with preoperative rapid intravenous infusion of taurine during myocardial revascularization, *Am.Heart J.* 123:339-345.
14. Newman, W.H., Frangakis, C.J., Grosso, D.S., and Bressler, R. 1977, A relation between myocardial taurine content and pulmonary wedge pressure in dogs with heart failure, *Physiol.Chem.Phys.* 9:259-263.
15. Penpargkul, S., Fein, F., Sonnenblick, E.H., and Scheuer, J. 1981, Depressed cardiac sarcoplasmic reticular function from diabetic rats, *J.Mol.Cell.Cardiol.* 13:303-309.
16. Peterson, M.B., Mead, R.J., and Welty, J.D. 1973, Free amino acids in congestive heart failure, *J.Mol.Cell.Cardiol.* 5:139-147.
17. Pierce, G.N., Ramjiawan, B., Dhalla, N.S., and Ferrari, R. 1990, Na^{+} - H^{+} exchange in cardiac sarcolemmal vesicles isolated from diabetic rats, *Am.J.Physiol.* 258:255-261.
18. Read, W.O. and Welty, J.D. 1963, Effect of taurine on epinephrine and digoxin induced irregularities of dog heart, *J.Pharmacol.Exptl.Ther.* 139:283-289.
19. Ruigrok, T.J.C., Boink, A.B.T.J., Spies, F., Blok, F.J., Maas, A.H.J., and Zimmerman, A.N.E. 1978, Energy dependence of the Ca^{2+} paradox, *J.Mol.Cell.Cardiol.* 10:991-1002.
20. Sordahl, L.A. and Stewart, M. 1980, Mechanism(s) of altered mitochondrial calcium transport in acutely ischemic canine hearts, *Circ.Res.* 47:814-820.
21. Stevens, M.J., Lattimer, S.A., Kamijo, M., Van Huysen, C., Sima, A.A.F., and Greene, D.A. 1993, Osmotically-induced nerve taurine depletion and the compatible osmolyte hypothesis in experimental diabetic neuropathy in the rat, *Diabetologia*, 36:608-614.
22. Tanaka, Y., Konno, N., and Kako, K.J. 1992, Mitochondrial dysfunction observed in situ in cardiomyocytes of rats in experimental diabetes, *Cardiovas.Res.* 26:409-414.
23. Tani, M. and Neely, J.R. 1988, Hearts from diabetic rats are more resistant to in vitro ischemia: possible role of altered Ca^{2+} metabolism, *Circ.Res.* 62:931-940.
24. Tatsumi, T. and Kako, K.J. 1993, Effects of hydrogen peroxide on mitochondrial enzyme function studied in situ in rat heart myocytes, *Basic Res.Cardiol.* 88:199-211.
25. Vary, T.C., Angelakos, E.T., and Schaffer, S.W. 1979, Relationship between adenine nucleotide metabolism and irreversible ischemic tissue damage in isolated perfused rat heart, *Circ.Res.*, 45:218-225.
26. Yamahara, Y., Asayama, J., Kobara, M., Ohta, B., Matsumoto, T., Miyazaki, H., Tatsumi, T., Ishibashi, K., Inoue, M., Inoue, D., and Nakagawa, M. 1993, Effects of ischemic preconditioning on the release of cardiac troponin T in isolated rat hearts, *Basic.Res.Cardiol.* 89:241-249.

REGULATION OF Ca^{2+} TRANSPORT BY INSULIN AND TAURINE

Interaction at the Level of the Na^{+}-Ca^{2+} Exchanger

Stephen W. Schaffer,[1] Khanh Nguyen,[1] Cherry Ballard,[1]
Nichelle Gardner,[1] and Junichi Azuma[2]

[1] Department of Pharmacology, School of Medicine
University of South Alabama
Mobile, Alabama
[2] The Third Department of Internal Medicine
Osaka University Medical School
2-2, Yamada-oka, Suita-City, Osaka 565, Japan

INTRODUCTION

In 1983 Lampson *et al.*[16] and Tokunaga *et al.*[31] introduced the concept that an interaction must exist between the actions of taurine and those of insulin. The initial studies examining this question focused on the metabolic actions of insulin. Tokunaga *et al.*[31] found that taurine suppressed insulin secretion in nondiabetic pancreatic islets, suggesting that taurine might serve as a physiological regulator of insulin secretion and plasma glucose levels. They also found that taurine levels increased significantly in diabetic islets, implying that taurine might contribute to the state of hypoinsulinemia and hyperglycemia. Lampson *et al.*[16] also concluded that taurine influenced plasma glucose levels, but their studies linked the hypoglycemic actions of taurine to a potentiation of insulin-mediated glucose uptake. Lampson *et al.*[16] also reported that taurine promoted several other insulin sensitive reactions, implying that taurine either facilitated the interaction of insulin with its receptor or increased the production of a putative second messenger of insulin. Subsequent studies revealed that taurine could interact directly with the insulin receptor[15,20]. This interaction appeared to be quite specific since insulin was capable of displacing taurine from its binding sites[20]. Based on these data, Maturo and Kulakowski[20] suggested that taurine might exhibit insulin-mimetic activity. However, this notion is inconsistent with a recent observation by Schaffer *et al.*[30] that insulin, but not taurine, could stimulate Na^{+}-Ca^{2+} exchanger activity. Yet, the studies by Schaffer *et al.*[30] did not address the possibility that taurine might function as a general modulator of insulin action.

Adequate characterization of the taurine-insulin interaction has been hampered by the complexity of the insulin signal transduction pathway, as well as the multiplicity of

Taurine 2, edited by Huxtable et al.
Plenum Press, New York, 1996

insulin and taurine actions. To simplify the problem, we have chosen to examine the taurine-insulin interaction using isolated membranes, which contain all of the components necessary for insulin action, but which lack the complexity of the intact cell. An important target of insulin in these membranes is the Na^+-Ca^{2+} exchanger[1]. Two mechanisms can be envisioned for the stimulation of the Na^+-Ca^{2+} exchanger by insulin. The hormone could either cause a change in the phospholipid microenvironment of the exchanger or activate a phosphorylation cascade, which would culminate in the direct activation of the exchanger. Interestingly, taurine could conceivably modulate both insulin-mediated pathways. Previous studies have shown that taurine is a physiological inhibitor of phospholipid N-methyltransferase, the enzyme catalyzing the conversion of phosphatidylethanolamine to phosphatidylcholine[12]. Since phospholipid N-methylation has a profound effect on membrane structure and function[5,8], taurine could alter the activity of an insulin-sensitive transporter, such as the Na^+-Ca^{2+} exchanger, by promoting a rise in the phosphatidylethanolamine/phosphatidylcholine ratio of the membrane. Alternatively, taurine could alter transporter activity by regulating its phosphorylation state. Lombardini and Liebowitz[18] and Schaffer *et al.*[29] have reported that taurine modifies the protein phosphorylation pattern of a number of cardiac proteins. In this report the interaction between taurine and insulin was examined by focusing on phospholipid N-methylation-dependent and independent changes in Na^+-Ca^{2+} exchanger activity.

METHODS

Isolated sarcolemma was prepared according to the method of Pitts[25]. Based on standard membrane markers, contamination of the membrane preparation by mitochondria and sarcoplasmic reticulum was minimal[12,30].

Phospholipid methyltransferase activity was assayed at all three catalytic sites[21]. Site 1, favoring the formation of phosphatidyl-N-monomethylethanolamine, was assayed by incubating 0.5 mg sarcolemmal protein in 50 mM Tris glycylglycine buffer (pH 8.0) containing 1 mM $MgCl_2$ and 0.1 μM S-adenosyl-L-[^{3}H-methyl]-methionine for 30 min at 37°C. Site 2 activity involving the methylation of phosphatidyl-N-monomethyl-ethanolamine, was determined by a 30 min incubation of 0.5 mg sarcolemmal protein with 50 mM phosphate buffer (pH 7.0, 37°C) containing 10 μM S-adenosyl-L-[^{3}H-methyl]-methionine but no magnesium. The final reaction favoring phosphatidylcholine formation occurs in 50 mM glycine buffer (pH 10.0, 37°C) containing 150 μM S-adenosyl-L-[^{3}H-methyl]-methionine and 0.5 mg sarcolemmal protein. When desired, the appropriate concentrations of insulin and/or taurine were added to the reaction mixture.

Na^+-Ca^{2+} exchanger activity was assayed according to a modification of the procedure of Reeves and Sutko[27]. Sarcolemmal vesicles were preloaded with 20 mM MOPS buffer (pH 7.4) containing 160 mM NaCl and 2 mM ATP. They were then incubated for 10 min with 20 mM MOPS buffer containing either 10 U/l insulin, 50 μM S-adenosylmethionine, 30 mM taurine or the appropriate combination of the three effectors. The incubated membrane vesicles (5 μg) were then suspended in 0.5 ml of NaCl or KCl containing buffer supplemented with 30 μM $^{45}CaCl_2$ and 0.5 μM valinomycin. The exchange process was terminated after 2 s by the addition of 3 ml ice-cold filtration buffer (160 mM KCl, 1 mM $LaCl_3$, pH 7.4) and rapidly filtered through a 0.2 μm filter. After washing, the ^{45}Ca content associated with the membranes was determined using a scintillation counter. Net Ca^{2+} uptake was calculated by subtracting ^{45}Ca uptake in NaCl buffer (no Na^+ gradient) from uptake in KCl buffer (outwardly directed Na^+ gradient).

RESULTS

Regulation of Phospholipid N-Methyltransferase by Taurine and Insulin

Phospholipid N-methyltransferase is an intrinsic membrane enzyme containing three catalytic sites (site 1, site 2 and site 3), each of which catalyzes one of the methylation steps involved in the conversion of phosphatidylethanolamine to phosphatidylcholine (Figure 1).

Although this enzyme is found in most biological membranes, in the myocardium the highest activity is associated with the sarcolemma and the sarcoplasmic reticulum[12,26]. In both membrane systems, taurine (30 mM) serves as an inhibitor of the enzyme. Interestingly, all three methylation steps catalyzed by phospholipid N-methyltransferase are equally inhibited by taurine in the two membrane systems (Table 1)[26].

Insulin also regulates phospholipid N-methyltransferase activity of the sarcolemmal membrane. However, in contrast to taurine, it only alters site 1 and site 3 methylation (Table 1). Another characteristic distinguishing insulin from taurine action is the degree of enzyme inhibition. While 30 mM taurine reduces site 1 and site 3 methylation approximately 30%, 10 U/liter insulin decreases the reactions at site 1 and site 3 by 48% and 43%, respectively.

Despite differences in the effects of taurine and insulin on phospholipid N-methylation, the two effectors appear to share some similarities, as evidenced by the failure of combined treatment to yield additive effects (Figure 2). In fact, exposure of sarcolemmal membranes to 1.0 U/liter insulin causes a larger decrease (37%) in site 1 phospholipid N-methylation than exposure to either 1.5 mM taurine alone (21%) or the combination of 1.5 mM taurine and 1.0 U/liter insulin (28%). All other taurine and insulin combinations examined, including a broad range of effector concentrations and combination ratios, also fail to yield additive responses.

Regulation of the Na^+-Ca^{2+} Exchanger by Phospholipid N-Methylation

The effects of phospholipid N-methylation on membrane function can be partially evaluated by treating membrane with the methyl donor, S-adenosylmethionine. A major effect of treating sarcolemma with 50 μM S-adenosylmethionine is a significant decrease in Na^+-Ca^{2+} exchanger activity (Figure 3). This effect is largely reversed by inclusion of either 10 U/liter insulin or 30 mM taurine in the reaction mixture. When both 10 U/liter insulin and 30 mM taurine are added to the reaction mixture simultaneously, the recovery of Na^+-Ca^{2+} exchange is comparable to the effect of adding either taurine or insulin alone.

Table 1. Effect of taurine and insulin on sarcolemmal phospholipid N-methyltransferase activity

Catalytic site	Control	With taurine (30 mM) (pmol/mg/30 min)	With insulin (10 U/liter)
Site 1	0.87 ± 0.06	0.61 ± 0.05*	0.45 ± 0.02*
Site 2	28 ± 3	18.2 ± 1.9*	22.1 ± 2.5
Site 3	237 ± 16	174 ± 18*	136 ± 11*

Activity of sarcolemmal phospholipid methyltransferase was assayed as described in the METHODS section. Catalytic site 1,2 and 3 refer to the reactions favoring the formation of phosphatidyl-N-monomethylethanolamine, phosphatidyl-N,N-dimethylethanolamine and phosphatidylcholine, respectively. Values shown are the means ± SEM of 3-5 different assays. Asterisks denote significant difference from the control values ($p < 0.05$).

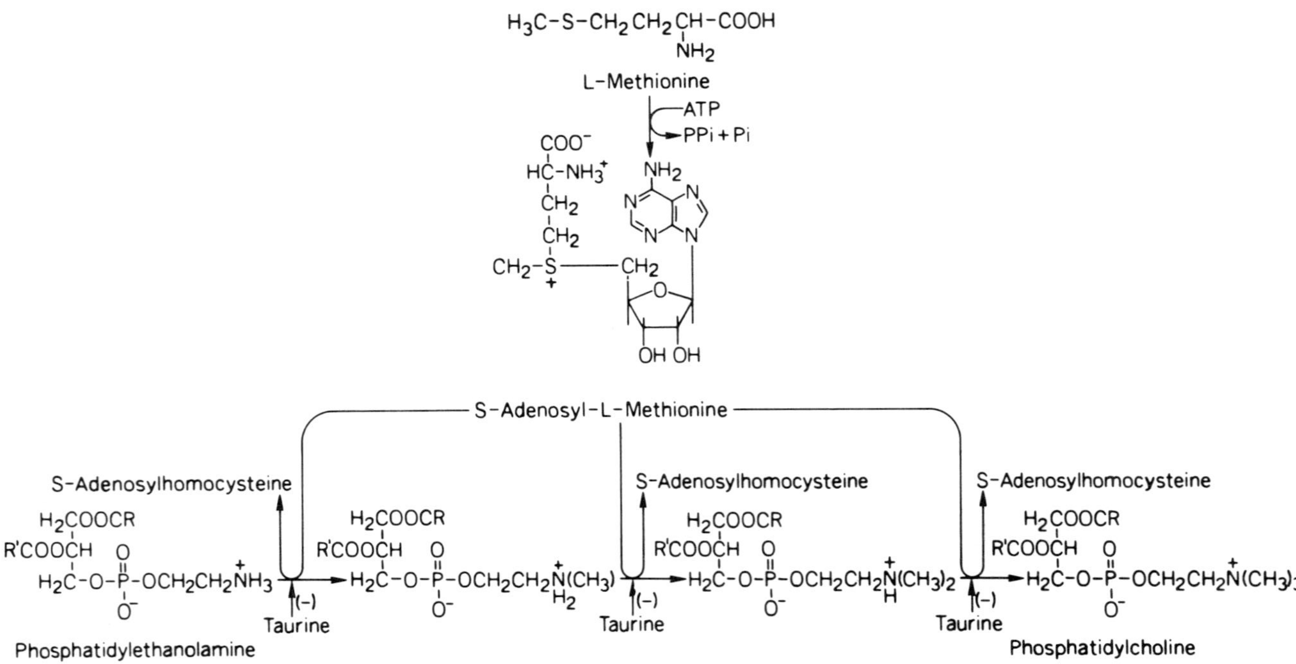

Figure 1. Three-step methylation reaction catalyzed by phospholipid N-methyltransferase. Sites 1, 2 and 3 favor the formation of phosphatidyl-N-monomethylethanolamine, phosphatidyl-N,N-dimethyl-ethanolamine and phosphatidylcholine, respectively.

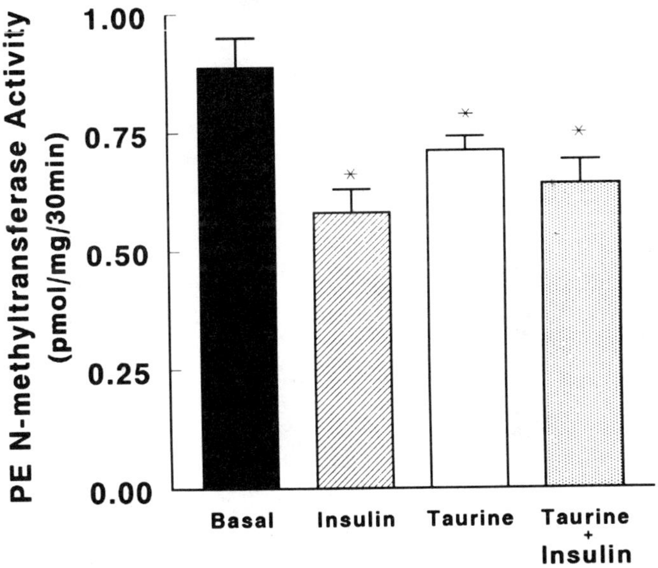

Figure 2. Effect of taurine and insulin on phosphatidylethanolamine N-methyltransferase activity. Isolated sarcolemma were exposed to either no effector, 1.0 U/liter insulin, 1.5 mM taurine or the combination of 1.0 U/liter insulin and 1.5 mM taurine during the site 1 phospholipid N-methyltransferase reaction. Activity of the enzyme was significantly reduced by insulin, taurine and the combination of insulin and taurine ($p < 0.05$). All data represent means ± SEM of 4-6 assays.

Regulation of the Na^+-Ca^{2+} Exchanger by Taurine and Insulin

In accordance with a previous study[30], we found that, in the absence of S-adenosylmethionine, taurine has no effect on sarcolemmal Na^+-Ca^{2+} exchanger activity. By contrast, insulin directly stimulates Na^+-Ca^{2+} exchanger activity in both the presence and absence of S-adenosylmethionine (Figure 4). In the absence of S-adenosylmethionine, 10 U/liter insulin increases Na^+-Ca^{2+} exchanger activity 85% compared to 94% in the presence of S-adenosylmethionine. Since the percent stimulation is comparable in the two conditions, it is unclear if insulin increases Na^+-Ca^{2+} exchange by reversing the effects of phospholipid N-methylation or by directly stimulating Na^+-Ca^{2+} exchanger activity.

Also unclear from the data is the nature of the interaction between taurine and insulin. To address the latter, the effect of taurine on insulin-mediated enhancement of Na^+-Ca^{2+} exchanger activity has been examined. As seen in Figure 4, addition of both 30 mM taurine and 10 U/liter insulin to the incubation medium leads to a 31% increase in Na^+-Ca^{2+} exchanger activity. Since insulin alone increases exchanger activity 85% while taurine alone has no effect, the results suggest that taurine inhibits the effects of insulin on this transporter.

DISCUSSION

Regulation of Phospholipid N-Methylation

Phospholipid N-methylation has a profound impact on membrane structure and function[5,8]. During the course of converting phosphatidylethanolamine to phosphatidyl-

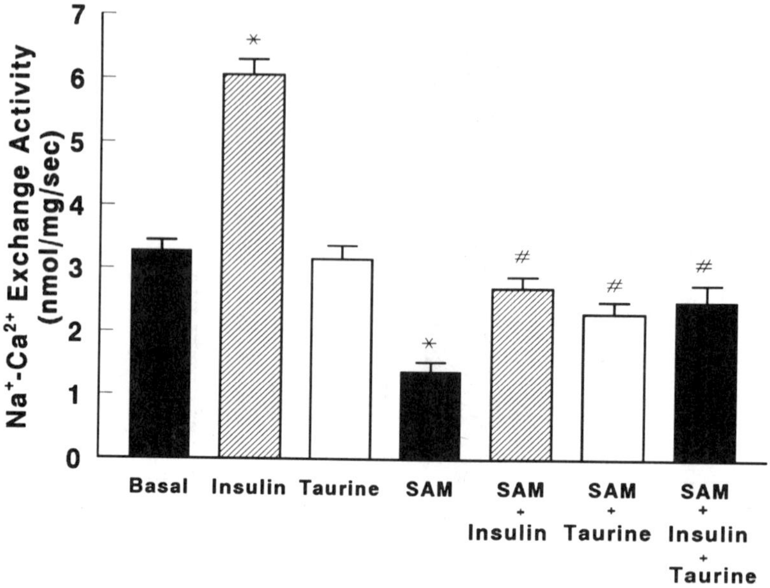

Figure 3. Effect of taurine, insulin and phospholipid N-methylation on Na^+-Ca^{2+} exchanger activity. Prior to assaying for Na^+-Ca^{2+} exchanger activity, isolated sarcolemma was incubated with buffer containing either no effector (BASAL), 10 U/liter insulin, 30 mM taurine, 50 μM S-adenosylmethionine (SAM) or the indicated combination of the effectors. Values shown represent means ± SEM of 4-6 assays. Asterisks and pound symbol denote significant difference from BASAL and SAM control values, respectively ($p < 0.05$). Phospholipid N-methylation reduced Na^+-Ca^{2+} exchanger activity but this effect was reversed by insulin, taurine or the combination of taurine and insulin. Insulin alone not only reversed the effects of phospholipid N-methylation, but also directly stimulated Na^+-Ca^{2+} exchanger activity.

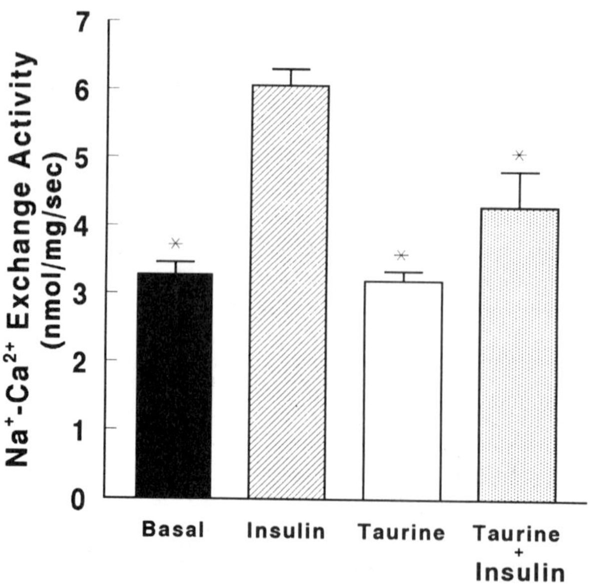

Figure 4. Effect of taurine and insulin on Na^+-Ca^{2+} exchanger activity. Prior to assaying for Na^+-Ca^{2+} exchanger activity, isolated sarcolemma was incubated with buffer containing either no effector (BASAL), 10 U/liter insulin, 30 mM taurine or the combination of the two effectors. Values shown represent means ± SEM of 4-6 assays. Asterisks denote significant difference from the insulin-treated group ($p < 0.05$). While taurine has no direct effect on Na^+-Ca^{2+} exchanger activity, it attenuates the significant increase in exchanger activity mediated by insulin.

choline, phospholipid is transferred from the inside or cytoplasmic side of the membrane to the exterior face of the membrane. This transfer reaction, catalyzed by aminophospholipid translocase, concentrates phosphatidylethanolamine in the inside leaflet of the membrane and phosphatidylcholine in the exterior leaflet. Among other effects, it has been suggested that the redistribution of membrane phospholipid between membrane leaflets is important in maintaining maximal Na^+-Ca^{2+} exchanger activity[13].

Equally important in altering membrane function is the change in polymorphism associated with the conversion of phosphatidylethanolamine to phosphatidylcholine. It is known that individual lipid species naturally adopt one of three structures in aqueous solution (micellar, bilayer or hexagonal). Phosphatidylcholine prefers the bilayer structure and is therefore considered a stabilizer of the phospholipid bilayer of biological membranes. By contrast, phosphatidylethanolamine induces hexagonal structures in mixed lipid systems. Although nonbilayer formers, such as phosphatidylethanolamine, render the bilayer membrane less stable, they play important roles in promoting certain membrane functions, including permeability, fusion, transport and enzyme activity[5,8]. Thus, it is not surprising that stimulators of phospholipid N-methylation cause dramatic decreases in Na^+-Ca^{2+} exchanger activity.

Modulation of phospholipid N-methyltransferase activity by taurine presumably involves competition between taurine and the substrate for the phospholipid headgroup binding site on the enzyme. This mechanism is based on the assumption that since taurine and the phospholipid headgroups are structurally similar, they will bind to the same site on the enzyme. Although enzyme kinetics usually provide information on the type of interaction which exists between a substrate and inhibitor, the complex kinetics of phospholipid N-methyltransferase makes this analysis virtually impossible. Thus, a more definitive explanation for the inhibitory effects of taurine must await the purification of the enzyme.

The mechanism underlying the inhibitory effect of insulin on phospholipid N-methylation also remains to be established, although a change in the phosphorylation state of phospholipid N-methyltransferase would be consistent with known actions of insulin[28]. Alternatively, insulin may restrict the availability of substrate for the N-methyltransferase reaction by promoting a remodeling of the phospholipid environment of the membrane[10]. This latter hypothesis is attractive because it provides a logical explanation for the nonadditivity of the interaction between taurine and insulin.

Regulation of the Na^+-Ca^{2+} Exchanger

Activity of the Na^+-Ca^{2+} exchanger is closely associated with the phospholipid microenvironment of the transporter. According to Philipson[22], Luciani *et al.*[19] and Hilgemann and Collins[13], changes in membrane surface charge, as well as perturbations in the hydrophobic region of the membrane, can alter Na^+-Ca^{2+} exchanger activity. The present study indicates that changes in the phosphatidylethanolamine/phosphatidylcholine ratio of the membrane can also lead to altered Na^+-Ca^{2+} exchanger activity. Thus, it is not surprising that the effects of both taurine and insulin on phospholipid N-methylation are translated into a change in Na^+-Ca^{2+} exchanger activity.

In addition to reversing the effects of phospholipid N-methylation on Na^+-Ca^{2+} exchanger activity, insulin also stimulates Na^+-Ca^{2+} exchanger activity by a mechanism unrelated to phospholipid N-methylation. This action of insulin is blocked by the protein kinase inhibitor, staurosporine[1], indicating involvement of the insulin-mediated phosphorylation cascade in the activation of the Na^+-Ca^{2+} exchanger. Although the Na^+-Ca^{2+} exchanger contains a phosphorylation consensus sequence for calmodulin-dependent kinase, there is considerable debate on whether protein phosphorylation alters Na^+-Ca^{2+} exchanger activity[2-4,6,17,32]. Caroni and Carafoli[3] initially introduced the idea that the

Na^+-Ca^{2+} exchanger is regulated by protein phosphorylation. However, Collins *et al.*[4] failed to detect inhibition of the stimulatory effect of ATP by peptide inhibitors of cAMP-dependent protein kinase, protein kinase C or calmodulin-dependent protein kinase. On the other hand, phorbol esters capable of activating protein kinase C have been reported to stimulate Na^+-Ca^{2+} activity[2,32]. Moreover, activation of the Na^+-Ca^{2+} exchanger by agonists linked to G_q proteins is blocked by the protein kinase C inhibitor, chelerythrine[2]. Thus, protein phosphorylation, in particular protein kinase C-mediated events, may play some role in the regulation of the Na^+-Ca^{2+} exchanger; the protein kinase involved in insulin action still remains to be established.

Insulin may also stimulate Na^+-Ca^{2+} exchange through the activation of phospholipase C. Although there has been considerable debate regarding which phospholipase C isoform is activated by insulin[7,9-11], the specificity of the insulin-sensitive phospholipase appears to be less important in determining Na^+-Ca^{2+} exchanger activity than the extent of phospholipid hydrolysis. Philipson *et al.*[23] have shown that cardiac vesicles exposed to a bacterial phospholipase C with preferential activity for phosphatidylcholine, phosphatidylethanolamine and sphingomyelin, exhibit enhanced Na^+-Ca^{2+} exchanger activity. Similarly, a phospholipase C isoform exhibiting high specificity for the hydrolysis of phosphatidylinositol has been shown to increase Na^+-Ca^{2+} exchanger activity[24]. These results raise the possibility that the insulin-sensitive phospholipase C isoform may either release a Na^+-Ca^{2+} exchange modulator from the membrane or cause a remodeling of the phospholipid microenvironment of the transporter.

One of the major findings of this study is the ability of taurine to inhibit insulin-mediated stimulation of the Na^+-Ca^{2+} exchanger. This observation is in apparent contradiction to earlier studies by Lampson *et al.*[16] showing potentiation of several insulin-mediated reactions by taurine. Although the basis for the conflicting effects is still poorly understood, the data suggest that taurine cannot interfere with an early step in the signal transduction pathway of insulin. According to this criteria, the interaction between taurine and insulin cannot occur at the level of the insulin receptor. This conclusion runs counter to the theory of Maturo and Kulakowski[20] that taurine alters insulin action through an effect on the insulin receptor. Also inconsistent with this observation is the possibility that taurine blocks the phosphorylation cascade initiated by insulin. However, the data do not rule out the possibility that taurine inhibits the phosphorylation of an insulin-sensitive protein by a mechanism unrelated to the phosphorylation cascade. Another logical explanation is that taurine may either alter insulin-mediated changes in membrane structure or the response of the transporter to insulin-mediated membrane remodeling. Huxtable[14] has advanced the idea that taurine is a membrane stabilizer while Hamaguchi *et al.*[12] has attributed the actions of taurine to changes in the phospholipid composition of the membrane. Since both the Na^+-Ca^{2+} exchanger and glucose transport are very sensitive to changes in the composition and structure of the phospholipid bilayer, the membrane modulating properties of taurine can account for the potentiation of insulin-mediated stimulation of glucose transport by taurine, as well as its attenuation of insulin-mediated stimulation of the Na^+-Ca^{2+} exchanger.

The modulation of Na^+-Ca^{2+} exchanger activity by taurine is consistent with its enantiostatic activity. On the one hand, taurine prevents a dramatic loss in basal Na^+-Ca^{2+} exchanger activity caused by phospholipid N-methylation. Yet, the stimulatory effects of taurine are limited by its attenuation of insulin-mediated activation of the exchanger. These opposite actions of taurine maintain exchanger activity within a narrow range. Presumably, this activity range is ideal for insuring a proper balance between Ca^{2+} influx via the Ca^{2+} channel and Ca^{2+} efflux via the Na^+-Ca^{2+} exchanger. Thus, the effect of taurine on Na^+-Ca^{2+} exchanger activity plays a crucial role in preventing heart failure due to either a decrease or elevation in myocardial $[Ca^{2+}]_i$.

REFERENCES

1. Ballard, C., Mozaffari, M., and Schaffer, S. 1994, Signal transduction mechanism for the stimulation of the sarcolemmal Na^+-Ca^{2+} exchanger by insulin, *Mol. Cell. Biochem.* 135:113-119.
2. Ballard, C., and Schaffer, S. 1995, Stimulation of the Na^+/Ca^{2+} exchanger by phenylephrine, angiotensin II and endothelin I, *J. Mol. Cell. Cardiol.*, in press.
3. Caroni, P., and Carafoli, E. 1983, The regulation of the Na^+-Ca^{2+} exchanger of heart sarcolemma, *Eur. J. Biochem.* 132:451-460.
4. Collins, A., Somlyo, A.V., and Hilgemann, D.W. 1992, The giant cardiac membrane path c method: Stimulation of outward Na^+-Ca^{2+} exchange current by MgATP, *J. Physiol.* 454:27-57.
5. Cullis, P.R., Hope, M.J., and Tilcock, C.P.S. 1986, Lipid polymorphism and the roles of lipids in membranes, *Chem. Physics Lipids* 40:127-144.
6. DiPolo, R., and Beauge, L. 1994, Effects of vanadate on MgATP stimulation of Na-Ca exchange support kinase-phosphatase modulation in squid axons, *Am. J. Physiol.* 266:C1382-C1391.
7. Donchenko, V., Zannetti, A., and Baldini, P.M. 1994, Insulin-stimulated hydrolysis of phosphatidylcholine by phospholipiase C and phospholipase D in cultured rat hepatocytes, *Biochim. Biophys. Acta* 1222:492-500.
8. Epand, R.M. 1990, Relationship of phospholipid hexagonal phases to biological phenomena, *Biochem. Cell Biol.* 68:17-23.
9. Etindi, R., and Fain, J.N. 1989, Insulin does not activate a phosphoinositide-specific phospholipase C in adipocytes, *Mol. Cell. Endocrinol.* 67:149-153.
10. Farese, R.V., and Cooper, D.R. 1989, Potential role of phospholipid-signaling systems in insulin action and states of clinical insulin resistance, *Diabetes/Metabol. Rev.* 5:455-474.
11. Fox, J.A., Soliz, N.M., and Saltiel, A.R. 1987, Purification of a phosphatidylinositol-glycan-specific phospholipase C from liver plasma membranes: A possible target of insulin action, *Proc. Natl. Acad. Sci. USA* 84:2663-2667.
12. Hamaguchi, T., Azuma, J., and Schaffer, S. 1991, Interaction of taurine with methionine: Inhibition of myocardial phospholipid methyltransferase, *J. Cardiovasc. Pharmacol.* 18:224-230.
13. Hilgemann, D.W., and Collins, A. 1992, Mechanism of cardiac Na^+-Ca^{2+} exchange current stimulation by MgATP: Possible involvement of aminophospholipid translocase, *J. Physiol.* 454:59-82.
14. Huxtable, R.J. 1992, The physiological actions of taurine, *Physiol. Rev.* 72:101-163.
15. Kulakowski, E.C., Maturo, J., and Schaffer, S.W. 1985, The low affinity taurine-binding protein may be related to the insulin receptor, *in*: "Taurine: Biological Actions and Clinical Perspectives", Oja, S.S., Ahtee, L., Kontro, P. and Paasonen, M.K., eds., Alan R. Liss, Inc., New York, pp. 127-136.
16. Lampson, W.G., Kramer, J.H., and Schaffer, S.W. 1983, Potentiation of the actions of insulin by taurine, *Can. J. Physiol. Pharmacol.* 61:457-463.
17. Lin, L.F., Kao, L-S, and Westhead, E.W. 1994, Agents that promote protein phosphorylation inhibit the activity of the $Na^+\backslash Ca^{2+}$ exchanger and prolong Ca^{2+} transients in bovine chromaffin cells, *J. Neurochem.* 63:1941-1947.
18. Lombardini, J.B., and Liebowitz, S.M. 1989, Taurine modifies calcium ion uptake and protein phosphorylation in rat heart, *in:* "Taurine and the Heart", Iwata, H., Lombardini, J.B. and Segawa, T., eds., Kluwer Academic Publishers, Boston/Dordrecht/London, pp. 117-137.
19. Luciani, S., Bova, S., Cargnelli, G., Cusinato, F., and Debetto, P. 1991, Modulation of sodium-calcium exchange by lipids, *Ann. N.Y. Acad. Sci.* 639:156-165.
20. Maturo, J., and Kulakowski, E.C. 1988, Taurine binding to the purified insulin receptor, *Biochem. Pharmacol.* 37:3755-3760.
21. Panagia, V., Makino, N., Ganguly, P.K., and Dhalla, N.S. 1987, Inhibition of Na^+-Ca^{2+} exchange in heart sarcolemmal vesicles by phosphatidylethanolamine N-methylation, *Eur. J. Biochem.* 166:597-603.
22. Philipson, K.D. 1990, The cardiac Na^+-Ca^{2+} exchanger: dependence on membrane environment, *Cell Biol. Int. Reports* 14:305-309.
23. Philipson, K.D., Frank, J.S., and Nishimoto, A.Y. 1983, Effects of phospholipase C on the Na^+-Ca^{2+} exchange and Ca^{2+} permeability of cardiac sarcolemmal vesicles, *J. Biol. Chem.* 258:5905-5910.
24. Pierce, G.N., and Panagia, V. 1989, Role of phosphatidylinositol in cardiac sarcolemmal membrane sodium-calcium exchange, *J. Biol. Chem.* 264:15344-15350.
25. Pitts, B.J.R. 1979, Stoichiometry of sodium-calcium exchange in cardiac sarcolemmal vesicles, *J. Biol. Chem.* 254:6232-6235.
26. Punna, S., Ballard, C., Hamaguchi, T., Azuma, J., and Schaffer, S. 1994, Effect of taurine and methionine on sarcoplasmic reticular Ca^{2+} transport and phospholipid methyltransferase activity, *J. Cardiovasc. Pharmacol.* 24:286-292.

27. Reeves, J.P., and Sutko, J.L. 1979, Sodium-calcium ion exchange in cardiac membrane vesicles, *Proc. Natl. Acad. Sci. USA* 76:590-594.
28. Saltiel, A.R. 1994, The paradoxical regulation of protein phosphorylation in insulin action, *FASEB J.* 8:1034-1040.
29. Schaffer, S.W., Allo, S., Harada, H., and Azuma, J. 1990, Regulation of calcium homeostasis by taurine: Role of calmodulin, *in:* "Taurine: Functional Neurochemistry, Physiology and Cardiology", Pasantes-Morales, H., Martin, D.L., Shain, W. and Martin del Rio, R., eds., Wiley-Liss, Inc., New York, pp. 217-225.
30. Schaffer, S.W., Punna, S., Duan, J., Harada, H., Hamaguchi, T., and Azuma, J. 1992, Mechanism underlying physiological modulation of myocardial contraction by taurine, *in:* "Taurine: Nutritional Value and Mechanisms of Action", Lombardini, B., Schaffer, S.W. and Azuma, J., eds., Plenum Press, New York, pp. 193-198.
31. Thakker, J.K., DiMarchi, R., MacDonald, K., and Caro, J.F. 1989, Effect of insulin and insulin-like growth factors I and II on phosphatidylinositol and phosphatidylinositol 4,5-bisphosphate breakdown in liver from humans with and without type II diabetes, *J. Biol. Chem.* 264:7169-7175.
32. Tokunaga, H., Yoneda, Y., and Kuriyama, K. 1983, Streptozotocin-induced elevation of pancreatic taurine content and suppressive effect of taurine on insulin secretion, *Eur. J. Pharmacol.* 87:237-243.
33. Vigne, P., Breittmayer, J-P, Duval, D., Frelin, C., and Lazdunski, M. 1988, The Na^+/Ca^{2+} antiporter in aortic smooth muscle cells, *J. Biol. Chem.* 263:8078-8083.

60

ANTIHYPERCHOLESTEROLEMIC ACTION OF TAURINE ON STREPTOZOTOCIN-DIABETIC RATS OR ON RATS FED A HIGH CHOLESTEROL DIET

K. Nanami,[1] H. Oda,[2] and H. Yokogoshi[1]

[1] School of Food and Nutritional Sciences
The University of Shizuoka
52-1 Yada, Shizuoka 422, Japan
[2] Department of Applied Biological Sciences
Nagoya University
Nagoya 464-01, Japan

INTRODUCTION

Taurine is especially abundant in excitable tissues, membrane-rich cells, and tissues producing oxidants and toxic substances including brain, retina, myocardium, skeletal muscle, liver, platelets, and leucocytes. Suggested biological actions of taurine include cell membrane stabilization, antioxidation, detoxification, osmoregulation, neuromodulation, and brain-retinal development. It also has antiarrhythmic, hypotensive, anticonvulsant, inotropic and thermoregulatory actions[2,4,6].

Cholesterol is important for the digestion and absorption of dietary fats via bile acid biosynthesis. It is also the precursor of steroid hormones and a component of biomembranes. However, raised serum cholesterol is a risk factor for coronary heart disease. Hypercholesterolemia is induced by many cases in physiological or unphysiological conditions. In this study, the effect of taurine on several kinds of hypercholesterolemia induced by feeding a high cholesterol diet, xenobiotic intake or diabetes was investigated.

MATERIALS AND METHODS

Male rats of the Wistar strain weighing about 100 g (SLC Inc., Hamamatsu, Japan) were housed in individual cages in our animal facility and were fed 20% casein diet as a basal diet. This diet contained 20% casein, appropriate amounts of minerals and vitamins, 5% corn oil, 5% cellulose and 64% carbohydrate mixtures (sucrose:starch; 1:2, w/w) (Table 1).

Taurine 2, edited by Huxtable et al.
Plenum Press, New York, 1996

Table 1. Composition of test diets (%)

	Basal (A)	A + taurine	High cholesterol (B)	B + taurine
Casein	20	20	20	20
Carbohydrate (starch)	42.57	41.90	41.74	38.40
Carbohydrate (sucrose)	21.28	20.95	20.86	19.20
Corn oil	5	5	5	5
AIN-76 vitamins	1	1	1	1
AIN-76A minerals	5	5	5	5
Choline Cl	0.15	0.15	0.15	0.15
Cellulose	5	5	5	5
Taurine		5		5
Cholesterol			1	1
Sodium cholate			0.25	0.25

Experiment 1. Rats were fed the basal diet or basal diet supplemented with 1% cholesterol and 0.25% sodium cholate for 14 days. Both diets were fed plus or minus taurine (5%). Blood was collected from a cervical cut, and livers were immediately removed, frozen and stored at -70°C. Serum lipids (triglyceride, total cholesterol, HDL-cholesterol, LDL/VLDL-cholesterol) and liver lipids were determined spectrophotometrically. The contents of lipoprotein mRNA (ApoA-I and ApoE) in liver were also determined.

Experiment 2. Animals were divided into three groups (six rats per group) and were fed basal diet and basal diet supplemented with 0.3% caffeine and/or 5% taurine for four

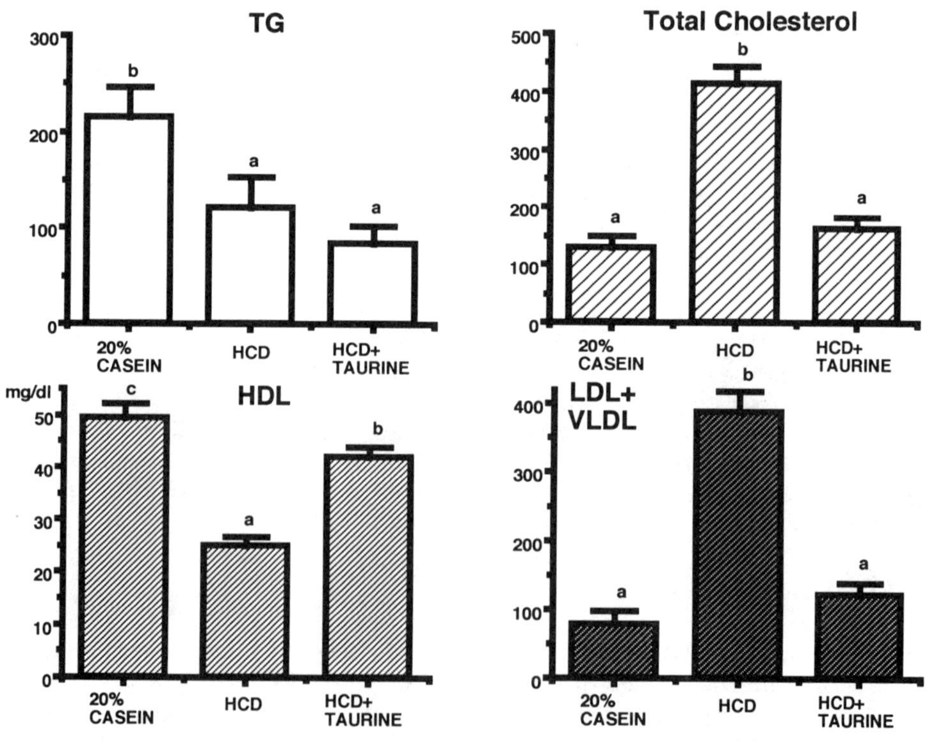

Figure 1. Effect of taurine on serum triglyceride and cholesterol of rats fed high a cholesterol diet (HCD).

weeks. The concentrations of lipids in serum and liver were determined by the same methods as in Experiment 1.

Experiment 3. Wistar rats were injected i.v. with streptozotocin (60 mg/kg), and were fed the commercial stock diet. Diabetic or nondiabetic rats drank a water or a taurine solution (1 g/day) for 90 days. Various lipids in serum or liver were determined by the same methods as in Experiment 1.

RESULTS AND DISCUSSION

Effect of Taurine on Hypercholesterolemia Induced by High Cholesterol Diet (Experiment 1)

When rats were fed the high cholesterol diet, both total cholesterol and LDL/VLDL-cholesterol in serum significantly increased and the concentrations of triglyceride and HDL-cholesterol significantly decreased (Fig. 1). Supplementation with taurine antagonized all these changes. On the other hand, hepatic content of total lipids, triglyceride, cholesterol and phospholipid were all increased by feeding the high cholesterol diet, and taurine decreased their enhanced concentrations (Fig. 2). ApoA-ImRNA content decreased on the high cholesterol diet, but taurine reversed this decrease.

There are numerous studies concerned with the effects of taurine on lipid metabolism, but the results are not always consistent, due to differences of the experimental condition.

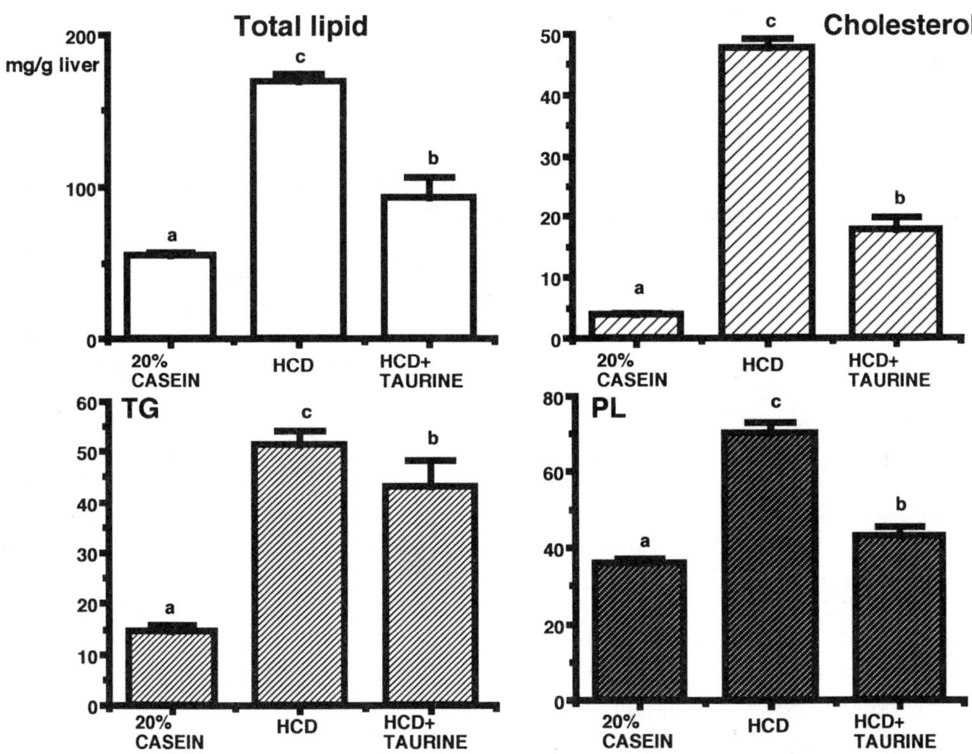

Figure 2. Effect of taurine on hepatic lipids of rats fed high cholesterol diet.

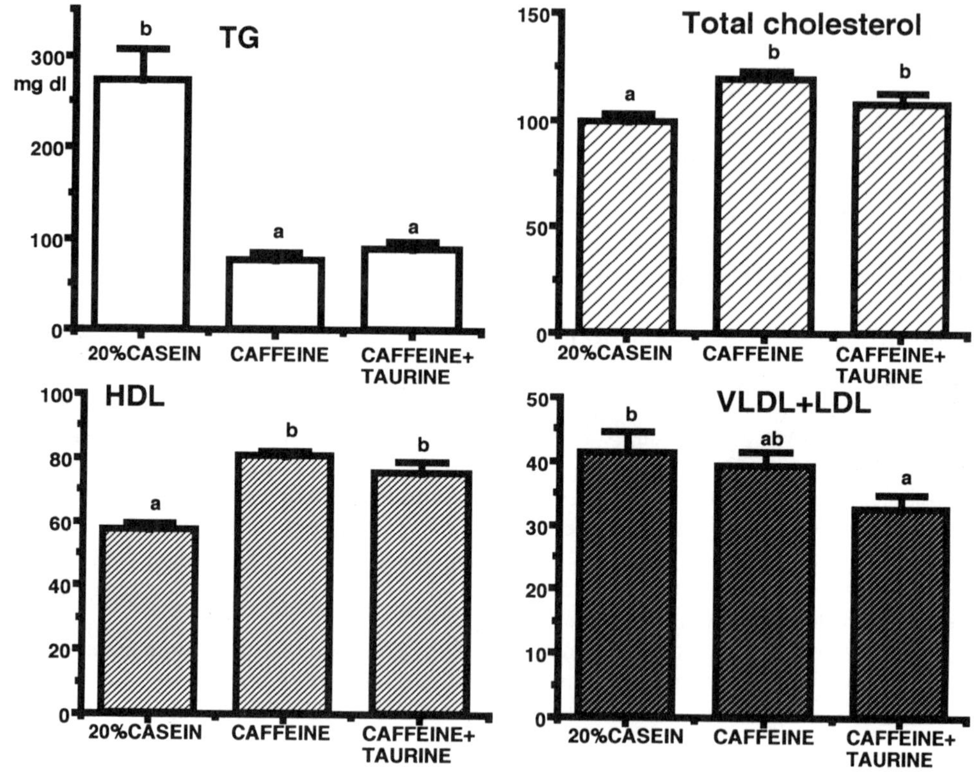

Figure 3. Effect of taurine on serum triglyceride and cholesterol of rats fed a caffeine-containing diet.

Venkatesan *et al.*[10] reported that taurine given i.p. for 10 days to Wistar strain rats did not affect serum lipid concentrations, but in liver the concentrations of triglyceride, cholesterol and total lipids tended to decrease. On the other hand, the concentration of total cholesterol in serum was significantly reduced by the supplementation of a 10% or 25% casein diet with taurine, while HDL-cholesterol increased[7]. Taurine supplementation also enhanced the activity of 7α-hydroxylase, the rate-limiting enzyme of bile acid synthesis[7], and the excretion of bile acids was also increased[1]. Therefore, in this experiment, supplementary taurine may enhance the cholesterol metabolism and bile acid excretion.

Effect of Taurine on Hypercholesterolemia Induced by Caffeine (Experiment 2)

When rats were fed a caffeine-containing diet with or without taurine, body weight gain significantly decreased as compared with that of rats fed the basal diet. By feeding the caffeine-containing diet, serum total cholesterol and HDL-cholesterol significantly increased, and the simultaneous supplementation of taurine did not affect these concentrations (Fig. 3). Serum triglycerides of rats fed the caffeine diet significantly decreased. Simultaneous supplementation of taurine did not affect triglyceride levels. On the other hand, the caffeine-containing diet decreased hepatic ApoA-ImRNA, total lipids, cholesterol and triglycerides, but the simultaneous supplementation of taurine did not affect these concentrations (Fig. 4). These phenomena are consistent with previous results[11]. Therefore, in the

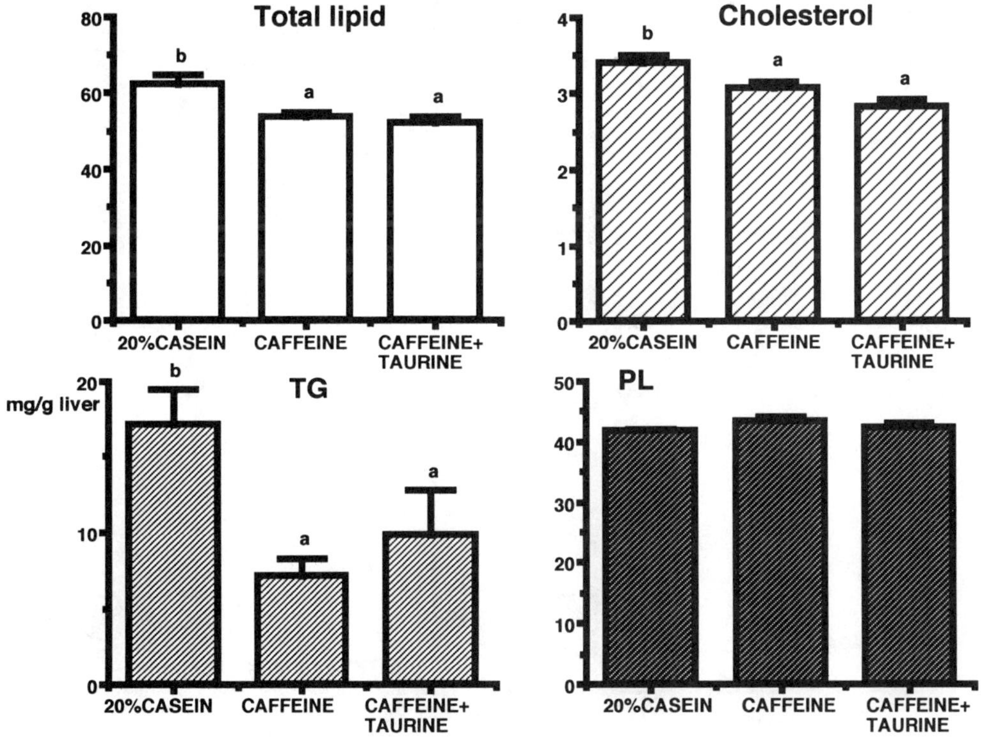

Figure 4. Effect of taurine on hepatic lipids of rats fed a caffeine-containing diet.

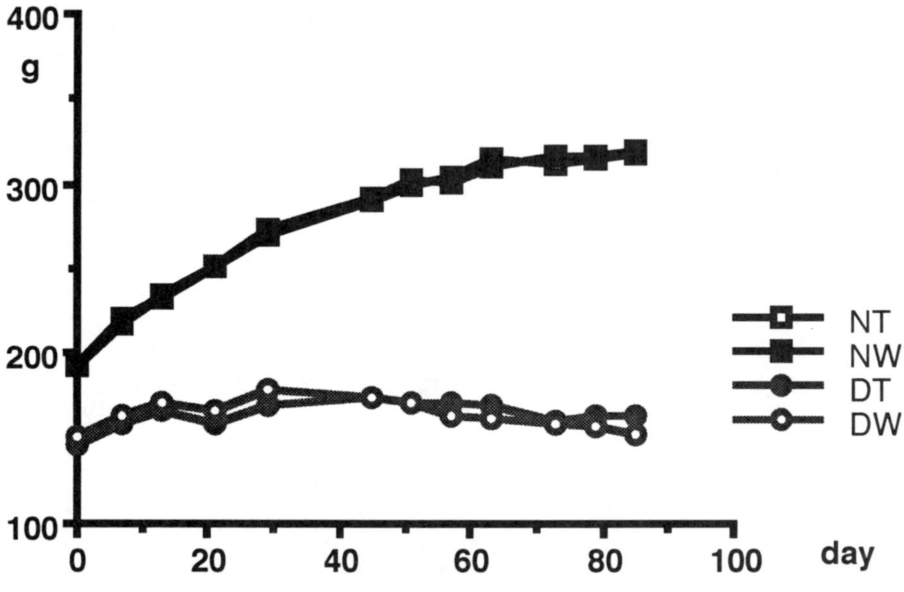

Figure 5. Changes in body weight of normal (N) or streptozotocin-induced diabetic (D) rats drinking water with (T) or without (W) taurine.

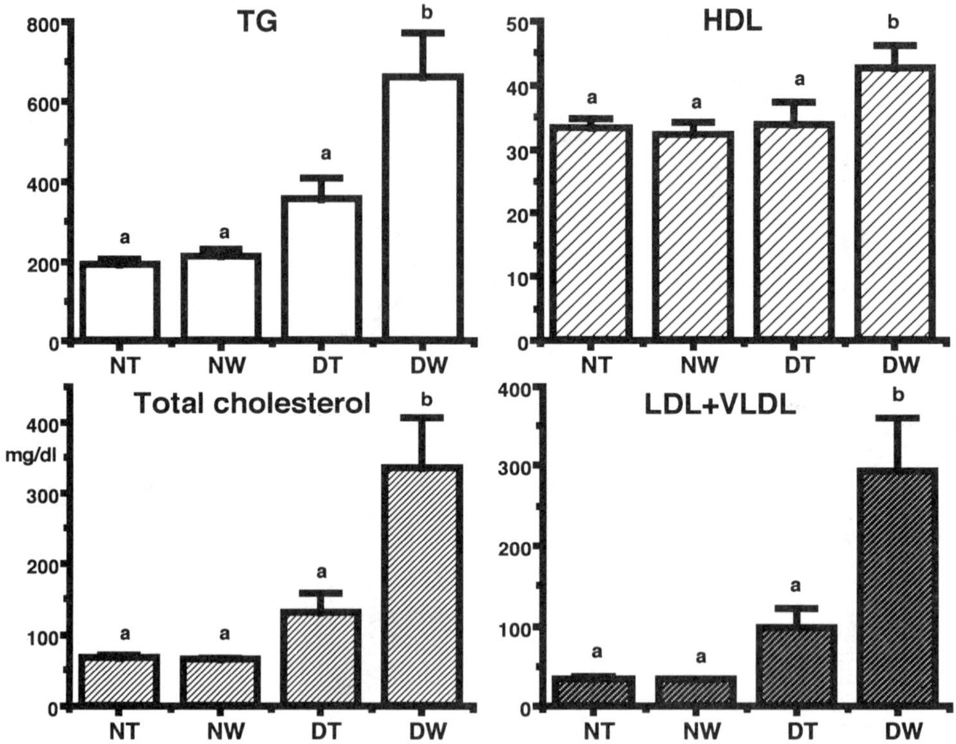

Figure 6. Effect of taurine on serum triglyceride and cholesterol of normal or diabetic rats. Abbreviations are for Fig. 5.

case of hypercholesterolemia caused by the increase of endogenous cholesterol synthesis resulting from a caffeine-containing diet, taurine did not affect cholesterol metabolism.

Effect of Taurine on Hypercholesterolemia in Streptozotocin-Diabetic Rats (Experiment 3)

In diabetic rats, body weight gain was smaller than that of normal rats regardless of taurine supplementation (Fig. 5). Serum levels of triglyceride, total cholesterol, HDL-cholesterol and LDL/VLDL-cholesterol in diabetic rats significantly increased compared with levels in normal rats. Taurine supplementation significantly reduced these increases (Fig. 6). Liver total cholesterol of diabetic rats significantly increased, and taurine decreased this increase (Fig. 7). The concentration of hepatic triglyceride in diabetic rats was significantly decreased, and taurine did not affect this level. The contents of ApoA-ImRNA tended to increase in diabetic rats. Taurine was without effect on ApoA-ImRNA.

It was considered that, in diabetic rats, hypercholesterolemia was induced by a decrease in lipoprotein lipase. In the streptozotocin-diabetic rat, taurine concentrations in pancreas promptly increased[8] and the shape of Langerhans's islands also changed[9]. In addition, enhancement of cholesterol metabolism, bile acid excretion and bile acid synthesis, and changes in bile acid composition have been reported[3,5]. It is considered, therefore, that, in diabetic rats, as the requirement of taurine and glycine for bile acid conjugation may be

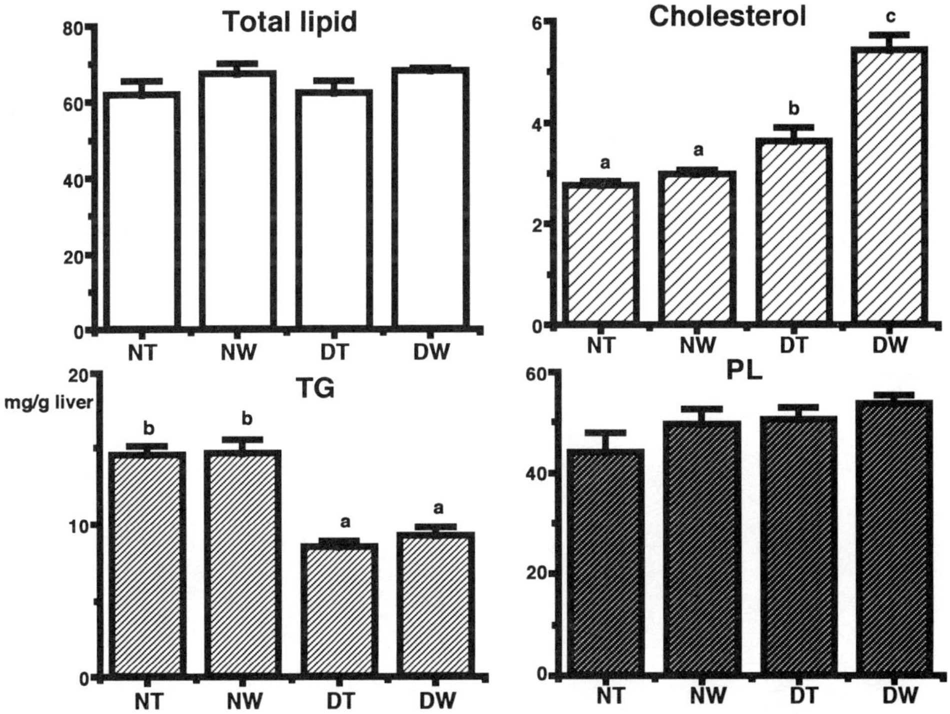

Figure 7. Effect of taurine on hepatic lipids of normal or diabetic rats. Abbreviations are for Fig. 5.

increased, supplemental taurine enhances bile acid metabolism and, in consequence, serum cholesterol decreases.

The overall results show that taurine has an antihypercholesterolemic action in exogenous hypercholesterolemia (induced by high cholesterol diet) and diabetes, but has no such action on endogenous hypercholesterolemia (induced by a caffeine-containing diet).

REFERENCES

1. Gandhi, V.M., Cherian, K.M., and Mulky, J.M. 1983, Hypolipidemic action of taurine in rats, *Ind.J.Exp.Biol.* 30:413-417.
2. Huxtable, R.J. 1992, Physiological actions of taurine, *Physiol.Rev.* 72:101-163.
3. Itoh, T. 1986, Effect of dietary cholesterol on serum cholesterol level and bile acid metabolism in streptozotocin-diabetic rats, *Yonago Acta Medica*, 29:34-48.
4. Kuriyama, K. 1980, Taurine as a neuromodulator, *Federation Proceedings*, 39:2680-2684.
5. Siow, Y., Schurr, A., and Vitale, G.C. 1991, Diabetes-induced bile acid composition changes in rat bile determined by high performance liquid chromatography, *Life Sci.* 49:1303-1308.
6. Sturman, J.A. 1993, Taurine in development, *Physiol.Rev.* 73:119-147.
7. Sugiyama, K., Ohishi, A., Ohnuma, Y., and Maramatsu, K. 1989, Comparison between the plasma cholesterol-lowering effect of glycine and taurine in rats fed on high cholesterol diets, *Agric.Biol.Chem.* 53:1647-1652.
8. Tokunaga, H., Yoneda, Y., and Kuriyama, K. 1979, Protective actions of taurine against streptozotocin-induced hyperglycemia, *Biochem.Pharmacol.* 28:2807-2811.
9. Tokunaga, H., Yoneda, Y., and Kuriyama, K. 1983, Streptozotocin-induced elevation of pancreatic taurine content and suppressive effect of taurine on insulin secretion, *Eur.J.Pharmacol.* 87:237-243.

10. Venkatesan, N., Rao, P.V., and Arumugam, V. 1993, Inhibitory effect of taurine on puromycin aminonucleoside-induced hyperlipidemia in rats, *Clin.J.Biochem.Nutr.* 15:203-210.
11. Yokogoshi, H., Mochizuki, S., Takahata, M., Quazi, S., and Yoshida, A. 1983, The hypercholesterolemic effect of caffeine-containing beverages and xanthine derivatives in rats, *Nutr.Rep.Int.* 28:805-814.

Part 8

TAURINE–CLINICAL STUDIES

61

TAURINE INTAKE OF KOREAN BREAST-FED INFANTS DURING LACTATION

Eul-Sang Kim,[1] Kum-Ho Cho,[1] Mee-Ah Park,[1] Kyu-Han Lee,[1] Jean Moon,[1] Young-Nam Lee,[2] and Hee-Kyung Ro[3]

[1] Department of Food Science and Nutrition
Dankook University
Seoul, Korea
[2] Department of Food and Nutrition
International College of Hotel Administration
Kyunghee University
Seoul, Korea
[3] Department of Food Science and Nutrition
Chosun University
Kwangju, Korea

INTRODUCTION

The roles of taurine in human nutrition and physiology have been intensively reviewed[1,11-13,21,24]. Recent interest in taurine metabolism and its nutritional requirement for humans has centered around the discovery that retinal degeneration and blindness develops in taurine-deficient cats[9,11,21], the observed depression in plasma and urine taurine concentrations in infants fed taurine-free formulas[5] and in children and adults maintained on long-term parenteral nutrition[6,7,23,26].

The enzyme cysteine sulfinic acid decarboxylase (CSAD) appears to be the rate-limiting step in taurine biosynthesis in many mammalian species[14]. The activities of CSAD in fetal liver and brain of humans, monkeys, rabbits, rats, guinea pigs and cats are lower than those of CSAD in adult tissues[21]. The CSAD activity is the lowest in man among six species. If taurine biosynthesis is limited in immature stages of various species, then a greater need for dietary or exogenous taurine may be necessary, because cystathionase and cysteine dioxygenase activity are also limited[24]. However, information regarding taurine intake in breast-feeding appears limited, in spite of the important role of taurine in infant nutrition. The objectives in this study were: (i) to find out any changes in taurine concentrations in human milk of nonvegetarians and lacto-ovovegetarians during the course of lactation; (ii) to find out daily taurine intakes of infants from nonvegetarians and lacto-ovovegetarians; and (iii) to provide basic data on taurine for Koreans in the field of obstetrics, pediatrics, and nutrition, and the infant formula industry.

Taurine 2, edited by Huxtable et al.
Plenum Press, New York, 1996

MATERIAL AND METHODS

Taurine concentrations were analyzed by amino acid autoanalyzer (LKB Alpha plus) from milk of nonvegetarians (n=22) in Seoul and lacto-ovovegetarians (n=23) from the Seventh Day Adventist Church in Seoul and its suburban areas. At 3, 5, 15, 30, 60, 90, 120, 150 days postpartum, milk intake of breast-fed infants was measured by the test weighing method[2,15]. Initial milk of 30 to 50 ml was collected by manual expression from 10 to 12 a.m. Taurine intakes of breast-fed infants per day were calculated by multiplying human milk intakes by taurine concentrations. Food intakes of mothers were measured by the weighing method. Statistical evaluation was performed by analysis of variance, t-test and Duncan's multiple range test.

RESULTS

Human Milk Intake

Human milk intakes in infants of nonvegetarians and lacto-ovovegetarians during the course of lactation are shown in Figure 1. The measurements of intake of infants at 3-5 days postpartum could not be done because of difficulty in test weighing by mothers during these days. As shown, infants of both groups consumed less at 15 days postpartum than any other times ($p < 0.05$). After 15 days, there were no significant day-to-day differences in human milk intake of infants within each group. No difference was shown between the two groups either.

Taurine Concentration of Human Milk

Figure 2 shows taurine concentration (nmol/ml) in nonvegetarians and lacto-ovovegetarians during the course of lactation. In both groups, the taurine concentrations of human milk decreased significantly ($p < 0.05$) during the course of lactation. Taurine concentrations in lacto-ovovegetarians after 90 days postpartum were significantly lower than those in nonvegetarians.

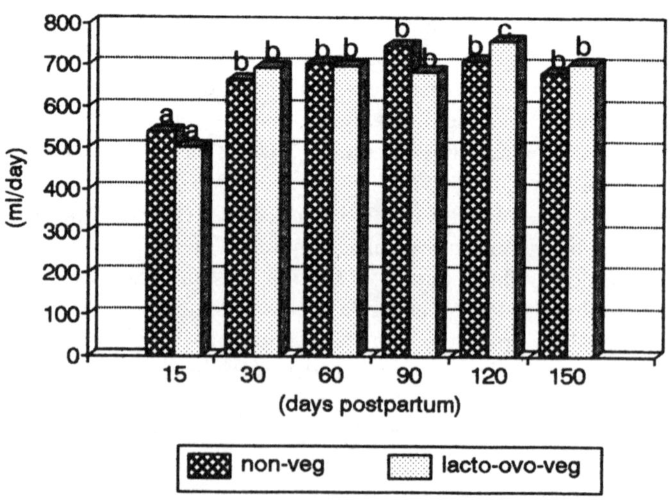

Figure 1. Human milk intakes in infants of nonvegetarians and lacto-ovovegetarians. Values shown are means. Values with a common superscript letter are not significantly different at the 5% level by Duncan's multiple range test.

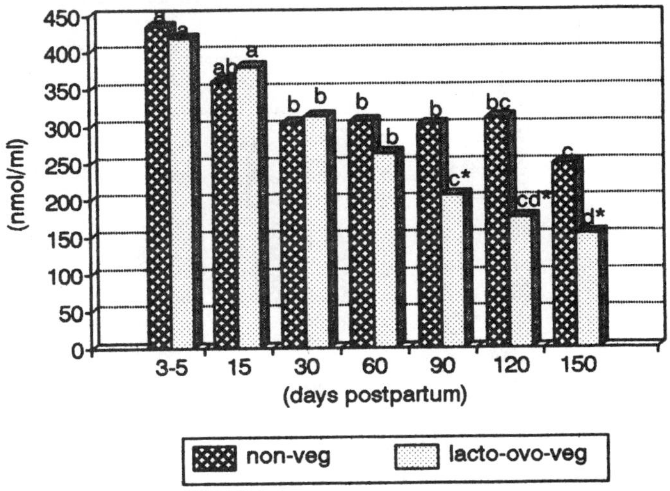

Figure 2. Taurine concentrations of human milk in nonvegetarian and lacto-ovovegetarians. Values shown are means. Values with a common superscript letter are not significantly different at the 5% level by multiple range test. *Taurine concentrations in nonvegetarians and lacto-ovovegetarians are statistically different at 5% level by t-test.

Taurine Intake of Breast-Fed Infants

Figure 3 shows taurine intake in infants of nonvegetarians and lacto-ovovegetarians. Taurine intakes in nonvegetarians were similar until 120 days. Taurine intakes in infants of lacto-ovovegetarians decreased gradually after 30 days. Taurine intakes in infants of lacto-ovovegetarians after 60 days postpartum were less than those in infants of nonvegetarians ($p < 0.05$).

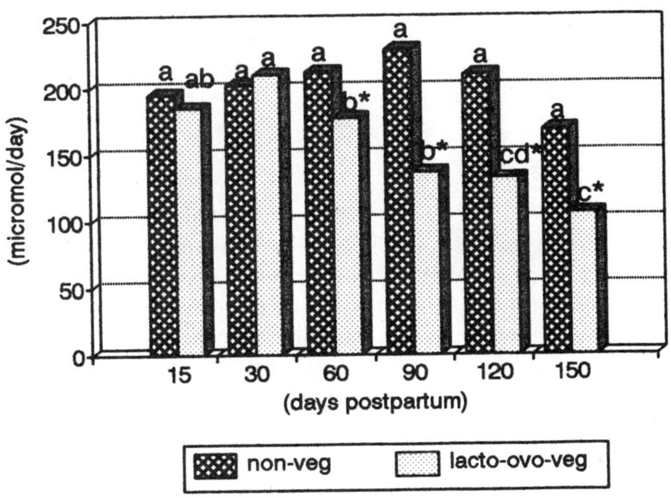

Figure 3. Taurine intakes in infants of nonvegetarians and lacto-ovovegetarians. Values shown are means. Values with a common superscript letter are not significantly different at the 5% level by Duncan's multiple range test. *Taurine concentrations in nonvegetarians and lacto-ovovegetarians are statistically different at 5% level by t-test.

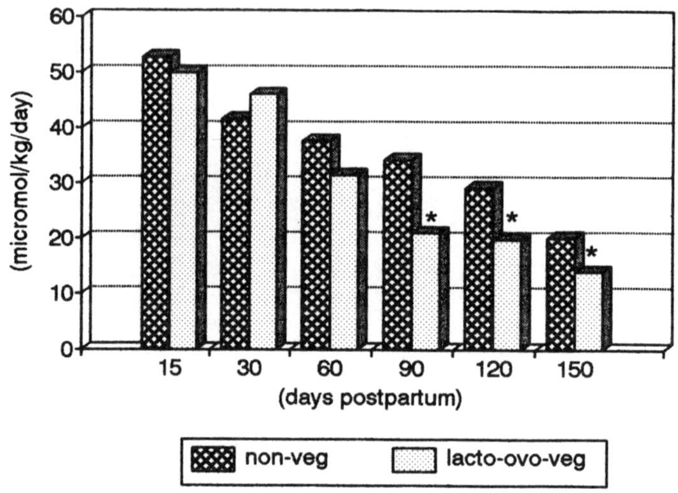

Figure 4. Taurine intakes per kg body weight of infants of nonvegetarians and lacto-ovovegetarians. *Taurine intakes in nonvegetarians and lacto-ovovegetarians are statistically different at 5% level by t-test.

Taurine Intake Per kg Body Weight of Breast-Fed Infants

Figure 4 shows taurine intake per kg body weight of infants of nonvegetarians and lacto-ovovegetarians. The intake per kg body weight during lactation decreased in both groups. However, taurine intake per kg body weight of infants of lacto-ovovegetarians after 90 days postpartum was less than that in infants of nonvegetarians ($p < 0.05$).

Maternal Intakes of Energy and Protein

Figures 5 and 6 show energy, protein and fat intakes of nonvegetarians and lacto-ovovegetarians during lactation. Energy intakes of non-vegetarians were higher at 30, 120 and 150 days postpartum than those of lacto-ovovegetarians. Protein intakes of lacto-ovovegetarians were higher than those of nonvegetarians, as we expected.

DISCUSSION

We have shown that taurine concentrations in milk of lacto-ovovegetarians are lower than those of nonvegetarians after 90 days postpartum. It is not clear why this only occurs after 90 days. However, it might be postulated that requirements for taurine can be met by metabolism of the methionine reserve in the body to cysteine and taurine for the first 90 days. After that period, the lower dietary intake of taurine by lacto-ovovegetarians might affect taurine concentrations in their human milk.

Rassin *et al.*[18] reported that, before 5 days postpartum, taurine content of human milk was 410 nmol/ml, but that 5 days postpartum it decreased to 340 nmol/ml. Shubat *et al.*[20] observed in their study that taurine content of colostrum in decreased gradually after attaining 400 nmol/ml. Furthermore, Erbersdobler *et al.*[3] reported that after 3-6 days postpartum, taurine content of human milk in 36 Germans was 400 nmol/ml. Rana and Sanders[17] observed that after 4-6 weeks postpartum, taurine content in milk of omnivores was 427 nmol/ml comparing with 277 nmol/ml in vegans in 14 British subjects. Taurine content in blood and

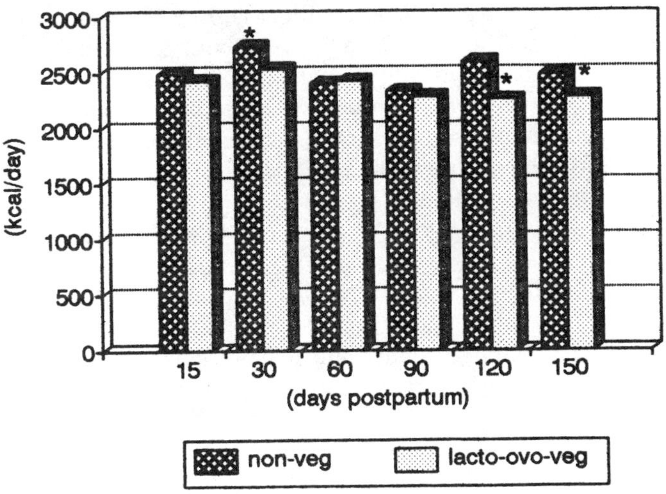

Figure 5. Energy intakes of nonvegetarians and lacto-ovovegetarians during lactation. Differences in energy intake between vegetarians and lacto-ovovegetarians are statistically different at the 5% level by t-test.

urine of vegans was lower than that of omnivores. These results are consistent with our findings in this study. However, other findings[8,16,19,22] are not in agreement with our findings. These differences might be due to limited sample size and the short duration of study period.

In the present study, taurine intakes in infants of lacto-ovovegetarians decreased significantly during lactation, in contrast to the pattern in nonvegetarians. Not much research has been done to determine taurine intakes of infant. More research using larger samples might be needed to determine such intakes.

Results from our studies showed that taurine intakes of infants was 13.3-26.3 mg/day in nonvegetarians and 21.1-28.6 mg/day in lacto-ovovegetarians. Thus, it seems that taurine content in formula (25 mg/100 g) in Korea might be insufficient for their needs for infants

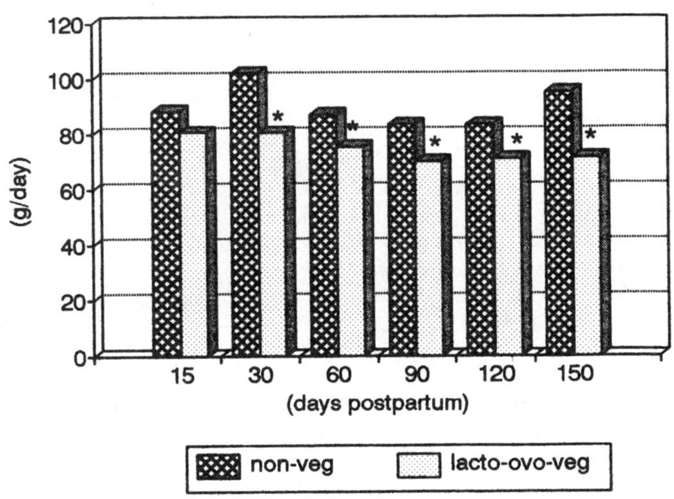

Figure 6. Protein intakes of nonvegetarians and lacto-ovovegetarians during lactation. Differences in protein intake between vegetarians and lacto-ovovegetarians are statistically different at the 5% level by t-test.

under 3 months of age. It has been suggested that taurine plays an important role in the development of brain[11,21] and in growth as a growth modulator[4,10]. It has also been reported that taurine deficiency is associated with abnormalities in the electroretinograms of patients maintained on long-term parenteral nutrition[7]. Because of the lack of enzyme for taurine biosynthesis in the neonatal period[21,24,25], precursor sulfur-containing amino acids can not serve to remedy dietary deficiencies in taurine. Special concern should be given to the area of formula diets for premature infants.

SUMMARY AND CONCLUSIONS

i. Taurine concentrations of human milk of nonvegetarians and lacto-ovovegetarians decreased significantly during the course of lactation. Taurine concentrations in lacto-ovovegetarians after 90 days postpartum were lower than those of nonvegetarians. Taurine concentrations of human milk by 150 days postpartum were: nonvegetarian 248-434 nmol/ml (31.0-54.4 mg/L); lacto-ovovegetarian 153-418 nmol/ml (19.1-52.3 mg/L).

ii. Taurine intakes in infants of lacto-ovovegetarians decreased significantly during lactation, unlike from the pattern in infants of nonvegetarians. Taurine intakes in infants of lacto-ovovegetarians at 90, 120 and 150 days postpartum were lower, compared with those in nonvegetarians. Taurine intakes in infants by 150 days postpartum were: Infants of nonvegetarians 169-229 μmol/day (21.1-28.6 mg/day); Infants of lacto-ovovegetarians 106-210 μmol/day (21.1-28.6 mg/day).

iii. This study showed that the taurine concentration in human milk and the taurine intake of infants were different during lactation between nonvegetarians and lacto-ovovegetarians, and that the intake per kg body weight decreased during lactation. Results suggest that the differences observed might be due to the effect of consumption of different types of food.

ACKNOWLEDGMENT

This work was supported by grants from Korea Research Foundation (KRF) and Pasteur Milk Company of Korea.

REFERENCES

1. Chesney, R. 1988, Taurine: Is it required for infant nutrition? *J.Nutr.* 118:6-10.
2. Choi, K.S. and Kim, E.S. 1991, A longitudinal study on human milk volume and lactational performance of Korean lacto-ovo-vegetarians, *Korn.J.Nutr.* 24:219-229.
3. Erbersdobler, H.F., Trautwein, E., and Greulich, H.G. 1984, Determination of taurine in milk and infant formula diets, *Eur.J.Ped.* 142:133-134.
4. Gaull, G.E. 1983, Taurine in human milk: growth modulator or conditionally essential amino acid? *Pediatr.Gastroenterol.Nutr.* 2:S266-S271.
5. Gaull, G.E., Rassin, D.K., Raiha, N.C.R., and Heinonen, K. 1977, Milk protein quantity and quality in low birth weight infants. III. Effects on sulfur-containing amino acids in plasma and urine, *J.Pediatr.* 90:348-355.
6. Geggel, H., S., Heckenlively, J.R., Martin, D.A., Ament, M.E., and Kollpe, J.D. 1982, Human retinal dysfunction and taurine deficiency, *Doc.Ophthalmol.Proc.Ser.* 31:199-207.
7. Geggel, H.S., Ament, M., Heckenlively, J.R., Martin, D.A., and Kopple, J.D. 1985, Nutritional requirement for taurine in patients receiving long-term parenteral nutrition, *N.Engl.J.Med.* 312:142-146.

8. Harzer, G., Franzke, V., and Birdels, J.G. 1984, Human milk nonprotein nitrogen components: changing patterns of free amino acids and urea in the course of early lactation, *Am.J.Clin.Nutr.* 40:303-309.
9. Hayes, K.C., Carey, R.E., and Schmidt, S.Y. 1975, Retinal degeneration associated with taurine deficiency in the cat, *Science*, 188:949-951.
10. Hayes, K.C., Stephan, Z.F., and Sturman, J.A. 1980, Growth depression in taurine-depleted infant monkeys, *J.Nutr.* 110:119-125.
11. Hayes, K.C. and Sturman, J.A. 1981, Taurine in metabolism, *Ann.Rev.Nutr.* 1:401-425.
12. Hayes, K.C. and Trautwein, E.A. 1994, Taurine, in: "Modern Nutrition in Health and Disease", 8th edition, Shils, M.E., Olson, J.A. and Shike, M. eds., Lea and Febiger, Philadelphia, pp. 477-485.
13. Huxtable, R.J. 1992, Physiological actions of taurine, *Physiol.Rev.* 72:101-163.
14. Jacobsen, J.G. and Smith, L.H. 1968, Biochemistry and physiology of taurine and taurine derivatives, *Physiol.Rev.* 48:424-511.
15. Lee, J.S. and Kim, E.S. 1991, A longitudinal study on human milk volume and lactational pattern, *Kor.J.Nutr.* 24:48-57.
16. Pamblanco, M., Portoles, M., Paredes, C., Ten, A., and Comin, J. 1989, Free amino acids in preterm milk from mothers delivering appropriate or small for gestational age infants, *Am.J.Clin.Nutr.* 50:778-781.
17. Rana, S.K. and Sanders, T.A.B. 1986, Taurine concentrations in the diet, plasma, urine and breast milk of vegans compared with omnivores, *Br.J.Nutr.* 56:17-28.
18. Rassin, D.K., Sturman, J.A., and Gaull, G.E. 1978, Taurine and other free amino acids in milk of man and other mammals, *Early Human Devel.* 2:1-13.
19. Sato, H., Tada, H., Yanagida, M., and Yamaguchi, K. 1983, The postnotal changes of trace nutrients in human breast milk, *Sulfur Amino Acids*, 6:297-300.
20. Shubat, P.J., Parker, D., and Huxtable, R.J. 1989, The effect of suckling and diurnal influences on the concentrations of taurine and other free amino acids in milk, *Eur.J.Clin.Nutr.* 43:675-680.
21. Sturman, J.A. and Hayes, K.C. 1980, The biology of taurine in nutrition and development, in: "Advances in Nutritional Research, Volume III", Draper, H.H. ed., Plenum Press, New York, pp. 231-299.
22. Svanberg, U.F., Gebre-Medhin, M., Ljungqvist, B., and Olsson, M. 1977, Breast milk composition in Ethiopian and Swedish mothers. III. Amino acids and other nitrogenous substances, *Am.J.Clin.Nutr.* 30:499-507.
23. Vinton, N.E., Laidlaw, S.A., Ament, M.E., and Kopple, J.D. 1987, Taurine concentrations in plasma, blood cells, and urine of children undergoing long-term total parenteral nutrition, *Pediat.Res.* 21:399-403.
24. Wright, C.E., Tallan, H.H., Lin, Y.Y., and Gaull, G.E. 1986, Taurine: biological update, *Ann.Rev.Biochem.* 55:427-453.
25. Yamaguchi, K. 1990, Nutrition and metabolism of sulfur amino acids, in: "Nutrition: Proteins and amino acids", Yoshida, A., Naito, H., Niiyama, Y. and Suguki, T. eds., Japan Sci. Soc. Press, Tokyo, pp. 165-183.
26. Zelikovic, I., Chesney, R.W., Friedman, A.L., and Ahlfors, C.E. 1990, Taurine depletion in very low birth weight infants receiving prolonged total parenteral nutrition: role of renal imaturitity, *J.Pediatr.* 116:301-306.

62

TAURINE AND DIABETES

Humans and Experimental Models

Flavia Franconi,[1] Mauro Miceli,[2] Alessandro Fazzini,[3] Giuseppe Seghieri,[2] Salvatore Caputo,[3] Mauro A. S. DiLeo,[3] Domenico Lepore,[3] and Giovanni Ghirlanda[3]

[1] Istituto di Chimica Biologica
Facoltà di Farmacia
Sassari, Italy
[2] Sezione Diabetologica
Spedali Riuniti
Pistoia, Italy
[3] Università Cattolica del Sacro Cuore
Roma, Italy

INTRODUCTION

Recent studies indicate that taurine levels are reduced both in plasma[3] and platelets obtained from patients with insulin-dependent diabetes mellitus (IDDM) in the absence of any complications of disease[3]. We have demonstrated that between plasma taurine levels and glycosylated hemoglobin levels (HbAlc) there is an inverse linear correlation indicating that taurine levels could be of some relevance in the control of metabolic state[3]. Moreover, when taurine levels are increased by taurine supplementation, the sensitivity of platelets obtained from IDDM patients to arachidonic acid is decreased[3].

The increased activity of platelets in patients with IDDM is considered to be a contributing factor to diabetic complications, including retinopathy[5,9]. It is well known that taurine deficiency is associated in cat, monkey, and human children with retinal alteration[4,6,11]. Therefore, we conducted experiments on human platelet-rich plasma (PRP) to evaluate the response of platelets to arachidonic acid at different levels of taurine. Taurine, *in vitro*, seems to potentiate the effect of anti-aggregating agents in platelets obtained from healthy controls. We also evaluated the response of platelets from IDDM patients to aspirin, a well known anti-aggregating agent. Moreover, in the streptozocin (STZ)-induced diabetic rat, retinal taurine uptake was studied at different times from the onset of diabetes.

Taurine 2, edited by Huxtable et al.
Plenum Press, New York, 1996

MATERIALS AND METHODS

Human Studies

Patients who fulfilled all criteria for IDDM as described by the National Data Group[1] and whose fasting baseline C peptide values were undetectable, were recruited as outpatients. The diet was assessed by means of a self-administered standard questionnaire. It remained relatively the same throughout the study period. IDDM patients were treated only with insulin and no alteration in serum cholesterol, HDL, LDL, triglycerides, serum or urinary uric acid, or plasma creatine levels was detected. The mean glycosylated haemoglobin (HbAlc) was 6.8 ± 0.03%, the mean glucose was 8.3 ± 0.4 mM, and the mean albumin excretion rate (AER) was < 20 mg/min.

Blood samples were obtained after overnight fasting. They were processed for taurine determination[8], platelet aggregation[2] and standard hematological and serum biochemical tests as described in Franconi *et al.*[3] at the beginning (basal value = B), after 3 months of taurine supplementation (T) and after 60 days washout (W). Taurine (1.5 g/day; p.o.) was administered three times a day (breakfast, lunch and dinner).

Arachidonic acid was used as aggregating agent for platelets and the dose-response curve was obtained. In some experiments the antiaggregating effect of aspirin was studied by preincubating PRP with different concentrations of aspirin before the addition of 1 mM arachidonic acid.

Animal Studies

Male Wistar rats, initially weighing 200 to 250 g, with free access to water were made diabetic with a single i.p. injection of streptozocin (60 mg/kg). Glycemia was monitored with chemstrips on 3 separate days, confirming the diabetic state on the day of experimentation. Normal adult male Wistar rats were used as control. After 15, 30, 60, 90 and 180 days of hyperglycemia, the STZ-D rats and those of a control group of the same size were anesthetized and killed with an overdose of pentobarbital and the bulbs were rapidly enucleated. After separation from the rest of the eyeball, the retinas were separated and incubated for 30 min in a rotary shaker with 4 ml of a buffered Krebs solution with a taurine concentration of either 0.1 mM or 0.01 mM. The solution also contained 20 μl of [^{3}H]taurine (32 mCi/mMol). Temperature (37°C) and light conditions were constant throughout all incubations. At 5-min intervals, 100 μl samples of the incubation medium were collected; radioactivity was measured as counts per minute (CPM) in each, using a liquid scintillation counter (460-C TRI-CARB, Packard Instruments). Reductions in levels of radioactivity in incubation medium were taken to reflect uptake of taurine by retinal tissue, although a portion of this "uptake" may be caused by simple diffusion of the marker into the extracellular space. The extracellular space estimation has been performed using [^{14}C]inulin as a marker: we found that [^{14}C]inulin counts in the medium remain constant, even while [^{3}H]taurine is taken up by the retina. At the end of the incubation period, the retinas were freeze-dried, weighed and fragmented. Radioactivity was then measured in the tissue homogenate itself, as described above. Results, expressed as nmol/mg dry weight, are the means ± SD of at least 10 incubations per group.

RESULTS AND DISCUSSION

The administration of taurine increased platelet and plasma taurine levels (Table 1). After 60 days of washout, taurine levels in the platelets are similar to basal levels. Plasma

Table 1. Taurine in plasma and platelets before (B) and at the end of taurine supplementation (T) and after 60 days of washout (W)

	Taurine Levels	
	Platelet (μmol/mg protein)	Plasma (μM)
B	0.64 ± 0.07	67.0 ± 0.1
T	0.99 ± 0.14*	126.8 ± 12.3*
W	0.58 ± 0.08	106.8 ± 6.7*

Values are means ± SE of 10 samples. *p < 0.001 versus basal.

Table 2. ED_{50} of arachidonic acid (mM) measured in PRP obtained from IDDM after taurine supplementation (1.5 g/day) and after 60 days washout

Basal	After taurine supplementation	Washout
0.44 ± 0.03	0.71 ± 0.06*	0.60 ± 0.05*

Values are means ± SE of 10 samples. *p < 0.001 versus basal.

taurine was higher than basal, although lower in comparison to levels at the end of supplementation. Platelets isolated from diabetics required less arachidonic acid to aggregate when taurine levels are low; taurine supplementation increased the ED_{50} of arachidonic acid (0.60 mM after 60 days washout) (Table 2).

When the mean ED_{50} of arachidonic acid was plotted against the mean plasma taurine levels, an inverse linear correlation was obtained ($y = -30.5+0.22x$; $r = 0.99$).

The dose-response of aspirin was measured in PRP obtained from patients after taurine supplementation and after 60 days of washout (Table 3). High plasma taurine levels increased platelet sensitivity to aspirin. In an experimental model of diabetes the retinal uptake of taurine was reduced. The reduction was a function of duration of diabetes (Fig. 1) and taurine concentration (data not shown). In IDDM patients both plasma and platelet taurine levels were decreased, as previously described[3]. This may increase platelet sensitivity to arachidonic acid. in fact, at higher taurine levels, platelets are less sensitive to arachidonic acid. Also, the action of aspirin seems to be affected by taurine levels, as is seen in healthy controls[10]. Moreover, Hayes *et al.*[7] demonstrated that in healthy subjects taurine supplementation decreased thromboxane formation.

Table 3. Dose-response curve of aspirin measured in PRP obtained from IDMM patients after taurine supplementation (1.5 g/day x 3 month) and after 60 days of washout

	Aspirin			
Taurine (μM)	0	1	10	50
126.8 ± 12.3	87.2 ± 3.9	48.9 ± 15.4	28.2 ± 14.3	2.0 ± 1.3
106.8 ± 6.7	94.7 ± 3.2	82.7 ± 8.3*	52.0 ± 17.9*	4.5 ± 2.1

Values (% of aggregation) are the means ± SE of at least 4 experiments. *p < 0.001

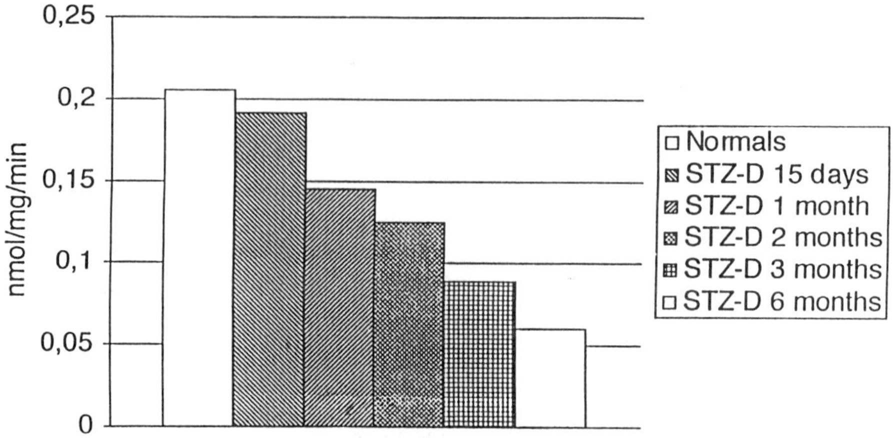

Figure 1. Taurine uptake in the retina of STZ-rats at different times from the injection of STZ. Values are the means of at least 5 experiments.

The decrease in platelet activity induced by taurine may lead to a reduction in diabetic complications such as micro- and macroangiopathies whose pathogenesis is associated with an increase in platelet aggregation[5]. In this context, taurine deficiency or decrease in taurine levels could lead to the emergence of retinopathy[4,6,11]. In our experimental model, the STZ rat, diabetic illness induces a decrease in taurine uptake in the retina. This could be important for diabetic retinal degeneration. Our findings encourage further investigation to focus on the role of taurine in IDDM patients.

REFERENCES

1. Born, G. V. R., and Cross, M. J., 1963, The aggregation of blood platelets, *J. Physiol.* **168**, 178.
2. Davi, G., Catalano, I., Arena, A., Notarbartolo, A., Strano, A., Ciabattoni, G., and Patrono, C., 1990, Thromboxane biosynthesis and platelet in type 2 diabetes mellitus, *N. Eng. J. Med.* **322**, 1769.
3. Franconi, F., Bennardini, F., Mattana, A., Miceli, M., Ciuti, M., Mian, M., Gironi, A., Bartolomei, G. C., Anichini, R., and Seghieri, G., 1995, Taurine deficiency in insulin-dependent diabetes mellitus (IDDM): Effect of taurine supplementation, *Am. J. Clin. Nutr.* **61**, 1115.
4. Geggel, H. S., Ament, M. E., Heckenlively, J. R., Martin, D. A., and Kapple, J. D., 1985, Nutritional requirement for taurine in patients receiving long-term parenteral nutrition, *N. Eng. J. Med.* **312**, 142.
5. Halushka, P. U., Rogers, R. C., Loadholt, C. B., and Colwell, J. A., 1981, Increased platelet thromboxane synthesis in diabetes mellitus, *J. Lab. Clin. Med.* **97**, 87.
6. Hayes, K. C., Carey, R. E., and Schmidt, S. Y., 1975, Retinal degeneration associated with taurine deficiency in the cat, *Science* **188**, 949.
7. Hayes, K. C., Pronezn, K. A., Addessa, A. E., and Stephan, Z. F., 1989, Taurine modulates platelet aggregation in cats and humans, *Am. J. Clin. Nutr.* **49**, 1211.
8. Jones, B. N., and Gilligan, J. P., 1983, O-phtalaldehyde pre-column derivatization and reversed-phase high performance liquid chromatography of polypeptide hydrolisates and physiological fluid, *J. Chromatogr.* **226**, 471.
9. Kaneti-Kaa, A., Goldberg, R. E., and Walsh, P. N., 1984, Platelet coagulant activities in diabetes mellitus: Evidences for platelet coagulant hyperactivity and platelet volume, *J. Lab. Clin. Med.* **103**, 82.
10. Miceli, M., Bennardini, F., Mattana, A., Cavarrubias, J., Saghieri, G., Franconi, F., 1992, Taurine modifies platelet sensitivity to aspirin, indomethacin, sodium nitroprusside and papaverine, *Platelets* **3**, 265.
11. Mustard, J. F., and Packman, M. A., 1984, Platelets and diabetes mellitus, *N. Eng. J. Med.* **311**, 665.
12. National Diabetes Data Group, 1979, Classification and diagnosis of diabetes mellitus and other categories of glucose intollerance, *Diabetes* **28**, 1039.
13. Sturman, J. A., Wen, G. Y., Wisniewski, H. M., and Neuringer, M. D., 1984, Retinal degeneration in primates raised on a synthetic human infant formula, *Int. Dev. Neurosci.* **2**, 121.

FIRST EXPERIMENTS IN TAURINE ADMINISTRATION FOR DIABETES MELLITUS

The Effect on Erythrocyte Membranes

E. P. Elizarova and L. V. Nedosugova

Cardiology Research Center
3-rd Cherepkovskaya st.
15-A, 121609, Moscow, Russia

INTRODUCTION

The hypoglycemic effect of taurine was discovered in 1935 by Ackerman and Heisen[1]. Dokshina *et al.*[3] found later in the seventies that taurine increased glucose consumption in diaphragm tissue and glycogen accumulation by leukocytes and liver. Kulakowski and Maturo[7] in the eighties also described insulin-like activity of taurine. Lampson, Kramer and Schaffer[5] found that taurine potentiated the effect of insulin and influenced activity of phosphorylase a and glycogen synthase I.

Usually in these articles effects of taurine were studied in acute experiments, and high doses of taurine were used. We have employed low doses of taurine to investigate its metabolic effects.

MATERIALS AND METHODS

Taurine was obtained from Kursk Medical Preparations Plant, Russia, and S-adenosyl-L-[methyl-^{3}H]methionine (15 Ci/mmol), from Amersham, UK.

Patient Characteristics

20 diabetic patients were under observation. All the patients were treated after clinical compensation. The first group (5 men and 5 women) consisted of patients with insulin-dependent diabetes (4 patients with severe diabetes implicated by nephropathy and diabetic retinopathy; 6 patients with diabetes of medium severity of damage). The second group (6 men and 4 women) consisted of patients with originally revealed diabetes. The patients from the first group were administered taurine tablets (0.5 g twice a day) against a background of

Taurine 2, edited by Huxtable et al.
Plenum Press, New York, 1996

insulin therapy. Average weight was 62 kg, average age 27 years old, and length of illness ranged from 6 months to 16 years. The median insulin dosage was 0.65 units per kg a day. Patients of the second group were administered taurine only. Average age of patients in this group was 52 years, and average weight was 72 kg. The sick ones had not strong vascular complications. Eight of 10 patients had atherosclerotic damage.

Preparation of Erythrocyte Ghosts

The methods of Steck and Kant[11] and Blostein and Chu[2] were used with modification for preparation of erythrocyte ghosts from blood of diabetes mellitus patients. Blood was centrifuged at 200 x g for 15 min for removal of leukocytes and platelets. Then erythrocytes were washed 3x with 10 volumes of 0.9% NaCl (15 min, 2000 x g). Ghosts were prepared by hemolysis of cells with 40 volumes of 0.5 mM Tris phosphate buffer, pH 8.0, containing 1 mM $MgCl_2$. Then ghosts were centrifuged for 50 min at 20000 x g. After centrifugation H_2O was added to the precipitate (1 vol ghosts to 1 vol H_2O) for 20 s. Ghosts were then diluted with 5 mM Tris phosphate buffer (pH 8.0) (1:25) and centrifuged for 50 min at 20000 x g. Then ghosts were washed 3x with 20 vol of 50 mM imidazole buffer, pH 6.5 (50 min at 20000 x g). After centrifugation the precipitate was suspended in imidazole buffer.

Assay of Phosphatide Methyltransferases

Phosphatide methyltransferase assays were performed the same day as membrane preparation. Assay medium consist of 100-200 μg protein; 10 mM $MgCl_2$; 0.2 μM S-adenosyl-L-[methyl-^{3}H]methionine; and 20 μg phosphatidylethanolamine (sonicated for 10 s). Where present, taurine (pH 6.5) was added in a concentration 12.5 mM. Incubation was carried out for 14-16 h at 4°C. Reaction was stopped by addition of 5 ml of chloroform/methanol/HCl (2:1:0.02, vol/vol). After addition of 2 ml of 0.1 M KCl in 50% methanol, the tube was centrifuged at 2000 x g for 10 min. The aqueous phase was aspirated, the chloroform phase was again washed with 2 ml of 0.1 M KCl in 50% methanol, and all the chloroform phase was transferred to a vial. After the solvent was evaporated to dryness at 80°C in N_2, 75 μl chloroform:methanol (1:2) was added. 30 μl was applied on plate (Kieselgel 60 F254). The chromatogram was developed in a solvent system of chloroform:propionic acid:n-propyl alcohol:water (13:13:19.5:6.5; vol/vol). Authentic phosphatidylcholine and its intermediates were chromatographed and the spots visualized by exposure to I_2 vapor or by spraying with 0.06% rhodamine 6G solution.

Triglycerides and cholesterol were determined by Technicon RA-XT, USA. Measurements of phospholipids were taken by the method of Kates[4].

Statistical analysis were performed by the Student's test: $p < 0.05$ was considered statistically significant.

RESULTS AND DISCUSSION

Clinical Data

A month after treatment the hypoglycemic profile changed, and average daily glycemia consequently decreased from 10.55 to 8.03 mM (Table 1). Amplitude of glycemic fluctuation also decreased. Glycosuria changed from 41 g to 8.5 g a day. Ideal compensation was reached in 50% of cases against a background of reduction of insulin dosage by 4-6 units. Unfortunately we did not test c-peptide. But we think that taurine did not stimulate insulin secretion, because most patients of this group had been ill for more than 2 years.

Table 1. Indicators of carbohydrate metabolism before and after course of taurine treatment in insulin-dependent diabetes

	Before treatment	After treatment
Average daily glycemia (mM)	10.55 ± 0.39	8.03 ± 0.04**
Amplitude of glycemic fluctuation (mM)	3.78 ± 0.48	2.81 ± 0.03*
Glycosuria (g/day)	41.08 ± 5.79	8.54 ± 2.73**

Data are for 10 patients. *$p < 0.05$; **$p < 0.001$

Table 2. Indicators of carbohydrate metabolism before and after course of taurine treatment in insulin-independent diabetes

	Before treatment	After treatment
Average daily glycemia (mM)	10.69 ± 0.58	8.01 ± 0.5*
Amplitude of glycemic fluctuation (mM)	5.16 ± 0.88	3.72 ± 0.81**
Glycosuria (g/day)	52.41 ± 8.87	18.28 ± 3.78*

Data are for 10 patients. *$p < 0.05$; **$p < 0.001$

Patients of the second group demonstrated similar performance of the preparation effect (Table 2). However, ideal compensation was achieved only in 2 patients.

Changes in triglycerides and cholesterol content of blood after a month's course of taurine treatment are represented on Fig. 1. Five patients were observed. Only triglyceride levels decreased reliably. Experiments in which triglycerides were estimated separately in the first and the second groups consisting of 9 patients supported significance of this data (date not presented). Changes in the one compensated patient were measured before treatment, and on day 14 and 30 (Fig. 1).

Lleu and Huxtable[6] have reported a negative correlation between phosphatidylethanolamine:lecithin ratio and taurine concentration in brain membranes. We investigated phospholipid content of erythrocyte membrane. After taurine treatment this ratio

n=5	Triglycerides	Total Chol.	Chol-HDL	Chol-LDL
Before	1.54±0.21	5.06±1.08	1.10±0.42	3.63±0.89
After	1.18±0.12	4.90±0.93	1.27±0.15	3.39±0.86

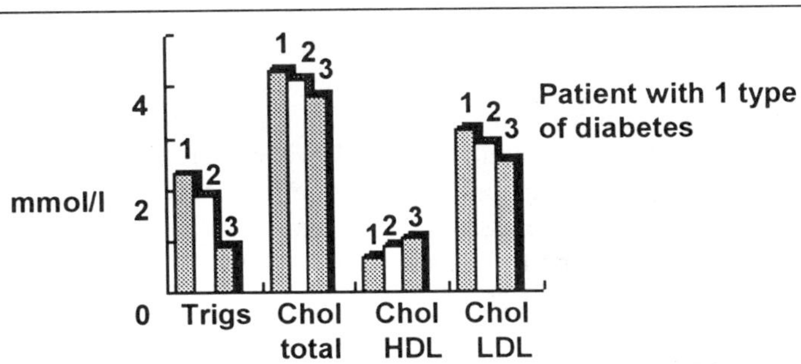

Figure 1. Effect of taurine on blood triglycerides and cholesterol in patients with diabetes mellitus before and after month course of taurine treatment. Bar graph shows data on one compensated patient, determined (1) before taurine treatment; (2) after 14 days treatment; and (3) after 30 days treatment.

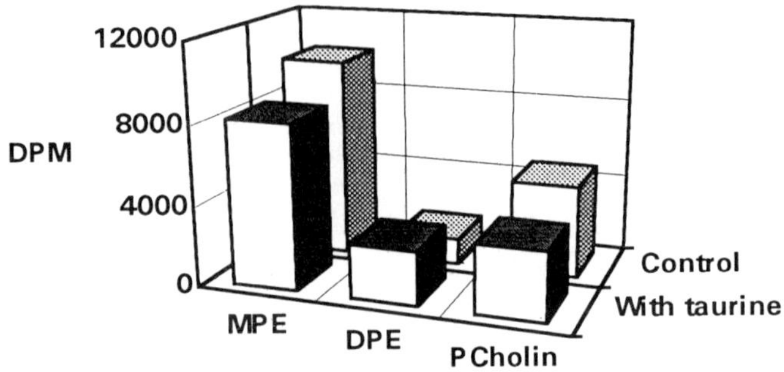

Figure 2. Effect of taurine on phospholipid methylation. MPE: Phosphatidylmonomethylethanolamine ($p < 0.01$); DPE: Phosphatidyldimethylethanolamine ($p < 0.01$); and PChol: Phosphatidylcholine ($p < 0.01$). 200 μg protein was incubated with S-adenosyl-L-[methyl-^{3}H]methionine and incorporation into each product determined. One of three representative experiments is shown.

decreased from 1.47 ± 0.29 (n=6) to 0.78 ± 0.36 (n=8). Thus, taurine must either decrease phosphatidylethanolamine level, or increase lecithin concentration.

Effects of Taurine on Phospholipid Methylation

Two methyltransferases exist in the membrane. This system catalyzes the conversion of phosphatidylethanolamine to phosphatidylcholine. There are three steps of methylation and each methylation requires optimal conditions to reach the highest rate of reaction. For example, Punna *et al.*[10] found taurine inhibited methylation at all three sites under three different conditions. But there are no ideal conditions in the cell for all enzymes (pH, concentration of S-adenosylmethionine, etc.). Furthermore, inhibition by taurine of the third stage of methylation (534 ± 31 pmol/mg/30 min without taurine and 319 ± 43 with taurine), may lead to the accumulation of the second product (dimethyl derivative of phosphatidylethanolamine).

We measured all three methylated products under similar conditions for the three sites. Taurine, under the experimental conditions described in Materials and Methods, produced a two-fold increase in dimethylethanolamine accumulation (Fig. 2). In these conditions taurine always decreased the level of phosphatidylmonomethylethanolamine and phosphatidylcholine, but sometimes did not influence or slightly decreased the dimethyl intermediate.

Some publications report the relation between phospholipid methylation and Ca^{2+}/Na^+ exchange. Dimethylethanolamine can inhibit Na^+/Ca^{2+} exchange[8,9]. Thus, taurine can either decrease or increase intracellular Ca^{2+} relative to the initial conditions depending on its effect on phospholipid methylation. However the question remains as to what way taurine increases phosphatidylcholine. As many effects of taurine depend on Ca^{2+}, we carried out the reaction in the presence of 2 mM Ca^{2+} (Table 3). Taurine increased not only dimethylethanolamine but phosphatidylcholine by more than 40%.

CONCLUSION

We propose that taurine must leave the cell in response to an appropriate stimulus. The released taurine influences the second stage of methylation (Fig. 3). As taurine may in

Table 3. Effect of taurine on the system of phospholipid methylation of the erythrocytes ghosts in the presence of calcium

	Products of Methylation (%)		
	MPhE	DPhE	PChol
Control	100%	100%	100%
Taurine	82.0 ± 6.0	137 ± 5.5	139 ± 5.3

Data are for 3 experiments with two parallel measurement in each experiment. Taurine 12 mM, Ca^{2+} 2 mM. MPhE: phosphatidylmonomethylethanolamine; DPhE: Phosphatidyldimethylethanolamine; PChol: Phosphatidylcholine.

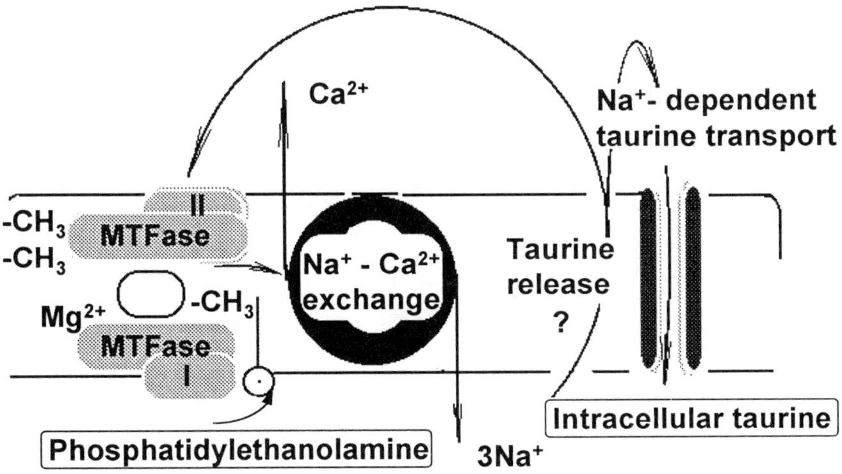

Figure 3. Proposed mechanism of taurine action. MTF: Phospholipid methyltransferase.

some conditions cause accumulation of phosphatidyldimethylethanolamine and under other conditions decrease its concentration, this may be part of one mechanism whereby taurine influences intracellular Ca^{2+} and hence other processes. This influence is therefore related to the membrane state. In congestive heart failure and diabetes, cellular release mechanisms for taurine may be impaired, leading to an accumulation of this substance.

REFERENCES

1. Ackermann, D. and Heinsen, H.A. 1935, Über die physiologische Wirkung des Asterubins und anderer zum Teil neu dargestellter, schwefelhaltiger Guanidinderivative, *Z.Physiol.Chem.* 235:115-121.
2. Blostein, R. and Chu, L. 1977, Sidedness of (sodium, potassium)-adenosine triphosphate of inside-out red cell membrane vesicles interaction with potassium, *J.Biol.Chem.* 252:3036-3043.
3. Dokshina, G.A., Yu-Silaeva, T., and Yartsev, E.I. 1976, Some insulin-like effects of taurine, *Vopr.Med.Khim.* 22:503-507.
4. Kates, M. 1988, "Techniques of Lipidology: Isolation, Analysis and Identification of Lipids", Elsevier, Amsterdam.
5. Lampson, W.G., Kramer, J.H., and Schaffer, S.W. 1983, Potentiation of the actions of insulin by taurine, *Can.J.Physiol.Pharmacol.* 61:457-463.
6. Lleu, P.-L. and Huxtable, R.J. 1992, Phospholipid methylation and taurine content of synaptosomes from cerebral cortex of developing rat, *Neurochem.Int.*, 21:109-118.

7. Maturo, J. and Kulakowski, E.C. 1987, Insulin-like activity of taurine, in: "The Biology of Taurine: Methods and Mechanisms", Huxtable, R.J., Franconi, F. and Giotti, A. eds., Plenum Press, New York, pp. 217-226.
8. McGivney, A., Crews, F.T., Hirata, F., Axelrod, J., and Siraganian, R.R. 1981, Rat basophilic leukemia cells defective in phospholipid methyltransferase enzyme, Ca^{2+} influx and histamine release: Reconstruction by hybridization, *Proc.Natl.Acad.Sci.USA*, 78:6176-6180.
9. Panagia, V., Makino, N., Ganguly, P.K., and Dhalla, N.S. 1987, Inhibition of Na^{+}-Ca^{2+} exchange in heart sarcolemmal vesicles by phosphatidylethanolamine N-methylation, *Eur.J.Biochem.* 166:597-603.
10. Punna, S., Ballard, C., Hamaguchi, T., Azuma, J., and Schaffer, S. 1994, Effect of taurine and methionine on sarcoplasmic reticular Ca^{2+} transport and phospholipid methyltransferase activity, *J.Cardiovasc.Pharmacol.* 24:286-292.
11. Steck, T.L. and Kant, J.A. 1974, Preparation of impermeable ghosts and inside-out vesicles from human erythrocyte membranes, in: "Methods in Enzymology, Vol. 31", Fleisner, S. and Parker, L. eds., Academic Press, New York, pp. 172-180.

64

EFFECTS OF TAURINE ON CALCIUM IN PLATELETS AND THEIR AGGREGATION

E. P. Elizarova

Cardiology Research Center
3-rd Cherepkovskaya Street
15-A, 121609 Moscow, Russia

INTRODUCTION

Taurine concentrations are 400-fold greater in platelets than in plasma[1,5]. The role of taurine in platelets has not been elucidated. For cardiologists, platelet taurine content is likely to have diagnostic value, since a decrease in taurine concentration in these cells indirectly reflects changes in taurine level in body tissues in general and in heart in particular[11]. Paasonen *et al.*[12] have reported that the taurine level in patients with hypertension does not differ from control values but increases in patients with myocardial infarction. In platelets, taurine transport decreases both in hypertension and in cardiovascular failure[9]. Marked losses of taurine from animals or humans affects the functional state of not only such organs as heart[13] and retina[8] but also of platelets. Taurine is known to model Ca^{2+} responses to PAF[3]. Taurine can inhibit PAF-induced aggregation of guinea pig platelets and aggregation of human platelets induced by ADP, serotonin and epinephrine[2]. Hayes *et al.*[9] reported the interrelation of taurine status and platelet aggregation in cat. They discovered that marked loss of taurine from platelets make them more sensitive to aggregation. These authors assume that this effect can be mediated via changes in GSH and altered thromboxane liberation. We suppose that this effect of taurine is not its only platelet action. Taurine may possess a homeostatic action that is revealed in certain pathological conditions. For this reason we have studied taurine effects in cells obtained from donors with different rhythm disturbances.

MATERIALS AND METHODS

Materials

The following reagents were used: taurine (Russia), bovine serum albumin, D-glucose, HEPES, PAF, ADP (Sigma, St Louis, MO, USA). Sources of other reagents are given in Avdonin *et al.*[4]

Taurine 2, edited by Huxtable et al.
Plenum Press, New York, 1996

Preparation of Platelets for Calcium Measurement

Venous blood samples were obtained from patients with coronary heart disease and/or arterial hypertension. The platelets were isolated and a fluorescent probe, quin-2, was incorporated into them[4,14]. Platelets were resuspended in medium containing (mM): 150 NaCl, 2.7 KCl, 0.37 NaH_2PO_4, 1 $CaCl_2$, 1 $MgCl_2$, 5 D-glucose, 10 HEPES (pH 7.4) and 0.35% bovine serum albumin. Fluorescence was measured by a Hitachi-850 spectrofluorimeter in a 1 ml thermostated cuvette at 30°C. The exciting wavelength was 340 nm and emission was 500 nm. Calcium concentration was calculated as described[10], and determined in the absence and presence of 10^{-7} M PAF or 10^{-5} M ADP. Stimulators induced nearly the same increase in cytosol Ca^{2+} in the cell (ΔCa^{2+}_{ind}). Taurine (10 mM) was added to the cuvette 1 min before addition of the inducer and its effect was assessed by the change of Ca^{2+} level in response to the taurine/inducer combination. Volumes added did not exceed 2% of the total volume of the medium.

ΔCa^{2+}_{taur} was defined by the difference in fluorescence levels ($\Delta Ca^{2+}_{taur+ind}$ - ΔCa^{2+}_{ind}). % Alteration of ΔCa^{2+}_{taur} was calculated using the formula:

$$\%\ \text{Change of}\ \Delta Ca^{2+}_{taur} = 100(\Delta Ca^{2+}_{taur}/\Delta Ca^{2+}_{ind})$$

Significance of the differences was determined using Student's t-test and correlative analysis.

Evaluation of Platelet Aggregation

Platelet aggregation was studied by a new optical method with the platelet aggregation analyser (PAA) (Biola Company) developed by the Cardiology Research Center, Moscow, Russia[6,7]. The method is based on the changes in light transmission fluctuations produced by changes in the number of platelets in a laser optical channel. PAA is a computer-controlled instrument unique in its ability to measure the average size of aggregate in real-time. All aggregation curves are stored and can be used later. Standard light transmission curves can be recorded simultaneously.

Blood from patients was obtained from the cubital vein by a butterfly needle (19 g). Blood was collected in 0.13 M sodium citrate (pH 7.4)[1,9]. Platelet-rich plasma (PRP) was prepared by centrifugation (200 x g, 10 min). Platelet-poor plasma (PPP) was prepared by recentrifugation of blood at 2000 x g for 20 min. The resulting supernatant contained 500 platelets/μl and was used to dilute autologous PRP.

The relative volume of platelets in PRP was determined by centrifugation in a special microcentrifuge cuvette with a graduated capillary at the bottom. 0.5 ml PRP was centrifuged with 0.05 ml of 100 mM EGTA at 2000 x g for 15 min. To exam platelet aggregation, relative platelet volume was adjusted to 0.25% with autologous PPP. All procedures were carried out at room temperature within 3 h of obtaining blood samples. Aggregation response was estimated at the upper point of the aggregation curve in relative units (r.u.).

We have investigated ADP-induced aggregation with final concentration of inducers of 0.5 μM and 5 μM[15,16]. Taurine (3 mM) was added before addition of ADP.

RESULTS

Modulation of Receptor-Dependent Increase of Ca^{2+} in Human Platelets by Taurine

Saturating concentrations of inducers led to an increase in concentration of cytosolic Ca^{2+} (ΔCa^{2+}_{ind}) in all 22 experiments. Increases ranged from 70 to 2510 nM (Table 1).

Table 1. Taurine effect on receptor-dependent increase of Ca^{2+} in human platelets

Number of Exps	ΔCa^{2+}_{ind} (nM)	$\Delta Ca^{2+}_{taur+ind}$ (nM)	% Change in platelet response in the presence of taurine[a]
PAF as inducer			
9	739.3 ± 224.2	476.8 ± 171.1	61.3 ± 6.7
8	289.6 ± 27.4	246.6 ± 28.9	88.0 ± 4.0
5	88.4 ± 14.2	170.2 ± 12.6	194.2 ± 14.4
ADP as inducer			
10			59.3 ± 8.5
8			89.5 ± 3.5
3			175.0 ± 9.6

Change in Ca $^{2+}$ concentration (nM) to PAF or ADP (ΔCa^{2+}_{ind}); [a]Change in platelet response to PAF (%) or ADP in the presence of 10 mM taurine; Data are means ± SEM.

Physiological concentrations of taurine caused no spontaneous changes in fluorescence of platelet suspensions. Cell preincubation with taurine led both to decreases (16 cases) and increases (6 cases) in response to inducers. The mean data of 22 independent experiments revealed no marked effect of taurine on inducer action. In the presence and absence of taurine, Ca^{2+} response of cell to stimulator was 427 ± 109 and 323 ± 23 nM, respectively ($p > 0.5$).

A more detailed examination revealed that the effect of taurine depended on the extent of Ca^{2+} increase in platelets in response to the inducer (Table 1). When the value was rather high (ΔCa^{2+}_{ind} >400 nM), Ca^{2+}-blocking effects of taurine were observed in all 9 experiments, the mean decrease of the response to the inducer being 262 ± 67 nM. When the initial response to the inducer was rather low (ΔCa^{2+}_{ind} < 150 nM), in all 5 cases the stimulating effect was found to enhance the response to the inductor by 81± 11 nM.

Taurine did not change substantially the cell response to stimulator if a rise in Ca^{2+} level induced by the inducer ranged from 150 to 400 nM.

Difference in alteration of $\%\Delta Ca^{2+}_{taur,}$ which defines the degree of taurine impact on inducer effect, depending on the initial response of the cell to hormone, are presented in Fig. 1. When platelet reaction to the inducer was high initially, taurine inhibited the increase in Ca^{2+} level by 39 ± 7% while at low reaction it enhanced this response by 94 ± 16% ($p < 0.001$). The results of all experiments gave a significant negative correlation between effect of taurine on the cell response to the inducer (both an increase and a decrease) and the level of the initial response of the platelets to the inducer (r 0.88; $p < 0.001$).

Modulation of Platelet Aggregation by Taurine

19 patients (7 women and 12 men) aged 23-63 were examined. All the patients suffered from cardiac rhythm disturbances. We introduced the term "spontaneous aggregation" and considered platelets to have aggregation if the slope of the curve describing the process of aggregation without ADP addition in the conditions of mixing was more than 0.1 and marked the platelet ability as (SA^+) (Fig. 2a). If the slope of the curve was less than 0.1 these cells were considered not have spontaneous aggregation and were marked as (SA^-) (Fig. 2b). The cells of nine of the patients failed to exhibit spontaneous aggregation. Platelets of the other 10 patients possessed spontaneous aggregation. One of the patients with hypertension showed a changed kinetics of aggregation with two bends. Responses to ADP varied markedly between patients (Table 2).

The effect of taurine on aggregation did not depend on the absolute value of platelet response to inducer (Table 3). Thus, taurine increased a small response by 0.5 μM ADP

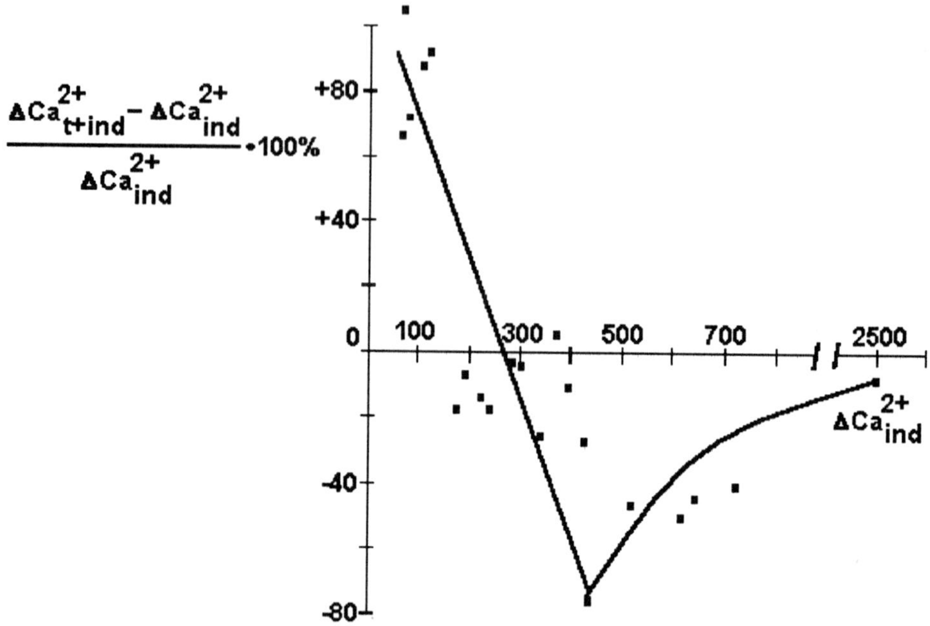

Figure 1. Correlation between Ca^{2+}-modulating effect of taurine and platelet response to PAF. Experimental conditions are described in Material and Methods. Change in Ca^{2+} concentration (nM) effected by PAF (ΔCa^{2+}_{ind}). Effect of taurine on response (%): $100([\Delta Ca^{2+}_{t+ind} - \Delta Ca^{2+}_{ind}]/\Delta Ca^{2+}_{ind})$.

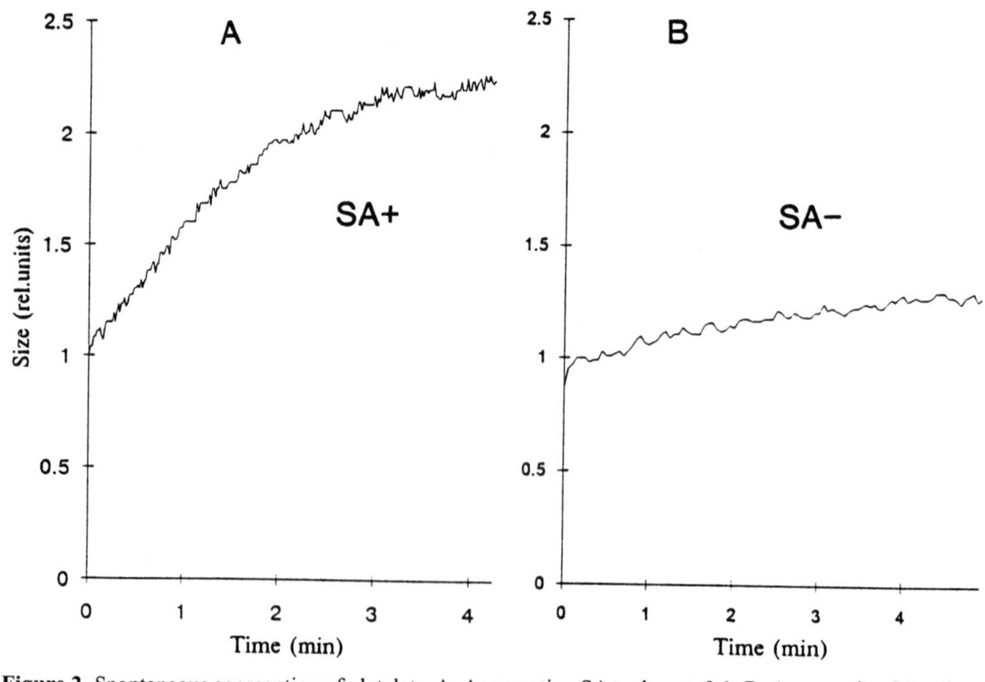

Figure 2. Spontaneous aggregation of platelets. A: Aggregation SA+, slope > 0.1; B: Aggregation SA-, slope < 0.1.

Table 2. ADP effect on platelet aggregation

	Relative Units of Aggregation		
ADP (μM)	Patient N1	Patient N2	Patient N3
5	9.7	14.2	22.6
2.5	12	15.9	21.9
1.0	9.9	13.9	21.3
0.5	2.2	3.4	10.9
0.1	1.1	0.9	4.5

Table 3. Taurine effect on ADP-stimulated response of platelets.

	Relative Units of Aggregation		
Conditions	Patient N1	Patient N7	Patient N8
ADP, 05 μM	2.2	10.9	2.8
ADP+ 3 mM taurine	4.3	7.2	1.3

(patient N8) and inhibited a substantial response (patient N7). The effect of taurine on the induced platelet response was also independent of the time of exposure (Table 4).

Addition of taurine (3 mM) to SA^- platelets increased ADP-stimulated aggregation, while addition of taurine to SA^+ platelets decreasedADP-stimulated aggregation. Data for the effect of taurine on platelet aggregation are reported for addition of 0.5 μM ADP. However, similar results were obtained with 5 μM ADP. In eight of the nine patients without spontaneous aggregation, addition of taurine induced a strengthened aggregation response to ADP. In eight patients of ten taurine decreased aggregational response of platelets to ADP (Fig. 3). It appeared that the effect of taurine correlated with either the presence or absence of spontaneous aggregation (r 0.75; $p < 0.001$).

DISCUSSION

The modification by taurine of the ADP- and PAF-induced Ca^{2+} responses of platelets implies that taurine could similarly change physiological activity typical of platelets (aggregation). Such effects have been demonstrated. Most likely, a single mechanism of action underlies both biochemical and physiological effects. The effect of taurine on platelet Ca^{2+} is reminiscent of the effect of taurine on the Ca^{2+} binding isotherm for cardiac ventricular sarcolemma[10]. Probably, the effects which were described in this article are mediated via the

Table 4. Effect of taurine on induced response of platelets depends on the exposure time

	Relative Units of Aggregation		
Time of incubation with taurine, min	Patient N4	Patient N5	Patient N6
0.5	6.9	2.1	9.5
1	6.5	2.5	8.9
3	9.7	1.7	9.4
5	11.5	1.7	9.6

Taurine was added 0.5-5 min prior to ADP in cuvette with platelets

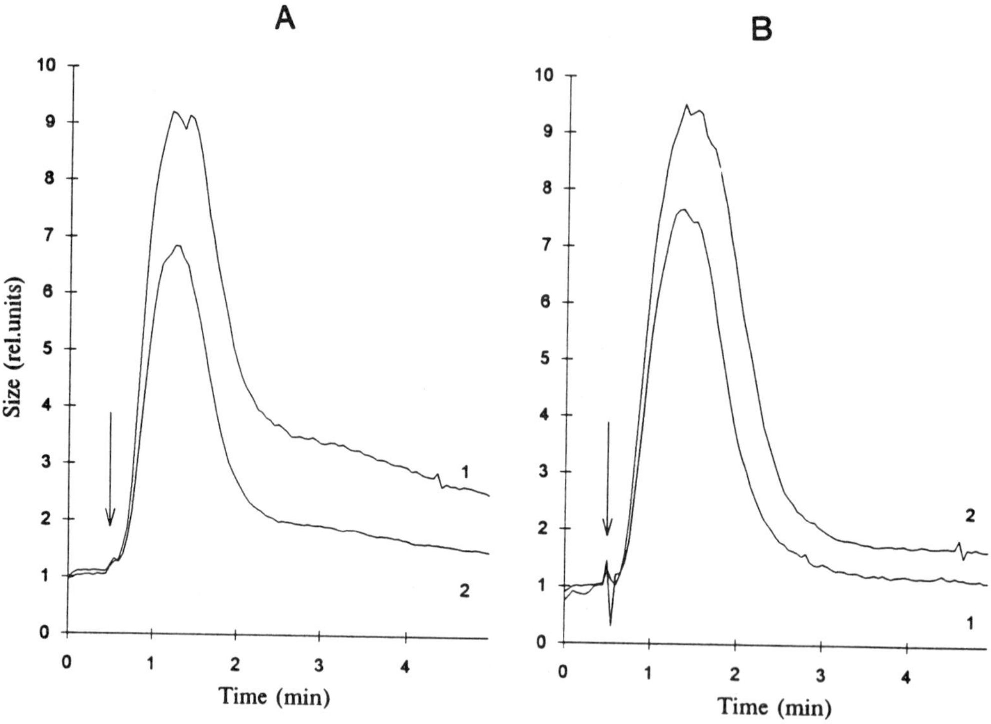

Figure 3. Effects of taurine on platelet aggregation. A: Platelets with spontaneous aggregation (SA+); B: Platelets without spontaneous aggregation (SA-). 1-0.5 μM ADP; 2-0.5 μM ADP + 3 mM taurine.

two membrane sites. Phospholipids may play an important role, as was reported in the article by Kulakowski *et al.*[10], because phospholipase C inhibited taurine binding. It is interesting to extend our researches with inhibitors of binding. Unfortunately, spontaneous aggregation has not been investigated closely enough to allow discussion of the functional and biochemical differences between platelets (SA^+) and (SA^-) and the relationship of this difference to the action of taurine. Platelets are convenient for studying membrane actions of taurine, as aggregation is a readily observed endpoint that is determined by the membrane state.

REFERENCES

1. Ahtee, L., Boullin, D.J., and Paasonen, M.K. 1974, Transport of taurine by normal human blood platelets, *Br.J.Pharm.* 52:245-251.
2. Almazov, V.A., Gurevich, V.S., Mikhailova, I.A., and Strelstova, E.N. 1986, Effect of taurine on Ca,Mg-ATPase activity and platelet aggregation in man, *Exp.Biol.Med.* 100:1314-1317.
3. Atahanov, S.E. and Elizarova, E.P. 1992, Modulation of receptor-dependent increase of calcium ions in human platelets by taurine, *Arzneimittelforschung*, 42:1311-1313.
4. Avdonin, P.V., Menchikov, M.Y., Orlov, S.N., Pokudin, N.I., and Tkachuk, V.A. 1985, Mechanisms of elevation of Ca^{2+} in platelet cytoplasm induced by aggregation factors, *Biochemistry (Russ.)*, 50:1055-1062.
5. Frendo, J., Koj, A., and Zgliczynski, J.M. 1959, Taurine in human blood platelets, *Nature*, 183:685-686.
6. Gabbasov, Z.A., Popov, E.G., Gavrilov, I.Y. and Pozin, A.Y. 1989, A new methodological approach to in vitro study aggregation, *Byull.Eksp.Biol.* N 10:437-440.

7. Gabbasov, Z.A., Popov, E.G., Gavrilov, I.Y. and Pozin, A.Y. 1989, Platelet aggregation: the use optical density fluctuations to study microaggregate formation in platelet suspension, *Thromb.Res.* 54:215-223.
8. Hayes, K.C., Carey, R.E., and Schmidt, S.Y. 1975, Retinal degeneration associated with taurine deficiency in the cat, *Science*, 188:949-951.
9. Hayes, K.C., Pronczuk, A., Addesa, A.E., and Stephan, Z.F. 1989, Taurine modulates platelet aggregation in cats and humans, *Am.J.Clin.Nutr.*, 49:1211-1216.
10. Kulakowski, E.C., Maturo, J., and Schaffer, S.W. 1978, The identification of taurine receptors from rat heart sarcolemma, *Biochem.Biophys.Res.Comm.*, 80:936-941.
11. Orlova, T.R., Yakushkin, V.V., and Elizarova, E.P. 1990, Taurine content in blood and myocardium of rabbits with experimental heart failure, *Physiol.J.*, 35:3-7.
12. Paasonen, M.K., Penttila, O., Himberg, J.J., and Solatunturi, E. 1980, Platelet taurine in patients with arterial hypertension, myocardial failure or infarction, *Acta Med.Scand.* 642(Suppl.):79-84.
13. Pion, P.D., Kittleson, M.D., Rogers, Q.R., and Morris, J.G. 1987, Myocardial failure in cats associated with low plasma taurine: a reversible cardiomyopathy, *Science*, 237:764-768.
14. Tsein, R.Y., Pozzan, T., and Rink, T.J. 1982, Calcium homeostasis in intact lymphocytes: Cytoplasmic free calcium monitored with a new intracellularly trapped fluorescent indicator, *J.Cell.Biol.* 94:325-334.
15. Yurenev, A., Balyakina, E., Dubov, P., Popov, E., Gabbasov, A.S. and Ataschanov, S. 1992, Laser aggregometry in hypertensives with and without left ventricular hypertrophy, *Am.J.Hypert.* 5:25a.
16. Yurenev, A.V., DeQuabbro, Dulov, P., Ostroumov, N., Niculin, I., Konyaeva, E., Balyakina, E., Atakhanov, S., Kosenka, A., Popov, E., Gabbasov, Z., and Gulyaev, V. 1992, Silent myocardial ischemia in patients with essential hypertension, *Am.J.Hypert.* 5:169s-174s.

PLASMA TAURINE IN LIVER CIRRHOSIS WITH PAINFUL MUSCLE CRAMPS

Shinichiro Yamamoto

Division of Gastroenterology
Kawasaki Medical School
577 Matsushima, Kurashiki 701-01, Japan

INTRODUCTION

Painful muscle cramps (MC) are frequently seen in liver cirrhosis (LC) patients. Matsuzaki *et al.*[5] reported that oral taurine administration helped to relieve MC in LC patients. We previously reported that in 34 out of 38 cases of LC (89.5%) patients oral taurine relieved MC[6].

Clinically there is little doubt about the effectiveness of taurine for the improvement of MC in LC. However, the precise mechanism of MC is unknown. In this study we examined the plasma taurine concentrations in LC patients with recurrent painful muscle cramps and compare them with those in LC patients without MC. Changes of plasma taurine concentrations following oral taurine ingestion was also compared in eight cases of LC with MC.

MATERIALS AND METHODS

Forty-eight patients with LC and twenty-two healthy controls were the subjects of this study. The LC patients were divided into two groups, 31 cases with MC and 17 cases without MC. Blood samples were collected for amino acid analysis. The plasma taurine concentration was determined using fluorescence detection with o-phthaldehyde after separation by high performance liquid chromatography (HPLC, Hitachi KLA 5)[3]. A generalized Wilcoxon test was used for statistical evaluation. The plasma taurine concentration was measured in eight LC patients with MC before and after oral administration of taurine in doses of 3 g/day for 4 weeks.

RESULTS

Plasma Taurine Concentration in LC with MC

As shown in Fig. 1, the plasma taurine concentration in LC cases with MC was 64.0 ± 26.2 μM. In LC patients without MC and normal controls, the concentrations were 73.0 ±

Taurine 2, edited by Huxtable et al.
Plenum Press, New York, 1996

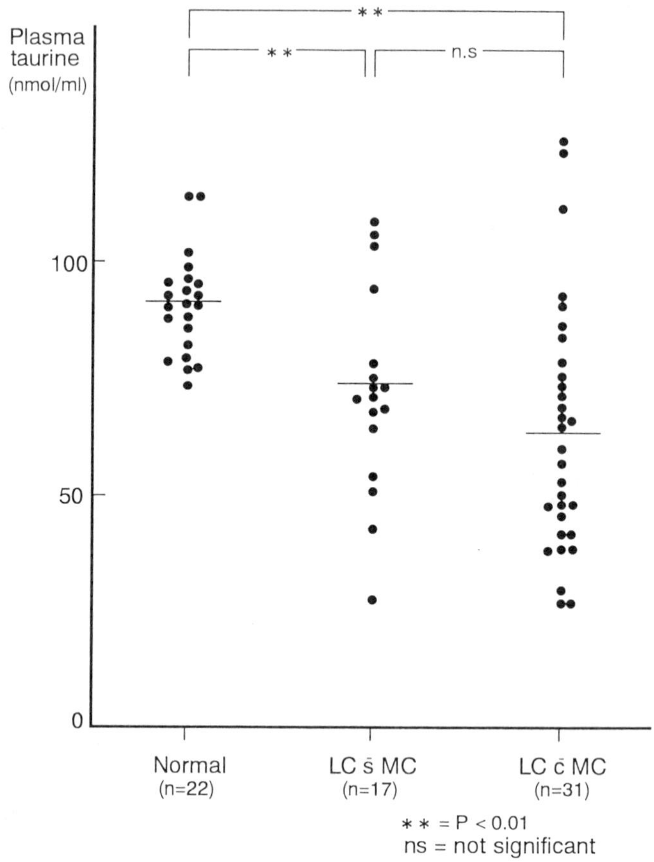

Figure 1. Plasma taurine concentrations in liver cirrhosis (LC) with or without muscle cramps (MC).

21.9 μM and 91.3 ± 10.6 μM, respectively. In LC patients with MC, the plasma taurine level was significantly lower than the controls. There was no significant difference between the MC(+) group and the MC(-) group.

Plasma Taurine Concentration after Oral Ingestion of Taurine

Taurine was administered orally to eight patients with MC in doses of 3 g/day for 4 weeks. Muscle cramps disappeared completely in five of these patients, while the frequency of MC decreased remarkably in the other three patients. Fig. 2 shows that the mean plasma taurine concentration for these patients was 54.1 ± 20.7 μM before ingestion of taurine and 125.1 ± 59.1 μM after ingestion.

DISCUSSION

There have been no reports, to our knowledge, concerning plasma taurine concentrations in LC patients with muscle cramps. Our data provide evidence that the plasma taurine concentration decreased significantly in LC patients with MC. There have been a few reports concerning the plasma taurine concentration in LC patients. Ganda *et al.*[2] reported that the

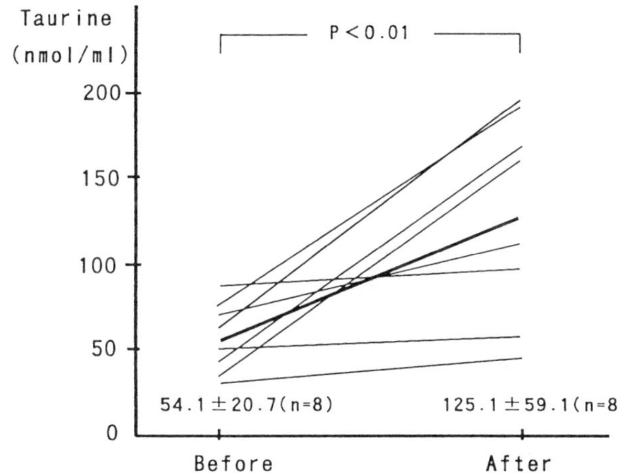

Figure 2. Changes of plasma taurine concentrations before and after oral taurine ingestion in 9 cases of LC with MC.

whole blood taurine level was increased in nine patients with decompensated liver cirrhosis. These authors stated that the increase in taurine might have occurred because they used whole blood rather than plasma, since the bulk of taurine in blood is located in red cells. Kawade *et al.*[4] examined free amino acids in 13 normal controls, ten compensated LC patients and 21 decompensated LC patients. The mean plasma taurine levels of the normal controls was 63 ± 20 μM, whereas those of the compensated LC and decompensated LC patients were 62 ± 16 μM and 61 ± 24 μM, respectively. The plasma taurine levels in normal controls differ from report to report. We obtained normal levels of 90.1 ± 12.1 μM, whereas those reported by Kawade *et al.* and Ganda *et al.* were 63 ± 20 μM and 154 ± 19 μM, respectively. These differences could be explained by differences in reagents, operational conditions or flow methods. Physiological fluctuations in the taurine concentration are small, but fasting is known to increase the plasma taurine level[4]. We also showed that oral administration of taurine relieved MC with an increase in the plasma taurine concentration. Durelli *et al.*[1] evaluated the oral administration of taurine to nine patients with myotonic dystrophy and noted subjective improvement in all these patients. The plasma taurine concentration increased from 53.4 ± 22.9 μM to 111.1 ± 42.4 μM after 6 months administration of taurine orally in doses of 100 to 150 mg/kg body weight. A stable antimyotonic effect may be maintained by a moderate increment of the plasma taurine concentration. A large-scale study is necessary, however, to evaluate whether there is a really significant decrease in the plasma taurine concentration in liver cirrhosis with recurrent painful muscle cramps.

REFERENCES

1. Durelli, L., Mutani, R., and Fassio, F. 1983, The treatment of myotonia - evaluation of chronic oral taurine therapy, *Neurology*, 33:599-603.
2. Ganda, O.P. and Ruderman, N.B. 1976, Muscle nitrogen metabolism in chronic hepatic insufficiency, *Metabolism*, 25:427-435.
3. Ganno, S. 1978, High speed amino acid analysis, *Kagakuno-Ryouiki*, 121:1-13.
4. Kawade, Y. and Takahashi, Y. 1979, Measurements of plasma free amino acids, *Nihon Rinsho*, 37:1708-1711.

5. Matsuzaki, Y., Tanaka, N., Yamaguchi, T., Chuganji, Y., Nishi, M., Chiba, T., Osuga, T., and Komatsu, Y. 1990, Effect of taurine administration for muscle cramp in chronic liver disease — a case of decompensated liver cirrhosis in which muscle cramp disappeared after oral administration of taurine, *Acta Hepatol.Jpn.* 31:1464-1469.
6. Yamamoto, S., Ohmoto, K., Saito, I., Ideguchi, S., Yamamoto, R., Mitsui, Y., Shimabara, M., Takatori, K., Ohumi, T., Hino, K., and Hirono, Y. 1993, Effect of taurine administration for muscle pain associated with liver cirrhosis and combined use of Arofuto, *Jpn.Pharmacol.Ther.* 21:1157-1161.

66

TAURINE IN HEPATIC ENCEPHALOPATHY

Roger F. Butterworth

Neuroscience Research Unit
Hopital Saint-Luc (University of Montreal)
Montreal, Quebec, Canada

ABSTRACT

Liver and brain are the major organs responsible for taurine synthesis. In both acute and chronic liver failure, brain taurine concentrations are decreased and, since taurine appears to be implicated in K^+ and Ca^{2+} homeostasis in brain, such losses could contribute to the pathophysiology of hepatic encephalopathy. Furthermore, taurine concentrations in cerebrospinal fluid in experimental acute liver failure are increased early in the progression of encephalopathy and prior to the onset of cerebral edema, a potentially fatal complication of acute liver failure. These findings suggest an osmoregulatory role for taurine in brain in acute liver failure. Monitoring of cerebrospinal fluid taurine may be of prognostic value in this severe, frequently fatal disorder.

INTRODUCTION

Hepatic Encephalopathy (HE) is a serious complication of acute and chronic liver diseases. Neurological symptoms progress from altered personality traits to neuromuscular incoordination, stupor and coma. There are two major sub-types of HE (Table 1). Acute liver failure resulting from viral infection or toxins results in a fulminating illness with a high (70-80%) mortality rate. In acute liver failure, HE progresses rapidly to coma within hours or days; death most commonly results from brain herniation as a consequence of increased intracranial pressure caused by cerebral edema. The most effective treatment for this condition is liver transplantation. The second, more common form of HE, known as portal-systemic encephalopathy (PSE), results from chronic liver disease, typically alcoholic cirrhosis. Progression of PSE is slow, mortality relatively low (~20%) and brain edema is a rare complication.

It has been suggested that HE in both acute and chronic liver failure results from an imbalance between excitatory and inhibitory neurotransmitter systems in the central nervous system. Modifications of glutamatergic[5] and monoaminergic[2,13] systems have been described.

Taurine 2, edited by Huxtable et al.
Plenum Press, New York, 1996

Table 1. Classification of the hepatic encephalopathies

HE in Acute Liver Failure	
	Results from viral infection or drug toxicity
	High mortality (70-80%)
	Brain edema a common finding
Portal-systemic Encephalopathy (PSE)	
	Occurs in chronic liver disease (alcoholic and viral)
	Precipitating factor common (eg: gastrointestinal bleeding)
	Brain edema uncommon

Taurine is one of the most abundant free amino acids within the brain and is localized in both neuronal and glial compartments. Taurine has been proposed as a classical neurotransmitter or neuromodulator. However, the concentration of taurine in the central nervous system is far in excess of that necessary for such functions; consequently, other putative functions for taurine have been proposed. Such functions include a membrane stabilizing action, and antioxidant osmoregulatory roles[29]. It is well established that taurine is important in osmoregulation in invertebrates. Changes in tissue taurine concentration have been described in numerous euryhaline invertebrate species (i.e: species that conform with the osmolarity of the surrounding environment[7]). Extracellular taurine concentrations in mammalian brain are increased in conditions characterized by cellular swelling. Such conditions include seizures[26], transient cerebral ischemia[1,9] and hypoglycaemic coma[23]. This report summarizes evidence that modifications of taurine may be implicated both in the pathogenesis of HE and of brain edema in acute liver failure.

TAURINE AND HEPATIC ENCEPHALOPATHY

Liver and brain are the major organs responsible for taurine synthesis. Taurine is transported from the liver to other organs (including brain) in order to maintain tissue levels. Liver taurine concentrations are in the 2-10 mM range and experimental cirrhosis resulting from CCl_4 hepatotoxicity results in a dose-dependent increase in urinary taurine[28]. Thus, acute or chronic liver disease could result in alterations of brain taurine concentrations and, in view of the neuromodulatory properties of taurine, contribute to HE. Recent reports have indeed described significant alterations of brain taurine in HE. Autopsied brain tissue from patients with chronic liver disease who died in hepatic coma contains significantly decreased taurine concentrations (Fig. 1) and brain taurine is reduced in some experimental models of acute liver failure[4,20,21]. Since it has been suggested that taurine, stored in astrocytes, may be important in K^+ and Ca^{2+} homeostasis in the CNS (27) and since movement of these ions is intimately involved in neural excitability, modifications of brain taurine concentrations could play a role in the pathogenesis of the neurological symptoms that characterize HE. For example, decreased brain taurine has been reported in both human and experimental epilepsy[14] and convulsions are encountered at preterminal stages of acute liver failure.

Pathogenesis of Brain Edema in Acute Liver Failure: Role of Ammonia and Taurine

Pathophysiologic mechanisms responsible for brain edema in acute liver failure are not completely understood. Recent studies suggest that ammonia-related mechanisms could

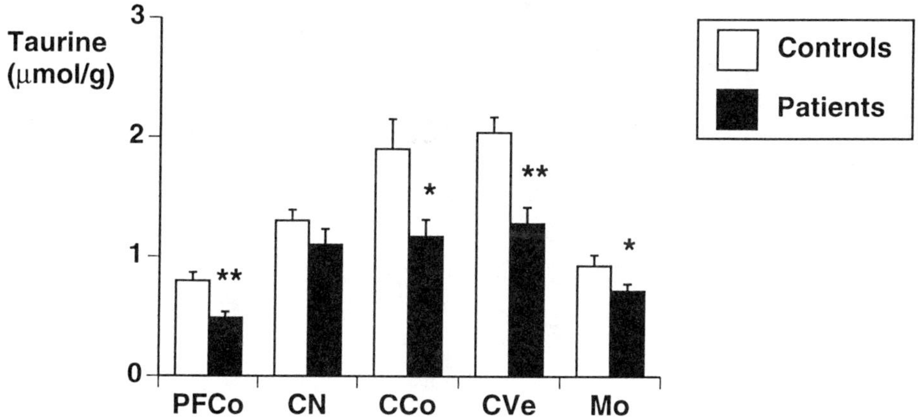

Figure 1. Taurine concentrations from 9 cirrhotic patients who died in hepatic coma and 9 age-matched controls. PFCo: prefrontal cortex; CN:caudate nucleus; CCo: cerebellar cortex; CVe: cerebellar vermis; Me: medulla oblangata. Values significantly different from control indicated by $*p < 0.05$; $**p < 0.01$ (Student's t test).

be implicated. In experimental acute liver failure, brain ammonia may reach mM concentrations, at which time there is significant brain edema (21). Addition of mM ammonia to rat brain cortical slices results in significant swelling[8]. Ammonia-induced swelling both *in vitro* and *in vivo* may be mediated by products of ammonia metabolism rather than by ammonia *per se*. Recent studies demonstrate a correlation between the rise in brain glutamine content and brain water concentrations in hyperammonemic animals[22], suggesting that ammonia-induced increases in brain water content are mediated by the osmotic effects of increased glutamine. In favour of this hypothesis, it was demonstrated that administration of methionine sulfoximine, an inhibitor of glutamine synthetase, the enzyme responsible for glutamine formation in brain, prevents both the increase in brain glutamine and in brain water in ammonia-treated animals[22]. In acute liver failure in humans, brain glutamine concentrations are increased[18], and in experimental acute liver failure, brain glutamine concentrations are increased 5-fold[20]. It has been suggested that, under normal physiological conditions, glutamine transport participates in the regulation of water movement in the brain[24]. Following up on these leads, a recent study has demonstrated a significant protective effect of inhibition of glutamine synthesis on brain edema formation in experimental liver failure[3].

In rats treated with ammonium acetate, brain taurine concentrations are decreased, concomitant with the development of brain edema (10).

Using an indwelling cisterna magna catheter to monitor CSF amino acids during the progression of HE in experimental acute liver failure, it was demonstrated that, in both ischemic and toxic models of the disorder, CSF taurine concentrations were significantly increased[21]. CSF taurine concentrations were significantly elevated early in the progression of HE in ischemic liver failure (Fig. 2). However, at later (coma) stages. CSF taurine concentrations attained a plateau. Such alterations of CSF taurine could relate to the osmosensitivity of this amino acid and could therefore be implicated in the pathogenesis of brain edema in acute liver failure. In favour of this possibility, extracellular taurine concentrations sampled by *in vivo* brain microdialysis are increased during cerebral ischemia[9] and in hypoglycaemic coma[23]. A common feature of these conditions is the occurrence of cellular swelling in the brain[19].

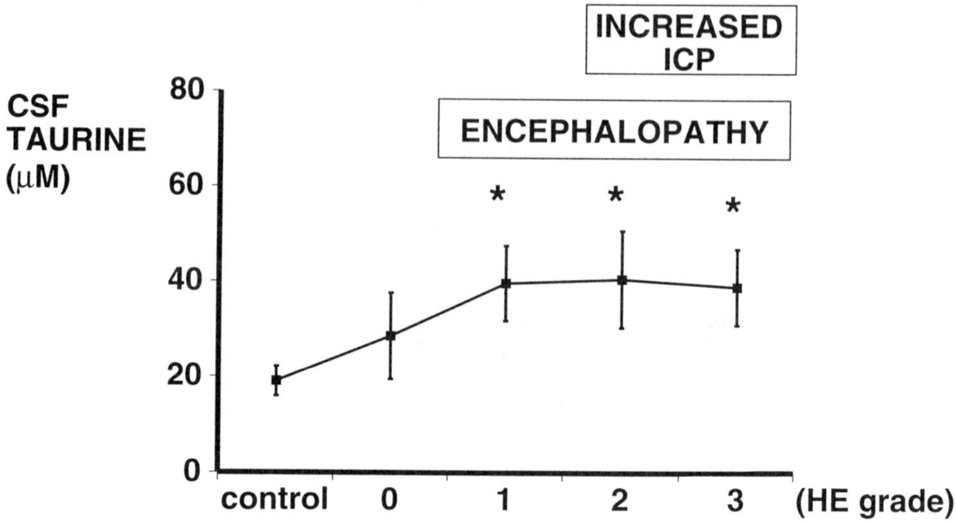

Figure 2. CSF taurine concentrations as a function of severity of hepatic encephalopathy and presence of increased intracranial pressure (ICP) hypertension in experimental acute ischemic liver failure. Data from Swain *et al.*[21]. Values significantly different from control by Analysis of Variance indicated by *$p < 0.02$. Increased ICP coincided with a plateau in CSF taurine concentrations.

Furthermore, K^+-stimulated release of taurine from striatal slices (15) or *in vivo* (using cerebral microdialysis)[4] is significantly increased in experimental animal models of acute liver failure. Whether this enhanced release of taurine results from increased size of the releasable pool or diminished astrocytic reuptake of the amino acid remains to be established. In contrast to acute liver failure, extracellular concentrations of taurine in brain in chronic liver failure are unchanged[17]. Brain edema is rarely a complication of chronic liver failure.

Studies using cultured astrocytes[16] and *in vivo* brain perfusion measurements[25] suggest that taurine is an important osmoeffector in astrocytes during volume regulatory changes. Release of taurine into the extracellular fluid could thus represent an important control mechanism for volume regulation in these cells. This viewpoint, however, has been challenged and it was proposed that taurine, although demonstrating osmotic sensitivity, was not present in sufficiently high concentrations in brain to play a major role in osmoregulation[12].

In summary, brain taurine concentrations are reduced in human HE. Furthermore, in acute liver failure, CSF taurine concentrations are increased until late (preterminal) stages at which brain edema and increased intracranial pressure are important features. In view of the osmosensitivity of taurine, monitoring of CSF taurine in acute liver failure could be of prognostic value in this severe, frequently fatal disorder.

ACKNOWLEDGMENTS

Studies from the author's laboratory were funded by a grant from The Medical Researcn Council of Canada (PG11118).

REFERENCES

1. Benveniste, H., Drejer, T., Schousboe, A. and Diemer, N.H. 1984, Elevation of the extracellular concentrations of glutamate and aspartate in rat hippocampus during transient cerebral ischemia monitored by intracerebral microdialysis. *J. Neurochem.* 43:1369-1374.
2. Bergeron, M., Pomier Layrargues, G. and Butterworth, R.F. 1989, Aromatic and branched-chain amino acids in autopsied brain tissue from cirrhotic patients with hepatic encephalopathy. Metab. *Brain Dis.* 4:169-176.
3. Blei, A.T., Olafsson, S., Therrien, G. and Butterworth, R.F. 1994, Ammonia-induced brain edema and intracranial hypertension in rats after portacaval anastomosis. *Hepatology* 19:1437-1444.
4. Bosman, D.K., Deutz, N.E.P., Maas, M.A.W., van Eijk, H.M.H., Smit, J.J.H., de Haan, J.G. and Chamuleau, R.A.F.M. 1992, Amino acid release from cerebral cortex in experimental acute liver failure, studied by *in vivo* cerebral cortex microdialysis. *J. Neurochem.* 59:91-599.
5. Butterworth, R.F. 1992, Evidence that hepatic encephalopathy results trom a defect of glutamatergic synaptic regulation. *Mol. Neuropharmacol.* 2:229-232.
6. de Knegt, R.J. and Schalm, S.W. 1995, Extracellular brain taurine during acute hepatic failure and acute hyperammonemia simulating acute hepatic failure, in rabbits using *in vivo* brain dialysis. In: "Advances in Hepatic Encephalopathy and Metabolic Nitrogen Exchange". CRC Press Inc., Boca Raton, Florida, pp. 259-264.
7. Florkin, M. and Schoffeniels, E. 1965, Euryhalinity and the concept of physiological radiation. In Studies in Comparative Biochemistry Munday K.A., ed., Pergamon Press, London, pp. 6-40.
8. Ganz R., Swain M., Traber P., DalCanto M., Butterworth, R.F. and Blei, A.T. 1989, Ammonia-induced swelling of rat cerebral cortical slices: implications for the pathogenesis of brain edema in acute hepatic failure. *Metab. Brain Dis.*. 4:213-223.
9. Hagberg, H., Lehmann, A., Sandberg, M., Nystrom, B., Jacobson, I. and Hamberger, A. 1985, Ischemia-induced shift of inhibitory and excitatory amino acids from intra- to extracellular compartments. *J. Cereb. Blood Flow Metab.* 5:413-419.
10. Hilgier, W. and Olson, J.E. 1994, Brain ion and amino acid contents during edema development in hepatic encephalopathy. *J. Neurochem.* 62:197-204.
11. Lavoie, J., Therrien, G., and Butterworth, R.F. 1995, Cerebrospinal fluid taurine in portal-systemic encephalopathy. In: "Advances in Hepatic Encephalopathy and Metabolic Nitrogen Exchange, vol. 42". CRC Press, Boca Raton, Florida, pp. 265-269.
12. Martin, D.L., Madelian, V. and Shain, W. 1995, Osmotic sensitivity of isoproterenol- and high$[K^+]_0$-stimulated taurine release by cultured astroglia. In: "Taurine: Functional Neurochemistry, Physiology, and Cardiology", Wiley-Liss Inc., pp. 349-356.
13. Mousseau, D.D., Perney, P., Pomier Layrargues, G. and Butterworth, R.F. 1993, Selective loss of pallidal dopamine D_2 receptor density in hepatic encephalopathy. *Neurosci. Lett.* 162:192-196.
14. Mutani, R., Bergamini, L. and Durelli, L. 1978, Taurine in experimental and human epilepsy. In: "Taurine and Neurological Disorders", Barbeau, A. and Huxtable, R.J. eds. Raven Press, New York, pp. 359-373.
15. Oja, S.S., Saransaari, P. and Wysmyk, U. 1990, Is taurine involved in the pathogenesis of hepatic encephalopathy? *Soc. Neurosci. Abs.* 454.10.
16. Pasantes-Morales, H. and Schousboe, A. 1988, Volume regulation in astrocytes: a role for taurine as an osmoeffector. *J. Neurosci. Res.* 20:505-509.
17. Raghavendra Rao, V.L., Audet, R.M. and Butterworth, R.F. 1996, Selective alterations of extracellular brain amino acids in relation to function in experimental portal-systemic encephalopathy: results of an *in vivo* microdialysis study. *J. Neurochem.* in press.
18. Record, C.O., Buxton, B., Chase, R.A., Curzon, G., Murray-Lyon, I.M. and Williams, R. 1976, Plasma and brain amino acids in fulminant hepatic failure and their relationship to hepatic encephalopathy. *J. Clin. Invest.* 6:387-394.
19. Siesjo, B.K. and Wielock, T. 1986, Epileptic brain damage: pathophysiology and neurochemical pathology. In: "Advances in Neurology, vol. 44", Delgado-Escueta, A.V., Ward, A.A.Jr., Woodbury, D.M. and Porter, R.J. eds., Raven Press, New York, pp. 813-847.
20. Swain, M.S., Bergeron, M., Audet, R., Blei, A.T. and Butterworth, R.F. 1992, Monitoring of neurotransmitter amino acids by means of an indwelling cisterna magna catheter: a comparison of two rodent models of fulminant liver failure. *Hepatology* 16:1028-1035.
21. Swain, M., Butterworth, R.F. and Blei, A.T. 1992, Ammonia and related amino acids in the pathogenesis of brain edema in acute ischemic liver failure in rats. *Hepatology* 15, 449-452.
22. Takahashi, H., Koehler, R.C., Brusilow, S.W., Traystman, R.J. 1991, Inhibition of brain glutamine accumulation prevents cerebral edema in hyperammonemic rats. *Am. J. Physiol.* 281:H826-H829.

23. Tossman, U., Wielock, T. and Ungerstedt, U. 1985, γ-Aminobutyric acid and taurine release in the striatum of the rat during hypoglycemic coma, studied by microdialysis. *Neurosci. Lett.* 62:231-235.
24. van Gelder, N.M. 1983, Metabolic interactions between neurons and astroglia: glutamine synthetase, carbonic anhydrase and water balance, in: "Basic Mechanisms of Neuronal Hyperexcitability", Ward, A. ed. A.R. Liss, New York, pp. 5-29.
25. Wade, J.V., Olson, P., Samson, F.E., Nelson, S.R. and Pazdernik, T.L. 1988, A possible role for taurine in osmoregulation within the brain. *J. Neurochem.* 51:740-745.
26. Wade, J.V., Samson, F.E., Nelson, S.R. and Pazdernik, T.L. 1987, Changes in extracellular amino acids during soman and kainic acid-induced seizures. *J. Neurochem.* 49:645-650.
27. Waltz, W. and Allen, A.F. 1987, Evaluation of the osmoregulatory function of taurine in brain cells. *Exp. Brain Res.* 68:290-298.
28. Waterfield, C.J., Turton, J.A., Scales, M.D.C. and Timbrell, J.A. 1991, Taurine, a possible urinary marker of liver damage: a study of taurine excretion in carbon tetrachloride-treated rats. *Arch. Toxicol.* 65:548-555.
29. Wright, C.E., Tallan, H.H. and Lin, Y.Y. 1986, Taurine: biological update. *Ann. Rev. Biochem.* 55:427-453.

67

EFFECT OF TAURINE ON THE FATTY LIVER OF CHILDREN WITH SIMPLE OBESITY

Kaoru Obinata, Takeshi Maruyama, Motohiko Hayashi, Toyohiko Watanabe, and Hiroshi Nittono

Department of Pediatrics
Koshigaya Municipal Hospital and
Juntendo University School of Medicine
10-47-1, Higashikoshigaya, Koshigaya, Saitama 343, Japan

SUMMARY

This study elucidated the effect of taurine on fatty liver in simple obesity. Taurine was orally administered to 10 children with fatty liver. During taurine administration, the CT numbers of the liver, which were low in the beginning, increased. Serum ALT levels were improved, especially in those children whose weight was well controlled. Even in those who failed in weight control, serum ALT levels were slightly recovered. Ratios of glycine/taurine-conjugated bile acids were decreased. Thus, taurine was effective in treating fatty liver of children with simple obesity regardless of the success/failure of weight control. Taurine administration is considered to be helpful as an adjuvant therapy for fatty liver.

INTRODUCTION

Obesity and its complications in children is of concern in Japan. Overnutritional fatty liver with obesity is commonly observed. It has been reported that fatty liver can progress to fatty hepatitis or fatty fibrosis[1]. Therefore, it is necessary to treat fatty liver as early as possible. The management of fatty liver consists mainly of weight control, which, however, is not very easy for children, especially those with severe obesity.

The role of taurine in pediatric preventive nutrition has been widely discussed. Taurine is now added to many infant formulas. It has been shown that taurine protects cell membranes by attenuating toxic compounds and improving lipid metabolism[2]. This study evaluated the efficacy of taurine as an adjuvant therapy for fatty liver of children with simple obesity.

Taurine 2, edited by Huxtable et al.
Plenum Press, New York, 1996

METHODS

Ten children (8-12 years) with fatty liver due to simple obesity (mean body mass index; 28.5) were given 2 - 6 g/day of taurine powder orally for a period of 6 months or longer. They were also put on a diet. Throughout the administration period, biochemical liver function and serum bile acids were measured. The degree of fatty liver was evaluated by CT scan images.

Fatty liver was diagnosed by the following criteria: (i) ratio of CT numbers between liver and spleen is below 0.85; (ii) serum transaminase elevates, with hepatomegaly; and (iii) viral hepatitis, EB virus and cytomegalovirus infections are excluded.

The degree of obesity (%) was calculated as:

$$100 \times (\text{present weight} - \text{standard weight})/(\text{standard weight}).$$

The subjects were divided into two groups: successful cases in weight control (more than 15% reduction in obesity degree); and unsuccessful cases in weight control (less than 15% reduction in obesity degree).

RESULTS

Weight control was achieved in 5 cases. Serum ALT levels were improved in these children after taurine administration (Fig. 1). Serum total cholesterol and triglyceride levels were decreased in two of the successful cases. Table 1 and Fig. 2 show the change in laboratory data and the changes of CT findings in a 12 year-old boy after taurine administration. Despite the low degree of obesity (19%), the CT number of the liver was O HU and the serum ALT level was 136 mU/ml before taurine administration. Eight weeks after taurine administration, the CT number of the liver increased to 10 HU and the ALT levels also

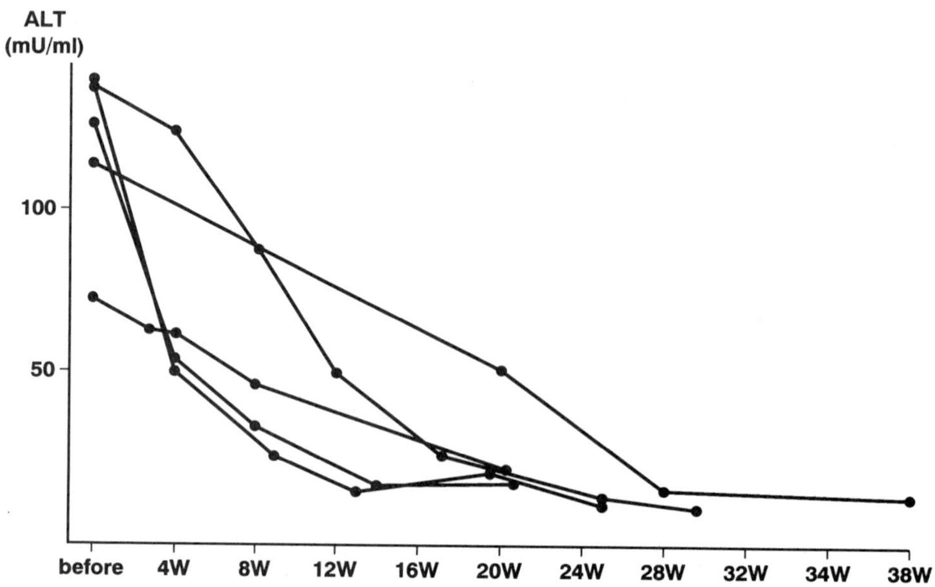

Figure 1. Changes in serum ALT levels in children who were successful in weight control during taurine administration.

Table 1. Changes of laboratory data and CT number of the liver during taurine administration (3g/day) and dietary therapy (a successful case in weight control)

	Before	4W	8W	12W	18W	25W	37W	45W	68W
AST(5-27mU/ml)	63	58	42	32	24	19	22	19	20
ALT(4-25mU/ml)	136	124	86	51	27	14	16	14	19
γ-GTP(9-37mU/ml)	40	27	29	21	17	14	14	12	14
Ch-E(0.6-1.2ΔpH)	1.8	1.5	1.5		1.4	1.3	1.2	1.3	1.3
T-Chol(120-230mg/dl)	221	178	186	167	142	137	129	124	135
HDL-Chol(40-71mg/dl)	16	17	17	28	26	28	24	31	33
TG(50-90mg/dl)	1,121	516	560	222	453	262	529	185	214
PL(130-235mg/dl)	327	244	259	190	212	174	197	175	181
TSBA(<3.6nmol/ml)	2.5				3.0	0.9	2.5	0.8	1.8
Degree of obesity	+19%		+18%	+19%	+14%		+3%	+1%	0%
CT number of the liver (HU)	0		10.4				65.4	56.3	57.0
Ratios of CT number between liver/spleen	0		0.2				1.1	1.1	1.0

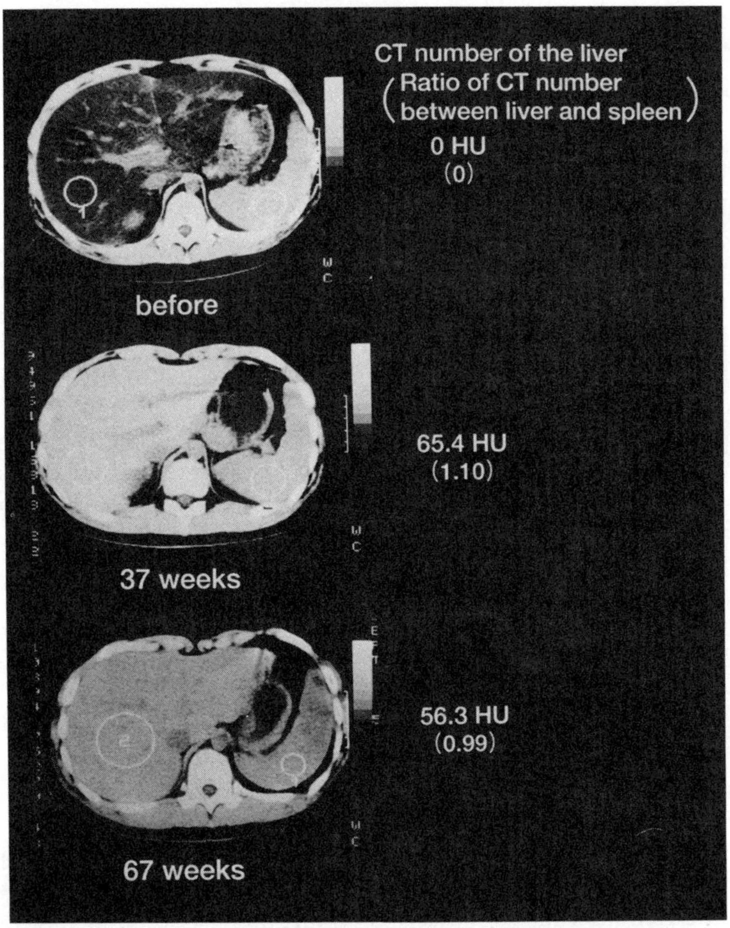

Figure 2. Changes in CT number of the liver during taurine administration in a 12 year-old boy who were successful in weight control.

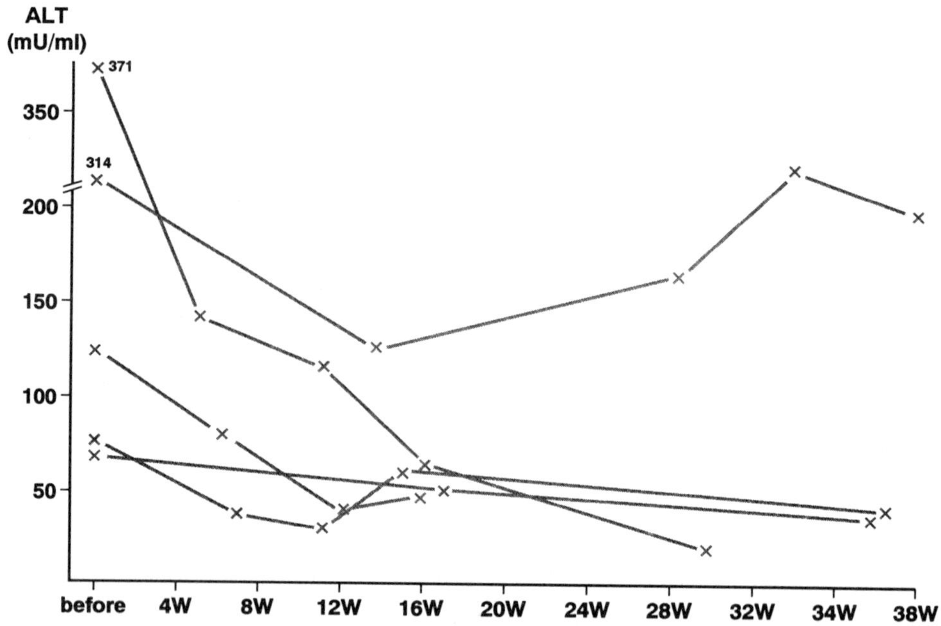

Figure 3. Changes in serum ALT levels in children who failed in weight control during taurine administration.

improved. There was, however, no change in the body weight. Thus the patient had to go on a restricted diet, whereupon obesity degree declined by 5%, and the ALT levels recovered to normal range. Thirty-seven weeks after taurine administration, the obesity degree became 3% and the CT number was 65.4 HU. Serum total cholesterol also recovered to the normal value.

Weight control failed in 5 cases. Serum ALT levels were slightly recovered (Fig. 3). There were, however, no significant changes in the serum lipids. Fig. 4 shows the changes in CT findings in a 13 year-old boy. Even in this case, the CT findings in the liver improved after 36 weeks of taurine administration; the CT number of the liver (4 HU) and the ratio of CT numbers between liver and spleen (0.07) increased to 26 HU and 0.45, respectively.

After taurine administration, the CT numbers of the liver which were low in the beginning increased even in those who failed in weight control (Fig. 5). Total bile acid levels and ratios of cholate/chenodeoxycholate seem to decrease to normal values. Ratios of glycine/taurine-conjugated bile acids increased after taurine administration in both groups. No remarkable side effects during administration period were recognized, and the compliance with taurine was good.

DISCUSSION

The hepatic lipid deposits consist largely of triglyceride with a little phospholipid and cholesterol. Most of the fatty acids taken up by the liver and esterified to triglycerides are originally derived from adipose tissue and the diet. In the case of fatty liver with simple obesity, increased levels of fatty acid synthesis and impaired secretion of lipoprotein from the liver may lead to increased accumulation of triglyceride.

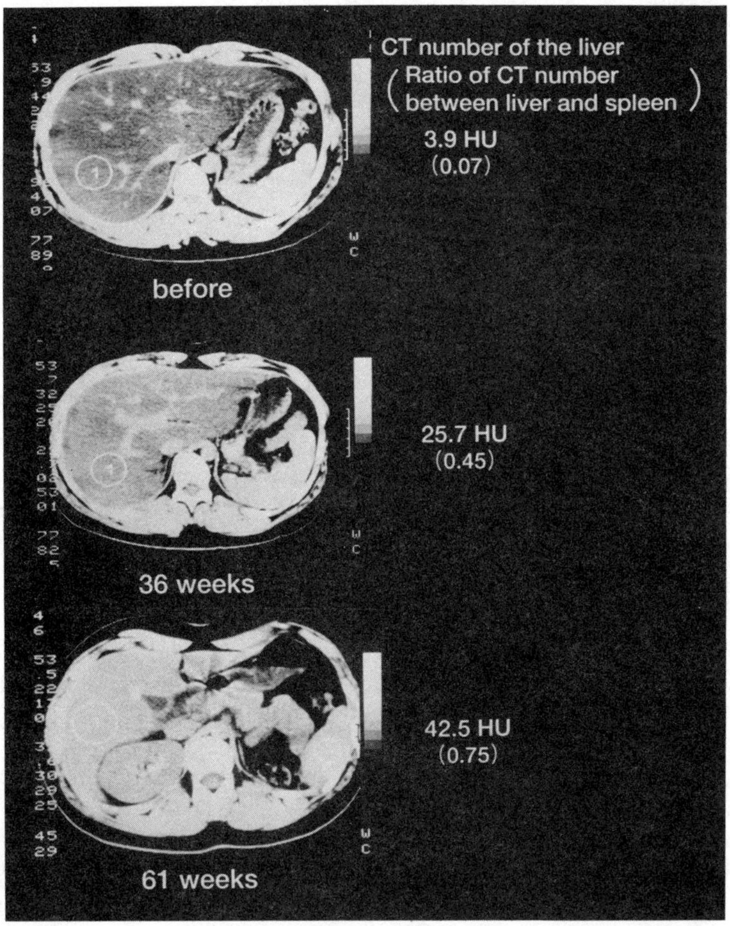

Figure 4. Changes in CT number of the liver during taurine administration in a 13 year-old boy who failed in weight control.

In fatty liver detected by CT images, lipids comprise more than 30% of total liver weight. The condition is detectable when the CT number of the liver falls below 50 HU and the ratio of CT numbers in the liver and spleen falls below 0.9[4]. The liver function tests show mild elevation of transaminases with an alanine aminotransferase (ALT)-dominant pattern. Liver dysfunction associated with virus infections such as viral hepatitis, EB virus and cytomegalovirus have to be excluded. When diagnostic uncertainly exists, needle biopsy of the liver is recommended. CT scan image is useful for evaluating and following up the degree of fatty infiltration in the liver.

Taurine plays an important role in the conjugation of bile acids which are final products of cholesterol catabolism. Bile acids are conjugated with either taurine or glycine before their secretion in bile. Taurine-conjugated bile acids are more soluble and less toxic than their glycine-conjugated counterparts. The glycine to taurine ratio of bile acids was reversed after taurine administration, suggesting that the hepatic taurine supply converts bile acids to taurine conjugates. Taurine supplementation increases the activity of cholesterol 7 α-hydroxylase, a promoting enzyme of hepatic cholesterol catabolism to cholate[7]. High concentration of bile acids and low cholesterol concentration in the duodenal juice in

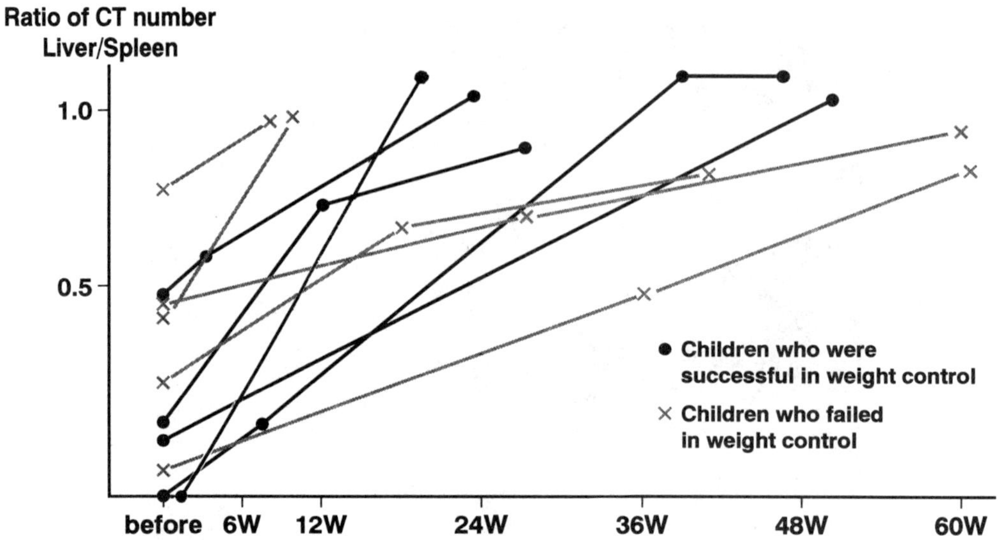

Figure 5. Changes in ratios of CT number between liver and spleen during taurine administration.

taurine-supplemented infants suggest a greater conversion of cholesterol to bile acids[9]. In patients with acute hepatitis, levels of serum bilirubin and total bile acids rapidly recovered in a taurine-administered group compared to the controls[6].

Taurine supplementation causes a significant decrease in hepatic concentration of triglyceride, that is probably due to inhibition of diacylglycerol:acyl CoA acyltransferase, the key enzyme in hepatic triglyceride synthesis[10]. Our co-worker Hayashi also reported the anti-fatty effect of taurine on experimental fatty degeneration in rat hepatocytes, and suggested that taurine has an accelerating effect on the secretion of very low density lipoprotein[3].

Taurine administration induces a protective effect against CCl_4-induced hepatic damage by suppressing lipid peroxide formation in hepatocytes[8]. Taurine may protect membranes by detoxification of destructive compounds and/or by directly preventing alterations in membrane permeability[5].

In the present study, the CT numbers and serum ALT levels were recovered even in those who failed in weight control. This suggests that taurine is effective not only for liver dysfunction but also for fatty infiltration in the liver of children with obesity. Taurine administration is recommended as an adjuvant form of therapy, especially for severe fatty liver.

REFERENCES

1. Adler, M. and Schaffner, F. 1979, Fatty liver hepatitis and cirrhosis in obese patients, *Am.J.Med.* 67:811-816.
2. Gaull, G.E. 1989, Taurine in pediatric nutrition: Review and update, *Pediatrics*, 83:433-442.
3. Hayashi, M. 1990, The effect of taurine on experimental fatty degeneration in primary cultured rat hepatocytes, *Act.Hepatol.Jpn.* 31:75-80.
4. Kato, K., Takayama, T., Katada, N., Nishimura, D., Shibata, T., and Takeichi, M. 1980, Evaluation of computed tomography in the diagnosis of liver diseases, *Acta.Hepatol.Jpn.* 21:58-69.
5. Kendler, B.S. 1989, Taurine: an overview of its role in preventive medicine, *Preventive Med.* 18:100

6. Matsuyama, Y., Morita, T., Higuchi, M., and Tsujii, T. 1983, The effect of taurine administration on patients with acute hepatitis, in: "Sulfur Amino Acids: Biochemical and Clinical Aspects", Kuriyama, K., Huxtable, R.J. and Iwata, H. eds., Alan R. Liss, Inc., New York, pp. 461-468.
7. Nakamura-Yamanaka, Y., Tsuji, K., and Ichikawa, T. 1987, Effect of dietary taurine on cholesterol 7α-hydroxylase activity in the liver of mice fed a lithogenic diet, *J.Nutr.Sci.Vitaminol.* 33:239-243.
8. Nakashima, T., Takino, T., and Kuriyama, K. 1983, Therapeutic and prophylactic effects of taurine administration on experimental liver injury, in: "Sulfur Amino Acids: Biochemical and Clinical Aspects", Kuriyama, K., Huxtable, R.J. and Iwata, H. eds., Alan R. Liss, Inc., New York, pp. 449-460.
9. Okamoto, E., Rassin, D.K., Zucker, C.L., Salen, G.S., and Heird, W.C. 1984, Role of taurine in feeding the low birth weight infant, *J.Peds.* 104:936-940.
10. Yan, C.C., Bravo, E., and Cantafora, A. 1993, Effect of taurine levels on liver lipid metabolism: an in vivo study in the rat, *Proc.Soc.Exp.Biol.Med.* 202:88-96.

68

EFFECTS OF ORAL TAURINE SUPPLEMENTATION ON LIPIDS AND SYMPATHETIC NERVE TONE

Shunsaku Mizushima,[1] Yasuo Nara,[1] Makoto Sawamura,[2] and Yukio Yamori[2]

[1] Otsuka Department of International Preventive Nutritional Medicine
[2] Otsuka Department of Life Science
Environmental Conservation and Development
Graduate School of Human and Environmental Studies
Kyoto University
Yoshida-Nihonmatsu-cho, Sakyo-ku, Kyoto 606, Japan

SUMMARY

Objectives

To assess effects of oral taurine supplementation on lipids and sympathetic nerve tone in healthy young men on experimental high fat and cholesterol diets.

Methods

Twenty-two healthy male volunteers, aged 18-29 years, were recruited for this randomized control trial after informed concent according to the Ethical Comittee of Shimane Medical University. Volunteers were randomly allocated into 2 study groups and given experimental diet of identical regimen [total calorie 2500 kcal, cholesterol 1000 mg, polyunsaturated fat/saturated fat (P/S) ratio 0.52, fat 40% of total energy intake (%E), protein 14%E, carbohydrate 46%E] to raise serum cholesterol (CHO) level for 3 weeks. Alcohol intake, smoking and strenuous physical activities were prohibited. Taurine powder (6 g/day) was supplied to one group (T-group, N=11) and placebo capsules to the other (C-group, N=11), by a single-blind approach. Blood samples and 24 h urine specimens were obtained once every week. Two men in the C-group dropped out due to upper respiratory infection. There were no difference in age, body mass index (BMI) or blood pressure (BP) between the groups. Statistical analysis was performed by analysis of variance (ANOVA, repeated measurement) and Student's t-test.

Taurine 2, edited by Huxtable et al.
Plenum Press, New York, 1996

Results

There were no changes in BMI and BP in either group during the period. Significant increases in total CHO (25.4 ± 17.5 mg/dl, mean ± SD), LDL-CHO (17.1 ± 14.5) and LDL (43.9 ± 37.6) were observed in C-group but were attenuated in the T-group. The T-group showed significant increases in VLDL-CHO, VLDL and TG. The T-group had significantly lower urinary norepinephrine excretion than the C-group in the last week.

Conclusion

Oral taurine supplementation attenuated increases in T-CHO, LDL-CHO and LDL in healthy men on high fat cholesterol diets but induced significant increases in VLDL-CHO, VLDL and TG, which could be explained by a possible effect of taurine on lipoprotein lipase. Significantly lower urinary norepinephrine excretion observed by the taurine administration implies the suppression of the sympathetic nervous system.

INTRODUCTION

Both genetics and environmental factors, including diets, are highly relevant to cardiovascular disease (CVD), including hypertension, coronary heart diseases and stroke.[6,10,18,28,29,32-34,36] Experimental data, especially those obtained from animal models for CVD, such as spontaneously hypertensive rats (SHR) and stroke-prone SHR (SHRSP), have shown that excess salt intake aggravates the development of hypertension and increases the incidence of stroke, whereas increased dietary intakes of K^+, Ca^{2+}, Mg^{2+}, unsaturated fatty acids, and proteins, especially those from fish (taurine), attenuate the development of hypertension and prevent stroke even in these genetically predisposed animal models.[18,28,29,32-34,36] Therefore, CVD are potentially preventable by dietary modifications. However, knowledge of appropriate diets for preventing CVD is still limited, in part because of the lack of appropriately designed and sufficiently wide studies, including epidemiological studies and radomized controlled trials, with special emphasis on CVD and related nutritional conditions.

In this study, to identify beneficial factors for prevention of CVD, the roles of taurine were investigated through a randomized controlled trial which assessed the effects of oral taurine supplementation on lipids and sympathetic nerve tone in healthy young males on experimental high fat and cholesterol diets.

SUBJECTS AND METHODS

Twenty-two healthy male volunteers, aged 18-29 years, were recruited for this randomized control trial after approval according to the ethics regulations of Shimane Medical University. Informed consent was obtained from each volunteer after detailed explanation of the procedures to be performed. Volunteers were randomly allocated into 2 study groups and given experimental diets of identical regimen [total calorie 2500 kcal, cholesterol 1000 mg, polyunsaturated fat/saturated fat (P/S) ratio 0.52, fat 40% of total energy intake (%E), protein 14%E, carbohydrate 46%E] to raise serum cholesterol level for 3 weeks. Alcohol intake, smoking and strenuous physical activities were prohibited. Taurine powder (6 g/day, Taisho Pharmaceutical Co. Ltd.) was supplied to one group (T-group, N=11) and placebo capsules to the other (C-group, N=11), by a single-blind approach. Blood samples and 24 h urine specimens were obtained once every week. Two men from the

C-group dropped out due to upper respiratory infection. There were no difference in age, body mass index (BMI) and blood pressure (BP) between the 2 groups. Neither obese nor hypertensive subjects were involved in this study. Health status of all subjects, at entry and during the trial, was routinely checked by a questionnaire on health conditions and by a physician. No particular signs and symptoms were found or reported, except a slight discomfort of bowel movement for a couple of days among 3 subjects, which was considered possibly due to the high-fat cholesterol of the diet. Blood testing and interview 2 weeks after the study did not indicate any lasting hyperlipidemia or any other problems among any subject. Statistical analysis was performed using analysis of variance (ANOVA, repeated measurement) and Student's t-test.

RESULTS

There were no significant changes in BMI and BP in either group. Lipid profiles, the values at entry into the study (pre) and on the 22nd day, and differences between these values are summarised in Table 1. Significant increases in total cholesterol (T-CHO), low density lipoprotein cholesterol (LDL-CHO) and low density lipoprotein (LDL) were observed in the C-group, while these increases were attenuated in the T-group. The average increase of T-CHO, expressed as % of T-CHO level at entry, was significant in C-group, whereas no significant change was observed in T-group (Fig. 1). T-group showed a significant increase of very low density lipoprotein cholesterol (VLDL-CHO), very low density lipoprotein (VLDL), and triglycerides (TG). T-group had significantly lower 24 h urinary norepinephrine excretion (77.3 μg/day) than the C-group (105.3) in the last week (Fig. 2).

DISCUSSION

Contemporary Japanese diet is characterized by major caloric intake from rice, complex carbohydrates, relatively low saturated fat intakes, and lots of soy protein and seafoods. Such a diet seems to have been causatively related to the low mortality and morbidity of coronary heart diseases in Japan, despite a high % of smokers (about 60% for men and about 15% for women according to the statistics of the Japan Tobacco Ltd.[2,23,24] Findings from our previous and ongoing epidemiological studies, as well as the experimental studies on SHRSP, clearly indicate the importance of nutritional factors in CVD prevention.

Table 1. Lipid profile (mg/dl)

	Control Group[a]			Taurine Group[b]		
	Pre	22nd Day	ΔChange	Pre	22nd Day	ΔChange
T-CHO	169.7 ± 26.3	195.1 ± 24.7**	25.4 ± 17.5	195.3 ± 37.7	217.4 ± 33.4	22.1 ± 33.8
VLDL-CHO	30.2 ± 6.2	34.8 ± 8.8	4.6 ± 4.0	24.9 ± 6.9	39.9 ± 11.6**	15.0 ± 12.7#
LDL-CHO	95.5 ± 25.6	112.6 ± 24.4**	17.1 ± 14.5	118.6 ± 29.6	125.3 ± 30.5	6.7 ± 21.4
HDL-CHO	52.4 ± 9.4	54.4 ± 11.4	2.0 ± 7.2	54.5 ± 2.2	53.1 ± 5.6	-1.4 ± 7.7
TG	87.6 ± 35.9	93.2 ± 45.2	5.7 ± 17.5	70.5 ± 19.2	109.5 ± 43.1	39.0 ± 42.8#
VLDL	97.8 ± 56.3	102.9 ± 58.0	5.1 ± 19.3	69.6 ± 31.8	126.1 ± 65.0*	56.5 ± 71.9#
LDL	346.1 ± 102.4	390.0 ± 105.5*	43.9 ± 37.6	433.2 ± 131.7	461.3 ± 124.6	28.1 ± 79.0

Data are means ± SD for [a]9 subjects; [b]11 subjects. T-CHO: total cholesterol; VLDL: very low density lipoprotein; LDL: low density lipoprotein; HDL: high density lipoprotein; TG: triglyceride. Significant difference from pre value: *p < 0.05; **p < 0.01. Significant difference in ΔChange in Control Group: #p < 0.05.

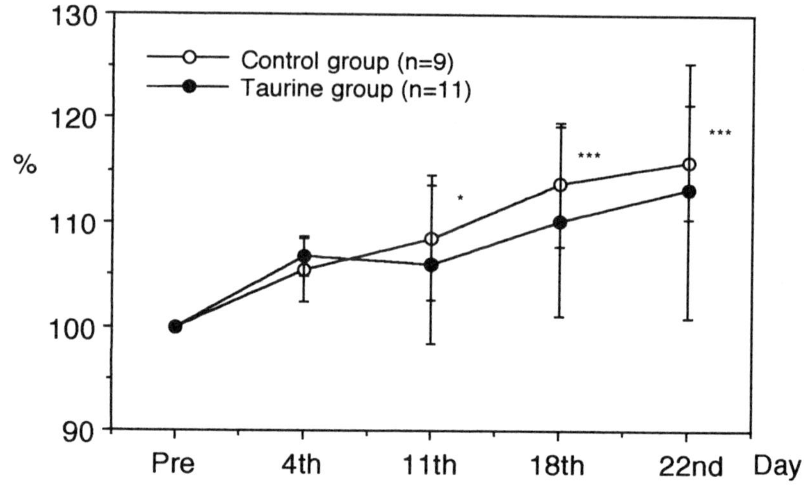

Figure 1. The effect of taurine on total cholesterol level during the high fat cholesterol diet. Data are means ± SD. Significant difference from the value at entry (pre as 100) by ANOVA (repeated measures), $*=p < 0.05$; $**p < 0.001$.

In particular, frequent seafood intakes showed protective effects on CVD in 3 approaches of investigation on the relationship between dietary factors and CVD: (i) Experimental studies with SHRSP of the effect of taurine supplementation on CVD[18,29,33,34,36]; (ii) The cross-sectional world-wide epidemiological study, Cardiovascular Diseases and Alimentary Comparison (CARDIAC) study, involving more than 10,000 middle-aged men and women;[16,19,26,27,30,31,35,37-41] and (iii) The study of 5 immigrant populations in Brazil, USA and Japan who have similar genetic background to populations in Okinawa and Shimane, well known regions for people's longevity in Japan.[3,11,14,15,17]

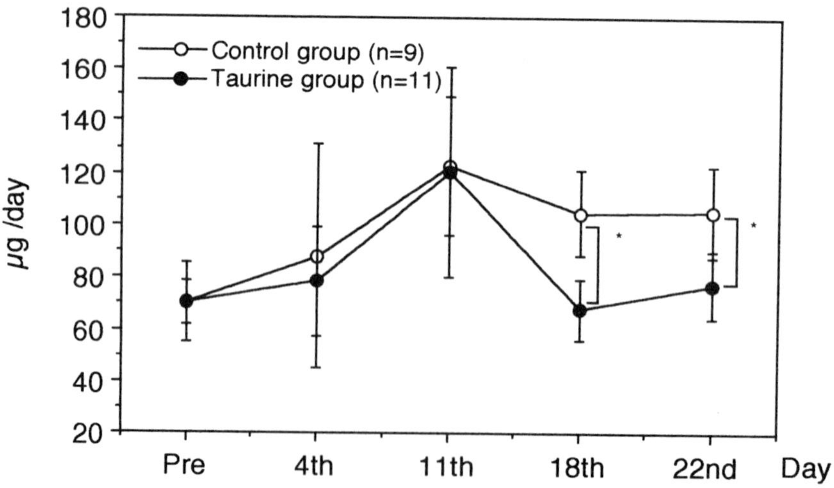

Figure 2. Urinary norepinephrine excretion during the high fat cholesterol diet. Data are Means ± SD. Significant difference between the the taurine and control groups, by Student's t-test, $\#p < 0.05$.

The effects of fish-protein-rich dietary constituents, especially taurine, on hypertension through central suppression of sympathetic tone, and on stroke prevention were first demonstrated by experimental studies on SHRSP;[18,28,29,33,34] fish-protein-rich diets without excess salt attenuated the development of severe hypertension and no stroke was observed.

In the CARDIAC study which has involved 57 populations in 25 countries as of 1995, 24-h urinary taurine excretion showed a tendency to a negative relationship with BP in the "Core study" on BP and dietary risk factors, and also an inverse association with coronary heart disease mortalities in the "Complete study" on CVD mortalities and dietary risk factors. A report of the CARDIAC study in 10 Chinese populations demonstrated significant cross-sectional inverse relationships between urinary taurine to urea nitrogen ratio and BP[19].

Immigrant studies are particularly useful for analysing interactions between genetics and environmental factors. We demonstrated various differences in dietary factors among 2 Japanese populations in Okinawa and Shimane, and 3 Japanese immigrant populations in Hilo (Hawaii), Sao Paulo and Campo Grande (Brazil), mainly reflected in differences in fish and meat intakes. These major differences were possibly associated with the accumulation of multiple CVD risk factors such as obesity, hypertension, hyperlipidemia, low n-3 polyunsaturated fatty acids, low taurine intakes and diabetes mellitus. Urinary taurine excretion was significantly higher in Japanese populations than in Japanese immigrants, reflecting their fish intake frequency.

Taurine is contained in many seafoods, including white fish, shellfish, octopus, squid, lobster, and shrimp.[4,9] A tight association between seafood intake and urinary taurine excretion was confirmed epidemiologically[13]. In addition to seafood, some foodstuffs such as dark meat of chicken and turkey, heart, liver and brain of mammals contain taurine to some degree[4,9]. Moderate levels of taurine excretion in Campo Grande, despite low fish intake (0.5 times/week), might be attributable to taurine derived from chicken and its precursor sulfur amino acids[15]. Furthermore, beneficial preventive effects on the process of atherosclerosis could be explained by activation of 7α-hydroxylase in the catabolism of cholesterol resulting in a decrease in the peripheral level and also by antioxidant activity[5,21].

The present study suggests the possible effect of taurine on lipoprotein lipase which hydrolyzes triacylglycerol of chylomicrons and VLDL being associated with decreased LDL and increased VLDL and TG. The mechanism is needs to be studied further. This study also showed a significant decrease in urinary norepinephrine excretion in the T-group, which might imply taurine suppress the central sympathetic nerve tension as proven experimentally[18,33].

An inverse relation between fish consumption and mortality from coronary heart diseases was found in some cohort studies[7,8,20,22]. These studies were conducted in populations who had relatively low fish intakes. In populations with relatively high fish intakes, such as Norway[25] and Hawaii[1], the relation was not clear. To reveal the causal relation between fish intakes and CVD, many aspects, including various biological markers related to fish, such as taurine, fatty acids, Mg^{2+}, etc. and dietary habits together with cooking style, need to be taken into account.

In conclusion, the present randomized controlled trial revealed that oral taurine supplementation attenuated increases in T-CHO, LDL-CHO and LDL, and suppressed urinary norepinephrine excretion in healthy young men on experimental high fat and cholesterol diets. Taurine could contribute to the prevention of cardiovascular disease, through beneficial effects on lipid metabolism and sympathetic nerve tone.

ACKNOWLEDGMENT

The study was supported by Taisho Pharmaceutical Co Ltd., Tokyo. Special gratitude is extended to Ms. Keiko Nogi (Nutritionist, Shimane Medical University Hospital) for her invaluable contribution to the preparation of daily diets during the trial.

REFERENCES

1. Curb, J.D. and Reed, D.M. 1985, Fish consumption and mortality from coronary heart disease, *N.Engl.J.Med.* 313:821-822.
2. Gonzalez-Biosca, M.D., Mizushima, S., Sawamura, M., Nara, Y., and Yamori, Y. 1992, The effect of nutritional prevention of cardiovascular diseases on longevity, *Nutrition Reviews*, 50:407-412.
3. Hekman, P.R.W., Mizushima, S., Gonzalez-Biosca, M.D., Moriguchi, E.H., Nara, Y., Moriguchi, Y., and Yamori, Y. 1993, Dietary habits as risk factors for cardiovascular diseases -comparative survey in Brazilian and Japanese immigrants living in Brazil, *The 3rd International conference on Preventive Cardiology*, 43 (Abstract).
4. Huxtable, R.J. 1992, Physiological actions of taurine, *Physiol.Rev.* 72:101-163.
5. Kalles, I. and Wikvall, K. 1981, Role of sulfhydryl groups in catalytic activity of purified cholesterol 7α-hydroxylase system from rabbit and rat liver microsomes, *Biochem.Biophys.Res.Commun.* 100:1361-1369.
6. Keys, A. 1970, Coronary heart disease in seven countries, *Circulation*, 41 (Suppl I):1-211.
7. Kromhout, D., Bosschieter, E.B., and Coulander, C.D.L. 1985, The inverse relation between fish consumption and 20-year mortality from coronary heart disease, *N.Engl.J.Med.* 312:1205-1209.
8. Kromhout, D., Feskens, E.J.M., and Bowles, C.H. 1995, The protective effect of a small amount of fish on coronary heart disease mortality in an elderly population, *Int.J.Epidemiol.* 24:340-345.
9. Laidlaw, S.A., Grosvenor, M. and Kopple, J.D. 1990, The taurine content of common foodstuffs, *J.Parent.Enter.Nutr.* 14:183-188.
10. Marmot, M. and Elliott, P. 1992, Coronary Heart Disease Epidemiology: From Aetiology to Public Health, Oxford University Press, Oxford.
11. Mizushima, S., Moriguchi, E.H., Nara, Y., Moriguchi, Y., and Yamori, Y. 1993, Impact of fish intake on relationship of dietary factors to cardiovascular disease among Japanese in Japan and Brazil, *The 3rd International conference on Preventive Cardiology*, 21 (Abstract).
12. Mizushima, S., Ueda, M., Sawamura, M., Nara, Y., and Yamori, Y. 1994, Diet and cardiovascular risk factors among Japanese middle-aged population in Japan, Hawaii and Brazil, *J.Hypertens.* 12:S61
13. Mizushima, S., Gonzalez-Biosca, M.D., Nara, Y., Sawamura, M., Horie, R., and Yamori, Y. 1991, Urinary taurine excretion as a biological index for dietary fish intake from cardiovascular diseases and alimentary comparison study, *International Epidemiological Association Regional Scientific Meeting*, 194 (Abstract).
14. Mizushima, S., Moriguchi, E.H., Gonzalez-Biosca, M.D., Nara, Y., Moriguchi, Y., and Yamori, Y. 1992, Immigrant study on relationship of dietary factors to cardiovascular disease among Japanese in Japan and Brazil, *J.Hypertens.* 10:S39.
15. Mizushima, S., Moriguchi, E.H., Nakada, Y., Gonzalez-Biosca, M.D., Nara, Y., Murakami, K., Horie, R., Moriguchi, Y., Mimura, G., and Yamori, Y. 1992, The relationship of dietary factors to caradiovascular diseases among Japanese in Okinawa and Japanese immigrants, orginally from Okinawa, in Brazil, *Hypertens.Res.* 15:45-55.
16. Mizushima, S., Yamori, Y., Sawamura, M., and Nara, Y. 1995, Ethnicity, environmental and salt-sensitivity in CARDIAC study: epidemiological implication for prevention, *Clin.Exp.Pharmacol.Physiol.* Suppl. 2:S412-S416.
17. Moriguchi, E.H., Mizushima, S., Gonzales-Biosca, M.D., Ishikawa, P.M., Heckman, P.R.W., Nara, Y., Moriguchi, Y., and Yamori, Y. 1993, Comparative cardiovascular diseases and alimentary comparison study (CARDIAC study) in Brazilians and Japanese immigrants in Brazil, *Canadian J.Cardiol.* 9:D37-D38.
18. Nara, Y., Yamori, Y., and Lovenberg, W. 1978, Effects of dietary taurine on blood pressure in spontaneously hypertensive rats, *Biochem.Pharmacol.* 27:2689-2692.

19. Nara, Y., Zhao, G.S., Huang, Z.D., Li, Y.H., Mizushima, S., Mano, M., Zhang, H.X., Sun, S.F., Sato, T., Horie, R., Zhang, M.X., He, B.S., Mori, C., Hatano, S., Liu, L.S., and Yamori, Y. 1990, Relationship between dietary factors and blood pressure in China, *J.Cardiovasc.Pharmacol.* 16 Suppl. 8:S40-S42.
20. Norel, S.E., Ahlbom, A., Feychting, M., and Pedersen, N. 1986, Fish consumption and mortality from coronary heart disease, *Br.Med.J.* 293:426
21. Petty, M.A., Kintz, J., and DiFrancesco, G.F. 1990, The effects of taurine on atherosclerosis development in cholesterol-fed rabbits, *Eur.J.Pharmacol.* 180:119-127.
22. Shekelle, R.B., Missell, L., Oglesby, P., Shryock, A.M., and Stamler, J. 1985, Fish consumption and mortality from coronary heart disease, *N.Engl.J.Med.* 313:820.
23. Tanaka, H., Yamaguchi, M., Date, C., Nakayama, T., Yamamoto, T., Yoshiike, N., Iwaya, M., Matsumura, Y., Yokoyama, T., Noji, A., Kushiro, W., and Cho, B.M. 1992, Nutrition and cardiovascular disease — a brief review of epidemiological studies in Japan, *Nutrition and Health*, 8:107-123.
24. Ueshima, H. 1990, Changes in dietary habits, cardiovascular risk factors and mortality in Japan, *Acta.Cardiol.* 45:311-327.
25. Vollset, S.E., Heuch, I., and Bjelke, E. 1985, Fish consumption and mortality from coronary heart disease, *N.Engl.J.Med.* 313:820-821.
26. WHO Collaborating Centre on Primary Prevention of Cardiovascular Diseases and Cardiovascular Diseases Unit 1986, CARDIAC study protocol and manual of operations, WHO Collaborating Centre on Primary Prevent of Cardiovascular Diseases and World Health Organization Izumo, Geneva.
27. Yamori, Y. 1989, Preliminary report of CARDIAC study: corss-sectional multicenter study on dietary factors of cardiovascular diseases, *Clin.Exp.Hypertens.* A11:957-972.
28. Yamori, Y., Horie, R., Nara, Y., Kihara, M., Fujiwara, K., Nabika, T., Mano, M., and Ikeda, K. 1992, Nutritional causation and prevention of cardiovascular diseases; experimental evidence in animal models and man, in: "Nutritional Prevention of Cardiovascular Diseases", Lovenberg, W. and Yamori, Y. eds., Academic Press, Orlando, FL, pp. 37-52.
29. Yamori, Y., Horie, R., Nara, Y., Tagami, M., Kihara, M., Mano, M., and Ishino, H. 1987, Pathogenesis and dietary prevention of cerebrovasular diseases in animal models and epidemiological evidence for the applicability in man, in: "Prevention of Cardiovascular Diseases, an Approach to Active Long Life", Yamori, Y. and Lenfant, C. eds., Elsevier, Amsterdam, pp. 163-177.
30. Yamori, Y., Mizushima, S., Sawamura, M., and Nara, Y. 1994, Nutritional factors for hypertension and major cardiovascular diseases: international cooperative studies for dietary prevention, *Deutsche Medizinische Wochenschrift*, 24:1825-1841.
31. Yamori, Y., Mizushima, S., Ueda, M., Sawamura, M., and Nara, Y. 1994, Inverse relationship of urinary magnesium and urea nitrogen excretion to blood pressure, *J.Hypertens.* 12:S61
32. Yamori, Y., Nagaoka, A., and Okamoto, K. 1974, Importance of genetic factors in stroke: an evidence obtained by slective breeding of stroke-prone and resistant SHR, *Jpn.Circ.J.* 38:1095-1100.
33. Yamori, Y., Nara, Y., Horie, R., Ooshima, A., and Lovenberg, W. 1981, Pathophysiological role of taurine in blood pressure regulation in stroke-prone spontaneously hypertensive rats (SHR), in: "The Effects of Taurine on Excitable Tissues", Schaffer, S.W., Baskin, S.I. and Kocsis, J.J. eds., Spectrum Publications, New York, pp. 391-403.
34. Yamori, Y., Nara, Y., Ikeda, K., Tsuchikura, S., Eguchi, T., Mano, M., Horie, R., and Sugawara, T. 1989, Recent advances in experimental studies on dietary prevention of cardiovascular diseases, in: "New Horizons in Preventing Cardiovascular Diseases", Yamori, Y. and Strasser, T. eds., Elsevier, Amsterdam, pp. 1-11.
35. Yamori, Y., Nara, Y., Mizushima, S., Gonzalez-Biosca, M.D., Sawamura, M., Horie, R., Mori, C., and Endo, J. 1992, Dietary risk factors for ischemic heart disease and stroke- a report on "complete" WHO CARDIAC study, *J.Hypertens.* 10:39.
36. Yamori, Y., Nara, Y., Mizushima, S., Ikeda, K., Sawamura, M., Nabika, T., and Horie, R. 1992, Gene-environmental interaction in hypertension, stroke and atherosclerosis in experimental models and supportive findings from a world-wide cross-sectional epidemiological survey: a WHO-CARDIAC study, *Clin.Exp.Pharmacol.Pysiol.* 19:43-52.
37. Yamori, Y., Nara, Y., Mizushima, S., Mano, M., Sawamura, M., and Horie, R. 1990, Risk factors for prevention in various populations with no frequent hypertensive treatment - a report from CARDIAC study, *J.Hypertens.* 8:S124.
38. Yamori, Y., Nara, Y., Mizushima, S., Mano, M., Sawamura, M., Horie, R., and Mori, C. 1992, Dietary risk factors for high blood pressure — a report on "core" WHO CARDIAC study, *J.Hypertens.* 10:39.
39. Yamori, Y., Nara, Y., Mizushima, S., Mano, M., Sawamura, M., Kihara, M., and Horie, R. 1990, International cooperative study on the relationship between dietary factors and blood pressure: a report

from the cardiovascular and alimentary comparison (CARDIAC) study, *J.Cardiovasc.Pharmacol.* 16:s43-s47.

40. Yamori, Y., Nara, Y., Mizushima, S., Mano, M., Sawamura, M., Kihara, M., and Horie, R. 1992, International cooperative study on the relationship between dietary factors and blood pressure: a preliminary report from the cardiovascular diseases and alimentary comparison (CARDIAC) study, *Nutrition and Health*, 8:77-90.
41. Yamori, Y., Nara, Y., Mizushima, S., Sawamura, M., and Horie, R. 1994, Nutritional factors for stroke and major cardiovascular diseases: international epidemiological comparison of dietary prevention, *Health Reports*, 6:181-188.

IS TAURINE A PREVENTIVE NUTRITIONAL FACTOR OF CARDIOVASCULAR DISEASES OR JUST A BIOLOGICAL MARKER OF NUTRITION?

Yukio Yamori,[1] Yasuo Nara,[2] Katsumi Ikeda,[2]
and Shunsaku Mizushima[2]

[1] Life Science, Environmental Conservation and Development
[2] Otsuka Department of International Preventive Nutritional Medicine
Graduate School of Human and Environmental Studies
Kyoto University, Kyoto, Japan 606

INTRODUCTION

For answering the question, "Is taurine a preventive nutritional factor of cardiovascular diseases (CVD) or just a biological marker of nutrition?" our experimental and world-wide cooperative studies on nutrition and CVD are reviewed. The data so far obtained indicates taurine is not only a good biological marker of sea food intakes but also an important nutrient for preventing hypertension and atherosclerosis.

EXPERIMENTAL STUDIES ON TAURINE AND CVD

Effect of Taurine on Hypertension and Stroke

Basic research on hypertension and stroke has made remarkable progress with the establishment of rat models of hypertension, particularly the spontaneously hypertensive rat (SHR)[8,10] and the stroke-prone SHR (SHRSP)[9,13], that develops severe hypertension and dies from hemorrhagic and ischemic strokes. These models are useful for studying the pathogenesis and prevention of hypertension and stroke.

Since we observed that fish protein-supplemented diet decreased blood pressure and stroke mortality in this model[12,15], we further proved that taurine, which is rich in fish, attenuated the development of hypertension and could prevent stroke[7]. We have recently demonstrated epidemiological evidence for the similar beneficial effect of fish protein intake on blood pressure in populations distributed world-wide[19].

Taurine 2, edited by Huxtable et al.
Plenum Press, New York, 1996

Effect of Taurine on Cholesterolemia and Atherogenesis

Taurine not only attenuates hypertension but also suppresses reactive hypercholesterolemia in SHRSP fed on high-fat cholesterol diets with or without 3% taurine supplementation[26]. This hypocholesterolemic effect is due to the increased excretion of cholesterol into bile acids by the activation of 7α-hydroxylase by taurine[5].

Such an attenuation of hypercholesterolemia was also demonstrated in healthy young volunteers given cholesterol-rich diets with or without taurine supplementation[4].

We further demonstrated 3% taurine-supplemented diets attenuated the activation of acetyl coenzyme A cholesterol acyltransferase, (ACAT) in the intestine, which was activated by cholesterol-rich diets[5]. Since ACAT of the intestinal microsomal fraction is important in cholesterol absorption from the intestine, taurine decreases cholesterol absorption to attenuate the reactive hypercholesterolemia.

When SHRSP are fed high-fat-cholesterol diets, abundant fat deposits are noted within a short period in mesenteric arteries. Taurine-supplemented diets not only attenuate hypercholesterolemia but also prevent fat deposition[19].

We further demonstrated cultured vascular smooth muscle cells (VSMC) obtained from the aorta of SHRSP incorporated more LDL cholesterol than those from normotensive Wistar-Kyoto control (WKY)[20,25] rats, and we noted the addition of catecholamins into the culture medium accelerated cholesterol incorporation into VSMC[16].

We previously reported catecholamine secretions induced by stress such as cold exposure were significantly greater in SHR and SHRSP, genetically predisposed to hypertension than in WKY[17]. We also found that diets could modify the neurohormonal response to stress. Exaggerated urinary norepinephrine excretion in stroke-prone SHR were significantly reduced by taurine-rich diets[2]. Therefore, taurine is expected to attenuate the norepinephrine-induced LDL incorporation into VSMC, thus preventing the process of atherosclerosis at the cellular level.

We can now conclude dietary components such as taurine are important and effective for preventing hypertension and atherosclerosis in our animal models.

EPIDEMIOLOGICAL STUDIES ON TAURINE AND CVD

Design of the Studies

In order to test the applicability of such effective dietary prevention to humans, we proposed a world-wide epidemiological study, the so-called CARDIAC Study[11], to WHO in 1982. CARDIAC Study is the abbreviation of Cardiovascular Diseases and Alimentary Comparison Epidemiological Study, which was designed to investigate the relation of biological markers of diet with hypertension in the "Core Study" and with CVD mortalities in the "Complete Study."[14,16,18,22-24]

We have carried out epidemiological surveys in 58 populations of 25 countries for the past 12 years, thanks to the great cooperation of leading international scientists. The target population consisted of 100 males and 100 females at the age range of 50 to 54 randomly selected from each study site. Participants numbered over 10,000 in total.

One characteristic of the CARDIAC Study was, first, standardized measurement of BP by an automated BP measurement system[1] which was used world-wide, even in the Masai, who used no salt at all, showed a low urinary NaCl excretion of around 2.5 g, and seldom had hypertension[14]. A second characteristic was the use of biological markers of 24 h urine and blood samples for objective nutritional estimation. For collecting 24 h urine easily, a new device, called 'Aliquot cups'[6,21] were introduced successfully, even to such inhabitants of developing regions such as Tibetan people who cooperatively collected 24 h urine samples, containing an average 18 g salt, nearly the highest in the world. All samples were centrally analyzed in our laboratory for the objective estimation of dietary intake of various nutrients by biological markers in urine and blood samples[11,14,16,18,22-24].

Nutrition and Blood Pressure

This analysis demonstrated for the first time on a world-wide basis that both systolic and diastolic BPs were positively related to 24 h urinary Na^+ excretion, indicating excess salt intake as a risk factor of hypertension[23].

Another risk factor of hypertension was proven to be obesity. In contrast, urinary Mg^{2+} excretion was inversely related to both systolic and diastolic BPs, indicating a beneficial effect of Mg^{2+} intake on blood pressure.

Nutrition and CVD

Our study further proved stroke mortalities were significantly positively related with urinary Na^+ excretion and also Na^+/K^+ ratios, indicating the adverse effect of Na^+ and the beneficial effect of K^+ on stroke[18].

Coronary heart disease mortalities were significantly positively related with serum cholesterol levels in various populations in the world. This was the confirmation of the previous "7 countries" study conducted by Keys and others[3].

What we found the most interesting among our abundant results was a significant inverse relationship between urinary taurine excretion and ischemic heart disease mortalities (Fig. 1).[18,22,24] These data suggest what we should eat for preventing ischemic heart diseases. Three populations with the highest taurine excretion and the lowest mortalities were all Japanese populations. Mediterranean populations followed the Japanese. Japanese now

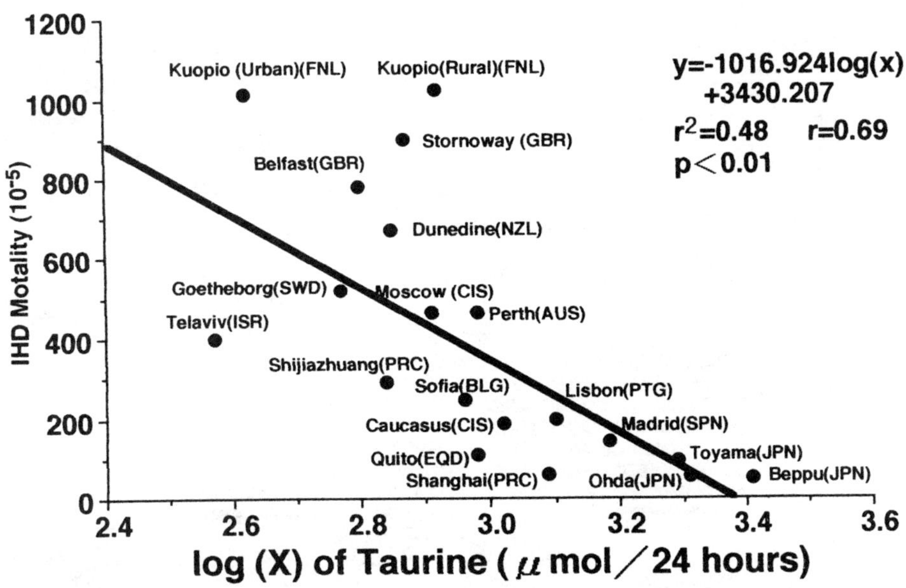

Figure 1. Correlation between urinary taurine excretion (μmol) for 24 h and age-adjusted mortality rates from ischemic heart disease (IHD) in 19 populations.

enjoy the longest life span in the world with the least mortality due to heart diseases. This may be ascribed to the high taurine intake from sea food. The dietary goal for preventing ischemic heart diseases in men appear to be taurine intake enough for maintaining 2000 μmol urinary taurine excretion per day. We also demonstrated a similar significant inverse relationship in females. This inverse relationship indicates coronary heart diseases might be reduced as low as in the Japanese who are fortunately maintaining urinary taurine excretion around 1500 μmol per day through their greater fish consumption.

Taurine Intake and CVD Prevention

Our CARDIAC Study also confirmed that the weekly frequency of fish intake was significantly positively related with urinary taurine excretion. In order to keep urinary taurine levels over 1500 μmol per day, fish should be eaten 5 days a week or more, as most Japanese and Mediterranean populations are eating.

If we combine our data of the fish intake frequency with the total sea food consumption per capita from OECD statistics, the intake of fish 5 days a week corresponds to 100 g of fish and shell fish intake a day on average. This suggests 100 g of fish intake per day on average or 5 days of fish weekly may enable us to keep high urinary taurine excretion in order to keep coronary heart disease mortality as low as in the Japanese.

Among 57 populations examined by our surveys, the 7 Japanese populations who were regularly eating fish 4 to 5 times a week showed the highest urinary taurine excretion (Fig. 2). Even the Japanese immigrants living in Sao Paulo, Brazil, showed a higher level than Mediterranean populations[18]. The lowest level in the world was the Tibetans, who never ate fish for religious reasons and were confirmed to have severe hypertension twice as frequently as in the Japanese living in north eastern Japan who had salt intakes close to the Tibetans.

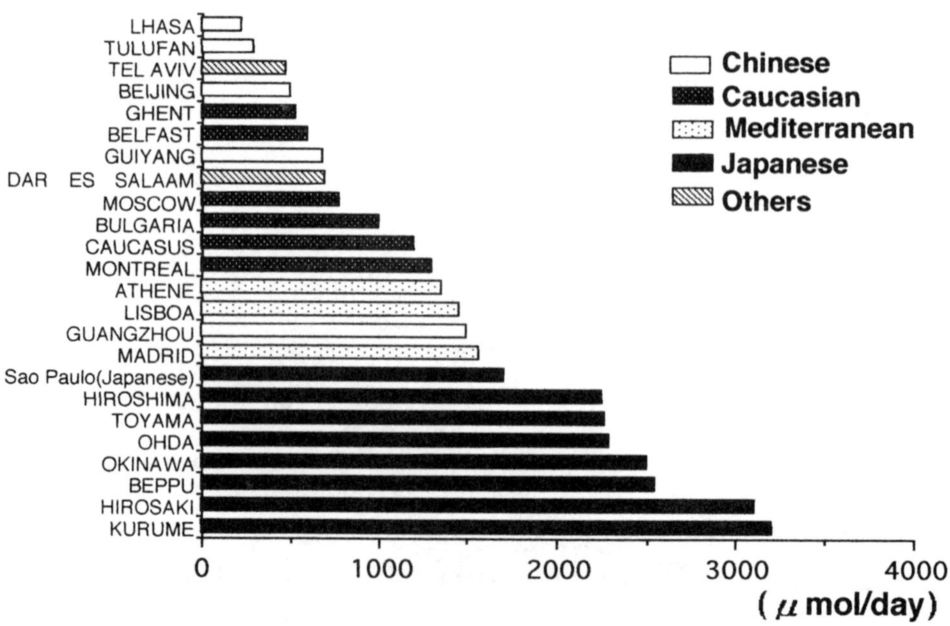

Figure 2. International comparison of urinary taurine excretion (μmol/day).

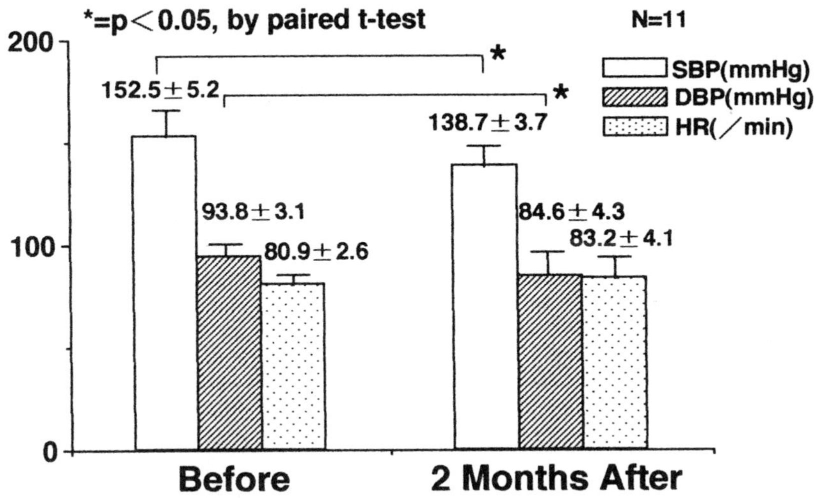

Figure 3. Effect of oral taurine intake (3 g/day) on blood pressure among mild or borderline hypertensives in Namche Bazar.

Clinical Trial of the Effect of Taurine on Blood Pressure

We did a clinical experiment to test the effect of taurine in Tibetans living in Namche Basar at the foot of Mt. Everest. Mild hypertensive and borderline hypertensive men were given 3 g of taurine per day for 2 months and their blood pressures were measured by an automated BP measurement system[1] to avoid observer bias. Both systolic and diastolic BPs were significantly decreased in 11 out of 17 who took taurine regularly, but not in the 6 who took taurine less than 10 days during 2-month period (Fig. 3). This experiment indicates that taurine administration may be more effective in the people whose daily taurine intake is not enough.

CONCLUSION

Among various biological markers of nutrition, taurine was inversely related to coronary heart disease mortality in our world-wide epidemiological survey. Taurine supplementation experimentally improves hypercholesterolemia by accelerating cholesterol excretion into bile acids and also by attenuating intestinal cholesterol absorption, and thus preventing atherosclerosis. Moreover, taurine suppresses the sympathetic nervous system to decrease BP and also attenuates stress-induced exaggerated catecholamine excretion to prevent LDL incorporation into VSMC. These and possibly other mechanisms may contribute to the prevention of coronary heart diseases through the direct or indirect improvement of hypertension, atherosclerotic and thrombotic processes. Therefore, taurine is not only a good marker of sea food intake but also an important contributory nutrient for cardiovascular disease prevention.

REFERENCES

1. Fukuoka, M. and Yamori, Y. 1987, A proposal for indirect and objective blood pressure measurement in adults, in: "Prevention of Cardiovascular Diseases: An Approach to Active Long Life", Yamori, Y. and Lenfant, C. eds., Elsevier, Amsterdam, pp. 127-137.

2. Ikeda, K., Nara, Y., Horie, R., and Yamori, Y. 1989, Diets modify pressor and catecholamine responses to stress in stroke-prone spontaneously hypertensive rats (SHRSP), in: "New Horizons in Preventing Cardiovascular Diseases", Yamori, Y. and Strasser, T. eds., Elsevier, Amsterdam, pp. 23-25.
3. Keys, A. 1980, " Seven Countries: A Multivariate Analysis of Death and Coronary Heart Disease", Harvard University, Boston.
4. Mizushima, S., Nara, Y., Sawamura, M., and Yamori, Y. 1996, Effects of oral taurine supplementation on lipids and sympathetic nerve tone, in: "Taurine: Basic and Clinical Aspects", Huxtable, R.J., Azuma, J., Kuriyama, K., Nakagawa, M. and Baba, A. eds., Plenum Press, New York.
5. Murakami, S., Yamagishi, I., Asami, Y., Ohta, Y., Toda, Y., Nara, Y., and Yamori, Y. 1996, Hypolipidemic effect of taurine in stroke-prone spontaneously hypertensive rats (SHRSP), *Pharmacology*, (in press).
6. Nara, Y., Kihara, M., Mano, M., Horie, R., and Yamori, Y. 1984, "Aliquot cups", simple method for collecting consecutive 24-hour urine samples for epidemiological and clinical studies, in: "Nutritional Prevention of Cardiovascular Disease", Lovenberg, W. and Yamori, Y. eds., Academic Press, Orlando, pp. 211-216.
7. Nara, Y., Yamori, Y., and Lovenberg, W. 1978, Effects of dietary taurine on blood pressure in spontaneously hypertensive rats, *Biochem.Pharmacol.* 27:2689-2692.
8. Okamoto, K. and Aoki, K. 1963, Development of a strain spontaneously hypertensive rats, *Jpn.Circ.J.* 27:202-293.
9. Okamoto, K., Yamori, Y., and Nagaoka, A. 1974, Establishment of the stroke-prone spontaneously hypertensive rat (SHR), *Circ.Res.* 34:1143-1153.
10. Okamoto, K., Yamori, Y., Ooshima, A., Park, C., Haebara, H., Matsumoto, M., Tanaka, T., Okuda, T., Hazama, F., and Kyogoku, M. 1972, Development and heredity: establishment of the inbred strain of the spontaneously hypertensive rat and genetic factors involved in hypertension, in: "Spontaneous Hypertension: Its Pathogenesis and Complications", Okamoto, K. ed., Igaku Shoin, Tokyo, pp. 1-8.
11. WHO and WHO Collaborating Centers 1986, " CARDIAC (cardiovascular diseases and alimentary comparison) study protocol", Shimane, Geneva.
12. Yamori, Y. 1981, Environmental influences on the development of hypertensive vascular diseases in SHR and related models, and their relation to human disease, in: "New Trends in Arterial Hypertension (INSERM Symposium No. 17)", Worcel, M., Bonvalet, J.P., Langer, S.Z., Menard, J. and Sassard, J. eds., Elsevier, Amsterdam, pp. 305-320.
13. Yamori, Y. 1984, The stroke-prone spontaneously hypertensive rat: contribution to risk factor analysis and prevention of hypertensive diseases, in: "Handbook of Hypertension, Vol. 4: Experimental and Genetic Models of Hypertension", de Jong, W. ed., Elsevier, Amsterdam, pp. 240-255.
14. Yamori, Y. 1989, Hypertension and biological dietary markers in urine and blood-a progress report from the CARDIAC study group, in: "New Horizons in Preventing Cardiovascular Diseases", Yamori, Y. and Strasser, T. eds., Elsevier, Amsterdam, pp. 111-126.
15. Yamori, Y., Horie, R., Ikeda, K., Nara, Y., and Lovenberg, W. 1979, Prophylactic effect of dietary protein on stroke and its mechanisms, in: "Prophylactic Approach to Hypertensive Diseases", Yamori, Y., Lovenberg, W. and Freis, E.D. eds., Raven Press, New York, pp. 497-504.
16. Yamori, Y., Horie, R., Nara, Y., Tagami, M., Kihara, M., Mano, M., and Ishino, H. 1987, Pathogenesis and dietary prevention of cerebrovasular diseases in animal models and epidemiological evidence for the applicability in man, in: "Prevention of Cardiovascular Diseases, an Approach to Active Long Life", Yamori, Y. and Lenfant, C. eds., Elsevier, Amsterdam, pp. 163-177.
17. Yamori, Y., Ikeda, K., Kulakowski, E.C., MaCarty, R., and Lovenberg, W. 1985, Enhanced sympathetic medullary response to cold exposure in spontaneously hypertensive rats, *J.Hypertens.* 3:63-66.
18. Yamori, Y., Mizushima, S., Sawamura, M., and Nara, Y. 1994, Nutrition factors for hypertension and major cardiovascular diseases: international cooperative studies for dietary prevention, *Deutsche Med.Woch.*, 15:1825-1841.
19. Yamori, Y., Mizushima, S., Ueda, M., Sawamura, M., and Nara, Y. 1996, Inverse relationship of urinary magnesium and urea nitrogen excretion to blood pressure, *WHO-Cardiac Study, Abstract of 15th Scientific Meeting of the International Society of Hypertension*, (in press).
20. Yamori, Y., Murakami, S., and Nara, Y. 1991, Activated acyl coenzyme A-cholesterol acyltransferase as a possible cause of rapid arterial fat deposition in genetic hypertensive rat models for atherosclerosis, *J.Vasc.Med.Biol.* 3:268-270.
21. Yamori, Y., Nara, Y., Kihara, M., Mano, M., and Horie, R. 1984, Simple method for sampling consecutive 24-hour urine for epidemiological and clinical studies, *Clin.Exp.Hypertens.* A6:1161-1167.
22. Yamori, Y., Nara, Y., Mizushima, S., Mano, M., Sawamura, M., Kihara, M., and Horie, R. 1992, International cooperative study on the relationship between dietary factors and blood pressure: a

preliminary report from the cardiovascular diseases and alimentary comparison (CARDIAC) study, *Nutrition and Health*, 8:77-90.
23. Yamori, Y., Nara, Y., Mizushima, S., Mano, M., Sawamura, M., Kihara, M., Horie, R., and Hatano, S. 1990, International cooperative study on the relationship between dietary factors and blood pressure: a report from the cardiovascular diseases and alimentary comparison study, *J.Cardiovasc.Pharmacol.* 16:S43-S47.
24. Yamori, Y., Nara, Y., Mizushima, S., Murakami, S., Ikeda, K., Sawamura, M., Nabika, T., and Horie, R. 1992, Gene-environment interaction in hyertension, stroke and atherosclerosis in experimental models and supportive findings from a world-wide cross-sectional epidemiological survey: a WHO-CARDIAC study, *Clin.Exp.Pharmacol.Physiol.* 19:43-52.
25. Yamori, Y., Nara, Y., Tagami, M., Mano, M., Kihara, M., Fujiwara, K., Horie, R., and Kamino, K. 1984, Common cellular disposition to hypertension and atherosclerosis, *J.Hypertens.* 2:S213-S215.
26. Yamori, Y., Wang, H., Ikeda, K., Kihara, M., Nara, Y., and Horie, R. 1983, Role of sulfur amino acids in the prevention and regression of cardiovascular diseases, in: "Sulfur Amino Acids, Biochemical and Clinical Aspects", Kuriyama, K., Huxtable, R.J. and Iwata, H. eds., A.R. Liss, New York, pp. 103-116.

THE INFLUENCE OF DIFFERENT TAURINE DIETS ON HEARING DEVELOPMENT IN NORMAL BABIES

A Preliminary Report

W. E. Davies,[1] P. C. Hopkins,[1] S. J. Rose,[2] and S. K. Dhillon[3]

[1] Department of Pharmacology
The Medical School, University of Birmingham
[2] Department of Paediatrics
Heartlands Hospital, Birmingham
[3] Hearing Services Centre
City Hospital, Birmingham

The cochleae of animals contain large quantities of taurine[1] but its function there remains uncertain[3,15]. It is well established that taurine levels of most organs are high at birth and decrease slowly during maturation but little is known about the clinical significance of this. Tyson *et al.*[13] suggest that dietary taurine levels influence hearing development in pre-term infants but whether this data justifies the routine supplementation of baby food is somewhat contentious[10,14]. Nevertheless, most baby milk powders are now supplemented with taurine at 30-40 mg/100 g. However, the decision to make this supplementation has been made on rather less than conclusive evidence. The major argument for supplementation is that human milk contains substantial quantities of taurine[11]. There are also a number of animal studies that indicate its essentiality, certainly in some species. There is little doubt that taurine is an essential dietary component in some species such as cat[4,12] and the evidence that taurine is essential for good development and function of the retina in monkey is substantial[6,9]. The evidence gathered from studies on babies, however, is much more equivocal. Studies on pre-term babies[13] and babies receiving long-term parenteral nutrition[2] suggest that taurine might be essential for good sensory development. Tyson *et al.*[13] used the auditory brain stem-evoked response (ABER) and electro-retinograms to monitor the development of their subjects and although they found no difference in the retinograms of babies fed the different diets, some differences in the ABERs were noted. Michalk *et al.*[8] were unable to detect any differences in the nervous or cardiovascular systems of term babies fed taurine-supplemented diets. Little further data are available on the influence of taurine on the development of normal babies.

We have consequently embarked on a study of normal term babies receiving different levels of dietary taurine using a developmental assessment of their hearing using ABERs

Taurine 2, edited by Huxtable et al.
Plenum Press, New York, 1996

Table 1. Selection criteria

No family history of congenital deafness
Born 38-42 weeks gestational age
No observed abnormalities
Birth weight ≥ 2.5kg
Apgar 1 min ≥ 7
Apgar 5 min ≥ 8
Bilirubin < 10mg/dL
Mother - Rubella immune
Negative VDRL

and transient evoked oto-acoustic emissions (EOAEs). The former was used to assess the integrity of the organ of Corti, the auditory nerve and some central neurotransmission processes, whilst the latter monitored the development of the outer hair cells, the attenuation/amplification gain system of the cochlea, and its efferent neural connections.

In an effort to get indirect evidence on the taurine relationship between mother and fetus, we have also investigated the taurine levels in mothers and cords at birth and also assayed taurine levels in the blood of women at different stages of pregnancy.

1. METHODS

Normal babies, selected by the criteria shown in Table 1, were randomly allocated to one of two casein/whey diets, one containing a relatively low level of taurine (2.5 mg/100 g powder) and one containing a relatively high level of taurine (34.5 mg/100 g powder). A breast-fed control group were also assessed in the same manner.

1.1 ABER Measurement

The acoustic stimulation consisted of 100μs clicks presented via TDH 39 earphones. ABERs were recorded with the infant in natural sleep. Two channel, ipsilateral and contralateral, recordings were made with monaural click presentation. The clicks were presented at a repetition rate of 19.1/s and at intensities of 75, 55, 35 and 25 dBnHL. At least one replication was recorded at each stimulus intensity. Four Ag/AgCl electrodes were attached to the high forehead and each mastoid, with a low forehead earth (ground). All electrode impedances were below 10 kΩ. A minimum of 1200 averaged stimuli were used with a filter bandwidth of 100 - 3000 Hz. The replicated responses were analysed off-line. Waves I through to V were identified and the interpeak intervals I-III, III-V and I-V calculated.

1.2 EOAE Measurement

Transient evoked otoacoustic emissions were recorded using the commercially available IL088 system (Otodynamics Ltd). Recordings were made with the child either still or asleep. The neonatal probe assembly was used to deliver click stimuli at supra threshold levels. Stimulus intensities within the individual ear canal ranged from 75 to 85 dBpk. A minimum of 50 and a maximum of 260 sweep averages were recorded from each ear. Testing was undertaken in non-soundproofed accommodation requiring the noise rejection to be set at 48-55 dB. Care was taken to monitor probe position throughout the recordings and each test was replicated. Responses were analysed off-line. Only responses with reproducibility greater than 60% were assessed.

Blood taken from a heel-prick was routinely assayed for whole blood taurine using the high pressure liquid chromatographic method of Hopkins *et al.*[5]. Auditory assessments were normally made two days after birth to allow the ears to clear of all fluid. Measurements were repeated at 2, 4, 8, 12 and 24 weeks after birth.

The blood samples from the pregnant mothers were all taken from rhesus-negative mothers and the cord and mother bloods were routine samples obtained from the Queen Elizabeth Maternity Hospital, Birmingham.

2. RESULTS

Figures 1 and 2 show a typical ABER (12 weeks) and EOAE (8 weeks) respectively.

Table 2 shows the latency of wave I, wave V and the I-V interpeak interval of the babies in each of the three groups. There is no significant difference to be seen in any of these values.

Table 3 shows the intensity of the otoacoustic emission at 8 weeks in the three groups of babies. Whilst there are some differences in the intensity of the full frequency spectrum intensities and the 1/3 octave centred around 4 KHz, these differences are not statistically significant at the 0.05 level.

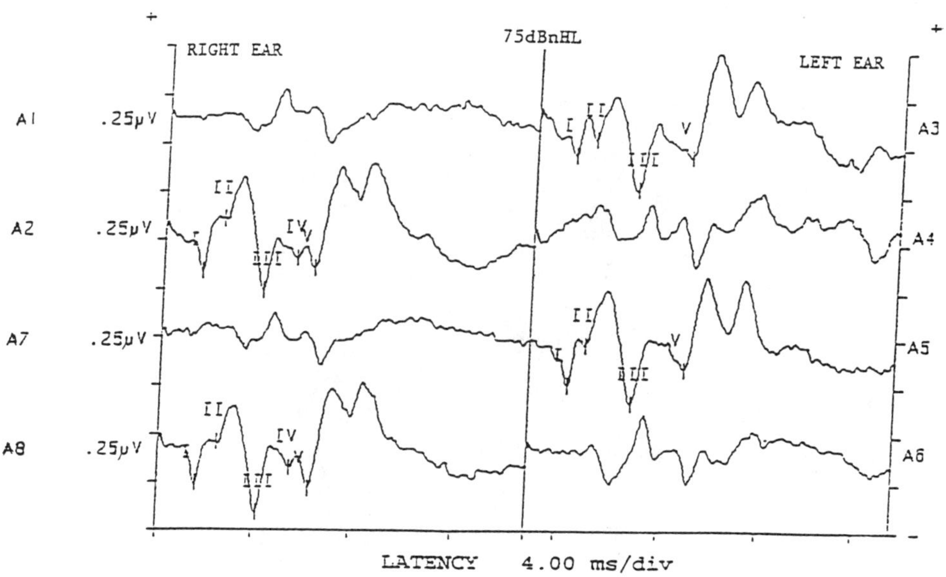

	Latencies (ms)					Interpeak Intervals (ms)		
	I	II	III	IV	V	I-III	III-V	I-V
A2	1.68	2.49	4.17	5.57	6.26	2.49	2.09	4.58
A3	1.51	2.26	3.94		6.03	2.43	2.09	4.52
A5	1.51	2.14	3.94		6.03	2.43	2.09	4.52
A8	1.68	2.49	4.17	5.51	6.32	2.49	2.15	4.64

Figure 1. Brain stem evoked responses recorded using click stimulation at 75 dBnHL showing ipsilateral and contralateral wave forms.

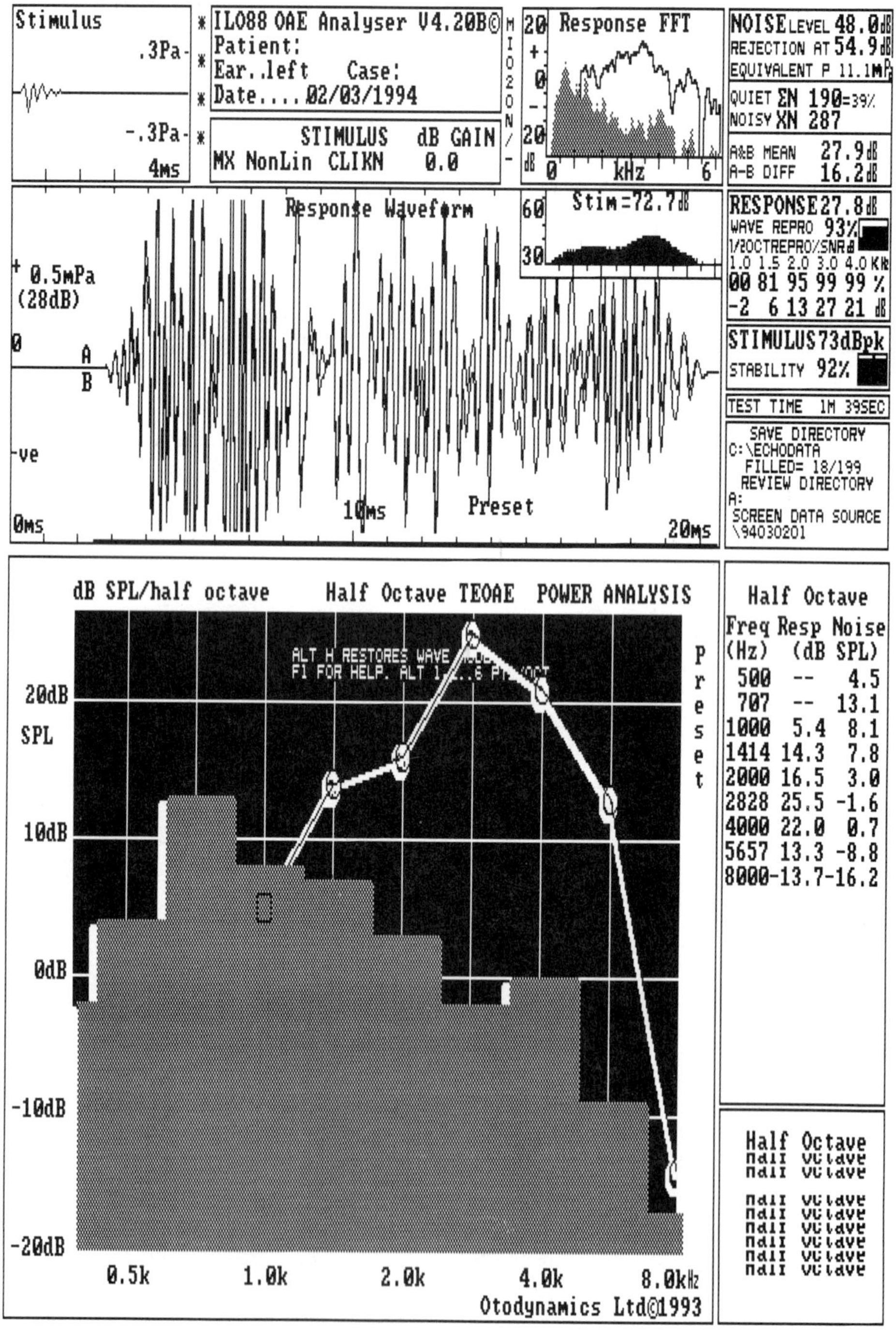

Figure 2. Various parameters measured when recording a transient evoked otoacoustic emission. On the "Response FFT" the hatched area represents the ambient noise and the plain area above the noise is the otoacoustic emission.

Table 2. ABERs 8 weeks post-partum

	Low taurine[1]	High taurine[1]	Breast-fed[2]
Wave I latency (ms)	1.605 ± 0.023	1.632 ± 0.050	1.687 ± 0.059
Wave V latency	6.414 ± 0.053	6.437 ± 0.086	6.577 ± 0.069
Wave I-V	4.809 ± 0.053	4.805 ± 0.097	4.890 ± 0.096

The stimulus intensity was 75 dBnHL. [1]n=10; [2]n=6. Data are means ± SEM

Table 3. EOAEs 8 weeks post-partum

	Low taurine[1]	High taurine[1]	Breast-fed[2]
Emission intensity (dB) (Full frequency spectrum)	22.01 ± 1.78	24.57 ± 2.16	22.38 ± 1.12
Emission intensity (dB) (at 4 kHz)	13.45 ± 2.27	18.41 ± 2.02	12.57 ± 1.85

[1]n=10; [2]n=6. Data are means ± SEM

Figure 3 shows no consistent difference between the taurine contents of whole blood of 53 mothers and their infants at the time of birth. This clearly shows that mothers' taurine levels at birth are neither consistently higher nor consistently lower than those of their infants at birth.

Figure 4 shows the whole blood taurine content of 44 pregnant mothers at various stages of pregnancy. There is clearly no correlation between the blood taurine and the pregnancy stage and, therefore, the mother does not appear to change her blood taurine content as pregnancy proceeds.

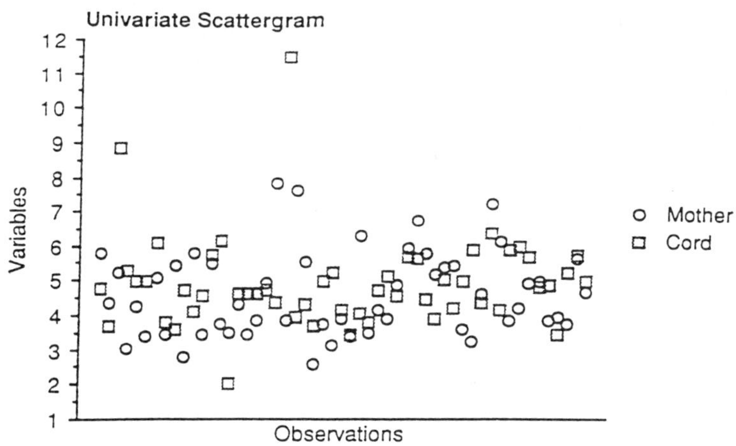

Paired Means Comparison
Hypothesized Difference = 0

	Mean Diff.	DF	t-Value	P-Value	95% Lower	95% Upper
Mother, Cord	-.306	52	-1.230	.2241	-.806	.193

Figure 3. Statistical analysis of the mother/cord taurine levels. Variables = taurine levels (mg/100 ml whole blood). Matched pair analysis showed that the taurine content of mother's blood was neither significantly higher nor significantly lower than her baby's blood taurine level.

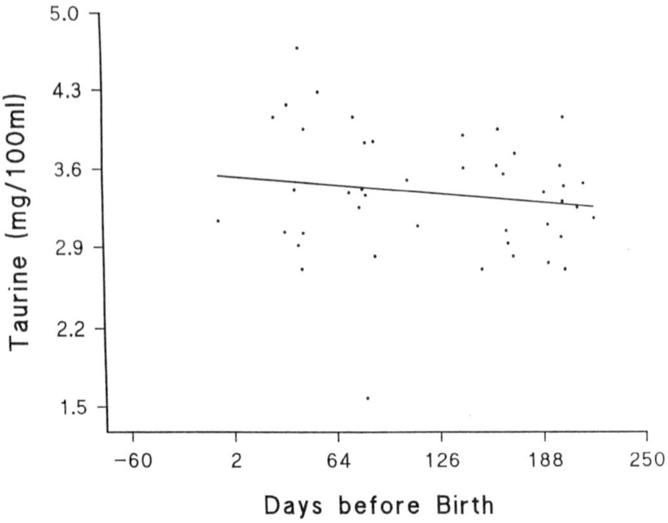

Figure 4. Correlation of whole blood taurine of pregnant mothers with length of pregnancy. With r = 0.150, there is no correlation between stage of pregnancy and blood taurine levels.

3. CONCLUSIONS

This study is still in progress and the complete data are not yet available. However, with 26 subjects having completed the 8-week post-partum stage, we have been unable to show any changes in our audiological assessment of the hearing ability of term infants. Thus we can provide no evidence to justify the supplementation of baby powders with the levels of taurine used at present (35 mg%). Whilst there may be a need for some dietary taurine in the human[7], supplementation at this level could, at best, be unnecessary. Normally, the kidney is able to stabilise the level of plasma taurine, but feeding babies with large quantities of taurine, before kidney function fully develops or where kidney function is impaired, could lead to toxic levels. Although it is generally accepted that taurine is nontoxic, it has been reported[16] that high taurine levels can be toxic to developing fertilised eggs. There is little doubt that one of the functions of taurine in the body is that of osmoregulation and cell volume control. It is, therefore, not inconceivable that during early development excess taurine could interfere with sensory end-organ function which is undergoing massive cellular/molecular development in response to sensory stimulation.

Our data on mother/cord blood taurine levels and the taurine levels of mothers during pregnancy suggest that mothers have no great metabolic concern for the taurine available to their babies. The fact that breast milk is particularly rich in taurine immediately after birth may only reflect that the mother is recovering from a period of stress and high metabolic activity with a concomitant increase in plasma by-products, including taurine. This could then lead to an elevated colostrum taurine level. With the possibility that the human may have an additional synthetic mechanism for the production of taurine[7] it is quite tenable that taurine in infant nutrition is not so vital as to justify the present taurine supplementation of baby food.

ACKNOWLEDGMENT

We gratefully acknowledge the financial support of Action Research and Cow and Gate Ltd. for the baby food.

REFERENCES

1. Davies, W.E. and Owen, C. 1985, The nature of neurotransmitters in the mammalian lower auditory system, in: "Auditory biochemistry", Drescher, D.G. ed., Thomas, Springfield,IL, pp. 244-257.
2. Geggel, H.S., Ament, M., Heckenlively, J.R., Martin, D.A., and Kopple, J.D. 1985, Nutritional requirement for taurine in patients receiving long-term parenteral nutrition, *N.Engl.J.Med.* 312:142-146.
3. Harding, N.J. and Davies, W.E. 1993, Cellular localisation of taurine in the organ of Corti, *Hearing Res.* 65:211-215.
4. Hayes, K.C., Carey, R.E., and Schmidt, S.Y. 1985, Retinal degeneration associated with taurine deficiency in the cat, *Nutr.Rev.* 43:84-86.
5. Hopkins, P.C., Kay, I.S., and Davies, W.E. 1989, A rapid method for the determination of taurine in biological tissue, *Neurochem.International*, 15:429-432.
6. Imaki, H., Moretz, R., Wisniewski, H., Neuringer, M., and Sturman, J. 1987, Retinal degeneration in three-month-old rhesus monkey infants fed a taurine-free human infant formula, *J.Neurosci.Res.* 18:602-614.
7. Irving, C.S., Marks, L., Klein, P.D., Foster, N., Gadde, P.L., Chase, T.N., and Samuel, D. 1986, New evidence for taurine biosynthesis in man obtained from $^{18}O_2$ inhalation studies, *Life Sci.* 38:491-495.
8. Michalk, D., Ringeisen, R., Tittor, F., Lauffer, H., Deeg, K., and Böhles, H. 1988, Development of the nervous and cardiovascular systems in low-birth-weight infants fed a taurine-supplemented formula, *Eur.J.Pediat.* 147:296-299.
9. Neuringer, M., Sturman, J.A., Wen, G.Y., and Wisniewski, H.M. 1985, Dietary taurine is necessary for normal retinal development in monkeys, in: "Taurine: biological actions and clinical perspectives", Oja, S.S., Ahtee, L., Kontro, P. and Paasonen, M.K. eds., Alan R. Liss, New York, pp. 53-62.
10. Perlman, M. 1989, Taurine and auditory system maturation, *Pediatrics*, 83:796-798.
11. Rassin, D.K., Sturman, J.A., and Gaull, G.E. 1978, Taurine and other free amino acids in milk of man and other mammals, *Early Human Devel.* 2:1-13.
12. Sturman, J.A. 1993, Taurine in development, *Physiol.Rev.* 73:119-147.
13. Tyson, J.E., Lasky, R., Flood, D., Mize, C., Picone, T., and Paule, C.L. 1989, Randomized trial of taurine supplementation for infants ≤1,300-gram birth weight: Effect on auditory brainstem-evoked responses, *Pediatrics*, 83:406-415.
14. Tyson, J.E., Mize, C., and Lasky, R. 1989, More on taurine, *Pediatrics*, 83:1072-1073.
15. Usami, S. and Ottersen, O.P. 1995, The localization of taurine-like immunoreactivity in the organ of Corti: A semiquantitative, post-embedding immuno-electron microscopic analysis in the rat with some observations in the guinea pig, *Brain Res.* 676:277-284.
16. Van Gelder, N.M. and Bélanger, F. 1988, Embryonic exposure to high taurine: A possible nutritional contribution to Friedreich's ataxia, *J.Neurosci.Res.* 20:383-389.

Part 9

SPECIAL LECTURE—TAURINE AND ENANTIOSTASIS

71

TAURINE

Past, Present, and Future

Ryan J. Huxtable

Department of Pharmacology
College of Medicine
University of Arizona
Tucson, Arizona 85724

INTRODUCTION

A major problem facing biomedical science in the Western nations is the loss of faith by the public in the conventional health care system. This has given rise to a pluralistic attitude in which the public simultaneously patronizes conventional health care delivery while also using "natural" remedies, over-the-counter preparations, herbs, vitamin and mineral supplements, and turning for treatment to those who practice any of a wide variety of so-called natural, holistic or traditional medicines. The latter may be eclectically derived from a variety of cultures, mixing elements of Traditional Chinese Medicine, Ayur-Vedic preparations, Japanese kampo and various European folk remedies. Accompanying this, at least in the U.S.A., has been the development of a common belief, articulated in a variety of ways, that being healthy is not so much a natural state, but one that must be maintained and protected by vigorous and continual intervention.

One result of these changed public perceptions has been an avid interest in taurine. Taurine is widely consumed in health drinks such as Lipovitan® and its analogs. A taurine-containing drink, Red Bull, is marketed in Europe as an aid to athletic performance[20]. Taurine-containing products are marketed in North America for the treatment of epilepsy, hyperactivity (Calms Kids®), substance abuse (Crave-Away), and other conditions. Yet this intriguing amino acid remains largely a mystery to science, despite the numerous phenomena with which it is associated. However, most of these phenomena have yet to be reduced to an understanding of mechanism. Taurine is established as a metabolic product of sulfur amino acid catabolism in many mammals, although apparently only to a limited extent in humans. It is a substrate for the biosynthesis of the bile salt, taurocholate. Apart from these functions, not much else has been unequivocally established as biochemical mechanisms for taurine, despite over 20 years of work by an international and interdisciplinary group of investigators.

Taurine 2, edited by Huxtable et al.
Plenum Press, New York, 1996

TAURINE: AN ENANTIOSTATIC AGENT

The dearth of scientific information on taurine thus seems to be at odds with its popular uses. Part of this discrepancy appears to be that scientists tend to be interested in what substances *do*, i.e. what effects are produced on a system when the substance is added. This contrasts with the public's intuitive sense that substances such as taurine need to be taken prophylactically to *prevent* something happening; i.e. that undesirable effects are produced on a system when a substance, such as an antioxidant vitamin, trace mineral, or taurine, is removed. This chapter discusses a reconciliation of these two viewpoints in terms of an interpretation of certain functions of taurine as enantiostatic actions.

The title of this talk was not chosen by me, but was suggested by Dr. Azuma. It is not an easy subject to talk on, considering the vast literature that has accumulated on taurine, the complexity and diversity of the phenomena associated with it, and the large number of scientists who have contributed to the present state of knowledge. For the past of taurine, we can turn to the proceedings of the various meetings that have been held over the last 20 years. I will start by considering a number of reports that taurine had a normalizing effect on function in the face of a biochemical or pharmacological disturbance. At the meeting in Mexico City in 1980, Flavia Franconi[15] reported that tension development in ventricular strips of the guinea pig heart depended both on the perfusate Ca^{2+} concentration and on the taurine concentration. The addition of taurine to the medium antagonized the effects of disturbances in Ca^{2+} concentration, and shifted tension development towards that seen at physiological Ca^{2+} concentration (Fig. 1). Taurine increased contractility at low Ca^{2+} concentrations and decreased it at high Ca^{2+} concentrations. Taurine, in other words, produces a "normalizing" effect on tension development even though the disturbance in homeostasis - an abnormal Ca^{2+} concentration - persists. This effect of taurine on the contractile response to Ca^{2+} is a general response in the heart[16-18,35,63].

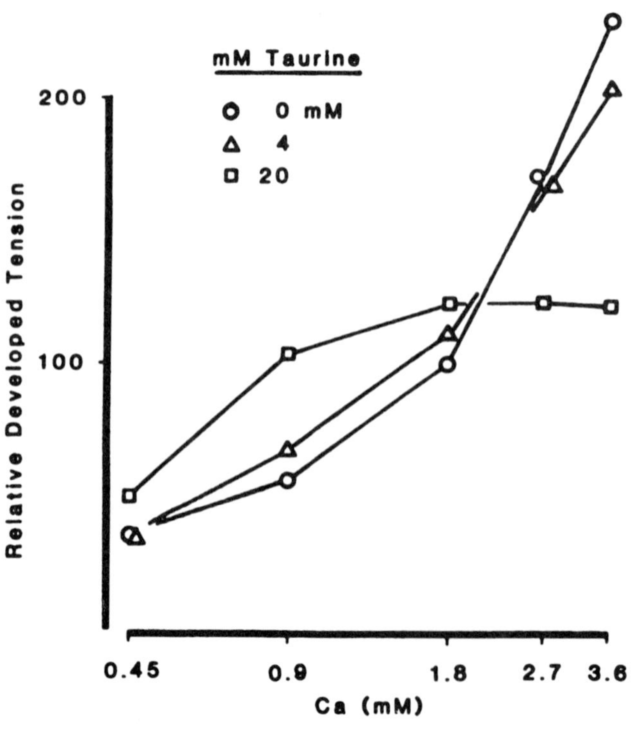

Figure 1. Taurine on tension development in ventricular strips, showing that taurine antagonizes both the increase in tension produced by raising Ca^{2+} concentrations above 1.8 mM and the decrease in tension produced by lowering Ca^{2+} concentrations below 1.8 mM. The data shown are recalculated from Franconi *et al.*[15,16]

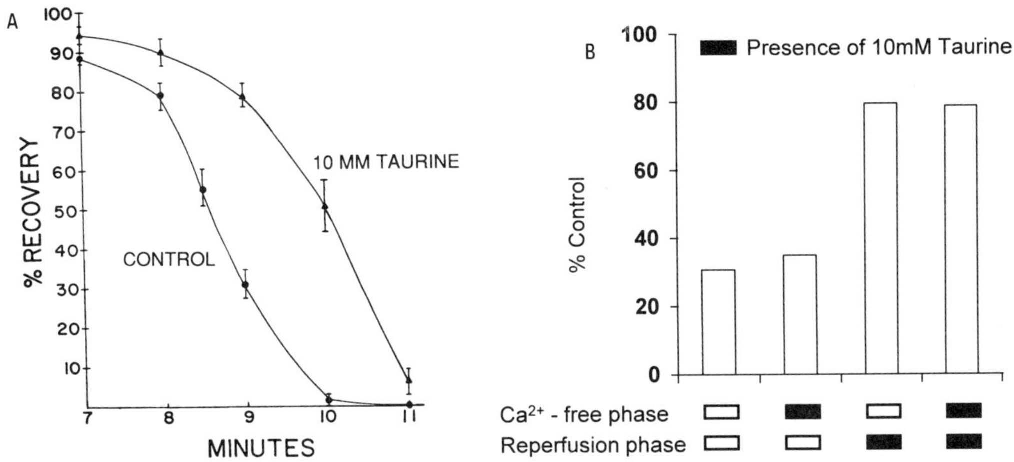

Figure 2. (a) Recovery of cardiac contractility in Ca^{2+} paradox. The longer the period of Ca^{2+}-free perfusion, the poorer the recovery of function on re-exposure to Ca^{2+}. Taurine increases the degree of recovery[52]; (b) The protective effect of taurine requires its presence during the recovery phase rather than the Ca^{2+}-free (insult) stage. Data from Kramer *et al.*[36]

A similar phenomenon is seen with the Ca^{2+} paradox. At the taurine meeting in Valley Forge, in 1979, Steven Schaffer *et al.*[52] reported that taurine increased the ability of the heart to recover function after a period of Ca^{2+}-free perfusion (Fig. 2a). They noted that taurine need only be present during reperfusion for this to occur; i.e. taurine did not alter the effects of Ca^{2+}-free conditions on the sarcolemmal membrane, but ameliorated the functional consequences of these effects (Fig. 2b).

Related effects of taurine are observed in other organ systems. At the taurine meeting in Tokyo in 1982, Kuriyama and his colleagues[48] showed that taurine, if given 12 h later, lessened the degree of hepatic lipid peroxidation induced by CCl_4. At the very first taurine meeting, in Tucson in 1975, Gruener showed that taurine stabilized isolated muscle fibers against loss of membrane potential[23].

In all of these examples, the protective effect of taurine is manifested without alteration in the environmental insult (altered Ca^{2+} levels, absence of O_2, presence of CCl_4). Taurine is maintaining function not by re-establishing homeostasis but by modifying the consequences of the insult in such a way as to oppose the deleterious changes produced in function. There are many similar examples. In the liver, taurine protects against radiation damage[10] and it lowers acetaldehyde levels in ethanol-loaded rats[37]. In the heart, taurine deficiency produces dilated cardiomyopathy[47,49], electrophysiological abnormalities[11,39,41] and myofibrillar loss[40]. Taurine supplemention protects against genetic cardiomyopathy[1]. Taurine has cardiotonic actions in both experimental[4,12] and clinical congestive heart failure[2,3]. In the lungs, taurine protects against nitric oxide-induced damage[22], and the toxic effects of ozone[5], bleomycin[60] amiodarone[61], nitrogen dioxide[42], cyclophosphamide[59] and paraquat[33].

A few years ago, the word "enantiostasis" was reintroduced to describe such phenomena[28]. An enantiostatic agent is one that antagonizes stress-induced changes on a system by producing other changes that result in maintenance of function without re-establishment of physicochemical homeostasis. This has been elegantly expressed as allowing "freedom from the simple consequences of environmental change". In this regard, for example, taurine

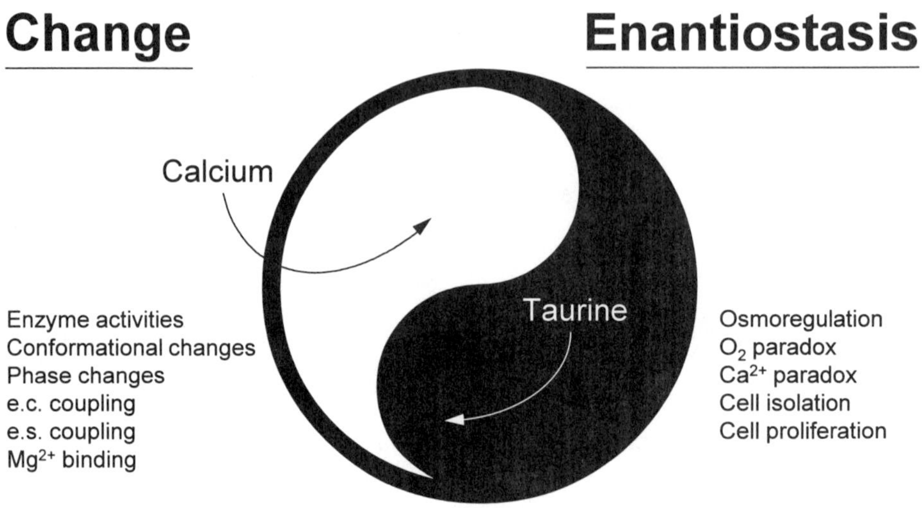

Figure 3. Enantiostasis as opposed to change. Due to the radius ratio, Ca^{2+} is a so-called "structure forming" element (discussed in Max, *Trends Pharmacol.Sci.* 7:342 (1976); 8:166 (1987). One consequence of this basic physicochemical property is that Ca^{2+} is an agent of change. Introduction of low concentrations of Ca^{2+} into the cell is accompanied by numerous changes that can function as signalling pathways or, with higher concentrations of Ca^{2+}, can produce toxicity and cell death. Some of the functions affected or altered by Ca^{2+} entry are listed on the lower left. Taurine, on the other hand, which is normally present in high concentrations in cells lacking cell walls, is an agent of stability, allowing cellular processes to continue functioning in the face of disruptive external influences. Some of the functions on which taurine has a stabilizing influence are listed on the lower right.

may be considered in opposition to Ca^{2+} ions, the introduction of which into the cell milieu typically catalyzes numerous changes in homeostasis. Many phenomena involving taurine can be rationalized in terms of enantiostasis (Fig. 3).

EFFECTS OF TAURINE ON TOXICITY OF PYRROLIZIDINE ALKALOIDS

We have recently added to the above phenomena by finding that taurine is protective against aspects of pyrrolizidine alkaloid (PA) toxicity[70,72]. PAs are widely distributed plant poisons. PAs are toxic to liver, heart and lung. Taurine is protective in all these organs. Many outbreaks of PA poisoning have been reported, particularly in the under-developed nations of the Caribbean, Africa, and Asia[26].

The actions of taurine appear to be enantiostatic, inasmuch as it protects against PA toxicity[70] without altering the metabolism of this PA or the delivery of the toxic pyrrolic metabolites to target tissues (Yan and Huxtable, unpublished observations). A particularly interesting epidemic of PA poisoning in which taurine status may have been of especial importance occurred in Uzbekistan, in Soviet Asia, in the 1950s[26,58]. Several hundred people were poisoned, and at least 44 died. This epidemic had two unusual aspects: the primary symptoms were neurological, and adults were more susceptible than children. Children are normally much more susceptible to PA poisoning than are adults. The same applies to other mammals, to the extent that PAs can be given to nursing goats, rats or primates in amounts that do not harm the nurse but which poison the suckling[21,34,53]. The poisoning agent in this central Asian outbreak was *Trichodesma incanum*, the seeds of which contaminated grain used for making bread. We arranged a collection of *T. incanum* seeds in order to investigate why trichodesmine, the major PA, was neurotoxic, and why developing brain was relatively protected. As reported by Sturman[55] at the first taurine meeting in 1975, high levels of taurine occur in mammalian brain, and the levels fall throughout development. Many similar observations have been made subsequently[30,43]. An obvious possibility is that taurine was involved in the protection, as it is high in developing brain relative to adult brain, and, as discused above, it had been shown in a number of other systems to protect from environmental insults.

GSH and cysteine are other sulfur-containing metabolites that ameliorate the consequences of exposure to PAs[64]. They do so by intercepting the toxic pyrrolic metabolites of PAs, the dehydroalkaloids formed by the action of hepatic cytochrome P450. The dehydroalkaloids are conjugated to form the much less toxic metabolite, 7-glutathionyl-6,7-dihydro-1-hydroxymethyl-5H-pyrrolizine (GSDHP)[38,45]. GSDHP is formed in common from hepatic metabolism of a number of PAs, as the esterifying acids that structurally differentiate the PAs are cleaved in the course of conjugation. The protective effect of GSH is a homeostatic one, inasmuch as conjugation of GSH with dehydroalkaloid reduces the amount of toxin to which the body is exposed.

In turn, the PAs also modify sulfur metabolism, diverting cysteine metabolism from oxidation to taurine towards GSH synthesis[65-67,69]. The PA, retrorsine, for example, markedly decreases hepatic taurine levels while increasing the levels of glutamate (a GSH precursor) and GSH[65,71]. In addition, there is increased excretion of taurine from the liver. Retrorsine has been implicated in various human intoxications, including the pediatric poisonings uncovered in Arizona in the 1970s[14,24,26,54]. Tissue taurine loss following an insult (e.g. X-ray or radiation exposure[9,13,50,62]; chemotherapy[8]; surgery[57]; myocardial infarction[7,44,56]) is common, and seems to be involved in the enantiostatic action. Enzymes involved in GSH synthesis and turnover show increased activity following PA exposure. Thus, the activities of γ-glutamylcysteine synthetase, the rate-limiting enzyme in GSH synthesis, and γ-glutamyltranspeptidase and GSH tranferase are increased [67,68]. These effects of PAs on sulfur

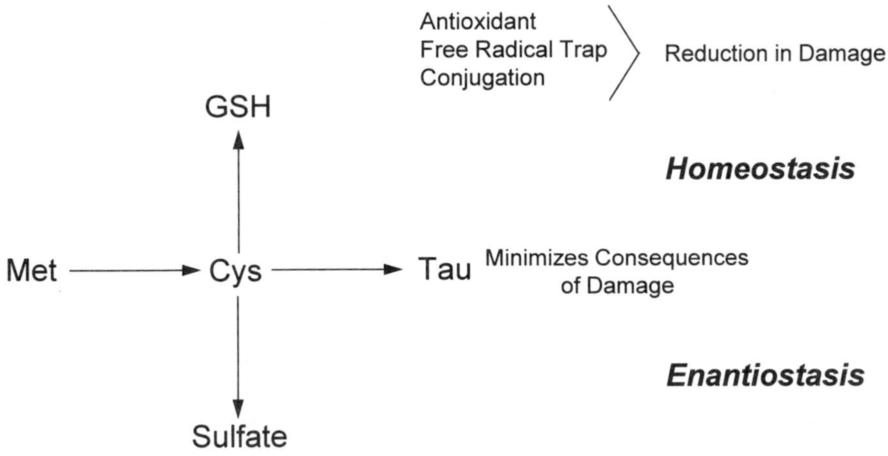

Figure 4. Protective mechanisms: Cysteine (Cys), obtained from the diet or formed from methionine (Met) by transsulfuration, is the substrate for both GSH and taurine biosynthesis. GSH functions as a homeostatic agent to reduce the levels of toxic xenobiotics and their metabolites to which the liver is exposed. Taurine functions as an enantiostatic agent to minimize damage or oppose the consequences of damage produced by toxic xenobiotics and their metabolites.

metabolism carry implications for the public health risks of PAs, in that toxicity at a given level of exposure may be conditionally dependent on the dietary sulfur amino acid status, and hence on the general state of protein nutrition.

In summary, sulfur metabolism is intimately involved in protecting the liver from xenobiotics. Two major prongs of sulfur metabolism are the path leading to GSH, present in all mammals, and the path leading to taurine[25,28]. This latter path is more significant in some species than others. The one arm protects by homeostatic mechanisms (Fig. 4). GSH functions as an antioxidant and a free radical trap. It also potentially toxic metabolites of xenobiotics by conjugation. These are all homeostatic mechanisms, tending to reduce toxicity by maintaining the intracellular milieu. Taurine, on the other hand, tends to reduce toxicity more by modifying the consequences of a disturbance in the milieu. This is an enantiostatic mechanism.

PUBLIC HEALTH IMPLICATIONS

The ameliorating effect of taurine on PA toxicity is an interesting observation, not only in terms of prophylaxis and treatment, but in terms of epidemiology. Taurine protects rats, even though this species has a high biosynthetic capacity for taurine[19,51]. Humans, however, do not[32,51]. In human liver, the activity of cysteine sulfinic acid decarboxylase, the enzyme involved in taurine biosynthesis, is only 0.1% of the activity in rat liver[19,51]. Humans, therefore, are dependent on dietary sources of taurine. The retrorsine findings[65,71] indicate that PAs cause a marked and immediate decrease in liver taurine content. If exogenous taurine is protective in the rat, therefore, one can predict that it should be even more protective in humans. It may be that differences in dietary factors such as taurine and the other sulfur amino acids underlie the variation in susceptibility to PA poisoning among exposed persons. Death can occur in humans consuming a cumulative dose of PAs amounting to about 2% of a rat LD_{50}[29]. Yet others consume far higher total doses without apparent harm. We can surmise, therefore, that exposed individuals on low taurine diets (vegetarians) or low sulfur amino acid diets may be at higher risk. Dietary factors may also underlie the high risk of populations in Third World countries. Epidemics of PA poison occur regularly in such nations. In the last 30 years there have been epidemics in India, Afghanistan, Uzbekistan, Tadjikistan, the Caribbean nations, and elsewhere[6,26,46]. The affected populations typically were consuming diets low in taurine, which is largely absent from plants, but found in meat and, to a lesser extent, dairy products.

In the underdeveloped countries, little can be done in the short term to make massive alterations in protein intake. However, there is a possibility that taurine could protect against the risk of PA poisoning. Taurine is a relatively low-priced, stable, easily transported substance. The use of taurine prophylaxis in PA-exposed populations may be justified, particularly in view of the absence of other useful interventions.

Another implication concerns the cases of PA poisoning regularly uncovered in developed countries[24,26,27,29]. The true incidence of PA poisoning is much higher than is realized because of the lack of adequate reporting. The cases of fatality we found in Arizona in the 1970s, for example, involved a herbal preparation that one wholesaler told us had been sold since the mid-1920s[14,24,54]. Despite the under-reporting, however, given the number of people who consume PA-containing herbs it is a mystery as to why PA poisoning is not more prevalent. In cases where the information is available, we have done careful studies of cumulative threshold doses required for toxicity or death[24,31]. These thresholds turn out to be low, and many consumers should easily exceed them. Why do not these people become ill? The suggestion from our sulfur amino acid studies is that PA poisoning, to an extent, is a conditional disease: conditional on low sulfur/low taurine intake. In other words, consumers of PA-containing herbs in developed countries who are also on low intake of sulfur amino acids/taurine (vegetarians, persons on certain dietary regimens, etc.) are at greater risk of PA intoxication. Nutritional status may underlie the puzzling variation in susceptibility to PA toxicity in exposed human populations in the United States.

CONCLUSION

The burgeoning interest in nutritional interventions for maintaining health may intersect with rising appreciation of the enantiostatic actions of taurine. This should result in increased application of this amino acid in prophylaxis for a variety of conditions in both the developed and under-developed nations.

ACKNOWLEDGMENTS

Work from the author's laboratory over the years has been supported by, among others, NIH Grant HL 25254, and the Taisho Pharmaceutical Co.

REFERENCES

1. Azari, J., Brumbaugh, P., and Huxtable, R.J. 1980, Prophylaxis by taurine in the hearts of cardiomyopathic hamsters, *J.Mol.Cell.Cardiol.* 12:1353-1366.
2. Azuma, J., Hasagawa, H., Awata, N., Sawamura, A., Harada, H., Ogura, K., Ohta, H., Yamauchi, K., and Kishimoto, S. 1983, Taurine for treatment of congestive heart failure in humans, in: "Sulfur Amino Acids: Biochemical and Clinical Aspects", Kuriyama, K., Huxtable, R.J. and Iwata, H. eds., A. R. Liss, New York, pp. 61-72.
3. Azuma, J., Sawamura, A., Awata, N., Ohta, H., Hamaguchi, T., Harada, H., Takihara, K., Hasegawa, H., Yamagami, T., Ishiyama, T., Iwata, H., and Kishimoto, S. 1985, Therapeutic effect of taurine in congestive heart failure: a double-blind crossover trial, *Clin.Cardiol.* 8:276-282.
4. Azuma, J., Takahisa, K., Awata, N., Ohta, H., Sawamura, A., Harada, H., and Kishimoto, S. 1984, Beneficial effect of taurine on congestive heart failure induced by chronic aortic regurgitation in rabbits, *Res.Comm.Chem.Pathol.Pharmacol.* 45:261-270.
5. Banks, M.A., Porter, D.W., Martin, W.G., and Castranova, V. 1991, Ozone-induced lipid peroxidation and membrane leakage in isolated rat alveolar macrophages: Protective effects of taurine, *J.Nutr.Biochem.* 2:308-313.
6. Chauvin, P., Dillon, J.-C., Moren, A., Talbak, S., and Barakaev, S. 1993, Heliotrope poisoning in Tadjikistan, *Lancet*, 341:1663
7. Cooper, M.W. and Lombardini, J.B. 1982, Elevated blood taurine levels after myocardial infarction or cardiovascular surgery: is there any significance? in: "Taurine in Nutrition and Neurology", Huxtable, R.J. and Pasantes-Morales, H. eds., Plenum Press, New York, pp. 191-206.
8. Desai, T.K., Maliakkal, J., Kinzie, J.L., Ehrinpreis, M.N., Luk, G.D., and Cejka, J. 1992, Taurine deficiency after intensive chemotherapy and/or radiation, *Am.J.Clin.Nutr.* 55:708-711.
9. Dilley, J.V. 1972, The origin of urinary taurine excretion during chronic radiation injury, *Radiation Res.* 50:191-196.
10. Dokshina, G.A., Yartsev, E.I., Kolesnikov, Y.A., and Kostesha, N.Y. 1974, Influence of taurine on the recovery of tissue respiration and oxidative phosphorylation in liver mitochondria in irradiated rats, *Radiology*, 14:44-48.
11. Eley, D.W., Lake, N., and ter Keurs, H.E.D.J. 1994, Taurine depletion and excitation-contraction coupling in rat myocardium, *Circ.Res.* 74:1210-1219.
12. Elizarova, E.P., Orlova, T.R., and Medvedeva, N.V. 1993, Effects on heart membranes after taurine treatment in rabbits with congestive heart failure, *Arzneimittelforschung*, 43:308-312.
13. Ewoane, C., Portemer, C., Bernard, P., and Chatagner, F. 1982, Effect of in vivo irradiation of the rat on the taurine concentration of the heart, *Experientia*, 38:1033

14. Fox, D.W., Hart, M.C., Bergeson, P.S., Jarrett, P.B., Stillman, A.E., and Huxtable, R.J. 1978, Pyrrolizidine (Senecio) intoxication mimicking Reye's syndrome, *J.Pediatr.* 93:980-982.
15. Franconi, F., Giotti, A., Manzini, S., Martini, F., Stendardi, I., and Zilletti, L. 1982, Observations on the action of taurine at arterial and cardiac levels, in: "Taurine in Nutrition and Neurology", Huxtable, R.J. and Pasantes-Morales, H. eds., Plenum Press, New York, pp. 181-190.
16. Franconi, F., Martini, F., Stendardi, I., Matucci, R., Zilletti, L., and Giotti, A. 1982, Effect of taurine on calcium levels and contractility in guinea pig ventricular strips, *Biochem.Pharmacol.* 31:3181-3186.
17. Franconi, F., Stendardi, I., Martini, F., Zilletti, L., and Giotti, A. 1982, Interaction between organic calcium-channel blockers and taurine in vitro and in vivo, *J.Pharm.Pharmacol.* 34:329-330.
18. Franconi, F., Stendardi, I., Matucci, R., Failli, P., Bennardini, F., Antonini, G., and Giotti, A. 1984, Inotropic effect of taurine in guinea-pig ventricular strips, *Eur.J.Pharmacol.* 102:511-514.
19. Gaull, G.E., Rassin, D.K., Raiha, N.C.R., and Heinonen, K. 1977, Milk protein quantity and quality in low birth weight infants. III. Effects on sulfur-containing amino acids in plasma and urine. *J.Pediatr.* 90:348-355.
20. Geiss, K.-R., Jester, I., Falke, W., Hamm, M., and WAAG, K.-L. 1994, The effects of a taurine-containing drink on performance in 10 endurance-athletes, *Amino Acids*, 7:45-56.
21. Goeger, D.E., Cheeke, P.R., Schmitz, J.A., and Buhler, D.R. 1982, Effect of feeding milk from goats fed tansy ragwort (Senecio jacobaea) to rats and calves, *Am.J.Vet.Res.* 43:1631-1633.
22. Gordon, R.E., Shaker, A.A., and Solano, D.F. 1986, Taurine protects hamster bronchioles from acute NO_2-induced alterations, *Am.J.Pathol.* 125:585-600.
23. Gruener, R., Bryant, H., Markovitz, D., Huxtable, R.J., and Bressler, R. 1976, Ionic actions of taurine on nerve and muscle membranes: electrophysiologic studies, in: "Taurine", Huxtable, R.J. and Barbeau, A. eds., Raven Press, New York, pp. 225-242.
24. Huxtable, R.J. 1980, Herbal teas and toxins: novel aspects of pyrrolizidine poisoning in the United States, *Perspect.Biol.Med.* 24:1-14.
25. Huxtable, R.J. 1989, Taurine in the central nervous system and the mammalian actions of taurine, *Prog.Neurobiol.* 32:471-533.
26. Huxtable, R.J. 1989, Human health implications of pyrrolizidine alkaloids and herbs containing them, in: "Toxicants of Plant Origin, Vol I: Alkaloids", Cheeke, P.R. ed., CRC Press, Boca Raton, Florida, pp. 41-86.
27. Huxtable, R.J. 1990, The harmful potential of herbal and other plant products, *Drug Safety*, 5 (Suppl. 1):126-136.
28. Huxtable, R.J. 1992, Physiological actions of taurine, *Physiol.Rev.* 72:101-163.
29. Huxtable, R.J. 1992, The toxicology of alkaloids in foods and herbs, in: "Handbook of Natural Toxins: Vol. 7 Food Poisoning", Tu, A.T. ed., Marcel Dekker, New York, pp. 237-262.
30. Huxtable, R.J., Crosswell, S., and Parker, D. 1989, Phospholipid composition and taurine content of synaptosomes in developing rat brain, *Neurochem.International*, 15:233-238.
31. Huxtable, R.J., Luthy, J., and Zweifel, V. 1986, Comfrey pepsin preparations - a warning, *N.Engl.J.Med.* 315:1095
32. Irving, C.S., Marks, L., Klein, P.D., Foster, N., Gadde, P.L., Chase, T.N., and Samuel, D. 1986, New evidence for taurine biosynthesis in man obtained from $^{18}O_2$ inhalation studies, *Life Sci.* 38:491-496.
33. Izumi, K., Nagata, R., Motoya, T., Yamashita, J., Hirokane, T., Nagata, T., Satoh, Y., Sawada, Y., Ishibashi, M., Yoshida, H., and Fukuda, T. 1989, Preventive effect of taurine against acute paraquat intoxication in beagles, *Jpn.J.Pharmacol.* 50:229-233.
34. Johnson, A.E. 1976, Changes in calves and rats consuming milk from cows fed chronic lethal doses of *Senecio jacobaea* (tansy ragwort), *Am.J.Vet.Res.* 37:107-110.
35. Khatter, J.C., Soni, P.L., Hoeschen, R.J., Alto, L.E., and Dhalla, N.S. 1981, Subcellular effects of taurine on guinea pig heart, in: "The Effects of Taurine on Excitable Tissues", Schaffer, S.W., Baskin, S.I. and Kocsis, J.J. eds., Spectrum Publications, New York, pp. 281-294.
36. Kramer, J.H., Chovan, J.P., and Schaffer, S.W. 1981, Effect of taurine on calcium paradox and ischemic heart failure, *Am.J.Physiol.* 240:H238-H246.
37. Kupryszewski, G., Ragnarsson, U., Rolka, K., and Wilusz, T. 1985, Lowering of liver acetaldehyde but not ethanol concentrations by pretreatment with taurine in ethanol-loaded rats, *Experientia*, 41:1421
38. Lafranconi, W.M., Ohkuma, S., and Huxtable, R.J. 1985, Biliary excretion of novel pneumotoxic metabolites of the pyrrolizidine alkaloid, monocrotaline, *Toxicon*, 23:983-992.
39. Lake, N. 1992, Effects of taurine deficiency on arrhythmogenesis and excitation-contraction coupling in cardiac tissue, *Adv.Exp.Med.Biol.* 315:173-179.
40. Lake, N. 1993, Loss of cardiac myofibrils: Mechanism of contractile deficits induced by taurine deficiency, *Am.J.Physiol.Heart Circ.Physiol.* 264:H1323-H1326.

41. Lake, N., Eley, D.W., and ter Keurs, H.E.D.J. 1991, The effects of taurine depletion on excitation-contraction coupling in rat cardiac trabeculae, *Biophys.J.* 59:64a
42. Lehnert, B.E., Archuleta, D.C., Ellis, T., Session, W.S., Lehnert, N.M., Gurley, L.R., and Stavert, D.M. 1994, Lung injury following exposure of rats to relatively high mass concentrations of nitrogen dioxide, *Toxicology*, 89:239-277.
43. Lleu, P.-L. and Huxtable, R.J. 1992, Phospholipid methylation and taurine content of synaptosomes from cerebral cortex of developing rat, *Neurochem.International*, 21:109-118.
44. Lombardini, J.B. and Bricker, D.L. 1981, Effects of cardiovascular surgery on blood concentrations of taurine and amino acids, *Proc.Soc.Exp.Biol.Med.* 167:498-505.
45. Mattocks, A.R., Croswell, S., Jukes, R., and Huxtable, R.J. 1991, Identity of a biliary metabolite formed from monocrotaline in isolated, perfused rat liver, *Toxicon*, 29:409-415.
46. Mayer, F. and Lüthy, J. 1993, Heliotrope poisoning in Tadjikistan, *Lancet*, 342:246-247.
47. Moise, N.S., Pacioretty, L.M., Kallfelz, F.A., Stipanuk, M.H., King, J.M., and Gilmour, R.F., Jr. 1991, Dietary taurine deficiency and dilated cardiomyopathy in the fox, *Am.Heart J.* 121:541-547.
48. Nakashima, T., Takino, T., and Kuriyama, K. 1983, Therapeutic and prophylactic effects of taurine administration on experimental liver injury, in: "Sulfur Amino Acids: Biochemical and Clinical Aspects", Kuriyama, K., Huxtable, R.J. and Iwata, H. eds., Alan R. Liss, Inc., New York, pp. 449-460.
49. Pion, P.D., Kittleson, M.D., Rogers, Q.R., and Morris, J.G. 1987, Myocardial failure in cats associated with low plasma taurine: a reversible cardiomyopathy, *Science*, 237:764-768.
50. Raghavan, K.G. and Nadkarni, G.B. 1970, Formation and excretion of taurine in X-irradiated rats, *Int.J.Radiat.Biol.* 18(1):41-49.
51. Rassin, D.K. 1982, Taurine, cysteine sulfinic acid decarboxylase and glutamic acid in brain, in: "Taurine in Nutrition and Neurology", Huxtable, R.J. and Pasantes-Morales, H. eds., Plenum Press, New York, pp. 257-268.
52. Schaffer, S.W., Chovan, J., Kramer, J., and Kulakowski, E. 1981, The role of taurine receptors in the heart, in: "The effects of taurine on excitable tissues", Schaffer, S.W., Baskin, S.I. and Kocsis, J.J. eds., Spectrum Publications, New York, pp. 261-279.
53. Schoental, R. 1959, Liver lesions in young rats suckled by mothers treated with the pyrrolizidine (*Senecio*) alkaloids, lasiocarpine, and retrorsine, *J.Path.Bact.* 77:485-495.
54. Stillman, A.E., Huxtable, R., Consroe, P., Kohnen, P., and Smith, S. 1977, Hepatic veno-occlusive disease due to pyrrolizidine poisoning in Arizona, *Gastroenterology*, 73:349-352.
55. Sturman, J.A. and Gaull, G.E. 1976, Taurine in the brain and liver of the developing human and rhesus monkey, in: "Taurine", Huxtable, R. and Barbeau, A. eds., Raven Press, New York, pp. 73-84.
56. Suleiman, M.-S., Fernando, H.C., Dihmis, W.C., Hutter, J.A., and Chapman, R.A. 1993, A loss of taurine and other amino acids from ventricles of patients undergoing bypass surgery, *Br.Heart J.* 69:241-245.
57. Turner, F.P. and Blum, V.G. 1964, The urinary excretion of free taurine in acute and chronic disease, following surgical trauma, and in patients with acute alcoholism, *J.Surg.Res.* 4:423-431.
58. United Nations Environment Programme, International Labour Organization, and World Health Organization 1988, "Environmental Health Criteria 80: Pyrrolizidine Alkaloids", World Health Organization, Geneva.
59. Venkatesan, N. and Chandrakasan, G. 1994, In vivo administration of taurine and niacin modulate cyclophosphamide-induced lung injury, *Eur.J.Pharmacol.Environ.Toxicol.Pharmacol.* 292:75-80.
60. Wang, Q., Giri, S.N., Hyde, D.M., and Li, C. 1991, Amelioration of bleomycin-induced pulmonary fibrosis in hamsters by combined treatment with taurine and niacin, *Biochem.Pharmacol.* 42:1115-1122.
61. Wang, Q., Hollinger, M.A., and Giri, S.N. 1992, Attenuation of amiodarone-induced lung fibrosis and phospholipidosis in hamsters by taurine and/or niacin treatment, *J.Pharmacol.Exp.Ther.* 262:127-132.
62. Watson, G.M. 1962, The origin of taurine excreted in the urine after whole body irradiation, *Inter.J.Rad.Biol.* 5:79-83.
63. Welty, J.D. and McBroom, M.J. 1985, Effects of verapamil and taurine administration on heart taurine and calcium in BIO 14.6 cardiomyopathic hamsters, *Res.Comm.Chem.Pathol.Pharmacol.* 49:141-144.
64. White, I.N.H. 1976, The role of liver glutathione in the acute toxicity of retrorsine to rats, *Chem.-Biol.Interact.* 13:333-342.
65. Yan, C.C. and Huxtable, R.J. 1995, Effects of the pyrrolizidine alkaloid, retrorsine, on sulfur metabolism in the liver, *Proc.Western Pharmacol.Soc.* 38:37-40.
66. Yan, C.C. and Huxtable, R.J. 1995, Effect of the pyrrolizidine alkaloid, monocrotaline, on bile composition of the isolated, perfused rat liver, *Life Sci.* 57:617-626.
67. Yan, C.C. and Huxtable, R.J. 1996, Effects of the pyrrolizidine alkaloid, monocrotaline, on glutathione metabolism in the rat, *Biochem.Pharmacol.* 51:375-379.

68. Yan, C.C. and Huxtable, R.J. 1995, The effect of the pyrrolizidine alkaloids, monocrotaline and trichodesmine, on hepatic glutathione synthesis in the rat, *Toxicon*, 33:627-634.
69. Yan, C.C. and Huxtable, R.J. 1995, The effect of the pyrrolizidine alkaloids, monocrotaline and trichodesmine, on tissue pyrrole binding and glutathione metabolism in the rat, *Toxicon*, 33:627-634.
70. Yan, C.C. and Huxtable, R.J. 1995, Effects of taurine and guanidinoethane sulfonate on right ventricular hypertrophy induced by the pyrrolizidine alkaloid, monocrotaline, *Biochem.Pharmacol.* 51:321-329.
71. Yan, C.C. and Huxtable, R.J. 1996, Effect of pyrrolizidine alkaloids on taurine and sulfur metabolism in the rat liver, in: "Taurine: Basic and Clinical Aspects", Huxtable, R.J., Azuma, J., Nakagawa, M., Kuriyama, K. and Baba, A. eds., Plenum, New York,
72. Yan, C.C. and Huxtable, R.J. 1996, Effect of taurine on toxicity of the pyrrolizidine alkaloid, monocrotaline, in rats, in: "Taurine: Basic and Clinical Aspects", Huxtable, R.J., Azuma, J., Nakagawa, M., Kuriyama, K. and Baba, A. eds., Plenum, New York,

INDEX